Random House Webster's Dictionary of *Scientists*

Random House Webster's Dictionary of *Scientists*

Random House
New York

This work was originally published in Great Britain under the title *The Hutchinson Dictionary of Scientists* by Helicon Publishing Limited.

Library of Congress Cataloging-in-Publication Data
Hutchinson dictionary of scientists.
Random House Webster's dictionary of scientists. — 1st ed.
p. cm.
Originally published : The Hutchinson dictionary of scientists.
Helicon Pub. Ltd., c1996.
ISBN 0-375-70057-9
1. Scientists—Biography—Dictionaries. 2. Science—History.
I. Title.
Q141.H88 1997 97-25159
509'.2'2—dc21 CIP

This book is available for special purchases in bulk by organizations and institutions, not for resale, at special discounts. Please direct your inquiries to the Random House Special Sales Department, toll-free 888-591-1200 or fax 212-572-4961.

Please address inquiries about electronic licensing of this division's products, for use on a network or in software or on CD-ROM, to the Subsidiary Rights Department, Random House Reference & Information Publishing, fax 212-940-7370.

Printed and bound in the United States of America.

Visit the Random House Web site at http://www.randomhouse.com/

First Edition
0 9 8 7 6 5 4 3 2 1
ISBN 0-375-70057-9

New York Toronto London Sydney Auckland

Contributors

Neil Ardley, Gareth Ashurst, Chris Backenist, Jim Bailey, Mary Basham, Alan Bishop, William Cooksey, David Cowey, Michael Darton, Sofie Fairbass, Keld Fenwick, Lorraine Ferguson, Judy Garlick, Richard Gulliver, Ian Harvey, Peter Lafferty, Nicholas Law, Claire Lewis, Lloyd Lindo, Robert Matthews, Nigel Morrison, Robert Mortimer, Diana Moule, Patricia Nash, Valerie Neal, Jon Osborne, Lucia Osborne, Adam Ostaszewski, Caroline Overy, Roy Porter, Helen Rapson, Peter Rodgers, Mary Sanders, Martin Sherwood, Robert Smith, E.M. Tansey, Christopher Tunney, Zusa Vrbova, David Ward

Editors

Managing Editor
Sara Jenkins-Jones

Project Editor
Shereen Karmali

Text Editors
Ingrid von Essen
Catherine Thompson
Karen Young

Proofreader
Catherine Thompson

Researchers
Anna Farkas
Justin Webb

Production
Tony Ballsdon

Design
Ann Dixon

Art Editor
Terence Caven

Contents

Preface

Astronomy

The origins of astronomy
Astronomy is the oldest of all sciences, its origins dating back at least several thousands of years. The word astronomy comes from the two Greek words *astron* 'star' and *nomos* 'law' and perhaps the first attempt at understanding the laws that governed the stars and their apparent movement across the night sky was prompted by the need to produce an accurate calendar.

In order to predict the flooding of the river Nile, and hence the time when the surrounding lands would be fertile enough for crops to be planted, the Egyptians made observations of the brightest star in the night sky, Sirius. It was discovered that the date when this star (called Sothis by the Egyptians) could first be seen in the dawn sky (the heliacal rising) enabled the date of the flooding to be calculated. This also enabled the length of the year to be calculated quite accurately; by 2780 BC the Egyptians knew that the time between successive heliacal risings was about 365 days. More accurate observations enabled them to show that the year was about $365\frac{1}{4}$ days long, with a slight difference of 20 minutes between the sidereal year (the time between successive appearances of a star in the same position in the sky) and the tropical year (the time between successive appearances of the Sun on the vernal equinox).

The making of calendars
Although evidence suggests that the grouping of stars into constellations was done before a fairly accurate calendar was drawn up, this latter achievement was probably the first scientific act carried out in the field of astronomy.

The prediction of phenomena, which is the fundamental activity of any science, was also being carried out by other ancient civilizations, such as the Chinese. Some historians suggest that the existence in Europe and elsewhere of megalithic sites such as Stonehenge in England and Carnac in France (some of which date back to almost 3000 BC) shows that even minor early civilizations could calculate their calendars and, possibly, predict basic astronomical events such as eclipses.

Certainly by 500 BC the prediction of eclipses had become quite accurate. Thales of Miletus (624–547 BC), a Greek philosopher, ended a war between the Medes and the Lydians by accurately predicting the occurrence of an eclipse of the Sun on 28 May 585 BC. His prediction was probably based on the Saros, a period of about 18 years, after which a particular sequence of solar (and lunar) eclipses recurs. This interval was known to Babylonian and Chaldean astronomers long before Thales' time.

The Greeks and early astronomy
The next 800 years of astronomy were dominated by the Greeks. Anaximander (610–546 BC), a pupil of Thales, helped bring the knowledge of the ancient Egyptians and others to Greece and introduced the sundial as a timekeeping device. He also pictured the sky as a sphere with the Earth floating in space at the centre.

Anaxagoras (500–428 BC) made great advances in astronomical thought, with his correct explanation of the cause of lunar and solar eclipses. In addition, he considered the

Moon to be illuminated by reflected light and all material in the heavens to be composed of the same material; a rocky meteorite falling on the Aegean coast in 468 BC may have brought him to this conclusion. All these thoughts are now known to be broadly correct.

Plato's geocentric theory

The Greek philosopher Plato (*c.* 420–340 BC) effectively cancelled out these advances with his insistence on the perfection of the heavens, which according to him implied that all heavenly bodies must follow the perfect curve (the circle) across the sky. This dogma cast its dark shadow over astronomical thought for the next 2,000 years. Both Eudoxus (408–355 BC) and Callippus (370–300 BC) tried to convert observations into proof for this idea: each planet was set into a sphere, so that the universe took on an onion-ring appearance, with the Earth at the centre. More spheres had to be added to make the theory even approximate to the observations, and by the time of Callippus there were 34 such spheres. Even so, the theory did not match observations.

It was Heraklides of Pontus (388–315 BC) who noted that the apparent motion of the stars during one night might be the result of the daily rotation of the Earth, not the stars, on its axis. He also maintained that Mercury and Venus revolved around the Sun, but he held to the geocentric belief that the Earth was the centre of the universe.

The heliocentric theory

Aristarchos of Samos (*c.* 320–240 BC) took these ideas one step further. In *c.* 260 BC he put forward the heliocentric theory of the Solar System, which puts the Sun at the centre of the system of the six planets then known, with the stars infinitely distant. This latter conclusion was based on the belief that the stars were motionless, their apparent motion resulting simply from the Earth's daily rotation. Aristarchos still maintained that the planets moved in Platonically perfect circles, however.

By noting that when the Moon was exactly half-illuminated it must lie at the right angle of a triangle formed by lines joining the Earth, Sun and Moon, Aristarchos was able to make estimates of the relative sizes and distances of the Sun and Moon. Unfortunately, although his theory was correct, the instruments he needed were not available, and his results were highly inaccurate. They were good enough, however, to show that the Sun was more distant and much larger than the Moon. Although this provided indirect support for the heliocentric theory, since it seemed logical for the small Earth to orbit the vast Sun, the geocentric dogma prevailed for another 1,800 years.

The final discoveries of the Greeks

Greek astronomy had its last success between 240 BC, when Eratosthenes (276–195 BC) made his calculation of the Earth's size, and *c.* AD 180, when Ptolemy died. In between those two events, Hipparchus (*c.* 146–127 BC) replaced the Eudoxian theory of concentric spheres with an even more contrived arrangement based on the ideas of Apollonius. The Sun and planets were all considered to revolve upon a small wheel, or epicycle, the centre of which revolved around the Earth, (the centre of the universe, on a larger circle (the deferent)). Using this theory, Ptolemy (or Claudius Ptolemaeus) was able to predict the position of the planets to within 1°, i.e. about two lunar diameters.

Having begun so well, Greek astronomy ended somewhat dismally, preferring dogma to observable evidence and so failing to come to terms with the heliocentric theory of Aristarchos.

The Arab influence

It was not until some 600 years after the death of Ptolemy that astronomy started once again to move forward. The lead was taken by the Arabs. Their great mathematical skill and ingenuity with instruments enabled them to refine the observations and theories of the Greeks and produce better star maps, which were becoming increasingly useful for navigation, one of the spurs to astronomical research for many centuries to come. During the Middle Ages, nevertheless, European astronomers did little more than tinker with Ptolemy's epicycles. The Roman Catholic Church decided that the geocentric theory was the only one compatible with the Scriptures, thus making anyone who attempted to put the Sun at the centre of the Solar System guilty of heresy, which was punishable by death.

It was, however, the publication of one book that made Europe the centre of astronomical development, after the science had effectively stood still for many centuries.

European astronomical thought

In 1543 the Polish astronomer Nicolaus Copernicus (1473–1543) published a book entitled *De revolutionibus orbium coelestium*, in which he demonstrated that by placing the Sun at the centre of the Solar System, with the planets orbiting about it, it was possible to account for the apparent motion of the planets in the sky much more neatly than by geocentric theory. To explain the phenomenon of retrogression, for example, cosmologists had been compelled to add complication upon complication to Ptolemy's theory. With the heliocentric theory it could be explained simply as the result of a planet's distances from the Sun.

The 'Copernican revolution' was far from a complete break with the past, nevertheless. Copernicus continued to believe that all celestial motion must be circular, so that his model retained the epicycles and deferents of Ptolemy. But whereas Aristarchos' heliocentric belief had been left to moulder for centuries, Copernicus' ideas were picked up by other cosmologists who, by the end of the 16th century, had far more accurate observations at their disposal.

The heliocentric theory is finally established

The supplier of these more accurate observations was the Danish astronomer Tycho Brahe (1546–1601), who, using only the naked eye, was able to make observations accurate to two minutes of arc, five times more precise than those of Ptolemy. The effect was enormous. Calendar reform took place, with the Gregorian calendar, now used throughout the Western world, being instituted in 1582. More important for astronomy, Brahe's observations were used by his German assistant, Johannes Kepler (1571–1630), to establish the heliocentric theory of the Solar System once and for all. Out went the complicated systems of epicycles and the dogma of the perfect circle; for Kepler's idea that the planets followed elliptical paths around the Sun, which itself sat at one focus of the ellipses, accorded nicely with the observations of Brahe.

In *Astronomia Nova*, published in 1609, Kepler enunciated the first two laws of planetary motion; the first stated that each planet moved in an elliptical orbit; the second that it did so in such a way that the line joining it to the Sun sweeps out equal areas in equal times. Kepler's third law, that the cube of the distances of a planet from the Sun is proportional to the time required for it to complete one orbit, was announced in 1618.

The invention of the telescope

At the same time as Kepler was making these theoretical breakthroughs, the invention of the telescope by the Dutch optician Hans Lippershey (1587–1619), in 1608 was effecting a revolution in observation. Galileo Galilei (1564–1642) quickly put the instrument to use. In the years 1609 to 1610 he discovered the phases of Venus (showing that planet to be orbiting inside the path of the Earth) and identified four satellites of Jupiter; he also established the stellar nature of the Milky Way. As if to underline the imperfect nature of the heavens, contrary to the Platonic dogma that had crippled astronomy for so long, Galileo also discovered spots on the Sun and mountains on the 'perfect' sphere of the Moon.

The discovery of the laws of gravity

The next major breakthrough was again a theoretical one, the discovery by Isaac Newton (1642–1727) of the law of universal gravitation in 1665. Gravity is the single most important force in astronomy and Newton's discovery enabled him to deduce all three of Kepler's laws.

By nature a somewhat reticent man, Newton did not publish his discoveries, in the *Philosophiae Naturalis Principia Mathematica*, until 1687, after the prompting of his friend Edmund Halley (1656–1742). Newton showed that his law could account even for small effects such as the precession of the equinoxes discovered by Hipparchus and that the slight deviations of the planets from their Keplerian orbits were the result of their mutual gravitational attraction. By applying this perturbations theory to the Earth–Sun–Moon system, Newton was able to solve problems about the various motions that had baffled Kepler and his predecessors.

Advances in observational astronomy

Newton also made a significant contribution to observational astronomy in 1668, when he built the first reflecting telescope, with optics that were free of some of the defects of the refractors then in use.

Seven years later, in 1675, Charles II founded the Royal Observatory at Greenwich, essentially to find an accurate way of determining longitude for British ships involved in overseas exploration and colonization. The line of zero longitude was set at Greenwich, and the first Astronomer Royal, John Flamsteed (1646–1719), drew up a new star catalogue with positions accurate to 20 seconds of arc. Published in 1725, it was the first map of the telescopic age.

Careful observer as Flamsteed was, he failed to notice anything strange about the star which he noted in his catalogue as 34 Tauri. Its true nature was discovered by William Herschel (1738–1822), using the best reflector then in existence, in 1781. The 'star' was in fact, a previously undiscovered planet, and was named Uranus. Its discovery doubled the dimensions of the known Solar System.

Beginnings of astrophysics

The opening of the 19th century marked the beginning of one of the most important branches of astronomical observation: spectroscopy. In 1802 William Wollaston (1766–1828) discovered that the Sun's spectrum was crossed by a number of dark lines, and in 1815 Joseph Fraunhofer (1787–1826) made a detailed map of these lines. Fraunhofer noticed that the spectra of stars were slightly different from that of the Sun, but the enormous significance of this observation was not grasped until half a century later by Gustav Kirchhoff (1824–1887). In the meantime, Fraunhofer's skill as a telescope-maker enabled Friedrich Bessel (1784–1846) to determine the distance to the star, 61 Cygni; he found it to lie at a distance of about six light years, a term that had then became commonly used as a measure of stellar distance.

Although Kepler's laws enabled the calculation of the relative sizes of the planetary orbits to be obtained, a figure in absolute terms for the mean Earth–Sun separation was still needed. In the 19th century much effort was expended on this task, using the method suggested by Kepler of observing transits of Venus from widely separated places in order to take parallax observations. By the middle of the century a figure within 2% of the correct value had been obtained.

The discovery of Neptune

The year 1846 saw another triumph for Newton's theory of gravitation, when John Adams (1819–1892) and Urbain Le Verrier (1811–1877) predicted the position of an as-yet-unseen planet. Their prediction was based on the observed discrepancies in the motion of Uranus – discrepancies which the two astronomers took to be caused by another, massive planet orbiting beyond the path of Uranus. By calculating the new planet's orbit and its position at certain times, Johann Galle (1812–1910) spotted it on 23 September 1846, less than two lunar diameters from the predicted spot. The new planet was later named Neptune.

Spectroscopy

Fraunhofer's study of the dark lines in the solar spectrum bore fruit in 1859 with Kirchoff's explanation of them. The lines were absorption lines and were the result of the presence in the Sun of certain chemical elements. Kirchoffs discovery made it possible to determine the chemical composition of the Sun from Earth-bound observations. His work was extended by William Huggins (1824–1910), who was able to show that the stars were made of similar elements to those found on the Earth, thus supporting the

2,300-year-old contention of Anaxagoras. Huggins used the new invention of photography to record the spectra. He also made the first measurement of stellar red-shift, to determine the relative motion of stars towards or away from the Earth. These developments were crucial to the future development of astronomy and astrophysics.

By the end of the 19th century, photography had begun to take over from naked-eye observations of the universe. Stars, comets, nebulae and the Andromeda galaxy had all been photographed by 1900. Spectroscopy had also been used on all these objects, and the composition, motion and distance of many of them determined. This latter achievement was possible because of the discovery by Ejnar Hertzsprung (1873–1967) of a relationship between the spectrum of a star and its intrinsic luminosity. The relation was also found by Henry Russell (1877–1957) in the USA and published nine years after Hertzsprung in 1914. As a result, the diagram plotting the luminosities of stars against their spectra is called the Hertzsprung–Russell diagram. Its importance lies in its ability to show how stars evolve and how distant a star of a given spectrum and apparent brightness is.

Hertzsprung was able, by means of his diagram, to calculate the distance of a Cepheid variable star. As Henrietta Leavitt (1868–1921) discovered in 1904, variable stars exhibit a relationship between their intrinsic luminosity and their period of variability, so that a measurement of the period above can give the Cepheid's distance. As such a measurement could be carried out even on Cepheids in other galaxies, the distances of these galaxies could now be calculated by means of Leavitt's period-luminosity law, published in 1912.

The origins of the universe

In 1916 Albert Einstein published his 'Foundation of the General Theory of Relativity' in *Annalen der Physik*. Essentially a theory of gravitation, it marked the greatest theoretical advance in our understanding of the universe since Newton's *Principia* and like Newton's theory it had far-reaching implications for astronomy.

Einstein's theory immediately cleared up a long-standing problem concerning the orbit of Mercury, which was slowly rotating at the rate of about 43 seconds of arc per century. Einstein's theory showed this to be a result of effects arising from the high orbital velocity of Mercury and the intense gravitational field of the Sun. The theory also made two predictions. First, that in the presence of an intense gravitational field, light should be red-shifted to longer wavelengths as it struggled to escape the field. In 1925 this was found by Walter Adams (1876–1956) to be true of the spectrum of the white-dwarf companion of Sirius. Second, that according to General Relativity, light should be bent by the space-time curvature pictured in the theory as being the cause of gravitation. Observations of a solar eclipse in 1919 showed the light of stars close to the position of the Sun was indeed bent by the amount predicted by Einstein.

The expanding universe

When he applied his theory to the entire universe, in an attempt to reach a universal understanding of dynamics, Einstein was dismayed to discover that in its pure form it would only apply to a universe that is in overall motion. Such a prediction was contrary to the contemporary belief that, apart from the individual motions of the stars within galaxies, the universe was static.

However, by applying spectroscopy to the light of distant galaxies, Vesto Slipher (1875–1969) was able to show that the galaxies were in a state of recession. Surprisingly, it was not until 1924 that proof that the galaxies were star-systems separate from our own galaxy was given by Edwin Hubble (1889–1953). Armed with this knowledge, Hubble was then able to show that the universe as a whole was expanding, and that the rate at which a galaxy appeared to be moving away from Earth was proportional to its distance from us (as determined by the Cepheid law). Thus, Einstein's theory was correct in predicting an expanding universe.

The Big Bang theory

By the end of the 1920s the idea that the universe was born in a 'Big Bang', as proposed by Abbé George Lemaître (1894–1966) in 1927, had become, as it still remains, the established dogma of cosmology. Estimates based on Hubble's law currently set the date of the creation at about 15,000 million years ago.

Will the universe expand forever? This depends upon the mean density of matter in the universe. If the density exceeds a certain critical value (roughly equal to three protons per 1 cu m/35 cu ft) the expansion will halt sometime in the future and the universe will collapse to a 'big crunch'. With this in mind, astronomers have tried to estimate the mean density of matter, but have been hampered by the fact that most of the material in the universe is invisible; this conclusion was reached in the 1980s by considering the gravitational pulls on galaxies. The hunt for the invisible 'dark matter' continues in the 1990s. Modern theories of particle physics have thrown up a wealth of possible dark matter candidates. These theories also allow cosmologists to discuss events in the early Big Bang, possibly as far back as 10^{-43} seconds after the initial event. One theory dealing with the very early universe was developed by US cosmologist Alan Guth, in which the universe is supposed to have undergone a brief phase of accelerated expansion (called inflation) around 10^{-35} seconds after its birth.

The discovery of Pluto

The year 1930 saw the discovery by photographic means of a ninth planet in the Solar System, Pluto. After a painstaking search involving millions of star images, Clyde Tombaugh (1906–) detected the tiny speck of light on plates taken at the Lowell observatory in Arizona.

The Sun's power source

In 1938 the long-standing problem of the power-source of the Sun and stars was finally solved by Hans Albrecht Bethe (1906–) and Carl Friedrich von Weizsäcker (1912–). They found that the vast outpourings of energy were attributable to the fusing of hydrogen nuclei deep within stars, the process being so efficient that luminosity could be sustained for thousands of millions of years.

The universe in a new light

Experiments in America by Karl Jansky (1905–1950) and Grote Reber (1911–) in the 1930s heralded the start of a new era in astronomical observation, marked by the use of wavelengths other than light, in particular radio waves. Solar radio emission was detected in 1942 and following a theoretical prediction by Hendrik van de Hulst (1918–) in the Netherlands in 1944, interstellar hydrogen radio emission was detected and this was used to produce a map of our Galaxy which was not limited to those regions not obscured by light-absorbing dust.

The steady-state theory

A radically different theory of the universe, which was free of the Big Bang and its attendant difficulties was proposed by Thomas Gold (1920–), Hermann Bondi (1919–), and Fred Hoyle (1915–) in 1948. Called the steady-state theory, it pictured the universe as being in a constant state of expansion, with new matter being created to compensate for the dilution caused by the Hubble recession. The theory aroused much criticism, but it was not until 1965 that the theory was considered by many to have been finally disproved.

In that year, experiments by Arno Penzias (1933–) and Robert Wilson (1936–) in America resulted in the discovery that the universe appears to contain an isotropic background of microwave radiation. On the Big-Bang picture, that can be interpreted as the red-shifted remnant of the radiation generated in the original Big Bang. Since it is difficult to reproduce the properties of this background on the steady-state theory, the theory was abandoned by most astronomers.

In 1992 the Cosmic Background Explorer (COBE) satellite detected slight variations in the strength of the background radiation. These are believed to mark the first stage in the formation of galaxies.

Significant discoveries made by radio telescopes
Radio astronomy, which had progressed far in attempts to discover the true nature of the universe by observations of distant galaxies, had two major successes in the 1960s. The first was the discovery of quasi-stellar objects, and their identification as extremely remote yet very powerful sources by Maarten Schmidt (1929–) in 1963. Their power-source remains enigmatic, but the current belief is that they are galactic centres which contain massive black holes, sucking material into themselves. The second discovery was made by Jocelyn Bell Burnell (1943–) at Cambridge in 1967. Rapid bursts of radio energy at extremely regular intervals were picked up and interpreted as being generated by a rapidly rotating magnetized neutron star, or pulsar. Such an object may be the result of a supernova explosion. In February 1987, a supernova SN1987A exploded in the Small Magellanic Cloud, the first naked-eye supernova since 1604. Observations of SN1987A allowed refinement of supernova theory. In particular, observation of neutrinos from the explosion confirmed that neutrinos carry off much of the energy released in the explosion.

Techniques have been developed in which the outputs from two or more radio telescopes are combined to allow better resolution than is possible with a single dish. In aperture synthesis, several dishes are linked together to simulate the performance of one very large single dish. This technique was pioneered by Martin Ryle (1918–) at Cambridge, England. Very long baseline interferometry uses radio telescopes spread across the world to resolve minute details in radio sources.

Satellite and space-probe astronomy
The late 1960s and early 1970s were marked by the advent of the crewed exploration of the Moon by the American Apollo mission, beginning in 1969 with the expedition of Apollo 11. Throughout the 1960s and 1970s, uncrewed probes to the planets revealed more about them than had been discovered in all the previous centuries of study combined. The missions were either fly-bys, beginning with the American *Mariner 2* probe of Venus in 1962, or landings, such as the descent of the Russian Venera 4 on to the surface of Venus in 1967.

Detailed maps of the four terrestrial planets, Mercury, Venus, Earth and Mars, have now been made, and American Pioneer and Voyager probes passed by Jupiter and Saturn, taking in much of their satellite systems, in the 1970s and early 1980s. *Voyager 2* flew past Uranus in 1986, and Neptune in 1989. Earth-bound observations, which have always been hampered by the interference of a turbulent and polluted atmosphere, are now being supplemented by orbiting observatories operating at a wide range of wavelengths. The first X-ray observatory, UHURU, was launched in 1971, and it detected many sources of X-rays both within and beyond our galaxy.

Other highlights of probe and satellite astronomy include the 1986 rendevous of the *Giotto* probe with Halley's comet, the 1990 launch of the Hubble Space Telescope, and the 1990 arrival of *Magellan* at Venus. In 1991 the *Galileo* probe flew to within 1,600 km/994 mi of the asteroid Gaspra. In 1992 *Giotto* flew at a speed of 14 km/8.7 mi per second to within 200 km/128 mi of comet Grigg–Skjellerup, at a distance of 240 million km/ 150 million mi from Earth (13 light minutes away). On 25 September 1992 the *Mars Observer* was launched from Cape Canaveral, the first US mission to Mars for 17 years. On 28 August 1993 *Galileo* flew past the asteroid Ida. *Galileo* completed its six-year journey to Jupiter Dec 1995. It released a small probe that parachuted into the planet's atmosphere and sent back data for 57 minutes. The main part of the probe went into orbit round the planet and relayed signals from the atmospheric probe back to Earth.

New telescopes
The 1990s saw the installation of an array of outstanding telescopes, incorporating radical new technology. The Keck telescope on Mauna Kea, Hawaii, with a mirror consisting of 36 hexagonal sections, came into operation. The New Technology Telescope, La Silla, Chile, has a 3.5 m/11 ft, computer-controlled flexible mirror system which adjusts its shape 100 times per second to achieve better viewing. The UKIT (United Kingdom

Infrared Telescope), Mauna Kea, Hawaii, has a single mirror 374 cm/147 in across. Its performance is so good that it is used for observing with visible light as well as infrared.

Great comet crash
Astronomers watched in awe as icy fragments of comet Shoemaker–Levy 9 slammed into Jupiter July 1994, the first time a planetary collision has been observed. Brilliant fireballs erupted in Jupiter's atmosphere as the fragments burned up, leaving dark blotches in the planet's clouds that were visible through telescopes for months afterwards.

Planets discovered
In 1996 astronomers from the San Francisco State University claimed the discovery of two planets orbiting stars 35 light years away that could support life. The two planets were detected by sophisticated techniques that analysed 'wobbles' in the period of the parent stars, 70 Virginis (in the constellation Virgo) and 47 UMa (in Ursa Major). Using similar techniques, astronomers had already claimed two other planet discoveries in other constellations, neither of which could sustain life. One planet is orbiting a pulsar or neutron star; the other, closely orbiting the star 51 Pegasi in the constellation of Pegasus, was spotted 1995 by Swiss astronomers.

David Abbott

Biology

Ancient times
Like so much else, the systematic study of living things began with the Greeks. Earlier cultures such as those in Egypt and Babylon in the Near East, and the early Indian and Chinese civilizations in Asia, had their own approaches to the study of Nature and its products. But it was the Greeks who fostered the attitudes of mind and identified the basic biological problems and methods from which modern biology has grown. Enquiry begun by Greek philosophers such as Alcmaeon (born *c.* 535 BC) and Empedocles (*c.* 492–*c.* 432 BC) culminated in the biological work of Aristotle (384–323 BC) and the medical writings of Hippocrates (*c.* 460–377 BC) and his followers.

Natural philosophy
Aristotle was an original thinker of enormous power and energy who wrote on physics, cosmology, logic, ethics, politics, and many other branches of knowledge. He also wrote several biological works which laid the foundations for comparative anatomy, taxonomy (classification), and embryology. He was particularly fascinated by sea creatures; he dissected many of these as well as studying them in their natural habitats. Aristotle's approach to anatomy was functional: he believed that questions about structure and function always go together and that each biological part has its own special uses. Nature, he insisted, does nothing in vain. He therefore thought it legitimate to enquire about the ultimate purposes of things. This teleological approach has persisted in biological work until the 20th century. In addition, Aristotle studied reproduction and embryological development, and he established many criteria by which animals could be classified. He believed that animals could be placed on a vertical, hierarchical scale ('scale of being'), extending from humans down through quadrupeds, birds, snakes, and fishes to insects, molluscs, and sponges. Hierarchical thinking (as reflected in the terms 'higher' and 'lower' organisms) is still present in biology. One of Aristotle's pupils, Theophrastus (*c.* 372–287 BC), founded for botany many of the fundamentals that Aristotle had established for zoology.

What Aristotle achieved for biology, Hippocrates and his followers contributed for medicine: they established a naturalistic framework for thinking about health and disease.

Unlike earlier priests and doctors, they did not regard illness as the result of sin, or as a divine punishment for misdeeds. They were keen observers whose most influential explanatory framework saw disease as the result of an imbalance in one of the four physiologically active humours (blood, phlegm, black bile, and yellow bile). The humours were schematically related to the four elements (earth, air, fire, and water) of Greek natural philosophy. Each person was supposed to have his or her own dominant humour, although different humours tended to predominate at different times of life (such as youth and old age) or seasons of the year. The therapy of Hippocrates aimed at restoring the ideal balance through diet, drugs, exercise, change of life style, and so on.

Galen's influence on medicine

After the classical period of Greek thought, the most important biomedical thinker was Galen (*c.* 129–*c.* 199), who combined Hippocratic humoralism with Aristotle's tendency to think about the ultimate purposes of the parts of the body. Galen should not be blamed for the fact that later doctors thought that he had discovered everything, so that they had no need to look at biology and medicine for themselves. In fact, Galen was a shrewd anatomist and the most brilliant experimental physiologist of antiquity. For most of that time, human dissection was prohibited and Galen learnt his anatomy from other animals such as pigs, elephants, and apes. It took more than a thousand years before it was discovered that some of the structures which he had accurately described in animals (such as the five-lobed liver and the rete mirabile network of veins in the brain) were not present in the human body. Like that of Aristotle, Galen's anatomy was functional, and often his tendency to speculate went further than sound observation would have permitted, as when he postulated invisible pores in the septum of the heart which were supposed to allow some blood to seep from the right ventricle to the left.

The Renaissance

Following Galen's death at the end of the 2nd century and the collapse of the Roman Empire, biology and medicine remained stagnant for a thousand years. Most Classical writings were lost to the European West, to be preserved and extended in Constantinople and other parts of the Islamic Empire. From the 12th century these texts began to be rediscovered in southern Europe – particularly in Italy, where universities were also established. For a while scholars were content merely to translate and comment on the works of men such as Aristotle and Galen, but eventually an independent spirit of enquiry arose in European biology and medicine. Human dissections were routinely performed from the 14th century and anatomy emerged as a mature science from the fervent activity of Andreas Vesalius (1514–1564), whose *De humani corporis fabrica* (1543; 'On the Fabric of the Human Body') is one of the masterpieces of the Scientific Revolution. His achievement was to examine the body itself rather than relying simply on Galen; the illustrations in his work are simultaneously objects of scientific originality and of artistic beauty. The rediscovery of the beauty of the human body by Renaissance artists encouraged the study of anatomy by geniuses such as Leonardo da Vinci (1452–1519). Shortly afterwards, William Harvey (1578–1657) discovered the circulation of the blood and established physiology on a scientific footing. His little book *De Motu Cordis* (1628; 'On the Motion of the Heart') was the first great work on experimental physiology since the time of Galen. The eccentric wandering doctor Paracelsus (1493–1541) had also deliberately set aside the teachings of Galen and other Ancients in favour of a fresh approach to Nature and medicine and to the search for new remedies for disease.

The age of discovery

While these achievements were happening in medicine, anatomy and physiology, other areas of biology were not stagnating. Voyages of exploration alerted naturalists to the existence of many previously unknown plants and animals and encouraged them to establish sound principles of classification, to create order out of the apparently haphazard profusion of Nature. Zoological and botanical gardens began to be established so that the curious could view wonderful creatures like the rhinoceros and the giraffe. And just as the

Great World (Macrocosm) was revealing plants and animals unknown in Europe only a short time before, so the invention of the microscope in the 17th century gave scientists the opportunity of exploring the secrets of the Little World (Microcosm). The microscope permitted Anton van Leeuwenhoek (1632–1723) to see bacteria, protozoa, and other tiny organisms; it enabled Robert Hooke (1635–1703) to observe in a thin slice of cork regular structures which he called 'cells'. And it aided Marcello Malpighi (1628–1664) to complete the circle of Harvey's concept of the circulation of the blood by first seeing it flowing through capillaries, the tiny vessels that connect the arterial and venous systems. Many of these microscopical discoveries were communicated to the Royal Society of London, one of several scientific societies established during the mid-17th century.

The full potential of the microscope as a biological tool had to wait for technical improvements effected in the early 19th century. But it also led scientists along some blind alleys of theory. Observations of sperm 'swimming' in seminal fluid provided some presumed evidence for a theory that was much debated during the 18th century, concerning the nature of embryological development. Aristotle had thought that the body's organs (heart, liver, stomach, and so on) only gradually appear once conception has initiated the growth of the embryo. Later scientists, including William Harvey, extended Aristotle's theory with new observations. But now the visualization of moving sperm suggested that some miniature, but fully formed organism was already present in the reproductive fluids of the male or female. The tiny *homunculi* were thought to be stimulated to growth by fertilization. If the homunculus was always there, it followed that its own reproductive parts contained all of its future offspring, which in turn contained its future offspring and so on, back to Adam and Eve (depending on whether the male or the female was postulated as the carrier). This doctrine, called preformationalism, was held by most 18th-century biologists, including Albrecht von Haller (1708–1777) and Lazzaro Spallanzani (1729–1799), two of the century's greatest scientists. Both men, like virtually all scientists of the period, were devout Christians, and preformationalism did not conflict with their belief that God established regular, uniform laws which governed the development and functions of living things. They did not believe that inert matter could join together by accident to make a living organism. They rejected, for instance, the possibility of spontaneous generation, and Spallanzani devised some ingenious experiments designed to show that maggots found in rotting meat or the teeming life discoverable after jars of water are left to stand did not spontaneously generate. Haller, Spallanzani, John Hunter (1728–1793), and most other great 18th-century experimentalists held that the actions of living things could not be understood simply in terms of the laws of physics and chemistry. They were Vitalists, who believed that unique characteristics separated living from non-living matter. The special attributes of humans were often ascribed to the soul, and lower animals and plants were thought to possess more primitive animal and vegetable souls which gave them basic biological capacities such as reproduction, digestion, movement, and so on.

Taxonomy and classification

Experimental biology was well established in the 18th century; another great area of 18th-century activity was classification. Again inspired by Aristotle, and drawing on the work of previous biologists such as John Ray (1627–1705), the Swedish botanist Carolus Linnaeus (1707–1778) spent a lifetime trying to bring order to the ever increasing number of plants and animals uncovered by continued exploration of the Earth and its oceans. His *Systema Naturae* (1735; 'System of Nature') was the first of many books in which he elaborated a philosophy of taxonomy and established the convention of binomial nomenclature still followed today. In this convention, all organisms are identified by their genus and species; thus human beings in the Linnaean system are *Homo sapiens*. Depending on the nature of the characteristics examined, however, plants and animals could be placed in a variety of groups, ranging from the kingdom at the highest level through phyla, classes, orders, and families and so on beyond the species to the variety and, finally, the individual. Naturalists had traditionally accepted that the species was the most significant taxonomic category, Christian doctrine generally holding that God had specially created each individual species. It was also assumed that the number of species existing was fixed during the Creation, as described in the Book of Genesis – no new species having been created and none becoming extinct. Linnaeus, however, believed that God had created genera and that it was possible that new species had emerged during the time since the original Creation.

Palaeontology and evolution

Some 18th-century naturalists such as Georges Buffon (1707–1788) began to suggest that the Earth and its inhabitants were far older than the 6,000 or so years inferred from the Bible. General acceptance of a vastly increased age of the Earth, and of the reality of biological extinction, awaited the work of early 19th-century scientists such as Georges Cuvier (1769–1832), whose reconstructions of the fossil remains of large vertebrates like the mastodon and dinosaurs found in the Paris basin and elsewhere so stirred both the popular and scientific imaginations of his day. Despite Cuvier's work on the existence of life on Earth for perhaps millions of years, he firmly opposed the notion that these extinct creatures might be the ancestors of animals alive today. Rather he believed that the extent to which any species might change (variability) was fixed and that species themselves could not change much over time. His contemporary and scientific opponent, Jean Baptiste Lamarck (1744–1829), argued however that species do change over time. He insisted that species never become extinct; instead they are capable of change as new environmental conditions and new needs arise. According to this argument, the ancestors of the giraffe need not have had such a long neck, which instead might have slowly developed as earlier giraffe-like creatures stretched their necks to feed on higher leaves. Lamarck believed that physical characteristics and habits acquired after birth could – particularly if repeated from generation to generation – become inherited and thus inborn in the organism's offspring. We still call the doctrine of the inheritance of acquired characteristics 'Lamarckianism', although most naturalists before Lamarck had already believed it. It continued to be generally accepted (for instance, by Charles Darwin) until late in the 19th century.

The debates between Cuvier and Lamarck were part of the new possibilities opened up by the revolution in thinking about the age of the Earth and of life on it. During the same fertile period, the systematic use of improved microscopes revolutionized the way in which biologists conceived organisms. In the closing years of the 18th century a French pathologist named Xavier Bichat (1771–1802), aided only with a hand lens, developed the idea that organs such as the heart and liver are not the ultimate functional units of animals. He postulated that the body can be divided into different kinds of tissues (such as nervous, fibrous, serous, and muscular tissue) which make up the organs. Increasingly, biologists and doctors began thinking in terms of smaller functional units, and microscopists such as Robert Brown (1773–1858) began noticing regular structures within these units, which we now recognize as cells. Brown called attention to the nucleus in the cells of plants in 1831 and by the end of the decade two German scientists, Matthias Schleiden (1804–1881) and Theodor Schwann (1810–1882) systematically developed the idea that all plants and animals are composed of cells. The cell theory was quickly established for adult organisms, but in certain situations – such as the earliest stages of embryological development or in the appearance of 'pus' cells in tissues after inflammation or injury – it appeared that new cells were actually crystallized out of an amorphous fluid which Schwann called the 'blastema'. The notion of continuity of cells was enlarged upon by the pathologist Rudolf Virchow (1821–1902), who summarized it in his famous slogan 'All cells from cells'. The cell theory gave biologists and physicians a new insight into the architecture and functions of the body in health and disease.

Microorganisms and disease

Concern with one-celled organisms also lay behind the work of Louis Pasteur (1822–1895), which helped to establish the germ theory of disease. Pasteur trained as a chemist, but his researches into everyday processes such as the souring of milk and the fermentation of beer and wine opened for him a new understanding of the importance of yeast, bacteria, and other microorganisms in our daily lives. It was Pasteur who finally convinced scientists that animals do not spontaneously generate on rotting meat or in infusions of straw; our skin, the air, and everything we come into contact with can be a source of these tiny creatures. After reading about Pasteur's work Joseph Lister (1827–1912) first conceived the idea that by keeping away these germs (as they were eventually called) from the wounds made during surgical operations, healing would be much faster and post-operative infection would be much less common. When in 1867 Lister published the first results using his new technique, antiseptic surgery was born. He spent much time developing the methods, which were taken up by other surgeons who soon realized that it was better to prevent infection altogether (asepsis) by carefully sterilizing their hands, instruments, and dressings. By the time Lister died, he was world famous and surgeons were performing operations that would have been impossible without his work.

After Lister drew attention to the importance of Pasteur's discoveries for medicine and surgery, Pasteur himself showed the way in which germs cause not only wound infections, but also many diseases. He first studied a disease of silkworms which was threatening the French silk industry; he then turned his attention to other diseases of farm animals and human beings. In the course of this research, he discovered that under certain conditions an organism could be grown which, instead of causing a disease, actually prevented it. He publicly demonstrated these discoveries for anthrax, then a common disease of sheep, goats and cattle which sometimes also affected human beings. He proposed to call this process of protection vaccination, in honour of Edward Jenner (1749–1823), who in 1796 had shown how inoculating a person with cowpox (vaccinia) can protect against the deadly smallpox. Pasteur's most dramatic success came with a vaccine against rabies, a much-dreaded disease occasionally contracted after a bite from an infected animal.

By the 1870s, other scientists were investigating the role played by germs in causing disease. Perhaps the most important of them was Robert Koch (1843–1910), who devised many key techniques for growing and studying bacteria, and who showed that tuberculosis and cholera – prevalent diseases of the time – were caused by bacteria. Immunology, the study of the body's natural defence mechanisms against invasion by foreign cells, was pioneered by another German, Paul Ehrlich (1854–1915), who also began looking for drugs that would kill disease-causing organisms without being too dangerous for the patient. His first success, a drug named Salvarsan, was effective in the treatment of syphilis. Ehrlich's hopes in this area were not fully realized immediately, and it was not until the 1930s that the synthetic sulfa drugs, also effective against some bacterial diseases, were developed by Gerhard Domagk (1895–1964) and others. Slightly earlier Alexander Fleming (1881–1955) had noticed that a mould called *Penicillium* inhibited the growth of bacteria on cultures. Fleming's observation was investigated during World War II by Howard Florey (1898–1968) and Ernst Chain (1906–1979), and since then many other antibiotics have been discovered or synthesized. But antibiotics are not effective against diseases caused by viruses; such infections can, however, often be prevented using vaccines. An example is poliomyelitis, vaccines against which were developed in the 1950s by Albert Sabin (1906–) and Jonas Salk (1914–).

Evolution and genetics

Many of these advances in modern medical science are a direct continuation of discoveries made in the 19th century, although of course we now know much more about bacteria and other pathogenic microorganisms than did Pasteur and Koch. Another 19th-century discovery which is still being intensively investigated is evolution. Charles Darwin (1809–1882) was not the first to suggest that biological species can change over time, but his book *The Origin of Species* (1859) first presented the idea in a scientifically plausible form. As a young man, Darwin spent five years (1831–36) on HMS *Beagle*, during which he studied fossils, animals, and geology in many parts of the world, particularly in South America. By 1837 he had come to believe the fact of evolution; in 1838 he hit upon its mechanism: natural selection. This principle makes use of the fact that organisms produce more offspring than can survive to maturity. In this struggle for existence, those offspring with characteristics best suited to their particular environment will tend to survive. In this way, Nature can work on the normal variation which plants and animals show and, under changing environmental conditions, significant change can occur through selective survival.

Darwin knew that his ideas would be controversial so he initially imparted them to only a few close friends, such as the geologist Charles Lyell (1797–1875) and the botanist Joseph Hooker (1817–1911). For 20 years he continued quietly to collect evidence favouring the notion of evolution by natural selection, until in 1858 he was surprised to receive a short essay from Alfred Russel Wallace (1823–1913), then in Malaya, perfectly describing natural selection. Friends arranged a joint Darwin–Wallace publication, and then Darwin abandoned a larger book he was writing on the subject to prepare instead *The Origin of Species.* In it he marshalled evidence from many sources, including palaeontology, embryology, geographical distribution, ecology (a word coined only later), and hereditary variation. Darwin did not have a very clear idea of how variations occur, but his work convinced a number of scientists, including Thomas Huxley (1825–1895), Ernst Haeckel (1834–1919), and Francis Galton (1822–1911), Darwin's cousin. Huxley became Darwin's chief publicist in Britain, Haeckel championed Darwin's ideas in Germany, and Galton quietly absorbed the evolutionary perspective into his own work in psychology, physical anthropology and the use of statistics and other forms of mathematics in the life sciences.

Meanwhile, unknown to Darwin (and largely unrecognized during his lifetime), a monk named Gregor Mendel (1822–1884) was elucidating the laws of modern genetics through his studies of inheritance patterns in pea plants and other common organisms. Mendel studied characteristics that were inherited as a unit; this enabled scientists to understand such phenomena as dominance and recessiveness in these units, called 'genes' in 1909 by the Danish biologist W L Johannsen (1859–1927). By 1900, when Hugo de Vries (1848–1935), William Bateson (1861–1926), and others were recognizing the importance of Mendel's pioneering work, much more was known about the microscopic appearances of cells both during adult division (mitosis) and reduction division (meiosis). In addition, August Weismann (1834–1914) had developed notions of the continuity of the inherited material (which he called the 'germ plasm') from generation to generation, thus suggesting that acquired characteristics are not inherited. Only a few scientists in the 20th century, such as the Soviet botanist Trofim Lysenko (1898–1976), have continued to believe in Lamarckianism, for modern genetics has accumulated overwhelming evidence that characteristics such as the loss of an arm or internal muscular development do not change the make-up of reproductive cells.

It is now believed that new inheritable variations occur when genes mutate. The study of this process and of the factors (such as X-rays and certain chemicals) that can make the occurrence of mutations likely was pioneered by geneticists such as Thomas Hunt Morgan (1866–1945) and Herman Muller (1890–1967). They did much of their work with fruit flies (*Drosophila*). While these and other scientists were showing that genes are located on chromosomes – strands of darkstaining material in the nuclei of cells – other researchers were trying to determine the exact nature of the hereditary substance itself. Originally it was thought to be a protein, but in 1953 Francis Crick (1916–) and James Watson (1928–), were able to show that it is dioxyribonucleic acid (DNA). Their work was an early triumph of molecular biology, a branch of the science that has grown enormously since the 1950s. Scientists now know a great deal about how DNA works. Among those who have contributed are George Beadle (1903–), Edward Tatum (1909–1975), Jacques Monod (1910–1976), Joshua Lederberg (1925–) and Maurice Wilkins (1916–). Molecular biologists and chemists have also been concerned with determining the structures of many other large biological molecules, such as the muscle protein myoglobin by John Kendrew (1917–) and Max Perutz (1914–). In their researches they often have to interpret the diffraction patterns produced when X-rays pass through these complex molecules, a technique pioneered by Dorothy Hodgkin (1910–1994).

Modern biological sciences

Molecular biology is only one of several new biological disciplines to be developed during the past century. The oldest of these, biochemistry, was established in Britain by Frederick Gowland Hopkins (1861–1947) who, along with Casimir Funk (1884–1967) and Elmer McCollum (1879–1967), is remembered for his fundamental work in the discovery of vitamins, substances that help to regulate many complex bodily processes. Other biochemists such as Carl Cori (1896–1984) and Gerty Cori (1896–1957) have studied the ways in which organisms make use of the energy gained when food is broken down. Many of these internal processes are also moderated by the action of hormones, one important example of which is insulin, discovered in the 1920s by Frederick Banting (1891–1941) and others.

Modern biologists also often use physics in their work, and biophysics is now an important discipline in its own right. Archibald Hill (1886–1977) and Otto Meyerhof (1884–1951) pioneered in this area with their work on the release of heat when muscles contract. More recently, Bernhard Katz (1911–) has used biophysical techniques in studying the events at the junctions between muscles and nerves, and at the junctions between pairs of nerves (synapses). The events at synapses are initiated by the release of chemical substances such as adrenaline and acetylcholine, as was demonstrated by Henry Dale (1875–1968) and Otto Loewi (1873–1961). The way in which nerve impulses move along the nerve axon has been investigated by Alan Hodgkin (1914–) and Andrew Huxley (1917–). For this work, they made use of the giant axon of the squid, an experimental preparation whose importance for biology was first shown by John Young (1907–). The complicated way in which the nervous system operates as a whole was first rigorously investigated by Charles Sherrington (1857–1952).

Another area of fundamental importance in modern biology and medicine is immunology. For instance, the discovery by Karl Landsteiner (1868–1943) of the major human blood group system (A, B, and O) permitted safe blood transfusions. The development of the immune system – and the way in which the body recognizes foreign substances ('self' and 'not-self') has been investigated by such scientists as Peter Medawar (1915–1987) and Frank Macfarlane Burnet (1899–1985). Much of this knowledge has been important to transplant surgery, pioneered for kidneys by Roy Calne (1930–) and for hearts by Christiaan Barnard (1922–).

Genetic engineering

Genetic engineering consists of a collection of methods used to manipulate genes. The first of these techniques dates back to 1952 when Joshua Lederberg found that bacteria exchange genetic material contained in a body he called a plasmid. The next year William Hayes established that plasmids were rings of DNA free from the main DNA in the chromosome of the bacteria. The next step was taken by Werner Arber (1929–), who studied viruses (called bacteriophages) which infect bacteria. He found that bacteria resist phages by splitting the phage DNA using enzymes. By 1968 Arber had discovered the enzymes produced by bacteria that split DNA at specific locations. In addition, he found that different genes that have been split at the same location by one of the restriction enzymes, as they are called, will recombine when placed together in the absence of the enzyme. The resulting product is called recombinant DNA.

In 1973 Stanley H Cohen (1922–) and Herbert W Brown combined restriction enzymes with plasmids in the first genetic engineering experiment. They cut a chunk out of a plasmid found in the bacterium *Eschericia coli* and inserted in the gap a gene created from a different bacterium. In 1976 Har Gobind Khorana (1922–) and co-workers constructed the first artificial gene to function naturally when inserted into a bacterial cell.

In 1995 a team at the Institute for Genomic Research in Gaitherburg, Maryland, USA, unveiled the first complete genetic blueprint for a free-living organism – a bacterium *Haemophilus influenza*. In theory, the blueprint, consisting of 1.8 million genetic instructions, would allow scientists to construct the bacterium from scratch. The achievement demonstrates the speed of the DNA decoding techniques used by genetic researchers.

Genetics and medicine
The application of genetic engineering to medicine has progressed rapidly. The Human Genome Organization was established 1988 in Washington, DC, with the aim of mapping the complete sequence of human DNA. In 1985 the first human cancer gene was isolated by US researchers. In 1993 the gene for Huntington's disease was discovered.

The first person to undergo gene therapy (in Sept 1990) was a four-year-old girl suffering from a rare enzyme (ADA) deficiency that cripples the immune system. A healthy ADA gene was inserted into a virus that had been rendered harmless. The virus was inserted into a blood-forming cell taken from the child's bone marrow. This cell reproduced, creating millions of cells containing the missing gene. Finally these cells were infused into the child's bloodstream, to be carried to the bone marrow where they produced healthy blood cells complete with the ADA gene.

New phylum discovered
In 1995 a new phylum (a group of creatures that share a distinct body plan) was discovered. The phylum, Cycliophora, is only the 36th ever described for all the 1.5 million or so named organisms. Most other phyla were described in the 1800s. A single species has so far been assigned to the phylum. *Symbion pandora* is a tiny creature found clinging to the mouthparts of Norwegian lobsters by zoologists at the University of Copenhagen.

David Abbott

Chemistry

The origins of chemistry
Chemistry seems to have originated in Egypt and Mesopotamia several thousand years before Christ. Certainly by about 3000 BC the Egyptians had produced the copper-tin alloy known as bronze, by heating the ores of copper and tin together, and this new material was soon common enough to be made into tools, ornaments, armour, and weapons. The Ancient Egyptians were also skilled at extracting juices and infusions from plants, and pigments from minerals, which they used in the embalming and preserving of their dead. By 600 BC the Greeks were also becoming a settled and prosperous people with leisure time in which to think. They began to turn their attention to the nature of the universe and to the structure of its materials. They were thus the first to study the subject we now call chemical theory. Aristotle (384–322 BC) proposed that there were four elements – earth, air, fire and water – and that everything was a combination of these four. They were thought to possess the following properties: earth was cold and dry, air was hot and moist, fire was hot and dry, and water was cold and moist. The idea of the four elements persisted for 2,000 years. The Greeks also worked out, at least hypothetically, that matter ultimately consisted of small indivisible particles, *atomos* – the origin of our word 'atom'.

From the Egyptians and the Greeks comes *khemeia*, alchemy and eventually chemistry as we know it today. The source of the word *khemeia* is debatable, but it is certainly the origin of the word chemistry. It may derive from the Egyptians' word for their country *Khem*, 'the black land'. It may come from the Greek word *khumos* (the juice of a plant), so that *khemeia* is 'the art of extracting juices'; or from the Greek *cheo* 'I pour or cast', which refers to the activities of the metal workers. Whatever its origin, the art of *khemeia* soon became akin to magic and was feared by the ordinary people. One of the greatest aims of the subject involved the attempts to transform base metals such as lead and copper into silver or gold. From the four-element theory, it seemed that it should be possible to perform any such change, if only the proper technique could be found.

The Arabs and alchemy

With the decline of the Greek empire *khemeia* was not pursued and little new was added to the subject until it was embraced by the increasingly powerful Arabs in the 7th century AD. Then for five centuries *al-kimiya*, or alchemy, was in their hands. The Arabs drew many ideas from the *khemeia* of the Greeks, but they were also in contact with the Chinese – for example, the idea that gold possessed healing powers came from China. They believed that 'medicine' had to be added to base metals to produce gold, and it was this medicine that was to become the philosopher's stone of the later European alchemists. The idea that not only could the philosopher's stone heal 'sick' or base metals, but that it could also act as the elixir of life, was also originally Chinese. The Arab alchemists discovered new classes of chemicals such as the caustic alkalis (from the Arabic *al-qalíy*) and they improved technical procedures such as distillation.

Western Europe had its first contact with the Islamic world as a result of the Crusades. Gradually the works of the Arabs – handed down from the Greeks – were translated into Latin and made available to European scholars in the 12th and 13th centuries. Many people spent their lives trying in vain to change base metals into gold; and many alchemists lost their heads for failing to supply the promised gold.

The beginning of modern chemistry

A new era in chemistry began with the researches of Robert Boyle (1627–1691), who carried out many experiments on air. These experiments were the beginning of a long struggle to find out what air had to do with burning and breathing. From Boyle's time onwards, alchemy became chemistry and it was realized that there was more to the subject than the search for the philosopher's stone.

Chemistry as an experimental science

During the 1700s the phlogiston theory gained popularity. It went back to the alchemists' idea that combustible bodies lost something when they burned. Metals were thought to be composed of a calx (different for each) combined with phlogiston, which was the same in all metals. When a candle burned in air, phlogiston was given off. It was believed that combustible objects were rich in phlogiston and what was left after combustion possessed no phlogiston and would therefore not burn. Thus wood possessed phlogiston but ash did not; when metals rusted, it was considered that the metals contained phlogiston but that its rust or calx did not. By 1780 this theory was almost universally accepted by chemists. Joseph Priestley (1733–1804) was a supporter of the theory and in 1774 he had succeeded in obtaining from mercuric oxide a new gas which was five or six times purer than ordinary air. It was, of course, oxygen but Priestley called it 'dephlogisticated air' because a smouldering splint of wood thrust into an atmosphere of this new gas burst into flames much more readily than it did in an ordinary atmosphere. He took this to mean that the gas must be without the usual content of phlogiston, and was therefore eager to accept a new supply.

It was Antoine Lavoisier (1743–1794) who put an end to the phlogiston theory by working out what was really happening in combustion. He repeated Priestley's experiments in 1775 and named the dephlogisticated air oxygen. He realized that air was not a single substance but a mixture of gases, made up of two different gases in the proportion of 1 to 4. He deduced that one-fifth of the air was Priestley's dephlogisticated air (oxygen), and that it was this part only that combined with rusting or burning materials and was essential to life. Oxygen means 'acid-producer' and Lavoisier thought, erroneously, that oxygen was an essential part of all acids. He was a careful experimenter and user of the balance, and from his time onwards experimental chemistry was concerned only with materials that could be weighed or otherwise measured. All the 'mystery' disappeared and Lavoisier went on to work out a logical system of chemical nomenclature, much of which has survived to the present day.

The 19th century

Early in the 19th century many well-known chemists were active. Claude Berthollet (1748–1822) worked on chemical change and composition, and Joseph Gay-Lussac

(1778–1850) studied the volumes of gases that take part in chemical reactions. Others included Berzelius, Cannizzaro, Avogadro, Davy, Dumas, Kolbe, Wöhler, and Kekulé. The era of modern chemistry was beginning.

Atomic theory and new elements

An English chemist, John Dalton (1766–1844), founded the atomic theory in 1803 and in so doing finally crushed the belief that the transmutation to gold was possible. He realized that the same two elements can combine with each other in more than one set of proportions, and that the variation in combining proportions gives rise to different compounds with different properties. For example, he determined that one part (by weight) of hydrogen combined with eight parts of oxygen to form water, and if it was assumed (incorrectly) that a molecule of water consisted of one atom of hydrogen and one atom of oxygen, then it was possible to set the mass of the hydrogen atom arbitrarily at 1 and call the mass of oxygen 8 (on the same scale). In this way Dalton set up the first table of atomic weights (now called relative atomic masses), and although this was probably his most important achievement, it contained many incorrect assumptions. These errors and anomalies were researched by Jöns Berzelius (1779–1848), who found that for many elements the atomic weights were not simple multiples of that of hydrogen. For many years, oxygen was made the standard and set at 16.000 until the mid-twentieth century, when carbon (= 12.000) was adopted. Berzelius suggested representing each element by a symbol consisting of the first one or two letters of the name of the element (sometimes in Latin) and these became the chemical symbols of the elements as still used today.

At about the same time, in 1808, Humphry Davy (1778–1829) was using an electric current to obtain from their oxides elements that had proved to be unisolatable by chemical means: potassium, sodium, magnesium, barium, and calcium. His assistant, Michael Faraday (1791–1867), was to become even better known in connection with this technique, electrolysis. By 1830, more than 50 elements had been isolated; chemistry had moved a long way from the four elements of the ancient Greeks, but their properties seemed to be random. In 1829 the German chemist Johann Döbereiner (1780–1849) thought that he had observed some slight degree of order. He wondered if it was just coincidence that the properties of the element bromine seemed to lie between those of chlorine and iodine, but he went on to notice a similar gradation of properties in the triplets calcium, strontium, and barium and with sulphur, selenium, and tellurium. In all of these examples, the atomic weight of the element in the middle of the set was about half-way between the atomic weights of the other two elements. He called these groups 'triads', but because he was unable to find any other such groups, most chemists remained unimpressed by his discovery. Then in 1864 John Newlands (1837–1898) arranged the elements in order of their increasing atomic weights and found that if he wrote them in horizontal rows, and started a new row with every eighth element, similar elements tended to fall in the same vertical columns. Döbereiner's three sets of triads were among them. Newlands called this his 'Law of Octaves' by analogy with the repeating octaves in music. Unfortunately there were many places in his chart where obviously dissimilar elements fell together and so it was generally felt that Newland's similarities were not significant but probably only coincidental. He did not have his work published.

In 1862 a German chemist, Julius Lothar Meyer (1830–1895), looked at the volumes of certain fixed weights of elements, and talked of atomic volumes. He plotted the values of these for each element against its atomic weight, and found that there were sharp peaks in the graph at the alkali metals – sodium, potassium, rubidium, and caesium. Each part of the graph between the peaks corresponded to a 'period' or horizontal row in the table of the elements, and it became obvious where Newlands had gone wrong. He had assumed that each period contained only seven elements; in fact the later periods had to be longer than the earlier ones. By the time Meyer published his findings, he had been anticipated by the Russian chemist Dmitri Mendeleyev (1834–1907), who in 1869 published his version of the periodic table, which was more or less as we have it today. He had the insight to leave gaps in his table for three elements which he postulated had not yet been discovered, and was even able to predict what their properties would be. Chemists were sceptical, but within 15 years all three of the 'missing' elements had been discovered and their properties were found to agree with Mendeleyev's predictions.

The beginnings of physical chemistry

Until the beginning of the 19th century, the areas covered by the subjects of chemistry and physics seemed well-defined and quite distinct. Chemistry studied changes where the molecular bonding structure of a substance was altered, and physics studied phenomena in which no such change occurred. Then in 1840 physics and chemistry merged in the work of Germain Hess (1802–1850). It had been realized that heat – a physical phenomenon – was produced by chemical reactions such as the burning of wood, coal and oil, and it was gradually becoming clear that all chemical reactions involved some sort of heat transfer. Hess showed that the quantity of heat produced or absorbed when one substance was changed into another was the same no matter by which chemical route the change occurred, and it seemed likely that the law of conservation of energy was equally applicable to chemistry and physics. Thermochemistry had been founded and work was able to begin on thermodynamics. Most of this research was done in Germany and it was Wilhelm Ostwald (1853–1932), towards the end of the 19th century, who was responsible for physical chemistry developing into a discipline in its own right. He worked on chemical kinetics and catalysis in particular, but was the last important scientist to refuse to accept that atoms were real – there was at that time still no direct evidence to prove that they existed. Other contemporary chemists working in the new field of physical chemistry included Jacobus van't Hoff (1852–1911) and Svante Arrhenius (1859–1927). Van't Hoff studied solutions and showed that molecules of dissolved substances behaved according to rules analogous to those that describe the behaviour of gases. Arrhenius carried on the work which had been begun by Davy and Faraday on solutions that could carry an electric current. Faraday had called the current-carrying particles 'ions', but nobody had worked out what they were. Arrhenius suggested that they were atoms or groups of atoms which bore either a positive or a negative electric charge. His theory of ionic dissociation was used to explain many of the phenomena in electrochemistry.

Gases

Towards the end of the 19th century, mainly as a result of the increasing interest in the physical side of chemistry, gases came under fresh scrutiny and some errors were found in the law that had been proposed three centuries earlier by Robert Boyle. Henri Regnault (1810–1878), James Clerk Maxwell (1831–1879) and Ludwig Boltzmann (1844–1906) had all worked on the behaviour of gases, and the kinetic theory of gases had been derived. Taking all their findings into account, Johannes van der Waals (1837–1923) arrived at an equation that related pressure, volume and temperature of gases and made due allowance for the sizes of the different gas molecules and the attractions between them. By the end of the century William Ramsay (1852–1916) had begun to discover a special group of gases – the inert or rare gases – which have a valency (oxidation state) of zero and which fit neatly into the periodic table between the halogens and the alkali metals.

Organic chemistry becomes a separate discipline

Meanwhile the separate branches of chemistry were emerging and organic substances were being distinguished from inorganic ones. In 1807 Berzelius had proposed that substances such as olive oil and sugar, which were products of living organisms, should be called organic, whereas sulphuric acid and salt should be termed inorganic. Chemists at that time had realized that organic substances were easily converted into inorganic substances by heating or in other ways, but it was thought to be impossible to reverse the process and convert inorganic substances into organic ones. They believed in Vitalism – that somehow life did not obey the same laws as did inanimate objects and that some special influence, a 'vital force', was needed to convert inorganic substances into organic ones. Then in 1828 Friedrich Wöhler (1800–1882) succeeded in converting ammonium cyanate (an inorganic compound) into urea. In 1845 Adolf Kolbe (1818–1884) synthesized acetic acid, squashing the Vitalism theory for ever. By the middle of the 19th century organic compounds were being synthesized in profusion; a new definition of organic compounds was clearly needed, and most organic chemists were working by trial and error. Nevertheless there was a teenage assistant of August von Hofmann (1818–1892), called William Perkin (1838–1907), who was able to retire at the age of only 35 because of a brilliant chance discovery. In 1856 he treated aniline with potassium chromate, added alcohol, and obtained a beautiful purple colour, which he suspected might be a dye (later called aniline purple or mauve). He left school and founded what became the synthetic dyestuffs industry.

Then in 1861 the German chemist Friedrich Kekulé (1829–1886) defined organic chemistry as the chemistry of carbon compounds and this definition has remained, although there are a few carbon compounds (such as carbonates) which are considered to be part of inorganic chemistry. Kekulé suggested that carbon had a valency of four, and proceeded to work out the structures of simple organic compounds on this basis. These representations of the structural formulae showed how organic molecules were generally larger and more complex than inorganic molecules. There was still the problem of the structure of the simple hydrocarbon benzene, C_6H_6, until 1865 when Kekulé suggested that rings of carbon atoms might be just as possible as straight chains. The idea that molecules might be three-dimensional came in 1874 when van't Hoff suggested that the four bonds of the carbon atom were arranged tetrahedrally. If these four bonds are connected to four different types of groups, the carbon atom is said to be asymmetric and the compound shows optical activity – its crystals or solutions rotate the plane of polarized light. Viktor Meyer (1848–1897) proposed that certain types of optical isomerism could be explained by bonds of nitrogen atoms. Alfred Werner (1866–1919) went on to demonstrate that this principle also applied to metals such as cobalt, chromium and rhodium, and succeeded in working out the necessary theory of molecular structure, known as coordination theory. This new approach allowed there to be structural relationships within certain fairly complex inorganic molecules, which were not restricted to bonds involving ordinary valencies. It was to be another 50 years before enough was known about valency for both Kekulé's theory and Werner's to be fully understood, but by 1900 the idea was universally accepted that molecular structure could be represented satisfactorily in three dimensions.

Modern synthetic organic chemistry

Kekulé's work gave the organic chemist scope to alter a structural formula stage by stage, to convert one molecule into another, and modern synthetic organic chemistry began. Richard Willstätter (1872–1942) was able to work out the structure of chlorophyll and Heinrich Wieland (1877–1957) determined the structures of steroids. Paul Karrer (1889–1971) elucidated the structures of the carotenoids and other vitamins and Robert Robinson (1886–1975) tackled the alkaloids – he worked out the structures of morphine and strychnine. The alkaloids have found medical use as drugs, as have many other organic compounds. The treatment of disease by the use of specific chemicals is known as chemotherapy and was founded by the bacteriologist Paul Ehrlich (1854–1915). The need for drugs to combat disease and infection during World War II spurred on research, and by 1945 the antibiotic penicillin, first isolated by Howard Florey (1898–1968) and Ernst Chain (1906–1979), was being produced in quantity. Other antibiotics such as streptomycin and the tetracyclines soon followed.

Some organic molecules contain thousands of atoms; some, such as rubber, are polymers and others, such as haemoglobin, are proteins. Synthetic polymers have been made which closely resemble natural rubber; the leader in this field was Wallace Carothers (1896–1937), who also invented nylon. Karl Ziegler (1898–1973) and Giulio Natta (1903–1979) worked out how to prevent branching during polymerization, so that plastics, films and fibres can now be made more or less to order. Work on the make-up of proteins had to wait for the development of chemical techniques such as chromatography (by

Mikhail Tswett (1872–1919) and by Archer Martin (1910–) and Richard Synge (1914–1994) and electrophoresis (Arne Tiselius (1902–1971). In the forefront of molecular biological research are Frederick Sanger (1918–), John Kendrew (1917–) and Max Perutz (1914–). One technique that has been essential for their work is X-ray diffraction, and for the background to this development we have to return to the area of research between chemistry and physics at the beginning of the present century.

Modern atomic theory

Ever since Faraday had proposed his laws of electrolysis, it had seemed likely that electricity might be carried by particles. The physicist Ernest Rutherford (1871–1937) decided that the unit of positive charge was a particle quite different from the electron, which was the unit of negative charge, and in 1920 he suggested that this fundamental positive particle be called the proton. In 1895 Wilhelm Röntgen (1845–1923) discovered X-rays, but other radiation components – alpha and beta rays – were found to be made up of protons and electrons. In about 1902 it was proved, contrary to all previous ideas, that radioactive elements changed into other elements, and by 1912 the complicated series of changes of these elements had been worked out. In the course of this research, Frederick Soddy (1877–1956) realized that there could be several atoms differing in mass but having the same properties. They were called isotopes and we now know that they differ in the number of neutrons which they possess, although the neutron was not to be discovered until 1932, by the physicist James Chadwick (1891–1974).

Rutherford evolved the theory of the nuclear atom, which suggested that sub-atomic particles made up the atom, which had until that time been considered to be indivisible. The question now was, how did the nuclear atom of one element differ from that of another? In 1909 Max von Laue (1879–1960) began a series of brilliant experiments. He established that crystals consist of atoms arranged in a geometric structure of regularly repeating layers, and that these layers scatter X-rays in a set pattern. In so doing, he had set the scene for X-ray crystallography to be used to help to work out the structures of large molecules for which chemists had not been able to determine formulae.

In 1913 the young scientist Henry Moseley (1887–1915) found that there were characteristic X-rays for each element and that there was an inverse relationship between the wavelength of the X-ray and the atomic weight of the element. This relationship depended on the size of the positive charge on the nucleus of the atom, and the size of this nuclear charge is called the atomic number. Mendeleyev had arranged his periodic table, by considering the valencies of the elements, in sequence of their atomic weights, but the proper periodic classification is by atomic numbers. It was now possible to predict exactly how many elements were still to be discovered. Since the proton is the only positively charged particle in the nucleus, the atomic number is equal to the number of protons; the neutrons contribute to the mass but not to the charge. For example, a sodium atom, with an atomic number of 11 and an atomic weight (relative atomic mass) of 23, has 11 protons and 12 neutrons in its nucleus.

Isotopes and biochemistry

The new electronic atom was also of great interest to organic chemists. It enabled theoreticians such as Christopher Ingold (1893–1970) to try to interpret organic reactions in terms of the movements of electrons from one point to another within a molecule. Physical chemical methods were being used in organic chemistry, founding physical organic chemistry as a separate discipline. Linus Pauling (1911–1994), a chemist who was to suggest in the 1950s that proteins and nucleic acids possessed a helical shape, worked on the wave properties of electrons, and established the theory of resonance. This idea was very useful in establishing that the structure of the benzene molecule possessed 'smeared out' electrons and was a resonance hybrid of the two alternating double bond/single bond structures. The concept of atomic number was clarified by Francis Aston (1877–1945) with the mass spectrograph. This instrument used electric and magnetic fields to deflect ions of identical charge by an extent that depended on their mass – the greater the mass of the ion, the less it was deflected. He found for instance that there were two kinds of neon atoms, one of mass 20 and one of mass 22. The neon-20 was ten times as common as the neon-22, and so it seemed reasonable that the atomic weight of the element was 20.2 – a weighted average of the individual atoms and not necessarily a whole number. In some cases, the weighted average (atomic weight) of a particular atom may be larger than that for an atom of higher atomic number. This explains the relative positions of iodine and tellurium in the periodic table, which Mendeleyev had placed correctly without knowing why.

In 1931 Harold Urey (1893–1981) discovered that hydrogen was made up of a pair of isotopes, and he named hydrogen-2 deuterium. In 1934 it occurred to the physicist Enrico Fermi (1901–1954) to bombard uranium (element number 92, the highest atomic number known at that time) in order to see whether he could produce any elements of higher atomic numbers. This approach was pursued by Glenn Seaborg (1912–) and the transuranium elements were discovered, going up from element 94 but becoming increasingly difficult to form and decomposing again more rapidly with increasing atomic number.

In Nov 1994, researchers working at the GSI heavy-ion cyclotron at Darmstadt, Germany produced element 110. The element, atomic mass 269, was produced when atoms of lead were bombarded with atoms of nickel. As is usual for super heavy atoms, the new element has a very short half-life; it decayed in less than a millisecond. A second element was discovered Dec 1994. Three atoms of element 111, atomic mass 272, were detected when bismuth-209 was bombarded with nickel atoms. It decayed into two previously unknown isotopes of elements 109 and 107 after about a millisecond. In Feb 1996 element 112 was discovered by the same team.

The boundaries between chemistry and other sciences

The area between physics and chemistry has been replaced by a common ground where atoms and molecules are studied together with the forces that influence them. A good example is the discovery in the early 1990s of a new form of carbon, with molecules called buckyballs, consisting of 60 carbon atoms arranged in 12 pentagons and 20 hexagons to form a perfect sphere.

The boundary between chemistry and biology has also become less well defined and is now a scene of intense activity, with the techniques of chemistry being applied successfully to biological problems. Electron diffraction, chromatography and radioactive tracers have all been used to help discover what living matter is composed of, although it is possible that these investigations in biology are only now at the stage that atomic physics was at the beginning of this century. It was Lavoisier who said that life is a chemical function, and perhaps the most important advance of all is towards understanding the chemistry of the cell. Biochemical successes of recent years include the synthesis of human hormones, the development of genetic fingerprinting, and the use of enzymes in synthesis. The entire field of genetic engineering is essentially biochemistry.

Retrosynthesis

In the 1960s, Elias J Corey (1928–) made a breakthrough in organic chemistry when he developed retrosynthesis, a powerful tool for building complex molecules from smaller, cheaper and more readily available ones. Retrosynthesis can be used to picture a molecule like a jigsaw, working backwards to find reactive components to complete the puzzle. Modern chemists use retrosynthesis to design everything from insect repellents to better drugs.

David Abbott

Engineering and Technology

Ancient times

Our existence today is powerful evidence of our ability to invent. Were it not for the invention of simple tools made from sharp-edged stones some two million years ago, it is doubtful whether our relatively weak and slow ancestors would have survived for long. Such tools enabled early humans to fight off predators and to hunt for food.

As well as being surrounded by potentially hostile animals, early humans were at the mercy of the climate. It was the second of the major inventions of prehistory, a means of creating fire, that enabled them to survive the Ice Ages. *Homo erectus* was using this to live through the second Ice Age some 400,000 years ago.

These two inventions served our ancestors well for an extremely long time. Not until the foundation of Jericho, the world's first walled town, *c.* 7000 BC does another major invention reveal itself, in the development of pottery. Copper, and the alloy made from copper and tin called bronze, also appeared at around that time. Behind the foundation of Jericho, were the beginnings of organized agriculture, and it was this that provided the stimulus for the development of increasingly sophisticated tools, such as the plough and the sickle.

The wheel

That most famous and significant of inventions of the ancient world, the wheel, first appeared around 3000 BC, in what is now southern Russia. These early wheels were solid, wooden and fixed to sleds which had previously been dragged across the ground. Although its first use was in transporting heavy loads, the wheel and axle combination later became a feature of milling devices, and irrigation systems. Sumerian and Assyrian engineers used wheel-driven water drawing devices in irrigation networks which are still in use today.

Greek influences

For several thousand years, these early inventions were enlarged and improved upon, without any major advances being made. By the time Archimedes (*c.* 287–*c.* 212 BC) was investigating the principle of the lever and producing his famous helical screw, the scene of the most significant developments had shifted from the Middle East to Greece. The measurement of time in particular remained a continuing challenge, and by the 2nd century BC, Ctesibius of Alexandria had developed the Egyptian *clepsydra* (water-clock) to give accuracy not surpassed until well into the Middle Ages.

Hero of Alexandria (*c.* AD 60) was the last of the Greek technologists, his most famous invention being the *aeolipile*, a primitive steam turbine that gave a hint, as early as the 1st century AD, of the potency of steam as a power-source.

Roman and Chinese influences

Despite their astonishing ability in geometry, physics, and mathematics, the Greeks were unable to make the advance which transformed architecture and civil engineering: the arch. The pre-Roman Etruscans used the semicircular arch as an architectural feature, but it was the Romans who put the arch to full use. Because of its ability to spread imposed stresses more evenly, the arch allowed greater spans in buildings than the Greeks' simple pillar-and-beam arrangements. Aqueducts comprising 6 m/20 ft wide arches were possible as early as 142 BC, as evidenced by the Pons Aemilius in Italy.

Combining the structural economy of the arch with the availability of good cement the Romans set up an infrastructure and communications network that gave them a standard of living hitherto unprecedented in history. It also enabled them to extend that standard, and maintain it by its armies, over a similarly unprecedented expanse of the world.

One major difficulty facing the Romans was the shortage of labour. As a result of the lack of an efficient means of harnessing animals to ploughs and other implements, the Romans had to use the weaker human to provide the power. It was the Chinese who first produced an efficient animal harness, freeing humans from such drudgery, and enabling animals to be used in tandem to haul great loads. The harness did not reach the West, however, until the 9th century, well after the fall of the Roman Empire. China was the birthplace of a number of other major inventions: of paper from pulp (by Ts'ai Lun *c.* AD 100), of the magnetic compass, and of gunpowder, *c.* 500 and 850 respectively.

The development of Western technology

It was also around this time that north-western Europe began its climb to ascendency in technology that it has held onto for centuries since. The poorer climate of this region, combined with the need to develop a new form of agriculture, was responsible for the emergence around the 8th century of the crop rotation methods still used today. The wind was put to use in both sea-going vessels and land-based mills. The region grew more populous and, by the 11th century, northern Europeans were moving their influence into the Mediterranean and Middle East.

As the societies grew, becoming more complex, the need for metals for housing, tools, equipment, and coinage increased concomitantly. This caused a renewed interest in the extraction and treatment of ores, which were also in increasing demand by rulers anxious to develop weapons capable of keeping their rivals at bay. Many of these rulers, notably in 15th-century Italy, employeed engineers to come up with new systems for both defence and attack. Undoubtedly the most famous such engineer was Leonardo da Vinci (1452–1519), military engineer to the Duke of Milan. Among the thousands of pages of da Vinci's notes are to be found an astonishing number of prescient plans for modern-day inventions: tanks, submarines, helicopters, and a whole range of firearms.

The printing press

The Renaissance was ushered in by one of the most influential inventions in history: the development of the movable type printing press by Johannes Gutenberg (*c.* 1397–1468) of Mainz, Germany. The Gutenberg Bible, the first book to be printed using this process, appeared in 1454. The Englishman William Caxton set up a press in England in 1476. As well as disseminating religious knowledge over a far wider scale, the invention enabled the speeding up of the transfer of technological advances from one country to another.

Steam power and the Industrial Revolution

The ever-increasing use of metals made mining the focus of much effort in the 15th century, one of the biggest problems being that of adequately draining the mines. Pumping water fast enough and in sufficient volume was also a problem facing those wanting to create more agricultural land from poorly drained areas, and to supply towns with their needs. By the mid-17th century, a number of patents had been granted to water pumps which used a new, remarkably versatile source of power: steam.

Thomas Savery (*c.* 1650–1715) demonstrated an engine for 'raising of water and occasioning motion to all sorts of mill works, by the impellant force of fire', to the Royal Society of London in 1699. He formed a partnership with Thomas Newcomen (1663–1729) and together they used the power of steam, in the form of a beam piston engine, to maintain the mines of Staffordshire, Cornwall and Newcastle in workable condition.

Watt's improvements to the steam engine

Newcomen steam engines grew in popularity during the 18th century, and it was while repairing a model one used on the physics course at Glasgow University in 1763 that the young James Watt (1736–1819) saw how major improvements could be made, improvements that led to Watt's condenser engine completely replacing the earlier model by 1800.

Inventions which revolutionized manufacturing

Through its predominance as a manufacturing market, Britain was able to reap rich rewards but competition from overseas urged on the hunt for greater efficiency of production. Technologists were at the forefront of this effort, with John Kay's (1704–1780) invention of the flying shuttle for the production of textiles leading to James Hargreaves' (1720–1778) spinning Jenny. Arkwright's (1732–1792) spinning machine was the centrepiece of the first cotton factory of 1771, and semi-automated mass production as a technique was born. The Industrial Revolution, which is arguably still under way, had begun.

The end of the 18th century saw Richard Trevithick (1771–1833) experimenting with the use of steam to provide motive power for boats, road vehicles and locomotives on steel rails. Dogged by bad luck, his ideas did not achieve the success or acclaim they deserved, and the world had to wait longer than it should have for the full exploitation of steam in such applications. French engineers had more success initially, with Nicholas Joseph Cugnot (1725–1804) developing a three-wheeled steam-powered tractor, *c.* 1770, capable of 6 km/h/3.5 mph. The Montgolfier brothers Joseph-Michael (1740–1810) and Jacques Etiènne (1745–1799) were responsible for the first sustained human flight, aboard a hot-air balloon in 1763.

The first passenger railway

By the time Trevithick left England for Peru, his high-pressure engine had shown that steam was amply capable of providing a mobile power source. He returned to find his place as the greatest engineer in the new technology usurped by others. Most famous of these was George Stephenson (1781–1848), whose steam locomotives were responsible for the setting up of the first practical passenger railway ever built, in 1825, between Stockton and Darlington.

Improved communications

Rapid communication, which remains a hallmark of an advanced technological society, became of increasing importance as the 19th century wore on, both in peacetime and in war. Thomas Telford (1757–1834) became famous for his canals and aqueduct building, enabling very large loads to be transported using little motive power. In France, the quality of the roads built by Pierre Trésaguet (1716–1796) made the rapid strikes of Napoleon's armies possible. By contrast, Britain's roads were in an appalling state through years of neglect and the operation of the Turnpike Trusts. The scientific approach to road construction devised by John McAdam (1756–1836) radically improved this aspect of Britain's infrastructure.

Steam power also found its way into ocean-going vessels. Robert Fulton (1765–1815) returned from Europe, where he had seen what steam engines were capable of, to set up the first regular steamship service, between New York and Albany, in 1807. Whether steam-powered ships were capable of a longer journey, in particular across the Atlantic Ocean, was a major debating point when Isambard Kingdom Brunel's (1806–1859) *Great Western* succeeded in sailing to New York without refuelling, in 1838. Brunel went on to design and launch the first iron ship (and the first to use a screw propellor, rather than paddle wheels) the *Great Britain*, and the colossal *Great Eastern*, whose steadiness and manoeuvrability were put to use in a key event in another field of technology altogether: the laying of the Atlantic telegraph cable in 1865.

Such very long-distance communication was made possible by advances in the understanding of electricity, and the translation of this into devices that could transmit and receive messages at the speed of light. The first successful system was brought out by Charles Wheatstone (1802–1875), in which electrical signals deflected magnetized needles indicating letters of the alphabet. The development of a communication system using a code of dots and dashes to represent the letters by Samuel Morse (1791–1872) proved so successful it is still in use today.

Developments brought about by war

The outbreak of the Crimean War involving the British, French and Turkish against the Russians led Henry Bessemer (1813–1898) to develop the method for removing impurities, in particular carbon, from molten iron, thus enabling steel to be produced cheaply in the quantities required by both military and civilian engineers. The war also resulted in the development of the first wrought-iron breech-loading gun by William Armstrong (1847–1908).

The wars affecting the United States also assisted the development of weapons on that side of the Atlantic. The war against Mexico, which broke out in 1846, accelerated the revolution in small arms manufacture initiated by Samuel Colt (1814–1862), who had produced the first revolver in 1836. The Civil War of 1861–65 prompted Richard Gatling (1818–1903) to develop the rapid-fire gun that bears his name.

The internal combustion engine

It was around the middle of the 19th century that attention began to shift from steam to gas and other combustible materials as a means of providing motive power. As early as 1833, an engine that ran on an inflammable mixture of gas and air had been described, and a number of the fundamental principles of the fuel-powered engine had been described by the time Jean Lenoir (1822–1900) began building engines using the system which operated smoothly in 1860. Together, Nikolaus Otto (1832–1891) and Eugen Langen managed by 1877 to solve the basic problems facing the development of the four-stroke internal combustion engine. This work led to the development of the modern motor car, and of powered flight. Gottlieb Daimler (1834–1900) was to join the pair as an engineer, leaving in 1883 to develop lighter, more efficient high-speed engines capable of driving cycles and boats, as well as automobiles.

Rudolph Diesel (1858–1913) experimented with internal combustion engines during the 1890s; by using the heat developed by compression of the fuel-air mixture, rather than a spark from an ignition system, to ignite the mixture, Diesel succeeded in producing an engine that could use cheaper fuels than the Otto cycle engines. However, the high pressures produced in the engines required the use of very heavy-gauge metal, with consequent weight–power ratio problems. Later advances in metallurgy enabled this disadvantage to be significantly reduced, with the result that the Diesel engine is still used in a wide range of vehicles today.

Flight

Work in the USA, as well as in Germany, succeeded in bringing the power of the internal combustion engine to bear on the problem of powered flight. Early work on the flow of air over gliders by Otto Lilienthal (1848–1896) and others established a body of knowledge needed to supplement the work of earlier enthusiasts such as George Cayley (1773–1857), who had defined the basic aerodynamic forces acting on a wing as early as 1799.

Internal combustion engines, coupled to balloons to form airships, were in use by the turn of the century. The first flight of a heavier-than-air machine powered by a light, efficient internal combustion engine was constructed by the Wright brothers, Orville (1871–1948) and Wilbur (1867–1912), on 17 December 1903. As well as finding a suitable engine for such a machine, the Wrights had succeeded in solving the problem of controlling the aeroplane in all three axes.

Electricity as a source of power

Despite the success of the internal combustion engine in powering a wide range of machines, the use of water and steam as power sources remained important in another major field of technology: the generation of electricity. Water had been used to drive turbines of increasing efficiency for many years when Charles Parsons (1854–1931) adapted the basic design of a water turbine to enable a jet of steam to impart its kinetic energy to a series of turbine blades which then rotate. By combining this rotation with the ability of a dynamo to convert rotary motion into electric power, the electric generator was born. The first ever turbine-powered generating station was set up in 1888, using four Parsons turbines each developing 75 kW/100 hp. Direct use of the mechanical power developed by steam was made in Parson's 44 tonne/44.7 ton Turbinia, whose turbine engine developed 1.5 MW/2,000 hp, enabling it to travel at 60 km/h/37 mph in 1897.

Work by Joseph Swan (1828–1914) in England and Thomas Edison (1847–1931) in the United States finally resulted in the creation of a long-lasting light source powered by electricity – the filament lamp – around 1880.

While Germany and the United States were quick to use electrical power to bring about a revolution in their industrial processes, the availability of cheap labour and concentration on waning industries based on traditional raw material inhibited the adoption of electrical power in Britain. Concentration on telegraphic technology and the generation of electric illumination did have an indirect advantage, however. The invention of the two-electrode electric valve, the diode, by John Fleming (1849–1945) provided a new outlet for the vacuum bulb technology. Such developments as radio communication, radar, television, and the computer all benefited from this.

Radio communications

Communication over long distances without the use of cables – 'wireless' communication – had been a practical possibility from the day when the electromagnetic wave physicist James Clerk Maxwell's (1831–1879) theory combining electrical and magnetic phenomena had been investigated by Heinrich Hertz (1857–1894) in 1888. Both transmitters and detectors of these radio waves were developed until Guglielmo Marconi (1874–1937) succeeded in transmitting messages over a few yards using electromagnetic waves in 1895. By 1901, he had succeeded in sending signals right across the Atlantic.

More sophisticated communication was made possible by Lee De Forest (1873–1961) and Reginald Fessenden (1866–1932), and their invention of the triode amplifier and amplitude modulation respectively. These advances enabled speech and sound to be transmitted over very long distances, and gave birth to modern communications.

Developments of the war years

World War I (1914–1918) saw the use of technology on an unprecedented scale. Although many of the advances then simply led to the deaths of hundreds of thousands of troops, many later found major applications in peacetime. An excellent example of this is provided by the development of the nitrogen fixation process to an industrial scale by Karl Bosch (1874–1940). This enabled the Germans to manufacture explosives such as TNT without relying on foreign imports of nitrogen-bearing materials, capable of being blockaded by the allies. In peacetime, the process allowed the cheap manufacture of fertilizers, equally vital to the survival of a country.

The war also had a profound effect on the aircraft industry. Starting the war as chiefly reconnaissance vehicles, the aircraft became directly involved in the fighting by the end, and mass production of tens of thousands became necessary. Governments spent money on research, accelerating advances in aerodynamics and power systems enormously. Civil aviation, begun by the Germans before the war, benefited, initially using modified military aircraft.

As with the automobile, engineers started to look at new power sources for the aircraft. The use of gas turbines was put forward in 1926, a suggestion turned into reality by Frank Whittle (1907–) in 1930. By combining a gas turbine with a centrifugal compressor, he created the jet engine.

Space travel

The inter-war years saw considerable advances in rocket technology: an area of engineering that was to enable humans to leave the planet of their birth. The Chinese had used solid-fuelled rockets in battles as early as 1232; their direct ancestors are still to be seen strapped to the central booster of the Space Shuttle. It was Konstantin Tsiolkovskii (1857–1935) who pointed out that liquid propellants had distinct advantages of power and controllability over solid fuels. The American astronautics pioneer Robert Goddard (1882–1949) succeeded in launching the first liquid-fuelled rocket in 1926. Just 35 years later, Soviet engineers used a liquid-filled booster to send the first human being into Earth orbit. Eight years after that a human being set foot on another celestial body for the first time.

Entertainment

By the 1920s, a number of devices born in research laboratories had become established as massively popular forms of entertainment. The work of George Eastman (1854–1932), Thomas Edison (1847–1931), and others brought photography and sound-and-motion 'movie' pictures to millions. A working system of television was devised by John Logie Baird (1888–1946) and shown in 1925, while the modern electronic system later adopted as standard for television was demonstrated by Vladimir Zworykin (1889–1982) in 1929. The British Broadcasting Corporation's forebear, the British Broadcasting Company, was formed in 1922, transmitting radio programmes to the public on a national scale. In 1936, experimental television broadcasts were made by the BBC from Alexandra Palace near London.

Atomic power

The growth in the use of electrical power put greater emphasis on ways of generating it cheaply. The United States in particular built many large storage dams, producing electricity by hydroelectric turbine technology. By 1920 some 40% of electricity in the USA was generated by this means. But developments in particle physics during these years were beginning to show that the fundamental constituents of matter would be capable of providing another, far more concentrated, form of energy: atomic power. By 1939 and the outbreak of World War II, a number of physicists had begun to appreciate the possibilities offered by 'chain reactions' involving the fission of unstable, chemical elements such as uranium.

The first electronic computers

The war itself again proved to be a sharp stimulus for the refinement of old ideas and the development of new ones. Radar, devised by Robert Watson-Watt (1892–1973) in 1935, was developed into a national defence system against aircraft that were growing ever faster and more deadly. More sophisticated weapons and ever more complex message-encoding systems resulted in the development of early electronic computers. These were needed to rapidly sift through data and perform arithmetical operations upon it, and also to carry out numerical integrations which were particularly useful in the precise calculation of the trajectories for artillery shells. The pioneering work on mechanical computing machines by Charles Babbage (1792–1871) in the early 1800s was transformed by the introduction of electronic devices by Vannevar Bush (1890–1974) and others during the war years.

The inter-war ideas of jet and rocket propulsion were used in the development of fighter aircraft and missiles such as the 'V' (*Vergeltung*) 1 and 2, powered by ram-jet and liquid fuel respectively. Most devastating of all was the use of the chain reaction of atomic energy in an uncontrolled explosive device against the Japanese in 1945. Although the use of the atomic bomb finally ended World War II, the world still lives under the threat of their use, in still more deadly form, to this day. The use of the first atomic bombs has tended to overshadow another event in the development of atomic power that took place during the war. This was the setting up of the first controlled atomic chain reaction in 1942 by a team of scientists at the University of Chicago, which paved the way for the peaceful use of atomic power to generate electricity. The first nuclear electricity was generated by the Experimental Breeder Reactor in Idaho, USA 1951. The world's first commercial scale nuclear power station entered service at Calder Hall, UK 1956.

The age of the microchip

While the demand for yet more electric power grew among industrial nations mass-producing cars, ships, and aircraft, ways of reducing the complexity and power consumption of electronic devices such as computers, radios and televisions were being sought. Most crucial of these was the invention of the transistor, by John Bardeen (1908–1991), William Shockley (1910–1989), and Walter Brattain (1902–1987) at Bell Laboratories in the USA, in 1948. These tiny semiconductor-based devices could achieve the rectification and amplification of the thermionic valves of the pre-war years at a fraction of the power consumption. The use of such devices in still smaller form allowed the miniaturization of electronic devices.

In 1958 the first integrated circuit, which contained the components of a complete circuit on a single piece of silicon, was built by Jack Kilby in the USA. The first microprocessor, a complete computer on a chip was designed by US computer engineer Ted Hoff 1971, containing 2,250 components. By 1990 memory chips capable of holding 4 million bits of information were being mass-produced in Japan. In 1993 the US chip manufacturer Intel introduced the Pentium microprocessor chip which was about five times more powerful than earlier processors and had 3.1 million transistors on a silicon square 15 mm/0.6 in across.

Computing power packed into smaller volumes has been the driving force of many areas of technology over the last 40 years. It made possible the era of human space flight, where keeping the mass of all components to a minimum is vital. Telecommunication satellites, such as *Telstar*, (which in 1962 transmitted the first live television pictures across the Atlantic), weather satellites and planetary probes were all made possible as a result of the semiconductor breakthrough.

Computer-aided design and manufacture have enabled new ideas in fields from architecture to aircraft manufacture to be tried out, tested and produced far more quickly and cheaply. The influence of the computer is felt in everyday life, from the diagnosis of disease by the CAT scanner invented by Godfrey Hounsfield (1919–) to the production of bank statements. The first home computers were introduced in the early 1980s. In the 1990s millions of home computers are linked to the worldwide Internet. Home computer users have also powered the market for CD-ROM, able to hold vast amounts of information.

Communications advances

The first optical fibre cable, capable of carrying digital signals, was installed in California 1977. In 1988 the International Services Digital Network (ISDN), an international system for sending signals in digital format along optical fibres and coaxial cable, was launched in Japan. The first transoceanic optical fibre cable, capable of carrying 40,000 simultaneous telephone conversations, was laid between Europe and the USA 1989. In 1992 videophones, made possible by advances in image compression and the development of ISDN, were introduced in the UK. The Japanese began broadcasting high-definition television 1989. All digital high-definition television was demonstrated in the USA 1992.

Lasers

The laser was another significant development, invented 1960 by Theodore Maiman (1927–), and often used in robotic welders. Holograms, or three-dimensional pictures, became practicable after the development of laser technology.

Power sources for the future

There is still considerable interest in finding ways of generating cheap power. Using nuclear fission has proved only a partial answer to the question of what will replace the burning of hydrocarbons such as coal to generate electricity. Public concern about both its inherent safety and the toxic waste produced has cast a shadow over the long-term future of fission-generated electricity.

Engineers and physicists in Europe, the United States, the former Soviet Union, and Japan are currently studying the generation of power by nuclear fusion. Using hydrogen and its isotopes derived from seawater, they hope to be able to mimic the reactions that have kept the Sun burning for thousands of millions of years. The engineering difficulties presented by trying to keep a plasma stable at a temperature of 100 million degrees are immense, but in 1991 fusion power came a step nearer when the Joint European Torus (JET) at Culham, England, produced fusion power for the first time. In 1994 the Tokamak Fusion Test Reactor at Princeton University produced 9 megawatts of power. It needed 33 megawatts to drive it and power production lasted for only 0.4 seconds. Nevertheless, the hot plasma of deuterium and tritium was so well behaved that researchers believe output can be boosted.

Others are turning to less exotic sources of energy, such as the wind, solar energy, and tides, to find better ways of exploiting them and generating cheap, clean power. The world's largest photovoltaic power station was plugged into the power grid at Davis, California, USA, in 1993. A wind farm in California's San Bernados Mountains, uses over 4,000 wind turbines to supply electricity to the Coachella Valley in S California.

David Abbott

Geology

Introduction
Throughout history humans have sought to control and understand their environment. Practical activities like agriculture and quarrying naturally lead to enhanced knowledge, and science suggests further ways of utilizing the Earth.

Growing interaction with the Earth has been important in the development of numerous sciences – not just geology but cosmogony and geophysics; alchemy and chemistry; mineralogy, and crystallography; meteorology, physical geography, topography, and oceanography; natural history, biology, and ecology. Distinct investigation of the Earth itself – geology – has been a recent development. Geology (literally 'Earth-knowledge') does not date back more than two hundred years.

Antiquity
Scientific thinking about the Earth grew out of traditions of thought which took shape in the Middle East and the Eastern Mediterranean. Early civilization needed to adapt to the seasons, to deserts and mountains, volcanoes and earthquakes. Yet inhabitants of Mesopotamia, the Nile Valley, and the Mediterranean littoral had experience of only a fraction of the Earth. Beyond lay *terra incognita*. Hence legendary alternative worlds were conjured up in myths of burning tropics, lost continents, and unknown realms where the gods lived.

The Greeks
The first Greek philosopher about whom much is known was Thales of Miletus (*c.* 640–546 BC). He postulated water as the primary ingredient of material nature. Thales' follower, Anaximander, believed the universe began as a seed which grew; and living things were generated by the interaction of moisture and the Sun. Xenophanes (*c.* 570–475 BC) is credited with a cyclic worldview: eventually the Earth would disintegrate, returning to a watery state.

Like many other Greek philosophers, Empedocles (*c.* 500–*c.* 430 BC) was concerned with change and stability, order and disorder, unity and plurality. The terrestrial order was dominated by strife. In the beginning, the Earth had brought forth living structures more or less at random. Some had died out. The survivors became the progenitors of modern species.

The greatest Greek thinker was Aristotle. He considered the world was eternal. Aristotle drew attention to natural processes continually changing its surface features. Earthquakes and volcanoes were due to the wind coursing about in underground caves. Rivers took their origin from rain. Fossils indicated that parts of the Earth had once been covered by water.

Ptolemy and Pliny
In the 2nd century AD, Ptolemy composed a geography that summed up the Ancients' learning. Ptolemy accepted that the equatorial zone was too torrid to support life, but he postulated an unknown land mass to the south, the *terra australis incognita*. Antiquity advanced a 'geocentric' and 'anthropocentric' view. The planet had been designed as a habitat for humans. A parallel may be seen in the Judaeo-Christian cosmogony.

The centuries from Antiquity to the Renaissance accumulated knowledge on minerals, gems, fossils, metals, crystals, useful chemicals and medicaments, expounded in encyclopedic natural histories by Pliny (23–79) and Isidore of Seville (560–636). The great Renaissance naturalists were still working within this 'encyclopedic' tradition. The most eminent was Konrad Gesner, whose *On Fossil Objects* was published in 1565, with superb illustrations. Gesner saw resemblances between 'fossil objects' and living sea creatures.

The Christian view

At the same time, comprehensive philosophies of the Earth were being elaborated, influenced by the Christian revelation of Creation as set out in 'Genesis'. This saw the Earth as recently created. Bishop Ussher (1581–1656) in his *Sacred Chronology* (1660), arrived at a creation date for the Earth of 4004 BC. In Christian eyes, time was directional, not cyclical. God had made the Earth perfect but, in response to Original Sin, he had been forced to send Noah's Flood to punish people by depositing them in a harsh environment, characterized by the niggardliness of Nature. This physical decline would continue until God had completed his purposes with humans.

The scientific revolution

The 16th and 17th centuries brought the discovery of the New World, massive European expansion and technological development. Scientific study of the Earth underwent significant change. Copernican astronomy sabotaged the old notion that the Earth was the centre of the system. The new mechanical philosophy (Descartes, Gassendi, Hobbes, Boyle, and Hooke) rejected traditional macrocosm–microcosm analogies and the idea that the Earth was alive. Christian scholars adopted a more rationalist stance on the relations between Scripture and scientific truth. The possibility that the Earth was extremely old arose in the work of 'savants' like Robert Hooke. For Enlightenment naturalists, the Earth came to be viewed as a machine, operating according to fundamental laws.

The old quarrel as to the nature of fossils was settled. Renaissance philosophies had stressed the living aspects of Nature. Similarities between fossils and living beings seemed to prove that the Earth was capable of growth. Exponents of the mechanical philosophy denied these generative powers. Fossils were petrified remains, rather like Roman coins, relics of the past, argued Hooke. Such views chimed with Hooke's concept of major terrestrial transformations and of a succession of faunas and floras now perished. Some species had been made extinct in great catastrophes.

The significance of fossils

This integrating of evidence from fossils and strata is evident in the work of Nicolaus Steno (1638–1686). He was struck by the similarity between shark's teeth and fossil *glossopetrae*. He concluded that the stones were petrified teeth. On this basis, he posited six successive periods of Earth's history. Steno's work is one of the earliest 'directional' accounts of the Earth's development that integrated the history of the globe and of life. Steno treated fossils as evidence for the origin of rocks.

The Enlightenment

Mining schools developed in Germany. German mineralogists sought an understanding of the order of rock formations which would be serviceable for prospecting purposes. Johann Gottlob Lehmann (1719–1776) set out his view that there were fundamental distinctions between the various *Ganggebergen* (masses formed of stratified rock). These distinctions represented different modes of origin, strata being found in historical sequence. Older strata had been chemically precipitated out of water, whereas more recent strata had been mechanically deposited.

Abraham Gottlob Werner (1749–1817) was appointed in 1775 to the Freiberg Akademie. He was the most influential teacher in the history of geology. Werner established a well-ordered, clear, practical, physically-based stratigraphy. He proposed a succession of the laying down of rocks, beginning with 'primary rocks' (precipitated from the water of a universal ocean), then passing through 'transition', 'flútz' (sedimentary), and finally 'recent' and 'volcanic'. The oldest rocks had been chemically deposited; they were therefore crystalline and without fossils. Later rocks had been mechanically deposited. Werner's approach linked strata to Earth history.

The development of stratigraphy

Thanks to the German school, but also to French observers like Guettard, Lavoisier and Dolomieu, to Italians such as Arduino, to Swedes like Bergman, stratigraphy was beginning to emerge in the 18th century.

Of course, there were many rival classifications and all were controversial. In particular, battle raged over the nature of basalt: was it of aqueous or igneous origin? The Wernerian, or Neptunist, school saw the Earth's crust precipitated out of aqueous solution. The other, culminating in Hutton, asserted the formation of rock types from the Earth's central heat.

The ideas of Buffon and Hutton

A pioneer of this school was Buffon (1707–1778). He stressed ceaseless transfigurations of the Earth's crust produced by exclusively 'natural' causes. In his *Epochs of Nature* (1779) he emphasized that the Earth had begun as a fragment thrown off the Sun by a collision with a comet. Buffon believed the Earth had taken at least 70,000 years to reach its present state. Extinction was a fact, caused by gradual cooling. The seven stages of the Earth explained successive forms of life, beginning with gigantic forms, now extinct, and ending with humans.

Though a critic of Buffon, James Hutton shared his ambitions. Hutton (1726–1797) was a scion of the Scottish Enlightenment, being friendly with Adam Smith and James Watt. In his *Theory of the Earth* (1795), Hutton demonstrated a steady-state Earth, in which natural causes had always been of the same kind as a present, acting with precisely the same intensity ('uniformitarianism'). There was 'no vestige of a beginning, no prospect of an end'. All continents were gradually eroded by rivers and weather. Debris accumulated on the sea bed, to be consolidated into strata and thrust upwards by the central heat to form new continents. Hutton thus postulated an eternal balance between uplift and erosion. All the Earth's processes were gradual. The Earth was incalculably old. His maxim was that 'the past is the key to the present'.

Hutton's theory was much attacked in its own day. Following the outbreak of the French revolution in 1789, conservatives saw all challenges to the authority of the Bible as socially subversive. Their writings led to ferocious 'Genesis versus Geology' controversies in England.

The 19th century

New ideas about the Earth brought momentous social, cultural and economic reverberations. Geology clashed with traditional religious dogma about Creation. Modern state-funded scientific education and research organizations emerged. German universities pioneered scientific education. The Geological Survey of Great Britain was founded, after Henry De la Beche (1796–1855) obtained state finance for a geological map of south-west England. De la Beche's career culminated in the establishment of a Mines Record Office and the opening in 1851 of the Museum of Practical Geology and the School of Mines in London.

Specialized societies were founded. The Geological Society of London dates from 1807. In the United States, the government promoted science. Various states established geological surveys, New York's being particularly productive. The US Geological Survey was founded in 1879, under Clarence King and later John Wesley Powell. In 1870 Congress appointed Powell to lead a survey of the natural resources of the Utah, Colorado, and Arizona area.

The stratigraphical column

Building on Werner, the great achievement of early 19th-century geology lay in the stratigraphical column. After 1800, it was perceived that mineralogy was not the master key. Fossils became regarded as the indices enabling rocks of comparable age of origin to be identified. Correlation of information from different areas would permit tabulation of sequences of rock formations, thereby displaying a comprehensive picture of previous geological epochs.

Britain

In Britain the pioneer was William 'Strata' Smith (1769–1839). Smith received little formal education and became a canal surveyor and mining prospector. By 1799 he set out a list of the secondary strata of England. This led him to the construction of geological maps. In 1815 he brought out *A Delineation of the Strata of England and Wales,* using a scale of five miles to the inch. Between 1816 and 1824 he published *Strata Identified by Organized Fossils,* which displayed the fossils characteristic of each formation.

France

Far more sophisticated were the French naturalists Georges Cuvier (1769–1832) and Alexandre Brongniart (1770–1837), who worked on the Paris basin. Cuvier's contribution lay in systematizing the laws of comparative anatomy and applying them to fossil vertebrates. He divided invertebrates into three phyla, and conducted notable investigations into fish and molluscs. In *Researches on the Fossil Bones of Quadrupeds* (1812), he reconstructed such extinct fossil quadrupeds as the mastodon, applying the principles of comparative anatomy. Cuvier was the most influential paleontologist of the 19th century.

Fossils, in Cuvier's and Brongniart's eyes, were the key to the identification of strata and Earth history. Cuvier argued for occasional wholesale extinctions caused by geological catastrophes, after which new flora and fauna appeared by migration or creation. Cuvier's *Discours sur les révolutions de la surface du globe* (1812) became the foundation text for catastrophist views.

Older rock types

Classification of older rock types was achieved by Adam Sedgwick (1785–1873) and Roderick Murchison (1792–1871). Sedgwick unravelled the stratigraphic sequence of fossil-bearing rocks in North Wales, naming the oldest of them the Cambrian period (now dated at 500–570 million years ago). Further south, Murchison delineated the Silurian system amongst the *grauwacke.* Above the Silurian, the Devonian was framed by Sedgwick, Murchison and De la Beche. Shortly afterwards, Charles Lapworth developed the Ordovician.

Uniformitarianism

Werner's retreating-ocean theory was quickly abandoned, as evidence accumulated that mountains had arisen not by evaporation of the ocean, but through processes causing elevation and depression of the surface. This posed the question of the rise and fall of continents. Supporters of 'catastrophes' argued that terrestrial upheavals had been sudden and violent. Opposing these views, Charles Lyell advocated a revised version of Hutton's gradualism. Lyellian uniformitarianism argued that both uplift and erosion occurred by natural forces.

Expansion of fieldwork undermined traditional theories based upon restricted local knowledge. The retreating-ocean theory collapsed as Werner's students travelled to terrains where proof of uplift was self-evident.

Geologists had to determine the earth movements that had uplifted mountain chains. Chemical theories of uplift yielded to the notion that the Earth's core was intensely hot, by consequence of the planet commencing as a molten ball. Many hypotheses were advanced. In 1829, Elie de Beaumont published *Researches on Some of the Revolutions of the Globe,* which linked a cooling Earth to sudden uplift: each major mountain chain represented a unique episode in the systematic crumpling of the crust. The Earth was like an apple whose skin wrinkled as the interior shrank through moisture loss. The idea of horizontal (lateral) folding was applied in America by James Dwight Dana to explain the complicated structure of the Appalachians. Such views were challenged by Charles Lyell in his bid to prove a steady-state theory. His classic *Principles of Geology* (1830–33) revived Hutton's vision of a uniform Earth that precluded cumulative, directional change in overall environment; Earth history proceeded like a cycle, not like an arrow. In *Principles of Geology,* Lyell thus attacked diluvialism and catastrophism by resuscitating Hutton's vision of an Earth subject only to changes currently discernible. Time replaced violence as the key to geomorphology.

Lyell discounted Cuvier's apparent evidence for the catastrophic destruction of fauna and flora populations. For over 30 years he opposed the transmutation of species, reluctantly conceding the point at last only in deference to his friend, Charles Darwin, and the cogency of Darwin's *Origin of Species* (1859).

Ice ages

Landforms presented a further critical difficulty. Geologists had long been baffled by beds of gravel and 'erratic boulders' strewn over much of Northern Europe and North America. Bold new theories in the 1830s attributed these phenomena to extended glaciation. Jean de Charpentier and Louis Agassiz contended that the 'diluvium' had been moved by vast ice sheets covering Europe during an 'ice age'. Agassiz's *Studies on Glaciers* (1840) postulated a catastrophic temperature drop, covering much of Europe with a thick covering of ice that had annihilated all terrestrial life.

The ice-age hypothesis met opposition but eventually found acceptance through James Geikie, James Croll and Albrecht Penck. Syntheses were required. The most impressive unifying attempt came from Eduard Suess. His *The Face of the Earth* (1885–1909) was a massive work devoted to analysing the physical agencies contributing to the Earth's geographical evolution. Suess offered an encyclopedic view of crustal movement, the structure and grouping of mountain chains, of sunken continents, and the history of the oceans. He made significant contributions to structural geology.

Suess disputed whether the division of the Earth's relief into continents and oceans was permanent, thus clearing the path for the theory of continental drift. Around 1900, the US geologist and cosmologist Thomas C Chamberlin (1843–1928) proposed a different synthesis: the Earth did not contract; its continents were permanent. Continents, Chamberlin argued, were gradually filling the oceans and thereby permitting the sea to overrun the land.

The 20th century

By 1900, study of the Earth had become fragmented into specialisms like stratigraphy, mineralogy, crystallography, sedimentology, petrography, and palaeobotany, and there was no univerally-accepted unifying research programme. Geophysics increasingly provided intellectual coherence. Geophysics emerged as a distinct discipline in the late 19th century. Study of the Earth's magnetic field came to early prominence. In 1919 the American Geophysical Union was formed, and 1957 was designated the International Geophysical Year. The modern term 'earth sciences', to some degree replacing geology, marks the triumph of geophysics.

Nineteenth-century fieldwork had set the agenda for an enduring tradition of stratigraphic surveying and investigation of landforms. These traditions continued to yield valuable harvests. Immensely influential was the US geomorphologist William Morris Davis (1850–1934). Davis developed the organizing concept of the cycle of erosion. He proposed a stage-by-stage life-cycle for a river valley, marked by youth (steep-sided V-shaped valleys), maturity (flood-plain floors), and old age, as the river valley was imperceptibly worn down into the rolling landscape he termed a 'peneplain'.

Use of radioactivity for dating purposes

Nineteenth-century geology was built on the idea of the cooling Earth. Lord Kelvin's estimates of the Earth's age suggested a relatively low antiquity, but this was soon challenged from within physics itself, for in 1896 the discovery of radioactivity revealed a new energy source unknown to Kelvin. In *The Age of the Earth* (1913), Arthur Holmes (1890–1965) pioneered the use of radioactive decay methods for rock-dating. By showing the Earth had cooled far more slowly than Kelvin asserted, the new physicists undermined the 'wrinkled apple' analogy.

Continental drift

Amidst such challenges Alfred Wegener (1880–1930) went further and declared that continental rafts might actually slither horizontally across the Earth's face. From 1910 Wegener developed a theory of continental drift. Empirical evidence for such displacement lay, he thought, in the close jigsaw-fit between coastlines on either side of the Atlantic, and notably in palaeontological similarities between Brazil and Africa. Wegener was also convinced that geophysical factors would corroborate wandering continents.

Wegener supposed that a united supercontinent, Pangaea, had existed in the Mesozoic. This had developed numerous fractures and had drifted apart, some 200 million years ago. During the Cretaceous, South America and Africa had largely been split, but not until the end of the Quaternary had North America and Europe finally separated. Australia had been severed from Antarctica during the Eocene.

The causes of continental drift

What had caused continental drift? Wegener offered a choice of possibilities. One was a westwards tidal force caused by the Moon. The other involved a centrifugal effect propelling continents away from the poles towards the equator (the 'flight from the pole'). In its early years, drift theory won few champions, and in the English-speaking world reactions were especially hostile. A few geologists were intrigued by drift, especially the South African, Alexander Du Toit (1878–1948), who adumbrated the similarities in the geologies of South America and South Africa, suggesting they had once been contiguous. In *Our Wandering Continents* (1937), Du Toit maintained that the southern continents had formed the supercontinent of Gondwanaland.

The most ingenious support for drift came, however, from the British geophysicist Arthur Holmes. Assuming radioactivity produced vast quantities of heat, Holmes argued for convection currents within the crust. Radioactive heating caused molten magma to rise to the surface, which then spread out in a horizontal current before descending back into the depths when chilled. Such currents provided a new mechanism for drift.

The real breakthrough required diverse kinds of evidence accumulating from the 1940s, especially through oceanography and palaeomagnetism. Advances in palaeomagnetism arose from controversies over origins of the Earth's magnetic field. The evidence for changing directions of the magnetic field recorded by the rocks was linked to a baffling anomaly: in many cases the direction of the field seemed to be reversed. This led geophysicists to suspect that the terrestrial magnetic field occasionally switched. Over millions of years, there would be intermittent reversal events in which the North and South magnetic poles would alternate. Remnant magnetization would record these events, and, if the rocks could be dated sufficiently precisely, a complete register of reversals could be traced against the geological record.

By 1960, US scientists had refined the radiometric technique for dating rocks, deploying especially the potassium–argon method. A group at Berkeley developed a timescale of reversals for the Pleistocene era; Australian scientists produced their own scale, based on the dating of Hawaiian lava flows. Oceanography was developing too. Here the work of William Maurice Ewing (1906–1974) was especially significant. Ewing ascertained that the crust under the ocean is much thinner than the continental shell. Ewing also demonstrated that mid-ocean ridges were common to all oceans. Ewing's work demonstrated that far from being ancient, ocean rocks were recent.

The US geophysicist Harry Hess (1906–1969) played a key role in promoting the new theories, viewing the oceans as the major centre of activity. The new crust was produced in the ridges, whereas trenches marked the sites where old crust was subtended into the depths, completing the convection current's cycle. Carried by the horizontal motion of the convection current, continents would glide across the surface. Constantly being formed and destroyed, the ocean floors were young; only continents – too light to be drawn down by the current – would preserve testimony of the remote geological past. Support came from J Tuzo Wilson (1908–1993), a Canadian geologist, who provided backing for the sea-floor spreading hypothesis. A dramatic new line of evidence, developed by Drummond Hoyle Matthews (1931–) and Fred Vine (1939–1988) of Cambridge University, confirmed sea-floor spreading.

The majority of Earth scientists accepted the new plate tectonics model with remarkable rapidity. In the mid-1960s, a full account of plate tectonics was expounded. The Earth's surface was divided into six major plates, the borders of which could be explained by way of the convection-current theory. Deep earthquakes were produced where one section of crust was driven beneath another, the same process also causing volcanic activity in zones like the Andes. Mountains on the western edge of the North and South American continents arose from the fact that the continental 'raft' is the leading edge of a plate, having to face the oncoming material from other plates being forced beneath them. The Alps and Himalayas are the outcome of collisions of continental areas, each driven by a different plate system.

Geologists of the late 1960s and 1970s undertook immense reinterpretation of their traditional doctrines. Well-established stratigraphical and geomorphological data had to be redefined in terms of the new forces operating in the crust. Tuzo Wilson's *A Revolution in Earth Science* (1967) was a persuasive account of the plate tectonics revolution.

Geology is remarkable for having undergone such a dramatic and comprehensive conceptual revolution within recent decades. The fact that the most compelling evidence for the new theory originated from the new discipline of ocean-based geophysics has involved considerable revaluing of skills and priorities within the profession. Above all, the ocean floor now appears to be the key to understanding the Earth's crust, in a way that Wegener never appreciated.

Satellite observations

In recent years Earth observation satellites have measured continental movements with unprecedented accuracy. The surface of the Earth can be measured using global positioning geodesy (detecting signals from satellites by Earth-based receivers), satellite laser ranging (in which satellites reflect signals from ground transmitters back to ground receivers), and very long-long-baseline interferometry, which compares signals received at ground-based receivers from distant extraterrestrial bodies. These techniques can measure distances of thousands of kilometres to accuracies of less than a centimetre. Movements of faults can be measured, as can the growth of tectonic plates. Previously, such speeds were calculated by averaging displacements measured over decades or centuries. The results show that in the oceanic crust, plate growth is steady: from 12 mm/0.05 in per year across the Mid-Atlantic Ridge to 160 mm/6.5 in per year across the East Pacific Rise. The major continental faults seem to be very irregular in their movement; the Great Rift Valley has remained stationary for 20 years, when long-term averages suggest that it would have opened up about 100 mm/4 in in that time.

Roy Porter

Mathematics

The beginnings of maths

Most ancient civilizations had the means to make accurate measurements, to record them in writing, and to use them in calculations involving elementary addition and subtraction. Apparently for many of them that was sufficient. It seems, for example, that the ancient Egyptians relied on simple addition and subtraction even for calculations of area and volume, although their number system was founded upon base 10, as ours is today. They certainly never thought of mathematics as a subject of potential interest or study for its own sake.

Not so the Babylonians. Contemporaries of the Egyptians, they nevertheless had a more practical form of numerical notation and were genuinely interested in improving their mathematical knowledge. (Perversely, however, their system used base 10 up to 59, after which 60 became a new base; one result of this is the way we now measure time and angles.) By about 1700 BC the Babylonians not only had the four elementary algorithms – the rules for addition, subtraction, multiplication and division – but also had made some progress in geometry. They knew what we now call Pythagoras' theorem, and had formulated further theorems concerning chords in circles. This even led to a rudimentary understanding of algebraic functions.

The ancient Greeks

Until the very end of their own civilization, the ancient Greeks had little use for algebra other than within a study of logic. After all, to them learning was as interdisciplinary as possible. Even Thales of Miletus (lived *c.* 585 BC), regarded as the first named mathematician, considered himself a philosopher in a school of philosophers; mathematics was peripheral. The Greeks' attitude of scientific curiosity was, however, to result in some notable advances in mathematics, especially in the endeavour to understand why and how algorithms worked, theorems were consistent, and calculations could be relied on. It led in particular to the notion of mathematical proof, in an elementary but no less factual way. Pythagoras (lived *c.* 530 BC), having proved the theorem now called after him, imbued mathematics with a kind of religious mystique on the basis of which he became a rather unsuccessful social reformer. Others became fascinated by solving problems using a ruler and compass, in which an outline of the concept of an irrational number (such as π) inevitably appeared. Further investigations of curves followed, and resulted in the first suggestions of what we now call integration. Such geometrical studies were often applied to astronomy. A corpus of various kinds of mathematical knowledge was beginning to accumulate.

The man who recorded much of it was Euclid (*c.* 330–260 BC). His work *The Elements* is intended as much as a history of mathematics as a compendium of knowledge, and was massive therefore in both scope and production. It contained many philosophical elements (as we would now define them) and astronomical hypotheses, but the exposition of the mathematical work was masterly, and became the style of presentation emulated virtually to this day. Euclid's geometry, especially, became the standard for millennia: mathematicians still distinguish between Euclidean and non-Euclidean geometry. He even included discussion and ideas on spherical geometry. Unfortunately, some of *The Elements* was lost, including the work on conic sections.

Conic sections seem to have been a source of fascination to many ancient Greek mathematicians. Archimedes (*c.* 287–212 BC), one of the most practical men of all time, used the principle of conic sections in an investigation into how to solve problems of an algebraic nature. A little later, Apollonius of Perga (lived *c.* 230 BC), wrote definitively of the subject, adducing a considerable number of associated theorems and including relevants proofs. The significance of part of this extra material was established only at the end of the 19th century.

The Romans

After about 150 BC, the study of astronomy dominated the scientific world. Consequently, for a while, little mathematical progress was made except in the context of the cosmological theories of the time. (There was accordingly some significant research into spherical geometry and spherical trigonometry.) It was then too that Roman civilization briefly flourished and began to recede – again with little effect on the status of mathematics. Surprisingly, however, after about 400 years, the Alexandrian Diophantus (lived *c.* AD 270–280) devised something of extreme originality: the algebraic variable, in which a symbol stands for an unknown quantity. Equations involving such indeterminates – Diophantus included one indeterminate per equation, needing thus only one symbol – are now commonly called Diophantine equations.

Alexandria thus became the centre for mathematical thought at the time. Very shortly afterwards, Pappus (lived *c.* 320) deemed it time again for a compilation of all known mathematical knowledge. In *The Collection* he revised, edited, and expanded the works of all the classic writers and added many of his own proofs and theorems, including some well-known problems that he left unsolved. It is this work more than any other that ensured the survival of the mathematics of the Greeks until the Renaissance about a thousand years later.

The Arabs

In the meantime the initiative was taken by the Arabs, whose main sphere of influence was, significantly, farther east. They were thus in contact with Persian and Indian scientific schools, and accustomed to translating learned texts. Both Greek and Babylonian precepts were assimilated and practised – the best known proponent was Al-Khwarizmi (lived *c.* 840), whose work was historically important to later mathematicians in Europe. The Arabs devised accurate trigonometrical tables (primarily for astronomical research) and continued the development of spherical trigonometry; they also made advances in descriptive geometry.

Mathematical knowledge returns to Europe

It was through his learning in the Arab markets of Algeria that the medieval merchant from Pisa Leonardo Fibonacci (or Leonardo Pisano; *c.* 1180–1250), brought much of contemporary mathematics back to Europe. It included – only then – the use of the 'Arabic' numerals 1 to 9 and the 'zephirum' (0), the innovation of partial numbers or fractions, and many other features of both geometry and algebra. From that time, hundreds of translators throughout Europe (especially in Spain) worked on Latin versions of Arab works and transcriptions. Only when Europe had regained all the knowledge and, so to speak, updated itself could genuine development take place. The effort took nearly 400 years before any truly outstanding advances were made – but may be said to be directly responsible for the overall updating and advance in science that then came about, known as the Renaissance.

The 16th century

One of the first instances of genuine progress in mathematics was the means of solving cubic equations, although acrimonious recriminations over priority surrounded its initial publication. One particularly charismatic contender – Niccolo Fontana (*c.* 1499–1557), usually known as Tartaglia – besides being a military physicist, remained an inspirational figure in the propagation of mathematics. The means of solving quartic equations was discovered soon afterwards.

Within another 20 years, the French mathematician François Vieta (1540–1603) was improving on the systematization of algebra in symbolic terms and expounding on mathematical (as opposed to astronomical) applications of trigonometry. It was he, if anyone, who initiated the study of number theory as an independent branch of mathematics. At the time of Vieta's death, Henry Briggs (1561–1630) in England was already professor of geometry; a decade later he combined with John Napier (1550–1617), the deviser of 'Napier's Bones', to produce the first logarithm tables using the number 10 as its base, a means of calculation commonly used until the late 1960s but now outmoded by the computer and pocket calculator. Simultaneously, the astronomer Johannes Kepler was publishing one of the first works to consider infinitesimals, a concept that would lead later to the formulation of the differential calculus.

The 17th century

It was in France that the scope of mathematics was then widened by a group of great mathematicians. Most of them met at the scientific discussions run by the director of the convent of Place Royale in Paris, Fr Marin Mersenne (1588–1648). To these discussions sometimes came the philosopher-mathematician René Descartes (1596–1675), the lawyer and magistrate Pierre de Fermat (1601–1675), the physicist and mathematician Blaise Pascal (1623–1662), and the architect and mathematician Gérard Desargues (1591–1661). Descartes was probably the foremost of these in terms of mathematical innovation, although it is thought that Fermat – for whom mathematics was an absorbing but part-time hobby – had a profound influence upon him. His greatest contribution to science was in virtually founding the discipline of analytical (coordinate) geometry, in which geometrical figures can be described by algebraic expressions. He applied the tenets of geometry to algebra, and was the first to do so, although the converse was not uncommon. Unfortunately, Descartes so much enjoyed the reputation his mathematical discoveries afforded him that he began to envy anyone who then also achieved any kind of mathematical distinction. He therefore regarded Desargues – who published a well-received work on conics – not only as competition but actually as retrogressive. When Pascal then publicly championed Desargues (whom Descartes had openly ridiculed), putting forward an equally accepted form of geometry now known as projective geometry, matters became more than merely unfriendly.

In the meantime, Fermat took no sides, studied both types of geometry, and was in contact with several other European mathematicians. In particular, he used Descartes' geometry to derive an evaluation of the slope of a tangent, finding a method by which to compute the derivative and thus being considered by many the actual formulator of the differential calculus. Part of his study was of tangents as limits of secants. With Pascal he investigated probability theory, and in number theory he independently devised many theorems, one of them now famous as Fermat's Last Theorem.

It is now known that at about this time in Japan, a mathematician called Seki Kowa (*c.* 1642–1708) was independently discovering many of the mathematical innovations also being formulated in the West. Even more remarkably, he managed to change the social order of his time in order to popularize the subject.

The discovery of calculus

Three years after Pascal died a religious recluse haunted by self-doubt, Isaac Newton (1642–1727), was obliged by the spread of the plague to his university college in Cambridge to return home to Woolsthorpe in Lincolnshire and there spend the next year and a half in scientific contemplation. One of his first discoveries was what is now called the binomial theorem, which led Newton to an investigation of infinite series, which in turn led to a study of integration and the notion that it might be achieved as the opposite of differentiation. He arrived at this conclusion in 1666, but did not publish it. More than seven years later, in Germany, Gottfried Leibniz (1646–1716) – who had possibly read the works of Pascal – arrived at exactly the same conclusion, and did publish it. He received considerable acclaim in Europe, much to Newton's annoyance, and a priority argument was very quickly in process. Naively, Leibniz submitted his claim for priority to a committee on which Newton was sitting, so the outcome was a surprise to no one else, but it was in fact Leibniz's notation system that was eventually universally adopted. It was not until 1687 that Newton's studies on calculus were published within his massive *Principia Mathematica*, which also included much of his investigations into physics and optics. Leibniz went on to try to develop a mathematical notation symbolizing logic, but although he made good initial progress it met with little general interest, and despite his energy and status he died a somewhat lonely and forlorn figure.

Another who died in even worse straits was an acquaintance of both Newton and Leibniz: Abraham de Moivre (1667–1754), a Huguenot persecuted for his religious background to the extent that he could find no professional position despite being a first-class and innovative mathematician. He met his end broken by poverty and drink – but not before he had formulated game theory, reconstituted probability theory, and set the business of life insurance on a firm statistical basis.

Leibniz's work on calculus was greatly admired in Europe, and particularly by the great Swiss mathematician family domiciled in Basle: the Bernoullis. The eldest of three brothers, Jakob (or Jacques; 1654–1705), actually corresponded with Leibniz; the youngest, Johan (or Jean; 1667–1748), was recommended by the physicist Christiaan Huygens to a professorship at Groningen. Both brothers were fascinated by investigating possible applications of the new calculus. Unhappily, their study of special curves (particularly cycloids) using polar coordinates proceeded independently along identical lines and resulted in considerable animosity between them. When Jakob died, however, Johan succeeded him at Basle, where he educated his son Daniel – also a brilliant mathematician – whose great friends were Leonhard Euler (1707–1783) and Gabriel Cramer (1704–1752).

The 18th century

Euler may have been the most prolific mathematical author ever. He had amazing energy, a virtually photographic memory and a gift for mental calculation that stood him in good stead late in life when he became totally blind. Not since Descartes had anyone contributed so innovatively to mathematical analysis – Euler's *Introduction* (1748) is considered practically to define in textbook fashion the modern understanding of analytical methodology, including especially the concept of a function. Other works introduced the calculus of variations and the now familiar symbols π, e and i, and systematized differential geometry. He also popularized the use of polar coordinates, and explained the use of graphs to represent elementary functions.

It was his friend Daniel Bernoulli (1700–1782) who had originally managed to secure a position for him in St Petersburg. When, in 1766, Euler returned there from a post at the Prussian Academy, his place in Berlin was taken by the Frenchman Joseph Lagrange (1736–1813) whose ideas ran almost parallel with Euler's. In many ways Lagrange was equally as formative in the popularizing of mathematical analysis, for although he might not have been as energetic or outrightly creative as Euler, he was far more concerned with exactitude and axiomatic rigour, and combined with this a strong desire to generalize. The publication of his studies of number theory and algebra were thus models of precise presentation, and his mathematical research into mechanics began a process of creative thought that has not ceased since. One immediate result of the latter was to inspire his friend and fellow-Frenchman Jean le Rond d'Alembert (1717–1783) to great achievements in dynamics and celestial mechanics. It was d'Alembert who first devised the theory of partial differential equations.

Towards the end of Lagrange's life, when he was already ailing, he became professor of mathematics at the institution which for the next 50 years at least was to exercise considerable influence over the progress of mathematics; the newly-established École Polytechnique in Paris. Two of his contemporaries there were Pierre Laplace (1749–1827) and Gaspard Monge (1746–1818). Laplace became famous for his astronomical calculations, Monge for his textbook on geometry; both were acquaintances of Napoleon Bonaparte – as was Joseph Fourier (1768–1830), the physicist who demonstrated that a function could be expanded in sines and cosines through a series now known as the Fourier series.

It was one of Gaspard Monge's pupils – Jean-Victor Poncelet (1788–1867) – who first popularized the notion of continuity and outlined contemporary thinking on the principle of duality. And it was one of Laplace's colleagues (whom he disliked), Adrien Legendre (1752–1833), who took over where Lagrange left off, and researched into elliptic functions for more than 40 years, eventually deriving the law of quadratic reciprocity and, in number theory, proving that π is irrational.

The 19th century

Legendre's investigations into elliptic integrals were outdated almost as soon as they were published by the work of the Norwegian Niels Abel (1802–1829) and the German Karl Jacobi (1804–1851). Jacobi went on to make important discoveries in the theory of determinants: he was a great interdisciplinarian. The tragically shortlived Abel has probably had the longer-lasting influence, in that he devised the functions now named after him. He was unlucky, too, in that his proof, that in general roots cannot be expressed in radicals was discovered simultaneously and independently by the equally tragic Evariste Galois (1811–1832), who only just had time before his violent death to initiate the theory of groups. Further progress in function theory was made by Augustin Cauchy (1789–1857), a prolific mathematical writer who in his works pioneered many modern mathematical methods, developing in particular the use of limits and continuity. He also originated the theory of complex variables, based at least partly on the work of Jean Argand (1767–1822), who had succeeded in representing complex numbers by means of a graph.

By this time, however, the centre of mathematics in Europe was undoubtedly Göttingen, where the great Karl Gauss (1777–1855) had long presided. Sometimes compared with Archimedes and Newton, Gauss was indisputably not only a mathematical genius who made a multitude of far-reaching discoveries – particularly in geometry and statistical probability – but was also an exceptionally inspirational teacher who inculcated in his pupils the need for meticulous attention to proofs. Late in his tenure at Göttingen, three of his pupils/colleagues were Lejeune Dirichlet (1805–1859), Bernhard Riemann (1826–1866) and Julius Dedekind (1831–1916). There could not have been a more influential quartet in the history of mathematics: the work of all four provides the basis for a major part of modern mathematical knowledge.

Non-Euclidean geometry

Gauss himself was most interested in geometry. Jakob Steiner (1796–1863) in Germany was trying to remove geometry from the 'taint' of analysis as propounded by the French, but Gauss went further and decided to investigate geometry outside the scope of that described by Euclid. It was a momentous decision – made almost simultaneously and quite independently by Nikolai Lobachevsky (1792–1856) and János Bolyai (1802–1860). Between them they thus derived non-Euclidean geometry. The ramifications of this were widespread and fast-moving. In Ireland William Hamilton (1805–1865) suggested the concept of *n*-dimensional space; in Germany Hermann Grassmann (1809–1877) not only defined it but went on to use a form of calculus based on it. But it was Gauss's own pupil, Riemann, who really became the archapostle of the subject. He invented elliptical hyperbolic geometries, introduced 'Riemann surfaces' and redefined conformal mapping (transformations) explaining his innovations with such enthusiasm and accuracy that the modern understanding of time and space now owes much to his work.

Boolean algebra

Meanwhile Dirichlet – who succeeded Gauss when the great man died and himself became an influential teacher – and Dedekind concentrated more on number theory. Dirichlet slanted his teaching of mathematics towards applications in physics, whereas Dedekind was determined to arrive at a philosophical interpretation of the concept of numbers. Such an interpretation was thought likely to be of use in the contemporary search for a mathematical basis for logic. George Boole (1815–1864) had already attempted to create a form of algebra intended to represent logic that, although not entirely successful, was stimulating to others.

Topology

As the study of geometry expanded rapidly, the importance of algebra also increased accordingly. Riemann was influential; Karl Weierstrass (1815–1897) provided important redefinitions in function theory; but in algebraic terms development was next most instigated by the Englishman Arthur Cayley (1821–1895) who discovered the theory of algebraic invariants even as he carried out research into *n*-dimensional geometry. The principles of topology were being established one by one even though the branch itself was not yet complete. Sophus Lie (1842–1899) made important contributions to geometry and to algebra – and indirectly to topology – with the concept of continuous groups and contact transformations, and Cayley went on to invent the theory of matrices. Gaston Darboux (1842–1917) revised popular thinking about surfaces. Felix Klein (1849–1925) – an influential figure in his time – unified all the geometries within his *Erlangen Programme* (1872). But it is Felix Hausdorff (1868–1942) who is actually credited with the formulation of topology.

Dedekind finally achieved his goal and axiomatized the concept of numbers – only for his axioms to be (albeit apologetically and acknowledgedly) 'stolen' from him by Giuseppe Peano (1858–1932). The axioms, however, may have inspired – among others – Hausdorff to conceive the idea of point sets in topology, and Georg Cantor (1843–1918) to define set theory (the basis on which most mathematics is taught in schools today) and transfinite numbers, and certainly caused a revival of interest in number theory generally. Immanuel Fuchs (1833–1902) reformulated much of function theory while attempting to refine Riemann's method for solving differential equations. His pupil, Henri Poincaré (1854–1912) – similarly fascinated by Riemann's work – made many conjectures that were later useful in the investigation of topology and of space and time, but less successfully spent years researching into what are now called integral equations, only to discover after they were finally axiomatized by Ivar Fredholm (1866–1927) that he had done all the work without perceiving the answer.

The 20th century

A different result of the Peano axioms was a renewal of the quest to find a relationship between mathematics and logic. Another system of symbolic logic had been devised by Gottlob Frege (1848–1925), whose pride was turned to ashes when Bertrand Russell (1872–1970) pointed out to him an internal, and fundamental, inconsistency. Russell, with his pupil and friend Alfred North Whitehead (1861–1947), attended lectures given by Peano; together they then published a large work on the foundations of mathematics, entitled *Principia Mathematica*. It had an immediate impact, and remained influential. Other prominent figures in the philosophy of mathematics at the time included Hermann Weyl (1885–1935) and Jacques Herbrand (1908–1931).

Maths and philosophy

The search for meaning in mathematics was not solely philosophical, however. One of Fredholm's pupils was David Hilbert (1862–1943), possibly the latest of the truly great mathematicians. A genuine polymath and an enthusiastic teacher, he expanded virtually all branches of mathematics, especially in the interpretation of geometric structures implied by infinite-dimensional space. He too was involved in the debate over the primary nature of mathematics, formal or intuitional. But all philosophical theories were dealt a heavy blow by the theorem formulated in 1930 by Kurt Gödel (1906–1978). This stated that the overall consistency (i.e. completeness) of mathematics cannot itself be proved mathematically – which means that the foundations of mathematics must forever remain impenetrable.

The days of debate were over; Hilbert went on with his work. Mathematics became gradually either more theoretical or more practical. Theoretically interest swung towards finding features in common between disparate mathematical structures. Henri Lebesgue (1875–1941) devised a concept of measure that contributed greatly to the theory of abstract spaces. Andrei Kolmogorov (1903–) and others not only related this to probability theory but thereby to problems of statistical mechanics and the clarification of the ergodic theorems provided by George Birkhoff (1884–1944) in 1932. In algebraic topology, René Thom (1923–) categorized surfaces. It is worthy of note that thereafter most modern mathematics has concerned itself with such abstract mathematical structures or concepts as fields, rings, or ideals.

Practical applications

The study of statistics and probability was also taken up with new enthusiasm for more practical applications. Karl Pearson (1857–1936) refined Gauss's ideas to derive the notion of standard deviation. Agner Erlang (1878–1929) used probability theory in a highly practical way to aid the efficiency of the circuitry of his capital's telephone system. Alonzo Church (1903–) defined a 'calculable function' and by so doing clarified the nature of algorithms. Following this, George Dantzig (1914–) was able to set up complex linear programmes for computers. Such progress is being maintained, sometimes now as a result of using the machines themselves to devise further advances.

Computer scientists have devised symbolic computation systems which manipulate algebraic expressions in the same way that a human mathematician would do, only faster and more accurately. The result might be called 'computer-assisted mathematics'. A good example is the proof in 1972 of the four-colour theorem by Kenneth Appel and Wolfgang Haken. In 1850 Francis Guthrie conjectured that no more than four colours need be used in order to ensure that no two adjacent colours on a map share the same colour. Mathematicians quickly proved that five colours would suffice, but had no success whatsoever in reducing the number to four. A direct attack by computer would not be possible, for how could a computer consider all possible maps? But Appel and Haken came up with a list of 1936 particular maps, and showed that if each had a rather complicated property, then the conjecture must be true. They then checked this property, case by case, on a computer, taking about 1,200 hours.

Chaos theory

Another modern computer-based development is chaos theory, a theory of non-linear dynamic systems. The central discovery, made in 1961 by US meteorologist Edward Lorenz, is that random behaviour can arise in systems whose mathematical description contains no hint whatever of randomness. The geometry of chaos can be explored using theoretical techniques such as topology but the most vivid pictures are obtained using computer graphics. The geometric structures of chaos are called fractals; they have the same detailed form on all scales of magnification. Frenchman Benoit Mandelbrot (1924–) produced the first fractal images in 1962, using a computer that repeated the same mathematical pattern over and over again. In 1975 US mathematician Mitchell Feigenbaum discovered a new universal constant (approximately 4.669201609103) which is important in chaos theory. Order and chaos, traditionally seen as opposites, are now viewed as two aspects of the same basic process, the evolution of a system in time.

In 1980 mathematicians completed the classification of all finite and simple groups. The classification has taken over a hundred mathematicians more than 35 years to complete, and covers over 14,000 pages in mathematical journals. In 1989 a team of US computer mathematicians discovered the highest known prime number (the number contains 65,087 digits).

In Oct 1994, Andrew Wiles, an English mathematician at Princeton University, announced that he had solved Fermat's Last Theorem. The theorem's proof had eluded mathematicians for over 300 years.

David Abbott

Physics

Introduction

Physics is a branch of science in which the theoretical and the practical are firmly intertwined. It has been so since ancient times, as physicists have striven to interpret observation or experiment in order to arrive at the fundamental laws that govern the behaviour of the universe. Physicists aim to explain the manifestations of matter and energy that characterize all things and processes, both living and inanimate, extending from the grandest of galaxies down to the most intimate recesses of the atom.

The history of physics has not been a straight and easy road to enlightenment. The exploration of new directions sometimes leads to dead ends. New ways of looking at things may result in the overthrow of a previously accepted system. Not Aristotle's system, nor Newton's, nor even Einstein's was 'true'; rather statements, or 'laws', in physics satisfy contemporary requirements or – in the existing state of knowledge – contemporary possibilities. The question that physicists ask is not so much 'Is it true?' as 'Does it work?'

Physics has many strands – such as mechanics, heat, light, sound, electricity, and magnetism – and, although they are often pursued separately, they are also all ultimately interdependent. To pursue the history of physics, therefore, it is necessary to follow several separate chains of discovery and then to find the links between them. The story is of frustration and missed opportunities as well as of genius and perseverence. But however complex it may appear, all physicists seek or have sought to play a part in the evolution of an ultimate explanation of all the effects that occur throughout the universe. That goal may be unattainable but the thrust towards it has kept physics as alive and vital today as it was when it originated in ancient times.

Force and motion

The development of an understanding of the nature of force and motion was a triumph for physics, one which marked the evolution of the scientific method. As in most other branches of physics, this development began in ancient Greece.

The earliest discovery in physics, apart from observations of effects like magnetism, was the relation between musical notes and the lengths of vibrating strings. Pythagoras (*c.* 582–*c.* 497 BC) found that harmonious sounds were given by strings whose lengths were in simple numerical ratios, such as 2:1, 3:2 and 4:3. From this discovery the belief grew that all explanations could be found in terms of numbers. This was developed by Plato (*c.* 427–*c.* 347 BC) into a conviction that the cause underlying any effect could be expressed in mathematical form. The motion of the heavenly bodies, Plato reasoned, must consist of circles, since these were the most perfect geometric forms.

Atoms

Reason also led Democritus (*c.* 470–*c.* 380 BC) to propose that everything consisted of minute indivisible particles called atoms. The properties of matter depend on the characteristics of the atoms of which it is composed, and the atoms combine in ways that are determined by unchanging fundamental laws of nature.

Aristotle's four elements

A third view of the nature of matter was given by Aristotle (384–322 BC), who endeavoured to interpret the world as he observed it, without recourse to abstractions such as atoms and mathematics. Aristotle reasoned that matter consisted of four elements – earth, water, air, and fire – with a fifth element, the ether, making up the heavens. Motion occurred when an object sought its rightful place in the order of elements, rocks falling through air and water to the earth, air rising through water as bubbles, and fire through air as smoke.

There was value in all these approaches and physics has absorbed them all to some degree. Plato was essentially correct; only his geometry was wrong, the planets following elliptical, not circular, orbits. Atoms do exist as Democritus foretold and they do explain the properties of matter. Aristotle's emphasis on observation (though not his reasoning) was to be a feature of physics and many other sciences, notably biology, of which he may be considered the founder.

Archimedes' scientific method

These ideas were, however, mainly deductions based solely on reason. Few of them were given the test of experiment to prove that they were right. Then came the achievements of Archimedes (*c.* 287–212 BC), who discovered the law of the lever and the principle of flotation by measuring the effects that occur and deduced general laws from his results. He was then able to apply his laws, building pulley systems and testing the purity of the gold in King Hieron's crown by a method involving immersion.

Archimedes thus gave physics the scientific method. All subsequent principal advances made by physicists were to take the form of mathematical interpretations of observations and experiments. Archimedes developed the method in founding the science of statics – how forces interact to produce equilibrium. But an understanding of motion lay a long way off. In the centuries following the collapse of Greek civilization in around AD 100, physics marked time. The Arabs kept the Greek achievements alive, but they made few advances in physics, while in Europe the scientific spirit was overshadowed by the 'Dark Ages'. Then in about 1200, the spirit of enquiry was rekindled in Europe by the import of Greek knowledge from the Arabs. Unfortunately, progress was hindered somewhat by the fact that Aristotle's ideas, particularly his views on motion, prevailed. Aristotle had assumed that a heavy object falls faster than a light object simply because it is heavier. He also argued that a stone continues to move when thrown because the air displaced by the stone closes behind it and pushes the stone. This explanation derived from Aristotle's conviction that nature abhors a vacuum (which is why he placed a fifth element in the heavens).

Gravity and Newton's laws of motion
Aristotle's ideas on falling bodies were probably first disproved by Simon Stevinus (1548–1620), who is believed to have dropped unequal weights from a height and found that they reached the ground together. At about the same time Galileo (1564–1642) measured the speeds of 'falling' bodies by rolling spheres down an inclined plane and discovered the laws that govern the motion of bodies under gravity. This work was brought to a brilliant climax by Isaac Newton (1642–1727), who in his three laws of motion achieved an understanding of force and motion, relating them to mass and recognizing the existence of inertia and momentum. Newton thus explained why a stone continues to move when thrown; and he showed the law of falling bodies to be a special case of his more general laws. Newton went on to derive from existing knowledge of the motion and dimensions of the Earth–Moon system a universal law of gravitation, one which provided a mathematical statement for the laws of planetary motion discovered empirically by Johann Kepler (1571–1630).

Newton's laws of motion and gravitation, which were published in 1687, were fundamental laws which sought to explain all observed effects of force and motion. This triumph of the scientific method heralded the Age of Reason – not the Greek kind of reasoning, but a belief that all could be explained by the deduction of fundamental laws upheld by observation or experiment. It was to result in an explosion of scientific discovery in physics that has continued to the present day. In the field of force and motion, important advances were made with the discovery of the law governing the pendulum and the principle of conservation of momentum by Christiaan Huygens (1629–1695) and the determination of the gravitational constant by Henry Cavendish (1731–1810).

The behaviour of matter
Physics is basically concerned with matter and energy, and investigation into the behaviour of matter also originated in ancient Greece with Archimedes' work concerning flotation. As with force and motion, Simon Stevinus made the first post-Greek advance with the discovery that the pressure of a liquid depends on its depth and area. This achievement was developed by Blaise Pascal (1623–1662), who found that pressure is transmitted throughout a liquid in a closed vessel, acting perpendicularly to the surface at any point. Pascal's principle is the basis of hydraulics. Pascal also investigated the mercury barometer invented in 1643 by Evangelista Torricelli (1608–1647) and showed that air pressure supports the mercury column and that there is a vacuum above it, thus disproving Aristotle's contention that a vacuum cannot exist. The immense pressure that the atmosphere can exert was subsequently demonstrated in several sensational experiments by Otto von Guericke (1602–1686).

Solid materials were also investigated. The fundamental law of elasticity was discovered by Robert Hooke (1635–1703) in 1678 when he found that the stress (force) exerted is proportional to the strain (elongation) produced. Thomas Young (1773–1829) later showed that a given material has a constant, known as Young's modulus, that defines the strain produced by a particular stress.

The effects that occur with fluids (liquids or gases) in motion were then explored. Daniel Bernoulli (1700–1782) established hydrodynamics with his discovery that the pressure of a fluid depends on its velocity. Bernoulli's principle explains how lift occurs and led eventually to the invention of heavier-than-air flying machines. It also looked forward to ideas of the conservation of energy and the kinetic theory of gases. Other important advances in our understanding of fluid flow were later made by George Stokes (1819–1903), who discovered the law that relates motion to viscosity, and Ernst Mach (1838–1916) and Ludwig Prandtl (1875–1953), who investigated the flow of fluids over surfaces and made discoveries vital to aerodynamics.

The effects of light

The Greeks were aware that light rays travel in straight lines, but they believed that the rays originate in the eyes and travel to the object that is seen. Euclid (*c.* 330–260 BC), Hero (lived AD 60) and Ptolemy (lived 2nd century AD) were of this opinion although, recognizing that optics is essentially a matter of geometry, they discovered the law of reflection and investigated refraction.

Optics made an immense stride forward with the work of Alhazen (*c.* 965–1038), who was probably the greatest scientist of the Middle Ages. Alhazen recognized that light rays are emitted by a luminous source and are then reflected by objects into the eyes. He studied images formed by curved mirrors and lenses and formulated the geometrical optics involved. Alhazen's discoveries took centuries to filter into Europe, where they were not surpassed until the 17th century. The refracting telescope was then invented in Holland in 1608 and quickly improved by Galileo and Kepler, and in 1621 Willebrord Snell (1580–1626) discovered the laws that govern refraction.

The discovery of the spectrum

The next major steps forward were taken by Newton, who not only invented the reflecting telescope in 1668, but a couple of years earlier found that white light is split into a spectrum of colours by a prism. Newton published his work in optics in 1704, provoking great controversy with his statement that light consists of a stream of particles. Huygens had put forward the view that light consists of a wave motion, an opinion reinforced by the discovery of diffraction by Francesco Grimaldi (1618–1663). Such was Newton's reputation, however, that the particulate theory held sway for the following century. In 1801 Young discovered the principle of interference, which could be explained only by assuming that light consisted of waves. This was confirmed in 1821, when Augustin Fresnel (1788–1827) showed from studies of polarized light, which had been discovered by Étienne Malus (1775–1812) in 1808, that light is made up of a transverse wave motion, not longitudinal as had previously been thought.

Newton's discovery of the spectrum remained little more than a curiosity until 1814, when Joseph von Fraunhofer (1787–1826) discovered that the Sun's spectrum is crossed by the dark lines now known as Fraunhofer lines. Fraunhofer was unable to explain the lines, but he did go on to invent the diffraction grating for the production of high-quality spectra and the spectroscope to study them. An explanation of the lines was provided by Gustav Kirchhoff (1824–1887), who in 1859 showed that they are caused by elements present in the Sun's atmosphere. With Robert Bunsen (1811–1899), Kirchhoff discovered that elements have unique spectra by which they can be identified, and several new elements were found in this way. In 1885, Johann Balmer (1825–1898) derived a mathematical relationship governing the frequencies of the lines in the spectrum of hydrogen. This later proved to be a crucial piece of evidence for revolutionary theories of the structure of the atom.

Meanwhile, several scientists investigated the phenomenon of colour, notably Young, Hermann von Helmholtz (1821–1894), and James Clerk Maxwell (1831–1879). Their research led to the establishment of the three-colour theory of light, which showed that the eye responds to varying amounts of red, green, and blue in light and mixes them to give particular colours. This led directly to colour photography and other methods of colour reproduction used today.

The speed of light

The velocity of light was first measured accurately in 1862 by Jean Foucault (1819–1868), who obtained a value within 1% of the correct value. This led to a famous experiment performed by Albert Michelson (1852–1931) and Edward Morley (1838–1923) in which the velocity of light was measured in two directions at right angles. Their purpose was to test the theory that a medium called the ether existed to carry light waves. If it did exist, then the two values obtained would be different. The Michelson–Morley experiment, performed in 1881 and then again in 1887, yielded a negative result both times (and on every occasion since), thus proving that the ether does not exist.

More important, the Michelson–Morley experiment showed that the velocity of light is constant regardless of the motion of the observer. From this result, and from the postulate that all motion is relative, Albert Einstein (1879–1955) derived the special theory of relativity in 1905. The principal conclusion of special relativity is that in a system moving relative to the observer, length, mass, and time vary with the velocity. The effects become noticeable only at velocities approaching light; at slower velocities, Newton's laws hold good. Special relativity was crucial to the formulation of new ideas of atomic structure and it also led to the idea that mass and energy are equivalent, an idea used later to explain the great power of nuclear reactions. In 1915 Einstein published his general theory of relativity, in which he showed that gravity distorts space. This explained an anomaly in the motion of Mercury, which does not quite obey Newton's laws, and it was dramatically confirmed in 1919 when a solar eclipse revealed that the Sun's gravity was bending light rays coming from stars.

Electricity and magnetism

The phenomena of electricity and magnetism are believed to have been first studied by the ancient Greek philosopher Thales (624–546 BC), who was considered by the Greeks to be the founder of their science. Thales found that a piece of amber picks up light objects when rubbed, the action of rubbing thus producing a charge of static electricity. The words 'electron' and 'electricity' came from this discovery, *elektron* being the Greek word for amber. Thales also studied the similar effect on each other of pieces of lodestone, a magnetic mineral found in the region of Magnesia. It is fitting that the study of electricity and magnetism originated together, for the later discovery that they are linked was one of the most important ever made in physics.

No further progress was made, however, for nearly 2,000 years. The strange behaviour of amber remained no more than a curiosity, though magnets were used to make compasses. From this, Petrus Peregrinus (lived 13th century) discovered the existence of north and south poles in magnets and realized that they attract or repel each other. William Gilbert (1544–1603) first explained the Earth's magnetism and also investigated electricity, finding other substances besides amber that produce attraction when rubbed.

Then Charles Du Fay (1698–1739) discovered that substances charged by rubbing may repel as well as attract in a similar way to magnetic poles and Benjamin Franklin (1706–1790) proposed that positive and negative charges are produced by the excess or deficiency of electricity. Charles Coulomb (1763–1806) measured the forces produced between magnetic poles and between electric charges and found that they both obey the same inverse square law.

The invention of the battery

A major step forward was taken in 1800, when Alessandro Volta (1745–1827) invented the battery. A source of current electricity was now available and in 1820 Hans Oersted (1777–1851) found that an electric current produces a magnetic field. This discovery of electromagnetism was immediately taken up by Michael Faraday (1791–1867), who realized that magnetic lines of force must surround a current. This concept led him to discover the principle of the electric motor in 1821 and electromagnetic induction in 1831, the phenomenon in which a changing magnetic field produces a current. This was independently discovered by Joseph Henry (1797–1878) at the same time.

Electrical theory

Meanwhile, important theoretical developments were taking place in the study of electricity. In 1827, André Ampère (1775–1836) discovered the laws relating magnetic force to electric current and also properly distinguished current from tension, or EMF. In the same year, Georg Ohm (1789–1854) published his famous law relating current, EMF and resistance. Kirchhoff later extended Ohm's law to networks, and he also unified static and current electricity by showing that electrostatic potential is identical to EMF.

In the 1830s, Carl Gauss (1777–1855) and Wilhelm Weber (1804–1891) defined a proper system of units for magnetism; later they did the same for electricity. In 1845

Faraday found that materials are paramagnetic or diamagnetic, and Lord Kelvin (1824–1907) developed Faraday's work into a full theory of magnetism. An explanation of the cause of magnetism was finally achieved in 1905 by Paul Langevin (1872–1946), who ascribed it to electron motion.

Electricity and magnetism were finally brought together in a brilliant theoretical synthesis by James Clerk Maxwell. From 1855 to 1873 Maxwell developed the theory of electromagnetism to show that electric and magnetic fields are propagated in a wave motion and that light consists of such an electromagnetic radiation. Maxwell predicted that other similar electromagnetic radiations must exist and, as a result, Heinrich Hertz (1857–1894) produced radio waves in 1888. X-rays and gamma rays were discovered accidentally soon after.

The nature of heat and energy

The first step towards measurement – and therefore an understanding – of heat was taken by Galileo, who constructed the first crude thermometer in 1593. Gradually these instruments improved and in 1714 Daniel Fahrenheit (1686–1736) invented the mercury thermometer and devised the Fahrenheit scale of temperature. This was replaced in physics by the Celsius or Centigrade scale proposed by Anders Celsius (1701–1744) in 1742.

At this time, heat was considered to be a fluid called caloric that flowed into or out of objects as they got hotter or colder, and even after 1798 when Count Rumford (1753–1814) showed the idea to be false by his observation of the boring of cannon, it persisted. Earlier Joseph Black (1728–1799) had correctly defined the quantity of heat in a body and the latent heat and specific heat of materials, and his values had been successfully applied to the improvement of steam engines. In 1824, Sadi Carnot (1796–1832), also a believer in the caloric theory, found that the amount of work that can be produced by an engine is related only to the temperature at which it operates.

Carnot's theorem, though not invalidated by the caloric theory, suggested that, since heat gives rise to work, it was likely that heat was a form of motion, not a fluid. The idea also grew that energy may be changed from one form to another (that is from heat to motion) without a change in the total amount of energy involved. The interconvertibility of energy and the principle of the conservation of energy were established in the 1840s by several physicists. Julius Mayer (1814–1878) first formulated the principle in general terms and obtained a theoretical value for the amount of work that may be obtained by the conversion of heat (the mechanical equivalent of heat). Helmholtz gave the principle a firmer scientific basis and James Joule (1818–1889) made an accurate experimental determination of the mechanical equivalent. Rudolf Clausius (1822–1888) and Kelvin developed the theory governing heat and work, thus founding the science of thermodynamics. This enabled Kelvin to propose the absolute scale of temperature that now bears his name.

The equivalence of heat and motion led to the kinetic theory of gases, which was developed by John Waterston (1811–1883), Clausius, Maxwell and Ludwig Boltzmann (1844–1906) between 1845 and 1868. It gave a theoretical description of all effects of heat in terms of the motion of molecules.

Heat radiation – quantum theory

During the 19th century it also came to be understood that heat may be transmitted by a form of radiation. Pioneering theoretical work on how bodies exchange heat had been carried out by Pierre Prévost (1751–1839) in 1791, and the Sun's heat radiation had been discovered to consist of infrared rays by William Herschel (1738–1822) in 1800. In 1862 Kirchhoff derived the concept of the perfect black body – one that absorbs and emits radiation at all frequencies. In 1879 Josef Stefan (1835–1893) discovered the law relating the amount of energy radiated by a black body to its temperature, but physicists were unable to relate the frequency distribution of the radiation to the temperature. This increases as the temperature is raised, causing an object to glow red, yellow and then white as it gets hotter. Lord Rayleigh (1842–1919) and Wilhelm Wien (1864–1928) derived incomplete theories of this effect, and then in 1900 Max Planck (1858–1947) showed that it could be explained only if radiation consisted of indivisible units, called quanta, whose energy was proportional to their frequency.

Planck's quantum theory revolutionized physics. It showed that heat radiation and other electromagnetic radiations including light must consist of indivisible particles of energy and not of waves as had previously been thought. In 1905 Einstein found a ready explanation of the photoelectric effect using quantum theory, and the theory was experimentally confirmed by James Franck (1882–1964) in the early 1920s.

Low temperature physics and superconductivity

Another advance in the study of heat that took place in the same period was the production of low temperatures. In 1852 Joule and Kelvin found the effect named after them is used to produce refrigeration by adiabatic expansion of a gas, and James Dewar (1842–1923) developed this effect into a practical method of liquefying gases from 1877 onwards. Heike Kamerlingh-Onnes (1853–1926) first produced temperatures within a degree of absolute zero and in 1911 he discovered superconductivity. A theoretical explanation of superconductivity had to await the work of John Bardeen (1908–1991), Leon Cooper (1930–), and John Schrieffer (1931–). Their ideas, the 'BCS theory', explained superconductivity as the result of electrons coupling in pairs, called Cooper pairs, that do not undergo scattering by collision with atoms in a conductor. In 1986 IBM researchers in Zurich, Georg Bednorz and Alex Muller, produced superconductivity in metalllic ceramics at relatively high temperatures, around 35K. The theoretical explanation of high-temperature superconductivity was still being developed in the early 1990s.

Sound

Sound is the one branch of physics that was well established by the Greeks, especially by Pythagoras. They surmised, correctly, that sound does not travel through a vacuum, a contention proved experimentally by Guericke in 1650. Measurements of the velocity of sound in air were made by Pierre Gassendi (1592–1655) and in other materials by August Kundt (1839–1894). Ernst Chladni (1756–1827) studied how the vibration of surfaces produces sound waves, and in 1845 Christian Doppler (1803–1853) discovered the effect relating the frequency (pitch) of sound to the relative motion of the source and observer. The Doppler effect is also produced by light and other wave motions and has proved to be particularly valuable in astronomy.

The structure of the atom

The existence of atoms was proved theoretically by chemists during the 19th century, but the first experimental demonstration of their existence and the first estimate of their dimensions was made by Jean Perrin (1870–1942) in 1909.

The principal direction taken in physics in this century has been to determine the inner structure of the atom. It began with the discovery of the electron in 1897 by J J Thomson (1856–1940), who showed that cathode rays consist of streams of minute indivisible electric particles. The charge and mass of the electron were then found by John Townsend (1868–1937) and Robert Millikan (1868–1953).

Radioactivity

Meanwhile, another important discovery had been made with the detection of radioactivity by Antoine Becquerel (1852–1908) in 1896. Three kinds of radioactivity were found; these were named alpha, beta, and gamma by Ernest Rutherford (1871–1937). Becquerel recognized in 1900 that beta particles are electrons. In 1903 Rutherford explained that radioactivity is caused by the breakdown of atoms. In 1908 he identified alpha particles as helium nuclei, and in association with Hans Geiger (1882–1945) produced the nuclear model of the atom in 1911, proposing that it consists of electrons orbiting a nucleus. Then in 1914 Rutherford identified the proton and in 1919 he produced the first artificial atomic disintegration by bombarding nitrogen with alpha particles.

Rutherford's pioneering elucidation of the basic structure of the atom was aided by developments in the use of X-rays, which had been discovered in 1895 by Wilhelm Röntgen (1845–1923). In 1912 Max von Laue (1879–1960) produced diffraction in X-rays by passing them through crystals, showing X-rays to be electromagnetic waves, and Lawrence Bragg (1890–1971) developed this method to determine the arrangement of atoms in crystals. His work influenced Henry Moseley (1887–1915), who in 1914 found by studying X-ray spectra that each element has a particular atomic number, equal to the number of protons in the nucleus and to the number of electrons orbiting it.

In 1913, Niels Bohr (1885–1962) achieved a brilliant synthesis of Rutherford's nuclear model of the atom and Planck's quantum theory. He showed that the electrons must move in orbits at particular energy levels around the nucleus. As an atom emits or absorbs radiation, it moves from one orbit to another and produces or gains a certain number of quanta of energy. In so doing, the quanta give rise to particular frequencies of radiation, producing certain lines in the spectrum of the radiation. Bohr's theory was able to explain the spectral lines of hydrogen and their relationship, found earlier by Balmer.

Wave-particle theory

These discoveries, made so quickly, seemed to achieve an astonishingly complete picture of the atom, but more was to come. In 1923, Louis de Broglie (1892–1987) described how electrons could behave as if they made up waves around the nucleus. This discovery was developed into a theoretical system of wave mechanics by Erwin Schrödinger (1887–1961) in 1926 and experimentally confirmed in the following year. It showed that electrons exist both as particles and waves. Furthermore it reconciled Planck's quantum theory with classical physics by indicating that electromagnetic quanta or photons, which were named and detected experimentally in X-rays by Arthur Compton (1892–1962) in 1923, could behave as waves as well as particles. A prominent figure in the study of atomic structure was Werner Heisenberg (1901–1976), who showed in 1927 that the position and momentum of the electron in the atom cannot be known precisely, but only found with a degree of probability or uncertainty. His uncertainty principle follows from wave-particle duality and it negates cause and effect, an uncomfortable idea in a science that strives to reach laws of universal application.

The next step was to investigate the nucleus. A series of discoveries of nuclear particles accompanying the proton were made, starting in 1932 with the discovery of the positron by Carl Anderson (1905–1991) and the neutron by James Chadwick (1891–1974).

This work was aided by the development of particle accelerators, beginning with the voltage multiplier built by John Cockcroft (1897–1967) and Ernest Walton (1903–1995), which achieved the first artifical nuclear transformation in 1932. It led to the discovery of nuclear fission by Otto Hahn (1879–1968) in 1939 and the production of nuclear power by Enrico Fermi (1901–1954) in 1942.

Particle physics

Much of modern physics has been concerned with the behaviour of elementry particles. The first major theory in this area was quantum electrodynamics (QED), developed by US physicists Richard Feynman (1918–1988) and Julian Schwinger (1918–1994), and by Japanese physicist Sin-Itiro Tomonaga (1906–1979). This theory describes the interaction of charged subatomic particles in electric and magnetic fields. It combines quantum theory and relativity and considers charged particles to interact by the exchange of photons. QED is remarkable for the accuracy of its predictions – for example, it has been used to calculate the value of some physical quantities to an accuracy of ten decimal places, a feat equivalent to calculating the distance between New York and Los Angeles to within the thickness of a hair.

By 1960 the existence of around 200 elementary particles had been established, some of which did not behave as theory predicted. They did not decay into other particles as quickly as theory predicted, for example. To explain these anomalies, US theoretical physicist Murray Gell-Mann (1929–) developed a classification for elementary particles, called the eightfold way. This scheme predicted the existence of previously undetected particles. The omega-minus particle found in 1964 confirmed the theory. In the same year Gell-Mann suggested that some elementary particles were made up of smaller particles called quarks which could have fractional electric charges. This idea explained the eightfold classification and now forms the basis of the standard model of elementary particles and their interactions.

The details of the standard model have been confirmed by experiment. In 1991 experiments at CERN, the European particle physics laboratory at Geneva, confirmed the existence of three generations of elementary particles, each with two quarks and leptons (light particles) as predicted by the standard model. In 1995 researchers at Fermilab discovered the top quark, the final piece of evidence in support of the standard model.

Quantum chromodynamics (QCD) is the mathematical theory, similar in many ways to quantun electrodynamics, which describes the interactions of quarks by the exchange of particles called gluons. The mathematics involved is very complex and although a number of successful predictions have been made, as yet the theory does not compare in accuracy with QED.

The success of the mathematical methods of quantum electrodynamics and quantum chromodynamics encouraged others to use these methods to unify the theory of the fundamental forces. Abdus Salam (1926–), Steven Weinberg (1933–), and Sheldon Glashow (1932–) demonstrated that at high energies the electromagnetic and weak nuclear force could be regarded as aspects of a single combined force, the electroweak force. This was confirmed in 1983 by the discovery of new particles predicted by the theory.

In the 1980s a mathematical theory called string theory was developed, in which the fundamental objects of the universe were not point-like particles but extremely small string-like objects. These objects exist in a universe of ten dimensions, although for reasons which are not yet understood, only three space dimensions and one time dimension are discernible. There are many unresolved difficulties, but some physicists think that string theory, or some variant of it, could develop into a 'theory of everything' that explains space–time, together with the elementary particles and their interactions, within one comprehensive framework.

New states of matter

In 1995 scientists at the University of Colorado discovered a new form of matter when they used lasers and magnetic fields to cool rubidium atoms to within 20 billionths of a degree of absolute zero. For a brief time, the atoms lost their individuality and behaved as if they were part of a single giant atom. This type of material is called a Bose–Einstein condensate.

The first atoms of antimatter were produced at CERN 1995 by combining antiprotons and positrons to make antihydrogen. Eleven antiatoms were created. Unfortunately they only lasted about 30 nanoseconds before being destroyed in the detectors.

David Abbott

Abbe Ernst 1840–1905. German physicist who, working with Carl Zeiss, greatly improved the design and quality of optical instruments, particularly the compound microscope. This enabled researchers to observe microorganisms and internal cellular structures for the first time.

Abbe was born in Eisenach, Thuringia, and studied physics at Jena and Göttingen, becoming a professor at Jena 1870 and director of the observatory 1878. Zeiss supplied optical instruments to the university and repaired them. Abbe became a partner in Zeiss's firm.

Abbe worked out how to overcome spherical aberration in lenses and why, contrary to expectation, the definition of a microscope decreases with a reduction in the aperture of the objective; he found that the loss in resolving power is a diffraction effect. In 1872 he developed the Abbe substage condenser for illuminating objects under high-power magnification.

Abel Frederick Augustus 1827–1902. British scientist and inventor who developed explosives. As a chemist to the War Department, he introduced a method of making guncotton and was joint inventor with James Dewar of cordite. He also invented the Abel close-test instrument for determining the flash point (ignition temperature) of petroleum. Baronet 1893.

Abel John Jacob 1857–1938. US biochemist, discoverer of adrenaline. He studied the chemical composition of body tissues, and this led, in 1898, to the discovery of adrenaline, the first hormone to be identified, which Abel called epinephrine. He later became the first to isolate amino acids from blood.

Abel Niels Henrik 1802–1829. Norwegian mathematician. He demonstrated that the general quintic equation $ax^5 + bx^4 + cx^3 + dx^2 + ex + f = 0$ could not be solved algebraically. Subsequent work covered elliptic functions, integral equations, infinite series, and the binomial theorem.

Abel was born at Finnöy, a small island near Stavanger, and studied at Oslo. In 1823 he provided the first solution in the history of mathematics of an integral equation and published a paper demonstrating that a radical expression to represent a solution to fifth- or higher-degree equations was impossible.

In 1825, Abel went to Berlin, where he met Leopold Crelle (1780–1855), a privy councillor and engineer much taken with problems in mathematics. Together they brought out the first issue of *Crelle's Journal*, which was to become the leading 19th-century German organ of mathematics. A year later Abel moved on to Paris, where he wrote 'Mémoiresur une propriété générale d'une classetrès-étendue de fonctions transcendantes'. It dealt with the sum of the integrals of a given algebraic function and presented the theorem that any such sum can be expressed as a fixed number of these integrals with integration arguments that are algebraic functions of the original arguments.

By studying the masters – not their pupils.

Niels Abel

when asked how he became
a great mathematician

Abel transformed the theory of elliptic integrals by introducing elliptic functions, and this generalization of trigonometric functions led eventually to the theory of complex multiplication, with its important implications for algebraic number theory. He also provided the first stringent proof of the binomial theorem. A number of useful concepts in modern mathematics, notably the Abelian group and the Abelian function, bear his name.

Abetti Giorgio 1882–1982. Italian astrophysicist who wrote a popular text on the Sun 1963. He participated in numerous expeditions to observe eclipses of the Sun, and led one such expedition to Siberia to observe the total solar eclipse of 19 June 1936.

Abetti was born in Padua and studied at the universities of Padua and Rome. He was a professor at the University of Florence 1921–57 and director of the Arcetri Observatory there 1921–52.

He wrote a handbook of astrophysics, published 1936, and a popular history of astronomy, which appeared in English 1952.

Abraham Edward Penley 1913– . British biochemist who isolated the antibiotic cephalosporin, capable of destroying penicillin-resistant bacteria. Knighted 1980.

Achard Franz Karl 1753–1821. German chemist who was largely responsible for developing the industrial process by which table sugar (sucrose) is extracted from sugar beet. He improved the quality of available beet and erected the first factory for the extraction of sugar in Silesia (now in Poland) 1802.

Achard was a student of Andreas Margraff (1709–1782), who isolated sugar from beet on an experimental scale in Germany 1747.

Adams John Couch 1819–1892. English astronomer who mathematically deduced the existence of the planet Neptune 1845 from the effects of its gravitational pull on the motion of Uranus, although it was not found until 1846 by J G Galle. Adams also studied the Moon's motion, the Leonid meteors, and terrestrial magnetism.

Adams was born in Landeast, Cornwall, and educated at Cambridge, where he spent virtually his entire career. He became professor of mathematics at the University of St Andrews, Fife 1858, Lowndean professor of astronomy and geometry at Cambridge 1859–92, and director of Cambridge observatory 1861–92.

The calculations to account for certain aberrations in the orbit of Uranus were taken up independently by Adams and French astronomer Urbain Leverrier.

By 1845 Adams had determined the position and certain characteristics of the hypothetical planet affecting the orbit, but a search for the new planet was not instigated for nearly a year at Cambridge. Meanwhile, Leverrier sent his figures to Galle at the Berlin Observatory, and Galle, having better maps, was able to find the planet within a few hours. The discovery of Neptune was credited to Leverrier.

Adams Roger 1889–1971. US organic chemist, known for his painstaking analytical work to determine the composition of naturally occurring substances such as complex vegetable oils and plant alkaloids.

Adams Walter Sydney 1876–1956. US astronomer who developed the use of spectroscopy in the study of stars and planets. He found that luminosity and the relative intensities of particular spectral lines could distinguish giant stars from dwarf stars. Spectra could also be used to study the physical properties, motions, and distances of stars.

Adams was born near Antioch, Syria, and studied celestial mechanics at the University of Chicago. His work on stellar spectroscopy began under George Hale at the Yerkes Observatory in Wisconsin. In 1904 he assisted Hale in the establishment of the Mount Wilson Observatory above Pasadena in California, becoming its director 1923.

Adams was involved in a long-term project with other astronomers to determine the absolute magnitudes of stars; together, they found the value for 6,000 stars. A second long-term collaborative project was the determination of the radial velocities of more than 7,000 stars.

In 1925 Adams made an observation of the gravitational field of Sirius B which corroborated Albert Einstein's theory of relativity.

Adams studied the atmosphere of Mars and Venus, reporting in 1932 the presence of carbon dioxide in the atmosphere of Venus and, in 1934, the occurrence of oxygen in concentrations of less than 0.1% on Mars. He was responsible for the design and installation of the 254-cm/100-in and 508-cm/200-in telescopes at Mounts Wilson and Palomar.

At Mount Wilson Adams was able to demonstrate that sunspots have a lower temperature than the rest of the solar disc. He also used Doppler displacements to study the rotation of the Sun.

Adamson Joy Friedericke Victoria (born Gessner) 1910–1985. German-born naturalist whose work with wildlife in Kenya, including the lioness Elsa, is described in the book *Born Free* 1960. She was murdered at her home in Kenya. She worked with her third husband, British game warden George Adamson (1906–1989), who was murdered by bandits.

Adanson Michel 1727–1806. French botanist who developed a classification system for plants which, unlike Linnaeus's system, was based upon many characteristics rather than just fructification, and brought the two systems into conflict. A genus of African savannah trees, *Adansonia* is named after him.

Adanson was born in Aix-en-Provence, France, but his family moved to Paris 1730 and he spent most of his childhood there. He studied at Plessis Sorbon, the Collége Royal and the Jardin du Roi. He travelled to Senegal for the Compagnie des Indes and collected numerous plant and animal specimens, many of which he later contributed to Antoine de Jussieu and also to Linnaeus. His classification scheme was outlined in *Familles des plantes* 1763–64, and he had a wide influence upon contemporary botanists. He was a member of the Legion of Honour and a foreign member of the Royal Society, London.

Addison Thomas 1793–1860. British physician who first recognized the condition known as Addison's disease in 1855. He was the first to correlate a collection of symptoms with pathological changes in an endocrine gland. He is also known for his discovery of what is now called pernicious (or Addison's) anaemia.

Addison was born in Longbenton, Northumberland, and studied medicine at Edinburgh and at Guy's Hospital, London, where he remained for the rest of his life. He gave a preliminary account of Addison's disease in 1849, and more fully in *On the Constitutional and Local Effects of Disease of the Suprarenal Capsules* 1855, differentiating it from pernicious anaemia.

Addison also described xanthoma (flat, soft spots that appear on the skin, usually on the eyelids) and wrote about other skin diseases, tuberculosis, pneumonia, and the anatomy of the lung. In collaboration with John Morgan (1797–1847), he wrote *An Essay on the Operation of Poisonous Agents Upon the Living Body* (1829), the first book on this subject to be published in English. And in 1839 appeared the first volume of *Elements of the Practice of Medicine*, written by Addison and Richard Bright, also at Guy's. In this volume Addison gave the first full description of appendicitis.

Ader Clément 1841–1926. French aviation pioneer and inventor. He demonstrated stereophonic sound transmission by telephone at the 1881 Paris Exhibition of Electricity. His steam-driven aeroplane, the *Eole*, made the first powered takeoff in history 1890, but it could not fly. In 1897, with his *Avion III*, he failed completely, despite false claims made later.

Whenever a child lies you will always find a severe parent. A lie would have no sense unless the truth were felt to be dangerous.

Alfred Adler

New York Times 1949

Adler Alfred 1870–1937. Austrian psychologist who saw the 'will to power' as more influential in accounting for human behaviour than the sexual drive. A dispute over this theory led to the dissolution of his ten-year collaboration with psychiatry's founder Sigmund Freud.

The concepts of inferiority complex and overcompensation originated with Adler.

Born and trained in Vienna, Adler was a general practitioner and nerve specialist there 1897–1927. By 1902 he had made contact

Adler *Alfred Adler was a prominent member of the circle of psychologists surrounding Sigmund Freud during the early 1900s. In 1911, after professional disagreement concerning Freud's theories, he left and developed his psychoanalytical theory of individual psychology.*

with Freud. He played a major part in the development of the psychoanalytical movement, and was president of the Vienna Psychoanalytical Society. But by 1907 he had shifted his theory away from Freud's emphasis on infantile sexuality towards power as the origin of neuroses; in 1911 Adler, and a number of others, left the Freudian circle and founded the Individual Psychology Movement. He moved to the USA 1935.

Adler held that much neurotic behaviour is a result of feelings of inadequacy or inferiority caused by, for instance, being the youngest in a family or being a child who is trying to compete in an adult world. In an attempt to overcome these feelings the patient overcompensates, often at the expense of normal social behaviour or, as Adler put it, 'social interest'. Adler's belief led on to his idea that a person can realize this ambition alone, which affects the way in which a psychiatrist helps a patient. Although his psychology made good sense, it lacked adequate definition and rigour of method.

Adler's works include *Study of Organ Inferiority and its Psychical Compensation* 1907, translated 1917, *The Practice and Theory of Individual Psychology* 1920, translated 1924, and *Understanding Human Nature* 1927. In his essentially most significant book, *The Neurotic Constitution* 1912, translated 1917, he insists that human character and actions must be explained teleologically.

Adrian Edgar Douglas, 1st Baron Adrian 1889–1977. English physiologist. He received the Nobel Prize for Physiology or Medicine 1932 for his work with Charles Sherrington in the field of nerve impulses and the function of the nerve cell. Adrian was also one of the first to study the electrical activity of the brain. Created a baron 1955.

Adrian was born in London and educated at Cambridge and St Bartholomew's Hospital, London. From 1919 he held academic posts at Cambridge.

Between 1925 and 1933 Adrian successfully recorded trains of nerve impulses travelling in single sensory or motor nerve fibres. He began to use thermionic valve amplifiers and found that in a single nerve fibre the electrical impulse does not change with the nature or strength of the stimulus. He also discovered that some sense organs, such as those concerned with touch, rapidly adapt to a steady stimulus whereas others, such as muscle spindles, adapt slowly or not at all.

Between 1933 and 1946 he worked on the ways in which the nervous system generates rhythmic electrical activity. He was one of the first scientists to use extensively the electroencephalogram (EEG). The last 20 years of his research life, from 1937 to 1959, were spent studying the sense of smell.

Adrian's works include *The Mechanism of Nervous Action* 1932 and *The Physical Background of Perception* 1947.

Every great scientific truth goes through three stages. First, people say it conflicts with the Bible. Next they say it had been discovered before. Lastly they say they always believed it.

Louis Agassiz

Attributed remark

Agassiz (Jean) Louis Rodolphe 1807–1873. Swiss-born US palaeontologist and geologist who developed the idea of the ice age. He established his name through his work on the classification of fossil fishes. Unlike Charles Darwin, he did not believe that individual species themselves changed, but that new species were created from time to time.

Agassiz was born in Motier and studied at various European universities. At Paris he adopted French anatomist Georges Cuvier's pioneering application of the techniques of comparative anatomy to palaeontology. In 1832 Agassiz became professor at Neuchâtel, Switzerland.

Moving to the USA 1846, he joined the faculty of Harvard.

Travelling in 1836 in the Alps, Agassiz developed the novel idea that glaciers, far from being static, were in a constant state of almost imperceptible motion. Finding rocks that had been shifted or abraded, presumably by glaciers, he inferred that in earlier times much of northern Europe had been covered with ice sheets. *Etudes sur les glaciers/Studies on Glaciers* 1840 developed the original concept of the ice age, which he viewed as a cause of extinction, demarcating past flora and fauna from those of the present.

His book *Researches on Fossil Fish* 1833–44 described and classified over 1,700 species. *Contributions to the Natural History of the United States* 1857–62 is an exhaustive study of the American natural environment.

Agricola Georgius 1494–1555. Latinized name of Georg Bauer. German mineralogist who pioneered mining technology. His book *De Re Metallica/On Metals* 1556, illustrated with woodcuts, became a standard text on smelting and mining.

Agricola was born at Glauchau in Saxony and trained as a physician. Involvement with the medicinal use of minerals sparked curiosity for the products of the Earth, and he quickly made himself an authority on mining, metal extraction, smelting, assaying, and related chemical processes. His *The Nature of Fossils* 1546 advances one of the first comprehensive classifications of minerals. He went on to explore the origins of rocks, mountains, and volcanoes.

Aiken Howard Hathaway 1900–1973. US mathematician and computer pioneer. In 1939, in conjunction with engineers from IBM, he started work on the design of an automatic calculator using standard business-machine components. In 1944 the team completed one of the first computers, the Automatic Sequence Controlled Calculator (known as the Harvard Mark I), a programmable computer controlled by punched paper tape and using punched cards.

Aiken was born in Hoboken, New Jersey, and studied engineering at the University of Wisconsin. His early research at Harvard in the 1930s was sponsored by the Navy Board of Ordnance and in 1939 he and three IBM engineers were placed under contract to develop a machine to produce mathematical tables and to assist the ballistics and gunnery divisions of the military.

The Harvard Mark I was principally a mechanical device, although it had a few electronic features; it was 15 m/49 ft long and 2.5 m/8 ft high, and weighed more than 30 tonnes. Addition took 0.3 sec, multiplication 4 sec. It was able to manipulate numbers of up to 23 decimal places and to store 72 of them. The Mark II, completed 1947, was a fully electronic machine, requiring only 0.2 sec for addition and 0.7 sec for multiplication. It could store 100 ten-digit figures and their signs.

Airy George Biddell 1801–1892. English astronomer. He installed a transit telescope at the Royal Observatory at Greenwich, England, and accurately measured Greenwich Mean Time by the stars as they crossed the meridian.

Airy became the seventh Astronomer Royal 1835. He began the distribution of Greenwich time signals by telegraph, and Greenwich Mean Time as measured by Airy's telescope was adopted as legal time in Britain 1880.

Airy was born in Alnwick, Northumberland, and studied mathematics at Cambridge, where he became professor of mathematics 1826 and of astronomy 1828.

As director of the Cambridge Observatory, he introduced a much improved system of meridian observations and set the example of reducing them in scale before publishing them. As Astronomer Royal, in 1847 he had erected the alt-azimuth (an instrument he devised to calculate altitude and azimuth) for observing the Moon in every part of the sky. Other innovations included photographic registration in 1848, transits timed by electricity in 1854, spectroscopic observations from 1868, and a daily round of sunspots using the Kew heliograph in 1873.

Airy's mathematical skills were used in establishing the border between Canada and the USA and the boundaries of the states of Oregon and Maine. His scientific expertise was also called on during the launch of the steamship *Great Eastern*, the laying of the transatlantic telegraph cable, and the construction of the chimes of the clock in the tower of the Houses of Parliament ('Big Ben').

His *Mathematical Tracts on Physical Astronomy* 1826 became a standard work.

Aitken Robert Grant 1864–1951. US astronomer who discovered and observed thousands of double stars. From 1895 he worked at the Lick Observatory on Mount Hamilton, California, and was its director 1930–35.

Aitken was born in Jackson, California. In 1891 he became professor of mathematics at the University of the Pacific.

He began a survey of double stars 1899, not finished until 1915. During the early years of the project he was assisted by W J Hussey (1862–1926), and between them they discovered nearly 4,500 new binary systems. Their primary tool was the 91-cm/36-in refractor. Aitken then carried out a thorough statistical examination of this vast amount of information, which he published as *Binary Stars* 1918 and, revised, as *New General Catalogue of Double Stars* 1935. His work lay not merely in the discovery of new binary stars, but also in determining their motions and orbits.

Albertus Magnus, St 1206–1280. German scholar of Christian theology, philosophy (especially Aristotelian), natural science, chemistry, and physics. He was known as 'doctor universalis' because of the breadth of his knowledge.

He studied at Bologna and Padua, and entered the Dominican order 1223. He taught at Cologne and lectured from 1245 at Paris University. St Thomas Aquinas was his pupil there, and followed him to Cologne 1248. He became provincial of the Dominicans in Germany 1254, and was made bishop of Ratisbon 1260. Two years later he resigned and eventually retired to his convent at Cologne. He tried to reconcile Aristotelian thought with Christian teachings.

Alder Kurt 1902–1958. German organic chemist who with Otto Diels developed the diene synthesis 1928, a fundamental process that has become known as the Diels–Alder reaction. It is used in organic chemistry to synthesize cyclic (ring) compounds, including many that can be made into plastics and others – which normally occur only in small quantities in plants and other natural sources – that are the starting materials for various drugs and dyes. Alder and Diels shared the 1950 Nobel Prize for Chemistry.

Alder was born in Königshütte in Upper Silesia (now Krolewska Huta in Poland). He studied at Berlin and at Kiel, where he worked under Otto Diels. Alder was director of the Chemical Institute at the University of Cologne from 1940.

The Diels–Alder reaction involves the adding of a conjugated diene (an organic compound with two double bonds separated by a single bond) to a dienophile (a compound with only one, activated double bond). The reaction is equally general with respect to dienophiles, provided that their double bonds are activated by a nearby group such as carboxyl, carbonyl, cyano, nitro, or ester. Many of the compounds studied were prepared for the first time in Alder's laboratory.

The diene synthesis stimulated and made easier the understanding of this important group of natural products. The ease with which the reaction takes place suggests that it may be the natural biosynthetic pathway.

Aldrin Edwin Eugene ('Buzz') 1930– . US astronaut who landed on the Moon with Neil Armstrong during the *Apollo 11* mission in July 1969, becoming the second person to set foot on the Moon.

Born at Montclair, New Jersey, he graduated from the US Military Academy, West Point, New York, and flew for the airforce in Korea and West Germany. He received a PhD for his thesis on orbital mechanics from the Massachusetts Institute of Technology 1963. During the *Gemini 12* flight with James Lovell 1966, Aldrin spent $5\frac{1}{2}$ hours in outer space without any ill effects. His 'walk' in space set a record for extravehicular activity, and proved that people could work outside an orbiting vehicle.

He published *Return to Earth* with Wayne Warga 1975.

Aleksandrov Pavel Sergeevich 1896–1982. Russian mathematician who was a leading expert in the field of topology and one of the founders of the theory of compact and bicompact spaces.

Aleksandrov was born in Bogorodsk (now Noginok), near Moscow, and studied at Moscow University, where he became professor of mathematics 1929.

Aleksandrov introduced many of the basic concepts of topology, such as the notion that an arbitrarily general topological space can be approximated to an arbitrary degree of accuracy by simple geometric figures such as polyhedrons. Of great importance, too, were his investigations into that branch of topology known as homology, which examines the relationships between the ways in which spatial structures are dissected. He formulated the theory of essential mappings and the homological theory of dimensionality, which led to a number of basic laws of duality relating to the topological properties of an additional part of space.

His passion for international cooperation led him to supervise the publication of an English–Russian dictionary of mathematical terminology 1962.

Alembert Jean le Rond d' 1717–1783. French mathematician, encyclopedist, and theoretical physicist. In association with Denis Diderot, he helped plan the great *Encyclopédie*, for which he also wrote the 'Discours préliminaire' 1751. He framed several theorems and principles – notably ***d'Alembert's principle*** – in dynamics and celestial mechanics, and devised the theory of partial differential equations.

The principle that now bears his name was first published in his *Traité de dynamique* 1743, and was an extension of the third of Isaac Newton's laws of motion. D'Alembert maintained that the law was valid not merely

Alembert *From scandalous beginnings – he was found on a church doorstep, the illegitimate son of a courtesan – Jean le Rond d'Alembert went on to become an eminent mathematician and philosopher. He discovered the calculus of partial differences, and, in 1743, developed the principle that now bears his name, by extending the Newtonian theory of dynamics to include mobile bodies.*

for a static body, but also for mobile bodies. Within a year he had found a means of applying the principle to the theory of equilibrium and the motion of fluids. Using also the theory of partial differential equations, he studied the properties of sound, and air compression, and also managed to relate his principle to an investigation of the motion of any body in a given figure.

A Paris foundling, d'Alembert had his education financed by a sponsor. First he studied law, then medicine, before deciding to devote his life to mathematics.

D'Alembert's first published mathematical work was a paper on integral calculus 1739. This and later papers were fundamental to the development of calculus. His mathematical treatises were collected in *Opuscules mathématiques* 1761–80.

From the early 1750s, together with other mathematicians such as Joseph Lagrange and Pierre Laplace, he applied calculus to celestial mechanics. In particular, d'Alembert worked out in 1754 the theory needed to set Newton's discovery of the precession of the equinoxes on a sound mathematical basis, and explained the phenomenon of the oscillation of the Earth's axis. He also gave accurate calculations of the perturbations in the orbits of the known planets.

For the *Encyclopédie* d'Alembert wrote on scientific topics, linking, especially, various branches of science. But when the Catholic church in France denounced the project, he resigned his editorship.

Alexander Frederick Matthias 1869–1955. Australian founder and teacher of the ***Alexander technique***, a method of relaxing the mind and body. At one time a professional reciter, he developed throat and voice trouble, and his experiments in curing himself led him to work out the system of mental and bodily control described in his book *Use of the Self* 1931.

By correcting bad habits of posture, breathing, and muscular tension, Alexander technique is effective in the prevention of disorders and is said to alleviate many conditions, such as back ache, migraine, asthma, and hypertension.

Alfvén Hannes Olof Gösta 1908–1995. Swedish astrophysicist who made fundamental contributions to plasma physics, particularly in the field of magnetohydrodynamics (MHD) – the study of plasmas in magnetic fields. He shared the 1970 Nobel Prize for Physics.

Alfvén was born in Norrköping, Sweden, and educated at the University of Uppsala. In 1940 he joined the Royal Institute of Technology, Stockholm, later dividing his academic career between that and the University of California, San Diego, from which he obtained a professorship in 1967. In 1972 he was among a group of Oxford scientists who appealed to governments to abandon fast-breeder nuclear reactors and concentrate efforts on nuclear fusion.

Alfvén formulated the frozen-in-flux theorem, according to which a plasma is – under certain conditions – bound to the magnetic lines of flux passing through it; later he used this theorem to explain the origin of cosmic rays. In 1939 he proposed a theory to explain aurorae and magnetic storms, which greatly influenced later ideas about the Earth's magnetosphere.

In 1942 Alfvén postulated that a form of electromagnetic wave would propagate through plasma; other scientists later observed this phenomenon in plasmas and liquid metals. Also in 1942 he developed a theory that the planets in the Solar System were formed from material captured by the Sun from an interstellar cloud of gas and dust.

Alhazen Ibn al-Haytham, (Abu Ali al-Hassan ibn al-Haytham) *c.* 965–1038. Arabian scientist, author of the *Kitab al-Manazir/Book of Optics*, translated into Latin as *Opticae thesaurus* (1572). For centuries it remained the most comprehensive and authoritative treatment of optics in both East and West.

Alhazen was born in Basra (now in Iraq). He made many contributions to optics, contesting the Greek view of Hero and Ptolemy that vision involves rays that emerge from the eye and are reflected by objects viewed. Alhazen postulated that light rays originate in a flame or in the Sun, strike objects, and are reflected by them into the eye. He studied lenses and mirrors, working out that the curvature of a lens accounts for its ability to focus light. He measured the refraction of light by lenses and its reflection by mirrors, and formulated the geometric optics of image formation by spherical and parabolic mirrors. He constructed a camera obscura. He also tried to account for the occurrence of rainbows, appreciating that they are formed in the atmosphere, which he estimated extended for about 15 km/9 mi above the ground.

Allbutt Thomas Clifford 1836–1925. British physician. He invented a compact medical thermometer, proved that angina is caused by narrowing of the coronary artery, and studied hydrophobia and tetanus. KCB 1907.

Alpher Ralph Asher 1921– . US scientist who carried out the first quantitative work on nucleosynthesis and in 1948 was the first to predict the existence of primordial background radiation, which is now regarded as one of the major pieces of evidence for the validity of the Big Bang model of the universe.

Alpher was born in Washington DC and educated there as a night-school student. Later he carried out research at Johns Hopkins University and, 1955–86, at the Central Electric Research Laboratory.

In 1948 Alpher and cosmologist George Gamow published the results of their work on nucleosynthesis in the early universe. They included the name of physicist Hans Bethe as a co-author of this paper, so that their new theory became popularly known as the alpha-beta-gamma theory.

Alpini Prospero 1553–1616. Italian botanist and physician, director of the botanical gardens at Padua, which was originally developed in order to grow plants for their medicinal uses. He studied plants for both their therapeutic uses and also because he was interested in their structure and function.

His *De plantis Aegypti* 1592 included the first European descriptions of the coffee bush (coffee arabica) and the banana tree. He also studied the flora of Crete. Such was his reputation as a botanist in the 16th century that Linnaeus named the genus *Alpinia* in his honour.

Alpini was born in Maroshica, Italy and studied medicine at Padua University, graduating 1578. In 1580, he went to Cairo for three years as the Venetian Consul's physician and then returned to Venice. In 1603, he was made the director of the botanical gardens at Padua, the earliest European botanical garden of which there are reliable records. The position was later occupied by his son Alpino. He died in Padua of a kidney infection.

Alter David 1807–1881. US inventor and physicist who in 1854 put forward the idea that each element has a characteristic spectrum, and that spectroscopic analysis of a substance can therefore be used to identify the elements present.

Alter was born in Westmoreland County, Pennsylvania, and studied medicine in New York. He spent the rest of his life experimenting and making inventions, working alone and using home-made apparatus. His inventions included a successful electric clock, a model for an electric locomotive (which was not put into production), a new process for purifying bromine, an electric telegraph that spelled out words with a pointer, and a method of extracting oil from coal (which was not put into commercial practice because of the discovery of oil in Pennsylvania).

Alter also investigated the Fraunhofer lines in the solar spectrum. The significance of his work in spectroscopy was not recognized at that time, but his idea was experimentally verified a few years later, and today spectroscopic analysis is extensively used in chemistry and astronomy.

Altman Sidney 1939– . Canadian-born US biochemist who shared the Nobel Prize for Chemistry in 1989 with Thomas Cech for his research on the catalytic activities of RNA (the nucleic acid involved in translating DNA into proteins).

Altman studied ribonuclease-P, an enzyme that catalyses the depolymerization (decoupling of molecules) of RNA in the formation of transfer RNA (*t*RNA). Ribonuclease-P is comprised of RNA and a protein. Altman showed that the RNA component is all that is required to catalyse the formation of *t*RNA with the protein playing no part in this process.

Born in Montréal, Altman received an education at the Massachusetts Institute of Technology and the University of Colorado, Boulder. After obtaining a PhD, he took a job as a teaching assistant at Columbia University in 1960. Later, in 1971, he moved to Yale and embarked on his research career, becoming professor of biology in 1980.

There is no democracy in physics. We can't say that some second-rate guy has as much right to opinion as Fermi.

Luis Alvarez

In D S Greenberg *The Politics of Pure Science* 1967

Alvarez Luis Walter 1911–1988. US physicist who led the research team that discovered the Ξ_0 subatomic particle 1959. He also made many other breakthroughs in fundamental physics, accelerators, and radar. He worked on the US atom bomb for two years, at Chicago and at Los Alamos, New Mexico, during World War II. Nobel prize 1968.

In 1980 Alvarez was responsible for the theory that dinosaurs disappeared because a meteorite crashed into Earth 70 million years ago, producing a dust cloud that blocked out the Sun for several years, causing dinosaurs and plants to die. The first half of the hypothesis is now widely accepted.

Alvarez was born in San Francisco and studied at Chicago. In 1945 he became professor at the University of California, working at the Lawrence Livermore Radiation Laboratory there 1954–59. During World War II he moved to the Massachusetts Institute of Technology, where he developed the VIXEN radar for the airborne detection of submarines, phased-array radars, and ground-controlled approach radar that enabled aircraft to land in conditions of poor visibility. He also participated in creating the atomic bomb dropped on Hiroshima, Japan.

Alvarez built the first practical linear accelerator and an accelerator for breeding plutonium, and invented the tandem electrostatic accelerator. He also devised, but never built, the microtron for accelerating electrons.

In 1953 Alvarez met Donald Glaser, inventor of the bubble-chamber detector for subatomic particles. Alvarez decided to build a much larger chamber than Glaser had used, and to fill it with liquid hydrogen. He also developed automatic scanning and measuring equipment whose output could be stored on punched cards and then analysed using computers. Alvarez and co-workers used the bubble chamber to discover a large number of new short-lived particles, including the K (the first meson) and the Ω meson. These experimental findings were crucial in the development of the 'eightfold way' model of elementary particles and, subsequently, the theory of quarks.

Alzheimer Alois 1864–1915. German neuropathologist. In 1906 he became the first to describe Alzheimer's disease, a degenerative illness affecting the nerve cells of the frontal and temporal lobes of the cerebrum of the brain, characterized by severe memory impairment. It is a major cause of presenile dementia.

Alzheimer was professor of psychiatry and neurology at Breslau University (now Wroclaw, Poland) from 1912.

Ambartsumian Viktor Amazaspovich 1908–1996. Soviet-Armenian astronomer who in 1955 proposed the manner in which enormous catastrophes might take place within stars and galaxies during their evolution.

Ambartsumian was born in Tiflis (now Tbilisi), Georgia, and studied at the University of Leningrad, where he later taught.

In 1944 he was appointed head of the Byurakan Observatory in Yerevan, Armenia.

The radio source in Cygnus had been associated with what appeared to be a closely connected pair of galaxies, and it was generally supposed that a galactic collision was taking place. If this were the case, such phenomena might account for many extragalactic radio sources. However,

Ambatsumian presented convincing evidence in 1955 of the errors of this theory. He suggested instead that vast explosions occur within the cores of galaxies, analogous to supernovae, but on a galactic scale.

His book *Theoretical Astrophysics* 1939 was influential.

Ames Adelbert 1880–1955. US scientist who studied optics and the psychology of visual perception. He concluded that much of what a person sees depends on what he or she expects to see, based (consciously or unconsciously) on previous experience.

Amici Giovanni Battista 1786–1863. Italian botanist and microscopist who in the 1820s made a series of observations clarifying the process by which pollen fertilizes the ovule in flowering plants.

Ampère André Marie 1775–1836. French physicist and mathematician who made many discoveries in electromagnetism and electrodynamics. He followed up the work of Hans Oersted on the interaction between magnets and electric currents, developing a rule for determining the direction of the magnetic field associated with an electric current. The unit of electric current, the ***ampere***, is named after him.

Ampère's law is an equation that relates the magnetic force produced by two parallel current-carrying conductors to the product of their currents and the distance between the conductors. Today Ampère's law is usually stated in the form of calculus: the line integral of the magnetic field around an arbitrarily chosen path is proportional to the net electric current enclosed by the path.

Ampère was born near Lyon and taught mathematics there until moving to Paris 1805. He held academic posts in mathematics, philosophy, astronomy, and experimental physics.

Ampère's first publication was an early contribution to probability theory, *Considérations sur la théorie mathématique de jeu/Considerations on the Mathematical Theory of Games* 1802, in which he discussed the inevitability of a player losing a gambling game of chance against an opponent with vastly greater financial resources.

Ampère *French mathematician, physicist, and chemist André Ampère, whose name is given to the basic SI unit of electric current. Although a brilliant scientist, Ampère's private life was tragic. His father was guillotined during the French Revolution in 1793, his first wife died 1804 four years after the birth of their only son, and he endured a disastrous second marriage until his death of pneumonia in 1836.*

In a series of papers beginning 1820 Ampère expounded the theory and basic laws of electromagnetism (which he called electrodynamics to differentiate it from the study of stationary electric forces, which he called electrostatics). He showed that two parallel wires carrying current in the same direction attract each other, whereas when the currents are in opposite directions, mutual repulsion results. He also predicted and demonstrated that a helical coil of wire (which he called a solenoid) behaves like a bar magnet while it is carrying an electric current.

Trying to explain electromagnetism, Ampère proposed that magnetism is merely electricity in motion. His suggestion that molecules are surrounded by a perpetual electric current may be regarded as a precursor of the electron-shell model.

He published *Mémoire sur la théorie mathématique des phénomènes électrodynamiques uniquement déduite de l'expérience/Notes on the Mathematical Theory of Electrodynamic Phenomena Deduced Solely from Experiment* 1827.

The future science of government should be called 'la cybernetique'.

André Ampère

1843

Amsler-Laffon Jakob, (born Amsler) 1823–1912. Swiss mathematical physicist who designed and manufactured precision instruments for use in engineering, including an improved tool to measure areas inside curves – the polar planimeter.

Amsler was born in Stalden bei Brugg, Valais, and studied mathematics and physics at Königsberg, Germany. He took up the design and manufacture of scientific instruments in Schaffhausen 1854.

Earlier models of tools to measure the surface of spheres had been based on Cartesian coordinates, but they were bulky and expensive. Amsler-Laffon's design, based on a polar coordinate system, was not only more delicate and flexible than its predecessors, but was also much cheaper to manufacture. It was particularly valuable to shipbuilders and railway engineers. By the time he died, his factory had produced more than 50,000 polar planimeters.

Anaximander *c.* 610–*c.* 546 BC. Greek astronomer and philosopher. He claimed that the Earth was a cylinder three times wider than it is deep, motionless at the centre of the universe, and that the celestial bodies were fire seen through holes in the hollow rims of wheels encircling the Earth. According to Anaximander, the first animals came into being from moisture and the first humans grew inside fish, emerging once fully developed.

He was born in Miletus, in what is now Turkey, and was a pupil of Thales. He is thought to have been the first to determine solstices and equinoxes, by means of a sundial, and he is credited with drawing the first geographical map of the whole known world. He believed that the universe originated as a formless mass containing within itself the contraries of hot and cold, and wet and dry, from which land, sea, and air were formed out of the union and separation of these opposites. Perpetual rotation in the universe created cosmic order by sorting heavier from lighter matter.

Overall, he seems to have shared the early Greek philosophical urge to explain the universe with a tiny number of general laws.

Anderson Carl David 1905–1991. US physicist who discovered the positive electron (positron) in 1932; he shared the Nobel Prize for Physics 1936. His discovery of another particle, the muon, in 1937 launched elementary-particle physics.

Anderson was born in New York and educated at the California Institute of Technology, where he spent his entire career.

Using a modified cloud chamber of his own devising, Anderson found that positive electrons, or positrons, were present in cosmic rays, energetic particles reaching Earth from outer space. For this discovery, Anderson shared the 1936 Nobel prize with Victor Hess, the discoverer of cosmic rays. In 1937, Anderson found a new particle in cosmic rays, one with a mass between that of an electron and a proton. The new particle was first called a mesotron and then a meson muon. The muon was the first elementary particle to be discovered beyond the constituents of ordinary matter (proton, neutron, and electron).

Anderson Elizabeth Garrett 1836–1917. English physician, the first English woman to qualify in medicine. Unable to attend medical school, Anderson studied privately and was licensed by the Society of Apothecaries in London 1865. She was physician to the Marylebone Dispensary for Women and Children (later renamed the Elizabeth Garrett Anderson Hospital), a London hospital now staffed by women and serving female patients.

She was the first woman member of the British Medical Association 1873 and the

Anderson *In 1860 Elizabeth Garrett spent a session as a medical student at the Middlesex Hospital, London, but was asked to leave when the male students objected to her presence. She was finally granted an MD by the Paris Faculty of Medicine.*

first woman mayor in Britain. She lectured at the London School of Medicine for Women 1875–97, and was its dean 1883–1903.

Anderson Philip Warren 1923– . US physicist who shared the 1977 Nobel Prize for Physics with his senior colleague John Van Vleck for his theoretical work on the behaviour of electrons in magnetic, noncrystalline solids.

Anderson was born in Indianapolis, Indiana, and educated at Harvard University. In 1949 he joined Bell Laboratories in New Jersey, and later took on additional academic posts.

In the early 1960s he investigated the interatomic effects that influence the magnetic properties of metals and alloys, devising a theoretical model (now called the Anderson model) to describe the effect of the presence of an impurity atom in a metal. He also described the movements of impurities within crystalline substances by a method now known as Anderson localization.

In addition, Anderson has studied the relationship between superconductivity, superfluidity, and laser action, and predicted the existence of resistance in superconductors. His studies of disordered glassy solids indicate that they could be used instead of the expensive crystalline semiconductors now used in many electronic devices, such as computer memories, electronic switches, and solar energy converters.

Andrews Thomas 1813–1885. Irish physical chemist, best known for postulating the idea of critical temperature and pressure from his experimental work on the liquefaction of gases, which demonstrated the continuity of the liquid and gaseous states. He also studied heats of chemical combination and was the first to establish the composition of ozone, proving it to be an allotrope of oxygen.

Andrews was born in Belfast and attended five universities, beginning at the age of 15. In 1835 he graduated from Edinburgh as a qualified doctor and surgeon. He was professor of chemistry at Queen's College, Belfast, 1849–79.

Studying chemical combination, Andrews succeeded in the direct determination of heats of neutralization and of formation of halides (chlorides, bromides, and iodides).

Many other scientists had tried to explain the relationship between gases and liquids. Andrews constructed elaborate equipment in which he initially investigated the liquefaction of carbon dioxide, and by 1869 he had concluded that it has a critical temperature (or critical point) of 30.9°C/87.6°F, above which it cannot be condensed into a liquid by any pressure.

Anfinsen Christian Boehmer 1916– . US biochemist who shared the Nobel Prize for Physiology or Medicine in 1972 with Stanford Moore and William Stein for his work on the shape and primary structure of ribonuclease (the enzyme that hydrolyses RNA). Specifically, he determined how ribonuclease kept its configuration after it had completed its enzymatic function.

Ribonuclease is made up of a single peptide (a molecule consisting of two or more amino acid molecules joined by a peptide bond) chain folded into a sphere bound together by four disulphide bonds. These bonds can be broken down so that the enzyme becomes denatured (collapses), losing all of its enzyme properties. Anfinsen found that its shape and consequently its enzymatic power could be restored, and concluded that ribonuclease must retain all of the information about its configuration within its amino acids. He went on to study the secondary and tertiary structures of ribonuclease.

Anfinsen was born in Monessen, Pennsylvania and was educated at the University of Pennsylvania and then Harvard, where he obtained his doctorate in 1943. From 1943 to 1950, he taught at Harvard Medical School and also collaborated with Axel Theorell at the Medical Nobel Institute. He then moved to the National Heart Institute in Bethesda, Maryland, where he was responsible for the laboratory of cellular physiology until he was appointed professor of biology at Johns Hopkins University in 1982.

Ångström Anders Jonas 1814–1874. Swedish astrophysicist who worked in spectroscopy and solar physics. In 1861 he identified the presence of hydrogen in the Sun. His outstanding *Recherches sur le spectre solaire* 1868 presented an atlas of the solar spectrum with measurements of 1,000 spectral lines expressed in units of one ten-millionth of a millimetre, the unit which later became the angstrom.

Ångström was educated at Uppsala and spent his entire academic career there. His main work was in spectroscopy, but he also investigated the conduction of heat and devised a method of determining thermal conductivity (1863). His 'Optical investigations' 1853 contains his principle of spectrum analysis, demonstrating that a hot gas emits light at the same frequency as it absorbs it when it is cooled. In 1867 he investigated the spectrum of the aurora borealis, the first person to do so.

Anokhin Piotre Kuzmich 1897–1974. Russian psychologist. He worked with V M Bechterev (1857–1927), and later with Ivan Pavlov, in examining the physiological bases of animal behaviour.

Anokhin proposed that behaviour is a system of functions, each relating to a definite goal, and suggested that, even in simple conditioning, it is regulated by its consequences rather than reflexively determined. His main ideas are collected in *Biology and Neurophysiology of the Conditioned Reflex and its Role in Adaptive Behaviour* 1974.

Antoniadi Eugène Marie 1870–1944. Turkish-born French astronomer who demolished the theory of canals on Mars. He became an expert also on the scientific achievements of ancient civilizations.

Antoniadi was born in Constantinople (now Istanbul). He began to make astronomical observations 1888, and in 1893 went to France to utilize better telescopes, first at the observatory at Juvisy-sur-Orge and then at Meudon. Detecting an apparent spot on the surface of Mars, he soon realized that it was merely an optical effect caused by the diffraction of light by the Earth's atmosphere. There was at that time widespread belief that there was an intricate pattern of canals on the surface of Mars suggestive of advanced technology, as proposed 1877 by Italian astronomer Giovanni Schiaparelli. Antoniadi suggested that the canals were also an optical illusion, produced by the eye's linking of many tiny surface details into an apparently meaningful pattern.

Antoniadi's later work included research into the behaviour and properties of Mercury (published in *La planète Mercure/The Planet Mercury* 1934). He then turned to a study of the history of astronomy and, in particular, the work of the ancient Greek and Egyptian astronomers.

Apollonius of Perga *c.* 245–*c.* 190 BC. Greek mathematician, called 'the Great Geometer'. In his work *Conica/The Conics* he showed that a plane intersecting a cone will generate an ellipse, a parabola, or a hyperbola, depending on the angle of intersection. In astronomy, he used a system of circles called epicycles and deferents to explain the motion of the planets; this system, as refined by Ptolemy, was used until the Renaissance.

It is thought that Apollonius may have studied at the school established by Euclid at Alexandria, especially since much of his work was built on Euclidean foundations.

Of his eight-volume treatise *Conica*, seven volumes are extant. The first four books consisted of an introduction and a statement of the state of mathematics provided by his predecessors. In the last four volumes Apollonius put forth his own important work on conic sections, the foundation of much of the geometry still used today in astronomy, ballistic science, and rocketry.

Appert Nicolas François *c.* 1750–1841. French pioneer of food preservation by canning. He devised a system of sealing food in

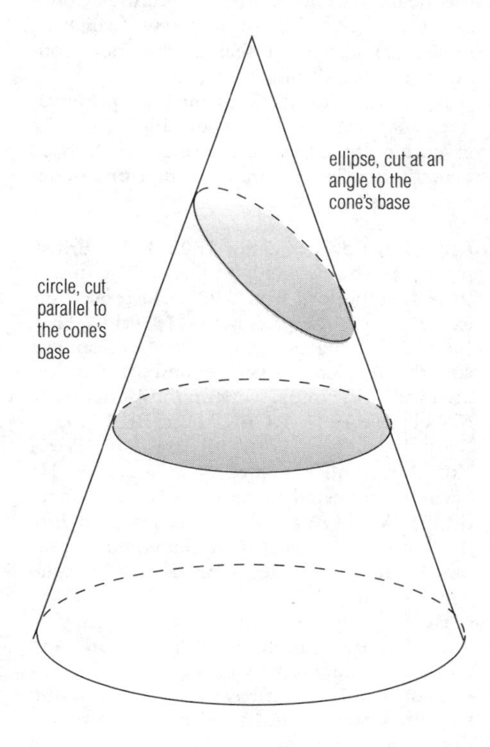

Apollonius of Perga *Conic sections were studied avidly by ancient Greek mathematicians. Archimedes in particular wrote several contemplative articles on the subject, but it was Apollonius whose work became the definitive description.*

glass bottles and subjecting it to heat, described in his book *L'Art de conserver les substances animales et végétales/The Art of Preserving all Kinds of Animal and Vegetable Substances for Several Years* 1810.

In 1822 he changed to using cylindrical tin-plated steel cans.

Appert was born near Paris and trained as a confectioner. By 1795, the feeding of Napoleon's troops had become a problem and the French Directory offered a prize for a practical method of preserving food. This encouraged Appert to begin a 14-year period of experimentation. In 1804 he opened the world's first canning factory – the House of Appert – in Massy, S of Paris. By 1809 he had succeeded in preserving certain foods in glass bottles that had been immersed in boiling water, and he was awarded the prize.

Since nothing was known of bacteriology and the causes of decay, Appert proceeded by trial and error. He based his methods on the heating of food to temperatures above 100°C/212°F, using an autoclave (which he perfected) and then sealing the food container to prevent putrefaction.

In addition to his work on food preservation, Appert was responsible for the invention of the bouillon cube and he devised a method of extracting gelatin from bones without using acid.

Appleton Edward Victor 1892–1965. British physicist who worked at Cambridge under Ernest Rutherford from 1920. He proved the existence of the Kennelly–Heaviside layer (now called the E layer) in the atmosphere, and the Appleton layer beyond it, and was involved in the initial work on the atom bomb. Nobel prize 1947. KCB 1941, GBE 1946.

Appleton was born in Bradford, West Yorkshire, and educated at Cambridge. He became interested in radio as signals officer during World War I, and his research into the atmosphere was of fundamental importance to the development of radio communications.

By periodically varying the frequency of the BBC transmitter at Bournemouth and measuring the intensity of the received transmission 100 km/62 mi away, Appleton found that there was a regular fading in and fading out of the signals at night but that this effect diminished considerably at dawn as the Kennelly–Heaviside layer broke up. Radio waves continued to be reflected by the atmosphere during the day but by a higher-level ionized layer. By 1926 this layer, which Appleton measured at about 250 km/155 mi above the Earth's surface (the first distance measurement made by means of radio), became generally known as the Appleton layer (it is now also known as the F layer).

Arago Dominique François Jean 1786–1853. French physicist and astronomer who made major contributions to the early study of electromagnetism. In 1820 he found that iron enclosed in a wire coil could be magnetized by the passage of an electric current. In 1824, he was the first to observe the ability of a floating copper disc to deflect a magnetic needle, the phenomenon of magnetic rotation.

From 1815, Arago worked with French physicist Augustin Fresnel on the polarization of light and was able to elucidate the fundamental laws governing it. Together they established the wave theory of light.

Arago was born near Perpignan, studied at the Ecole Polytechnique in Paris, and was appointed to the Bureau of Longitudes. He was professor of analytical geometry at the Ecole Polytechnique 1809–30, when he became permanent secretary to the Academy of Sciences, director of the Paris Observatory, and deputy for Pyrénées Orientales until 1852.

To get to know, to discover, to publish – this is the destiny of a scientist.

Dominique Arago

Attributed

His political affiliation was with the extreme left, and the revolution of 1848 saw him elected to a ministerial position in the provisional government. It was under his administration that slavery was abolished in the French colonies.

Arago also investigated the compressibility, density, diffraction, and dispersion of gases; the speed of sound, which he found to be 331.2 m/1,087 ft per second; lightning, of which he found four different types; and heat. His studies in astronomy included investigations of the solar corona and chromosphere, measurements of the diameters of the planets, and a theory that light interference is responsible for the twinkling of stars.

Arber Agnes (born Robertson) 1879–1960. English botanist and plant morphologist who researched into gymnosperms (cone-bearing plants) and monocotyledons (seed-bearing plants with a single seed leaf).

Arber's work on the comparative anatomy of monocotyledon plants culminated in the book *Monocotyledons: a Morphological Study* 1925. She also produced a popular book on herbals, *Herbals, Their Origin and Evolution*, which she illustrated herself.

Robertson was born in London and inherited her love of botany from her mother and her artistic ability from her father. She attended University College, London and after graduating, was admitted to Newnham College, Cambridge where she studied Natural Science. Following her graduation from Cambridge she went to work with Ethel Sargent in Reigate, and in 1908, was awarded a lectureship in botany at University College. In 1909 she married Edward Alexander Newell Arber, a university demonstrator in palaeobotany at Cambridge and continued her research after her marriage, first at the Balfour Laboratory and then at home. She was elected to the Royal Society 1946 and won the Linnaean Society's Gold medal 1948. Upon her death in Cambridge on 22 March she was buried at Girton College, Cambridge.

Arber Werner 1929– . Swiss biochemist who won the Nobel Prize for Physiology or Medicine 1978 for his discovery of restriction enzymes (bacterial enzymes that can break a chain of DNA in two at a specific point). Restriction enzymes are used in genetic engineering.

Arber was born in Granichen, Switzerland and studied at the Swiss Federal Institute of Technology. In 1949, he left for Geneva to work on bacteriophages (viruses that attack and grow in bacteria). By 1962, Arber had conducted a series of experiments to show the genetic basis of 'host-induced' variation (the phenomenon by which a bacteriophage adapts to the particular strain of bacteria that it grows in).

In Arber's theory, certain bacterial strains were postulated to contain restriction enzymes which were able to cleave unprotected (bacteriophage) DNA. Furthermore, these restriction enzymes must have the ability to recognize a specific sequence of nucleotides within a bacteriophage DNA molecule in order not to destroy the bacteria's own DNA. These enzymes would protect bacteria from infection since bacteriophage DNA would be broken before it could replicate and destroy a cell.

Arber went on to purify and characterize a sequence-specific restriction enzyme. Today such enzymes are routinely used by molecular biologists in genetic engineering to create pieces of DNA of a specified length.

Archimedes *c.* 287–212 BC. Greek mathematician who made major discoveries in geometry, hydrostatics, and mechanics. He formulated a law of fluid displacement (Archimedes' principle), and is credited with the invention of the Archimedes screw, a cylindrical device for raising water.

Archimedes was born in Syracuse, Sicily. He designed engines of war for the defence of Syracuse, and was killed when the Romans seized the town. It is alleged that Archimedes' principle was discovered when he stepped into the public bath and saw the water overflow. He was so delighted that he rushed home naked, crying 'Eureka! Eureka!' ('I have found it! I have found it!'). He used his discovery to prove that the goldsmith of Hieron II, King of Syracuse, had adulterated a gold crown with silver.

Archimedes' method of finding mathematical proof to substantiate experiment and observation would become the method of modern science in the High Renaissance. For example, the lever had been used by other scientists, but it was Archimedes who demonstrated mathematically that the ratio of the effort applied to the load raised is equal to the inverse ratio of the distances of the effort and load from the pivot, or fulcrum, of the lever.

Give me but one firm place on which to stand, and I will move the earth.

Archimedes

On the lever, quoted in *Pappus Alexander*

Among Archimedes' inventions was a design for a model planetarium able to show the movement of the Sun, Moon, planets, and possibly constellations across the sky.

Archimedes wrote many mathematical treatises, some of which still exist in altered forms in Arabic translation. Among the areas he investigated were the value for π, which he approximated very closely. He also examined the expression of very large numbers, using a special notation. Archimedes also evolved methods to solve cubic equations and to determine square roots by approximation, as

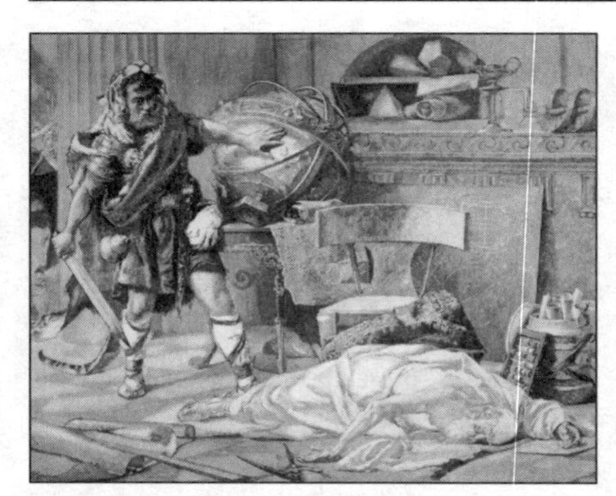

Archimedes *The death of Archimedes during the capture of Syracuse. Engrossed in a complex mathematical problem, Archimedes was killed by an impatient Roman soldier – despite orders that he be taken alive – after supposedly ignoring a challenge to surrender.*

well as formulas for the determination of the surface areas and volumes of curved surfaces and solids.

Argand Jean-Robert 1768–1822. Swiss mathematician who 1806 invented a method of geometrically representing complex numbers and their operations – the ***Argand diagram***.

Argand was born in Geneva and later moved to Paris. He appears to have been entirely self-taught as a mathematician.

Argand adopted Descartes's practice of calling all multiples of $\sqrt{-1}$ 'imaginary'. He demonstrated that real and imaginary parts of a complex number could be represented as rectangular coordinates.

The diagram is a graphic representation of complex numbers of the form $a + b$i, in which a and b are real numbers and i is $\sqrt{-1}$. One axis represents the pure imaginary numbers (those belonging to the bi category) and the other the real numbers (those belonging to the a category); it is thus possible to plot a complex number as a set of coordinates in the field defined by the two axes.

Argand's book *Essai sur une manière de représenter les quantités imaginaires dans les constructions géométriques* 1806 was published anonymously and it was not until 1813 that he became known as the author.

Argelander Friedrich Wilhelm August 1799–1875. Prussian astronomer who made a catalogue of 324,198 stars in the northern hemisphere, the *Bonner Durchmusterung/Bonn Survey* 1859–62. His *Uranometrica nova* 1843 introduced the 'estimation by steps' method for determining stellar magnitudes with the naked eye.

Argelander was born in Memel, East Prussia (now Klaipeda in Lithuania), and studied at Königsberg under German astronomer Friedrich Bessel. In 1823 he went to Finland and worked as an astronomical observer in Åbo (Turku) until the observatory was destroyed by fire 1827, and then as professor and director of the observatory at the University of Helsinki. In 1836 he became professor at Bonn and had a new observatory constructed there.

In Åbo, Argelander studied the proper motion of more than 500 stars and published the most accurate catalogue of the day on the subject.

Next, he studied the movement of the Sun through the cosmos, continuing on and confirming work done in England by William Herschel 1783.

An extension of Bessel's study of stars in the northern sky eventually resulted in the publication of the *Bonner Durchmusterung/ Bonn Survey*. This catalogued the position and brightness of nearly 324,000 stars, and although it was the last major catalogue to be produced without the aid of photography, its value is such that it was reprinted as recently as 1950.

Aristarchus of Samos *c.* 320–*c.* 250 BC. Greek astronomer. The first to argue that the Earth moves around the Sun, he was ridiculed for his beliefs. He was also the first astronomer to estimate (quite inaccurately) the sizes of the Sun and Moon and their distances from the Earth.

Aristarchus was born on Samos and may have studied in Alexandria, where he died.

His only surviving work is *Magnitudes and Distances of the Sun and Moon*. He produced

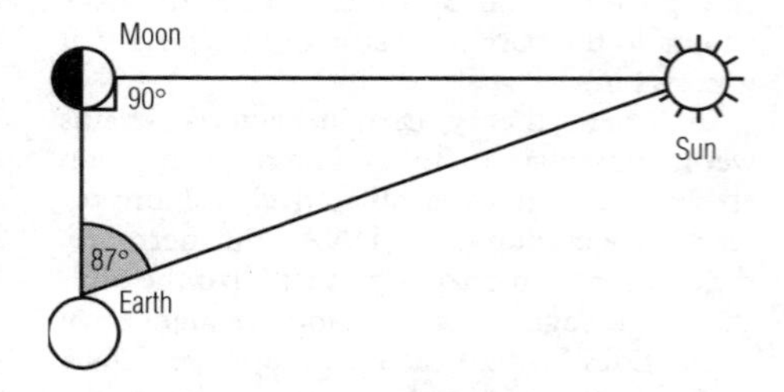

Aristarchus *Aristarchus calculated the distance between the Earth and the Sun in terms of the Earth–Moon distance by measuring the angle to the Sun when the Moon was exactly half full (and its angle to the Sun was 90°).*

methods for finding the relative distances of the Sun and Moon that were geometrically correct but rendered useless by inaccuracies in observation.

Aristarchus' model of the universe described the Sun and the fixed stars as stationary in the cosmos, and the planets – including the Earth – as travelling in circular orbits around the Sun. He stated that the apparent daily rotation of the sphere of stars is due to the Earth's rotation on its axis as it travels along its orbit, and that the reason no stellar parallax (change in position of the stars) was observed from one extreme of the orbit to the other is that even the diameter of the Earth's orbit is insignificant in relation to the vast dimensions of the universe.

Aristotle 384–322 BC. Greek philosopher who advocated reason and moderation. He maintained that sense experience is our only source of knowledge, and that by reasoning we can discover the essences of things, that is, their distinguishing qualities. In his works on ethics and politics, he suggested that human happiness consists in living in conformity with nature. He derived his political theory from the recognition that mutual aid is natural to humankind, and refused to set up any one constitution as universally ideal. Of Aristotle's works some 22 treatises survive, dealing with logic, metaphysics, physics, astronomy, meteorology, biology, psychology, ethics, politics, and literary criticism.

What we have to learn to do, we learn by doing.

Aristotle

Nicomachaean Ethics Bk II

Aristotle was born in Stagira in Thrace and studied in Athens, where he became a distinguished member of the Academy founded by Plato. He then opened a school at Assos. At this time he regarded himself as a Platonist, but his subsequent thought led him further from the traditions that had formed his early background and he was later critical of Plato. In about 344 he moved to Mytilene in Lesvos, and devoted the next two years to the study of natural history. Meanwhile, during his residence at Assos, he had married Pythias, niece and adopted daughter of Hermeias, ruler of Atarneus.

In 342 he accepted an invitation from Philip II of Macedon to go to Pella as tutor to Philip's son Alexander the Great. In 335 he opened a school in the Lyceum (grove sacred to Apollo) in Athens. It became known as the 'peripatetic school' because he walked up and down as he talked, and his works are a collection of his lecture notes. When Alexander died 323, Aristotle was forced to flee to Chalcis, where he died.

Among his many contributions to political thought were the first systematic attempts to distinguish between different forms of government, ideas about the role of law in the state, and the conception of a science of politics.

In the *Poetics*, Aristotle defines tragic drama as an imitation (mimesis) of the actions of human beings, with character subordinated to plot. The audience is affected by pity and fear, but experiences a purgation (catharsis) of these emotions through watching the play. The second book of the *Poetics*, on comedy, is lost. The three books of the *Rhetoric* form the earliest analytical discussion of the techniques of persuasion, and the last presents a theory of the emotions to which a speaker must appeal.

In the Middle Ages, Aristotle's philosophy first became the foundation of Islamic philosophy, and was then incorporated into Christian theology; medieval scholars tended

Aristotle *Until the scientific revolution of the 16th and 17th centuries, Aristotle's impact on Western philosophy and science was so great that he was known simply as 'the philosopher'. Much of Aristotle's teachings were lost following the collapse of the Roman Empire, but were resurrected by Arab and Jewish scholars in medieval times.*

to accept his vast output without question. Aristotle held that all matter consisted of a single 'prime matter', which was always determined by some form. The simplest kinds of matter were the four elements – earth, water, air, and fire – which in varying proportions constituted all things. According to Aristotle's laws of motion, bodies moved upwards or downwards in straight lines. Earth and water fell, air and fire rose. To explain the motion of the heavenly spheres, Aristotle introduced a fifth element, ether, whose natural movement was circular.

His major writings on cosmology, or astronomy, are brought together in the four-volume *De caelo/On the Heavens*. Aristotle rejected the notion of infinity and the notion of a vacuum. A vacuum he held to be impossible because an object moving in it would meet no resistance and would therefore attain infinite velocity. Space could not be infinite, because in Aristotle's view, the universe consisted of a series of concentric spheres which rotated around the centrally placed, stationary Earth. If the outermost sphere were an infinite distance from the Earth, it would be unable to complete its rotation within a finite period of time, in particular within the 24-hour period in which the stars, fixed, as Aristotle believed, to the sphere, rotated around the Earth.

Aristotle's work in astronomy also included proving that the Earth was spherical. He observed that the Earth cast a circular shadow on the Moon during an eclipse and he pointed out that as one travelled north or south, the stars changed their positions. Aristotle overestimated the Earth's diameter by only 50%.

Aristotle saw nature as always striving to perfect itself. The principle of life he termed a soul, which he regarded as the form of the living creature, not as a substance separable from it. The intellect, he believed, can discover in sense impressions the universal, and since the soul thus transcends matter, it must be immortal. Art embodies nature, but in a more perfect fashion, its end being the purifying and ennobling of the affections. The essence of beauty is order and symmetry. Aristotle also first classified organisms into species and genera.

Arkwright Richard 1732–1792. English inventor and manufacturing pioneer who developed a machine for spinning cotton (he called it a 'spinning frame') 1768. He set up a water-powered spinning factory 1771 and installed steam power in a Nottingham factory 1790. Knighted 1786.

Arkwright was born in Preston, Lancashire, and experimented in machine designing with a watchmaker, John Kay, until, with Kay and John Smalley (died 1782), he set up the spinning frame, the first machine capable of producing sufficiently strong cotton thread to be used as warp. Soon afterwards he moved to Nottingham to escape the fury of the spinners, who feared that their handicraft skills would become redundant. In 1771 he went into partnership with Jebediah Strutt (1726–1797), a Derby man who had improved the stocking frame, and Samuel Need (died 1781), and built a water-powered factory at Cromford in Derbyshire.

In 1773 Arkwright produced the first cloth made entirely from cotton; previously, the warp had been of linen and only the weft was cotton. A special act of Parliament was passed 1774 to exempt Arkwright's fabric from the double duty imposed on cottons by an act of 1736. By 1782 Arkwright employed 5,000 workers.

Armstrong Edwin Howard 1890–1954. US radio engineer who developed a system known as superheterodyne tuning for reception over a very wide spectrum of radio frequencies and frequency modulation (FM) radio transmission for static-free reception.

Armstrong was born and educated in New York, and became a professor at Columbia University. He was involved in much litigation over patents and eventually killed himself.

The superheterodyne receiving circuit was developed by Armstrong during World War I in an attempt to make a receiver that could detect the presence of enemy aircraft by means of the electromagnetic (radio) waves given off by the sparking of the ignition systems of their engines.

At that time, radio broadcasting used amplitude modulation (AM), which is susceptible to static interference. In 1933 Armstrong, with US physicist Michael Pupin (1858–1935), developed and patented the FM method of radio broadcasting, in which the transmitted signal is made to modulate the frequency of the carrier wave over a wide waveband. FM is unaffected by static and is capable of high-fidelity sound reproduction; it remains the basis of quality radio, television, microwave, and satellite transmissions, although the high frequencies used are generally limited to line-of-sight distances. FM broadcasting did not gain ground until after World War II.

Armstrong William George 1810–1900. English engineer. He developed hydraulic equipment and a revolutionary method of making gun barrels 1855, by building a breech-loading artillery piece with a steel and wrought-iron barrel (previous guns were muzzle-loaded and had cast-bronze barrels). By 1880 the 150-mm/16-in Armstrong gun was the standard for all British ordnance. Created a baron 1887.

Armstrong was born in Newcastle-upon-Tyne and studied law in London. In 1839 he constructed an overshot water wheel and soon afterwards he designed a hydraulic crane. This depended simply on the pressure of water acting directly on a piston in a cylinder. The resulting movement of the piston produced a corresponding movement through suitable gears. The first example was erected on the quay in Newcastle 1846, the pressure being obtained from the water mains of the town. Abandoning his law practice 1847, he founded an engineering works at Elswick to specialize in building hydraulic cranes.

In 1850 Armstrong invented the hydraulic pressure accumulator. It consisted of a large cylinder containing a piston that could be loaded to any desired pressure, the water being pumped in below it by a steam engine or other prime mover. This device made possible the installation of hydraulic power in almost any situation, and it was particularly used for the manipulation of heavy naval guns.

In 1854 Armstrong designed submarine mines for use in the Crimean War. A year later he designed a gun with a barrel made of wrought iron wrapped round an inner steel tube. It was rifled, and threw not a round ball but an elongated shell. The British Army first adopted the Armstrong gun 1859, but then temporarily reverted to muzzle-loaders. By 1880 he had completed the steel-wire-coiled design that was the prototype of all subsequent artillery.

Arnold Joseph 1782–1818. English naturalist and explorer who, while on a trip to Sumatra, discovered the largest flower ever described. It measured about 1 m/3 ft across and weighed over 30 kg/66 lb; Arnold named the plant (a parasite), *Rafflesia Arnoldii*.

Born in Beccles, Suffolk, Arnold showed a childhood interest in botany and published his findings in various magazines. As a young man he was apprenticed to a local surgeon and in 1807 graduated with an MD from Edinburgh University. He then joined the British Navy and, in 1815, sailed to Botany Bay as the doctor on a ship of female convicts. During this trip he made an extensive collection of insects, especially those from South America and Australia, although many were destroyed by a fire on board his ship, *The Indefatigable,* on its return journey. In 1818 he went to Sumatra where he was employed by Stamford Raffles as a naturalist; he died there of a fever in July of the same year and his collection of shells and fossils was donated to the Linnaean Society.

Arp Halton Christian 1927– . US astronomer who has worked particularly on the identification of galaxies. He published *Atlas of Peculiar Galaxies* 1965. He also carried out the first photometric work on the Magellanic Clouds – the nearest extragalactic system.

Arp was born in New York and educated at Harvard and the California Institute of Technology. In 1953 he began research at the observatory on Mount Palomar, and after working at a number of other institutions returned there in 1969. During the 1980s he put forward some unorthodox views on the red shifts of quasars. In the ensuing controversy he left the Palomar Observatory and went to the Max Planck Institute for Astrophysics, near Munich, in Germany.

During his research on globular clusters, globular-cluster variable stars, novae, Cepheid variables, extragalactic nebulae, and so on, Arp has attempted to relate the listings of galaxies to radio sources; the optical identification of these sources can now be done fairly accurately. He worked with other astronomers on the question of whether the red shifts in the spectrum of quasars are due to the general expansion of the universe.

Arrhenius Svante August 1859–1927. Swedish scientist, the founder of physical chemistry. For his study of electrolysis, he received the Nobel Prize for Chemistry 1903. In 1905 he predicted global warming as a result of carbon dioxide emission from burning fossil fuels.

Arrhenius explained that in an electrolyte the dissolved substance is dissociated into electrically charged ions. The electrolyte conducts electricity because the ions migrate through the solution.

Arrhenius was born near Uppsala, and studied at Uppsala and Stockholm, specializing in solutions and electrolytes. He was

Arrhenius *Svante August Arrhenius's doctoral thesis was almost rejected by the University of Uppsala, because of his innovative and novel ideas. He was fortunate enough to be highly regarded by several prominent chemists, such as Ostwald and van't Hoff, and was able to continue his pioneering work in Europe on a travelling scholarship. In 1887, by consolidating and extending his doctoral thesis, Arrhenius published his theory of electrolytic dissociation, which is still largely accepted today. In 1903, he was awarded the Nobel Prize for Chemistry.*

soon offered academic posts throughout Europe but chose to stay in Stockholm as professor of physics at the university from 1891. In 1905 he was appointed director of the Nobel Institute for Physical Chemistry and held that post until shortly before his death.

After 1905 Arrhenius applied the laws of theoretical chemistry to physiological problems (particularly immunology); he also published papers on cosmic physics concerning the aurora borealis and the transport of living matter ('spores') through space from one planet to another.

Artin Emil 1898–1962. Austrian mathematician who made important contributions to the development of class field theory and the theory of hypercomplex numbers. He was one of the creators of modern algebra.

Artin was born in Vienna and educated there and at Leipzig, Germany. From 1923 to 1937 he lectured at Hamburg, emigrating to the USA 1937. There Artin lectured at the University of Indiana 1938–46 and Princeton 1946–58. In 1958 he returned to Hamburg.

Artin's early work was concentrated on the analytical and arithmetical theory of quadratic number fields. In his doctoral thesis of 1921 he formulated the analogue of the Riemann hypothesis about the zeros of the classical zeta function, studying the quadratic extension of the field of rational functions of one variable over finite constant fields, by applying the arithmetical and analytical theory of quadratic numbers over the field of natural numbers. Then in 1923, in the most important discovery of his career, he derived a functional equation for his new-type L-series. The proof of this he published 1927, thereby providing, by the use of the theory of formal real fields, the solution to Hilbert's problem of definite functions. The proof produced the general law of reciprocity – Artin's phrase – which included all previously known laws of reciprocity and became the fundamental theorem in class field theory.

Artin made two other important theoretical advances. His theory of braids, given in 1925, was a major contribution to the study of nodes in three-dimensional space. A year later, in collaboration with Schrier, he succeeded in treating real algebra in an abstract manner, defining a field as real-closed if it itself was real but none of its algebraic extensions were. He was then able to demonstrate that a real-closed field could be ordered in an exact manner and that, in it, typical laws of algebra were valid.

In 1944 his discovery of rings with minimum conditions for right ideals – now known as Artin rings – was a fertile addition to the theory of associative ring algebras.

Class Field Theory 1961 is a summation of his life's work.

Ashburner Michael 1942– . English geneticist and one of the world's leading authorities on the genetics of the fruit fly *Drosophila melanogaster*. His work has advanced the study of genes involved in the development of living organisms, including the human body, since many of the *Drosophila* genes are also present and play a similar part in human cells.

He is known for his detailed studies of the structure and function of genes. In his studies he chose to use 'heat shock genes' (genes that are activated by increased temperature) in *Drosophila*, as a model for gene regulation. He also cloned many genes involved in the inheritance of interesting characteristics of *Drosophila* flies.

Ashburner studied at Cambridge University in the 1960s and went on to join the staff. He was elected a Fellow of the Royal Society 1990 in recognition of his contribution to the field of genetics. In 1991 he became professor of biology at Cambridge.

Asimov Isaac 1920–1992. Russian-born US author and editor of science fiction and nonfiction. He published more than 400 books, including his science-fiction novels *I, Robot* 1950 and the *Foundation* trilogy 1951–53, continued in *Foundation's Edge* 1983. His two-volume work *The Intelligent Man's Guide to Science* 1960 gained critical acclaim.

Asimov received a PhD in biochemistry from Columbia University 1948 and joined the faculty of the Boston University Medical School.

Asimov saw his first science-fiction story published 1939, but did not become a full-time writer until 1958, when he largely turned from fiction to popular science. His 'three laws of robotics', which became widely accepted in science fiction, first appeared in a 1941 story. They specify that a robot must not harm a human being; must obey orders from humans unless they conflict with the first law; and must preserve its own existence unless this conflicts with the first or second law.

Asimov *Science-fiction author Isaac Asimov. Asimov published more than 200 books and gained much of his recognition through his adeptness at presenting science to the general public. He won the Hugo award for his* Foundation *trilogy published 1951–53.*

If my doctor told me I only had six months to live, I wouldn't brood. I'd type a little faster.

Isaac Asimov

Life

Aston Francis William 1877–1945. English physicist who developed the mass spectrometer, which separates isotopes by projecting their ions (charged atoms) through a magnetic field. For his contribution to analytic chemistry and the study of atomic theory he was awarded the 1922 Nobel Prize for Chemistry.

Aston was born and educated in Birmingham. From 1910 he worked in the Cavendish Laboratory, Cambridge, where J J Thomson was investigating positive rays from gaseous discharge tubes. Thomson and Aston examined the effects of electric and magnetic fields on positive rays, showing that the rays were deflected depending on their mass. The deflected rays were made to reveal their positions by aiming them at a photographic plate. The image produced on the photographic plate became known as a mass spectrum, and the instrument itself as a mass spectrometer. This became an essential tool in the study of nuclear physics and later found application in the determination of the structures of organic compounds.

Aston first examined neon gas and found that it consists of two isotopes. Over the next few years he examined the isotopic composition of more than 50 elements, and published *Isotopes* 1922.

Atkinson Robert D'escourt 1898–1982. Welsh astronomer and inventor. His research was in the field of atomic synthesis, stellar energy, and positional astronomy.

His contributions were fundamental to our basic understanding of how stars like the Sun work and how they evolve. He was also involved in instrument design.

Atkinson was born in Rhayader, Wales, and educated at Oxford and Göttingen, Germany. After early research at Oxford and

Berlin, he moved to the USA, ending his career at Rutgers University, New Jersey.

In 1924 British astrophysicist Arthur Eddington, to explain the radiation from the Sun, suggested a process whereby atoms were broken down inside the central core of a star, converting matter into energy. He was supported in this view by Atkinson, who in 1932 was working at the Royal Greenwich Observatory. In that year physicists had for the first time succeeded in splitting the nucleus of an atom. Atkinson was able to work out a theoretical model of the way in which matter could be annihilated. Not only did he determine the amount of energy released from atomic reactions within stars, but he was also able to suggest the kinds of reactions necessary to produce the vast quantities of radiation required.

Attenborough David Frederick 1926– . English traveller and zoologist who has made numerous wildlife films for television. He was the writer and presenter of the television series *Life on Earth* 1979, *The Living Planet* 1983, *The Trials of Life* 1990, and *The Private Life of Plants* 1995. Knighted 1985.

Attenborough studied zoology at Cambridge. In 1952 he joined the BBC Television Service as a trainee producer. Between 1954 and 1964 he made annual trips to film and study wildlife and human cultures in remote parts of the world; these expeditions were recorded in the *Zoo Quest* series of TV programmes and books. He was director of programmes for BBC Television 1969–72 and a member of its board of management. *Life on Earth* set out to illustrate evolution; *The Living Planet* dealt with ecology and the environment; *The Trials of Life* described life cycles; and *The Private Life of Plants* comprehensively covered the plant kingdom.

He is the brother of the actor and director Richard Attenborough.

Audubon John James 1785–1851. US naturalist and artist. In 1827, after extensive travels and observations of birds, he published the first part of his *Birds of North America*, with a remarkable series of colour plates. Later he produced a similar work on North American quadrupeds.

The ***National Audubon Society*** (founded 1886) has branches throughout the USA and Canada for the study and protection of birds.

Audubon was born in Santo Domingo (now Haiti) and educated in Paris, emigrating to the USA at 18. He travelled throughout the country collecting and painting the wildlife around him; he also painted portraits and even street signs. By 1825 he had compiled his beautiful set of bird paintings, but US publishers were not interested and he moved to the UK.

Audubon *US biologist and artist John James Audubon. Audubon published artworks depicting the wildlife of North America in the early nineteenth century. The majority of his paintings are of birds.*

Before Audubon most painters of birds used stylized techniques; stuffed birds were often used as subjects. Audubon painted from life and his compositions were startling, his detail minute. *The Birds of America* was published in Britain in 87 parts 1827–38. On his return to the USA in 1839 he published a bound edition of the plates with additions. He illustrated *Viviparous Quadrupeds of North America* 1845–48, compiling the text 1846–54 with his sons and John Bachman (1790–1874).

Auer Carl, Austrian chemist and engineer, see Baron von ⇨Welsbach.

Austin Herbert, 1st Baron Austin 1866–1941. English industrialist who began manufacturing cars 1905 in Northfield, Birmingham, notably the Austin Seven 1921. KBE 1917, Baron 1936.

Avery Oswald Theodore 1877–1955. Canadian-born US bacteriologist whose work on transformation in bacteria established 1944 that DNA (deoxyribonucleic acid) is responsible for the transmission of

Avery *Bacteriologist Oswald Theodore Avery. Oswald purified a molecule from heat-killed pathogenic smooth pneumococci bacteria extracts that could transform nonpathogenic mutant rough pneumococci into the smooth form in vitro. This molecule was deoxyribonucleic acid, DNA.*

heritable characteristics. He also proved that polysaccharides play an important part in immunity.

Avery was born in Halifax, Nova Scotia, but spent most of his life in New York. He studied medicine at Columbia University, and worked at the Rockefeller Institute Hospital 1913–48.

Avery's work on transformation – a process by which heritable characteristics of one species are incorporated into another species – was stimulated by the research of F Griffith (1877–1941), who 1928 published the results of his studies on *Diplococcus pneumoniae*, a bacterium that causes pneumonia in mice. Avery proved conclusively that DNA was the transforming principle responsible for the development of polysaccharide capsules in unencapsulated bacteria that had been in contact with dead, encapsulated bacteria. This implicated DNA as the basic genetic material of the cell.

Avogadro Amedeo, Conte di Quaregna 1776–1856. Italian physicist, one of the founders of physical chemistry, who proposed ***Avogadro's hypothesis*** on gases 1811. His work enabled scientists to calculate ***Avogadro's number***, and still has relevance for atomic studies.

Avogadro made it clear that the gas particles need not be individual atoms but might consist of molecules, the term he introduced to describe combinations of atoms. No previous scientists had made this fundamental distinction between the atoms of a substance and its molecules.

Avogadro was born in Turin, where he spent his whole academic career. He based most of his findings on a mathematical approach.

In 1809 Joseph Gay-Lussac had discovered that all gases, when subjected to an equal rise in temperature, expand by the same amount. From this Avogadro deduced his hypothesis.

Avogadro's hypothesis states that equal volumes of all gases, when at the same temperature and pressure, have the same numbers of molecules. It follows from this law that a mole of any substance contains the same number of molecules. This quantity is now known as ***Avogadro's number*** or constant.

Leading chemists of the day paid little attention to Avogadro's work, with the result that the confusion between atoms and molecules and between atomic weights and molecular weights continued for nearly 50 years.

Axelrod Julius 1912– . US neuropharmacologist who shared the 1970 Nobel Prize for Physiology or Medicine with the biophysicists Bernard Katz and Ulf von Euler (1905–1983) for his work on neurotransmitters (the chemical messengers of the brain).

Axelrod wanted to know why the messengers, once transmitted, should stop operating. Through his studies he found a number of specific enzymes that rapidly degraded the neurotransmitters.

Ayrton William Edward 1847–1908. English physicist and electrical engineer who invented many of the prototypes of modern electrical measuring instruments. He also created the world's first laboratory for teaching applied electricity, in Tokyo, Japan 1873.

Ayrton was born in London and studied mathematics there at University College. He joined the Indian Telegraph Service and was sent to Bombay 1868. In 1873 he became professor of physics and telegraphy at the new Imperial Engineering College in Tokyo, at that time the world's largest technical university. Returning to London, from 1879 he held various professorships in applied physics and electrical engineering.

In 1881 Ayrton and his colleague John Perry (1850–1920) invented the surface-contact system for electric railways, and they brought out the first electric tricycle 1882. There followed a series of portable electrical measuring instruments, including the ammeter (so named by its inventors), an electric power meter, various forms of improved voltmeters, and an instrument used for measuring self and mutual induction. In this, great use was made of an ingeniously devised flat spiral spring which yields a relatively large rotation for a small axial elongation.

Ayrton *English electrical engineer and inventor William Edward Ayrton. As well as establishing the world's first laboratory devoted to the teaching of applied electricity, he co-invented the ammeter and the world's first electric tricycle.*

Baade (Wilhelm Heinrich) Walter 1893–1960. German-born US astronomer who made observations that doubled the distance, scale, and age of the universe. He discovered that stars are in two distinct populations according to their age, known as Population I (the younger) and Population II (the older). Later, he found that Cepheid variable stars of Population I are brighter than had been supposed and that distances calculated from them were wrong.

Baade, born in Shröttinghausen, studied at Münster and Göttingen. He emigrated to the USA 1931, working at Mount Wilson Observatory, Pasadena, California, until 1948 and nearby Mount Palomar Observatory until 1958, when he returned to Germany.

During World War II, in 1943, Baade made use of the blackout darkness to study the Andromeda galaxy and was able to observe, for the first time, some of the stars in the inner regions of the galaxy. He found that the most luminous stars towards the centre are not blue-white but reddish, and proposed that these have differing structures and origins. Population I stars, bluish, are young and formed from the dusty material of the spiral arms – hydrogen, helium, and heavier elements; Population II stars, reddish, are old, were created near the nucleus and contain fewer heavy elements.

Babbage Charles 1792–1871. English mathematician who devised a precursor of the computer. He designed an analytical engine, a general-purpose mechanical computing device for performing different calculations according to a program input on punched cards (an idea borrowed from the Jacquard loom). This device was never built, but it embodied many of the principles on which digital computers are based.

Babbage was born in Totnes, Devon. As a student at Cambridge, he assisted John Herschel with his astronomical calculations and thought they could be better done by machines. His mechanical calculator, or difference engine, begun 1822, which could compute squares to six places of decimals, got him a commission from the British Admiralty for an expanded version. But this project was abandoned in favour of the analytical engine, which he worked on for the rest of his life. The difference engine could perform only one function, once it was set up. The analytical engine was intended to perform many functions; it was to store numbers and be capable of working to a program.

A Swede, Pehr Georg Scheutz, built a simpler version of the difference engine 1854, which worked and was bought by the Dudley Observatory, Albany, New York (and is now in the Smithsonian Institution).

The whole of the developments and operations of analysis are now capable of being executed by machinery ... As soon as an Analytical Engine exists, it will necessarily guide the future course of science.

Charles Babbage

Passages from the Life of a Philosopher 1864

Babbage was a founder member of the Royal Astronomical Society, the British Association, the Cambridge Philosophical Society, and the Statistical Society of London. He was elected Fellow of the Royal Society 1816. His book *On the Economy of Machinery and Manufactures* 1832 is an analysis of industrial production systems and their economics.

In 1991, the British Science Museum completed Babbage's second difference engine (to demonstrate that it would have been possible with the materials then available). It evaluates polynomials up to the seventh power, with 30-figure accuracy.

Babcock George Herman 1832–1893. US co-inventor of the first polychromatic printing press. He devised the Babcock–Wilcox steam boiler 1867 with his partner, Stephen Wilcox.

Babcock was born near Otego, New York. He went to work with his father, Asher Babcock, in daguerreotype and job printing for newpapers. Some time before 1854 Babcock and his father invented the polychromatic printing press. The father-and-son team also invented a job printing press which was still manufactured in the 1980s.

Employed at the Hope Iron Works, Providence, Rhode Island, Babcock and Stephen Wilcox designed a sectionally headed steam boiler with automatic cut-off, based on a safety water tube patented by Wilcox 1856. This boiler was able to withstand very high pressures and ensured a high standard of protection against explosions. It was first manufactured in Providence and then in New York, where the firm of Babcock and Wilcox was incorporated 1881. More than 100 years later, the firm was still manufacturing high-quality steam boilers.

Babcock Harold Delos 1882–1968. US astronomer and physicist. He measured the Sun's general magnetic field 1948 and studied the the relationship between sunspots and local magnetic fields. He also did important work in spectroscopy.

Babcock was born in Edgerton, Wisconsin, and studied electrical engineering at the University of California in Berkeley. In 1909 George Hale invited him to work at the Mount Wilson Observatory near Los Angeles, where he remained until 1948, with breaks during the two world wars.

Babcock made an investigation of the Zeeman effect (whereby a magnetic field causes a substance's spectral lines to be split) in chromium and vanadium – important elements in the solar spectrum. He then produced a revised table of wavelengths for the solar spectrum, published 1928 and including 22,000 spectral lines (extended 1947 and 1948).

It was in collaboration with his son Horace Welcome Babcock (1912–) that he succeeded in measuring the solar magnetic field. They used an instrument of their own design, the 'solar magnometer', which exploited the Zeeman effect to produce a continuously changing record of the Sun's local magnetic fields.

Bachelard Gaston 1884–1962. French philosopher and scientist who argued for a creative interplay between reason and experience. He attacked both Cartesian and positivist positions, insisting that science was derived neither from first principles nor directly from experience.

Balzac said that bachelors replace feelings by habits. In the same way, academics replace research by teaching.

Gaston Bachelard

La formation de l'ésprit scientifique 1938

Bacon Francis, 1st Baron Verulam and Viscount St Albans 1561–1626. English politician, philosopher, and essayist. He became Lord Chancellor 1618, and the same year confessed to bribe-taking, was fined £40,000 (which was later remitted by the king), and spent four days in the Tower of London. His works include *Essays* 1597, characterized by pith and brevity; *The Advancement of Learning* 1605, a seminal work discussing scientific method; *Novum organum* 1620, in which he redefined the task of natural science, seeing it as a means of

Bacon *After a bribery scandal that left his political career in ruins, Francis Bacon devoted the rest of his life to the advancement of science. His own scientific work was generally behind the times, but it was his ideas and philosophies, particularly his insistence of the importance of experiment over deduction, that made him such an influential figure.*

empirical discovery and a method of increasing human power over nature; and *The New Atlantis* 1626, describing a utopian state in which scientific knowledge is systematically sought and exploited.

Truth comes out of error more easily than out of confusion.

Francis Bacon

Quoted in R L Weber, *A Random Walk in Science*

Bacon was born in London, studied law at Cambridge from 1573, was part of the embassy in France until 1579, and became a member of Parliament 1584. He was the nephew of Queen Elizabeth's adviser Lord Burghley, but turned against him when he failed to provide Bacon with patronage. He helped secure the execution of the Earl of Essex as a traitor 1601, after formerly being his follower. Bacon was accused of ingratitude, but he defended himself in *Apology* 1604.

Satirist Alexander Pope called Bacon 'the wisest, brightest, and meanest of mankind'. Knighted on the accession of James I 1603, he became Baron Verulam 1618 and Viscount St Albans 1621. His writings helped to inspire the founding of the Royal Society. The ***Baconian Theory***, originated by James Willmot 1785, suggesting that the works of Shakespeare were written by Bacon, is not taken seriously by scholars.

He died after catching a cold while stuffing a chicken with snow in an early experiment in refrigeration.

I have taken all knowledge to be my province.

Francis Bacon

Letter to Lord Burleigh, 1592

Bacon Roger *c.* 1214–1294. English philosopher and scientist. He was interested in alchemy, the biological and physical sciences, and magic. Many discoveries have been credited to him, including the magnifying lens. He foresaw the extensive use of gunpowder and mechanical cars, boats, and planes.

In 1266, at the invitation of his friend Pope Clement IV, he began his *Opus majus/Great Work*, a compendium of all branches of knowledge. In 1268 he sent this with his *Opus minus/Lesser Work* and other writings to the pope. In 1277 Bacon was condemned and imprisoned by the Christian church for 'certain novelties' (heresy) and not released until 1292.

Bacon wrote in Latin and his works include *On Mirrors*, *Metaphysical*, and *On the Multiplication of Species*. He followed the maxim 'Cease to be ruled by dogmas and authorities; look at the world!'

Bacon was born in Somerset and educated at Oxford and Paris. He became a Franciscan monk and lectured in Paris about 1241–47, then at Oxford University. He described a hypothetical diving apparatus and some of the properties of gunpowder. He promoted the use of latitude and longitude in mapmaking, and suggested the changes necessary to improve the Western calendar that were carried out by Pope Gregory XIII in 1582.

Mathematics is the door and the key to the sciences.

Roger Bacon

Opus Majus part 4 *Distinctia Prima* cap 1, 1267 transl Robert Belle Burke, 1928

Baekeland Leo Hendrik 1863–1944. Belgian-born US chemist. He invented Bakelite, the first commercial plastic, made from formaldehyde and phenol. He also made a photographic paper, Velox, which could be developed in artificial light.

Baekeland was born in Ghent and educated there and at Charlottenburg Technische Hochschule in Germany. In 1889 he moved to the USA, setting up as a consultant in his own laboratory in New York. He began making Velox photographic paper 1893, and sold the invention and the manufacturing company to Kodak for $1 million 1899. In 1909 he founded the General Bakelite Corporation, later to become part of the Union Carbide and Carbon Company.

Baer Karl Ernst von 1792–1876. Estonian embryologist who discovered the mammalian ovum 1827. He made a significant contribution to the systematic study of the development of animals, and showed that an embryo develops from simple to complex, from a homogeneous to a heterogeneous stage.

Baer was born in Piep and studied at Dorpat (Tartu); at Vienna, Austria; and in

Baer *Using a colleague's pet dog as a subject for dissection, von Baer was the first embryologist to show that mammalian reproduction involved the fusion of a female ovum with male sperm, instead of, as previously thought, a mingling of mutual seminal fluids. He also showed that during embryonic development, general characteristics appeared before species-dependent ones: this concept eventually became known as the 'biogenetic law'.*

Germany at Würzburg. He taught at Königsberg 1817–34, then moved to St Petersburg, Russia. In 1837 he led the first of many expeditions into Novaya Zemlya, in Arctic Russia, where he was the first naturalist to collect plant and animal specimens. He later led expeditions to Lapland, the Caucasus, and the Caspian Sea. He was professor at the Medico-Chirurgical Academy in St Petersburg 1846–62.

Baer conceived that the goal of early development is the formation of three layers in the vertebrate embryo – the ectoderm, endoderm, and mesoderm – out of which all later organs are formed. He also suggested that the younger the embryos of various species are, the stronger is the resemblance between them.

In his observations of the embryo, von Baer discovered the extraembryonic membranes – the chorion, amnion, and allantois – and described their functions. He also identified for the first time the notochord and revealed the neural folds.

Baeyer Johann Friedrich Wilhelm Adolf von 1835–1917. German organic chemist who synthesized the dye indigo 1880. He discovered barbituric acid 1863, later to become the parent substance of a major class of hypnotic drugs. In 1888 he carried out the first synthesis of a terpene. Nobel Prize for Chemistry 1905.

Baeyer was born in Berlin and studied there and at Heidelberg.

He became professor of chemistry at the University of Strasbourg 1872 and three years later at Munich, where he stayed for the rest of his career.

Baeyer discovered fluorescein 1871. He also found the resinous condensation product of phenol and formaldehyde (methanal), which Leo Baekeland later developed into the first thermosetting plastic Bakelite.

His work with ring compounds and the highly unstable polyacetylenes led him to consider the effects of carbon–carbon bond angles on the stability of organic compounds. He concluded that the more a bond is deformed away from the ideal tetrahedral angle, the more unstable it is; this is known as ***Baeyer's strain theory***. It explains why rings with five or six atoms are much more common, and stable, than those with fewer or more atoms.

Bailey Donald Coleman 1901–1985. English engineer, inventor in World War II of the portable ***Bailey bridge***, made of interlocking, interchangeable, adjustable, and easily transportable units. Knighted 1946.

Bailey Liberty Hyde 1858–1954. US horticulturist and botanist who advised President Roosevelt on his agricultural policy and ran Roosevelt's Country Life Commission 1908, which aimed to improve the standard of life and living conditions of rural communities. His own research included work on *Carex* and *Cucurbita* (plants of the gourd family).

Bailey was born in South Haven Township, Michigan. During his childhood he became interested in natural history and geology. As a young man of nineteen he was given a place at Michigan State Agricultural College, where he first encountered Charles Darwin's work on evolution. In 1882, after obtaining his BSc, he became a reporter in Illinois, but he quickly obtained the position of assistant curator of the Harvard University herbarium. In 1885, he was made the professor of horticulture and landscape gardening at Michigan State Agricultural College and was made professor of practical and experimental horticulture at Cornell University 1888.

As a botanist and horticulturist, he had a particular interest in agriculture and was appointed the first dean of the New York State College of Agriculture 1904. He was then invited by the US Government to run Roosevelt's Country Life Commission 1908.

It is time for me to enjoy another pinch of snuff. Tomorrow my hands will be bound, so as to make it impossible.

Jean Bailly

H Hoffmeister *Anekdotenschatz*. Remark on the evening before his execution.

Bailly Jean Sylvain 1736–1793. French astronomer who wrote about the satellites of Jupiter and the history of astronomy. Early in the French Revolution he was president of the national assembly and mayor of Paris, but resigned in 1791; he was guillotined during the Reign of Terror.

Baily Francis 1774–1844. British astronomer who described 1836 the light effect called Baily's beads, observable round the edge of the Moon, immediately before and after the Sun's disappearance during a total eclipse. These are caused by sunlight shining between the mountains on the Moon's horizon.

Baily was born in Newbury, Berkshire, and apprenticed to a firm of merchant bankers in London; then he set out to explore unsettled parts of North America. On his return to England in 1798 he became a stockbroker, but retired 1825 in favour of full-time astronomy. Baily travelled to Italy 1842 and was again able to see his beads during a solar eclipse. He was not the first to have noticed the beads, but his description of the 1836 eclipse was so exciting that it sparked a renewed and lasting interest in eclipses.

Baily began to publish his astronomical observations 1811. He was the author of an accurate revised star catalogue in which he plotted the positions of nearly 3,000 stars. He also measured the Earth's elliptical shape.

Bainbridge Kenneth Tompkins 1904– . US physicist who was director of the first atomic-bomb test at Alamogordo, New Mexico, 1945. He also made important innovations in the mass spectrometer.

Bainbridge was born in Cooperstown, New York, and educated at the Massachusetts Institute of Technology (MIT) and Princeton University. He worked at the Cavendish Laboratory, Cambridge, England, in the early 1930s, and held academic posts at Harvard from 1934; he also worked in the radiation laboratory at MIT 1940–43 and then in the Los Alamos laboratory on the atomic bomb until 1945. He also carried out research in radar.

The mass spectrometer invented by English physicist Francis Aston focused ion beams of varying velocity but not varying direction. In 1936 Bainbridge developed a machine in which ion beams that are non-uniform in both direction and velocity can be brought to a focus.

Baird John Logie 1888–1946. Scottish electrical engineer who pioneered television. In 1925 he gave the first public demonstration of television and in 1926 pioneered fibre optics, radar (in advance of Robert Watson-Watt), and 'noctovision', a system for seeing at night by using infrared rays.

Born at Helensburgh, Scotland, Baird studied electrical engineering in Glasgow at what is now the University of Strathclyde, at the same time serving several practical apprenticeships. He was working on television possibly as early as 1912, and he took out his first pro-

Baird *Scottish electrical engineer John Logie Baird, who gave the first demonstration of a televised image. Baird suffered from persistant ill-health and was forced to retire from his job as a salesman in 1923. On his retirement, he invested all his time into the development of television and, with very little money, constructed the first working set from scrap materials in his attic.*

visional patent 1923. He also developed video recording on both wax records and magnetic steel discs (1926–27), colour TV (1925–28), 3-D colour TV (1925–46), and transatlantic TV (1928). In 1936 his mechanically scanned 240-line system competed with EMI-Marconi's 405-line, but the latter was preferred for the BBC service from 1937, partly because it used electronic scanning and partly because it handled live indoor scenes with smaller, more manoeuvrable cameras. In 1944 he developed facsimile television, the forerunner of Ceefax, and demonstrated the world's first all-electronic colour and 3-D colour receiver (500 lines).

Baker Alan 1939– . English mathematician whose chief work has been devoted to the study of transcendental numbers (numbers that cannot be expressed as roots or as the solution of an algebraic equation with rational coefficients).

Baker was born in London and studied mathematics there and at Cambridge. He remained at Cambridge, except for many visiting professorships abroad, becoming professor 1974.

In 1966 he extended French mathematician Joseph Liouville's original proof of the existence of transcendental numbers by means of continued fractions, by obtaining a result on linear forms in the logarithms of algebraic numbers. This solution opened the way to the resolution of a wide range of Diophantine problems and in 1967 Baker used his results to provide the first useful theorems concerning the theory of these problems.

Apart from individual papers, Baker's most important publication is *Transcendental Number Theory* 1975.

Baker Benjamin 1840–1907. English engineer who designed, with English engineer John Fowler (1817–1898), London's first underground railway (the Metropolitan and District) 1869; the Forth Rail Bridge, Scotland, 1890; and the original Aswan Dam on the river Nile, Egypt. KCMG 1890.

Baker was born near Frome, Somerset, and at 16 was apprenticed at Neath Abbey ironworks. In 1862 he joined the staff of John Fowler, becoming his partner 1875. In the construction of the Central Line of the London Underground, Baker incorporated an ingenious energy-conservation measure: he dipped the line between stations to reduce the need both for braking to a halt and for the increase in power required to accelerate away.

The Forth Bridge was built just after the

Baker *British civil engineer Benjamin Baker 1896. Baker worked in the latter half of the nineteenth century and is famous for designing the Forth Rail Bridge.*

collapse of the Tay Bridge 1879, and made Baker's name internationally. It is a cantilever structure of mild steel, which had just become available through the new Siemens open-hearth process. The two main spans are each of 521 m/1,710 ft.

Balfour Eve 1898–1990. English agriculturalist and pioneer of modern organic farming. She established the Haughley Experiment, a farm research project at New Bells Farm near Haughley, Suffolk, to demonstrate that a more sustainable agricultural alternative existed. The experiment ran for almost 30 years, comparing organic and chemical farming systems. The wide-ranging support it attracted led to the formation of the Soil Association 1946.

Balfour John Hutton 1808–1884. Scottish botanist who was an inspired teacher and introduced microscopy into undergraduate botany courses.

Balfour was born in Edinburgh, the son of an army surgeon and educated at the Edinburgh High School and St Andrews University. He intended to be ordained into the Church of Scotland, but gave this up and graduated with an MD from Edinburgh University 1832. After continuing his medical studies in Paris he set up his own practice in Edinburgh 1834. He had an abiding love of botany however, and helped to set up the Botany Society of Edinburgh 1836 and the Edinburgh Botany Club 1838. In 1841 he gave up medicine in order to become the professor of botany at Glasgow University, and in 1845 he was made professor of botany at Edinburgh and the Regis Keeper of the

Royal Botanical Garden in Edinburgh, as well as the Queen's Botanist for Scotland. He was an inspired teacher and, while professor at Edinburgh, led his students on regular botanical excursions to every part of Scotland and established the use of light microscopes as teaching aids within the department. He wrote several textbooks, but did not do much original research.

Balmer Johann Jakob 1825–1898. Swiss physicist and mathematician who developed a formula in 1884 that gave the wavelengths of the light emitted by the hydrogen atom (the hydrogen spectrum). This simple formula played a central role in the development of spectral and atomic theory.

Balmer was born near Basel, where he taught school for 40 years. Although a mathematician and not trained in physics, he became interested in spectroscopy. In 1885 he published an equation that described the four visible spectral lines of hydrogen (all that were then known) and also predicted the existence of a fifth line at the limit of the visible spectrum, which was soon detected and measured. He further predicted the existence of other hydrogen spectral lines beyond the visible spectrum. The five lines in the visible part of the hydrogen spectrum are now known as the Balmer series.

Baltimore David 1938– . US virologist who shared the Nobel Prize for Physiology or Medicine in 1975 with Renato Dulbecco and Howard Temin for their discovery that certain viruses contain an enzyme, called reverse transcriptase, that makes deoxyribonucleic acid (DNA) from ribonucleic acid (RNA).

In 1970, Baltimore and Temin independently discovered that viruses called retroviruses, which contain their genetic information in the form of RNA, could transfer their genes into the DNA of a cell infected with the virus. Baltimore and his wife, Alice Huang, discovered that the virus that causes vesicular stomatitis reproduces itself using an enzyme, reverse transcriptase, to copy its own RNA.

This work was a surprise to the scientific community, which had accepted Francis Crick's earlier 'Central Dogma' of molecular biology indicating that genetic information flows in the opposite direction – from DNA on the chromosomes into RNA and from that RNA the production of protein in the cytoplasm of the cell.

Baltimore was born on 7 March 1938 in New York City. He studied at Swarthmore College in Pennsylvania and Rockefeller University in New York before being asked to join Renato Dulbecco's research team at the Salk Institute in La Jolla, California 1965 to start his work on viral genetics. In 1972 he was appointed to a staff position at the Massachusetts Institute of Technology (MIT), where he currently works as a professor.

In 1990, Baltimore was appointed president of Rockefeller University but resigned this post a year later due to a public dispute over a fraudulent research paper that he had published jointly with a colleague at MIT. Baltimore himself was not included in any charges of misconduct.

Banks Joseph 1743–1820. British naturalist and explorer. He accompanied Capt James Cook on his voyage around the world 1768–71 and brought back 3,600 plants, 1,400 of them never before classified. The *Banksia* genus of shrubs is named after him. Created a baronet 1781.

Banks was born in London and educated at Oxford. Inheriting a fortune, he made his first voyage 1766, to Labrador and Newfoundland. In 1768 Banks obtained the position of naturalist on an expedition to the southern hemisphere in the *Endeavour*, commanded by Capt James Cook. The expedition explored the coasts of New Zealand and Australia. Banks's plant-collecting activities at the first landing place in Australia (near present-day Sydney) gave rise to the name of the area – Botany Bay. He also studied the Australian fauna.

Returning to England 1771, he brought back a vast number of plant specimens, more than 800 of which were previously unknown. As a result of the friendship between Banks and George III, the Royal Botanic Gardens at Kew – of which Banks was the honorary director – became a focus of botanical research.

In 1772 Banks went on his last expedition, to Iceland, where he studied geysers. He was instrumental in establishing the first colony at Botany Bay in 1788.

Banneker Benjamin 1731–1806. American astronomer, surveyor, and mathematician who published almanacs 1792–97. He took part in the survey that prepared the establishment of the US capital, Washington DC.

Banneker was born near the Patapsco River in Baltimore County. He was self-taught and worked most of his life on a tobacco farm. He engaged in a long correspondence with politician Thomas Jefferson, defending the mental capacities of black people and urging the abolition of slavery.

In 1753, at the age of 21, having studied only a pocket watch, Banneker constructed a striking clock. This was the first clock of its kind in America and operated for more than 40 years.

Having taught himself to calculate an ephemeris (table showing future positions of planets, comets, and so on) and to make projections for lunar and solar eclipses, Banneker compiled an ephemeris for each year 1791–1804, though not all were published. He also wrote a dissertation on bees and did a study of locust plague cycles.

Banting Frederick Grant 1891–1941. Canadian physician who discovered a technique for isolating the hormone insulin 1921 when he and his colleague Charles Best tied off the ducts of the pancreas to determine the function of the cells known as the islets of Langerhans. This made possible the treatment of diabetes. Banting and John J R Macleod (1876–1935), his mentor, shared the 1923 Nobel Prize for Physiology or Medicine, and Banting divided his prize with Best. KBE 1934.

Banting *With his assistant Charles Best, Frederick Banting developed the first practical method for the commercial preparation of insulin. Insulin is a pancreatic hormone which reduces glucose levels in blood, and is effective in the treatment of diabetes. For his discovery, Banting was awarded the 1923 Nobel Prize for Physiology or Medicine.*

Banting was born in Alliston, Ontario, and studied medicine at the University of Toronto, where from 1921 he carried out research into diabetes. It had been suggested that insulin might be concerned in glucose metabolism and that its source might be the islets of Langerhans. Banting reasoned that if the pancreas were destroyed but the islets of Langerhans were retained, the absence of digestive enzymes would allow them to isolate insulin. With Charles Best, one of his undergraduate students, he experimented on dogs. Next, he obtained fetal pancreatic material from an abattoir. Eventually reasonably pure insulin was produced and commercial production of the hormone started.

Banu Musa (Arabic 'sons of Musa') three brothers, Muhammad (died 873), Ahmad, and al-Hasan, who lived in Baghdad. They compiled an important mathematical work on the measurement of plane and spherical figures, and one of the earliest works on mechanical engineering, *Kitab al-Hiyal/The Book of Ingenious Devices*.

The Banu Musa were educated at a scientific academy, and commissioned Arabic translations of Greek scientific and medical works. They calculated the circumference of the Earth by trigonometry, and Muhammad and Ahmad engaged in large engineering projects, such as a canal and a new town, while al-Hasan devoted himself to theoretical study. *The Book of Ingenious Devices* contains 100 designs ranging from a siphon to a dredging machine.

Bardeen John 1908–1991. US physicist who won a Nobel prize 1956, with Walter Brattain and William Shockley, for the development of the transistor 1948. In 1972 he became the first double winner of a Nobel prize in the same subject (with Leon Cooper and Robert Schrieffer (1931–)) for his work on superconductivity.

At the Bell Telephone laboratory, New Jersey, 1945–51, in a team with Shockley and Brattain, Bardeen studied semiconductors, especially germanium, used in radar receivers in the same way that crystals had been used in the earliest radio sets. The work led to the development 1956 of the transistor.

The second Nobel prize was won for explaining superconductivity, the total loss of electrical resistance by some metals when cooled within a few degrees of absolute zero. The theory developed in 1957 by Bardeen, Schrieffer, and Cooper states that super-conductivity arises when electrons travelling through a metal interact with the vibrating atoms of the metal.

Barker Herbert Atkinson 1869–1950. British manipulative surgeon, whose work established the popular standing of orthopaedics (the study and treatment of disorders of the spine and joints), but who was never recognized by the world of orthodox medicine.

Barkla Charles Glover 1877–1944. English physicist who studied the phenomenon of X-ray scattering. He found that X-ray emissions are a form of transverse electromagnetic radiation, like visible light, and monochromatic. Nobel Prize for Physics 1917.

Barkla was born in Widnes, Lancashire, and studied in Liverpool and at Cambridge. He was professor of physics at King's College, London, 1909–13, when he became professor of natural philosophy at Edinburgh University.

In 1903 Barkla published his first paper on secondary radiation – the effect whereby a substance subjected to X-rays re-emits secondary X-radiation. He found that the more massive an atom, the more charged particles it contains, and it is these charged particles that are responsible for the X-ray scattering. Barkla was one of the first to emphasize the importance of the amount of charge in an atom (rather than merely its atomic mass) in determining an element's position in the periodic table.

Between 1904 and 1907 Barkla found that, unlike the low atomic mass elements, the heavy elements produced secondary radiation of a longer wavelength than that of the primary X-ray beam, and that the radiation from the heavier elements is of two characteristic types. Barkla named the two types of characteristic emissions the K-series (for the more penetrating emissions) and the L-series (for the less penetrating emissions). He later predicted that an M-series and a J-series of emissions with different penetrances might exist, and an M-series was subsequently discovered.

Barnard Christiaan Neethling 1922– . South African surgeon who performed the first human heart transplant 1967 at Groote Schuur Hospital in Cape Town. The 54-year-old patient lived for 18 days.

Barnard also discovered that intestinal artresia – a congenital deformity in the form of a hole in the small intestine – is the result of an insufficient supply of blood to the fetus during pregnancy. It was a fatal defect before he developed the corrective surgery.

Barnard was born in Beaufort West, Cape Province, and studied at the University of Cape Town. He held academic and research posts at that university and Groote Schuur Hospital until 1983.

It is infinitely better to transplant a heart than to bury it to be eaten by worms.

Christiaan Barnard

Time 31 Oct 1969

Barnard's early research involved experiments with heart transplants in dogs, and in Dec 1967 he first applied the technique to a human patient. Surgically this transplant was a success, but the patient died from double pneumonia – probably contracted as a result of the immunosuppressive drugs administered to him to prevent his body rejecting the new heart.

Open-heart surgery was first introduced in South Africa by Barnard, and he further developed cardiothoracic surgery by new designs for artificial heart valves.

Barnard Edward Emerson 1857–1923. US observational astronomer who discovered the fifth satellite of Jupiter 1892 and Barnard's star 1916. He was the first to realize that the apparent voids in the Milky Way are in fact dark nebulae of dust and gas.

Barnard's star is the second closest star to the Sun, six light years away in the constellation Ophiuchus. A faint red dwarf, it is visible only through a telescope. It came to be called 'Barnard's run-away star', because it has a proper motion of 10 seconds of arc per year – faster than any star known until 1968.

Barnard discovered his first comet 1881 and by 1892 he had found 16. He also investigated the surface features of Jupiter.

Barnard was born in Nashville, Tennessee, and from the age of nine worked as an assistant in a photographic studio. A fascination with astronomy led him to take a job in the observatory at Vanderbilt University, where he spent most of his time using the telescopes. He went to California to work at the Lick Observatory when it opened 1888. In 1895 he took up the chair of practical astronomy at the University of Chicago and became astronomer at the Yerkes Observatory, Wisconsin. He participated in an expedition to Sumatra, Indonesia, to observe the solar eclipse of 1901.

Barr Murray Llewellyn 1908– . Canadian anatomist and geneticist who carried out research into defects of the human reproductive system, and simplified diagnostic tests for chromosomal defects.

Barr was born in Belmont, Ontario, and attended the University of Western Ontario, where he spent his entire academic career.

In 1949 Barr noticed that the nuclei of nerve cells in females have a mass of chromatin (the nucleoprotein of chromosomes, which stains strongly with basic dyes), whereas those in males do not. He found that this sex difference occurs in the cells of most mammals. From his investigations, the sex chromatin (called the Barr body) is now known to be one of the two X-chromosomes in the cells of females; it is more condensed than the other chromosomes and is genetically inactive. The other X-chromosome in females is attenuated and genetically active in resting cells. Stained sex chromatin offered a much needed investigative and diagnostic procedure for patients with developmental anomalies of the reproductive system.

Barr and his colleagues also devised a buccal smear test by rubbing the lining of the patient's mouth (the buccal cavity) and examining the cells obtained for chromosomal defects. This test is now used extensively to screen patients, including newborn babies.

Barrow Isaac 1630–1677. British mathematician, theologian, and classicist. His *Lectiones geometricae* 1670 contains the essence of the theory of calculus, which was later expanded by Isaac Newton and Gottfried Leibniz.

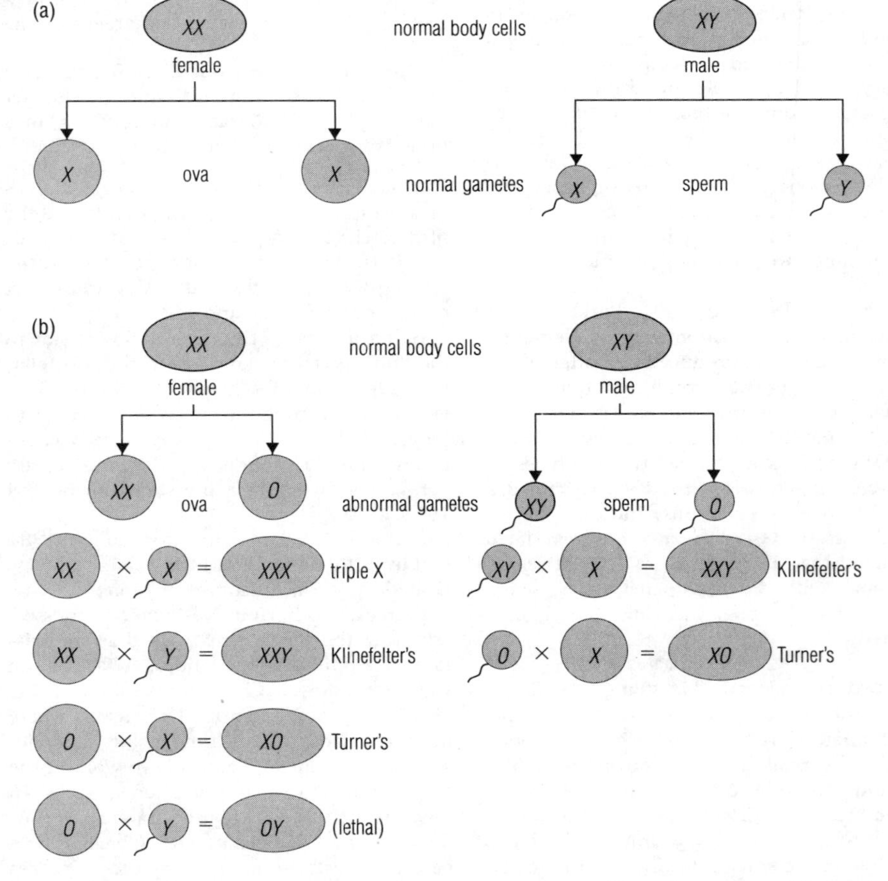

Barr *(a) Meiosis with the formation of normal gametes; (b) chromosomal defects arise from fertilization of abnormal gametes by normal gametes.*

Barrow was born in London and studied at Cambridge, where he was professor of mathematics 1663–69. Isaac Newton attended his lectures and was inspired by Barrow's work in the field of optics. To Barrow is due the credit for two original contributions: the method of finding the point of refraction at a plane interface, and his point construction of the diacaustic of a spherical interface.

Barrow's mathematical importance is slight, the *Lectiones Mathematicae* 1669 being marred by his insistence that algebra be separated from geometry and his desire to relegate algebra to a subsidiary branch of logic.

Bartlett Frederic Charles 1886–1969. English psychologist. He put forward the view of sensory and memory processes as the expression of a dynamic integration of an organism's past experience with its current situation and needs. The results of his extensive researches, centred on perception, recognition, and recall processes, are collected in his book *Remembering: A Study of Experimental and Social Psychology* 1932. Knighted 1948.

Bartlett, born in Gloucestershire, studied at London and Cambridge universities and, with C S Myers (1873–1946) and W H R Rivers (1864–1922), founded the Psychological Laboratory at Cambridge. He went on to become the first professor of experimental psychology at Cambridge 1931.

He also carried out influential work in applied and industrial psychology on problems associated with submarine detection, the design and control of aircraft, and the training of air-force personnel during both world wars.

Bartlett Neil 1932– . British-born chemist who in 1962 prepared the first compound of one of the inert gases, which were previously thought to be incapable of reacting with anything.

Bartlett was born in Newcastle-upon-Tyne and attended the University of Durham. He took an appointment at the University of British Columbia, Canada, 1958, became professor of chemistry at Princeton, USA, 1966, and at the University of California, Berkeley, 1969.

In Canada Bartlett was working with platinum hexafluoride, PtF_6, and found that it is extremely reactive. When he reacted it with xenon, the heaviest of the stable inert gases, he obtained xenon platinofluoride (xenon fluoroplatinate, $XePtF_6$), the first chemical compound of an inert gas. Other compounds of xenon followed, including xenon fluoride (XeF_4) and xenon oxyfluoride ($XeOF_4$).

Barton Derek Harold Richard 1918– . English organic chemist who investigated the stereochemistry of natural compounds. He showed that their biological activity often depends on the shapes of their molecules and the positions and orientations of key functional groups. He shared the 1969 Nobel Prize for Chemistry.

Barton was born in Gravesend, Kent, and studied at Imperial College, London. He has held various professorships in the UK, beginning at Birkbeck College, London, 1953 and ending at the University of London from 1978, the same year that he was appointed director of the Institute for the Chemistry of Natural Substances in France.

While lecturing in the USA at Harvard 1949–50, Barton studied the different rates of reaction of certain steroids and their triterpenoid isomers (substances with the same composition but differing in the way their atoms are joined and arranged in space). He deduced that the difference in the spatial orientation of their functional groups accounts for their behaviour, and so developed a new field in organic chemistry which became known as conformational analysis.

Barton went on to examine many natural products, concluding that the structures of many phenols and alkaloids could be explained and predicted.

Bartram John 1699–1777. US botanist described by Linnaeus as the greatest 'natural botanist' in the world. He made many expeditions in order to study the native flora and its environment, and his research included the first plant hybridizing experiments to be carried out in America.

Bartram was born in Marple, Pennsylvania and became a farmer and collector of North American plants. He corresponded with the botanist and businessman, Peter Collinson, in England and sold his plants through Collinson to botanists in Europe. In 1728, he was responsible for the formation of the Botanical Garden at Kingsessing on the banks of the Schuylkill river. He travelled to West Virginia and the Blue Ridge Mountains 1738, to the Catskills 1755, and to the Carolinas 1760. By 1765 he had been appointed royal botanist and in this capacity he travelled from Charleston, South Carolina to St Augustine, Florida. As part of this expedition, he explored the St John's River by canoe. His published work includes *Observations on the Inhabitants, Climate, Soil, etc .... made by John Bartram in his travels from Pennsylvania to Lake Ontario.*

Basov Nikolai Gennadievich 1922– . Soviet physicist who, with his compatriot Aleksandr Prokhorov, developed the microwave amplifier called a maser. They were both awarded the Nobel Prize for Physics 1964, which they shared with Charles Townes of the USA.

Bassi Agostino 1773–1856. Italian microbiologist who was the first to demonstrate that microscopic organisms could cause certain infectious diseases. His work preceded that of the Louis Pasteur, who formulated 1868 the theory that germs might cause some diseases.

At the beginning of the 19th century Bassi studied the silkworm disease muscardine. He discovered 1807 that it was caused by a minute parasitic fungus (the fungus was later named *Botrytis bassiana* after its discoverer) that was transmitted by infected food and from animal to animal by contact. He went on to describe methods for treating fungally infected worms, which was of considerable interest at the time, as muscardine was causing financial losses to those working in the European silk trade.

Bassi was born 25 September 1773 in a village near Lodi in what was then part of the Austrian Empire but is now a part of Italy. He graduated in law and worked as a civil servant in Italy while devoting much of his spare time to the study of living organisms using an early version of the microscope.

Although Anton van Leeuwenhoek first discovered and described such minute microorganisms as bacteria 1663, the link between these tiny organisms and the induction of infectious diseases was not recognized for another two hundred years. Bassi was the first to understand this link.

Bates H(enry) W(alter) 1825–1892. English naturalist and explorer, who spent 11 years collecting animals and plants in South America and identified 8,000 new species of insects. He made a special study of camouflage in animals, and his observation of insect imitation of species that are unpleasant to predators is known as 'Batesian mimicry'.

Bates was born in Leicester and left school at 13, but studied natural history in his spare time. In 1844 he met English naturalist Alfred Russel Wallace, and together they travelled to the Amazon region of South America 1848 to study and collect its flora and fauna. Wallace returned to England 1852 but Bates remained until 1859. He returned with a vast number of specimens, including more than 14,000 species of insects.

Bates *During a two-year insect hunting expedition to the Amazon, H W Bates collected over 14,000 specimens, including almost 8,000 that were new to science. On his return, he published a paper describing the phenomenon of insect mimicry, now known as Batesian mimicry. His theory proposes that edible species of insects may gradually evolve markings which are very similar to poisonous species in an attempt to deceive predators.*

In 1861 Bates presented a paper entitled 'Contributions to an insect fauna of the Amazon Valley', in which he outlined his observations of mimicry. He had discovered that several different species of butterflies have almost identical patterns of colours on their wings, and that some are distasteful to bird predators whereas others are not. He suggested that the latter types, influenced by natural selection, mimic the distasteful species and thus increase their chances of survival.

In *The Naturalist on the River Amazon* 1863, Bates described both his explorations and his scientific findings.

Bateson William 1861–1926. English geneticist who was one of the founders of the science of genetics (a term he introduced), and a leading proponent of Austrian biologist Gregor Mendel's work on heredity. Bateson also made contributions to embryology.

Bateson was born in Whitby, Yorkshire, and educated at Cambridge, where in 1908 he became the first professor of genetics. He was director of the John Innes Horticultural Institution in Surrey 1910–26.

Doing embryological research in the USA in the 1880s, Bateson discovered evidence that chordates had evolved from echinoderms – a theory now widely accepted. He spent the next years investigating the fauna of the salt lakes of Europe, central Asia, and Egypt. In his book *Material for the Study for Variation* 1894, he put forward his theory of discontinuity to explain the long process of evolution. According to this theory, species do not develop in a predictable sequence of very gradual changes but instead evolve in a series of discontinuous jumps. Mendel's work, which he translated and championed, provided him with supportive evidence.

...knowledge grows by solid increments, definite, predictable discoveries of fact. Rarely is any piece of interpretation an event of equal consequence.

William Bateson

Bateson also carried out breeding experiments, described in *Mendel's Principles of Heredity* 1908. He showed that certain traits are consistently inherited together; this phenomenon (called linkage) is now known to result from genes being situated close together on the same chromosome.

Batten Jean 1909–1982. New Zealand aviator who made the first return solo flight by a woman Australia–Britain 1935, and established speed records.

Bauer Ferdinand Lucas 1760–1826. Austrian botanist and illustrator who contributed to the replacement of the Linnaean classification system for plants during a trip to Australia 1801 with Robert Brown.

Bauer was born in Feldsberg, Lower Austria and later became known for his expert illustration of John Sibthorp's *Flora Graeca.* He was educated with his two brothers Joseph and Franz by monks in Feldsberg and helped to illustrate Father Boccius's 14-volume *Hotus Botanicus* while he studied under him. Then he went to Vienna to work for Nikolamus von Jacquin and it was there that he met John Sibthorp, who was professor of botany at Oxford University.

He travelled in Greece and the Levant with Sibthorp in 1786 and 1787, painting over 1,000 watercolours, mainly of plants. In 1801 he went to Australia with Matthew Flinders and Robert Brown. The work which he produced in Australia with Brown was important because it helped to overthrow Linnaeus's taxonomic system. Bauer was able to produce large numbers of watercolour illustrations because he sketched the plants in the field and painted them later.

Bauer Franz Andreas 1758–1840. Austrian botanist who introduced an early begonia hybrid to the Britain, worked on orchids, and helped to establish the Botanical Gardens in Edinburgh.

Bauer was born in Feldsberg, Lower Austria and educated with his two brothers, Ferdinand and Joseph, by monks in Feldsberg. He trained as a botanical illustrator and helped to illustrate Father Boccius's 14-volume *Hotus Botanicus* while studying under him. He then went to Vienna to work for Nikolamus von Jacquin and in 1788 travelled to London in the company of Jacquin's son. While he was in London he studied in Joseph Banks's library and he was subsequently employed by Banks at the Royal Garden at Kew for the rest of his life. His work at Kew did not keep him fully occupied and he was able to carry out his own research. His publications include *Delineations of Exotic Plants Cultivated in the Royal Garden at Kew* 1796 and the paper 'Microscopical Observations on the Red Snow' 1819.

Bauhin Gaspard Casper 1560–1624. Swiss botanist and physician who developed an important early plant classification system. In *Pinax theatri botanica/Illustrated Exposition of Plants* 1623, he attempted to classify all known species of plant, naming 6,000 species.

Bauhin's classification system was used by both Linnaeus and John Ray, and was to some extent based upon the system which the Italian Andrea Cesalpino had outlined earlier in the 16th century. Bauhin also wrote several textbooks of human anatomy, including *Theatrum anatomicum* 1605.

Bauhin was born in Basel, Switzerland and studied medicine at the University of Padua in Italy 1577–88, where he was a student of Fabricius ab Aquapendente. He then returned to the University of Basel, where he obtained his MD 1581. He was appointed professor of Greek the following year, professor of anatomy and botany 1588, and professor of medicine 1614. Bauhin died in Basel on 5 December.

Baxter George 1804–1867. English engraver and printmaker; inventor 1834 of a special process for printing in oil colours, which he applied successfully in book illustrations.

Bayes Thomas 1702–1761. English mathematician whose investigations into probability led to what is now known as Bayes' theorem.

Bayes' theorem relates the probability of particular events taking place to the probability that events conditional upon them have occurred.

Bayley Isaac 1853–1922. Scottish botanist whose main interests were rhododendrons and primulas; he also studied plant propagation and germination.

Bayley was born in Edinburgh and educated first at the Edinburgh Academy and then the University, where he graduated with a BSc 1873. He then went to Wurzburg University where he studied under the influential naturalist, Julius von Sachs.

In 1875 he returned to Edinburgh University to become an assistant to the professor of natural history. Whilst working here he participated in the transit of Venus expedition to Rodriguez Island. He was appointed to the chair of botany at Glasgow University 1879. He then moved to Oxford to take up the post of Sherardian professor of botany at Oxford in 1884. He was also made the king's botanist in Scotland and the keeper of the royal botanical gardens. He was elected Fellow of the Royal Society 1884 and knighted in 1920.

Bayliss William Maddock 1860–1924. English physiologist who discovered the digestive hormone secretin, the first hormone to be found, with Ernest Starling 1902. During World War I, Bayliss introduced the use of saline (salt water) injections to help the injured recover from shock. Knighted 1922.

Bayliss was born in Wolverhampton, Staffordshire, and studied at University College, London, and at Oxford, becoming professor at University College.

By experimenting with the inner lining of the duodenum (first part of the intestines), Bayliss concluded that as hydrochloric acid from the stomach's digestive juices passes into the duodenum during the normal digestive process, the duodenal lining releases a chemical into the bloodstream which, in turn, makes the pancreas secrete its juices. This is the hormone secretin.

Bayliss went on to study the activation of enzymes, particularly the pancreatic enzyme trypsin. Bayliss and Starling also investigated the peristaltic movements of the intestines and their nerve supply, and pressures within the venous and arterial systems.

Bayliss's *Principles of General Physiology* 1915 became a standard work.

Beadle George Wells 1903–1989. US biologist. In 1958 he shared a Nobel prize with Edward L Tatum and Joshua Lederberg for his work in biochemical genetics, forming the 'one-gene–one-enzyme' hypothesis (a single gene codes for a single kind of enzyme).

Beadle was born in Wahoo, Nebraska. In 1931, he went to the California Institute of Technology, where he researched into the genetics of the fruit fly *Drosophila melanogaster*. From 1937 to 1946 he was professor at Stanford University, California, and it was during this period that he collaborated with Tatum.

Earlier, Beadle had shown that the eye colour of *Drosophila* is a result of a series of chemical reactions under genetic control. At Stanford, he used the red bread mould *Neurospora crassa*, which is a simpler organism than *Drosophila*. He subjected colonies of *Neurospora* to X-rays and studied the changes in the nutritional requirements of, and therefore enzymes formed by, the mutant *Neurospora* produced by the irradiation. By repeating the experiment with various mutant strains and culture mediums, Beadle and Tatum deduced that the formation of each individual enzyme is controlled by a single, specific gene. This concept found wide applications in biology and virtually created the science of biochemical genetics.

Beaufort Francis 1774–1857. British admiral, hydrographer to the Royal Navy from 1829; the Beaufort scale and the Beaufort Sea in the Arctic Ocean are named after him. KCB 1848.

Beaufort was born in County Meath, Ireland. In 1790 he enlisted in the Royal Navy. He did major surveying work, especially around the Turkish coast in 1812. As hydrographer to the navy, he promoted voyages of discovery such as that of Joseph Hooker with the *Erebus*.

Drawing up the scale named after him, Beaufort specified the amount of sail that a full-rigged ship should carry under the various wind conditions. It was officially adopted by the Admiralty in 1838. Modifications

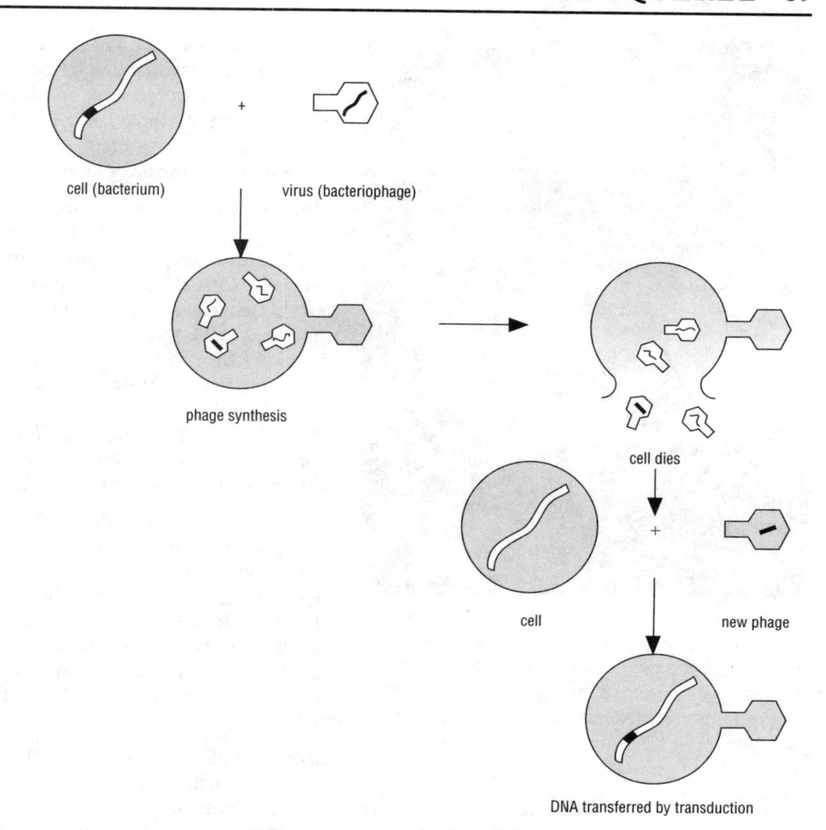

Beadle *Beadle showed how a bacteriophage can bring about the transfer of DNA by transduction.*

were made to the scale when sail gave way to steam.

Beaumont William 1785–1853. US surgeon who conducted pioneering experiments on the digestive system. In 1822 he saved the life of a Canadian trapper wounded in the side by a gun blast; the wound only partially healed, and through an opening in the stomach wall, Beaumont was able to observe the workings of the stomach. His *Experiments and Observations on the Gastric Juice and the Physiology of Digestion* was published 1833.

Beaumont was born in Lebanon, Connecticut, and joined the army as a surgeon 1812. Between 1825 and 1833 he carried out experiments on the Canadian patient. He extracted and analysed gastric juice and stomach contents at various stages of digestion, observed changes in secretions, and noted the muscular movements of the stomach. From 1834 he worked in St Louis, Missouri, where he became professor of surgery.

Becker Lydia Ernestine 1827–1890. English botanist and campaigner for women's rights. She established the Manchester Ladies' Literary Society 1865 as a forum for women to study scientific subjects. In 1867 she co-founded and became secretary of the National Society for Women's Suffrage. In 1870 she founded a monthly newsletter, the *Women's Suffrage Journal*.

Becquerel (Antoine) Henri 1852–1908. French physicist who discovered penetrating radiation coming from uranium salts, the first indication of radioactivity, and shared a Nobel prize with Marie and Pierre Curie 1903.

Becquerel was born and educated in Paris. In 1875 he began private scientific research, investigating the behaviour of polarized light in magnetic fields and in crystals. The discovery of X-rays 1896 prompted Becquerel to investigate fluorescent crystals for the emission of X-rays, and in so doing he accidentally discovered radioactivity in uranium

Becquerel *In 1896, Henri Becquerel discovered radioactivity while investigating the X-ray emission of fluorescent uranium salts using photographic plates. His accidental discovery marked the beginning of the nuclear age and in 1903 he was awarded the Nobel Prize for Physics – an award he shared with Marie and Pierre Curie.*

salts in the same year. Marie and Pierre Curie then searched for other radioactive materials, which led them to the discovery of polonium and radium in 1898.

Becquerel subsequently investigated the radioactivity of radium, and showed in 1900 that it consists of a stream of electrons. In the same year, Becquerel also obtained evidence that radioactivity causes the transformation of one element into another.

Bednorz Johannes Georg 1950– . German physicist who, with Alexander Müller discovered high-temperature superconductivity in ceramic materials. The discovery of these materials contributed towards the use of superconductors in computers, magnetic levitation trains, and the more efficient generation and distribution of electricity. Nobel Prize for Physics 1987.

Bednorz and Müller showed 1986 that a ceramic oxide of lanthanum, barium and copper became superconducting at temperatures above 30 K, much hotter than for any previously known superconductor.

Bednorz was born in Neuenkirkchen, North-Rhine Westphalia, in the Federal Republic of Germany. He studied chemistry at the University of Mü. In 1972 he worked as a summer student at IBM Zürich under Mü. In 1977 he started his doctoral studies at the Swiss Federal Insitute of Technology in Zürich. He joined the IBM Zürich Research Laboratories at Rüschlikon 1982.

Beebe Charles 1877–1962. US naturalist, explorer, and writer. His interest in deep-sea exploration led to a collaboration with the engineer Otis Barton and the development of a spherical diving vessel, the bathysphere. On 24 Aug 1934 the two men made a record-breaking dive to 923 m/3,028 ft.

Beebe was curator of birds for the New York Zoological Society 1899–1952. He wrote the comprehensive *Monograph of the Pheasants* 1918–22, and his expeditions are described in a series of memoirs.

Beeching Richard, Baron Beeching 1913–1985. British scientist and administrator. He was chair of the British Railways Board 1963–65, producing the controversial *Beeching Report* 1963, which advocated concentrating resources on intercity passenger traffic and freight, at the cost of closing many rural and branch lines. Created a baron 1965.

Behring Emil von 1854–1917. German physician who discovered that the body produces antitoxins, substances able to counteract poisons released by bacteria. Using this knowledge, he developed new treatments for such diseases as diphtheria. He won the first Nobel Prize for Physiology or Medicine, in 1901.

Behring discovered the diphtheria antitoxin and developed serum therapy together with Japanese bacteriologist Shibasaburō Kitasato, and they went on to apply the technique to tetanus. Behring also introduced early vaccination techniques against diphtheria and tuberculosis.

Behring was born in Prussia and educated in Berlin, where he was assistant to bacteriologist Robert Koch before becoming professor of hygiene at Halle and Marburg.

Investigating tuberculosis, Behring recognized that the bovine and human forms of the disease were caused by the same micro-organism, thus identifying the danger to humans of drinking contaminated milk.

Beijerinck Martinus Willem 1851–1931. Dutch bacteriologist who in 1898 published his finding that an agent smaller than bacteria could cause diseases, an agent that he called a virus.

Beijerinck was born in Amsterdam and studied chemical engineering at the Delft Polytechnic, where, after working as a bacteriologist with an industrial company, he taught and carried out research from 1895.

In the 1880s and 1890s Beijerinck studied the disease that stunts the growth of tobacco plants and mottles their leaves (now called the tobacco mosaic virus disease). Trying to isolate a causative agent, he pressed out the juice of infected tobacco leaves and found that the juice alone was able to infect healthy plants, even after he had passed it through a filter that removed bacteria. Beijerinck was also certain that the causative agent was not a toxin because he could infect a healthy plant and from that plant infect another healthy plant, continuing this process indefinitely – therefore the infective agent had to be capable of reproduction. He believed that the filtered juice of the infected plants was itself alive, and he called the causative agent a filterable virus (the Latin word for poison).

Békésy Georg von 1899–1972. Hungarian-born US scientist who resolved a long-standing controversy on how the inner ear functions. For his discovery concerning the mechanism of stimulation within the cochlea, he received the 1961 Nobel Prize for Physiology or Medicine.

Békésy *Scientist Georg von Békésy. Békésy worked in telecommunications holding both academic and industrial positions. He won the 1961 Nobel Prize for Physiology or Medicine after discovering how the inner ear works. Fascinated by the senses, he also investigated the mechanisms underlying sight and touch.*

Békésy was born in Budapest and studied there and at Bern, Switzerland. From 1923 to 1946 he worked in the laboratories of the Hungarian Telephone System and of Siemens and Halske in Germany, as well as holding academic posts. He emigrated to Sweden 1946, and then to the USA 1947. He was on the staff of Harvard 1947–66, when he became professor of sensory sciences at the University of Hawaii.

While working as a telecommunications engineer, Békésy investigated how the human ear actually receives sound. He constructed models of the cochlea and also worked with cadavers, whose auditory mechanisms he stimulated electrically. By substituting a saline solution containing fine aluminium particles for the fluid in the cochlea and by using stroboscopic illumination, he was able to observe how sound vibrations are transmitted to a membrane in the form of travelling waves. Each wave causes maximum vibration at different sections of the membrane according to its frequency.

Mr Watson, come here; I want you.

Alexander Graham Bell

First complete sentence spoken over the telephone March 1876

Bell Alexander Graham 1847–1922. Scottish-born US scientist and inventor, the first person ever to transmit speech from one point to another by electrical means. This invention – the telephone – was made 1876. Later Bell experimented with a type of phonograph and, in aeronautics, invented the tricycle undercarriage.

Bell also invented a photophone, which used selenium crystals to apply the telephone principle to transmitting words in a beam of light. He thus achieved the first wireless transmission of speech.

Bell was born in Edinburgh, where his grandfather was a speech tutor. As a boy he constructed an automaton simulating the human organs of speech, using rubber, cotton, and a bellows. He was educated at the universities of Edinburgh and London, and in 1870 went first to Canada and then to the USA, where he opened a school for teachers of the deaf in Boston and in 1873 became professor of vocal physiology at the university. With the money he had made from his telephone system, Bell set up a laboratory in

Bell, Alexander Graham *A drawing of Alexander Graham Bell testing his second telephone in the lecture hall of Boston University. Bell's initial attempts to market his new invention led to failure: Western Union considered it a 'very interesting novelty' but saw no future for Bell's 'electric toy'. Other companies were not so short-sighted and within four years of its invention, America had 60,000 telephones.*

Nova Scotia, Canada; in 1880 he established, in addition, the Volta Laboratory in Washington DC.

There, Bell patented the gramophone and wax recording cylinder, which were commercially successful improvements on Thomas Edison's first phonograph and cylinders of metal foil. The laboratory also experimented with flat disc records, electroplating records, and impressing permanent magnetic fields on records – the embryonic tape recorder.

In 1881, Bell developed two telephonic devices for locating metallic masses (usually bullets) in the human body. One, an induction balance method, was first tried out on President Garfield, who was assassinated that year, while the other, a probe, was widely used until the advent of X-rays. Bell also built hydrofoil speedboats and tetrahedral kites capable of carrying a person. He invented an air-cooling system, a way of desalinating sea water, the forerunner of the iron lung, and a sorting machine for punch-coded census cards.

Bell Charles 1774–1842. Scottish anatomist and surgeon who carried out pioneering research on the human nervous system. He gave his name to Bell's palsy, an extracranial paralysis of the facial nerve, and to the long thoracic nerve of Bell, which supplies a muscle in the chest wall. Knighted 1829.

Bell was born in Edinburgh and became a surgeon at the Edinburgh Royal Infirmary, having learned from his surgeon brother ***John Bell*** (1763–1820). Charles went to London 1804, held various academic posts, and returned to Edinburgh 1836 as professor of surgery.

Bell discovered that nerves are composite structures, each with separate fibres for sensory and motor functions. His findings first appeared 1811; his main written work was *The Nervous System of the Human Body* 1830.

Bell John 1928–1990. British physicist who in 1964 devised a test to verify a point in quantum theory: whether two particles that were once connected are always afterwards interconnected even if they become widely separated. As well as investigating fundamental problems in theoretical physics, Bell contributed to the design of particle accelerators.

Bell worked for 30 years at CERN, the European research laboratory near Geneva, Switzerland. He demonstrated how to measure the continued interconnection of particles that had once been closely connected, and put forward mathematical criteria that had to be obeyed if such a connection existed, as required by quantum theory. In the early 1980s, a French team tested Bell's criteria, and a connection between widely separated particles was detected.

Bell Patrick 1799–1869. Scottish inventor of a reaping machine, developed around 1828.

It was pushed by two horses and used a rotating cylinder of horizontal bars to bend the standing corn on to a reciprocating cutter that was driven off the machine's wheels (in much the same way as on a combine harvester).

Bell was born near Dundee and became a cleric. While still at St Andrews University, he constructed the reaping machine. He started trials in deep secrecy inside a barn on a crop which had been planted by hand, stalk by stalk. In 1828, he and his brother carried out night-time trials which were a success, leading them to exhibit the machine the following year. In the years to 1832 at least 20 machines were produced, and later, since Bell did not take out a patent, the design was widely copied and improved on, until mechanical harvesting became the norm.

Bell Burnell (Susan) Jocelyn 1943– . Irish astronomer. In 1967 she discovered the first pulsar (rapidly flashing star) with Antony Hewish and colleagues at Cambridge University, England.

Jocelyn Bell was born in Belfast, near the Armagh Observatory, where she spent much time as a child. She was educated at Glasgow and Cambridge universities. It was while a research student at Cambridge that the discovery of pulsars was made. Between 1968 and 1982 she did research in gamma-ray astronomy at the University of Southampton and in X-ray astronomy at the Mullard Space Science Laboratory, University College London. Then she worked on infrared and optical astronomy at the Royal Observatory, Edinburgh. In 1991 she was appointed professor of physics at the Open University, Milton Keynes.

Nobel prizes are based on long-standing research, not on a flash-in-the-pan observation of a research student. The award to me would have debased the prize.

Jocelyn Bell Burnell

Bell spent her first two years in Cambridge building a radio telescope that was specially designed to track quasars. The telescope had the ability to record rapid variations in signals. In 1967 she noticed an unusual signal, which turned out to be composed of a rapid set of pulses that occurred precisely every 1.337 sec. One attempted explanation of this curious phenomenon was that it emanated from an interstellar beacon, so initially it was nicknamed LGM, for Little Green Men. Within a few months, however, Bell located three other similar sources. They too pulsed at an extremely regular rate but their periods varied over a few fractions of a second and they all originated from widely spaced locations in our Galaxy. Thus it seemed that a more likely explanation of the signals was that they were being emitted by a special kind of star – a pulsar.

Bell Burnell *Jocelyn Bell Burnell, the discoverer of the pulsar. Since her discovery in 1967 over 500 radio pulsars have been found.*

Beltrami Eugenio 1835–1899. Italian mathematician who pioneered modern non-Euclidean geometry. His work ranged over almost the whole field of pure and applied mathematics, but especially theories of surfaces and space of constant curvature.

Beltrami was born in Cremona, Lombardy, and studied mathematics at Pavia. He held academic posts at Bologna, Pisa, Pavia, and Rome.

In 1862 he published his first paper, an analysis of the differential geometry of curves, to which he would return in his most important paper, 'Saggio di interpretazione della geometria non-Euclidia' 1868. This advanced a theory of hyperbolic space that laid the analytical base for the development of non-Euclidean geometry.

Beltrami demonstrated that the concepts and formulae of Russian mathematician Nikolai Lobachevsky's geometry are realized for geodesics on surfaces of constant negative curvature. He showed also that there are rotation surfaces of this kind – he called them 'pseudospherical surfaces'. He also demonstrated the usefulness of employing differential parameters in surface theory, thereby beginning the use of invariant methods in differential geometry. After 1872 Beltrami switched his attention to questions of applied mathematics, especially problems in elasticity and electromagnetism.

Benacerraf Baruj 1920– . Venezuelan-born US immunologist who shared the 1980 Nobel Prize for Physiology or Medicine with Jean Dausset and George D Snell for their discovery of the immune response genes and major histocompatability complex (MHC) (gene complex involved in the immune response). They studied a gene region within one chromosome, which influences the immune system in various ways. This gene region is the MHC and is of great medical and biological significance because of the part it plays in the rejection of transplants made between incompatible individuals.

The action of immune cells to defend the body from potentially harmful substances depends upon their ability to recognize unusual molecules (antigens) on the surface of invading cells, such as bacteria. These antigens are then engulfed by macrophages (a type of white blood cell) and small fragments of antigen presented on its surface in association with proteins encoded by the major histocompatability complex. This process is called antigen presentation and is crucial for lymphocyte (a type of white blood cell) activation, and the production of antibodies to the antigen.

In rare cases, however, a foreign protein will have no peptides with a suitable motif for binding to any of the MHC molecules encoded by an individual. In his mouse studies, Benacerraf showed that there was a genetic basis for this difference between individuals in their response to antigen and that involved a group of genes he called immune response genes.

Baruj Benacerraf was born in Caracas, Venezula. His family moved to New York 1940. Benacerraf was educated at Columbia University, New York, and the Medical College of Virginia, Richmond. He became a naturalized US citizen 1943. During World War II he served with the US Army Medical Corps in Europe. In 1948 he began research in immunology at the Neurological Institute, Columbia University. In 1949 he moved to the Brossais Hospital, Paris, France. He returned to the US 1956 as assistant professor of pathology at New York University School of Medicine. In 1968 he became director of the Laboratory of Immunology of the National Institute of Allergy and Infectious Disease in Bethesda, Maryland. In 1970 he was appointed to the chair of pathology at the Harvard Medical School, Boston, Massachusetts.

Benz *Karl Benz's* Motorenveloziped *drove on the streets of Mannheim for the first time in 1886 and is generally regarded as one of the world's first vehicles to be successfully propelled by an internal combustion engine.*

Benz Karl Friedrich 1844–1929. German automobile engineer who produced the world's first petrol-driven motor vehicle. He built his first model engine 1878 and the petrol-driven car 1885.

Benz made his first four-wheeled prototype in 1891 and by 1895, he was building a range of four-wheeled vehicles that were light, strong, inexpensive, and simple to operate. These automobiles ran at speeds of about 24 kph/15 mph. In 1926, the thriving company merged with the German firm of Daimler to form Daimler-Benz.

Benz was born and educated in Karlsruhe, and worked for mechanical and engineering companies before setting up on his own in Mannheim 1871. He produced a two-stroke engine of his own design 1878, and in 1885, the first vehicle successfully propelled by an internal-combustion engine. It achieved a speed of up to 5 kph/3 mph. Benz, who for his invention drew on experimental work by engineers in many different fields, believed that this vehicle would be a completely new system and not simply a carriage with a motor replacing the horse. The engine had a massive fly-wheel and was mounted horizontally in the rear, using electric ignition by coil and battery. The cooling system consisted simply of a cylinder jacket in which the water boiled away, being topped up as necessary. The production model Tri-car appeared in 1886–87 and had a 1 kW/1.5 hp single-cylinder engine.

Berg Paul 1926– . US molecular biologist. In 1972, using gene-splicing techniques developed by others, Berg spliced and combined into a single hybrid the DNA from an animal tumour virus (SV40) and the DNA from a bacterial virus. For his work on recombinant DNA, he shared the 1980 Nobel Prize for Chemistry.

Berg was born in New York and educated at Pennsylvania State University and Case Western Reserve University. Between 1955 and 1974 he held several positions at Washington University.

In 1956 Berg identified an RNA molecule (later known as a transfer RNA) that is specific to the amino acid methionine. He then perfected a method for making bacteria accept genes from other bacteria. This genetic engineering can be extremely useful for creating strains of bacteria to manufacture specific substances, such as interferon. But there are also dangers: a new, highly virulent pathogenic microorganism might accidentally be created, for example. Berg has therefore advocated restrictions on genetic engineering research.

Berger Hans 1873–1941. German psychiatrist and philosopher of science. He first described the human electroencephalogram (EEG) 1929. The differential patterns of cortical electrical activity he observed in alert and relaxed subjects led him to attempt the application of EEG to the study of psychophysical relationships and of conscious processes in general. He saw EEG as a key to the mind–body problem, a problem with which he was preoccupied for much of his life.

Latterly he proposed a confusing form of psychophysical parallelism, allowing for interaction between physical and mental domains through the transfer and transformation of physical and psychic energy respectively.

Bergius Friedrich Karl Rudolph 1884–1949. German research chemist who invented processes for converting coal into oil and wood into sugar. He shared a Nobel prize 1931 with Carl Bosch for his part in inventing and developing high-pressure industrial methods.

Bergius was born near Breslau, Silesia (now in Poland), the son of the owner of a chemical factory. He studied chemistry at the universities of Breslau and Leipzig, and did research at Karlsruhe Technische Hochschule with German chemist Fritz Haber, who introduced him to high-pressure reactions. Bergius worked in industry 1914–45, then left Germany and eventually settled in Argentina 1948, as a technical adviser to the government.

In 1912 Bergius worked out a pilot scheme for using high pressure, high temperature, and a catalyst to hydrogenate coal dust or heavy oil to produce paraffins (alkanes) such as petrol and kerosene. Yielding nearly 1 tonne of petrol from 4.5 tonnes of coal, the process became important to Germany during World War II as an alternative source of supply of petrol and aviation fuel. He also discovered a method of producing sugar and alcohol from simple substances made by breaking down the complex molecules in wood; he continued this work in Argentina, and found a way of making fermentable sugars and thus cattle food from wood.

Bergstrom Sune Karl 1916– . Swedish biochemist who shared the Nobel Prize for Physiology or Medicine 1982 with John Vane and Bengt Samuelsson for the purification of prostaglandins, chemical messengers that regulate many processes in the body – such as blood pressure, blood platelet aggregation, and inflammatory response – and therefore have many clinical applications.

A century before Bergstrom began his studies in Stockholm, the German physiologist Johannes Peter Müller had discovered that human semen and extracts of sheep seminal vesicular glands had peculiar properties. Both substances caused contraction of smooth muscle in vitro (in an artificial environment) and sharp decreases in the blood pressure of experimental animals. Müller called the active agents in these substances prostaglandins, as they are primarily made in the prostate gland.

Prostaglandins are synthesized in all cells that have a nucleus and provide a general regulatory mechanism in mammalian physiology. The purification of the prostaglandins was complicated by the very low amounts present in the seminal substances and their extremely short half lives. In 1957, Bergstrom and his student Bengt Samuelsson managed to obtain crystals from two prostaglandins, alprostadil (PGE1) and PGF1a. They characterized these two chemicals 1962 and went on to unravel their function.

Bergstrom was born in Stockholm and graduated in medicine and chemistry from the Karolinska Institute, where he was to return to work 1948.

The synthesis of prostaglandins is inhibited by aspirin and other anti-inflammatory drugs. Prostaglandins control the aggregation of blood platelets and therefore can be used to prevent heart disease and stroke. Prostaglandins can also function homeostatically, lowering blood pressure or constricting bronchial airways. They have inflammatory actions that are more potent than those of histamine and are known to cause contraction of intestinal and uterine muscles. Due to all these functions, prostaglandins now have many clinical applications.

Bernard Claude 1813–1878. French physiologist and founder of experimental medicine. Bernard first demonstrated that digestion is not restricted to the stomach, but takes place throughout the small intestine. He discovered the digestive input of the pancreas, several functions of the liver, and the vasomotor nerves which dilate and contract the blood vessels and thus regulate body temperature. This led him to the concept of the *milieu intérieur* ('internal environment') whose stability is essential to good health.

Bernard was born in St-Julien, Rhoñe-Alpes region, and studied medicine in Paris. He never practised but devoted himself to research. He was made professor 1854 at the Faculty of Sciences in Paris, and 1855 at the Collège de France.

Bernard performed a series of important experiments on the physiology of digestion, showing, for example, that pancreatic secretions were important in fat metabolism, and investigating the mechanisms of nervous control of gastric secretion. He also investigated the physiology of fetal tissues and the nutritive role of the placenta; and the role of drugs such as curare and opium alkaloids and their effects on the nervous system.

Bernard *French physiologist Claude Bernard discovered the role of the small intestine and pancreas in the digestive process. He later served in the French Senate.*

He summed up his work in *The Introduction to the Study of Experimental Medicine* 1865.

Bernays Paul Isaak 1888–1977. British-born Swiss mathematician whose theory of sets and classes is now widely believed to be the most useful arrangement, and a major contribution to the modern development of logic.

Bernays was born in London but grew up in Berlin and studied mathematics, philosophy, and theoretical physics at the universities of Berlin and Göttingen. He researched at Zürich, Switzerland, 1912–17, and then became German mathematician David Hilbert's assistant at Göttingen. In 1933, when the Nazis came to power in Germany, Bernays moved back to Zürich, where he eventually became a professor at the Technical High School.

Bernays became interested in axiomatic thoughts, and presented his principles of axiomatization most fully in lectures at the Princeton Institute for Advanced Study in 1935–36.

In Bernays' set theory there are two kinds of individuals: sets and classes. A set is a multitude forming a real mathematical object, whereas a class is a predicate to be regarded only with respect to its extension.

Bernoulli Daniel 1700–1782. Swiss mathematical physicist who made important contributions to trigonometry and differential equations (differentiation). In hydrodynamics he proposed Bernoulli's principle, an early formulation of the idea of conservation of energy.

Bernoulli was born in Groningen in the Netherlands, the son of mathematician Johann Bernoulli. Having studied philosophy, logic, and medicine in Basel, Switzerland, he became professor of mathematics at the St Petersburg Academy, Russia, 1725–32, and professor of anatomy and botany at the University of Basel from 1733. During his career he won ten prizes from the French Academy, for papers on subjects which included marine technology, oceanology, astronomy, and magnetism.

Bernoulli's *Hydrodynamica* 1738 is both a theoretical and practical study of equilibrium, pressure, and velocity in fluids. Bernoulli's principle states that the pressure of a moving fluid decreases the faster it flows (which explains the origin of lift on the aerofoil of an aircraft's wing). *Hydrodynamica* also contains the first attempt at a thorough mathematical explanation of the behaviour of gases by assuming they are composed of tiny particles, producing an equation of state that enabled Bernoulli to relate atmospheric pressure to altitude, for example. This was the first step towards the kinetic theory of gases achieved a century later.

Among his achievements in mathematics, Bernoulli demonstrated how the differential calculus could be used in problems of probability. He did pioneering work in trigonometrical series and the computation of trigonometrical functions. Bernoulli also showed the shape of the curve known as the lemniscate.

Bernoulli Jakob 1654–1705. Swiss mathematician who with his brother Johann pioneered German mathematician Gottfried Leibniz's calculus. Jakob used calculus to study the forms of many curves arising in practical situations, and studied mathematical probability (*Ars conjectandi* 1713); ***Bernoulli numbers*** are named after him.

Jakob Bernoulli's papers on transcendental curves (1696) and isoperimetry (1700, 1701) contain the first principles of the calculus of variations. It is probable that these papers owed something to collaboration with Johann. His other great achievement was his treatise on probability, *Ars conjectandi*, which contained both the Bernoulli numbers (a series of complex fractions) and the Bernoulli theorem.

Jakob Bernoulli was born in Basel. On a trip to England 1676 he met Irish physicist Robert Boyle and other leading scientists, and decided to devote himself to science. He became particularly interested in comets (which he explained by an erroneous theory 1681) and in 1682 began to lecture in mechanics and natural philosophy at the University of Basel. During the next few years he came to know the work of Leibniz and began a correspondence with him. In 1687 he was made professor of mathematics at Basel.

Bernoulli Johann 1667–1748. Swiss mathematician who with his brother Jakob Bernoulli pioneered German mathematician Gottfried Leibniz's calculus. He was the father of Daniel Bernoulli.

Bernoulli, *Johann Bernoulli was an early pioneer of differential calculus, working particularly on the nature and mathematical properties of curves.*

Johann also contributed to many areas of applied mathematics, including the problem of a particle moving in a gravitational field. He found the equation of the catenary 1690 and developed exponential calculus 1691.

Bernoulli was born in Basel and studied medicine, but became professor of mathematics at Groningen, the Netherlands, 1694–1705, and then at Basel. Both Johann and Jakob wrote papers on a wide variety of mathematical and physical subjects, and it is often difficult to separate their work, although they never published together.

Bernstein Jeremy 1929– . US mathematical physicist who has written many articles and books on various topics of pure and applied science for the nonspecialist reader. He has also sought to give a mathematical analysis and description of the behaviour of elementary particles.

Bernstein was born in Rochester, New York, and educated at Harvard. He held academic posts there and at New York University before becoming professor of physics at the Stevens Institute of Technology in Hoboken, New Jersey, 1967. He has also been a consultant to the Rand Corporation and the General Atomic Company.

In 1962 Bernstein joined the staff of the urbane magazine the *New Yorker*. His lengthy articles for that magazine include 'The analytical engine: computers, past, present and future'. He published a general survey of the historical progress of scientific knowledge,

Ascent, 1965, and a biography of Albert Einstein 1973.

Bert Paul 1833–1886. French physiologist and politician who discovered the physiological effects of altitude and atmospheric pressure on the composition of gases in the bloodstream.

In 1878, he published his work *La Pression Barometrique*, which described the effects of abnormally high or low atmospheric pressures, the composition of gases in the air, and altitude on the respiration of animals and himself. Bert correctly concluded that the physiological effects of oxygen and other gases in the blood are due to their partial pressures (fractions of total gas pressure that are exerted by each of the chemical components that comprise the gas), and not their proportions as had previously been thought. His research had widespread implications for the effects of diving, climbing and aviation on respiration and paved the way for the development of various forms of specially adapted breathing apparatus to meet the needs of such activities.

Bert was born in Auxerre, France and obtained qualifications in law and medicine from the Sorbonne in Paris. He was later appointed professor of physiology at the Sorbonne 1869. His career in science was interrupted by the Franco-Prussian war, after which Bert became interested in politics and was elected 1872 to the Chamber of Deputies. He featured prominently in forming a widespread educational structure in France.

Berthelot Pierre Eugène Marcellin 1827–1907. French chemist and politician who carried out research into dyes and explosives, proving that hydrocarbons and other organic compounds can be synthesized from inorganic materials.

Berthelot was born in Paris, where he studied and became professor of organic chemistry. In 1870–71, during the siege of Paris in the Franco-Prussian War, he was consulted about the defence of the capital and supervised the manufacture of guns and explosives. Thereafter he took an increasing part in politics, becoming a senator 1881, minister for public instruction 1886, and foreign minister 1895–96.

Berthelot first studied alcohols, showing in 1854 that glycerol is a triatomic alcohol; he combined it with fatty (aliphatic) acids to make fats, including fats that do not occur naturally. This work provided increasing justification for the view that organic chemistry deals with all the compounds of carbon and not just compounds formed and found in nature. He continued his research by investigating sugars, which he identified as being both alcohols and aldehydes. Using crude but effective methods, he also synthesized many simple organic compounds. His work during the 1850s was summed up in his book *Chimie organique fondée sur la synthèse* 1860.

Berthelot began his studies of thermochemistry 1864. He measured the heat changes during chemical reactions, inventing the bomb calorimeter to do so and to study the speeds of explosive reactions. He introduced the term exothermic to describe a reaction that evolves heat, and endothermic for a reaction that absorbs heat. He published *Mécanique chimique* 1878 and *Thermochimie* 1897.

Berthollet Claude Louis, Count 1748–1822. French chemist who carried out research into dyes and bleaches (introducing the use of chlorine as a bleach) and determined the composition of ammonia. Modern chemical nomenclature is based on a system worked out by Berthollet and Antoine Lavoisier.

Berthollet was born in the then Italian region of Savoy. He qualified as a physician at the University of Turin, moving to Paris to study chemistry. As private physician in the household of the duke of Orléans, he carried out research in the laboratory at the Palais Royale. He was appointed inspector of dyeworks and director of the Gobelins tapestry factory 1784. He taught chemistry to Napoleon and went with him to Egypt 1798. There he observed the high concentration of sodium carbonate (soda) by Lake Natron on the edge of the desert. He reasoned that, under the prevailing physical conditions, sodium chloride in the upper layer of soil had reacted with calcium carbonate from nearby limestone hills – the beginning of his theory that chemical affinities are affected by physical conditions, in this case the heat and high concentration of calcium carbonate. In 1804 he became a senator but ten years later voted for the deposition of Napoleon.

Berthoud Ferdinand 1727–1807. Swiss clockmaker and maker of scientific instruments. He improved the work of John Harrison, devoting 30 years to the perfection of the marine chronometer, giving it practically its modern form.

Berthoud was born near Couvet, Neuchâtel, and was apprenticed to his clockmaker brother Jean-Henri at the age of 14. In 1745 he went to Paris and in 1764 was appointed horologer to the navy. He made over 70 chronometers using a wide variety of mechanisms, and wrote ten volumes on the subject.

If navigators did not know the time at the zero meridian, it was impossible for them to plot their precise position, so accurate chronometers were vital. Clocks made by Berthoud were tested at sea in 1768 and 1769 and showed variations in their working of about 5–20 seconds a day.

Berzelius Jöns Jakob 1779–1848. Swedish chemist who accurately determined more than 2,000 relative atomic and molecular masses. He devised (1813–14) the system of chemical symbols and formulae now in use and proposed oxygen as a reference standard for atomic masses. His discoveries include the elements cerium (1804), selenium (1817), and thorium (1828); he was the first to prepare silicon in its amorphous form and to isolate zirconium. The words 'isomerism', 'allotropy', and 'protein' were coined by him.

God knows what happens to your time once you have begun to get old. You are busy all the time, you do important things, you work, and yet when you sum it all up the result is nothing.

Jöns Berzelius

Berzelius noted that some reactions appeared to work faster in the presence of another substance which itself did not appear to change, and postulated that such a substance contained a ***catalytic force***. Platinum, for example, was capable of speeding up reactions between gases. Although he appreciated the nature of catalysis, he was unable to give any real explanation of the mechanism.

Berzelius, born in Östergötland, studied natural sciences and medicine at Uppsala University and began to experiment in chemistry. In 1807 he was appointed professor of medicine and pharmacy at the College of Medicine, Stockholm. Papers he published 1810–16 describe the preparation, purification, and analysis of about 2,000 chemical compounds. In the course of this work he improved many existing methods and developed new techniques. Quantitative analysis on this scale established beyond doubt British chemist John Dalton's atomic theory and French chemist J L Proust's law of definite proportions. It also laid the foundation of Berzelius's determination of the atomic weights of the 40 elements known at that time.

In the early 19th century it became apparent that elements could be grouped by similar chemical properties. Chlorine, bromine, and iodine formed such a grouping. Each of these elements could be found as salts in sea water, so Berzelius coined the name 'halogens' (salt formers) to describe the family collectively.

His *Textbook of Chemistry* 1803 was soon accepted as the definitive work for the time.

Bessel Friedrich Wilhelm 1784–1846. German astronomer and mathematician, the first person to find the approximate distance to a star by direct methods when he measured the parallax (annual displacement) of the star 61 Cygni in 1838. In mathematics, he introduced the series of functions now known as ***Bessel functions***.

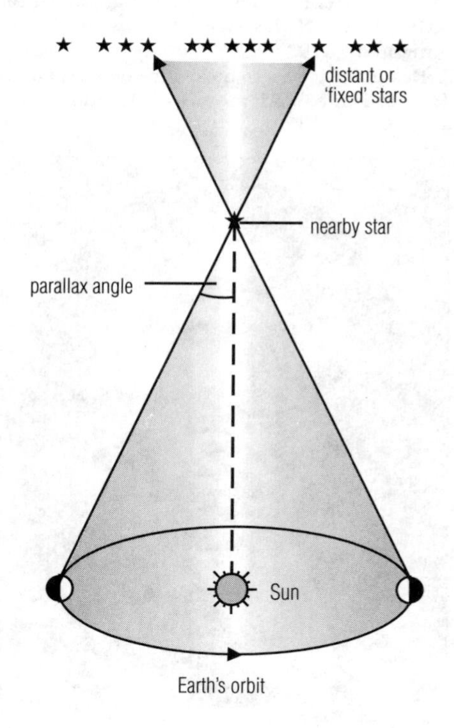

Bessel *Bessel developed the technique of stellar parallax, by means of which distances to nearby stars can be calculated by observing their apparent change in position when viewed from opposite ends of a long baseline such as the diameter of the Earth's orbit*

Bessel was born in Minden, NW Germany. As an amateur, he wrote a paper on Halley's comet 1804 which got him a post as an assistant at Lilienthal Observatory. After only four years the Prussian government commissioned Bessel to construct the first large German observatory at Königsberg; this was completed 1813 and Bessel spent his life as its director.

Bessel's work laid the foundations of a more accurate calculation of the scale of the universe and the sizes of stars, galaxies, and clusters of galaxies. In addition, he made a fundamental contribution to positional astronomy (the exact measurement of the position of celestial bodies), to celestial mechanics (the movements of stars), and to geodesy (the study of the Earth's size and shape). In 1840 he predicted the existence of Neptune.

Bessemer Henry 1813–1898. British engineer and inventor who developed a method of converting molten pig iron into steel (the ***Bessemer process***) 1856. Knighted 1879.

Bessemer was born near Hitchin, Hertfordshire, and moved to London. His early inventions included a typesetting machine, new ways of making gold paint and lead pencils, and machinery for sugar refining.

Bessemer *While trying to devise a method for manufacturing stronger rifle barrels for use in the Crimean War, Henry Bessemer discovered a process in which molten pig-iron could be turned directly into steel. This process used a special furnace, a Bessemer converter, and was much cheaper than previous methods.*

During the Crimean War of the early 1850s Bessemer turned to the problem of high gas pressures causing guns to explode. The British military commanders showed no interest, but Napoleon III of France encouraged Bessemer in his experiments. By modifying the standard process, he found a way to produce steel without an intermediate wrought-iron stage, reducing its cost dramatically. However, to obtain high-quality steel, phosphorus-free ore was required. In 1860 Bessemer erected his own steel works in Sheffield, importing phosphorus-free iron ore from Sweden.

Best Charles H(erbert) 1899–1978. Canadian physiologist, one of the team of Canadian scientists including Frederick Banting whose research resulted in 1922 in the discovery of insulin as a treatment for diabetes.

Best also discovered the vitamin choline and the enzyme histaminase, and introduced the use of the anticoagulant heparin.

Best was born in Maine, USA. As one of Banting's undergraduate students at the University of Toronto, he took part in the experiments to isolate insulin. They tied off the pancreatic duct in a group of dogs, which caused atrophy of the pancreas except for the part known as the islets of Langerhans. This eliminated the digestive enzymes normally produced by the pancreas, and left only insulin, produced by the islets of Langerhans. An extract of this was injected into another group of dogs, whose pancreas had been entirely removed so that they had developed diabetes. Gradually, these dogs' condition improved with the injections.

A Banting–Best Department of Medical Research was founded in Toronto, and Best was its director 1941–67.

Bethe Hans Albrecht 1906– . German-born US physicist who in 1938 worked out the details of how nuclear mechanisms power stars. He also worked on the first atom bomb but later became a peace campaigner. Nobel prize 1967.

Bethe was born in Strasbourg (now in France), and educated at the universities of Frankfurt and Munich. He left Germany 1933, moving first to England and in 1935 to the USA, where he became professor of theoretical physics at Cornell University; his research was interrupted by World War II and by his appointment as head of the theoretical division of the Los Alamos atom-bomb project. He has been a leading voice in emphasizing the social responsibility of the scientist and opposed the US government's Strategic Defense Initiative (Star Wars) programme.

Betti Enrico 1823–1892. Italian mathematician who was the first to provide a thorough exposition and development of the theory of equations formulated by French mathematician Evariste Galois. This greatly advanced the transition from classical to abstract algebra.

Betti was born near Pistoia, Tuscany, and studied physical and mathematical sciences at Pisa, where he was professor from 1856. Betti had fought against Austria during the first wars of Italian independence and in 1862 he became a member of the new Italian parliament. He entered the government as undersecretary of state for education 1874 and served in the senate after 1884.

Betti's most important work was on algebra and the theory of equations. In papers published 1852 and 1855 he gave proofs of most of Galois's major theorems. In so doing he became the first mathematician to resolve integral functions of a complex variable into their primary factors. He also developed the theory of elliptical functions.

Betti turned to mathematical physics 1863. In 1878 he gave the law of reciprocity in elasticity theory which became known as ***Betti's theorem***. Along the way, conducting research into 'analysis situs' in hyperspace in 1871, he also did valuable work on numbers characterizing the connection of a variety, these later becoming known as ***Betti numbers***.

Bhabha Homi Jehangir 1909–1966. Indian theoretical physicist who made several important explanations of the behaviour of subatomic particles. He was also responsible for the development of research and teaching of advanced physics in India, and for the establishment and direction of the nuclear-power programme in India.

Bhabha was born in Bombay and studied in the UK at Cambridge, doing research there until 1939. He was in India when World War II broke out. At the Bangalore Institute of Science he was put in charge of a department investigating cosmic rays. The Tata Institute of Fundamental Research was established in Bombay 1945 with Bhabha as director.

Bhabha made major contributions to the early development of quantum electrodynamics, a part of high-energy physics. A primary gamma ray dissipates its energy in the formation of electron showers. In 1935, Bhabha became the first person to determine the cross-section (and thus the probability) of electrons scattering positrons. This phenomenon is now known as ***Bhabha scattering***.

Bichat Marie François Xavier 1771–1802. French physician and founder of histology, the study of tissues. He studied the organs of the body, their structure, and the ways in which they are affected by disease.

This led to his discovery and naming of 'tissue', a basic biological and medical concept; he identified 21 types. He argued that disease does not affect the whole organ but only certain of its constituent tissues.

Bichat was born in the Jura region. His medical studies were interrupted by military service in the French Revolution. From 1801 he was physician at the Hôtel-Dieu hospital for the poor in Paris. He distrusted the microscope and made little use of it; his analysis of tissues consequently did not include any understanding of their cellular structure.

Bickford William 1774–1834. English inventor of the miner's safety fuse 1831. He made a major contribution to safety and productivity in mines and quarries.

Bickford, born in Devonshire, set up as a leather merchant in Cornwall. Later he started a company to construct the machinery for fuse production. This firm was eventually merged into what would become Imperial Chemical Industries (ICI).

When used for blasting in mines, gunpowder was put into a borehole and lit by a fuse that was generally made of goose quills. Quill fuses frequently seemed to fail but then rekindled so that the miner who went to inspect the apparently extinct fuse was injured in the blast. Bickford came up with a safety fuse that virtually eliminated this danger. A given length burned in a given time, and it had better resistance to water. Several holes could be fired at a time. Fuses based on this design were still being used in the 1950s.

Biffen Rowland Harry 1874–1949. English botanist and geneticist who produced several wheat hybrids that are resistant to disease. In the development of better hybrids of wheat, he applied Mendel's laws and demonstrated for the first time that resistance to yellow rust disease of wheat was inherited.

Biffen was born in Cheltenham and educated at Emmanuel College, Cambridge, graduating with a first in natural sciences 1896. Whilst studying botany at Gonville and Caius College, he became interested in agriculture. In 1899, he was made a lecturer in the recently established school of agriculture and was appointed the first chair in agricultural botany 1908. Although Biffen did not enjoy teaching, he was an inspirational head of department. He was elected as a Fellow of the Royal Society 1914, awarded the Darwin medal 1920, and knighted 1925. Biffen continued to live in Cambridge until his death 12 July.

Bigelow Erastus Brigham 1814–1879. US inventor and industrialist who devised power looms for weaving various patterned fabrics and carpets, such as Wilton and Brussels carpets. Carpets could now be made of virtually any colour, to virtually any pattern.

Bigelow was born in West Boylston, Massachusetts, and forced to go out to work at the age of ten. Nonetheless he invented a loom for weaving lace to trim stagecoach upholstery 1837, and in 1843 set up the Clinton Company to manufacture ginghams. His power loom for weaving Brussels and Wilton carpets was developed between 1845 and 1851. Together with his other looms, this made the company so successful that the town of Clinton, Massachusetts, grew up around the plant. Bigelow helped found the Massachusetts Institute of Technology 1861.

Bigelow Jacob 1787–1879. US physician and botanist. His *Florula Bostoniensis* 1814 was a study of the plants growing within a ten mile radius of Boston. He later extended this study to cover many of the flora in other parts of New England and produced a standard manual of E American botany which is still popular today.

Bigelow was born in Sudbury, Massachusetts and studied at Harvard University, graduating 1806. He then studied medicine at the University of Pennsylvania and obtained his MD 1810. He decided to work as a physician in Boston but also found time to lecture on botany. He was made professor of Materia Medica at Harvard Medical School 1816 and became widely respected throughout the American scientific community for both his botanical and medical work, his work having far reaching effects upon the effective practise of medicine in America. His three-volume series *American Medical Botany* was published 1817, 1818, and 1820, and *Discourse on Self-limited Diseases* appeared 1835.

Binet Alfred 1857–1911. French psychologist who introduced the first intelligence tests 1905. They were standardized so that the last of a set of graded tests the child could successfully complete gave the level described as 'mental age'. If the test was passed by most children over 12, for instance, but failed by those younger, it was said to show a mental age of 12. Binet published these in collaboration with Théodore Simon.

Binet was born in Nice and studied neurology and psychology in Paris, becoming director of the physiological psychology laboratory at the Sorbonne 1895.

Binet wrote several books on mental processes and reasoning ability, including *L'Etude experimentale de l'intelligence/ Experimental Study of Intelligence* 1903, a study of his two daughters. In addition he devised several tests that involved interpreting a subject's response to visual stimuli such as ink blots and pictures, the forerunners of some modern personality tests.

Binnig Gerd 1947– . German physicist who was involved in the invention of the scanning tunnelling electron microscope (STM), an ultra-powerful microscope capable of imaging individual atoms. Nobel Prize for Physics 1986 (shared with Ernst Ruska and Heinrich Rohrer).

The scanning tunnelling electron microscope produces a magnified image by using a tiny tungsten probe, with a tip so fine that it may consist of a single atom, which moves across a specimen. The probe tip moves so close to the specimen surface that electrons jump (or tunnel) across the gap between the tip and the surface. The magnitude of the electron flow (current) depends upon the distance from the tip to the surface, and so by measuring the current the contours of the surface can be determined. The contours can be used to form an image on a computer screen of the surface.

Binnig was born in Frankfurt, Germany, and educated there and in Offenbach, a nearby town. He joined the IBM Zürich Research Laboratories at Rüschlikon, Switzerland 1978. The first STM was built at IBM with Heinrich Rohrer 1981.

Biot Jean Baptiste 1774–1862. French physicist who studied the polarization of light. In 1804 he made a balloon ascent to a height of 5 km/3 mi, in an early investigation of the Earth's atmosphere.

Bird Adrian Peter 1947– . English molecular biologist who demonstrated that the process of DNA methylation provides a switching mechanism by which a cell can control the activity of specific genes in its nucleus. This process is thought to be particularly important for cell differentiation (the process by which cells become increasingly more specialized).

His research showed that the activity of genes studded along the length of chromosomes in living cells can be blocked by methylation of specific sites (called 'CpG sites') on the DNA of those genes. Thus, when these sites in a gene are methylated, the gene is inactive and when they are demethylated the gene has the potential to become active and influence the functioning of the cell. The exact site where the DNA is methylated and its degree of methylation also affects how stable the gene is when it is inactive.

Bird's research is particularly important to the study of the development of the embryo when the activity of different genes has to be tightly controlled to induce the development of specific parts of the body in the right place at the right time.

Adrian Bird graduated from the University of Sussex and obtained his PhD from the University of Edinburgh 1971. He was elected a Fellow of the Royal Society 1989 and the Buchanan professor of genetics at the University of Edinburgh 1990.

Birdseye Clarence 1886–1956. US inventor who pioneered food refrigeration processes. While working as a fur trader in Labrador 1912–16 he was struck by the ease with which food could be preserved in an Arctic climate. Back in the USA he found that the same effect could be obtained by rapidly freezing prepared food between two refrigerated metal plates. To market his products he founded the General Sea Foods Company 1924, which he sold to General Foods 1929.

Birkhoff George David 1884–1944. US mathematicians who made fundamental contributions to the study of dynamical systems such as the Solar System. He formulated the 'weak form' of the ergodic theorem.

Birkhoff was born at Overisel, Michigan, and studied at the University of Chicago and at Harvard. From 1912 he taught at Harvard, and was made professor 1919.

Birkhoff developed a system of differential equations which is still inspiring research. His work on difference equations was notable for the prominence he gave to the use of matrix algebra.

Birkhoff notably investigated the theory of dynamical systems and Jules Poincaré's celestial mechanics. He began to examine the motion of bodies in the light of his work on asymptotic expansions and boundary value problems of linear differential equations. In 1913 he proved Poincaré's last geometric theorem on the three-body problem.

With John Von Neumann, Birkhoff was chiefly responsible for establishing, in the 1930s, the modern science of ergodics. He arrived at the statement of his 'positive ergodic theorem', or what is known as the 'weak form' of ergodic theory, just before Von Neumann published his 'strong form' of it. Birkhoff also transformed the Maxwell–Boltzmann hypothesis of the kinetic theory of gases, which was undermined by the number of exceptions found to it, into a vigorous principle.

Throughout his life Birkhoff argued that Einstein's general relativity was an unhelpful theory.

Biró Lazlo 1900–1985. Hungarian-born Argentine who invented a ballpoint pen 1944. His name became generic for ballpoint pens in the UK.

Bishop (John) Michael 1936– . US virologist and molecular biologist who won the Nobel Prize in Physiology or Medicine 1989 with Harold Varmus for their discovery of oncogenes (cancer-causing genes) which, when over-activated or damaged, trigger a normal cell to divide in an uncontrolled fashion.

Together with his colleague Harold Varmus Bishop demonstrated that a gene (called the *src* gene) which is present in some viruses could transform mammalian cells into cells resembling cancer cells. The *src* gene was the first oncogene to be identified. This gene is present, for example, in the avian sarcoma virus, which produces highly malignant tumours when injected into chickens.

He also demonstrated that the viral *src* gene is present naturally in the nucleus of some mammalian cells (under these circumstances it is called the cellular gene, *c-src*), where it plays a part in the normal growth and differentiation (specialization) of cells.

When *c-src* genes become overactive or faulty, as in the presence of a virus, they produce an enzyme (called tyrosine kinase) that stimulates a cascade of growth signals to be produced by the cell, causing it to divide rapidly.

Bishop was born in York, Pennsylvania and studied at Gettysburg College and Harvard Medical School. Since his discovery of the *src* gene many more oncogenes have been found and it is hoped that cancer specialists will be able to use knowledge of these proto-oncogenes to develop drugs which will stop them being switched on, thus preventing cancer in the future.

Bishop Ronald Eric 1903–1989. British aircraft designer. He joined the de Havilland Aircraft Company 1931 as an apprentice, and designed the Mosquito bomber, the Vampire fighter, and the Comet jet airliner.

Bjerknes Vilhelm Firman Koren 1862–1951. Norwegian scientist whose theory of polar fronts formed the basis of all modern weather forecasting and meteorological studies. He also developed hydrodynamic models of the oceans and the atmosphere and showed how weather prediction could be carried out on a statistical basis, dependent on the use of mathematical models.

Bjerknes was professor at Stockholm, Sweden, and Leipzig, Germany, before returning to Norway and founding the Bergen Geophysical Institute 1917.

During World War I, Bjerknes instituted a network of weather stations throughout Norway; coordination of the findings from such stations led him and his co-workers to the theory of polar fronts, based on the discovery that the atmosphere is made up of discrete air masses displaying dissimilar features.

He coined the word 'front' to denote the boundary between such air masses.

Black Davidson 1884–1934. Canadian anatomist. In 1927, when professor of anatomy at the Union Medical College, Peking (Beijing), he unearthed the remains of Peking man, an example of one of our human ancestors.

Black James Whyte 1924– . British physiologist, director of therapeutic research at Wellcome Laboratories (near London) from 1978. He was active in the development of beta-blockers (which reduce the rate of heartbeat) and anti-ulcer drugs. He shared the Nobel Prize for Physiology or Medicine 1988 with US scientists George Hitchings and Gertrude Elion (1918–). Knighted 1981.

Black Joseph 1728–1799. Scottish physicist and chemist who in 1754 discovered carbon dioxide (which he called 'fixed air'). By his investigations in 1761 of latent heat and specific heat, he laid the foundation for the work of his pupil James Watt.

In 1756 Black described how carbonates become more alkaline when they lose carbon dioxide, whereas the taking-up of carbon dioxide reconverts them. He discovered that carbon dioxide behaves like an acid, is produced by fermentation, respiration, and the combustion of carbon, and guessed that it is present in the atmosphere. He also discovered the bicarbonates (hydrogen carbonates).

Born in Bordeaux, France, Black qualified as a doctor in Edinburgh, where he became professor of chemistry.

Black noticed that when ice melts it absorbs heat from its surroundings without itself undergoing a change in temperature, from which he argued that the heat must have combined with the ice particles and become latent. He also observed that equal masses of different substances require different quantities of heat to change their temperatures by the same amount, an observation that established the concept of specific heat capacity.

Black *Prior to Joseph Black's discovery of carbon dioxide, chemists of the time had thought that no other gases apart from air existed. He called this new gas 'fixed air' and showed that it could support neither life nor combustion.*

Black Max 1909– . Azeri-born US philosopher and mathematician. Investigating the question 'What is mathematics?', he divided the answers into three schools: the logical, the formalist, and the intuitional.

Black, born in Baku, studied philosophy at Cambridge and London universities. Moving to the USA 1940, he worked from 1946 at Cornell, where he was professor of philosophy 1954–77.

Black described mathematics as the study of all structures whose form may be expressed in symbols. Within that broad spectrum are the three main schools. The ***logical*** considers that all mathematical concepts, such as numbers or differential coefficients, are capable of purely logical definition. The ***formalist*** concerns itself with the structural properties of symbols, independent of their meaning. The formalist approach has been especially fruitful in its application to geometry. The ***intuitional*** considers mathematics to be grounded on the basic intuition of the possibility of constructing an infinite series of numbers.

This approach has had most influence in the theory of sets of points.

Black's works include *The Nature of Mathematics* 1950 and *Problems of Analysis* 1954.

Blackburn James 1803–1854. English-born Australian engineer and architect, transported to Tasmania in 1833 for forgery and granted a free pardon in 1841. Among the churches he designed in Tasmania are the Gothic Revival Holy Trinity, Hobart, 1840–47, and the Norman style St Mark's, Pontville, 1839. He also surveyed and designed the Yan Yean water-supply works in Victoria 1850–51.

Blackett Patrick Maynard Stuart. Baron Blackett 1897–1974. British physicist. He was awarded a Nobel prize 1948 for work in cosmic radiation and his perfection of the cloud chamber, an apparatus for tracking ionized particles, with which he confirmed the existence of positrons.

Blackett was born in Croydon, Surrey, and joined the navy 1912; after World War I, he studied science at Cambridge. He held posts at various British academic institutions.

In 1924, working under physicist Ernest Rutherford at Cambridge, Blackett made the first photograph of an atomic transmutation, which was of nitrogen into an oxygen isotope. He continued to develop the cloud chamber and 1932 designed one where photographs of cosmic rays were taken automatically. Later he discovered particles with a lifespan of 10^{-10} sec, which became known as strange particles.

In the 1950s he turned to the study of rock magnetism. He was made a life peer 1969.

A first-rate laboratory is one in which mediocre scientists can produce outstanding work.

Patrick Blackett

Quoted by M G K Menon in his commemoration lecture on H J Bhabha, Royal Institution, 1967

Blackman Frederick Frost 1866–1947. English botanist after whom the Blackman reactions of photosynthesis are named. Leading a successful research group, he worked initially upon respiration in plants and showed that the exchange of CO_2 between the leaves and the air occurred via the stomata (the pores in the epidermis of a plant). His later work continued to apply physiochemical concepts to biology.

Blackman was born in Lambeth and educated in medicine at St Bartholomew's Hospital, obtaining a BSc 1885. However, in 1887, he abandoned his study of medicine and went to St John's College, Cambridge, where he obtained a first in natural sciences. Following graduation he was appointed as a demonstrator in the Cambridge School of Botany. He was promoted to reader 1904 and remained in this position until he retired 1936.

He was involved in planning the extension of the botany school in 1933 and the general administration of the department. He was elected a Fellow of the Royal Society 1906 and was awarded its Royal Medal 1921.

Blackwell Elizabeth 1821–1910. English-born US physician, the first woman to qualify in medicine in the USA 1849, and the first woman to be recognized as a qualified physician in the UK 1869.

Her example inspired Elizabeth Garrett Anderson and many other aspiring female doctors.

Elizabeth Blackwell was taken to the USA as a child. She studied at Geneva Medical School in New York State and then opened a private clinic in New York City. On her return to the UK 1869, she became the first woman to appear in the Medical Register.

She was professor of gynaecology at the London School of Medicine for Women 1875–1907.

Blakemore Colin Brian 1944– . English physiologist who has made advanced studies of how the brain works. In his book *Mechanics of the Mind* 1977 he explains the mechanics of sensation, sleep, memory, and thought, and discusses a number of philosophical questions.

Blakemore was born in Stratford-upon-Avon and studied at Cambridge and the Neurosensory Laboratory at the University of California in Berkeley. He was on the staff of Cambridge University 1967–79, when he became professor of physiology at Oxford.

In his experimental work, Blakemore has shown that cells in the visual cortex of the brain of a newborn kitten are able to detect visual outlines. But if the kitten is kept at a critical period in an environment with only, say, vertical lines, it will later prove to have in the cortex only cells that can recognize these patterns and not others. Blakemore suggests that it is possible that the inherited DNA of genes already contains the capacity to synthesize RNA, the protein that is involved in the storage of any new remembrance.

Blakeslee Albert Francis 1874–1954. US botanist who was the first to artificially produce polyploid plants (plants having three or more sets of chromosomes per cell in cases where the normal number is two sets). He developed the technique using the chemical colchicine to increase the number of chromosomes.

Blakeslee was born in Geneseo, New York, and graduated 1896 from Wesleyan University in Middletown, Connecticut. Until 1899 he taught mathematics and science in a preparatory school, but then went to Harvard to do graduate research, obtaining his PhD 1904. He demonstrated that lower fungi can reproduce sexually in his PhD thesis which was on *Sexual Reproduction in the Mucorineae*. He went to Connecticut Agricultural College, Storrs, where he was a professor until 1914.

In 1915, he moved to the Carnegie Institution experimental laboratories at Cold Spring Harbor, Long Island, New York, where he continued his research into plant genetics. In 1936, he was promoted to director of the institute. He was then made professor of botany at Smith College, Northampton, where he founded a Genetics Experimental Station.

Blériot Louis 1872–1936. French aviator who, in a 24-horsepower monoplane of his own construction, made the first flight across the English Channel on 25 July 1909.

Bloch Felix 1905–1983. Swiss-US physicist who invented the analytical technique of nuclear magnetic resonance (NMR) spectroscopy 1946. For this work he shared the Nobel Prize for Physics 1952 with US physicist Edward Purcell (1912–).

Bloch Konrad 1912– . German-born US chemist whose research concerned cholesterol. Making use of the radioisotope carbon-14 (the radioactive form of carbon), Bloch was able to follow the complex steps by which the body chemically transforms acetic acid into cholesterol. He shared the 1964 Nobel Prize for Physiology or Medicine with his collaborator in Germany, Feodor Lynen.

Bloch was born in Neisse (now Nysa, Poland) and graduated as a chemical engineer from Munich Technische Hochschule; he emigrated to the USA 1936 and studied biochemistry at Columbia University. In 1954 he became a professor at Harvard.

Bloch demonstrated that carbon atoms of carbon-labelled acetate (ethanoate) fed to rats was incorporated into cholesterol in the animals' livers. Using acetic acid (ethanoic acid) labelled with deuterium he showed for the first time that this acid, a compound having only two carbon atoms, is the major precursor of cholesterol. This discovery was the first of a long series that elucidated the biological synthesis of the steroid.

Bloembergen Nicolaas 1920– . Dutch-born US physicist who worked on laser spectroscopy using the laser as a tool to excite or break chemical bonds during the study of chemical reactions. Nobel Prize for Physics 1981 (shared with Arthur Schawlow and Kai Siegbahn).

In the mid 1950s Bloembergen developed a type of maser, called the three-level or multi-level maser, that could produce a continuous output of microwave radiation – earlier masers could only work intermittently. Bloembergen has also worked on non-linear optics, studying the effects produced by high-intensity radiation.

Bloembergen was born in Dordrecht, Netherlands, and educated at the universities of Utrecht and Leiden, where he gained his doctorate 1948. He emigrated to the United States soon after and joined Harvard

University, Cambridge, Massachusetts, 1949. He was appointed Gordon McKay Professor of Applied Physics 1957, Rumford Professor of Physics 1974, and Gerhard Gade university professor 1980.

Bode Johann Elert 1747–1826. German astronomer and mathematician who contributed greatly to the popularization of astronomy. He published the first atlas of all stars visible to the naked eye, *Uranographia* 1801, and devised Bode's law.

Bode's law is a numerical sequence that gives the approximate distances, in astronomical units (distance between Earth and Sun = one astronomical unit), of the planets from the Sun by adding 4 to each term of the series 0, 3, 6, 12, 24, ... and then dividing by 10. Bode's law predicted the existence of a planet between Mars and Jupiter, which led to the discovery of the asteroids. The relationship was first noted 1772 by German mathematician Johann Titius (1729–1796) and is also known as the Titius–Bode law.

Bode, born in Hamburg, taught himself astronomy. He was director of the Berlin observatory 1786–1825. Bode worked on the compilation of two atlases, *Vorstellung der Gestirne* and *Uranographia*, which described the positions of more than 17,000 stars and included for the first time some of the celestial bodies discovered by William Herschel. It was Bode who named Herschel's new planet Uranus.

Bodmer Walter Fred 1936– . German-born English geneticist who was a pioneer of research into the genetics of the HLA histocompatability system, which helps the immune system to distinguish between the body's own cells and foreign, potentially harmful cells, like bacteria, that need to be destroyed and cleared from the body.

The HLA system, which consists of a set of genes and their corresponding antigens presented on the surface of cells, is also important in the way that cells recognize one another and work together during embryonic development. Only when there is overlapping expression of these HLA antigens and genes by two individuals, can transplantation of tissues, organs, or limbs take place between them without rejection by the immune system of the recipient.

Bodmer was born in Frankfurt, Germany just before World War II and studied at Cambridge University in England. After working at Stanford University 1961–70, Bodmer was professor of genetics at Oxford University 1971–79, and director of research then director general at the Imperial Cancer Research Fund in the UK 1979–95, during which time he was elected a Fellow of the Royal Society and knighted.

Boerhaave Hermann 1668–1738. Dutch physician and chemist who made Leiden a European centre of medical knowledge. He re-established the technique of clinical teaching, taking his students to the bedsides of his patients. His *Elementia chemiae* 1732 presented a clear and precise approach to the chemistry of the day.

Boerhaave was born near Leiden, and studied there and at Haderwijk. He returned to Leiden as a physician, and also began teaching, eventually becoming professor of medicine, botany, 'physic', and chemistry.

Boerhaave described the structure and function of the sweat glands, and was the first to realize that smallpox is spread by contact.

Boerhaave's works remained authoritative for nearly a century.

They include a physiology textbook, *Institutiones medicae* (a classification of diseases with their causes and treatment) 1708, *Index plantarum* 1710, and *Historia plantarum*, a collection of his botanical lectures compiled by his ex-students.

Bohm David Joseph 1917–1992. US-born British physicist who specialized in quantum mechanics but also worked on plasmas, metals, and liquid helium. In 1959 he and his student Yakir Aharanov discovered the Aharanov–Bohm effect, showing that the motions of charged particles can be affected by magnetic fields even if they never enter the regions to which those fields are confined.

There are no things, only processes.

David Bohm

In C H Waddington *The Evolution of an Evolutionist* 1975

Bohm left the USA 1951 after an accusation of un-American activities during the McCarthy era made it impossible for him to find university employment. He settled permanently in London 1961, where he was professor of theoretical physics at Birkbeck College until 1983.

In 1952 he produced a radical alternative version of quantum theory which stated that particles are particles at all times, even when not observed, and that their wavelike behaviour is determined by a hidden field or 'pilot wave'. When Bohm published his ideas, they were immediately rejected, but in 1982 French physicists carried out Bohm's experiment and demonstrated that such a theory was possible.

Later in life Bohm developed an interest in philosophy and wrote several books on physics, philosophy, and the nature of consciousness.

Bohr Aage Niels 1922– . Danish physicist who produced a new model of the nucleus of the atom 1952, known as the collective model. For this work, he shared the 1975 Nobel Prize for Physics. He was the son of Niels Bohr.

Aage Bohr was born and educated in Copenhagen. During the German occupation of Denmark in World War II, he worked in London 1943–45. From 1946 he worked at his father's Institute of Theoretical Physics in Copenhagen, and became professor of physics at the University of Copenhagen 1956. He was director of the Niels Bohr Institute (formerly the Institute of Theoretical Physics) 1963–70 and director of Nordita (Nordic Institute for Theoretical Atomic Physics) 1975–81.

Aage Bohr, Ben Mottelson, and James Rainwater shared the 1975 Nobel prize for 'the discovery of the connection between the collective motion and the particle motion in atomic nuclei and the development of the theory of the structure of the atomic nucleus, based on this connection'. Rainwater had suggested the theory 1950 and Bohr and Mottelson obtained experimental results that proved it.

The work of these three people paved the way for nuclear fusion.

Bohr Niels Henrik David 1885–1962. Danish physicist. His theoretical work produced a new model of atomic structure, now called the Bohr model, and helped establish the validity of quantum theory. He also explained the process of nuclear fission. Nobel Prize for Physics 1922.

Bohr's first model of the atom was developed working with Ernest Rutherford at Manchester, UK. He was director of the Institute of Theoretical Physics in Copenhagen from 1920. During World War II he took part in work on the atomic bomb in the USA. In 1952 he helped to set up CERN, the European nuclear research organization in Geneva. He proposed the doctrine of ***complementarity***: that a fundamental particle is neither a wave nor a particle, because these are complementary modes of description.

Bohr, Niels *Danish physicist Niels Bohr, who recieved the Nobel Prize for Physics 1922 for his theory of the structure of the atom. Bohr escaped from German-occupied Denmark in a fishing boat during World War II and assisted in the development of the atom bomb in the United States. After the war he returned to Denmark and campaigned publicly against the spread of nuclear weapons.*

Bohr was born and educated in Copenhagen. In 1911 he went to the UK to study at the Cambridge atomic-research laboratory under J J Thomson, but moved 1912 to Manchester to work with New Zealand physicist Rutherford. Bohr developed models of the atom in which electrons are disposed in rings around the nucleus, a first step towards an explanation of atomic structure.

In 1913 Bohr developed his theory of atomic structure by applying quantum theory to the observations of radiation emitted by atoms. The authorities in Denmark made him a professor 1916 and then built the Institute of Theoretical Physics for him. Leading physicists from all over the world developed Bohr's work there, resulting in the theories of quantum and wave mechanics that more fully explain the behaviour of electrons within atoms. Bohr's atomic theory was validated 1922 by the discovery of an element he had predicted, which was given the name hafnium.

In 1939 Bohr proposed his liquid-droplet model for the nucleus, which explained why a heavy nucleus could undergo fission following the capture of a neutron. Working from experimental results, Bohr was able to show that only the isotope uranium-235 would undergo fission with slow neutrons.

An expert is a man who has made all the mistakes which can be made in a very narrow field.

Niels Bohr

Quoted in Mackay
The Harvest of a Quiet Eye

When Denmark was invaded by Nazi Germany, Bohr took an active part in the resistance movement. In 1943, he escaped to Sweden and on to the USA. After working on the atomic bomb, he became a passionate advocate for the control of nuclear weapons.

Bok Bart (Jan) 1906–1983. Dutch-born US astrophysicist who discovered small, circular dark spots in nebulae (now ***Bok's globules***). His work broadened our understanding of the formation of stars.

Bok, born in Hoorn, was educated at Leiden and Groningen in the Netherlands and in the USA at Harvard, where he remained and became professor 1947. In 1957 he went to Australia as head of the Department of Astronomy at the Australian National University as well as director of the Mount Stromlo Observatory near Canberra. He ended his career as professor at the University of Arizona 1966–1974.

Photographs had shown that the Milky Way was dotted with dark patches or nebulae. Bok also discovered small, circular dark spots, which were best observed against a bright background. Measurements of their dimensions and opacity suggested that their masses were similar to that of the Sun. Bok suggested that the globules were clouds of gas in the process of condensation and that stars might be in the early stages of formation there.

Boksenberg Alexander 1936– . English astronomer and physicist who devised a light-detecting system that can be attached to telescopes, vastly improving their optical powers. His image photon-counting system (IPCS) revolutionized observational astronomy, enabling Boksenberg and others to study distant quasars.

Boksenberg studied at London University. He became professor of physics 1978 and director of the Royal Greenwich Observatory 1981.

In the early 1960s Boksenberg became interested in the instrumentation carried aboard space vehicles, and in image-detecting systems. The IPCS, rather than recording light with a photographic emulsion, uses a television camera and a computer to detect and store the locations of individual photons of light collected by a telescope from a faint astronomical object, and to present the incoming results as an instantaneous picture. Successfully tested 1973 at Mount Palomar, California, the invention was subsequently installed on all modern telescopes.

Boksenberg also designed instruments for ultraviolet astronomy, for use on high-altitude balloon-borne platforms and on satellites, particularly the European observatory satellite TD-1A 1972 and the International Ultraviolet Explorer (IUE) satellite observatory 1978.

Boksenberg's study of the absorption lines in the spectra of quasars has shown that these are not a manifestation of the quasar itself but a reflection of the state of the universe – galaxies and intergalactic gas – that exists between the quasar and the Earth. They can thus provide direct information on the nature and evolution of the universe.

Boltzmann Ludwig Eduard 1844–1906. Austrian physicist who studied the kinetic theory of gases, which explains the properties of gases by reference to the motion of their constituent atoms and molecules. He established the branch of physics now known as statistical mechanics.

He derived a formula, the ***Boltzmann distribution***, which gives the number of atoms or molecules with a given energy at a specific temperature. The constant in the formula is called the ***Boltzmann constant***.

Boltzmann was born in Vienna, educated at Linz and Vienna, and held a number of academic posts in Vienna, Graz, Munich, and Leipzig. Opposition to his theories about the structure of atoms made him depressed and he committed suicide just before they became widely accepted.

A paper 1868 on thermal equilibrium in gases gave the Boltzmann distribution formula and included the constant, which has become a fundamental part of virtually every mathematical formulation of a statistical nature in both classical and quantum physics.

Boltzmann's work in the fields of gases, electromagnetism, and thermodynamics led him to consider phenomena in terms of probability theory and atomic events, which led to the establishment of statistical mechanics. This discipline holds that macroscopic properties of matter, such as conductivity and viscosity, can be understood and are determined by the cumulative properties of the constituent atoms. Boltzmann held that the second law of thermodynamics should be considered from this viewpoint.

Bolyai (Farkas) Wolfgang 1775–1856. Hungarian mathematician, father of János Bolyai; their work on geometry is closely interlinked.

Bolyai was born in Nagyszeben, Hungary (now Sibiu, Romania), and studied mathematics at Göttingen, Germany, where he made friends with German mathematician Karl Gauss. Bolyai taught at colleges first in Nagyszeben and then in Marosvásárhely (now Tirgu Mures, Romania) until 1853.

His son's growing interest in higher mathematics inspired Bolyai to write the book *Tentamen juventutem/Attempt to Introduce Studious Youth into the Elements of Pure Mathematics* 1832–33, a survey of mathematics. He also tried to find a proof for Euclid's fifth postulate – that parallel lines do not meet – and János caught his enthusiasm.

Detest it just as much as lewd intercourse; it can deprive you of all your leisure, your health, your rest, and the whole happiness of life.

Wolfgang Bolyai

To his son János, warning him to give up his attempts to prove the Euclidean postulate on parallels. Attributed

---

Bolyai János 1802–1860. Hungarian mathematician, one of the founders of non-Euclidean geometry. He was the son of Wolfgang Bolyai.

Bolyai was born in Koloszvár, Hungary (now Cluj, Romania), and was taught mathematics by his father. In 1818 he entered the Royal College of Engineers in Vienna, and on graduation joined the army, but retired 1833.

By about 1820, János Bolyai had become convinced that a proof of Euclid's postulate about parallel lines was impossible; he began instead to construct a geometry which did not depend upon Euclid's axiom. This was a theory of absolute space in which several lines pass through the point P without intersecting the line L. He developed his formula relating the angle of parallelism of two lines with a term characterizing the line. In his new theory Euclidean space was simply a limiting case of the new space, and János introduced his formula to express what later became known as the space constant.

János described his new geometry in a paper of 1823 called 'The absolute true science of space'. Wolfgang sent it to German mathematician Karl Gauss, who replied that he had been thinking along the same lines for more than 25 years. János's paper was printed as an appendix to his father's *Tentamen juventutem* 1832–33. Its publication received little attention and he subsequently discovered that Nikolai Lobachevsky had published an account of a very similar geometry (also ignored) 1829.

Bolzano Bernardus Placidus Johann Nepomuk 1781–1848. Czech philosopher and mathematician who formulated a theory of real functions and introduced the non-differentiable ***Bolzano function***. He was also able to prove the existence and define the properties of infinite sets.

Bolzano was born and educated in Prague, where he became professor of mathematics 1804 and of philosophy 1805. For the next 14 years he lectured mainly on ethical and social questions, although also on the links between mathematics and philosophy. His liberal and Czech nationalist views brought him into disfavour with the Austro-Hungarian authorities and in 1819 he was suspended from his professorship and forbidden to publish. In 1824 Bolzano resigned his chair. Owing to the opposition of the imperial authorities, most of his work remained unpublished until 1962.

Bolzano formulated a proof of the binomial theorem and, in one of his few works published in his lifetime (1817), attempted to lay down a rigorous foundation of analysis. One of the most interesting parts of the book was his definition of continuous functions. During the 1830s Bolzano concentrated on the study of real numbers.

Bond George Phillips 1825–1865. US astronomer who developed astronomical photography and in 1850 took the first photograph of a star (Vega). His research was carried out at the Harvard Observatory together with his father, William Cranch Bond.

Bond was born in Dorchester, Massachusetts, and educated at Harvard, remaining there at the observatory under his father and succeeding him as director when he died 1859.

During the late 1840s the Bonds worked on developing photographic techniques for astronomy. In 1857 George Bond became the first person to photograph a double star, Mizar, with the aid of wet collodion plates. He suggested that a star's magnitude could be quantitatively determined by measuring the size of the image it made. A bright star would affect a greater area of silver grains.

Bond also made numerous studies of comets. He discovered 11 new comets and made calculations on the factors affecting their orbits.

Bond William Cranch 1789–1859. US astronomer who established the Harvard College Observatory as a centre of astronomical research. Much of his work as done in collaboration with his son George Phillips Bond. He also designed chronometers.

Bond was born in Falmouth, Maine, and worked as a watchmaker. An amateur astronomer, he was one of the independent observers who discovered the comet of 1811. He was commissioned by Harvard College to investigate the equipment at observatories in England during a trip he made there in 1815, and in 1839 Harvard invited him to move his private observatory into their premises. Bond thus became the first director of the Harvard College Observatory.

William and George Bond discovered Hyperion (the eighth satellite of Saturn) 1848 and the Crêpe Ring (a faint ring inside two bright rings) around Saturn 1850. Their observation that stars could be seen through the Crêpe Ring led to their conclusion that the rings of Saturn are not solid.

The two Bonds also collaborated on the development of photographic techniques for astronomy. They took superior photographs of the Moon, and the first photographs of stars.

Bondi Hermann 1919– . Austrian-born British cosmologist. In 1948 he joined with Fred Hoyle and Thomas Gold in developing the steady-state theory of cosmology 1948, which suggested that matter is continuously created in the universe. KCB 1973.

Bondi was born in Vienna and studied mathematics at Cambridge University. While interned 1940 as an 'enemy alien' in the UK, during World War II, he met Thomas Gold. Bondi returned to Cambridge 1941 and began to do naval radar work for the British Admiralty 1942; through this work he met Fred Hoyle. Gold soon joined them and the three discussed cosmology and related subjects. Bondi's academic career after the war was spent mostly at Cambridge and King's College, London.

[Science doesn't deal with facts; indeed] fact is an emotion-loaded word for which there is little place in scientific debate. Science is above all a cooperative enterprise.

Hermann Bondi

Nature 1977

The steady-state model stimulated much debate for, while its ideas were revolutionary, it was fully compatible with existing knowledge. However, evidence that the universe had once been denser and hotter emerged in the 1950s and 1960s, and the theory was abandoned by most scientists.

Bondi also described the likely characteristics and physical properties of gravitational waves, and demonstrated that such waves are compatible with and are indeed a necessary consequence of the general theory of relativity.

Bonestell Chesley 1888–1986. US artist who specialized in such realistic-looking astronomical illustrations that many believe they were instrumental in persuading the US government that space exploration was possible.

Bonestell was born in San Francisco, trained as an architect, and took part in the design of the Golden Gate Bridge. He then became a Hollywood special-effects painter, working on films ranging from *Citizen Kane* 1941 to *The War of the Worlds* 1953.

Bonestell began illustrating spacescapes for magazines and science-fiction books in the 1940s. From 1951 he collaborated closely with rocket engineer Wernher von Braun. The articles they published 1952–54 gripped the imagination of the US public and pressure mounted for the government to invest in space exploration.

The Conquest of Space 1949, written by German-born rocket scientist Willy Ley

(1906–1969), was the first popular astronomy book to carry Bonestell's illustrations, many in colour.

Bonpland Aime Jacques Alexandre 1773–1858. French botanist and explorer of South America who, with Alexander von Humboldt, travelled 9,650 km/6,301 mi in South America and collected over 6,000 new species of plants. Their travels and discoveries are described in 23 volumes published 1805–34: *Voyage de Humboldt et Bonpland aux regions quinoxiales du nouveau continent, fait en 1799–1804*.

Bonpland was born in Rochelle, France. He met Alexander von Humboldt in Paris and, in 1799, they set out on an expedition to Central and South America together. Over five years, they explored both the Orinoco and Amazon rivers, visited the Andes and Cuba, and climbed several volcanoes in Ecuador. They each had peaks in the Sierra Nevada National Park in NW Venezuela named in their honour.

Bonpland contracted malaria 1800 during his trips with Humboldt, a condition which plagued him with illness and lethargy for the rest of his life. He was arrested 1816 in South America on a trip in Paraguay and kept in prison for nine years, during which time he continued to correspond with Humboldt.

Boole George 1815–1864. English mathematician whose work *The Mathematical Analysis of Logic* 1847 established the basis of modern mathematical logic, and whose ***Boolean algebra*** can be used in designing computers.

Boole's system is essentially two-valued. By subdividing objects into separate classes, each with a given property, his algebra makes it possible to treat different classes according to the presence or absence of the same property. Hence it involves just two numbers, 0 and 1 – the binary system used in the computer.

Boole was born in Lincoln and was largely self-taught. In 1849 he was appointed professor of mathematics at Queen's College in Cork, Ireland.

In 1847 Boole announced that logic was more closely allied to mathematics than to philosophy. He argued not only that there was a close analogy between algebraic symbols and those that represented logical forms but also that symbols of quantity could be separated from symbols of operation. These ideas received fuller treatment in *An Investigation of the Laws of Thought on which are Founded the Mathematical Theories of Logic and Probabilities* 1854.

Boole *George Boole was the founder of the modern science of symbolic logic and the first mathematician to consider logic algebraically. Boolean algebra, applied to binary addition and subtraction, is important today in the design and construction of circuits and computers.*

Booth Hubert Cecil 1871–1955. English mechanical and civil engineer who invented the vacuum cleaner 1901.

Booth, born in Gloucester, studied at the City and Guilds Institute. He formed an engineering consultancy 1901 and his British Vacuum Cleaner Company 1903.

Booth conceived the principle of his vacuum cleaner after witnessing the cleaning of a railway carriage by means of compressed air which simply blew a great cloud of dust around.

In his machine, one end of a tube was connected to an air pump, while the other, with nozzle attached, was pushed over the surface being cleaned. The cleaner incorporated an air filter.

Because of the large size and high price of early vacuum cleaners and the fact that few houses had mains electricity, Booth initially offered cleaning services rather than machine sales. The large vacuum cleaner, powered by petrol or electric engine and mounted on a four-wheeled horse carriage, was parked in the street outside a house while large cleaning tubes were passed in through the windows. The machine was such a novelty that society hostesses held special parties at which guests watched operatives cleaning carpets or furniture. Transparent tubes were provided so that the dust could be seen departing down them. Booth's machines received a great popular boost when they were used to clean the blue pile carpets laid in Westminster Abbey for the coronation of King Edward VII in 1902.

Smaller, more compact indoors vacuum cleaners followed, but until the first electrically powered model appeared 1905, two people were required to operate them – one to work the pump by bellows or a plunger, the other to handle the cleaning tube.

Having worked on the design of engines for two Royal Navy battleships, Booth was commissioned 1894 to work on a Ferris wheel at Earl's Court, London, and afterwards on similar structures in Blackpool, Vienna, and Paris. In 1902 Booth directed the erection of the Connel Ferry Bridge over Loch Etive, Scotland.

Bordet Jules Jean Baptiste Vincent 1870–1961. Belgian bacteriologist and immunologist who researched the role of blood serum in the human immune response. He was the first to isolate 1906 the whooping-cough bacillus.

Borel Emile Félix Edouard Justin 1871–1956. French mathematician who rationalized the theory of functions of real variables.

Borel was born in St-Affrique and studied in Paris. In 1893 he was appointed to the faculty of mathematics at the University of Lille. The Sorbonne created a chair in function theory for him, which he held 1909–40. From 1910 he was also in charge of science at the Ecole Normale Supérieure. He was a Radical-Socialist member of the national chamber of deputies 1924–36, serving as minister of the navy 1925.

In the 1890s Borel did his most important work: on probability, the infinitesimal calculus, divergent series, and, most influential of all, the theory of measure. He provided a proof of Charles Emile Picard's theorem 1896. In the 1920s he wrote on the subject of game theory, before John Von Neumann (generally credited with being the founder of the subject) first wrote on it in 1928. Borel's theory of integral functions and his analysis of measure theory and divergent series established him, alongside French mathematician Henri Lebesgue, as one of the founders of the theory of functions of real variables.

Borelli Giovanni Alfonso 1608–1679. Italian scientist who explored the links between physics and medicine and showed how mechanical principles could be applied to animal physiology. This approach, known as iatrophysics, has proved basic to understanding how the mammalian body works.

Born Max 1882–1970. German-born British physicist who received a Nobel prize 1954 for fundamental work on the quantum theory, especially his 1926 discovery that the wave function of an electron is linked to the probability that the electron is to be found at any point.

In 1924 Born coined the term 'quantum mechanics'. Born made Göttingen a leading centre for theoretical physics and together with his students and collaborators – notably Werner Heisenberg – he devised 1925 a system called matrix mechanics that accounted mathematically for the position and momentum of the electron in the atom. He also devised a technique, called the Born approximation method, for computing the behaviour of subatomic particles, which is of great use in high-energy physics.

I am now convinced that theoretical physics is actual philosophy.

Max Born

Autobiography

Born was born in Breslau (now Wroclaw, Poland) and studied at Breslau and Göttingen. He was professor of physics at Frankfurt-am-Main 1919–21 and at Göttingen 1921–33.

With the rise to power of the Nazis, Born left Germany for the UK, and in 1936 he became professor of natural philosophy at Edinburgh. In 1953 he retired to Germany.

Encouraged by German chemist Fritz Haber to study the lattice energies of crystals, Born was at Frankfurt able to determine the energies involved in lattice formation, from which the properties of crystals may be derived, and thus laid one of the foundations of solid-state physics.

Born was inspired by Danish physicist Niels Bohr to seek a mathematical explanation for Bohr's discovery that the quantum theory applies to the behaviour of electrons in atoms. This led to matrix mechanics. But in 1926, Erwin Schrödinger expressed the same theory in terms of wave mechanics.

Born used statistical probability to reconcile the two systems.

Bosch Carl 1874–1940. German metallurgist and chemist. He developed the Haber process from a small-scale technique for the production of ammonia into an industrial high-pressure process that made use of water gas as a source of hydrogen. He shared the Nobel Prize for Chemistry 1931 with Friedrich Bergius.

The Haber process, now also called the Haber–Bosch process, combines nitrogen and hydrogen at 400–500°C/750–930°F and at 200 atmospheres pressure. The two gases, in the proportions 1:3 by volume, are passed over a catalyst of finely divided iron. Around 10% of the reactants combine to form ammonia, and the unused gases are recycled. The ammonia is separated either by being dissolved in water or by being cooled to liquid form.

Bosch was born in Cologne and studied at Leipzig. He took a job with Badische Anilin und Sodafabrik (BASF), and by 1902 was working on methods of fixing the nitrogen present in the Earth's atmosphere. At that time, the only large sources of nitrogen compounds essential for the production of fertilizers and explosives were in the natural deposits of nitrates in Chile. Learning of the Haber process 1908, Bosch set up a team of chemists and engineers to improve on it and reproduce it on a large scale. Heavy demand from the military during World War I caused BASF to expand. From 1925 Bosch was chair of the vast industrial conglomerate IG Farbenindustrie AG after its formation from the merger of BASF with other German industrial concerns.

Bosch's other scientific work was on the synthesis of methyl alcohol and of petrol from coal tar.

Boscovich Ruggero Giuseppe 1711–1787. Croatian-born Italian scientist. An early supporter of Newton, he developed a theory, popular in the 19th century, of the atom as a single point with surrounding fields of repulsive and attractive forces.

Bose Jagadis Chunder 1858–1937. Indian physicist and plant physiologist. He was professor of physical science at Calcutta 1885–1915, and studied the growth and minute movements of plants and their reaction to electrical stimuli. He founded the Bose Research Institute, Calcutta. Knighted 1917.

Bose Satyendra Nath 1894–1974. Indian physicist who formulated the Bose–Einstein law of quantum theory with Einstein. He was professor of physics at the University of Calcutta 1945–58. The boson particle is named after him.

Bothe Walther Wilhelm Georg 1891–1957. German physicist who showed 1929 that the cosmic rays bombarding the Earth are composed not of photons but of more massive particles. Nobel Prize for Physics 1954.

Boucher de Crèvecoeur de Perthes Jacques 1788–1868. French geologist whose discovery of palaeolithic hand axes 1837 challenged the accepted view of human history dating only from 4004 BC, as proclaimed by the calculations of Bishop James Ussher.

Boulton Matthew 1728–1809. English factory owner who helped to finance James Watt's development of the steam engine.

Boulton had an engineering works near Birmingham. He went into partnership with Watt 1775 to develop engines to power factory machines that had previously been driven by water.

Boulton, born in Birmingham, first joined forces with Watt 1769. They established the steam engine by erecting pumps in machines to drain the Cornish tin mines. Boulton foresaw a great industrial demand for steam power and urged Watt to develop the double-action rotative engine patented 1782 and the Watt engine 1788 to drive the lapping machines in his factory.

I sell here, Sir, what all the world desire to have – POWER.

Matthew Boulton

To Boswell, of his engineering works 1776

In 1786 Boulton applied steam power to coining machines, obtaining a patent 1790. So successful was the process that as well as his home market Boulton supplied coins to foreign governments and to the East India Company. In 1797 he was commissioned to reform the copper currency of the realm. Boulton's machines for the Royal Mint continued in efficient operation until 1882.

Bourdon Eugène 1808–1884. French engineer and instrumentmaker who invented the pressure gauge that bears his name 1849.

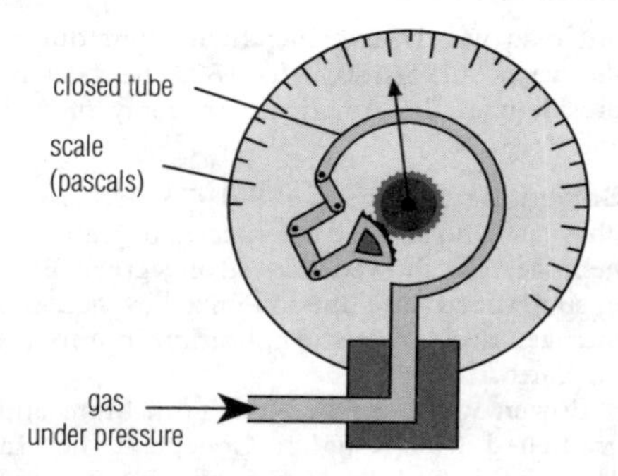

Bourdon *The most common form of Bourdon gauge is the C-shaped tube. However, in high-pressure gauges spiral tubes are used; the spiral rotates as pressure increases and the tip screws forwards.*

The Bourdon gauge contains a C-shaped tube, closed at one end. When the pressure inside the tube increases, the tube uncurls slightly, causing a small movement at its closed end. A system of levers and gears magnifies this movement and turns a pointer, which indicates the pressure on a circular scale.

Bourdon was born in Paris, where he set up an instrument and machine shop. He took an early interest in steam engines, of which he made more than 200, mostly small ones for demonstration purposes.

In developing the steam engine, Scottish engineer James Watt had used relatively low-pressure systems, but at higher pressures there was a need for a compact, accurate pressure gauge. Bourdon's solution to the problem was simple and ingenious, and remains the most widely used gauge for measuring a wide range of fluid pressures.

Boveri Theodor Heinrich 1862–1915. German biologist who showed that it is not a specific number but a specific assortment of chromosomes that is responsible for normal development, indicating that individual chromosomes possess different qualities.

Boveri was born in Bamberg, Bavaria, and studied at Munich. From 1893 he was professor of zoology and comparative anatomy as well as director of the Zoological-Zootomical Institute in Würzburg.

In 1889 Boveri experimented on sea urchins' eggs, fertilizing nucleated and non-nucleated fragments, and found that both types could develop normally; occasional nonfertilized fragments containing only a nucleus were also able to develop normally. He showed that at fertilization the egg and the spermatozoon incorporate the same number of chromosomes each in the creation of the new individual. Eventually he was able to demonstrate that cytoplasm (the part of the cell outside the nucleus) plays an important part in development.

He also studied tumours, believing that malignancy may result from abnormal chromosome numbers. He was the first to realize tumours formed when cells malfunctioned.

Bovet Daniel 1907–1992. Swiss physiologist. He pioneered research into antihistamine drugs used in the treatment of nettle rash and hay fever, and was awarded a Nobel Prize for Physiology or Medicine 1957 for his production of a synthetic form of curare, used as a muscle relaxant in anaesthesia.

Bowden Frank Philip 1903–1968. Australian physicist and chemist who worked mainly in Britain.

He studied friction, lubricants, and surface erosion. For example, he realized that the thin layer of water between a ski or an ice skate and the snow or ice is produced not by pressure due to the weight on them but by friction-induced heat caused by irregularities in the sliding surfaces.

Bowden was born in Hobart, Tasmania, and studied at the University of Tasmania. In 1927 he went to Cambridge University, England, where he remained except during World War II. In 1939 he headed a lubricants and bearings section at Melbourne University, and when the war ended he set up a similar research group at Cambridge. There, he was director of the laboratory for physics and chemistry of solids 1946–65, and a chair in surface physics was created for him 1966.

Bowden began to publish papers on friction 1931, and demonstrated that sliding produces friction over only a small fraction of the total area of the surface, but can produce exceedingly high temperatures and even induce melting in hot spots. An implication of this work is that a lubricant might be decomposed by heat at exactly the place where it is most needed.

His war research covered a broad range of areas of military significance: machine and tool lubricants, flame-throwing fuels, the accurate measurement of shell velocities for gun calibration, and the casting of aircraft bearings.

Bowditch Henry Pickering 1840–1911. US physiologist who investigated the control of the contraction of cardiac muscle, confirming that the cardiac muscle has an inherent rhythmical power that acts automatically (the myogenic theory) and determining that a certain minimum amount of stimulus is needed to induce a contraction of the cardiac muscle.

He worked for may years on how heart muscles contract using frog cardiac muscle as an experimental model. Since a frogs heart continued to beat for some time after it was extracted from the animal, Albrecht von Haller had earlier proposed the myogenic theory, suggesting that cardiac muscle possesses an inherent rhythmical power that acts automatically.

Bowditch demonstrated that passing fluid under pressure into the apex of a frogs heart produces rhythmical activity even in the absence of its usual nervous stimulation. He could obtain the same result by applying a constant electrical current to the muscle. These results supported the myogenic theory and led Bowditch to propose the 'all-or-nothing' rule governing the contraction of cardiac muscle. This rule specifies that a certain minimum amount of stimulus is needed in order to induce a nerve impulse and subsequent muscle contraction. Any stimulus below this set amount has no effect, no matter how long it is applied to the muscle.

Bowditch was born in Boston and studied at Harvard Medical School. He obtained his medical degree 1868 after serving in the American Civil War, in which he fought and was wounded. After graduating, he went to work with eminent physiologists in France and Germany before returning to Harvard to teach physiology. He became dean of the physiology department at Harvard 1883 and founded the prestigious American Physiological Society 1887.

He also worked on how spinal reflexes in the body, such as the knee-jerk reflex, work and the role of nutrition and other environmental factors on growth in children.

Bowditch Nathaniel 1773–1838. US astronomer. He wrote *The New American Practical Navigator* 1802, having discovered many inaccuracies in the standard navigation guide of the day. *Celestial Mechanics* 1829–39 was his translation of the first four volumes of French astronomer Pierre Laplace's *Traité de mécanique céleste* 1799–1825.

Bowditch was born in Salem, Massachusetts. He had little formal education but read widely as a merchant sailor during the years 1795–1803. In 1829 he became president of the American Academy of Arts and Sciences.

Bowen Ira Sprague 1898–1973. US astrophysicist who studied the spectra of planetary nebulae. He showed that strong green lines in such spectra are due to ionized oxygen and nitrogen under extreme conditions not found on Earth.

Bowen was born in New York State and graduated from Oberlin College, Ohio. In 1921 he joined the California Institute of Technology, becoming a professor 1931. He was director of the Mount Wilson and Palomar Observatories 1946–64.

A spectral line is produced when an electron in an atom transfers itself from one energy level to another. Spectral analysis can determine the energy levels between which the electrons are moving, since strong lines are produced where it takes place easily ('permitted' transitions) and weak lines where it takes place with difficulty ('forbidden' transitions).

Bowen showed that the strong green lines in the spectra of planetary nebulae are caused by forbidden transitions in known elements under conditions not produced in the laboratory. The lines had previously been attributed to hypothetical undiscovered elements.

Bowen Norman Levi 1887–1956. Canadian geologist whose work helped found modern petrology. He demonstrated the principles governing the formation of magma by partial melting, and the fractional crystallization of magma.

Born in Kingston, Ontario, Bowen was educated at the local Queen's University, then moved to the recently founded Geophysical Laboratory in Washington DC, USA. His findings on the experimental melting and crystallization behaviour of silicates and similar mineral substances were published from 1912 onwards. In *The Evolution of Igneous Rocks* 1928 he dealt particularly with magma, becoming known as the head of the 'magmatist school' of Canadian geology.

Bower Frederick Orpen 1855–1948. English botanist responsible for building the first Botanical Institute at Glasgow University. His research included a study of the evolutionary morphology (physical structure and form) of the *Pteridophyta* (a simple type of vascular plant that does not bear seeds) and he wrote several books on primitive land plants.

Frederick Bower was born in Ripon. While he was at school at Repton, he began to study botany and decided to make it his life's work. When he read natural sciences at Trinity College, Cambridge, he was able to study botany, which had recently been introduced by S H Vines. He obtained a first 1877 and went to Wurzburg, where he learned laboratory techniques from Julius Sachs. In 1879, he went to study in Strasbourg, returning to London 1880. He became a lecturer 1882 under T H Huxley.

In 1885, he was appointed to the chair of botany in Glasgow University. While he was there he built the first British botanical Institute. He was elected a Fellow of the Royal Society 1891 and was awarded its Gold medal 1910 and the Darwin Medal 1935. He was president of the Royal Society of Edinburgh 1919–24. He never married, and lived to the age of 92.

Boyle Charles, 4th Earl of Orrery 1676–1731. Irish soldier and diplomat. The ***orrery***, a mechanical model of the Solar System in which the planets move at the correct relative velocities, is named after him. Succeeded to earldom 1703.

Boyle Robert 1627–1691. Irish chemist and physicist who published the seminal *The Sceptical Chymist* 1661. He formulated ***Boyle's law*** 1662. He was a pioneer in the use of experiment and scientific method.

Boyle's law states that the volume of a given mass of gas at a constant temperature is inversely proportional to its pressure.

Boyle questioned the alchemical basis of the chemical theory of his day and taught that the proper object of chemistry was to determine the compositions of substances. The term 'analysis' was coined by Boyle and many of the reactions still used in qualitative work were known to him. He introduced certain plant extracts, notably litmus, for the indication of acids and bases. He was also the first chemist to collect a sample of gas.

Boyle was born in Lismore, County Waterford. He lived in Oxford, England, 1654–68, and subsequently in London. As a student he joined a group whose aim was to cultivate the 'new philosophy'; this became the Royal Society 1662.

He learned Hebrew, Greek, and Syriac to further his studies of the Bible, and endowed the Boyle Lectures for the defence of Christianity.

Boyle pointed out that bodies alter in weight according to the varying buoyancy of the atmosphere. These findings were published in *The Spring of Air* 1660.

In *The Sceptical Chymist* he advanced towards the view that matter is ultimately composed of 'corpuscles' of various sorts and sizes, capable of arranging themselves into groups, and that each group constitutes a chemical substance. He successfully distinguished between mixtures and compounds and showed that a compound can have very different qualities from those of its constituents.

In 1667 Boyle was the first to study the phenomenon of bioluminescence, when he showed that fungi and bacteria require air (oxygen) for luminescence, becoming dark in a vacuum and luminescing again when air is readmitted. He can also be credited with the invention of the first match 1680.

Although his main importance is in chemistry, Boyle accomplished much important work in physics, with Boyle's law, the role of air in propagating sound, the expansive force of freezing water, the refractive powers of crystals, the density of liquids, electricity, colour, hydrostatics, and so on.

Boys Charles Vernon 1855–1944. English inventor and physicist who designed several scientific instruments, including a very sensitive torsion balance used 1895 to determine Isaac Newton's gravitational constant and the mean density of the Earth. Knighted 1935.

Boys was born in Wing, Rutland, and studied at the Royal School of Mines, London. For a time he was on the staff of the Royal College of Science, London, then worked for the Metropolitan Gas Board 1897–39.

Boys elaborated on the Cavendish experiment to determine the gravitational constant and the mean density of the Earth. In Boys's apparatus copper wire was replaced by an extremely fine quartz fibre – made by firing an arrow to the end of which was attached molten quartz.

Boys's other work included the invention of a 'radio-micrometer' 1890 to measure the heat radiated by the planets in the Solar System. He also developed a special camera to record fast-moving objects such as bullets and lightning flashes, and devised a calorimeter which became the standard instrument used to measure the calorific value of fuel gas in Britain. In addition, he performed a series of experiments on soap bubbles, which increased knowledge of surface tension and of the properties of thin films.

Bradley James 1693–1762. English astronomer who in 1728 discovered the aberration of starlight. From the amount of aberration in star positions, he was able to calculate the speed of light. In 1748, he announced the discovery of nutation (variation in the Earth's axial tilt).

Bradley was born in Sherborne, Dorset, and studied theology at Oxford. His keen interest in astronomy caused him to resign a clerical post 1721 to become professor of astronomy at Oxford.

As Astronomer Royal from 1742 he sought to modernize the observatory in Greenwich, and embarked on an extensive programme of stellar observation.

The determination of stellar parallax was the goal of many astronomers of Bradley's day because it would confirm Copernicus' hypothesis that the Earth moved around the Sun. When Bradley did find a displacement, it was not only too large, but was in an unexpected direction. Eventually Bradley realized that the displacement was simply a consequence of observing a stationary object from a moving one, namely the Earth. Bradley called this effect the 'aberration of light'.

This discovery allowed Bradley to produce more accurate tables of stellar positions, but he found that his observations on the distances of stars were still variable. He studied the distribution of these variations and deduced that they were caused by the oscillation of the Earth's axis, which in turn was caused by the gravitational interaction between the Moon and the Earth's equatorial bulge, so that the orbit of the Moon was sometimes above the ecliptic and sometimes below it. Bradley named this 'nutation'.

Bradshaw George 1801–1853. British publisher who brought out the first railway timetable in 1839. Thereafter *Bradshaw's Railway Companion* appeared at regular intervals.

He was apprenticed to an engraver on leaving school, and set up his own printing and engraving business in the 1820s, beginning in 1827 with an engraved map of Lancashire.

Bragg (William) Lawrence 1890–1971. Australian-born British physicist. In 1915 he shared with his father William Bragg the Nobel Prize for Physics for their research work on X-rays and crystals. Knighted 1941.

Bragg was born in Adelaide and studied mathematics there and at Cambridge, then switched to physics. He was professor of physics at the University of Manchester 1919–38 and at Cambridge 1938–54.

I sometimes feel it necessary to remind young research students that we are not writing our papers for consideration only by God and a committee of archangels.

Lawrence Bragg

Attributed

He became interested in the work of Max von Laue, who claimed to have observed X-ray diffraction in crystals. Bragg was able to determine an equation now known as ***Bragg's law*** that enabled both him and his father to deduce the structure of crystals such as diamond, using the X-ray spectrometer built by his father. For this, he was the youngest person (at age 25) ever to receive the Nobel prize.

Lawrence Bragg then went on to determine the structures of such inorganic substances as silicates.

Bragg, Lawrence *Lawrence Bragg was still at university when, in 1912, he discovered the famous law, describing the condition for X-ray diffraction in crystals, which now bears his name. In 1915, at the age of 25, he became the youngest person ever to be awarded the Nobel Prize for Physics – an award he shared with his father, William.*

Bragg William Henry 1862–1942. English physicist. In 1915 he shared with his son Lawrence Bragg the Nobel Prize for Physics for their research work on X-rays and crystals. KBE 1920.

Crystallography had not previously been concerned with the internal arrangement of atoms but only with the shape and number of crystal surfaces. The Braggs' work gave a method of determining the positions of atoms in the lattices making up the crystals, and for accurate determination of X-ray wavelengths. This led to an understanding of the ways in which atoms combine with each other and revolutionized mineralogy and later molecular biology, in which X-ray diffraction was crucial to the elucidation of the structure of DNA.

Bragg was born in Westward, Cumberland. He obtained a first-class degree in mathematics from Cambridge 1885 and was immediately appointed professor of mathematics and physics at the University of Adelaide, South Australia. In 1909 he returned to the UK as professor at Leeds; from 1915 he was professor at University College, London.

Bragg became convinced that X-rays behave as an electromagnetic wave motion. He constructed the first X-ray spectrometer 1913. He and his son used it to determine the structures of various crystals on the basis that X-rays passing through the crystals are diffracted by the regular array of atoms within the crystal.

Bragg, William Henry *In 1913, after several years work investigating the electromagnetic behaviour of X-rays, William Bragg constructed the world's first X-ray spectrometer. Working in conjunction with his son Lawrence Bragg, he developed X-ray crystallography: an experimental technique for analysing the lattice structure of crystals.*

Physicists use the wave theory on Mondays, Wednesdays and Fridays, and the particle theory on Tuesdays, Thursdays and Saturdays.

William Bragg

Attributed remark

Brahe Tycho 1546–1601. Danish astronomer whose accurate observations of the planets enabled German astronomer and mathematician Johannes Kepler to prove that planets orbit the Sun in ellipses. Brahe's discovery and report of the 1572 supernova brought him recognition, and his observations of the comet of 1577 proved that it moved in an orbit among the planets, thus disproving the Greek view that comets were in the Earth's atmosphere.

And when statesmen or others worry him [the scientist] too much, then he should leave with his possessions. With a firm and steadfast mind one should hold under all conditions, that everywhere the earth is below and the sky above, and to the energetic man, every region is his fatherland.

Tycho Brahe

Attributed 1597

Brahe was a colourful figure who wore a silver nose after his own was cut off in a duel, and who took an interest in alchemy. In 1576 Frederick II of Denmark gave him the island of Hven, where he set up an observatory. Brahe was the greatest observer in the days before telescopes, making the most accurate measurements of the positions of stars and planets. He moved to Prague as imperial mathematician in 1599, where he was joined by Kepler, who inherited his observations when he died.

Brahe was born in Skåne (then under Danish rule). He studied at Copenhagen and in Germany at Wittenberg and Rostock.

Observing the 1577 comet, he came to the conclusion that its orbit must be elongated, which conflicted with the belief in planetary spheres. Brahe, the last great astronomer to reject the heliocentric theory of Copernicus, tried to compromise, suggesting that, with the exception of the Earth, all the planets revolved around the Sun.

He prepared tables of the motion of the Sun and determined the length of a year to within less than a second, necessitating the calendar reform of 1582.

Bramah Joseph, adopted name of Joe Brammer 1748–1814. British inventor of a flushing water closet 1778, an 'unpickable' lock 1784, and the hydraulic press 1795. The press made use of Blaise Pascal's regulation (that pressure in fluid contained in a vessel is evenly distributed) and employed water as the hydraulic fluid; it enabled the 19th-century bridge builders to lift massive girders.

Bramah took out patents for 18 inventions, but his training of a whole generation of engineers in the craft of precision engineering at the dawn of the Industrial Revolution was probably an even greater legacy.

Bramah was born near Barnsley, Yorkshire. On completing his apprenticeship with a cabinetmaker, he set up his own business in London. One of his assistants was Henry Maudslay.

In 1785 he suggested the locomotion of ships by means of screws; in 1790 and 1793 he constructed the hydraulic transmission of power. Among Bramah's other inventions were a beer pump and machines for numbering banknotes, for making paper, and for the manufacture of aerated water and pen nibs.

Branly Edouard Eugène Désiré 1844–1940. French physicist and inventor who in 1890 demonstrated the possibility of detecting radio waves; the apparatus he devised (the coherer) was soon used in the invention of wireless telegraphy and radio.

Branly was born in Amiens and studied in Paris at the Ecole Normale Supérieure and the Sorbonne. In 1876 he was appointed professor of physics at the Ecole Supérieure de Sciences of the Catholic Institute in Paris, then worked as a medical professor of electrotherapy 1897–1916.

Branly researched in various subjects, notably electricity, electrostatics, magnetism, and electrical dynamics. His most important work – on wireless telegraphy – was performed 1890, when he demonstrated the coherer, an invention of his that enabled radio waves from a distant transmitter to be detected. Once he had established the principle, however, he did not develop it further and the practical points were later taken up by Italian engineer Guglielmo Marconi.

Brattain Walter Houser 1902–1987. US physicist. In 1956 he was awarded a Nobel prize jointly with William Shockley and John Bardeen for their work on the development of the transistor, which replaced the comparatively costly and clumsy vacuum tube in electronics.

He was born in Amoy, China, the son of a teacher. From 1929 to 1967 he was on the staff of Bell Telephone Laboratories.

Braun Emma Lucy 1889–1971. US botanist, an early pioneer in recognizing the importance of plant ecology and conservation. Her book *Deciduous Forests of Eastern North America* 1950 describes the evolution of forest communities and their survival during periods of glaciation.

Born in Cincinnati, Braun studied geology and botany at the University of Cincinnati. She remained in academic positions at the university until 1948, becoming professor of plant ecology in 1946. She lived with her sister Annette Braun (1884–1978), an entomologist, and continued research work until the end of her life, the two setting up a home laboratory and an experimental garden.

Braun's work in ecology concentrated on the vegetation of a selected variety of habitats in Ohio and Kentucky. An early taxonomic study provided a detailed catalogue of the flora of the Cincinnati region, which she then compared with that of the same region a century earlier. This approach became very influential for analysing regional changes in flora over a period of time.

Braun also wrote and campaigned to save natural areas and to create nature reserves.

Braun Karl Ferdinand 1850–1918. German physicist who made improvements to Guglielmo Marconi's system of wireless telegraphy; they shared the 1909 Nobel Prize for Physics. Braun also discovered crystal rectifiers (used in early radios), and invented the oscilloscope 1895.

Braun was born in Fulda, Hesse, and educated at Marburg and Berlin. He held academic posts at a number of German universities, ending his career as professor and from 1895 director of the Institute of Physics at Strasbourg.

In an attempt to increase the radio transmitter range to more than 15 km/9 mi, Braun devised a system in which the power from the transmitter was magnetically coupled (using electromagnetic induction) to the antenna circuit. He patented this invention 1899, and the principle of magnetic coupling has since been applied to all similar transmission systems. Later Braun developed directional antennas.

In 1874 Braun discovered that some mineral metal sulphides conduct electricity in one direction only. These were later used in the crystal radio receivers that preceded valve circuits.

Braun's oscilloscope was an adaptation of the cathode-ray tube. A laboratory instrument to study high-frequency alternating currents, it was the forerunner of television and radar display tubes.

Braun Wernher von 1912–1977. German scientist responsible for Germany's rocket development programme in World War II. He was technical director of the army rocket research centre at Peenemunde and designed a number of rockets including the V2. He was taken to the USA with his research team 1945 and became a prominent figure in the NASA space programme.

We can lick gravity, but sometimes the paperwork is overwhelming.

Wernher von Braun

Bredig Georg 1868–1944. German physical chemist who devised a method of preparing colloidal solutions 1898. He studied the catalytic action of colloidal platinum and the 'poisoning' of catalysts by impurities.

Bredig was born in Glogau, Lower Silesia (now Glogow, Poland).

He went to work as an assistant in the laboratory of German chemist Wilhelm Ostwald in Leipzig, and later held academic appointments in physical chemistry in Germany and Switzerland; from 1911 at the Karlsruhe Hochschule, Germany.

Bredig devised his method of preparing colloidal solutions (lyophobic sols) using an electric arc. The colloidal particles are thought to be produced mainly by rapid condensation of the vapour of the arc, and they may be in the form of the metal or its oxide.

Brenner Sidney 1927– . South African scientist, one of the pioneers of genetic engineering. Brenner discovered messenger

Brenner *South African-born biochemist Sidney Brenner. Brenner discovered messenger RNA and the ribosomes which function to synthesize proteins encoded by DNA. He succeeded Max Perutz as director of the Laboratory for Molecular Biology in Cambridge in 1980, a post he held until 1986. He then became professor of genetic medicine at Cambridge University. In 1996 he became head of the Molecular Sciences Institute, La Jolla, California.*

RNA (a link between DNA and the ribosomes in which proteins are synthesized) 1960.

Brenner became engaged in one of the most elaborate efforts in anatomy ever attempted: investigating the nervous system of nematode worms and comparing the nervous systems of different mutant forms of the animal. About 100 genes are involved in constructing the nervous system of a nematode and most of the mutations that occur affect the overall design of a section of the nervous system.

Progress in science depends on new techniques, new discoveries and new ideas, probably in that order.

Sidney Brenner

Nature May 1980

Brenner was born in Germiston, near Johannesburg, and studied at the University of the Witwatersrand and in the UK at Oxford. From 1957 he researched in the Molecular Biology Laboratory of the Medical Research Council, Cambridge, and in 1980 was appointed its director.

Brenner is also interested in tumour biology and in the use of genetic engineering for purifying proteins, cloning genes, and synthesizing amino acids.

Brewster David 1781–1868. Scottish physicist who made discoveries about the diffraction and polarization of light, and invented the kaleidoscope 1816. Knighted 1831.

Brewster was born in Jedburgh. Although he never took his degree, he was made principal of Edinburgh University 1859.

And why does England thus persecute the votaries of her science? Why does she depress them to the level of her hewers of wood and her drawers of water? It is because science flatters no courtier, mingles in no political strife.

David Brewster

Quarterly Review 1830

In 1813 Brewster was able to demonstrate, by studying the polarization of light passing through a succession of glass plates, that the index of refraction of a particular medium determines the tangent of the angle of polarization for light that transverses it. Brewster then sought an expression for the polarization of light by reflection and found, in 1815, that the polarization of a beam of reflected light is greatest when the reflected and refracted rays are at right angles to each other. This is known as ***Brewster's law***: the tangent of the angle of polarization is numerically equal to the refractive index of the reflecting medium when polarization is maximum.

Brewster *Although trained as a priest, David Brewster was never ordained and spent most of his scientific career investigating the polarization of light. In 1815 he discovered the optical law, which now bears his name, that relates the angle of maximum polarization for light reflecting off a surface to the refractive index of the reflecting material.*

Brewster then worked on the polarization of light reflected by metals, and established the new field of optical mineralogy.

Bridgewater Francis Egerton, 3rd Duke of Bridgewater 1736–1803. Pioneer of British inland navigation. With James Brindley as his engineer, he constructed 1762–72 the Bridgewater canal from Worsley to Manchester and on to the Mersey, a distance of 67.5 km/42 mi. Initially built to carry coal, the canal crosses the Irwell valley on an aqueduct. Succeeded as Duke 1748.

Bridgman Percy Williams 1882–1961. US physicist. His research into machinery producing high pressure led in 1955 to the creation of synthetic diamonds by General Electric. He was awarded the Nobel Prize for Physics 1946.

Born in Cambridge, Massachusetts, he was educated at Harvard, where he spent his entire academic career.

Bridgman's experimental work on static high pressure began in 1908, and because this field of research had not been explored before, he had to invent much of his own equipment; for example, a seal in which the pressure in the gasket always exceeds that in the pressurized fluid. The result is that the closure is self-sealing. His discoveries included new, high-pressure forms of ice.

His technique for synthesizing diamonds was used to synthesize many more minerals and a new school of geology developed, based on experimental work at high pressure and temperature. Because the pressures and temperatures that Bridgman achieved simulated those deep below the ground, his discoveries gave an insight into the geophysical processes that take place within the Earth. His book *Physics of High Pressure* 1931 still remains a basic work.

Briggs Henry 1561–1630. English mathematician, with John Napier one of the founders of calculation by logarithms. Briggs's tables remain the basis of those used to this day.

Briggs was born in Halifax, Yorkshire, and studied at Cambridge, where he became a lecturer. He was professor of geometry at the newly established Gresham College,

London, 1596–1620, and then became professor of geometry at Oxford.

In 1614 Scottish mathematician John Napier had published his discovery of logarithms, and Briggs went to Edinburgh to meet him 1616. On this and subsequent visits the two men worked together to simplify Napier's original logarithms. It seems most probable that the idea of having a table of logarithms with 10 for their base was originally conceived by Briggs; the first such tables were published by him 1617, under the title *Logarithmorum Chilias Prima*, and were followed in 1624 by the *Arithmetica Logarithmica*, in which the tables were given to 14 significant figures.

The logarithms of Briggs (and Napier) were logarithms of sines, a reflection of their interest in astronomy and navigation. For that reason Briggs's logarithms were 10^9 times 'larger' than those in modern tables.

Bright Richard 1789–1858. British physician who described many conditions and linked oedema to kidney disease. ***Bright's disease***, an acute inflammation of the kidneys is named after him.

Bright was born in Bristol and studied medicine at Edinburgh University. He was on the staff of Guy's Hospital, London, from 1820.

Bright initiated the use of biochemical studies by working with chemists to demonstrate that urea is retained in the body in kidney failure. He also correlated symptoms in patients with the pathological changes he later found in postmortem examinations of the same people. In this way he found that the presence of the protein albumin in the urine and oedema (accumulation of fluid in the body) are associated with pathological changes in the kidneys.

Brindley James 1716–1772. British canal builder, the first to employ tunnels and aqueducts extensively, in order to reduce the number of locks on a direct-route canal. His 580 km/360 mi of canals included the Bridgewater (Manchester–Liverpool) and Grand Union (Manchester–Potteries) canals.

Brindley was born near Buxton, Derbyshire. He set up a machine shop in Staffordshire and began constructing flint and silk mills. He was virtually illiterate and made all calculations in his head.

In 1759 Brindley was engaged by the Duke of Bridgewater to construct a canal to transport coal to Manchester from the duke's mines at Worsley. Brindley's revolutionary scheme for this included a subterranean channel and an aqueduct over the river Irwell. He constructed impervious banks by puddling clay, and the canal simultaneously acted as a mine drain. The success of this project established him as the leading canal builder in the UK.

Brindley *A portrait of the engineer James Brindley. In the left-hand corner of the picture is the Barton Aqueduct, which carries the Bridgewater Canal over the river Irwell at a height of 12 m/40 ft. Brindley's ingenious design established him as the leading canal builder in England.*

Brinell Johan August 1849–1925. Swedish engineer who devised the Brinell hardness test, for measuring the hardness of substances, in 1900.

Brinell was born in Småland and received a technical education in Borås. As chief engineer at an ironworks, he became interested in metallurgy. While studying the internal composition of steel during heating and cooling, he devised the hardness test, which was put on trial at the Paris Exhibition 1900. It is based on the idea that a material's response to a load placed on one small point is related to its ability to deform permanently. Brinell also carried out investigations into the abrasion resistance of selected materials.

Brisbane Thomas Makdougall 1773–1860. Scottish soldier, colonial administrator, and astronomer. After serving in the Napoleonic Wars under Wellington, he was governor of New South Wales 1821–25. Brisbane in Queensland is named after him. He catalogued over 7,000 stars. KCB 1814, Baronet 1836.

Britton Nathaniel Lord 1859–1934. US botanist who set up the New York Botanical Gardens. His own research focussed on taxonomy (the classification of living organisms). His most famous book was *Illustrated Flora of the Northern United States, Canada, and the British Possessions* 1896.

Britton was born in Staten Island, New York and educated at the School of Mines of Columbia College, where he graduated as an engineer 1879. He worked initially as an assistant in geology at Columbia, but then became a botanist and the assistant geologist for the Geological Survey of New Jersey. He returned to Columbia 1887 to teach botany and geology and was appointed professor of botany. Following a visit to Kew Gardens in London in 1885, he initiated the development of the New York Botanical Gardens 1896, obtaining and developing 250 acres of land in Bronx Park.

Brockhouse Bertram N 1918– . Canadian physicist who, with Clifford Shull, developed neutron diffraction techniques used for studying the structure and properties of matter. Brockhouse designed ingenious instruments with which he recorded the energy of neutrons scattered from various materials. Nobel Prize for Physics 1994.

Brockhouse was born in Lethbridge, Alberta, Canada. The family moved to Chicago 1935 where Brockhouse completed his high school education. The family returned to Canada 1938. After World War II, Brockhouse attended the University of British Columbia, studying physics and mathematics. He moved to the University of Toronto to complete his studies.

In 1950 he became research officer at Chalk River Nuclear Laboratory in Ontario. Prior to that time, all nuclear reactors had been employed for a single purpose: to produce the first atomic bomb. A major new tool was placed in the hands of researchers when nuclear reactors were released from the war effort and could be used for tasks other than splitting atomic nuclei. Brockhouse and others took full advantage of the new tool. He started by studying how the energy of neutrons from the Chalk River nuclear reactor changed after passing through a material. The results of the first neutron scattering experiments were published in 1951.

In 1956 he built the first triple axis neutron spectrometer, an instrument which greatly improved the results of scattering experiments. Brockhouse was able to measure the frequency of atomic vibrations in solids and deduce the forces between the atoms. He later carried out similar experiments on the magnetic properties of atoms. In 1962 he became professor of physics at McMaster University, in Hamilton, Ontario, Canada.

Broglie (Louis César Victor) Maurice de, 6th Duc de Broglie 1875–1960. French physicist. He worked on X-rays and gamma rays, and helped to establish the Einsteinian description of light in terms of photons. He was the brother of Louis de Broglie.

Maurice, after a short naval career, equipped an extensive private laboratory at the family home, where he was assisted by his brother Louis. Maurice pioneered the study of X-ray spectra.

Broglie Louis Victor Pierre Raymond de, 7th Duc de Broglie 1892–1987. French theoretical physicist. He established that all subatomic particles can be described either by particle equations or by wave equations, thus laying the foundations of wave mechanics. He was awarded the 1929 Nobel Prize for Physics. Succeeded as Duke 1960.

Two seemingly incompatible conceptions can each represent an aspect of the truth .... They may serve in turn to represent the facts without ever entering into direct conflict.

Louis de Broglie

Dialectica I

De Broglie's discovery of wave–particle duality enabled physicists to view Einstein's conviction that matter and energy are interconvertible as being fundamental to the structure of matter. The study of matter waves led not only to a much deeper understanding of the nature of the atom but also to explanations of chemical bonds and the practical application of electron waves in electron microscopes.

De Broglie was born in Dieppe and educated at the Sorbonne, where he stayed on until 1928. He was professor at the Henri Poincaré Institute 1932–62. From 1946, he was a senior adviser on the development of atomic energy in France.

If particles could be described as waves, then they must satisfy a partial differential equation known as a wave equation. De Broglie developed such an equation in 1926, but found it in a form which did not offer useful information when it was solved.

Throughout his life, de Broglie was concerned with the philosophical issues of physics and he was the author of a number of books on this subject.

Brongniart Alexandre 1770–1847. French naturalist and geologist who first used fossils to date strata of rock and was the first scientist to arrange the geological formations of the Tertiary period in chronological order. He also wrote a classification of reptiles *Essay on the Classification of Reptiles* 1800, splitting them into four orders: Batrachia (now a separate class, Amphibia), Chelonia, Ophidia, and Sauria.

Brongniart introduced the idea of geological dating according to the distinctive fossils found in each stratum. From 1804–11, Brongniart and Georges Cuvier studied the fossils deposited in the Paris basin and showed that the fossils had been laid down during alternate fresh and salt water conditions.

Brongniart was born in Paris and was appointed professor of natural history at the Ecole Centrale des Quatre-Nations, Paris 1797. From 1800 until his death 1847, he was the director of the Sèvres porcelain factory and under his directorship the factory became one of the world leaders in enamelling. In 1822, he became professor of mineralogy at the National Museum of Natural History, Paris.

Bronowski Jacob 1908–1974. Polish-born British scientist, broadcaster, and writer, who enthusiastically popularized scientific knowledge in several books and in the 13-part BBC television documentary *The Ascent of Man*, issued as a book 1973.

Bronowski fled with his family to Germany when, in World War I, Russia occupied his native Poland. Moving to the UK 1920, he studied mathematics at Cambridge. He lectured at University College, Hull, 1934–42, and then did military research during World War II, remaining a government official until 1963. His last appointment was at the Salk Institute for Biological Studies in California, from 1964.

His book *The Common Sense of Science* 1951 displayed the history and workings of science around three central notions: cause, chance, and order. *Science and Human Values* 1958 collected newspaper articles written for the *New York Times* about nuclear science and the morality of nuclear weapons. *The Western Intellectual Tradition* 1960 is an illuminating survey of the growth of political, philosophical, and scientific knowledge from the Renaissance to the 19th century, written with Bruce Mazlish. Bronowski also wrote about literature; for example, *William Blake and the Age of Revolution* 1965.

You will die but the carbon will not; its career does not end with you ... it will return to the soil, and there a plant may take it up again in time, sending it once more on a cycle of plant and animal life.

Jacob Bronowski

'Biography of an Atom – and the Universe' *New York Times* 13 Oct 1968

Brønsted Johannes Nicolaus 1879–1947. Danish physical chemist whose work in solution chemistry, particularly electrolytes, resulted in a new theory of acids and bases.

Brønsted was born in Varde, Jutland, and studied at Copenhagen, where he became professor of physical and inorganic chemistry 1908. In his later years he turned to politics, being elected to the Danish parliament in 1947.

Brønsted laid the foundations of the theory of the infrared spectra of polyatomic molecules by introducing the so-called valency force field 1914. He also applied the newly developed quantum theory of specific heat capacities to gases, and published papers about the factors that determine the pH and fertility of soils.

In 1923 Brønsted published a new theory of acidity which had certain important advantages over that proposed by Swedish chemist Svante Arrhenius 1887. Brønsted defined an acid as a proton donor and a base as a proton acceptor. The definition applies to all solvents, not just water. It also explains the different behaviour of pure acids and acids in solution. In Brønsted's scheme, every acid is related to a conjugate base, and every base to a conjugate acid.

Brouwer Luitzen Egbertus Jan 1881–1966. Dutch mathematician. He worked on the nature and foundation of mathematics, and was the founder of the intuitionist school of mathematics, or intuitionism. He held that the foundation of mathematics is a fundamental intuition of temporal sequence – the counting of moments of time – and that numbers and mathematical entities were constructible from this intuition.

He was opposed to the derivation of mathematics from logic (as Bertrand Russell and Alfred North Whitehead had attempted), and from geometry (as in the formalism of David Hilbert). He also worked in the field of topology (the study of geometric surfaces).

Brouwer, born in Overschie, was professor of mathematics at Amsterdam until 1951. His principal work is *Intuitionism and Formalism* 1912.

Brown Ernest William 1866–1938. English mathematician who studied the effect of gravity on the motions of the planets and smaller members of the Solar System. He published extremely accurate tables of the Moon's movements.

Brown was born in Hull and educated at Cambridge. In 1891 he went to the USA to teach mathematics at Haverford College in Pennsylvania. He was professor of mathematics at Haverford 1893–1907, and then at Yale University.

Brown suggested that the fluctuations in the Moon's mean longitude arose as a consequence of a variable rate in the rotation of the Earth. He was also interested in the asteroid belt, rejecting the theory that the asteroids had at one time have been part of a planet. During the later years of his career he calculated the gravitational effect exerted by the planet Pluto on the orbits of its nearest neighbours, Uranus and Neptune.

Brown Herbert Charles 1912– . US inorganic chemist who is noted for his research on boron compounds. He manufactured sodium borohydride, a reduction agent (a substance that reduces another substance), and developed a simple technique to synthesize diborane. He also created a new class of compounds, the organoboranes, by reacting diborane with alkenes (unsaturated hydrocarbons containing one or more double bonds), and received the Nobel Prize for Chemistry 1979.

Born to Russian émigrés in London, his family moved to Chicago when he was two. When his father died, Brown took on the responsibility of running the family hardware store and looking after his mother and three sisters. But he was determined to fulfil his potential and managed to attend junior college, where he became interested in chemistry. He gained a place at Chicago University, obtained his doctorate, and took a job at Wayne University, Detroit in 1943. He move to Purdue University, Indiana 1947, when he was offered a professorship of inorganic chemistry. He retired in 1978.

Brown Michael Stuart 1941– . US geneticist who shared the Nobel Prize for Physiology or Medicine 1985 with Joseph Goldstein following their work on how the body metabolizes cholesterol. They discovered that individuals with inherited high cholesterol levels have either low levels or deficient forms of the low-density lipoprotein receptor (LDL-receptor) involved in the removal of cholesterol from the blood. The discovery lead to the development of new drugs that lower blood cholesterol levels and reduce the risk of heart disease.

Brown and Goldstein studied why some individuals have high levels of cholesterol in their blood and found that this susceptibility to hypercholesterolaemia can be inherited in some families. They showed that cholesterol is normally removed from the blood by the binding of cholesterol-carrying molecules, called low-density lipoproteins (LDLs) to specific receptors (LDL-receptors) on the surface of cells in the body. The resulting LDL-receptor complexes are then absorbed by the cells. They further demonstrated that this uptake inhibits the cells' production of new LDL-receptors, which explains why a diet of high-cholesterol foods can overwhelm the body's natural capacity for removing cholesterol from the blood.

Brown was born in New York in the USA. He received his medical degree from the University of Pennsylvania 1966 and worked as a junior doctor at the Massachusetts General Hospital in Boston, where he met and became friends with Goldstein. They were reunited and started to work together when Brown was appointed assistant professor at the Southwestern Medical School in Dallas, Texas 1971.

Brown Robert 1773–1858. Scottish botanist who in 1827 discovered ***Brownian motion***. Brownian motion is the continuous random motion of particles in a fluid medium (gas or liquid) as they are subjected to impact from the molecules of the medium. As a botanist, his more lasting work was in the field of plant morphology. He was the first to establish the real basis for the distinction between gymnosperms (pines) and angiosperms (flowering plants).

On an expedition to Australia 1801–05 Brown collected 4,000 plant species and later classified them using the 'natural' system of

Brown, Robert *Scottish botanist Robert Brown. He described the Brownian motion of particles in liquids and gases in 1827. His principal interest was always botany, however, and he held the position of Keeper of Botany at the British Museum in the early part of the nineteenth century.*

Bernard de Jussieu (1699–1777) rather than relying upon the system of Carolus Linnaeus.

Brown was born in Montrose, Forfarshire (now Tayside). He studied medicine at Edinburgh. In the late 1790s he was introduced to English botanist Joseph Banks, and served as his librarian 1810–20, after the Australian voyage.

The concept of Brownian motion arose from his observation that very fine pollen grains suspended in water move about in a continuously agitated manner. He was able to establish that inorganic materials such as carbon and various metals are equally subject to it, but he could not find the cause of the movement (now explained by kinetic theory).

Brown also described the organs and mode of reproduction in orchids. In 1831, he discovered that a small body that is fundamental in the creation of plant tissues occurs regularly in plant cells – he called it a 'nucleus', a name that is still used.

Browne Thomas 1605–1682. English author and physician. Born in London, he travelled widely in Europe before settling in Norwich 1637. His works display a richness of style as in *Religio medici/The Religion of a Doctor* 1643, a justification of his profession; *Vulgar Errors* 1646, an examination of popular legend and superstition; *Urn Burial* and *The Garden of Cyrus* 1658; and *Christian Morals*, published posthumously 1717. Knighted 1671.

Brunel Isambard Kingdom 1806–1859. British engineer and inventor. In 1833 he became engineer to the Great Western Railway, which adopted the 2.1-m/7-ft gauge on his advice. He built the Clifton Suspension Bridge over the river Avon at Bristol and the Saltash Bridge over the river Tamar near Plymouth. His shipbuilding designs include the *Great Western* 1837, the first steamship to cross the Atlantic regularly; the *Great Britain* 1843, the first large iron ship to have a screw propeller; and the *Great Eastern* 1858, which laid the first transatlantic telegraph cable.

The son of Marc Brunel, he made major contributions in shipbuilding and bridge construction, and assisted his father in the Thames tunnel project. Brunel University in Uxbridge, London, is named after both father and son.

Brunel was born in Portsmouth and educated partly in France.

In 1833 he was appointed to carry out improvements on the Bristol docks, and while working on this project his interest in the potential of railways was fired. In all, Brunel was responsible for building more than 2,600 km/1,600 mi of the permanent railway of the west of England, the Midlands, and South Wales. He also constructed two railway lines in Italy, acted as adviser on the construction of the Victoria line in Australia and on the East Bengal railway in India.

Brunel's last ship, the *Great Eastern*, was to remain the largest ship in service until the end of the 19th century.

With over ten times the tonnage of his first ship, it was the first ship to be built with a double iron hull. It was driven by both paddles and a screw propeller. A report of an explosion on board brought on his death.

Brunel Marc Isambard 1769–1849. French-born British engineer and inventor, father of Isambard Kingdom Brunel. He constructed the tunnel under the river Thames in London from Wapping to Rotherhithe 1825–43. Knighted 1841.

Brunel fled to the USA 1793 to escape the French Revolution. He became chief engineer in New York. In 1799 he moved to

England to mass-produce pulley blocks, which were needed by the navy. Brunel demonstrated that with specially designed machine tools 10 men could do the work of 100, more quickly, more cheaply, and yield a better product. Cheating partners and fire damage to his factory caused the business to fail and he was imprisoned for debt 1821. He spent the latter part of his life working on the Rotherhithe tunnel.

Buchner Eduard 1860–1917. German chemist who researched the process of fermentation. In 1897 he observed that fermentation could be produced mechanically, by cell-free extracts. Buchner argued that it was not the whole yeast cell that produced fermentation, but only the presence of the enzyme he named zymase. Nobel Prize for Chemistry 1907.

Buchner was born and educated in Munich, and held a number of academic posts in Germany from 1888. He was killed by a grenade in World War I.

Buchner had been interested in the problems of alcoholic fermentation since the 1880s, and showed 1886 that the absence of oxygen was not necessary for fermentation, but his discovery of zymase came about by accident. In 1893 Buchner and his brother Hans Buchner (1850–1902), a bacteriologist, found a way to make a cell-free liquid extract of microorganisms. They were using a yeast extract for pharmaceutical studies and added a thick sugar syrup to stop any bacterial action. Buchner fully expected the sugar to act as a preservative, but it had the opposite effect and carbon dioxide was produced. The sugar had fermented, producing carbon dioxide and alcohol, in the same way as if whole yeast cells had been present.

It was soon realized that the conversion of sugar into alcohol by means of yeast juice is a series of stepwise reactions, and that zymase is really a mixture of several enzymes.

Buckland William 1784–1856. English geologist and palaeontologist, a principal pioneer of British geology. He contributed to the descriptive and historical stratigraphy of the British Isles, inferring from the vertical succession of the strata a stage-by-stage temporal development of the Earth's crust.

Buckland was born in Axminster, Devon, and studied at Oxford, where he became reader in mineralogy 1813 and in geology 1818.

He was also a cleric and in 1845 became dean of Westminster.

Using the comparative anatomy of French palaeontologist Georges Cuvier, Buckland reconstructed *Megalosaurus* and, in his book *Relics of the Deluge* 1823, explored the geological history of Kirkdale Cavern, a hyena cave den in Yorkshire. His interest in catastrophic transformations of the Earth's surface in the geologically recent past, as indicated by such features as fossil bones and erratic boulders, led him to become an early British exponent of the glacial theory of Louis Agassiz, once he abandoned his assertion (set out in *Geology Vindicated* 1819) that these were caused by the biblical Flood.

Buffon Georges Louis Leclerc, Comte de Buffon 1707–1788. French naturalist and author of the 18th century's most significant work of natural history, the 44-volume *Histoire naturelle génerale et particulière* 1749–67. In *The Epochs of Nature*, one of the volumes, he questioned biblical chronology for the first time, and raised the Earth's age from the traditional figure of 6,000 years to the seemingly colossal estimate of 75,000 years. Became Count 1773.

Buffon was born in Montbard, Burgundy, and educated in Dijon. He travelled a great deal and spent some time in England, where science was undergoing a renaissance. He translated the works of Isaac Newton and Stephen Hales into French. Buffon was director of the Jardin du Roi botanical gardens from 1739.

Genius is only a great aptitude for patience.

Georges Buffon

Attributed remark

Buffon's encyclopedia was the first work to cover the whole of natural history and it was extremely popular. In considering the age of the Earth, he observed that some animals retain parts that are vestigial and no longer useful, suggesting that they have evolved rather than having been spontaneously generated. This upset the authorities because it conflicted with a literal reading of the Bible, and Buffon was forced to recant.

Bullard Edward Crisp 1907–1980. English geophysicist who, with US geologist Maurice Ewing, is generally considered to have founded the discipline of marine geophysics. He pioneered the application of the seismic method to study the sea floor. He also studied continental drift before the theory became generally accepted. Knighted 1953.

Bullard was born in Norwich and educated at Cambridge. During World War II he did military research, and he continued to advise the Ministry of Defence for several years after the war. He was professor of geophysics at the University of Toronto, Canada, 1948–50; director of the National Physical Laboratory at Teddington, Middlesex, 1950–57; and head of geodesy and geophysics at Cambridge 1957–74. He was also a professor at the University of California from 1963, and advised the US government on nuclear-waste disposal.

I should be prepared to argue that Rutherford was a disaster. He started the 'something for nothing' tradition which I was brought up in and had some difficulty freeing myself from – the notion that research can always be done on the cheap, the notion that things don't cost what they do cost. It is wrong. The war taught us differently. If you want quick and effective results you must put the money in.

Edward Bullard

In P Grosvenor and J McMillan *The British Genius* 1973

Bullard's earliest work was to devise a technique (involving timing the swings of an invariant pendulum) to measure minute gravitational variations in the East African Rift Valley. He then investigated the rate of efflux (outflow) of the Earth's interior heat through the land surface; later he devised apparatus for measuring the flow of heat through the deep sea floor.

While at Toronto University, Bullard developed his 'dynamo' theory of geomagnetism, according to which the Earth's magnetic field results from convective movements of molten material within the Earth's core.

Bunsen Robert Wilhelm 1811–1899. German chemist credited with the invention of the ***Bunsen burner***. The Bunsen burner, used in laboratories, consists of a vertical metal tube through which a fine jet of fuel gas is directed. Air is drawn in through holes near the base of the tube and the mixture is ignited and burns at the tube's upper opening. His name is also given to the carbon–zinc electric cell, which he invented 1841 for use in arc lamps. In 1860 he discovered two new elements, caesium and rubidium.

Bunsen *Robert Wilhelm Bunsen, German chemist and physicist. As well as the Bunsen burner, he invented a grease-spot photometer (a means of comparing the intensities of two light sources), an electric cell and, with Gustav Kirchhoff, the method of spectrum analysis which facilitated the discovery of new elements.*

Bunsen was born in Göttingen. He studied chemistry there and at Paris, Berlin, and Vienna. He ended his academic career as professor of experimental chemistry at Heidelberg from 1852.

Bunsen was working on cacodyl compounds, unpleasant and dangerous organic compounds of arsenic, until a laboratory explosion cost him the sight of one eye and he nearly died of arsenic poisoning. In 1844 he invented a grease-spot photometer to measure brightness.

His first work in inorganic chemistry made use of the Bunsen cell. Using electrolysis, he was the first to isolate metallic magnesium and demonstrate the intense light it produces when burned in air. The Bunsen burner was probably used to heat metal salts for spectroscopic analysis, a technique which he pioneered, together with physicist Gustav Kirchhoff, and by which he discovered the new elements.

Burali Forti Cesare 1861–1931. Italian mathematician who worked in the field of vector analysis, especially on the linear transformations of vectors.

He also framed the ***Burali Forti paradox*** 1897, which contradicted the notion that mathematics (or at least its foundations) could be adequately expressed in purely logical terms.

Burali Forti was born in Arezzo, Tuscany, studied at Pisa, and became a professor at the Academia Militare di Artiglieria e Genio in Turin.

Burali Forti's paradox states: 'To every class of ordinal numbers there corresponds an ordinal number which is greater than any element of the class.' In 1902 English philosopher Bertrand Russell demonstrated that this contradiction was of a fundamental logical character and could not be overcome by minor changes in the theory of infinite ordinal numbers.

Much of Burali Forti's work in the field of vector analysis was done in collaboration with Roberto Marcolongo. In 1904 they published a comprehensive analysis of existing systems of vector notation, producing 1909 their own proposals for a unified system. Burali Forti simplified the foundations of vector analysis by the introduction of the notion of the derivative of a vector with respect to a point.

In 1912–13 Burali Forti published more volumes on linear transformations and demonstrated their application to such things as the theory of mechanics of continuous bodies, hydrodynamics, optics, and some problems of mechanics. His last contribution, a paper on differential projective geometry, was finished 1930.

Burbidge Geoffrey 1925– . British astrophysicist who, with his wife Margaret Burbidge, made discoveries relating to nucleosynthesis – the creation of elements in space – and studied quasars and galaxies.

Burbidge was born in Chipping Norton, Oxfordshire. He studied physics at Bristol and University College, London, before going to the USA, first to Harvard and then the University of Chicago. During 1953–55 he worked at the Cavendish Laboratories, Cambridge, but then returned to the USA. He was professor of physics at the University of California, San Diego, 1963–78, and director of the Kitt Peak National Observatory, Arizona, 1978–84.

A paper published 1957 by the Burbidges, US astrophysicist William Fowler, and English astronomer Fred Hoyle began with the premise that at first stars consisted mainly of hydrogen and that most of the stars now visible are in the process of producing helium from hydrogen and releasing energy as starlight. They then suggested that as stars age some of their helium is 'burned' to form other elements, and described several ways in which this was likely to happen, relating it to red giants and supernovae.

In 1970, using evidence gained from observations, Geoffrey Burbidge calculated that the stars emitting light in elliptical galaxies could not account for more than 25% of the mass. He argued that black holes are the most likely source of the missing mass.

Burbidge (Eleanor) Margaret, (born Peachey) 1919– . British astrophysicist who, with her husband Geoffrey Burbidge, discovered processes by which elements are formed in the nuclei of stars. Together they published *Quasi-Stellar Objects* 1967, based on her research. Later, they suggested that quasars and galaxies are linked in some way.

Peachey was born in Davenport and studied at University College, London, and at the University of London Observatory, where she worked 1948–51 before going to the USA. She was alternately at Yerkes Observatory, University of Chicago, and at the California Institute of Technology, then moved to the University of California 1962, becoming professor 1964, as well as director of the Royal Greenwich Observatory 1972–73.

In addition to the work done jointly with her husband, Margaret Burbidge measured the red shifts of several objects suspected of being quasars, and in the process she found that some quasars do not give off any radio radiation.

Burkitt Denis Parsons 1911–1993. British surgeon who first described the childhood tumour named after him, ***Burkitt's lymphoma***. He also pioneered the trend towards high-fibre diets.

Burkitt was born in Enniskillen, Northern Ireland, and educated at Dublin and Edinburgh. He joined the Colonial Service 1946 and worked for the Ministry of Health in Kampala, Uganda. It was in Kampala 1957 that he realized that all the young children brought to him with facial tumours came from the same part of the country. Burkitt devoted himself to finding out where such cases occurred, and made a 15,000-km/9,300-mi safari with two other doctors. The resulting evidence pointed to a 'lymphoma belt' across Africa. It existed only in the hot, wet parts of Africa where malaria was endemic. Infected children therefore already had weakened immune systems. Using material provided by Burkitt, virologist Tony Epstein identified the virus that caused Burkitt's lymphoma. The virus, which ordinarily led to glandular fever, made lymphoid cells malignant in young malaria patients. Surgery was little use, so Burkitt experimented with chemotherapy. Remarkably, a single course of treatment caused the lymphoma to 'melt away'.

Geography also played a role in Burkitt's second major discovery. Now working in London, he knew that few African patients suffered from appendicitis, haemorrhoids, gallstones, or other diseases prevalent in the West. He also knew that the food in rural Africa was high in fibre. He conducted a wide-ranging survey in rural areas of the Third World, and found that where sugar and white flour were eaten, so were the diseases of the West found rampant. This led Burkitt to become convinced that a diet high in roughage prevents many ailments, and compelled scientists and the public to think differently about nutrition.

Burnell Jocelyn Bell, British astronomer. See ◊Bell Burnell.

Burnet (Frank) Macfarlane 1899–1985. Australian physician, an authority on immunology and viral diseases such as influenza, poliomyelitis, and cholera. He shared the 1960 Nobel Prize for Physiology or Medicine with immunologist Peter Medawar for his work on skin grafting. KBE 1969.

Burnet was born in E Victoria and studied at Melbourne and London universities. From 1927 he was associated with the Walter and Eliza Hall Institute for Medical Research in Melbourne, becoming its director 1944.

The idea of man as a dominant animal of the earth whose whole behaviour tends to be dominated by his own desire for dominance gripped me. It seemed to explain almost everything.

Macfarlane Burnet

Dominant Manual 1970

Burnet was the first to investigate the multiplication mechanism of bacteriophages (viruses that attack bacteria) and devised a method for identifying bacteria by the bacteriophages that attack them. In 1932 he developed a technique for growing and isolating viruses in chick embryos; this was to be used as a standard laboratory procedure for more than 20 years.

In 1949 Burnet predicted that an individual's ability to produce a particular antibody to a particular antigen was not innate, but developed during the individual's life. In 1951 Medawar carried out the experiments that confirmed this theory.

Burnet's second major contribution to immunology was made in 1957 – his 'clonal selection' theory of antibody formation, which explains why a particular antigen stimulates the production of its own specific antibody.

Burnside William 1852–1927. English mathematician who made advances in automorphic functions, group theory, and the theory of probability.

Burnside was born in London and educated at Cambridge, where he became a lecturer. In 1885 he left Cambridge to take up the chair of mathematics at the Royal Naval College in Greenwich, London; he retired 1919.

Burnside's study of elliptic functions led him, over the years, to study the functions of real variables and the theory of functions in general. One of his most influential papers, written 1892, was a development of some work by French mathematician Henri Poincaré on automorphic functions.

His study of automorphic functions in turn brought him to the field of group theory, particularly the theory of the discontinuous group of finite order. In 1897 he published the first book on group theory to appear in English; the revised edition 1911 is considered a standard work. A nearly completed manuscript on probability theory was published after his death.

Burroughs William Seward 1855–1898. US industrialist who invented the first hand-operated adding machine to give printed results.

Bush Vannevar 1890–1974. US electrical engineer and scientist who in the 1920s and 1930s developed several mechanical and mechanical-electrical analogue computers which were highly effective in the solution of differential equations. The standard electricity meter is based on one of his designs.

Bush was born in Everett, Massachusetts, and studied at Tufts College and the Massachusetts Institute of Technology (MIT). From 1932 he held senior positions at MIT. During World War I he worked on a magnetic device for detecting submarines. During World War II, as scientific adviser to President F D Roosevelt, Bush was one of the initiators of the atomic-bomb project. After the war he took part in the setting-up and running of the Office of Scientific Research and Development and its successor, the Research and Development Board.

In about 1925 Bush began to construct what he called the product integraph. It contained a linkage to form the product of two algebraic functions and represent it mechanically. He also devised a watt-hour meter, a direct ancestor of the electricity meter, which integrated the product of electric current and voltage. By similarly using current and voltage to represent equations, his machine was able to evaluate integrals involved in the solution of differential equations.

The product integraph was limited to the solution of first-order differential equations. To build a device capable of the solution of second-order equations, Bush coupled one of the watt-hour meters to a mechanical device known as a Kelvin integrator.

In 1931 Bush began work on an almost totally mechanical machine known as the differential analyser. This had six integrators, three input tables, and an output table, which showed the graphical solution of an equation. The Bush analyser was the model for developments in the mechanical world, and many similar machines were built.

Bush built several analysers designed to solve specific problems, and another large analyser, which had many of the operations electrified and a tape input that reduced the time taken for setting up a problem from days to minutes.

Butenandt Adolf Friedrich Johann 1903–1995. German biochemist who isolated the first sex hormones (oestrone, androsterone, and progesterone), and determined their structure. He shared the 1939 Nobel Prize for Chemistry with Leopold Ruzicka (1887–1976).

Butenandt was born in Lehe, near Bremerhaven and studied biology and chemistry in Marburg and Göttingen. He started to work on sex hormones whilst in Göttingen, and continued his research as professor of chemistry at the Danzig Institute 1933. In 1936 he became head of the Kaiser Wilhelm Institute of Biochemistry in Berlin-Dahlem, and transferred his attentions to the study of gene action, mainly that controlling eye colour in insects.

Having determined that eye pigmentation was caused by a substance, kynurenine, formed under genetic control, Butenandt concluded in 1940 that 'genes act by providing an enzyme system that oxidizes tryptophane to kynurene.2'. This is essentially the 'one gene – one enzyme' rule usually ascribed to George Beadle. Butenandt succeeded in isolating the moulting hormone in insects; the first time an insect hormone had been isolated. He then began working on the sex attractant of the female silk moth, and thus isolated the first pheromone.

In 1960 Butenandt became President of the Max Planck Institute, and in 1972, when his presidency ended, he was elected Honorary President. He remained active with the Institute for many years.

Byron (Augusta) Ada, Countess of Lovelace 1815–1852. English mathematician, a pioneer in writing programs for Charles Babbage's analytical engine. In 1983 a new, high-level computer language, Ada, was named after her.

She was the daughter of the poet Lord Byron.

Cailletet Louis Paul 1832–1913. French physicist and inventor who in 1877–78 was the first to liquefy oxygen, hydrogen, nitrogen, and air. He did it by cooling them below their critical temperatures, first compressing the gas, then cooling it, then allowing it to expand to cool it still further.

Cailletet was born in Chatillon-sur-Seine and educated in Paris at the Ecole des Mines, after which he returned to Chatillon to manage his father's ironworks.

Investigating the causes of accidents that occurred during the tempering of incompletely forged iron, Cailletet found that many were due to the highly unstable state of the iron while it was hot and had gases dissolved in it. He also analysed the gases from blast furnaces. As a result of these and other metallurgical studies, Cailletet developed a unified concept of the role of heat in changes of state of metals.

Cailletet's other achievements included the installation of a 300-m/985-ft high manometer on the Eiffel Tower; an investigation of air resistance on falling bodies; a study of a liquid-oxygen breathing apparatus for high-altitude ascents; and the construction of numerous devices, including automatic cameras, an altimeter, and air-sample collectors for sounding-balloon studies of the upper atmosphere.

Cairns Hugh John Forster 1922– . English virologist whose research has focused on cancer and influenza. In 1959 he succeeded in carrying out genetic mapping of an animal virus for the first time.

Cairns was born in Oxford and studied medicine there. He worked at the Viruses Research Institute, Entebbe, Uganda, 1951–63 and was director of the Cold Spring Harbor Laboratory of Quantitative Biology in New York 1963–68. He then took professorships at the State University of New York and with the American Cancer Society. From 1973 to 1981 he was in charge of the Mill Hill laboratories of the Imperial Cancer Research Fund, London. In 1982 he moved to the Harvard School of Public Health, Boston, USA.

The influenza virus, Cairns discovered 1952–53, is not released from the infected cell in a burst but in a slow trickle, and is completed as it is released through the cell surface.

Comparing the rates of replication of DNA in mammals with those in the bacterium *Escherichia coli*, he found that mammalian DNA is replicated more slowly than that of *E. coli*, but is replicated simultaneously at many points.

Cairns's later work studied the link between DNA and cancer, some forms of which may be caused by the alkylation of bases in the DNA. He showed that bacteria are able to inhibit the alkylation mechanism in their own cells, and later demonstrated this ability in mammalian cells.

Cajori Florian 1859–1930. Swiss-born US historian of mathematics. His books dealt with the history of both elementary and advanced mathematics, as well as the teaching of mathematics. His two-volume *History of Mathematical Notations* 1928–29 is a standard reference text. He also wrote biographies of eminent mathematicians.

Cajori was born near Thusis in Graubünden. At the age of 16 he emigrated to the USA, where he took up studies at the University of Wisconsin. He was professor of physics at Colorado College 1889–98, professor of mathematics at Tulane University, New Orleans, 1898–1918, and from 1918 professor of history of mathematics at the University of California at Berkeley.

Cajori's influence on the modern perception of the development of mathematics was profound, and his works are frequently quoted to this day. His reputation is founded mainly on his many books on the history of mathematics, although a number of his works – notably his edited version of Isaac Newton's *Principia* (published posthumously) – have been subject to some criticism for their interpretation of historical material. He also compiled *A History of Physics* 1899.

Callendar Hugh Longbourne 1863–1930. English physicist and engineer who carried out fundamental investigations into the behaviour of steam. One of the results was the compilation of reliable steam tables that enabled engineers to design advanced steam machinery.

Callendar, born in Gloucestershire, studied classics, mathematics, and physics at Cambridge, then medicine and law. He was professor of physics at the Royal Holloway College, Egham, Surrey, 1888–93, at McGill College, Montreal, 1893–98, at University College, London, 1898–1901, and at the Royal College of Science, London, from 1901. During World War I Callender was a consultant to the Board of Inventions, which received more than 100,000 'war-winning' ideas.

While he was at Cambridge, Callendar's main research was on the platinum resistance thermometer, with which he obtained an accuracy of 0.1°C in 1,000°C – about 100 times better than previous results. In 1928 the method was adopted as an international standard. This work led to recording temperatures on a moving chart, a principle now fundamental to any branch of science or industry that requires a continuous record of temperature.

Callendar's research topics were varied, most of them connected with thermodynamics. He carried out experiments on the flow of steam through nozzles, producing much information of great value to steam turbine designers. He also worked on antiknock additives for fuels.

In 1902 he published *The Properties of Steam and Thermodynamic Theory of Turbines*.

Calmette (Léon Charles) Albert 1863–1933. French bacteriologist. A student of Pasteur, he developed, with Camille Guérin, the BCG vaccine against tuberculosis 1921.

Calne Roy Yorke 1930– . British surgeon who developed the technique of organ transplants in human patients and pioneered kidney-transplant surgery in the UK. Knighted 1986.

Calne was educated at Guy's Hospital Medical School, London. He became professor of surgery at Cambridge University 1965.

Calvin Melvin 1911– . US chemist who, using radioactive carbon-14 as a tracer, determined the biochemical processes of photosynthesis, in which green plants use chlorophyll to convert carbon dioxide and water into sugar and oxygen. Nobel prize 1961.

Calvin was born in St Paul, Minnesota, and studied at Michigan College of Mining and Technology and the University of Minnesota. From 1937 he was on the staff of the University of California, becoming professor 1947.

Calvin began work on photosynthesis 1949, studying how carbon dioxide and water combine to form carbohydrates such as sugar and starch in a single-celled green alga, *Chlorella*. He showed that there is in fact a cycle of reactions (now called the Calvin cycle) involving an enzyme as a catalyst.

Cameron Alastair Graham Walter 1925– . Canadian-born US astrophysicist responsible for theories regarding the formation of the unstable element technetium within the core of red giant stars and of the disappearance of Earth's original atmosphere.

Cameron was born in Winnipeg, Manitoba, and educated at Manitoba and Saskatchewan. He emigrated to the USA 1959 and held successive posts at the California Institute of Technology; the Goddard Institute for Space Studies, New York; and Yeshiva University, New York. In 1973 he became professor of astronomy at Harvard.

Spectral lines denote the presence in red giants of technetium, an element too unstable to have existed for as long as the giants themselves. This means the element is created in the stellar core. Cameron suggested that technetium-97 (half-life 2.6×10^6 years) might result from the decay of a nucleus of molybdenum-97, a usually stable nuclide that becomes unstable when it absorbs an X-ray photon at high temperatures.

Cameron also suggested that the Earth's original atmosphere was blown off into space by the early solar 'gale' – as opposed to the present weak solar 'breeze' – with its associated magnetic fields.

Campbell Donald Malcolm 1921–1967. British car and speedboat enthusiast, son of Malcolm Campbell, who simultaneously held the land-speed and water-speed records. In 1964 he set the world water-speed record of 444.57 kph/276.3 mph on Lake Dumbleyung, Australia, with the turbojet hydroplane *Bluebird*, and achieved the land-speed record of 648.7 kph/403.1 mph at Lake Eyre salt flats, Australia. He was killed in an attempt to raise his water-speed record on Coniston Water, England.

Campbell William Wallace 1862–1938. US astronomer and mathematician who published a catalogue of nearly 3,000 radial velocities of stars 1928. His spectroscopic observations of Nova Auriga 1892 enabled him to describe the changes in its spectral pattern with time.

Campbell was born in Hancock, Ohio, and studied at the University of Michigan. From 1891 he worked at the newly established Lick Observatory, California, becoming director of the observatory 1901–30. There he was responsible for much of the spectroscopic work undertaken and was an active participant in and organizer of seven eclipse expeditions to many parts of the world.

In 1896 Campbell initiated his lengthiest project, the compilation of a vast amount of data on radial velocities. Useful for the determination of the motion of the Sun relative to other stars, the programme also led to the discovery of many binary systems, and the data were later used in the study of galactic rotation.

Campbell also confirmed the work done in 1919 by English astronomer Arthur Eddington on the deflection of light during an eclipse, which supported the general theory of relativity. The positive result Campbell obtained in 1922 was arrived at only after two previous attempts (1914 and 1918) had been frustrated by poor weather conditions and inadequate equipment.

Campbell went against the popular opinion of the time in reporting his observations on the absence of sufficient oxygen or water vapour in the Martian atmosphere to support life as found on Earth.

Candolle Alphonse Louis Pierre Pyrame de 1806–1893. Swiss botanist who developed a new classification system for plants based on a broad number of their morphological features. He worked extensively on the effects of climatic and other physical variables on the development of distinct species of flora, and published several books including *Introduction á l'étude de la botanique* 1835, *Géographie botanique raisonnée* 1855, and *Historie des sciences et ses savants depuis deux siècles* 1873.

Candolle was born in Geneva, the son of the eminent botanist Augustin Pyrame de Candolle. He graduated 1825 from Geneva University and, despite his love of botany, decided to study law, obtaining a doctorate 1829. However, he later returned to botany, working for his father until he was made director of the university's botanical gardens 1835 and then replaced him as professor of botany at Geneva University 1842.

He introduced postage stamps into Switzerland and was involved in public affairs throughout his lifetime.

Candolle Augustin Pyrame de 1778–1841. Swiss botanist who coined the term 'taxonomy' to mean the classification of plants on the basis of their gross anatomy in his book *Théorie élémentaire de la botanique* 1813. He posited that plant relationships can be determined by the symmetry of their sexual organs and introduced the concept of homologous parts, the idea that an organ or structure possessed by different plants may indicate a common ancestor.

Candolle was born in Geneva and kept detailed botanical notebooks from the age of 14, which included descriptions of the flora of the Swiss Plateau and the Jura mountains. In 1794, he began to study medicine at the College de Calvin, but transferred to Paris 1796 to study natural sciences. He was commissioned by the French government to make a botanical and agricultural survey of France 1806–12. In 1808, he was made professor of botany at the Ecole de Médecine and Faculté des Sciences in Montpellier.

In 1816, he became professor of natural history at the Academy of Geneva, where he was responsible for reorganizing its botanical gardens. He was made rector of the academy 1831–32, retiring 1835. His research included not only botany, but also pharmacology, and he published several textbooks, including *Cours de botanique* 1828 and his twenty volume work *Plantorum historia succulentarum.*

Cannizzaro Stanislao 1826–1910. Italian chemist who revived interest in the work of Avogadro that had, in 1811, revealed the difference between atoms and molecules, and so established atomic and molecular weights as the basis of chemical calculations.

Cannizzaro also worked in aromatic organic chemistry. In 1853 he discovered reactions (named after him) that make benzyl alcohol and benzoic acid from benzaldehyde.

Cannizzaro was born in Palermo, Sicily, and studied at Palermo, Naples, and Pisa. In 1848 he fought in the Sicilian Revolution, and was condemned to death, but in 1849 escaped to Marseille and went on to Paris. He was a professor at the Technical Institute of Alessandria, Piedmont, 1851–55, followed by professorships at Genoa, Palermo, and Rome. He became a senator 1871 and eventually vice president.

Cannizzaro's reaction involves the treatment of an aromatic aldehyde with an alcoholic solution of potassium hydroxide. The aldehyde undergoes simultaneous oxidation and reduction to form an alcohol and a carboxylic acid. It is an example of a disproportionation reaction, and finds many uses in synthetic organic chemistry.

Reviving Avogadro's hypothesis 1858, Cannizzaro pointed out that once the molecular weight of a volatile compound had been determined from a measurement of its vapour density, it was necessary only to estimate, within limits, the atomic weight of one of its elemental components. Then in investigating a sufficient number of compounds of that element, the chances were that at least one of them would contain only one atom of the element concerned, so that its equivalent weight (atomic weight divided by valency) would correlate with its atomic weight.

Cannon Annie Jump 1863–1941. US astronomer who carried out revolutionary work on the classification of stars by examining their spectra. Her system, still used today, has spectra arranged according to temperature into categories labelled O, B, A, F, G, K, M, R, N, and S. O-type stars are the hottest, with surface temperatures of over 25,000 K.

Studying photographs, Cannon discovered 300 new variable stars. In 1901 she published a catalogue of the spectra of more than 1,000 stars, using her new classification system. She went on to classify the spectra of over 300,000 stars. Most of this work was published in a ten-volume set which was completed 1924. It described almost all stars with magnitudes greater than nine. Her later work included classification of the spectra of fainter stars.

Cannon was born in Dover, Delaware, and studied at Wellesley and Radcliffe colleges. She spent her career at the Harvard College Observatory, as assistant 1896–1911, curator of astronomical photographs 1911–38, and astronomer and curator 1938–40.

A system had been established 1890 for classifying stellar spectra into categories labelled alphabetically A–Q. Cannon reformed this system. Stars in the O, B, A group are white or bluish, those in the F, G group yellow, those in the K group orange, and those in the M, R, N, S group red. Our Sun is yellow so its spectrum places it in the G group.

Cantor Georg Ferdinand Ludwig Philipp 1845–1918. German mathematician who followed his work on number theory and trigonometry by considering the foundations of mathematics. He defined real numbers and produced a treatment of irrational numbers using a series of transfinite numbers. Cantor's set theory has been used in the development of topology and real function theory.

Cantor was born in St Petersburg, Russia, but went to school in Germany and attended the universities of Zürich and Berlin. From 1869 he was on the staff at Halle University, as professor from 1879. His work gained little recognition, which may have contributed to his depression and mental illness in later life.

Investigating sets of the points of convergence of the Fourier series (which enables functions to be represented by trigonometric series), Cantor derived the theory of sets that is the basis of modern mathematical analysis. His work contains many definitions and theorems in topology. For the theory of sets, he had to arrive at a definition of infinity, and also therefore consider the transfinite; for this he used the ancient term 'continuum'. He showed that within the infinite there are countable sets and there are sets having the power of a continuum, and proved that for every set there is another set of a higher power.

Cantor considered metaphysics and astrology to be a science into which mathematics, and especially set theory, could be integrated.

Carathéodory Constantin 1873–1950. German mathematician who made significant advances to the calculus of variations and to function theory. His work also covered theory of measure and applied mathematics.

Carathéodory was born in Berlin and studied there and at Göttingen University. He held professorships at various universities in Germany and, 1920–24, in Greece; from 1924 he was at the University of Munich.

His first major contribution to the calculus of variations was his proposal of a theory of discontinuous curves. From his work on field theory he established links with partial differential calculus, and in 1937 he published a book on the application to geometrical optics of the results of his investigations into the calculus of variations.

One of Carathéodory's most significant achievements, the subject of a book in 1932, was a simplification of the proof of one of the central theorems of conformal representation. It formed part of his work on function theory.

Carathéodory's interest also extended beyond pure mathematics into the applications of the subject, particularly to mechanics, thermodynamics, and relativity theory.

Cardano Girolamo 1501–1576. Italian physician, mathematician, philosopher, astrologer, and gambler. He is remembered for his theory of chance, his use of algebra, and many medical publications, notably the first clinical description of typhus fever.

Born in Pavia, he became professor of medicine there 1543, and wrote two works on physics and natural science, *De subtilitate rerum* 1551 and *De varietate rerum* 1557.

Cardozo William Warrick 1905–1962. US physician and paediatrician who made pioneering investigations into sickle-cell anaemia. He concluded in 1937 that the disease was inherited following Mendelian law and almost always occurred in black people or people of African descent; not all persons with sickle cells were necessarily anaemic and not all patients died of the disease.

Cardozo studied at Ohio State University. In 1937 he started private practice in Washington DC, and was appointed part-time instructor in paediatrics at Howard University College of Medicine and Freedmen's Hospital, later being promoted to associate professor.

Cardozo's investigations were published in the *Archives of Internal Medicine* as 'Immunologic studies in sickle cell anaemia'.

Carlson Chester Floyd 1906–1968. US scientist who invented xerography. A research worker with Bell Telephone, he lost his job 1930 during the Depression and set to work on his own to develop an efficient copying machine. By 1938 he had invented the Xerox photocopier.

Carnegie Andrew 1835–1919. US industrialist and philanthropist, born in Scotland, who developed the Pittsburgh iron and steel industries, making the USA the world's leading producer. He endowed public libraries, education, and various research trusts.

Carnegie invested successfully in railways, land, and oil. From 1873 he engaged in steelmaking, adopting new techniques. Having built up a vast empire, he disposed of it to the US Steel Trust 1901. After his death the Carnegie trusts continued his philanthropic activities.

Carnegie Hall in New York, opened 1891 as the Music Hall, was renamed 1898 because of his large contribution to its construction.

Carnegie, born in Dunfermline, was taken by his parents to the USA 1848. He began work at 14, in Pittsburgh, and was largely self-educated. As a railway employee, he introduced sleeping cars. He saved some capital and bought railway shares, and made a small fortune in oil. On business trips to Europe, he became acquainted with first the Bessemer process and then the open-hearth process of steelmaking, which he introduced to the USA.

Carnegie was attacked by some as an exploiter of labour and an unscrupulous business competitor. However, on retirement he moved to Skibo Castle in Sutherland, Scotland, and used his wealth to endow libraries and universities, the Carnegie Endowment for International Peace, and other good causes.

Carnot (Nicolas Leonard) Sadi 1796–1832. French scientist and military engineer who founded the science of thermodynamics. His pioneering work was *Reflexions sur la puissance motrice du feu/On the Motive Power of Fire*, which considered the changes that would take place in an idealized, frictionless steam engine.

Carnot's theorem showed that the amount of work that an engine can produce depends only on the temperature difference that occurs in the engine. He described the maximum amount of heat convertible into work by the formula $(T_1 - T_2)/T_2$, where T_1 is the temperature of the hottest part of the machine and T_2 is the coldest part.

Carnot was born in Paris and educated there and at the Ecole Genie in Metz. He became an army engineer, at first inspecting and reporting on fortifications and from 1819 based in Paris.

In formulating his theorem, Carnot considered the case of an ideal heat engine following a reversible sequence known as the

Carnot cycle. This cycle consists of the isothermal expansion and adiabatic expansion of a quantity of gas, producing work and consuming heat, followed by isothermal compression and adiabatic compression, consuming work and producing heat to restore the gas to its original state of pressure, volume, and temperature. Carnot's law states that no engine is more efficient than a reversible engine working between the same temperatures. The Carnot cycle differs from that of any practical engine in that heat is consumed at a constant temperature and produced at another constant temperature; no work is done in overcoming friction at any stage; and no heat is lost to the surroundings.

At the time he wrote *Reflexions*, Carnot believed that heat was a form of fluid. But notes discovered 1878 indicate that he later arrived at the idea that heat is essentially work, or rather work that has changed its form. He had calculated a conversion constant for heat and work and showed he believed that the total quantity of work in the universe is constant – the first law of thermodynamics.

Carothers Wallace Hume 1896–1937. US chemist who carried out research into polymerization. He discovered that some polymers were fibre-forming, and in 1931 he produced nylon and neoprene, one of the first synthetic rubbers.

Carothers was born in Burlington, Iowa, and studied at Illinois and Harvard. In 1928 he became head of organic chemistry research at the Du Pont research laboratory in Wilmington, Delaware. He committed suicide by swallowing cyanide, some years later.

Much of Carothers's research effort was directed at producing a polymer that could be drawn out into a fibre. His first successful experiments involved polyesters, but for finer fibres with enough strength to emulate silk, he turned to polyamides. Nylon is a linear chain polymer which can be cold-drawn after extrusion through spinnerets to orientate the molecules parallel to each other so that lateral hydrogen bonding takes place.

Carothers also worked on synthetic rubbers. Neoprene, first produced commercially 1932, is resistant to heat, light, and most solvents.

Carr Emma Perry 1880–1972. US chemist who in the USA pioneered techniques to synthesize and analyse the structure of complex organic molecules using absorption spectroscopy. She also did research into unsaturated hydrocarbons and far ultraviolet vacuum spectroscopy.

Carr was born in Holmesville, Ohio, and studied at Mount Holyoke and the University of Chicago. From 1910 she was on the staff of Mount Holyoke, as professor of chemistry 1913–46.

Carrel Alexis 1873–1944. US surgeon born in France, whose experiments paved the way for organ transplantation. Working at the Rockefeller Institute, New York City, he devised a way of joining blood vessels end to end (anastomosing). This was a key move in the development of transplant surgery, as was his work on keeping organs viable outside the body, for which he was awarded the Nobel Prize for Physiology or Medicine 1912.

Carrington Richard Christopher 1826–1875. English astronomer who by studying sunspots established the Sun's axis and rotation. He was the first to record the observation of a solar flare, in 1859.

Carrington was born in London and educated at Cambridge. He became an

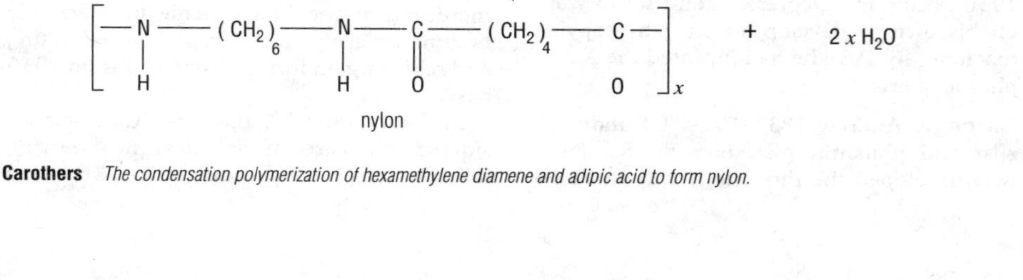

hexamethylene diamine
(hexan-1, 6-diamine)

adipic acid
(hexan-1, 6-dioic acid)

nylon

Carothers *The condensation polymerization of hexamethylene diamene and adipic acid to form nylon.*

astronomical observer at Durham University. In 1853 he set up his own observatory at Redhill, Surrey; in 1865 he moved to Churt, near Farnham, Surrey, where he built another observatory.

Sunspot activity manifests itself in an 11-year cycle, and Carrington observed them for seven years, plotting their positions and movements by a method of his own devising. The principal results of this extended work were, first, to determine the position of the Sun's axis and, second, to show that the Sun's rotation is differential, that is, that it does not rotate as a solid body, but turns faster at the equator than at the poles. Carrington also derived a useful expression for the rotation of a spot in terms of heliographical latitude.

Carrington's *Catalogue of 3,735 Circumpolar Stars* 1857 was so highly regarded that it was printed by the Admiralty at public expense. An extensive account of all the sunspot observations was published 1863.

Carroll James 1854–1907. English-born US physician who with US bacteriologist Walter Reed established that yellow fever is transmitted by mosquitoes.

Carroll and Reed were members of the Yellow Fever Commission established 1900 by the US government to investigate the causes of this disease, as thousands of US troops were dying of yellow fever in Cuba while fighting the Spanish. Reed discovered that yellow fever was transmitted via the mosquito parasite *Aedes aegypti*, and that the infectious agent had to develop within the mosquito for a number of days. He proved his theories with a series of controlled experiments in which diseased and healthy human volunteers were infected by being bitten by infected mosquitoes. Carroll allowed himself to be bitten by a mosquito that had fed on a patient with the disease twelve days earlier. He became very ill, but made a good recovery; a colleague was not so fortunate and later died from the disease.

Carroll was born in Woolwich in the UK but emigrated to North America with his family as a child. He trained as a doctor before working as a surgeon with the US Army.

Carroll Lewis, pen name of Charles Lutwidge Dodgson 1832–1898. English author of the children's classics *Alice's Adventures in Wonderland* 1865 and its sequel *Through the Looking-Glass* 1872. Among later works was the mock-heroic narrative poem *The Hunting of the Snark* 1876. He was fascinated by the limits and paradoxes of language and thought, the exploration of which leads to the apparent nonsense of Alice's adventures. He was an Oxford don and also published mathematical works.

Dodgson was a mathematics lecturer at Oxford 1855–81. There he first told the fantasy stories to Alice Liddell and her sisters, daughters of the dean of Christ Church. Dodgson was a prolific letter writer and one of the pioneers of portrait photography. He was also responsible, in his publication of mathematical games and problems requiring the use of logic, for a general upsurge of interest in such pastimes. He is said to be, after Shakespeare, the most quoted writer in the English language.

'Why', said the Dodo, 'the best way to explain it is to do it.'

Lewis Carroll

Alice in Wonderland p 33

Dodgson was born in Daresbury, Cheshire, and studied mathematics and classics at Oxford. He was ordained a deacon 1861. In 1867 he visited Russia.

His other major works of fiction, *Sylvie and Bruno* 1889 and *Sylvie and Bruno Concluded* 1893, were unsuccessful.

Interested in the use of number games that called for general intelligence to solve the problems, rather than specialized knowledge, he saw their potential as teaching aids. His books in this field include *A Tangled Tale* 1885, *The Game of Logic* 1886, and *Pillow Problems* 1893.

The chessboard featured in some of these games. Several of his books of puzzles suggest an awareness of the theory of sets – the basis on which most modern mathematical teaching is founded – which was then only just being formulated by German mathematician Georg Cantor.

Dodgson also wrote mathematical textbooks for the general syllabus, quite a few books on historical mathematics (particularly on Euclid and his geometry), and a number of specialized papers (such as 'Condensation of determinants').

Carson Rachel Louise 1907–1964. US biologist, writer, and conservationist. Her book *Silent Spring* 1962, attacking the indiscriminate use of pesticides, inspired the creation of the modern environmental movement.

Carson was born in Springdale, Philadelphia, and educated at Johns Hopkins University. She worked first at the University of Maryland and the Woods Hole Marine Biological Laboratory in Massachusetts, and then as an aquatic biologist with the US Fish and Wildlife Service 1936–49, becoming its editor-in-chief until 1952.

As cruel a weapon as the cave man's club, the chemical barrage has been hurled against the fabric of life.

Rachel Carson

The Silent Spring

Her first book, *The Sea Around Us* 1951, was a best-seller and won several literary awards. It was followed by *The Edge of the Sea* 1955, an ecological exploration of the seashore. *Silent Spring* was a powerful denunciation of the effects of the chemical poisons, especially DDT, with which humans were destroying the earth, sea, and sky. While writing about broad scientific issues of pollution and ecological exploitation, she also raised important issues about the reckless squandering of natural resources by an industrial world.

Carter Herbert James 1858–1940. Australian entomologist, born in England. He migrated to Australia in 1881 and worked as school teacher. He became interested in entomology, particularly beetles, and was an avid collector and classifier, describing over 1,000 new species. He was joint editor of the first *Australian Encyclopaedia* 1925–27, supervising the science articles. His published work includes 65 papers and the book *Gulliver in the Bush* 1933, recording his field experiences in Australia.

Cartwright Edmund 1743–1823. British inventor. He patented the power loom 1785, built a weaving mill 1787, and patented a wool-combing machine 1789.

Cartwright was born in Nottinghamshire and studied at Oxford. He became rector of Goadby Marwood, Leicestershire, 1779 and was prebendary of Lincoln from 1786. He set up a factory in Doncaster, Yorkshire, for weaving and spinning, went bankrupt 1793, but was awarded £10,000 by the government 1809.

Visiting the spinning mills of manufacturing pioneer Richard Arkwright inspired Cartwright to try to invent a weaving mill.

He patented his first, water-driven power loom 1785, and gradually improved it. It was followed by the wool-combing machine, which did the work of 20 hand-combers. The wool-combers – some 50,000 in number – organized a protest, but nothing came of it.

Cartwright's Doncaster factory was enlarged when a steam engine was erected to power it, and in 1799 a Manchester firm contracted with Cartwright for the use of 400 of his power looms and built a mill where some of these were powered by steam. The Manchester mill was burned to the ground, probably by workers who feared to lose their jobs, and this prevented other manufacturers from repeating the experiment.

Carver George Washington 1860–1943. US agricultural chemist. He devoted his life to improving the economy of the US South and the condition of blacks. He advocated the diversification of crops, promoted peanut production, and was a pioneer in the field of plastics.

At the Tuskegee Institute, Alabama, Carver demonstrated the need for crop rotation and the use of leguminous plants, especially the peanut. Following his advice, farmers were soon making more money from the peanut and its 325 by-products (including milk, cheese, face powder, printer's ink, shampoo, and dyes) which were developed by Carver, than from tobacco and cotton.

Born a slave in Missouri, he was kidnapped and raised by his former owner, Moses Carver. He won a scholarship to Highland University, Kansas, but was then rejected on account of his race. Instead he graduated from Iowa Agricultural College, where he began teaching agriculture and bacterial botany as well as conducting experiments into plant pathology. In 1897 he transferred to the Tuskegee Institute, becoming director of agriculture and of a research station.

Carver made peanuts the principal crop in the farming belt running from Montgomery to the Florida border. He also discovered 118 products which could be made from the sweet potato and 75 products from the pecan nut. Carver's other work included developing a plastic material from soya beans which Henry Ford later used in part of his automobile, and extracting dyes and paints from the clays of Alabama.

Cassegrain *c.* 1650–1700. French inventor of the system of mirrors used within many

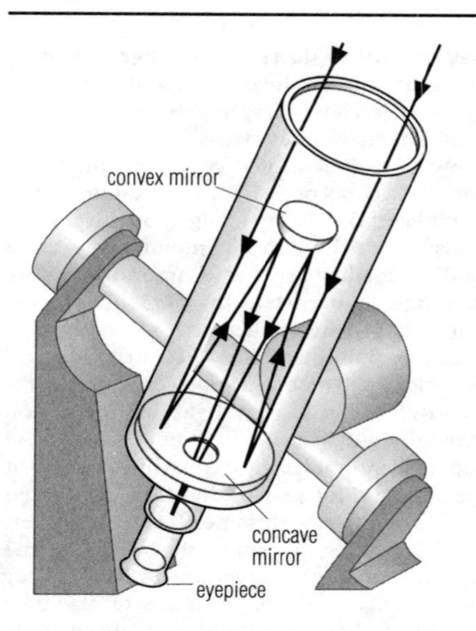

Cassegrain *In a Cassegrain reflecting telescope, a hole in the centre of the concave main mirror allows light reflected by the convex secondary mirror to reach the eyepiece (or a camera)*

modern reflecting telescopes and sometimes used in large refraction telescopes.

Nothing is known for certain about Cassegrain's life – not even his first name. Believed to have been a professor at the College of Chartres, he is variously credited with having been an astronomer, a physician, and a sculptor at the court of Louis XIV.

Cassegrain's telescope, an improvement on a design by English scientist Isaac Newton, used an auxiliary convex mirror to reflect the image through a hole in the objective – that is, through the end of the telescope itself. One intention behind this innovation was further to increase the angular magnification. A century later it was noted that this also partly cancelled out the spherical aberration. Cassegrain also submitted a scientific paper concerning the megaphone to the Academy of Sciences in Paris.

Cassini Giovanni Domenico 1625–1712. Italian-born French astronomer who discovered four moons of Saturn and the gap in the rings of Saturn now called the ***Cassini division***.

Cassini was born near Nice (then in Italy). Having assisted two astronomers at an observatory near Bologna, Cassini was made professor of astronomy at the University of Bologna at the age of 25. In 1669 he departed for France at the invitation of King Louis XIV, to construct and run the Paris Observatory. When he went blind 1710, his son ***Jacques Cassini*** (1677–1756) succeeded him.

Cassini refused to accept the Copernican cosmological model and rejected the concept of a finite speed of light, although its proof was demonstrated by Danish astronomer Ole Römer using Cassini's own data.

During 1664–67 Cassini determined the rotation periods of Mars, Jupiter, and Venus. In 1675 he distinguished two zones within what was thought to be the single ring around Saturn. Cassini correctly suggested that the rings were composed of myriads of tiny satellites.

In the 1670s Cassini made many observations of details on the lunar surface. He also took advantage of a good opposition of Mars 1672 to determine the distance between the Earth and that planet. He arranged for Jean Richer (1630–1696) to make measurements from his base in Cayenne, on the NE coast of South America, while Cassini made simultaneous measurements in Paris, which permitted them to make a triangulation of Mars. From the result, Cassini was able to deduce many other astronomical distances.

Cauchy Augustin Louis, Baron de 1789–1857. French mathematician who employed rigorous methods of analysis. His prolific output included work on complex functions, determinants, and probability, and on the convergence of infinite series. In calculus, he refined the concepts of the limit and the definite integral.

Cauchy has the credit for 16 fundamental concepts and theorems in mathematics and mathematical physics, more than any other mathematician. His work provided a basis for the calculus. He provided the first comprehensive theory of complex numbers, which contributed to the development of mathematical physics and, in particular, aeronautics.

Cauchy was born in Paris, studied engineering there and worked for a time in construction, then became a professor at the Ecole Polytechnique 1816 and later at the Collège de France. In 1830 Charles X was overthrown, and, refusing to take the new oath of allegiance, Cauchy went into exile. He became professor of mathematical physics at the University of Turin, and from 1833 he was tutor to Charles X's son in Prague, returning to Paris 1838 to resume his professorship at the Ecole Polytechnique. In

1843 he published a defence of academic freedom of thought, which was instrumental in the abolition of the oath of allegiance soon after the fall of Louis Philippe 1848. From 1848 to 1852 Cauchy was a professor at the Sorbonne.

In 1805 Cauchy provided a simple solution to the problem of Apollonius, namely to describe a circle touching three given circles, and in 1816 he published a paper on wave modulation. In mechanics he substituted the concept of the continuity of geometrical displacements for the principle of the continuity of matter and in astronomy he described the motion of the asteroid Pallas.

His main work was published in three treatises: *Cours d'analyse de l'Ecole Polytechnique* 1821, *Résumé des leçons sur le calcul infinitésimal* 1823, and *Leçons sur les applications de calcul infinitésimal à la géométrie* 1826–28.

Cavendish Henry 1731–1810. English physicist and chemist. He discovered hydrogen (which he called 'inflammable air') 1766, and determined the compositions of water and of nitric acid. The Cavendish experiment 1798 enabled him to discover the mass and density of the Earth. The Cavendish experiment was the measurement of the gravitational attraction between lead and gold spheres.

Cavendish demonstrated 1784 that water is produced when hydrogen burns in air, thus proving that water is a compound and not an element. He also worked on the production of heat and determined the freezing points for many materials, including mercury.

Cavendish was born in Nice, France, and left Cambridge without a degree. He spent the rest of his life in seclusion in London. Most of his work, especially his experiments with electricity, was unknown for 100 years or more. He believed electricity to be an elastic fluid.

In the late 1760s Cavendish began experimenting with 'facticious airs' (gases that can be produced by the chemical treatment of solids or liquids). He studied 'fixed air' (carbon dioxide) produced by mixing acids and bases; 'inflammable air' (hydrogen) generated by the action of acids on metals; and the 'airs' produced during decay and fermentation. He measured the specific gravities of hydrogen and carbon dioxide, comparing them with that of 'common' (atmospheric) air. In 1783 he found that the composition of the atmosphere is the same in different locations and at different times.

Cayley Arthur 1821–1895. English mathematician who developed matrix algebra, used by Werner Heisenberg in his elucidation of quantum mechanics. He also developed the study of *n*-dimensional geometry, introducing the concept of the 'absolute', and formulated the theory of algebraic invariants.

Cayley was born in Richmond, Surrey, and studied mathematics at Cambridge before becoming a barrister. In 1863 he became professor of pure mathematics at Cambridge.

Cayley published about 900 mathematical notes and papers on nearly every pure mathematical subject, as well as on theoretical dynamics and astronomy. Some 300 of these papers were published during his 14 years at the Bar, and for part of that time he worked in collaboration with James Joseph Sylvester, another lawyer. Together they founded the algebraic theory of invariants 1843. Cayley clarified many of the theorems of algebraic geometry that had previously been only hinted at, and he was one of the first to realize how many different areas of mathematics were drawn together by the theory of groups.

Cayley George 1773–1857. English aviation pioneer, inventor of the first piloted glider 1853, and the caterpillar tractor. Succeeded as 6th baronet 1792.

Cayley was born in Brompton, Yorkshire. In later life he helped to found London's Regent Street Polytechnic.

Cayley spent much of his life experimenting with kites and gliders, but also worked on other aspects of flight, including helicopters, streamlining, parachutes, and the idea of biplanes and triplanes. The 1853 glider was a triplane, in which his reluctant coach driver travelled 275 m/900 ft across a small valley – the first recorded flight by a person in an aircraft. Although delighted with the result, Cayley realized that control of flight could not be achieved until a lightweight engine was developed to give the thrust and lift required.

Cech Thomas 1947– . US biochemist who discovered the catalytic activity of RNA whilst working on introns, regions of genes that do not contain genetic information. He was jointly awarded the Nobel Prize for Chemistry in 1989 with Sidney Altman.

It was thought that DNA served as templates for messenger RNA (mRNA) molecules, and that these mRNA molecules were themselves direct templates for protein synthesis. However, before the protein could be translated from the messenger RNA, the introns (non-coding regions) had to be spliced out. Cech believed that an enzyme must be involved in this splicing.

He took nuclei from the ciliate *Tetrahymena thermophilia* and mixed them together with some unspliced RNA. As he expected the RNA introns were snipped away, but surprisingly none of the nuclear enzymes had been used. He deduced that the protein-free precursor ribosomal RNA mediates its own cleavage and splicing. Cech called this and other catalytic forms of RNA ribozymes.

Cech was born in Chicago and educated at the universities of California and Chicago. He worked briefly at the Massachusetts Institute of Technology before moving to become a professor at the University of Colarado, Boulder 1983.

Cech also postulated that RNA splicing was at the forefront of creation and that it initiated evolution. His work continued, encompassing the action of telomerase enzymes, which use an inbuilt template to add short repeat sections of DNA to chromosomal DNA.

Celestial Police group of astronomers in Germany 1800–15, who set out to discover a supposed missing planet thought to be orbiting the Sun between Mars and Jupiter, a region now known to be occupied by types of asteroid. Although they did not discover the first asteroid (found 1801), they discovered the second, Pallas (1802), third, Juno (1804), and fourth, Vesta (1807).

The group was called together by the German Johann Schroter in 1800 at his observatory in Lilienthal, Germany. They included Hungarian Franz Xaver von Zach, German Heinrich Olbers and Karl Harding.

They decided to split the zodiac into several zones, each of which would be searched by more than one astronomer. The group was disbanded in 1815.

Celsius Anders 1701–1744. Swedish astronomer, physicist, and mathematician who introduced the Celsius scale of temperature. His other scientifc works include a paper on accurately determining the shape and size of the Earth, some of the first attempts to gauge the magnitude of the stars in the constellation Aries, and a study of the falling water level of the Baltic Sea.

Celsius was born in Uppsala, where he succeeded his father as professor of astronomy in 1730. He travelled extensively in Europe, visiting astronomers and observatories in particular. In Uppsala he built Sweden's first observatory.

On his travels Celsius observed the aurora borealis; he published some of the first scientific documents on the phenomenon 1733. An expedition to Lapland confirmed the theory propounded by Isaac Newton that the Earth is flattened at the poles.

In 1742 Celsius presented a proposal to the Swedish Academy of Sciences that all scientific measurements of temperature should be made on a fixed scale based on two invariable (generally speaking) and naturally occurring points. His scale defined 0° as the temperature at which water boils, and 100° as that at which water freezes. This scale, in an inverted form devised eight years later by his pupil Martin Strömer, has since been used in almost all scientific work.

Cesalpino Andrea 1519–1603. Italian botanist who showed that plants could be and should be classified by their anatomy and structure. In *De plantis* 1583, Cesalpino offered the first remotely modern classification of plants. Before this plants were classed by their location – for example marsh plants, moorland plants, and even foreign plants.

Cesaro Ernesto 1859–1906. Italian mathematician who made important contributions to intrinsic geometry. He first defined ***Cesaro's curves*** 1896.

Cesaro was born in Naples and studied at the Ecole des Mines in Liège, Belgium, and at the University of Rome. He was professor of higher algebra at the University of Palermo 1886–91, and then professor of mathematical analysis at Naples.

In *Lezione di geometrica intrinsica/Lessons in Intrinsic Geometry* 1896, Cesaro simplified the analytical expression and made it independent of extrinsic coordinate systems. He stressed the intrinsic qualities of the objects. He also described the curves which now bear his name. The monograph also deals with the theory of surfaces and multidimensional spaces in general.

Cesaro's other work covered topics ranging from elementary geometrical principles to the application of mathematical analysis; from the theory of numbers to symbolic algebra; and from the theory of probability to differential geometry. He also made notable interpretations of James Clerk Maxwell's work in theoretical physics.

Chadwick James 1891–1974. English physicist. In 1932 he discovered the particle in the nucleus of an atom that became known as the ***neutron*** because it has no electric charge. Nobel Prize for Physics 1935. Knighted 1945.

Chadwick established the equivalence of atomic number and atomic charge. During World War II, he was closely involved with the atomic bomb, and from 1943 he led the British team working on the Manhattan Project in the USA.

Chadwick was born in Cheshire and studied at Manchester under Ernest Rutherford, investigating the emission of gamma rays from radioactive materials. In 1913 he went to Berlin to work with German physicist Hans Geiger, inventor of the Geiger counter. There he was interned as an enemy alien during World War I. After the war, he joined Rutherford at Cambridge. Chadwick was professor of physics at Liverpool 1935–48 and master of Gonville and Caius College, Cambridge, 1948–58.

Chadwick discovered the continuous nature of the energy spectrum and investigated beta particles emitted during radioactive decay. With Rutherford, he produced artificial disintegration of some of the lighter elements by alpha-particle bombardment. He returned to this at Liverpool, where he ordered the building of a cyclotron and developed a research school in nuclear physics.

Chain Ernst Boris 1906–1979. German-born British biochemist. After the discovery of penicillin by Alexander Fleming, Chain worked to isolate and purify it. For this work, he shared the 1945 Nobel Prize for Medicine with Fleming and Howard Florey. Chain also discovered penicillinase, an enzyme that destroys penicillin. Knighted 1969.

Born and educated in Berlin, Chain fled to Britain from the Nazis 1933. He worked at Cambridge University 1933–35, and then with Florey at Oxford. In 1949 Chain was invited to the Istituto Superiore di Sanità in Rome; he stayed there as professor until 1961, when he returned to the UK as professor of biochemistry at Imperial College, London.

At Oxford, Chain initially investigated the observation first made by Fleming 1924 that tears, nasal secretion, and egg white destroy bacteria. Chain showed that these substances contain an enzyme, lysozyme, which digests the outer cell wall of bacteria. In 1937 Chain

Chain *Nobel laureate biochemist Ernst Chain. Chain won the prize for Physiology or Medicine 1945 after purifying penicillin in collaboration with Howard Florey at the William Dunn School of Pathology in Oxford. They shared the prize with Alexander Fleming, who identified the drug.*

found another observation of Fleming's, that the mould *Penicillium notatum* inhibits bacterial growth. In collaboration with Florey, Chain isolated and identified the antibacterial factor in the mould.

Chain then elucidated the chemical structure of crystalline penicillin, finding that there are four types, each differing in its relative elemental constituents.

Chain also studied snake venoms and found that the neurotoxic effect of these venoms is caused by their destroying an essential intracellular respiratory coenzyme.

Challis James 1803–1882. English astronomer who failed to take the advice of John Couch Adams on where to search for the planet Neptune, leaving its discovery to French and German astronomers.

Challis was born in Braintree, Essex, and educated at Cambridge. An Anglican cleric, he was rector at Papworth Everard, Cambridgeshire, 1830–52, as well as professor of astronomy and director of the observatory at Cambridge from 1836.

In 1844, astronomer and mathematician Adams obtained via Challis data from the Greenwich Observatory regarding the known deviations in the orbit of the planet Uranus. These indicated the gravitational influence of a planet even farther out. In Sept 1845 Adams supplied Challis and Astronomer Royal George Airy with an estimated orbital path for the unknown planet and a prediction for its likely position on 1 Oct 1845. But Challis did not take the calculations seriously and Airy did not even see them until the following year.

By that time, the new planet had been discovered from the Berlin Observatory. Challis admitted that if he had indeed conducted a search at Adams's predicted position for 1 Oct 1845 he would have been within 2° of the planet's actual position and would almost certainly have spotted it.

Chamberlain Owen 1920– . US physicist whose graduate studies were interrupted by work on the Manhattan Project at Los Alamos. After World War II, working with Italian physicist Emilio Segrè, he discovered the existence of the antiproton. Both scientists were awarded the Nobel Prize for Physics 1959.

Chamberlin Thomas Chrowder 1843–1928. US geophysicist who asserted that the Earth was far older than then believed. He developed the planetesimal hypothesis for the origin of the Earth and other planetary bodies – that they had been formed gradually by accretion of particles.

Chamberlin, *US geophysicist Thomas Chamberlin is best known for his work on the geology of the Solar System and the development of the planetesimal hypothesis. This states that, during their early formation, planetary bodies attract material through gravitational contraction, and not, as previously thought, that they cool steadily from a large molten mass.*

Chamberlin was born in Illinois and brought up in Wisconsin. Partly self-taught in science, he joined the Wisconsin Geological Survey 1873, and rose to become its chief geologist, publishing *Geology of Wisconsin* 1877–83. He went on to work for the US Geological Survey before becoming professor at Chicago 1892–1918.

Irish physicist Lord Kelvin had postulated that the Earth was less than 100 million years old, basing his views on the assumption, derived from the nebular hypothesis, that the Earth had steadily cooled from a molten mass. Chamberlin countered this with the planetesimal hypothesis, and believed geological evidence in any case suggested the Earth to be older than Kelvin had estimated.

Chamisso Adelbert von 1781–1838. French-born German biologist and writer, best remembered for his fairy tale *Peter Schlemihl's Remarkable Story*. One of the most prominent German Romanticists, he published books on biology, a novel, and works of poetry. Some of his poems were set to music by Schumann. He was also a keen zoologist and was the first to report the peculiar sexual cycle of some forms of molluscs.

Chamisso was born in Champagne, France, but fled with his family when he was nine to Berlin, Germany to escape the horrors of the French Revolution. He adopted German as his first language and, following a career in the Prussian army, sailed around the world on the Russian vessel *Rurik*. On this voyage, which lasted from 1815 to 1818, he kept a detailed diary of events and catalogued many new species of plants in his capacity as ship's botanist. Upon his return he became the curator of the botanical gardens in Berlin.

His book *Peter Schlemihl's Remarkable Story* 1814, is about a man who sells his shadow to the Devil and roams the world trying to find peace of mind. The story is thought to represent his own feelings about his life as an individual without a nationality.

Chandrasekhar Subrahmanyan 1910–1995. Indian-born US astrophysicist who made pioneering studies of the structure and evolution of stars. The ***Chandrasekhar limit*** is the maximum mass of a white dwarf before it turns into a neutron star. Nobel Prize for Physics 1983.

Chandrasekhar has also investigated the transfer of energy in stellar atmospheres by radiation and convection, and the polarization of light emitted from particular stars.

Chandrasekhar was born in Lahore (now in Pakistan) and studied in Madras, India, and at Cambridge, UK, before joining the staff of the University of Chicago, Illinois, 1936. He became a professor there 1952.

A certain modesty toward understanding nature is a precondition to the continued pursuit of science.

Subrahmanyan Chandrasekhar

Interview 1984

The evolution of white dwarfs is explained in his *Introduction to the Study of Stellar Structure* 1939. He calculated that stellar masses below 1.44 times that of the Sun would form stable white dwarfs, but those above this limit would not evolve into white dwarfs; the limit is now believed to be about 1.2 solar masses. Stars with masses above the Chandrasekhar limit are likely to explode into supernovae; the mass remaining after the explosion may form a white dwarf if the conditions are suitable, but is more likely to form a neutron star.

Charcot Jean-Martin 1825–1893. French neurologist who studied hysteria, sclerosis, locomotor ataxia, and senile diseases. Among his pupils was the founder of psychiatry, Sigmund Freud.

One of the most influential neurologists of his day, Charcot exhibited hysterical women at weekly public lectures, which became fashionable events. He was also fascinated by the relations between hysteria and hypnotic phenomena.

Charcot was born and educated in Paris and worked at the Salpétrière hospital there. He was convinced that all psychiatric conditions followed natural laws, and studied the way certain mental illnesses correlate with physical changes in the brain. He published the results in a series of memoirs.

Chardonnet (Louis-Marie) Hilaire Bernigaud, Comte de 1839–1924. French chemist who developed artificial silk 1883, the first artificial fibre. He also worked on cellulose nitrate.

Chardonnet was born in Besançon, Franche-Comté. He trained first as a civil engineer in Paris, and then went to work under Louis Pasteur, who was studying diseases in silkworms. This inspired Chardonnet to seek an artificial replacement for silk. He opened a factory in Besançon 1889 and another in Hungary 1904.

Chardonnet's starting point was mulberry leaves, the food of silkworms; he turned them into a cellulose pulp with nitric and sulphuric acids and stretched it into fibres. The original fibre was highly flammable, but by 1889 he had eliminated this and developed rayon.

Chargaff Erwin 1905– . Czech-born US biochemist, best known for his work on the base composition of deoxyribonucleic acid (DNA). In 1950 he demonstrated a simple mathematical relationship between the proportions of nitrogenous bases in DNA. He worked in many fields of biochemistry, publishing on topics as diverse as how the body metabolizes fats to how blood coagulates.

Science is wonderfully equipped to answer the question 'How?' but it gets terribly confused when you ask the question 'Why?'

Erwin Chargaff

Columbia Forum 1969

His most influential work was his calculations of the chemical composition of DNA. In 1950, Chargaff published a simple mathematical relationship between the proportions of nitrogenous bases (molecules that connect the two strands of the DNA molecule like the rungs of a twisted ladder). He showed that the number of adenine molecules equals the number of thymidine molecules, and that the number of guanine molecules equals the number of cytosine molecules.

Chargaff was born in Czernowitz, which was then part of Austria but is now part of the Ukraine. He studied in Vienna and at Yale University in the USA. When Watson and Crick published their seminal work on DNA structure a few years later 1953, they claimed to have been unaware of Chargaff's work.

Although Chargaff made an important contribution to the early study of genetics, he went on to become a stern critic of molecular biology stating that 'by its claim to be able to explain everything, it actually impedes the flow of scientific explanation'.

What counts … in science is to be not so much the first as the last.

Erwin Chargaff

Science 1971

Charles Jacques Alexandre César 1746–1823. French physicist who studied gases and made the first ascent in a hydrogen-filled balloon 1783. His work on the expansion of gases led to the formulation of Charles's law.

Hearing of the hot-air balloons of the Montgolfier brothers, Charles and his brothers began experimenting with hydrogen balloons and made their ascent only ten days after the Montgolfiers' first flight. In later flights Charles ascended to an altitude of 3,000 m/10,000 ft.

Charles was born in Loiret and worked as a clerk at the Ministry of Finance in Paris. Stimulated by American scientist Benjamin Franklin's experiments with lightning and electricity, he constructed a range of apparatus which he demonstrated at public lectures. He went on to become professor of physics at the Paris Conservatoire des Arts et Métiers.

In about 1787 Charles experimented with hydrogen, oxygen, and nitrogen, and found that a gas expands by $\frac{1}{273}$ of its volume at 0°C for each degree rise in temperature. He communicated this to physical chemist Joseph Gay-Lussac, who repeated the experiments and published the result. Charles's law is therefore sometimes known as Gay-Lussac's law.

Charnley John 1911–1982. English orthopaedic surgeon who applied engineering principles to the practice of orthopaedics. He worked on degenerative hip disease and developed a new technique, the total hip replacement, or arthroplasty. He also successfully pioneered arthrodeses (fusing joint surfaces) for the knee and hip. Knighted 1977.

Charnley was born in Bury, Lancashire, and studied at Manchester University. He worked at Manchester Royal Infirmary from 1947 until the mid-1960s, and then became director of the Centre of Hip Surgery at Wrightington Hospital, Lancashire, turning it into the primary unit for hip replacement in the world.

By research and experiment, Charnley found in 1962 a low-friction, high-density polythene suitable for artificial hip joints. He also pioneered the use of methyl methacrylate cement for holding the metal prosthesis or implant to the shaft of the femur.

For the treatment of rheumatoid arthritis Charnley devised a system for surgically fusing joint surfaces (arthrodesis) to immobilize the knee joint, using an external compression device which bears his name. This leaves the joint immobile but pain-free.

Throughout his career Charnley developed a series of highly practical and successful surgical instruments. In his fight against postoperative infection he used air 'tents' which allowed the surgeon and the wound to be kept in a sterile atmosphere during the operation.

Charpak Georges 1924– . Polish-born French physicist who invented and developed particle detectors, in particular the multiwire proportional chamber (consisting of an array of wires, each at a high voltage, that produce an electrical signal as a particle passes through). Nobel Prize for Physics 1992.

Modern elementary particle physicists look deep inside matter using accelerators as microscopes. These accelerators produce showers of particles, like the sparks from fireworks, which can be discharged a hundred million times each second. Inside the particle shower there is information about the smallest constituents of matter – the elementary particles – and the forces with which they interact. However, it takes a special kind of detector to extract this information. A suitable detector must be able to react quickly, track the paths of particles and send its observations direct to a computer. These requirements are met by the multiwire proportional chamber which Charpak invented 1968.

In a multiwire chamber, thousands of wires cross the space traversed by the particles. When a particle passes through the chamber, electrical signals are produced in the wires. A computer connected to the wires then maps the path of each particle and displays it on a monitor screen. This type of detector is used, in some form or other, in almost all elementary particle physics experiment today. It is also used in medicine and industry.

Charpak was born at Dabrovica, Poland and naturalized French 1946. He was educated in Paris and received his doctorate 1954 for work in experimental nuclear physics. From 1959 he worked at CERN, the European particle physics laboratory near Geneva, Switzerland.

Chase Mary Agnes, (born Meara) 1869–1963. US botanist and suffragist who made outstanding contributions to the study of grasses. During the course of several research expeditions she collected many plants previously unknown to science, and her work provided much important information about naturally occuring cereals and other food crops.

Meara was born in Iroquois County, Illinois, and was self-educated. From 1903 she worked in Washington DC for the US Department of Agriculture Bureau of Plant Industry and Exploration, and became the principal scientist for agrostology (study of grasses). She was politically active in various reform movements, especially those for female suffrage, and on this account was jailed and forcibly fed during World War I.

Chase was particularly responsible for work in modernizing and extending the national grass herbarium. She travelled widely, collecting plants from several regions of North and South America, and also visiting European research institutes and herbaria during the 1920s. Altogether she collected more than 12,000 plants for the herbarium.

Chase's publications include the authoritative *Manual of the Grasses of the United States* 1950.

Châtelet (Gabrielle) Emilie de Breteuil, Marquise du 1706–1749. French scientific writer and translator into French of Isaac Newton's *Principia*.

Her marriage to the Marquis du Châtelet 1725 gave her the leisure to study physics and mathematics. She met controversial French writer Voltaire 1733, and settled with him at her husband's estate at Cirey, in the Duchy of Lorraine. Her study of Newton, with whom she collaborated on various scientific works, influenced Voltaire's work. She independently produced the first (and only) French translation of Newton's *Principia* (published posthumously in 1759).

Cherenkov Pavel Alexeevich 1904–1990. Soviet physicist. In 1934 he discovered ***Cherenkov radiation***; this occurs as a bluish light when charged atomic particles pass through water or other media at a speed in excess of that of light. He shared a Nobel prize 1958 with his colleagues Ilya Frank and Igor Tamm for work resulting in a cosmic-ray counter.

Cherenkov discovered that this effect was independent of any medium and depended for its production on the passage of high velocity electrons.

Cherwell Frederick Alexander Lindemann, 1st Viscount Cherwell 1886–1957. British physicist, scientific adviser to the government during World War II. He served as director of the Clarendon Laboratory, Oxford, 1919–56, and oversaw its transformation into a major research institute. Baron 1941, Viscount 1956.

Lindemann was born in Baden-Baden, Germany, and studied at Berlin, leaving on the outbreak of World War I 1914 to become director of the Royal Air Force Physical Laboratory, where he was concerned with aircraft stability. From 1919 he was professor of experimental philosophy at Oxford. He was also a member of the government as paymaster-general 1942–45 and 1951–53, helping to direct scientific research during World War II and to create the Atomic Energy Authority afterwards.

The perfectly designed machine is one in which all its working parts wear out simultaneously. I am that machine.

Frederick Lindemann, 1st Viscount Cherwell

Letter from Lindemann to Lord De L'Isle 1957, shortly before he died.

Together with German chemist Hermann Nernst, Lindemann made an advance in quantum theory in 1911 by constructing a special calorimeter and measuring specific heats at very low temperatures. They confirmed the specific heat equation proposed by Albert Einstein in 1907, in which he used the quantum theory to predict that the specific heats of solids would become zero at absolute zero. Lindemann also derived a formula that relates the melting point of a crystalline solid to the amplitude of vibration of its atoms.

Chevreul Michel-Eugène 1786–1889. French chemist who studied the composition of fats and identified a number of fatty acids, including 'margaric acid', which became the basis of margarine. He also studied sugars and dyes.

Chevreul was born in Angers, Maine-et-Loire, and studied in Paris. From 1824 he was director of dyeing at the Gobelins tapestry factory. He became professor of chemistry at the Museum of Natural History, and its director 1864.

By treating soaps with hydrochloric acid, Chevreul obtained and identified various fatty acids, including stearic, palmitic, oleic, caproic, and valeric acids. He realized that the soapmaking process is the treatment of a glyceryl ester of fatty acids with an alkali to form fatty acid salts (soap) and glycerol.

In 1825 Chevreul and Joseph Gay-Lussac patented a process for making candles from stearin (crude stearic acid), providing a cleaner and less odorous alternative to tallow candles. Chevreul determined the purity of fatty acids by measuring their melting points, and constancy of melting point soon became a criterion of purity throughout preparative and analytical organic chemistry.

At the Gobelins dyeworks he made various chemical discoveries, and his interest in the creation of the illusion of continuous colour gradation by using massed small monochromatic dots (as in an embroidery or tapestry) later influenced the Pointillist and Impressionist painters.

Cheyne John 1777–1836. Scottish physician who, with William Stokes, gave his name to ***Cheyne–Stokes breathing***, or periodic respiration.

Cheyne was born in Leith and apprenticed to his physician father at the age of 13. In 1809 he settled in Dublin. He took the first professorial chair in medicine at the Royal College of Surgeons of Ireland 1813.

In 1818 Cheyne described the periodic respiration that occurs in patients with intracranial disease or cardiac disease. His paper described the breathing that would cease entirely for a quarter of a minute or more, then would become perceptible and increase by degrees to quick, heaving breaths which gradually subside again.

Child Charles Manning 1869–1954. US zoologist who developed a theory of how the various cells and tissues in organisms are organized – by a gradation in the rate of physiological processes leading to relationships of dominance and subordination. Although not now thought to be correct, it was an important early contribution to the problem of functional organization within living organisms.

Child was born in Ypsilanti, Michigan, and studied zoology at the Wesleyan University in Connecticut and at Leipzig, Germany. He spent his academic career at the University of Chicago, becoming professor 1916.

In 1900, Child began a series of experiments on regeneration in coelenterates and flatworms. In 1910 he perceived that there is a gradation in the rate of physiological processes along the longitudinal axis of organisms. According to his gradient theory, developed 1911, each part of an organism dominates the region behind and is dominated by that in front. In general, the region of the highest rate of activity in eggs, embryos, and other reproductive parts becomes the head of the larval form; in plants it becomes the growing tip of the shoot or primary root. Regeneration follows the same principle.

In 1915 Child demonstrated that the parts of an organism that have the highest metabolic rates are most susceptible to poisonous substances, but also have the greatest powers of recovery after damage.

Chladni *By combining music with experimental physics, Ernst Chladni helped establish the science of acoustics. His most famous acoustic experiment used geometrically shaped metal or glass plates covered with a fine layer of sand. When the plates were struck, or even played with a bow, the sand moved to form patterns known as 'Chladni figures'.*

Chladni Ernst Florens Friedrich 1756–1827. German physicist, a pioneer in the field of acoustics. He developed an experimental technique whereby sand is vibrated on a metal plate and settles into regular and symmetric patterns (***Chladni's figures***), indicating the nodes of the vibration's wave pattern.

Chladni was born in Wittenberg, Saxony, and studied law at Leipzig. His interest in sound stemmed from his love of music.

In 1786 he began studying sound waves and worked out mathematical formulas that describe their transmission. In 1809 he demonstrated Chladni's figures to a group of scientists in Paris.

He also measured the velocity of sound in various gases by measuring the change in pitch of an organ pipe filled with a gas other than air (the pitch, or sound frequency, varies depending on the molecular composition of the gas). He invented various musical instruments, including one he called the euphonium, composed of rods of glass and metal that were made to vibrate by being rubbed with a moistened finger; he demonstrated it at lectures throughout Europe.

Christoffel Elwin Bruno 1829–1900. German mathematician who made a fundamental contribution to the differential geometry of surfaces, carried out some of the first investigations that later resulted in the theory of shock waves, and introduced what are now known as the ***Christoffel symbols*** into the theory of invariants.

Christoffel was born in Montjoie (now Monschau), near Aachen, and studied at Berlin. His first professorship was at the Polytechnicum in Zürich, Switzerland, 1862–69; the second at Berlin; and from 1872 he was at the University of Strasbourg.

In *Allgemeine Theorie der geodätischen Dreiecke* 1868, he presented a trigonometry of triangles formed by geodesics on an arbitrary surface.

Christoffel's paper 'Über die Transformation der homogen Differential-ausdrücke zweiten Grades' 1869 introduced the symbols that later became known as Christoffel symbols of the first and second order. The series of other symbols of more than three indices, including the four-index symbols already introduced by Bernhard Riemann, is now known as the Riemann–Christoffel symbols.

Christoffel formulated a theorem that also bears his name, concerning the reduction of a quadrilateral form.

In 1877, Christoffel published a paper on the propagation of plane waves in media with a surface discontinuity, and thus made an early contribution to shock-wave theory.

In *Vollständige Theorie der Riemannschen θ-Function* (published posthumously), Christoffel gave an independent interpretation of Riemann's work on surface geometry.

Chrysler Walter Percy 1875–1940. US industrialist. After World War I, he became president of the independent Maxwell Motor Company and went on to found the Chrysler Corporation 1925. By 1928 he had acquired Dodge and Plymouth, making Chrysler Corporation one of the largest US motor-vehicle manufacturers.

Chrysler was born in Wamego, Kansas. He worked first as a railway machinist, rising through the ranks to become manager of the American Locomotive Company in Pittsburgh 1912. Shifting to the car industry, he was hired by General Motors and was appointed president of the Buick division 1916.

Church Alonzo 1903– . US mathematician who in 1936 published the first precise definition of a calculable function, and so contributed enormously to the systematic development of the theory of algorithms.

Church was educated at Princeton and remained there for 40 years, becoming professor of mathematics and philosophy. In 1967 he moved to the University of California in Los Angeles.

The solving of algorithmic problems involves the construction of an algorithm capable of solving a given set with respect to some other set, and if such an algorithm cannot be constructed, it signifies that the problem is unsolvable. Theorems establishing the unsolvability of such problems are among the most important in the theory of algorithms, and Church's theorem was the first of this kind. From English mathematician Alan Turing's thesis, Church proved that there were no algorithms for a class of quite elementary arithmetical questions. He also established the unsolvability of the solution problem for the set of all true propositions of the logic of prediction.

Cierva Juan de la 1895–1936. Spanish engineer. In trying to produce an aircraft that would not stall and could fly slowly, he invented the autogiro 1923, the forerunner of the helicopter but differing from it in having unpowered rotors that revolve freely.

Cierva was born in Murcia and studied engineering in Madrid. He was twice elected to the Cortes (parliament), in 1919 and 1922. In 1925 he founded the Cierva Autogyro Company in the UK. Test-flying his own aircraft, he was killed in a crash in Croydon, S London.

Cierva's first aircraft was a biplane bomber which stalled and crashed when tested 1919. It occurred to him that a machine with a rotating wing would be less vulnerable to engine failure, and he developed the autogiro. Commercial production began 1925, and in 1928 he flew one of his company's aircraft across the English Channel. He then flew one all the way to Spain.

Clark Wilfrid Edward Le Gros 1895–1971. English anatomist and surgeon whose research made a major contribution to the understanding of the structural anatomy of the brain. By emphasizing the importance of relating structure to function, he had a profound influence on the teaching of anatomy. Knighted 1955.

Clark was born in Hemel Hempstead, Hertfordshire. He trained at St Thomas's Hospital, London, and served in the Royal Army Medical Corps during World War I. As principal medical officer in Sarawak, Borneo, he began research into the evolution of primitive primates. Clark returned to the UK 1923 tattooed on the shoulders with the insignia of the Sea Dyaks as a mark of their esteem. He became professor of anatomy first at St Bartholomew's Hospital, London, then at St Thomas's, and 1934–62 at Oxford, where he created a new department of anatomy.

Clark's research was directed mainly towards the brain, and the relationship of the thalamus to the cerebral cortex. He also carried out studies of the hypothalamus. His work on the sensory (largely visual) projections of the brain remains the basis of contemporary knowledge of this aspect of neuroanatomy.

His publications include *Morphological Aspects of the Hypothalamus* 1938, *The Tissues of the Body* 1939, and *History of the Primates* 1949.

Clarke Arthur C(harles) 1917– . English science-fiction and nonfiction writer who originated the plan for a system of communications satellites in geostationary orbit 1945. His works include the short story 'The Sentinel' 1951 (filmed 1968 by Stanley Kubrick as *2001: A Space Odyssey*), *Childhood's End* 1953, and *2010: Odyssey Two* 1982.

Any sufficiently advanced technology is indistinguishable from magic.

Arthur C Clarke

The Lost Worlds of 2001

Clarke was born in Minehead, Somerset, served in the Royal Air Force during World War II, and then studied at King's College, London. He became chair of the British Interplanetary Society 1946, the year his first story was published. In 1956 he moved to Sri Lanka. His popular-science books generally concern space exploration; his fiction is marked by an optimistic belief in the potential of science and technology.

Clarke Charles Baron 1832–1906. English botanist who collected and described vast numbers of botanical specimens indigenous to the Indian subcontinent, and donated 25,000 specimens to the herbarium at Kew.

Clarke was born in Hampshire and graduated from King's College School, London. He went up to Trinity College, Cambridge 1852, and, in 1856, became a fellow at Queen's College, lecturing in mathematics 1858–65. A man of many talents, he was also called to the Bar at Lincoln's Inn 1858. He left England 1865 to become an inspector of schools for the Civil Service of Bengal, India.

While in India, he continued his interest in alpine climbing and plant collecting. By 1868, he had collected 700 specimens, which were later to be lost at sea in a ship wreck. He went on collecting specimens undeterred, and upon his return to England 1877, he was able to donate 25,000 specimens to the herbarium at Kew. He returned to India 1883 until his retirement 1887, when he settled in Kew. He became a fellow of the Linnaean Society 1867 and was president 1894–96. He was also elected a Fellow of the Royal Society and a fellow of the Geological Society. He died 25 August following injuries he sustained in a bicycling accident.

Claude Georges 1870–1960. French industrial chemist, responsible for inventing neon signs. He discovered 1896 that acetylene, normally explosive, could be safely transported when dissolved in acetone. He later demonstrated that neon gas could be used to provide a bright red light in signs. These were displayed publicly for the first time at the Paris Motor Show 1910. As an old man, Claude spent the period 1945–49 in prison as a collaborator.

Clausius Rudolf Julius Emanuel 1822–1888. German physicist, one of the founders of the science of thermodynamics. In 1850 he enunciated its second law: heat cannot pass from a colder to a hotter body.

According to Clausius, there are two types of entropy: the conversion of heat into work, and the transfer of heat from high to low temperature. He concluded that entropy must inevitably increase in the universe.

Clausius was born in Pomerania (now in Poland) and educated at Berlin and Halle universities. He became professor of physics at Zürich 1855, returning to Germany 1867 for similar posts first at Würtzburg and then at Bonn.

Clausius also improved the mathematical treatment of the first law of thermodynamics, and studied the relationship between thermodynamics and kinetic theory. From 1857 onwards, he did important work on the kinetic theory of gases as well as on the theory of electrolysis.

Cleve Per Teodor 1840–1905. Swedish chemist and geologist who discovered the elements holmium and hulium 1879. He also demonstrated that the substance didymium, previously supposed to be an element, was in fact two elements, now known as neodymium and praseodymium.

Towards the end of his life he developed a method for identifying the age of glacial and postglacial deposits from the diatom fossils found in them.

Clifford William Kingdon 1845–1879. English mathematician and scientific philosopher who developed the theory of biquaternions and made advances in non-Euclidean geometry. ***Clifford parallels*** and ***Clifford surfaces*** are named after him.

Clifford was born in Exeter, Devon, and educated at Cambridge, where he spent his academic career until 1871, when he was appointed professor of applied mathematics at University College, London. From 1876 he lived in the Mediterranean region.

It was through a generalization of the quaternions (themselves a generalization of complex numbers) that Clifford derived his theory of biquaternions, associating them specifically with linear algebra. In this way representing motions in three-dimensional non-Euclidean space, and together with his suggestion in 1870 that matter itself was a kind of curvature of space, Clifford may be seen to have foreshadowed in some respects Albert Einstein's general theory of relativity.

Clifford continued his studies in non-Euclidean geometry, with reference particularly to Riemann surfaces, which he proved to be topologically equivalent to a box with holes in it. He further investigated the consequences of adjusting the definitions of parallelism, and found that parallels not in the same place can exist only in a Riemann space – and he proved that they do exist. He showed how three parallels define a ruled second-order surface that has a number of interesting properties.

Clusius alternative name for French botanist Charles de ⇨L'Ecluse.

Cockcroft John Douglas 1897–1967. British physicist. In 1932 he and Irish physicist Ernest Walton succeeded in splitting the nucleus of an atom for the first time. For this they were jointly awarded a Nobel prize 1951. Knighted 1948.

The voltage multiplier built by Cockcroft and Walton to accelerate protons was the first particle accelerator. They used it to bombard lithium, artificially transforming it into helium. The production of the helium nuclei was confirmed by observing their tracks in a cloud chamber. They then worked on the artificial disintegration of other elements, such as boron.

Cockcroft was born in Todmorden, W Yorkshire, and studied at Manchester and Cambridge, where he took up research work under Ernest Rutherford at the Cavendish Laboratory. Having been in charge of the construction of the first nuclear-power station in Canada during World War II, he returned to the UK to be director of Harwell Atomic Energy Research Establishment 1946–58, and in 1959 became first Master of Churchill College, Cambridge.

Cockerell Christopher Sydney 1910– . English engineer who invented the hovercraft in the 1950s.

Cockerell tested various ways of maintaining the air cushion. In 1957 he came up with the idea of a flexible skirt, which gave rise to much derision because nobody could believe that a piece of fabric could be made to support a large vessel. Knighted 1969.

Cockerell was born and educated at Cambridge. Employed by the Marconi Company 1935–50, he made a major contribution to aircraft radio navigation and communications. During this period he filed 36 patents.

Trained as a development engineer, he set himself the task of trying to make a boat go faster. First he experimented with air lubrication of the hull, but concluded that a major reduction in drag could be obtained only if the hull could be supported over the water by a really thick air cushion. In 1953 he began work on the hovercraft, carrying out his early experiments on Oulton Broad, Norfolk, and filing his first hovercraft patent 1955. Finally in 1958 he found commercial backing; the first full-size hovercraft was built and crossed

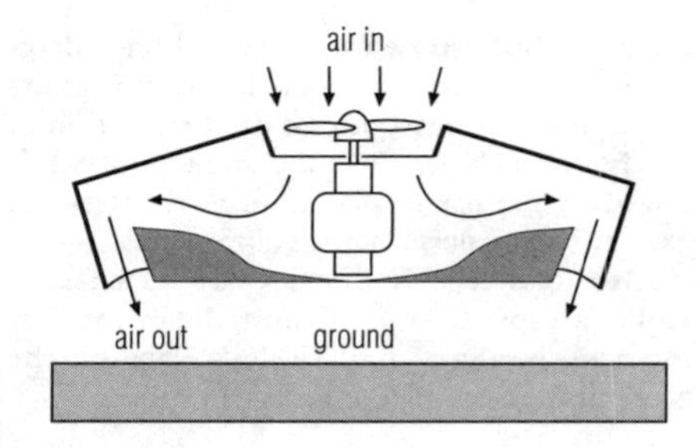

Cockerell *Cockerell's original hovercraft design made use of a peripheral jet of compressed air to achieve lift.*

the English Channel with the inventor on deck.

In the 1970s he began to interest himself in the generation of energy by wave power.

Cockerill William 1759–1832. English engineer who is generally regarded as the founder of the European textile-machinery industry. He was mainly active in Russia and Belgium.

Cockerill was born in Lancashire. His working career began with the building of spinning jennies and flying shuttles. In 1794 he went to St Petersburg, Russia, and enjoyed the patronage of Catherine II. Her successor, however, imprisoned Cockerill for failing to complete a contract within the given time. Eventually he escaped via Sweden to Belgium 1799, where he established himself as a manufacturer of textile machinery, first in Verviers and from 1807 in nearby Liège. There, together with his three sons William, Charles, and John, he made rotary carding machines, spinning frames, and looms for the French woollen industry.

Cody Samuel Franklin 1862–1913. US-born British aviation pioneer. He made his first powered flight on 16 Oct 1908 at Farnborough, England, in a machine of his own design. He was killed in a flying accident.

Born in Texas, USA, he took British nationality in 1909. He spent his early days with a cowboy stage and circus act, and made kites capable of lifting people.

Cohen Stanley 1922– . US biochemist who won the Nobel Prize for Physiology or Medicine 1986 jointly with Rita Levi-Montalcini for their work to isolate and characterize ***growth factors***, small proteins that regulate the growth of specific types of cells.

Cohen helped to purify and characterize nerve growth factor, a small protein produced in the male salivary gland that regulates the growth of small nerves and affects the development of the sensory and sympathetic neurons. He went on to discover another growth factor, called epidermal growth factor, that affects epithelial cell growth, tooth eruption, and eyelid opening. He then laboured to link epidermal growth factor to the regulation of embryonic growth. Subsequent studies by other scientists have shown that this growth factor also plays a crucial part in the exaggerated growth rate of some cancer cells.

Nature [is] that lovely lady to whom we owe polio, leprosy, smallpox, syphilis, tuberculosis, cancer.

Stanley Cohen

Attributed

Cohen was born in Brooklyn, New York City and studied at Brooklyn College and the Universities of Ohio and Michigan. While working at Vanderbilt University in the early 1950s, he became aware of the Levi-Montalcini discovery of the first growth factor, nerve growth factor.

Cohn Ferdinand Julius 1828–1898. German botanist and bacteriologist who showed that bacteria are not able to generate spontaneously but can develop from spores, and developed a system of classification for bacteria. Working primarily on algae, he also demonstrated that some primordial plant cells lack cellulose (complex carbohydrate that is the principle constituent of higher plants) cell walls.

In 1872, he developed a system for the classification of bacteria based upon their morphology, grouping them into four classes: Sphaerobacteria, Microbacteria, Desmobacteria, and Spirobacteria. In 1876, he demonstrated that *Bacillus subtilis* had thermoresistant (heat-resistant) spores and was not able to generate spontaneously.

Cohn was born in Breslau in Lower Silesia and is reported to have read by the age of two and to have known the basics of natural history by three. However, at the age of ten his hearing became impaired, slowing his amazing academic progress. In 1842, he went to the University of Breslau in order to study philosophy and soon became interested in botany, but because he was a Jew he was unable to obtain a degree. This prompted him to transfer to the University of Berlin, where he obtained a doctorate in botany 1847 at the age of nineteen. In 1849, he returned to Breslau, where he was made extraordinary professor of botany 1859 and ordinary professor of botany 1872.

Cohnheim Julius Friedrich 1839–1884. German pathologist who devised new ways of looking at specimens of human tissue under the microscope and worked out many of the early cellular events that occur in inflammation.

Cohnheim devised a variety of innovative methods for viewing slices of human tissue under the microscope to show that the essence of the inflammatory response is the migration of white blood cells to a wound. The inflammatory response results in an increased flow of blood to the site of injury and an increased permeability of the endothelium (the thin tissue that comprises the walls of capillaries). Both of these effects assist the migration of phagocytic (germ-engulfing) white blood cells from the blood to the interstitial fluid. Here the white blood cells can begin engulfing debris and any infecting microorganisms. Cohnheim also demonstrated that when inflammation subsides, the remaining pus consists largely of dead white blood cells.

Cohnheim was born in Demmin, Poland (now Germany). He studied in Berlin, where he graduated in medicine and studied for a year with the eminent cellular pathologist Rudolf Virchow.

Cohnheim held chairs in pathology at Breslau and Leipzig. His textbook of pathology *Lectures on General Pathology*, 1877–1880, runs to two volumes and was one of the first textbooks to combine pathological anatomy and morbid histology.

Colles Abraham 1773–1843. Irish surgeon who in 1814 observed and described a common fracture of the wrist, now named after him.

Colles was born in County Kilkenny and educated at Dublin and Edinburgh. He set up in practice in Dublin 1797 and began to teach anatomy and surgery. At the age of 29 he became president of the Royal College of Surgeons, where he was a professor 1804–36.

In 1814 the paper on Colles's fracture was published describing the fracture of the distal (carpel) end of the radius bone in the forearm. This causes deformity and swelling of the wrist, but can be easily treated once diagnosed. Colles advocated the use of tin splints to stabilize the wrist after closed reduction of the fracture. Nowadays the reduction is followed by the use of plaster of Paris casts but exactly the same principles apply.

Also named after him are Colles's fascia, Colles's space, the Colles ligament (of inguinal hernia), and Colles's law of the communication of (congenital) syphilis.

Colombo Matteo Realdo *c.* 1516–1559. Italian anatomist who discovered pulmonary circulation, the process of blood circulating from the heart to the lungs and back.

This showed that Galen's teachings were wrong, and was of help to William Harvey in his work on the heart and circulation. Colombo was a pupil of Andreas Vesalius and became his successor at the University of Padua.

Colt Samuel 1814–1862. US gunsmith who invented the revolver 1835 that bears his name. With its rotating breech which turned, locked, and unlocked by cocking the hammer, the Colt was superior to other revolving pistols, and it revolutionized military tactics.

Colt built a large factory in Hartford, Connecticut 1854. During the Crimean War 1853–56 he also manufactured arms in Pimlico, London. By 1855 he had the largest private armoury in the world. When the American Civil War broke out 1861, he supplied thousands of guns to the US government.

Colt, born in Hartford, Connecticut, made up his mind as a boy to become an inventor. One of his discoveries was how to fire gunpowder using an electric current. After a public demonstration at a mine, which covered all the spectators with mud, he was sent off to Amherst Academy. As a result of a fire caused by another of his experiments, he was asked to leave there as well. He then became apprenticed as a sailor. Watching the helm during a voyage 1830, Colt noticed that whichever way the wheel turned, each of its spokes always lined up with a clutch that locked it into position. He realized that this mechanism could be applied to a firearm.

By 1835, Colt had perfected his revolver and patented it. In 1836 he set up a company in Paterson, New Jersey, where he produced a five-shot revolver and rifles and shotguns based on the same revolving principle. These models were revolutionary, yet unreliable, and Colt's company failed 1842.

Colt *US inventor Samuel Colt shown holding a Colt revolver. In a time when firearms were notoriously unreliable, Colt's designs were revolutionary and were quickly adopted by many fighting forces. His first revolver, the 'Colt Patterson', fired five shots and was used extensively by Texas Rangers in the Mexican–American War. He also supplied weaponry to both the Unionist North in the American Civil War and the British in Crimea.*

But in 1846, the Mexican–American War broke out and, with a big government order, Colt was back in business. The pioneers then starting to open up the West soon adopted the Colt to use against Native Americans and wildlife.

Compton Arthur Holly 1892–1962. US physicist who in 1923 found that X-rays scattered by such light elements as carbon increased their wavelengths. He concluded from this unexpected result that the X-rays were displaying both wavelike and particle-like properties, since named the ***Compton effect***. He shared a Nobel prize 1927 with Scottish physicist Charles Wilson. Compton was also a principal contributor to the development of the atomic bomb.

The behaviour of the X-ray, previously considered only as a wave, is explained best by considering that it acts as a corpuscle or particle – as a photon (Compton's term) of electromagnetic radiation. Quantum mechanics benefited greatly from this interpretation. Further confirmation came from experiments using a cloud chamber in which collisions between X-rays and electrons were photographed and analysed.

Compton was born in Wooster, Ohio, studied at Princeton, and worked 1919–20 in the UK with nuclear physicist Ernest Rutherford at Cambridge. His academic career in the USA was spent at Washington University, St Louis, 1920–23 and 1945–61, and at the University of Chicago 1923–45.

To determine whether cosmic rays consist of particles or electromagnetic radiation, Compton made measurements at various latitudes of comparative cosmic-ray intensities, using ionization chambers. By 1938 he had collated the results and demonstrated that the rays are deflected into curved paths by the Earth's magnetic field, proving that at least some component of cosmic rays consists of charged particles.

During World War II, Chicago University was the prime location of the Manhattan Project, the effort to produce the first atomic bomb, and in 1942 Compton became one of its leaders. He organized research into methods of isolating fissionable plutonium and worked with Italian physicist Enrico Fermi on producing a self-sustaining nuclear chain reaction.

Cook James 1728–1779. British naval explorer. After surveying the St Lawrence River in North America 1759, he made three voyages: 1768–71 to Tahiti, New Zealand, and Australia; 1772–75 to the South Pacific; and 1776–79 to the South and North Pacific, attempting to find the Northwest Passage and charting the Siberian coast. He was killed in Hawaii.

In 1768 Cook was given command of an expedition to the South Pacific to witness Venus eclipsing the Sun. He sailed in the *Endeavour* with Joseph Banks and other scientists, reaching Tahiti in April 1769. He then sailed around New Zealand and made a detailed survey of the east coast of Australia, naming New South Wales and Botany Bay. He returned to England 12 June 1771.

Now a commander, Cook set out 1772 with the *Resolution* and *Adventure* to search for the southern continent. The location of Easter Island was determined, and the Marquesas and Tonga Islands plotted. He also went to New Caledonia and Norfolk Island. Cook returned 25 July 1775, having sailed 100,000 km/60,000 mi in three years.

On 25 June 1776, he began his third and last voyage with the *Resolution* and *Discovery*. On the way to New Zealand, he visited several of the Cook or Hervey Islands and revisited the Hawaiian or Sandwich Islands. The ships sighted the North American coast at latitude 45° N and sailed north hoping to discover the Northwest Passage. He made a continuous survey as far as the Bering Strait, where the way was blocked by ice. Cook then surveyed the opposite coast of the strait (Siberia), and returned to Hawaii early 1779, where he was killed in a scuffle with islanders.

Cook was born in Marton, Yorkshire, and joined the navy 1755. His expeditions set new standards of survey work, cartography, hydrography, and other scientific research.

Coolidge Julian Lowell 1873–1954. US geometrician who wrote many mathematical textbooks, in which he not only reported his results but also described the historical background, together with contemporary developments.

Coolidge was born in Brookline, Massachusetts. He studied at Harvard and at several universities in Europe, including Oxford. From 1900 he taught at Harvard, where he became professor 1918 and remained until 1940.

Coolidge's first book (1909) was on non-Euclidean geometry. He was especially interested in the use of geometry in the investigation of complex numbers.

Coolidge also wrote his first paper on probability theory 1909, in which he examined certain problems in game theory. Together with several later studies on statistics, this work was included in his 1925 book on probability – one of the first on the subject to be published in English.

His first work on the algebraic theory of curves appeared 1915, and a book 1931. Work on two classical geometrical figures – the circle and the sphere – also led to the writing of a book 1916. His last book was a historical text 1949.

Cooper Leon Niels 1930– . US physicist who in 1955 began work on the phenomenon of superconductivity. He proposed that at low temperatures electrons would be bound in pairs (since known as ***Cooper pairs***) and in this state electrical resistance to their flow through solids would disappear. He shared the 1972 Nobel Prize for Physics with John Bardeen and J Robert Schrieffer (1931–).

Cooper was born in New York, where he attended Columbia University, specializing in quantum field theory – the interaction of particles and fields in subatomic systems. His work with John Bardeen was carried out at the University of Illinois. In 1958 Cooper moved to Brown University, Rhode Island, and in 1978 became director of the Centre for Neural Science at Brown.

Whereas the decrease in resistance is gradual in most metals, the resistance of superconductors suddenly disappears below a certain temperature. Experiments had shown that this temperature was inversely related to the mass of the nuclei. Cooper showed that an electron moving through the lattice attracts positive ions, slightly deforming the lattice. This leads to a momentary concentration of positive charge that attracts a second electron. This is the Cooper pair. Although the electrons in the pair are only weakly bound to each other, Bardeen, Cooper, and Schrieffer were able to show that they all form a single quantum state with a single momentum. The scattering of individual electrons does not affect this momentum, and this leads to zero resistance. Cooper pairs cannot be formed above the critical temperature, and superconductivity breaks down.

Developing a theory of the central nervous system, Cooper has also worked on distributed memory and character recognition.

He published *The Meaning and Structure of Physics* 1968.

Copernicus Nicolaus 1473–1543. Polish astronomer who believed that the Sun, not the Earth, is at the centre of the Solar System, thus defying the Christian church doctrine of the time. For 30 years he worked on the hypothesis that the rotation and the orbital motion of the Earth were responsible for the apparent movement of the heavenly bodies. His great work *De revolutionibus orbium coelestium/On the Revolutions of the Heavenly Spheres* was not published until the year of his death.

Copernicus relegated the Earth from being the centre of the universe to being merely a planet (the centre only of its own gravity and the orbit of its solitary Moon). This forced a fundamental revision of the anthropocentric view of the universe and came as an enormous psychological shock to European culture. Copernicus's model could not be proved right, because it contained several

Finally we shall place the Sun himself at the centre of the Universe.

Nicolaus Copernicus

De revolutionibus orbium coelestium

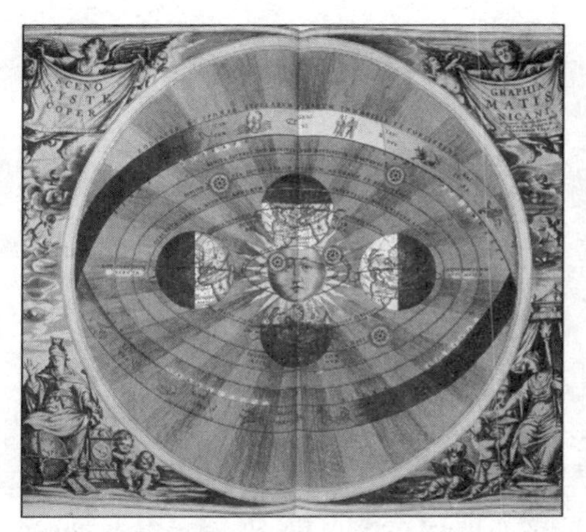

Copernicus *A representation of the heliocentric nature of the Solar System proposed by Copernicus. The orbit of the planets around the Sun, rather than* vice versa, is depicted.

fundamental flaws, but it was the important first step to the more accurate picture built up by later astronomers.

Copernicus was born in Toruné on the Vistula. He studied mathematics, astronomy, classics, law, and medicine at Kraków and various universities in Italy. On his return to Poland 1506 he became physician to his uncle, the bishop of Varmia, who had also got him the post of canon at Frombork, enabling him to intersperse astronomical work with the duties of various civil offices.

Copernicus began to make astronomical observations in 1497, although he relied mainly on data accumulated by others. In about 1513 he wrote a brief, anonymous text entitled *Commentariolus*, in which he outlined the material he later discussed in *De revolutionibus*.

Unable to free himself entirely from the constraints of classical thinking, Copernicus was able to imagine only circular planetary orbits. This forced him to retain the cumbersome system of epicycles, with the Earth revolving around a centre which revolved around another centre which in turn orbited the Sun. It was the work of Johannes Kepler, who introduced the concept of elliptical orbits, that rescued the Copernican model. Copernicus also held to the notion of spheres, in which the planets were supposed to travel. It was Tycho Brahe who rid astronomy of that archaic concept.

A Lutheran minister oversaw the publication of *De revolutionibus* and inserted a preface (without Copernicus's permission) stating that the theory was intended merely as an aid to the calculation of planetary positions, not as a statement of reality. This served to compromise the value of the text in the eyes of many astronomers, but it also saved the book from instant condemnation by the Roman Catholic Church. *De revolutionibus* was not placed on the index of forbidden books until 1616 (it was removed from the list 1835).

Corey Elias James 1928– . US organic chemist who received the Nobel Prize for Chemistry in 1990 for the development of ***retrosynthetic analysis***, a method of synthesizing complex substances. A prolific worker, Corey has synthesized more than 100 substances for the first time, including terpenes (a family of hydrocarbons found in plant oils) and ginkgolide B (an extract from the ginkgo tree used to control asthma).

Retrosynthetic analysis involves the breakdown of an organic compound in stages, with each step being tested for reversability. The starting point is a list of a compound's features and structure, such as how the carbon atoms are bonded together and whether they are linked together in chains, rings, or branches. The molecule is then simplified by unlinking the chains and breaking the bonds. From this process emerges a set of rules leading from compound to reactants and back to the compound again.

Corey was born in Methuen, Massachusetts, USA and studied at the Massachusetts Institute of Technology, where he originally planned a career as an electrical engineer. However, an enlightening lecture by an organic chemist inspired him to change his studies to chemistry. After obtaining a doctorate, he took a job as an instructor at the University of Illinois 1951, which led to a professorship. In 1959, he moved on to Harvard and was appointed Sheldon Emery professor 1965.

Cori Carl Ferdinand (1896–1984) and Gerty (Theresa, born Radnitz) (1896–1957) US biochemists born in Austro-Hungary who, together with Argentine physiologist Bernardo Houssay (1887–1971), received a Nobel prize 1947 for their discovery of how glycogen (animal starch) – a derivative of glucose – is broken down and resynthesized in the body, for use as a store and source of energy.

Both were born in Prague and married while studying at the medical school there.

They emigrated to the USA 1922, and in 1931 Carl Cori was appointed professor of biochemistry at Washington University School of Medicine in St Louis, Missouri. Gerty Cori also worked there, becoming professor 1947. Carl Cori remained at St Louis until 1967, when he moved to Harvard Medical School.

Glycogen is broken down in the muscles into lactic acid, which, when the muscles rest, is reconverted to glycogen. In the 1930s the Coris set out to determine exactly how these changes take place. Gerty Cori found a new substance in muscle tissue, glucose-1-phosphate, now known as ***Cori ester***. Its formation from glycogen involves only a small amount of energy change, so that the balance between the two substances can easily be shifted in either direction. The second step in the reaction chain involves the conversion of glucose-1-phosphate into glucose-6-phosphate. Finally this second phosphate is changed to fructose-1,6-diphosphate, which is eventually converted to lactic acid. The first set of reactions from glycogen to glucose-6-phosphate is now termed glycogenolysis; the second set, from glucose-6-phosphate to lactic acid, is referred to as glycolysis.

Coriolis Gaspard Gustave de 1792–1843. French physicist who in 1835 discovered the ***Coriolis effect***, which governs the movements of winds in the atmosphere and currents in the ocean. Coriolis was also the first to derive formulas expressing kinetic energy and mechanical work.

Coriolis was born in Paris and studied there, graduating in highway engineering. He was professor of mechanics at the Ecole Centrale des Arts et Manufactures 1829–36 and then at the Ecole des Ponts et Chaussées. In 1838 he became director of studies at the Ecole Polytechnique.

From 1829, Coriolis was concerned that proper terms and definitions should be introduced into mechanics. He succeeded in establishing the use of the word 'work' as a technical term in mechanics, defining it in terms of the displacement of force through a certain distance.

Investigating the movements of moving parts in machines and other systems relative to the fixed parts, Coriolis explained how the rotation of the Earth causes objects moving freely over the surface to follow a curved path relative to the surface – the Coriolis effect.

Corliss George Henry 1817–1888. US engineer and inventor of many improvements to steam engines, particularly the ***Corliss valve*** for controlling the flow of steam to and through cylinders.

Corliss was born in Easton, New York. He took out the first of his many patents in 1849, for the Corliss valve. In 1856 he founded the Corliss Engine Company in Providence, Rhode Island. This company designed and built the largest steam engine then in existence to power all the exhibits in the Machinery Hall at the 1876 Philadelphia Centennial Exposition.

To reduce heat loss and wear on moving parts, the Corliss valve had a separate inlet and exhaust port for the steam at each end of the cylinder. There was also a spring-loaded action to make the valve open and close as quickly as possible. Corliss's design continued in use for as long as large steam engines were manufactured.

Cormack Allen MacLeod 1924– . South-African born US physicist who shared the Nobel Prize for Physiology or Medicine 1979 with Godfrey Hounsfield for his invention of the X-ray diagnostic technique computer-assisted tomography (CAT scan).

While working on the administration of isotopes and film badge calibration at Groote Shuur Hospital he wondered what would happen if the attenuation (amount of beam weakening) of X-rays could be measured following their passage through the body. He thought that such information would contain enough information, if attained from enough different angles, to allow the construction of an image of the body's internal structure, which could be diagnostically useful.

On sabbatical at Harvard University he began work on a mathematical model for construction of such an image. He continued his endeavours with laboratory tests upon his return to South Africa 1957. Godfrey Hounsfield, working independently in England, had realized that a topographic, or slice by slice, approach was required.

Cormack was born in Johannesburg. After graduating from the University of Cape Town, he left South Africa to work in Cambridge for his PhD. He returned to South Africa briefly as a medical physicist at the University of Cape Town before moving to the USA to join the Physics Department at Tufts University in Medford, Massachusetts, where he was made a professor of physics 1964.

Following a great deal of work to refine and extend the studies of Cormack and Hounsfield, the first clinical machine for head (CAT) scanning was installed at the Atkinson Morleys Hospital, Wimbledon 1971.

Cornforth John Warcup 1917– . Australian chemist. Using radioisotopes as markers, he found out how cholesterol is manufactured in the living cell and how enzymes synthesize chemicals that are mirror images of each other (optical isomers).

He shared a Nobel prize 1975 with Swiss chemist Vladimir Prelog. Knighted 1977.

For him [the scientist], truth is so seldom the sudden light that shows new order and beauty; more often, truth is the uncharted rock that sinks his ship in the dark.

John Cornforth

Nobel prize address 1975

Cornforth was born in Sydney and educated there and at Oxford University. He settled in the UK 1941, and worked for the British Medical Research Council 1946–62, when he became director of the Milstead Laboratory of Chemical Enzymology, Shell Research Ltd. He remained there until 1975, when he accepted a professorship at the University of Sussex.

In his researches, Cornforth studied enzymes, trying to determine specifically which group of hydrogen atoms in a biologically active compound is replaced by an enzyme to bring about a given effect. By using the element's three isotopes, he was able to identify precisely which hydrogen atom was affected by enzyme action. He was able, for example, to establish the orientation of all the hydrogen atoms in the cholesterol molecule.

Correns Carl Franz Joseph Erich 1864–1933. German botanist and geneticist who is credited with rediscovering Mendel's laws (Hugo De Vries is similarly credited).

His work on the role of pollen in influencing characteristics of fruit and seed produced led him to discover ratios like those found by Mendel. He predicted that sex must be inherited in a Mendelian fashion and in 1907 he was able to demonstrate that this was true using experiments on *Bryonia*.

Correns was born in Berlin and educated at the University of Munich before studying in Berlin and Tübingen. He became assistant professor at Pfeffer's Institute in Leipzig and then professor at Munster 1909, and first director of the Kaiser Wilhelm Institut für Biologie in Berlin 1913.

Cort Henry 1740–1800. English iron manufacturer. For the manufacture of wrought iron, he invented the puddling process and developed the rolling mill (shaping the iron into bars), both of which were significant in the Industrial Revolution.

Cort's work meant that Britain no longer had to rely on imported iron and could become self-sufficient. His method of manufacture combined previously separate actions into one process, removing the impurities of pig iron and producing high-class metal relatively cheaply and quickly.

Cort, born in Lancaster, became an arms purchaser for the navy. All the best metal suitable for arms manufacture was imported from abroad, mainly from Russia, Sweden, and North America. Seeking a method of making this high-grade metal in England, Cort in 1775 set up his own forge near Farnham, Surrey.

By 1784 he was in a position to apply for a patent on his process of puddling and rolling, which allowed bar iron to be produced on a large scale and of a high quality. With the advent of the Napoleonic Wars, demand for iron rocketed. But Cort was cheated by his financial backer and, having handed over the rights to his patent as security for the capital, was left bankrupt, to watch the ironmasters of England grow wealthy on his hard work.

Coster Laurens Janszoon *c.* 1370–1440. Dutch printer. According to some sources, he invented movable type, but after his death an apprentice ran off to Mainz with the blocks and, taking Johann Gutenberg into his confidence, began a printing business with him.

Cotton William 1786–1866. English inventor, financier, and philanthropist who in 1864 invented a knitting machine for the production of hosiery. This machine had a straight-bar frame which automatically made fully fashioned stockings knitted flat and sewn up the back.

Cotton was born in Leyton, Essex, and self-educated. In 1807 he was admitted as a partner in the firm of Huddart and Company in E London, which manufactured a cordage-making machine designed by hydrographer and inventor Joseph Huddart. Cotton published a memoir of Huddart with an account of his inventions 1855, and gave money for the building of schools, churches, and lodging houses in the East End of London.

In 1821 Cotton was elected a director of the Bank of England, and was its governor 1843–46. He invented an automatic weighing machine for sovereigns.

Coulomb Charles Augustin de 1736–1806. French scientist, inventor of the torsion balance for measuring the force of electric and magnetic attraction. The coulomb was named after him. In the fields of structural engineering and friction, Coulomb greatly influenced and helped to develop engineering in the 19th century.

Coulomb's law of 1787 states that the force between two electric charges is proportional to the product of the charges and inversely proportional to the square of the distance between them.

Coulomb was born in Angoulême and trained as a military engineer. He was posted to Martinique to undertake construction work 1764–72. From 1781 he was in Paris, resigning from the army during the French Revolution.

Coulomb took full advantage of his various postings to pursue a variety of studies, including structural mechanics, friction in machinery, and the elasticity of metal and silk fibres. In his 1781 study of friction, he extended knowledge of the effects of friction caused by such factors as lubrication and differences in materials and loads, producing a classic work that was not surpassed for 150 years. He also carried out fundamental research in ergonomics.

The torsion balance Coulomb invented is described in a paper of 1777, and in a paper of 1785 he discussed its adaptation for electrical studies. He went on to investigate the distribution of electric charge over a body and found that it is located only on the surface of a charged body and not in its interior.

Coulson Charles Alfred 1910–1974. English theoretical chemist who developed a molecular orbital theory and the concept of partial valency. He developed many mathematical techniques for solving chemical and physical problems.

Coulson was born in Dudley, Yorkshire, and studied at Cambridge. He became professor of theoretical physics at King's College, London, 1947, professor of mathematics at Oxford 1952, and later Oxford's first professor of theoretical chemistry. He was chair of the charity Oxfam 1965–71.

The molecular orbital theory that Coulson developed is an extension of atomic quantum theory and deals with 'allowed' states of electrons in association with two or more atomic nuclei, treating a molecule as a whole. He was thus able to explain properly phenomena such as the structure of benzene and other conjugated systems, and invoked what he called partial valency to account for the bonding in such compounds as diborane.

Coulson also contributed significantly to the understanding of the solid state (particularly metals), such as the structure of graphite and its 'compounds'.

He wrote three best-selling books: *Waves* 1941, *Electricity* 1948, and *Valence* 1952.

Courant Richard 1888–1972. German-born US mathematician who wrote several textbooks that became standard reference works, and founded three highly influential mathematical institutes, one at Göttingen and two in New York.

Courant was born in Lublintz, Upper Silesia (now in Poland), and educated at Göttingen, where he became assistant to German mathematician David Hilbert. During World War I, he interested the military authorities in a device for sending electromagnetic radiation through the earth to carry messages. He was professor at Göttingen until the rise of the Nazis, and emigrated to the USA 1934, joining the teaching staff at New York University. He was director of the Institute of Mathematical Sciences of New York University 1953–58, and a new institute opened 1965 as the Courant Institute of Mathematical Sciences.

In *Methoden der mathematischen Physik/Methods of Mathematical Physics* 1924–27, Courant treated many of the subjects developed by David Hilbert; the book is now universally known as Courant–Hilbert. Other books include *Differential and Integral Calculus*, a university textbook.

Cournand Andre Frederic 1895–1988. French-born US physician who shared with Werner Forssman and Dickinson Richards the Nobel Prize for Physiology or Medicine 1956 for developing a method to catheterize the heart. This method paved the way for the development of the discipline of modern cardiology, as it meant that studies of the human heart could be performed without the need for surgery.

In 1929, the German surgeon Werner Forssman placed a hollow tube into a vein in his arm and advanced the tube until its tip entered the right atrium of his heart. This procedure, called cardiac catheterization, was developed further by Richards and Cournand so that the tube could be inserted into the right side of the heart and the pulmonary artery in order to study congenital heart disease.

They also succeeded in the catheterization of the left side of the heart by passing the tube through the septum, which separates the two sides of the heart. The procedure made it possible to measure the pressures in the various chambers of the heart, information that proved enormously valuable for the development of surgery for congenital heart disease and narrowed arteries.

Cournand was born in Paris and educated at the Sorbonne. He emigrated to the USA 1930 and joined the staff of Columbia University 1934.

Courtauld Samuel 1793–1881. British industrialist who developed the production of viscose rayon and other synthetic fibres from 1904. He founded the firm of Courtauld's 1816 in Bocking, Essex, and at first specialized in silk and crepe manufacture.

His great-nephew ***Samuel Courtauld*** (1876–1947) was chair of the firm from 1921, and in 1931 gave his house and art collection to the University of London as the Courtauld Institute.

The institute is based in Somerset House, London.

Cousteau Jacques Yves 1910– . French oceanographer who pioneered the invention of the aqualung 1943 and techniques in underwater filming. In 1951 he began the first of many research voyages in the ship *Calypso*. His film and television documentaries and books established him as a household name.

Cousteau was born in the Gironde. He joined the navy and worked for naval intelligence during the Nazi occupation. From 1936 he experimented with diving techniques. The compressed air cylinder had been invented 1933 but restricted the diver to very short periods of time beneath the surface. Testing new breathing equipment, Cousteau was several times nearly killed.

In 1942 Cousteau met Emile Gagnan, an expert on industrial gas equipment. Gagnan had designed an experimental demand valve for feeding gas to car engines. Together, Gagnan and Cousteau developed a self-contained compressed air 'lung', the aqualung. In June 1943 Cousteau made his first dive with it, achieving a depth of 18 m/ 60 ft.

I am not a scientist. I am, rather, an impresario of scientists.

Jacques Cousteau

On his role as a filmmaker. *Christian Science Monitor* 21 July 1971

Cowling Thomas George 1906–1990. English applied mathematician and physicist who contributed significantly to modern research into stellar energy, with special reference to the Sun.

Cowling was born in Walthamstow, Essex, and studied at Oxford.

He was professor of mathematics at University College, Bangor, 1945–48, and at Leeds University 1948–70.

Cowling was responsible for demonstrating the existence of a convective core in stars, suggesting that the Sun may behave like a giant dynamo whose rotation, internal circulation, and convection produce the immensely powerful electric currents and magnetic fields associated with sunspots. With Swedish physicist Hannes Alfvén, Cowling showed that such currents and fields are likely to have existed since the Sun was first formed.

Craik Kenneth John William 1914–1945. Scottish psychologist. Initially involved in the study of vision, Craik became increasingly interested in the nature of cognition, believing that the brain has mechanisms similar in principle to devices used, for example, in calculating machines and anti-aircraft predictors. In several aspects of his work he anticipated later ideas in cybernetics and artificial intelligence and, were it not for his early death, would have made important contributions to these fields.

Craik's views were presented in *The Nature of Explanation* 1943 and developed in a posthumously published collection of essays, notes, and papers titled *The Nature of Psychology* 1966. He was educated at the University of Edinburgh and later studied under Frederic Charles Bartlett at Cambridge. During World War II he applied an interest in servomechanisms and control principles to human factors in the design of tank and anti-aircraft gunnery and, in 1944, was appointed first director of the Applied Psychology Unit in Cambridge.

Cram Donald James 1919– . US chemist who shared the 1987 Nobel Prize for Chemistry with Jean-Marie Lehn and Charles J Pedersen for their work on molecules (groups of two or more atoms bonded together) with highly selective structure-specific interactions. The work has importance in the synthesis of organic compounds, analytic chemistry, and biochemistry.

Cram designed and produced complex 'host' molecules that could selectively recognize and bind with other 'guest' molecules and ions. The molecular recognition occurs because the structure or shape of the host molecule matches that of the guest molecule. Suitable host molecules can be used to catalyze (trigger) various types of chemical reactions or to transport ions through biological barriers, such as cell membranes.

Cram was born at Chester, Vermont, USA, and educated at Rollins College, Florida, and the University of Nebraska, Canada. During World War II, he worked for Merck and Company on penicillin. In 1945, he went to Harvard University, obtaining his doctorate 1947. After three months with the Massachusetts Institute of Technology, he joined the University of California at Los Angeles. He was appointed professor of chemistry 1956.

Cramer Gabriel 1704–1752. Swiss mathematician who introduced ***Cramer's rule*** 1750, a method for the solution of linear equations which revived interest in the use of determinants; ***Cramer's paradox***; and the concept of utility in mathematics.

Cramer was born and educated in Geneva, where from the age of 20 he was professor of mathematics at the Académie de la Rive. In 1750 he was made professor of philosophy. Cramer travelled in Europe and met leading mathematicians.

Cramer's paradox revolves around a theorem formulated by Scottish mathematician Colin Maclaurin, who stated that two different cubic curves intersect at nine points. Cramer pointed out that the definition of a cubic curve – a single curve – is that it is determined itself by nine points.

Cramer's concept of utility now provides a connection between the theory of probability and mathematical economics.

His major work is *Introduction à l'analyse des lignes courbes algébriques* 1750, in which he set out Cramer's rule.

Cray Seymour Roger 1925–1996. US computer scientist and pioneer in the field of supercomputing. He designed one of the earliest computers to contain transistors 1960. In 1972 he formed Cray Research to build the first supercomputer, the Cray-1, released 1976. Its success led to the production of further supercomputers, including the Cray-2 1985, the Cray Y-MP, a multiprocessor design 1988, and the Cray-3 1989.

Creed Frederick George 1871–1957. Canadian inventor who developed the teleprinter. He perfected the Creed telegraphy system (teleprinter), first used in Fleet Street, the headquarters of the British press, 1912 and subsequently, usually under the name Telex, in offices throughout the world.

If you want to understand function, study structure.

Francis Crick

What Mad Pursuit 1988 p 150

Crick Francis Harry Compton 1916– . English molecular biologist. From 1949 he researched the molecular structure of DNA, and the means whereby characteristics are transmitted from one generation to another. For this work he was awarded a Nobel prize (with Maurice Wilkins and James Watson) 1962.

Using Wilkins's and others' discoveries, Crick and Watson postulated that DNA consists of a double helix consisting of two parallel chains of alternate sugar and phosphate groups linked by pairs of organic bases. They built molecular models which also explained how genetic information could be coded – in the sequence of organic bases. Crick and Watson published their work on the proposed structure of DNA in 1953. Their model is now generally accepted as correct.

Crick was born in Northampton and studied physics at University College, London. During World War II he worked on the development of radar. He then went to do biological research at Cambridge. In 1977 he became a professor at the Salk Institute in San Diego, California.

Later Crick, this time working with South

Crick *English microbiologist Francis Crick. Crick shared the Nobel Prize for Physiology or Medicine with James Watson and Maurice Wilkins 1962, after discovering the molecular structure of the genetic material DNA, at the Cavendish Laboratory in Cambridge. He later turned his attention to investigating the nature of human consciousness.*

African Sydney Brenner, demonstrated that each group of three adjacent bases (he called a set of three bases a codon) on a single DNA strand codes for one specific amino acid. He also helped to determine codons that code for each of the 20 main amino acids. Furthermore, he formulated the adaptor hypothesis, according to which adaptor molecules mediate between messenger RNA and amino acids. These adaptor molecules are now known as transfer RNAs.

Crompton Rookes Evelyn Bell 1845–1940. English engineer who pioneered the dynamo, electric lighting, and road transport. He also contributed to the development of industry standards, both electrical and mechanical, and was involved in the founding of the National Physical Laboratory and what is now the British Standards Institution.

Crompton was born near Thirsk in Yorkshire. During his school holidays he built a steam-driven road locomotive. In India as an army officer 1864–75 he continued to develop his road vehicles. Their success was undermined by the poor quality of roads, the cheapness of other forms of transport, and the developing railway system.

On his return from service he began importing dynamos from France and set up his own company to develop and manufacture generating systems for lighting town halls, railway stations, and small residential areas. Direct-current electricity of about 400 volts was generated and used with large storage batteries (accumulators). This competed, in the end unsuccessfully, with the alternating-current system of Sebastian Ferranti.

During the Boer War, Crompton served in South Africa as commandant of the Electrical Engineers' Royal Engineers Volunteer Corps. He then returned to road transport, and contributed both to the principles of automobile engineering and the maintenance and design of roads. During World War I he was an adviser on the design and production of military tanks.

Crompton Samuel 1753–1827. British inventor at the time of the Industrial Revolution. He invented the 'spinning mule' 1779, combining the ideas of Richard Arkwright and James Hargreaves. This span a fine, continuous yarn and revolutionized the production of high-quality cotton textiles.

Crompton's invention was called the mule because it was a hybrid. It used the best from the spinning jenny and from Richard Arkwright's water frame of 1768. The strong, even yarn it produced was so fine that it could be used to weave delicate fabrics such as muslin, which became fashionable among the middle and upper classes, creating a new market for the British cotton trade. Spinning was taken out of the home and into the factories.

Crompton was born near Bolton, Lancashire. He developed the spinning mule for use in his own home and had no means of patenting it, so he sold it for a small fee to local manufacturers.

Cronin James Watson 1931– . US physicist who shared the 1980 Nobel Prize for Physics with Val Fitch for their work in particle physics. They showed for the first time that left–right asymmetry is not always preserved when some particles are changed in state from matter to antimatter.

Cronin was born in Chicago, Illinois, and educated at the Southern Methodist University and the University of Chicago. He was at Princeton 1958–71, as professor from 1965, and became professor at the University of Chicago 1971.

The discovery for which Fitch and Cronin received the Nobel prize was first published in 1964, and at that time was regarded as a bombshell in the field of particle physics. Their findings have had impact on an outstanding controversy about the symmetry of nature.

Crookes William 1832–1919. English scientist whose many chemical and physical discoveries include the metallic element thallium 1861, the radiometer 1875, and the Crookes high-vacuum tube used in X-ray techniques. Knighted 1897.

The radiometer consists of a four-bladed paddle wheel mounted horizontally on a pinpoint bearing inside an evacuated glass globe. Each vane of the wheel is black on one side (making it a good absorber of heat) and silvered on the other side (making it a good reflector). When the radiometer is put in strong sunlight, the paddle wheel spins round.

Crookes was born in London and studied at the Royal College of Chemistry. Financially independent, he carried out most of his researches in his own laboratory. In 1859 he founded the weekly *Chemical News*, which he edited until 1906.

During the 1870s Crookes's studies concerned the passage of an electric current through glass 'vacuum' tubes containing rarified gases; such discharge tubes became known as Crookes tubes. The ionized gas in a Crookes tube gives out light – as in a neon sign – and Crookes observed near the cathode a light-free gap in the discharge, now called the ***Crookes dark space***. He named the ion stream 'molecular rays' and demonstrated how they are deflected in a magnetic field and how they can cast shadows, proving that they travel in straight lines.

Topics covered by his publications included chemical analysis; the manufacture of sugar from sugar beet; dyeing and printing of textiles; oxidation of platinum, iridium, and rhodium; use of carbolic acid (phenol) as an antiseptic in the treatment of diseases in cattle; the origin and formation of diamonds in South Africa; and the use of artificial fertilizers and their manufacture from atmospheric nitrogen. He also published papers on spiritualism.

Crookes *A cartoon published in* Vanity Fair *in 1903 of English chemist William Crookes, shown holding the discharge tube which carries his name. The Crookes tube was used to investigate the passage of cathode rays through rarified gases in a vacuum. Crooke's experimental work using high-voltage discharge tubes was one of the key steps towards the discovery of X-rays and the electron.*

Crutzen Paul 1933– . Dutch meteorologist who shared the 1995 Nobel Prize for Chemistry with Mexican chemist Mario Molina and US chemist F Sherwood Rowland for their work in atmospheric chemistry, particularly concerning the formation and decomposition of ozone. They explained the chemical reactions which are destroying the ozone layer.

Crutzen, while working at Stockholm University in 1970, discovered that the nitrogen oxides NO and NO_2 speed up the breakdown of atmospheric ozone into molecular oxygen. These gases are produced in the atmosphere from nitrous oxide N_2O which is released by microorganisms in the soil. He showed that this process is the main natural method of ozone breakdown. Crutzen also discovered that ozone-depleting chemical reactions occur on the surface of cloud particles in the stratosphere.

Crutzen was born in Amsterdam. He received his doctor's degree in meteorology from Stockholm University 1973. He is currently at the Max Planck Institute for Chemistry in Mainz, Germany.

Cugnot Nicolas-Joseph 1725–1804. French engineer who produced the first high-pressure steam engine and, in 1769, the first self-propelled road vehicle. Although it proved the viability of steam-powered traction, the problems of water supply and pressure maintenance severely handicapped the vehicle.

While serving in the army, Cugnot was asked to design a steam-operated gun carriage. After several years, he produced a three-wheeled, high-pressure carriage capable of carrying 1,800 litres/400 gallons of water and four passengers at a speed of 5 kph/3 mph. Although he worked further on the carriage, the political upheavals of the French revolutionary era obstructed progress and his invention was ignored.

Cugnot *Cugnot's steam tractor.*

Cugnot was born in Void, Meuse. As a young soldier, he invented a new kind of rifle. After serving in the Seven Years' War, Cugnot returned to Paris in 1763 as a military instructor. He also devoted his time to writing military treatises and exploring a number of inventions he had conceived during his campaigning.

Culshaw John Royds 1924–1980. British record producer who developed recording techniques. Managing classical recordings for the Decca record company in the 1950s and 1960s, he introduced echo chambers and the speeding and slowing of tapes to achieve effects not possible in live performance. He produced the first complete recordings of Wagner's *Ring* cycle.

Cunningham Allan 1791–1839. English explorer and botanist who explored South America, Australia, and New Zealand, but is best known for exploring the Darling Downs in Australia.

Cunningham was born in Wimbledon, Surrey and started his career as a botanist as clerk to the curator of Kew Gardens. Between 1814 and 1831, he was employed by Kew Gardens to collect botanical specimens, and travelled to Rio in this capacity 1814–16, then Sydney, Australia, where he remained 1816–26. In 1826, he visited New Zealand before returning to England 1831. He was elected a fellow of the Linnaean Society 1832. When his brother Richard Cunningham, then colonial botanist for New South Wales, was killed by aborigines in 1835, he accepted this post himself. However, he found the work rather menial and boring, describing it as running the 'Government Cabbage Garden'. In 1838, he made another journey to New Zealand but had to return to Australia due to ill health six months later. He died in Sydney 27 June. Most of his specimens remain at Kew to this day.

Curie Marie (born Manya Sklodowska) 1867–1934. Polish scientist. In 1898 she and her husband Pierre Curie discovered two new radioactive elements in pitchblende ores: polonium and radium. They isolated the pure elements 1902. Both scientists refused to take out a patent on their discovery and were jointly awarded the Nobel Prize for Physics 1903, with Henri Becquerel. Marie Curie wrote a *Treatise on Radioactivity* 1910, and was awarded the Nobel Prize for Chemistry 1911.

From 1896 the Curies worked together on radioactivity, building on the results of Wilhelm Röntgen (who had discovered X-rays) and Becquerel (who had discovered that similar rays are emitted by uranium salts). They took no precautions against radioactivity and Marie Curie died of radiation poisoning. Her notebooks, even today, are too contaminated to handle.

Maria Sklodowska was born in Warsaw, then under Russian domination. She studied

Curie *Marie and Pierre Curie with their eldest daughter Irène 1904. All three members of the family won Nobel prizes – Marie and Pierre won the Nobel Prize for Physics for their work on radioactivity in 1903, Marie won the Nobel Prize for Chemistry for the discovery of polonium and radium in 1911, and Irène Joliet-Curie won the Nobel Prize for Chemistry in 1935 for her work on preparing the first artificial isotope.*

in Paris from 1891 and married in 1895. In 1906 she succeeded her husband as professor of physics at the Sorbonne; she was the first woman to teach there.

Marie Curie discovered that thorium also emits radiation and found that the mineral pitchblende was even more radioactive than could be accounted for by any uranium and thorium content. In July 1898 the Curies announced the discovery of polonium, followed in Dec of that year with the discovery of radium. They eventually prepared 1 g/ 0.04 oz of pure radium chloride – from 8 tonnes of waste pitchblende from Austria. They also established that beta rays (now known to consist of electrons) are negatively charged particles.

In 1910 with André Debierne (1874–1949), who in 1899 had discovered actinium in pitchblende, Marie Curie isolated pure radium metal.

At the outbreak of World War I in 1914 Curie helped to equip ambulances with X-ray equipment, which she drove to the front lines. The International Red Cross made her head of its Radiological Service.

It would be impossible, it would against the scientific spirit ... Physicists should always publish their researches completely. If our discovery has a commercial future that is a circumstance from which we should not profit. If radium is to be used in the treatment of disease, it is impossible for us to take advantage of that.

Marie Curie

Eve Curie *The discovery of radium* in *Marie Curie* transl V Sheean, 1938 [On the patenting of radium. Discussion with her husband, Pierre]

Curie Pierre 1859–1906. French scientist who shared the Nobel Prize for Physics 1903 with his wife Marie Curie and Henri Becquerel. From 1896 the Curies had worked together on radioactivity, discovering two radioactive elements.

Pierre Curie was born in Paris and educated at the Sorbonne, becoming an assistant there 1878. He discovered the piezoelectric effect and, after being appointed head of the laboratory of the Ecole de Physique et Chimie, went on to study magnetism and formulate ***Curie's law***, which states that magnetic susceptibility is inversely proportional to absolute temperature. In 1895 he discovered the ***Curie point***, the critical temperature at which a paramagnetic substance become ferromagnetic. In 1904 he became professor of physics at the Sorbonne.

Curtis Heber Doust 1872–1942. US astronomer who deduced that spiral nebulae were galaxies that produced a cloud of debris which accumulated in the plane of the galaxy.

Curtis was born in Muskegan, Michigan, and studied classics at the University of Michigan. At the age of 22, he became professor of Latin at Napa College, California. Access to the small observatory there changed his career, and in 1897 he became professor of mathematics and astronomy at the University of the Pacific. He subsequently worked at a number of US observatories and 1906–09 in Chile, but most of his research was done at the Lick Observatory, Mount Hamilton, California, between 1898 and 1920.

Curtis worked on a programme for the measurement of stellar radial velocities 1902–09. For the next 11 years he concentrated his efforts on spiral nebulae, trying to establish whether they were distant star clusters or clouds of debris. From studying photographs he concluded that they were both. If such a cloud of debris had also gathered outside our own Galaxy, this would explain why none appeared in the plane of the Milky Way. Spiral nebulae in that position would simply be obscured by dust.

Curtiss Glenn Hammond 1878–1930. US aeronautical inventor, pioneer aviator, and aircraft designer. In 1908 he made the first public flights in the USA, including the one-mile flight. He belonged to Alexander Graham Bell's Aerial Experiment Association 1907–1909 and established the first flying school 1909. In 1910 Curtiss staged his sensational flight down the Hudson River from Albany to New York City.

In 1916 he founded the Curtiss Aeroplane and Motor Corp, based on his invention of ailerons, which he designed for the first seaplanes 1911. He designed and constructed many planes for the Allied nations during World War I. After the end of the war 1918 he continued to improve plane and motor designs.

A physician is obligated to consider more than a diseased organ, more even than the whole man – he must view the man in his world.

Harvey Cushing

Man Adapting by René Dubos

Cushing Harvey Williams 1869–1939. US neurologist who pioneered neurosurgery. He developed a range of techniques for the surgical treatment of brain tumours, and also studied the link between the pituitary gland and conditions such as dwarfism. He first described the chronic wasting disease now known as ***Cushing's syndrome***.

Cushing was born in Cleveland, Ohio. He studied medicine at Yale and Harvard, and later in Switzerland and the UK. From 1912 to 1932 he was professor of surgery at the Harvard Medical School; in 1933 he became professor of neurology at Yale.

As a result of experimenting on the effect of artificially increasing intercranial pressure in animals, Cushing developed new methods of controlling blood pressure and bleeding during surgery on human beings.

Cushing wrote a description of the stages in the development of different types of intercranial tumours, classified such tumours, and published a definitive account of acoustic nerve tumours 1917.

In 1908 Cushing began studying the pituitary gland and, after experimenting on animals, discovered a way of gaining access to this gland surgically. As a result of this discovery it became possible to treat cases of blindness caused by tumours in that region.

Cuvier Georges Léopold Chrétien Frédéric Dagobert. Baron Cuvier 1769–1832. French comparative anatomist, the founder of palaeontology. In 1799 he showed that some species have become extinct by reconstructing extinct giant animals that he believed were destroyed in a series of giant deluges. These ideas are expressed in *Recherches sur les ossiments fossiles de quadrupèdes/Researches on the Fossil Bones of Quadrupeds* 1812 and *Discours sur les révolutions de la surface du globe* 1825.

In 1798 Cuvier produced *Tableau élémentaire de l'histoire naturelle des animaux*, in which his scheme of classification is outlined. He was the first to relate the structure of fossil animals to that of their living relatives. His great work *Le Règne animal/The Animal Kingdom* 1817 is a systematic survey.

Cuvier was born at Montebéliard in the principality of Württemburg, and studied natural history at Stuttgart, before becoming a private tutor in Normandy. He held academic posts in Paris from 1795, becoming professor at the Collège de France 1799 and at the Jardin des Plantes from 1802.

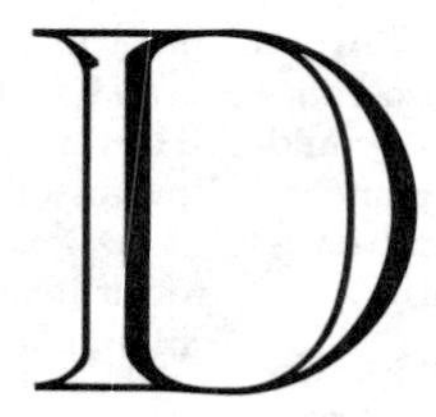

Daguerre Louis Jacques Mandé 1787–1851. French pioneer of photography. Together with Joseph Niépce, he is credited with the invention of photography (though others were reaching the same point simultaneously). In 1838 he invented the daguerreotype, a single image process superseded ten years later by Fox Talbot's negative/positive process.

Daimler Gottlieb Wilhelm 1834–1900. German engineer who pioneered the car and the internal-combustion engine together with Wilhelm Maybach. In 1885 he produced a motor bicycle and in 1889 his first four-wheeled motor vehicle. He combined the vaporization of fuel with the high-speed four-stroke petrol engine.

Daimler's work on the internal-combustion engine began in earnest in 1872 when he teamed up with Nikolaus Otto at a gas-engine works; Maybach was the chief designer. Daimler built his first petrol engines 1883. The Daimler Motoren Gesellschaft was founded 1890, and Daimler engines were also manufactured under licence; a Daimler-powered car won the first international car race: Paris to Rouen 1894.

Daimler was born near Stuttgart and trained as a mechanical engineer at the Stuttgart Polytechnic. He gained factory experience in various engineering works, including that of English engineer Joseph Whitworth in Manchester. In France he many have seen J J E Lenoir's newly developed gas engine.

Daimler's first working petrol-fuelled unit was an air-cooled, single-cylinder engine with a large cast-iron flywheel running at 900 rpm. The second engine was fitted to a bicycle in Nov 1885 (possibly even earlier). Daimler went on to try his engine as the power source for a boat.

Dale Henry Hallett 1875–1968. British physiologist who in 1936 shared the Nobel Prize for Physiology or Medicine with Otto Loewi for proving that chemical substances are involved in the transmission of nerve impulses.

Dale was born in London and studied at Cambridge; St Bartholomew's Hospital, London; University College, London; and Frankfurt, Germany. He was director of the Wellcome Physiological Research Laboratories 1906–14; worked at the Medical Research Council 1914–28, and was director of the National Institute for Medical Research 1928–42.

In 1910, investigating the chemical composition and effects of ergot (a fungus that infects cereals and other grasses), Dale identified the substance now known as histamine. In 1914 he isolated acetylcholine from biological material. With Loewi, he later showed it to be produced at the nerve endings of parasympathetic nerves.

d'Alembert see ◊Alembert, French mathematician.

Dalén Nils Gustav 1869–1937. Swedish industrial engineer who invented the light-controlled valve. This allowed lighthouses to operate automatically and won him the 1912 Nobel Prize for Physics.

Dalton John 1766–1844. English chemist who proposed the theory of atoms, which he considered to be the smallest parts of matter. He produced the first list of relative atomic masses in 'Absorption of Gases' 1805 and put forward the law of partial pressures of gases (***Dalton's law***).

From experiments with gases, Dalton noted that the proportions of two components combining to form another gas were always constant. He suggested that if substances combine in simple numerical ratios, then the macroscopic weight proportions represent the relative atomic masses of those substances. He also propounded the law of partial pressures, stating that for a mixture of gases the total pressure is the sum of the pressures that would be developed by each individual gas if it were the only one present.

Dalton was born near Cockermouth in Cumbria. As a Quaker, he was precluded from attending Oxford or Cambridge (then open only to members of the Church of England), and was self-taught. For 57 years he kept a diary of observations about the weather, and this led to his interest in gases. He was one of the first scientists to note and record colour blindness.

Dalton set out his atomic theory in *New System of Chemical Philosophy* 1808. The chemical symbols he introduced would soon be superseded by the system created by Swedish chemist Jöns Berzelius.

Dam Carl Peter Henrik 1895–1976. Danish biochemist who discovered vitamin K. For his success in this field he shared the 1943 Nobel Prize for Physiology or Medicine with US biochemist Edward Doisy (1893–1986).

In 1928 Dam began a series of experiments to see if chickens could live on a cholesterol-free diet. The birds, it turned out, were able to metabolize their own supply. Yet they continued to die from spontaneous haemorrhages. Dam concluded that their diet lacked an unknown essential ingredient to control coagulation, which he eventually found in abundance in green leaves. Dam named the new compound vitamin K.

Dancer John Benjamin 1812–1887. British optician and instrumentmaker who pioneered microphotography. By 1840 he had developed a method of taking photographs of microscopic objects, using silver plates. The photographic image was capable of magnification up to 20 times before clarity was lost. By 1859 he was showing microscope slides which carried portraits or whole book pages.

Dancer was born in London but grew up in Liverpool. He trained in the family business of manufacturing optical and other scientific instruments and giving public lectures at the Liverpool Mechanical Institution.

Dancer improved many of the standard laboratory practices of the period. He introduced unglazed porous jars in voltaic cells to separate the electrodes. Dancer improved on the Daniell cell by crimping or corrugating its copper plates to increase the electrode surface area. He also constructed the apparatus with which James Joule determined the mechanical equivalent of heat.

Dancer began a series of experiments in 1839 based on Daguerre's and Fox Talbot's photography techniques, and adopted the collodion method in the 1850s. By 1856 Dancer had prepared hundreds of microphotographs and his work was exhibited throughout Europe. They included photographs of distinguished scientists of the era and portraits of the British royal family.

Daniell John Frederic 1790–1845. British chemist and meteorologist who invented a primary electrical cell 1836. The ***Daniell cell*** consists of a central zinc cathode dipping into a porous pot containing zinc sulphate solution. The porous pot is, in turn, immersed in a solution of copper sulphate contained in a copper can, which acts as the cell's anode. The use of a porous barrier prevents polarization (the covering of the anode with small bubbles of hydrogen gas) and allows the cell to generate a continuous current of electricity.

The Daniell cell was the first reliable source of direct-current electricity.

Daniell was born in London and privately educated. He was the first professor of chemistry at King's College, London, 1831–45.

Daniell's other work included the development of improved processes for sugar manufacturing; investigations into gas generation by the distillation of resin dissolved in turpentine; and inventing a new type of dew-point hygrometer for measuring humidity (1820) and a pyrometer for measuring the temperatures of furnaces (1830). He also studied the behaviour of the Earth's atmosphere; gave an explanation of trade winds; researched into the meteorological effects of solar radiation and of the cooling of the Earth; suggested improvements for several meteorological instruments; and pointed out

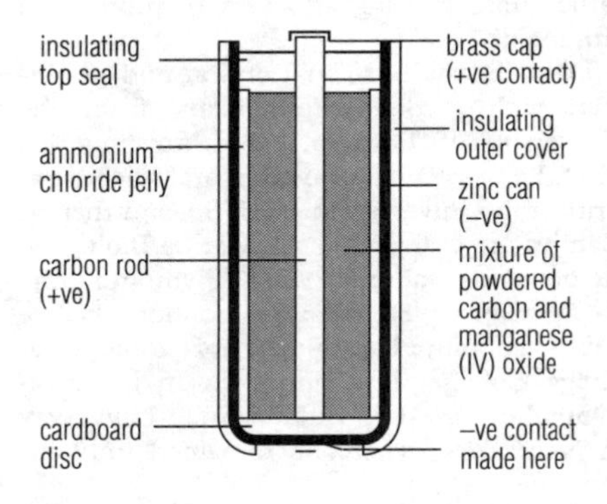

Daniell *A Daniell cell has a zinc cathode immersed in zinc sulphate solution contained in a porous pot, surrounded by copper sulphate solution. The whole is contained in a copper can, which also forms the anode.*

the importance of humidity in the management of greenhouses.

Danielli James Frederic 1911–1984. British cell biologist who hypothesized that the molecular structure of the cell membrane was a sandwich of two layers of proteins. In 1943, Danielli and Hugh Davson published their seminal theory on transport of substances across cell membranes in *The Permeability of Natural Membranes*. Their work provided a framework for future physiologists and cell biologists working on the role of the membrane in different cellular activities.

Danielli's major work was on the structure and physical properties of cell membranes. With Davson he proposed that the cell membrane was a sandwich of two juxtaposed phospholipid (lipid that contain a phosphate group) layers between two layers of globular proteins (proteins that have a globular three-dimensional structure). Phospholipids are amphipathic, that is they have both hydrophobic (water-hating) and hydrophilic (water-loving) regions. In the Davson–Danielli model, the hydrophobic parts of the phospholipid molecules are on the inside of a bilayer, while the hydrophilic parts form the outsides of the membrane.

In the 1950s, cell membranes were imaged using the electron microscope and it was revealed that they were too thin to possess coatings of globular proteins, as suggested by Danielli and Davson. Staining cell membranes with heavy metals, however, revealed a structure consistent with a Danielli and Davson's proposed phospholipid bilayer, and in 1972, Singer and William Nicholson revised the Davson–Danielli model to include proteins that spanned the membrane rather than existing as layers of protein on either side.

Danielli was born in London and graduated with a BSc in chemistry from the University of London. After finishing his PhD, he worked as a postdoctoral scientist at Princeton University in the USA and then at Cambridge University, the Marine Biological Associations laboratory in Plymouth, the Royal Cancer Hospital in London, before being appointed professor of zoology at Kings College, London 1948. In the mid-1950s he moved to Buffalo State University in New York, where he remained until his retirement 1980.

Dantzig George Bernard 1914– . US mathematician, an expert in linear computer programming and operations research. His work is fundamental to many university courses in business studies, industrial engineering, and managerial sciences. Dantzig has been involved in all the main areas of mathematical programming.

Dantzig was born in Portland, Oregon, and educated at the universities of Maryland and Michigan. During World War II he became attached to the Statistical Control Headquarters of the US Air Force. He was a research mathematician with the Rand Corporation at Santa Monica, California, 1952–60, and then became a professor at the University of California at Berkeley, moving 1966 to Stanford University, Palo Alto, California.

A fundamental problem in economics is the optimum allocation of scarce resources among competing activities – a problem that can be expressed in mathematical form. In 1947 Dantzig discovered that many such planning problems could be formulated as linear computer programs. He also devised an algorithm, known as the ***simplex method***, which was widely adopted for the purpose.

Darboux (Jean) Gaston 1842–1917. French mathematician who contributed immensely to the differential geometry of his time, and to the theory of surfaces. In defining the Riemann integral 1879, he derived the ***Darboux sums*** and used the ***Darboux integrals***.

Darboux was born in Nîmes and educated at the Ecole Normale Supérieure and at the Sorbonne in Paris, where he became a professor 1873 and assistant to mathematician Joseph Liouville.

Leçons sur la théorie générale des surfaces et les applications géométriques du calcul infinitésimal 1887–96 described all his work to date, but dealt mainly with the application of analysis to curves and surfaces, and the study of minimal surfaces and geodesics. In *Leçons sur les systèmes orthogonaux et les coordonnées curvilignes* 1898, Darboux applied the theorem on algebraic integrals to orthogonal systems in *n* dimensions. Important among Darboux's other work were his papers on the theory of integrations, proof of the existence of integrals of continuous functions, and the theory of analytical functions.

Darby Abraham 1677–1717. English iron manufacturer who developed a process for smelting iron ore using coke instead of the more expensive charcoal 1709.

He employed the cheaper iron to cast

strong thin pots for domestic use, and after his death it was used for the huge cylinders required by the new steam pumping-engines. In 1779 his grandson Abraham Darby (1750–1791) constructed the world's first iron bridge, over the river Severn at Coalbrookdale, Shropshire.

Darby was born near Dudley, Worcestershire, and trained in engineering, setting up his own business 1698. In about 1704 he visited Holland and brought back with him some Dutch brass founders, establishing them in Bristol, later moving to Coalbrookdale. They experimented with substituting cast iron for brass in some products, and in 1708 Darby took out a patent for a new way of casting iron pots and other ironware in sand only, without loam or clay. This process cheapened utensils much used by poorer people and at that time largely imported from abroad.

Dart Raymond Arthur 1893–1988. Australian-born South African palaeontologist and anthropologist who in 1924 discovered the first fossil remains of the australopithecenes, early hominids, near Taungs in Botswana.

Dart named them *Australopithecus africanus*, and spent many years trying to prove that they were early humans rather than apes. In the 1950s and 1960s, the Leakey family found more fossils of this type and of related types in the Olduvai Gorge of E Africa, establishing that australopithecines were hominids, walked erect, made tools, and lived as early as 5.5 million years ago. After further discoveries in the 1980s, they are today classified as *Homo sapiens australopithecus*, and Dart's assertions have been validated.

Dart was born in Brisbane and studied at the University of Sydney. He was professor of anatomy at the University of Witwatersrand, Johannesburg, South Africa, 1922–58.

Darwin Charles Robert 1809–1882. English scientist who developed the modern theory of evolution and proposed, with Alfred Russel Wallace, the principle of natural selection. After research in South America and the Galápagos Islands as naturalist on HMS *Beagle* 1831–36, Darwin published *On the Origin of Species by Means of Natural Selection or the Preservation of Favoured Races in the Struggle for Life* 1859. This explained the evolutionary process through the principles of natural and sexual selection. It aroused bitter controversy because it disagreed with the literal interpretation of the Book of Genesis in the Bible.

The theory of natural selection concerned the variation existing between members of a sexually reproducing population. According to Darwin, those members with variations better fitted to the environment would be more likely to survive and breed, subsequently passing on these favourable characteristics to their offspring.

On the Origin of Species also refuted earlier evolutionary theories, such as those of French naturalist J B de Lamarck. Darwin himself played little part in the debates, but his *Descent of Man* 1871 added fuel to the theological discussion, in which English scientist T H Huxley and German zoologist Ernst Haeckel took leading parts.

I see no good reasons why the views given in this volume should shock the religious feelings of anyone.

Charles Darwin

On the Origin of Species 1859

Darwin also made important discoveries in many other areas, including the fertilization mechanisms of plants, the classification of barnacles, and the formation of coral reefs.

Darwin was born in Shrewsbury, the grandson of Erasmus Darwin, and studied medicine at Edinburgh and theology at Cambridge. His first book was *Journal of Researches into the Geology and Natural History of the Various Countries Visited by HMS* Beagle 1839. By 1844 he had enlarged his sketch of ideas to an essay of his conclusions, but then left his theory for eight years while he studied barnacles. In 1858 he was forced into action by the receipt of a memoir from A R Wallace, embodying the same theory.

Darwinism alone is not enough to explain the evolution of sterile worker bees, or altruism. Neo-Darwinism, the current theory of evolution, is a synthesis of Darwin and genetics based on the work of Austrian scientist Gregor Mendel.

Darwin Erasmus 1731–1802. British poet, physician, and naturalist; he was the grandfather of Charles Darwin. He wrote *The Botanic Garden* 1792, which included a versification of the Linnaean system entitled *The Loves of the Plants*, and *Zoonomia* 1794–96, which anticipated aspects of evolutionary theory, but tended to French naturalist J B de Lamarck's interpretation.

Dausset Jean Baptiste Gadriel Joachim 1916– . French immunologist who was an early pioneer in the study of the human major histocompatability complex (MHC), the system of genes and their corresponding antigens on the surface of cells that enables the body to recognize its own cells.

Dausset's experience in World War II led him to question why individuals responded differently to blood transfusions and prompted him to study how the immune system recognizes its own cells and distinguishes them from foreign ones. He discovered that a set of genes called the human leucocyte antigen (HLA) genes cause different recognition molecules to be expressed on the surface of cells in different individuals. He extended these studies to show the phenomenon of tissue-typing by grafting small pieces of skin between individuals of the same family, demonstrating the need for matching of HLA expression in individuals prior to tissue or organ transplantation.

Dausset was born in Toulouse and studied medicine at the University of Paris. His studies were delayed by the outbreak of World War II, during which he was a member of the French Medical Corps. He received his medical degree 1945.

David Armand 1826–1900. French naturalist and Lazarist missionary who discovered a new species of deer 1865, now known as Père David's deer, *Elaphurus davidianus*.

Born in Espelette France in 1826, David developed a keen interest in both natural sciences and religion as a child, and later became a missionary in China. He discovered the deer in a game park near Beijing belonging to the Emperor. The deer were classified by Henri Milne-Edwards 1866 and several of them were brought to Europe. In 1895 a flood destroyed the majority of the Chinese herd, and the rest were killed during the Boxer Rebellion 1900. The deer in European zoos were the only surviving specimens.

da Vinci see ◊Leonardo da Vinci, Italian Renaissance artist.

Davis William Morris 1850–1934. US physical geographer who analysed landforms. In the 1890s he developed the organizing concept of a regular cycle of erosion, a theory that dominated geomorphology and physical geography for half a century.

Davis was born in Philadelphia and studied science at Harvard, where he taught 1877–1912.

Davis proposed a standard stage-by-stage life cycle for a river valley, marked by youth (steep-sided V-shaped valleys), maturity (flood-plain floors), and old age, as the river valley was imperceptibly worn down into the rolling landscape which he termed a 'peneplain'. On occasion these developments, which Davis believed followed from the principles of Scottish geologist Charles Lyell, could be punctuated by upthrust, which would rejuvenate the river and initiate new cycles.

Davisson Clinton Joseph 1881–1958. US physicist who in 1927 made the first experimental observation of the wave nature of electrons. George Thomson carried through the same research independently, and in 1937 they shared the Nobel Prize for Physics.

Davisson was born in Bloomington, Illinois, and studied at the University of Chicago. He worked for the Western Electric Company (later Bell Telephone) in New York 1917–46.

With Lester Germer (1896–1971), Davisson discovered that electrons can undergo diffraction, so proving French physicist Louis de Broglie's theory that electrons, and therefore all matter, can show wavelike structure.

Davy Humphry 1778–1829. English chemist. He discovered, by electrolysis, the metallic elements sodium and potassium in 1807, and calcium, boron, magnesium, strontium, and barium in 1808. In addition, he established that chlorine is an element and proposed that hydrogen is present in all acids. He invented the safety lamp for use in mines where methane was present, enabling miners to work in previously unsafe conditions. Knighted 1812, baronet 1818.

Davy's experiments on electrolysis of aqueous solutions from 1800 led him to suggest its large-scale use in the alkali industry. He theorized that the mechanism of electrolysis could be explained in terms of species

The eternal laws Preserve one glorious wise design; Order amidst confusion flows, And all the system is divine.

Humphry Davy

From Davy's notebooks at the Royal Institution

that have opposite electric charges, which could be arranged on a scale of relative affinities – the foundation of the modern electrochemical series. His study of the alkali metals provided proof of French chemist Antoine Lavoisier's idea that all alkalis contain oxygen.

Davy was born in Penzance, Cornwall, and apprenticed to an apothecary. As a laboratory assistant in Bristol in 1799, he discovered the respiratory effects of laughing gas (nitrous oxide). He moved to the Royal Institution, London, 1801. In 1813 he took on Michael Faraday as a laboratory assistant.

Davy introduced a chemical approach to agriculture, the tanning industry, and mineralogy; he designed an arc lamp for illumination and an electrolytic process for the desalination of sea water.

We are survival machines – robot vehicles blindly programmed to preserve the selfish molecules known as genes. This is a truth which still fills me with astonishment.

Richard Dawkins

The Selfish Gene Preface

Dawkins (Clinton) Richard 1941– . British zoologist whose book *The Selfish Gene* 1976 popularized the theories of sociobiology (social behaviour in humans and animals in the context of evolution). In *The Blind Watchmaker* 1986 he explained the modern theory of evolution.

Dawkins was born in Nairobi, Kenya, and educated at Oxford, where from 1975 he held academic posts.

In *The Selfish Gene* he argued that genes – not individuals, populations, or species – are the driving force of evolution. He suggested an analogous system of cultural transmission in human societies, and proposed the term 'mimeme', abbreviated to 'meme', as the unit of such a scheme. He considered the idea of God to be a meme with a high survival value. His contentions were further developed in *The Extended Phenotype* 1982, primarily an academic work.

De Bary Heinrich Anton 1831–1888. German botanist and founder of mycology. He demonstrated that fungi were not spontaneously generated as many scientists believed, and showed that rusts and smuts were the cause and not the result of disease in cereals. He provided evidence that the spores from the potato blight fungus could cause potato blight in previously healthy plants and was the first to use the term 'symbiosis' 1879 to describe the close and mutually beneficial relationship between some plant species.

De Bary was born in Frankfurt-am-Main, Germany. He was educated at the Gymnasium at Frankfurt and then went to Marburg University to study medicine. In 1850 he went to Berlin to study botany under Alexander Braun. He graduated 1853 with a degree in medicine, with a dissertation entitled 'De plantorum generatione sexuali'. He began practising medicine in Frankfurt, but soon gave this up, to work as Hugo von Mohl's assistant in the botany department at Tübingen University. This led to him setting up his own Botanical Laboratory in Freiburg im Breisgau 1855.

De Beer Gavin Rylands 1899–1972. English zoologist who made important contributions to embryology and evolution. He disproved the germ-layer theory and developed the concept of paedomorphism (the retention of juvenile characteristics of ancestors in mature adults). Knighted 1954.

De Beer was born in London, studied at Oxford and lectured there. In 1945 he became professor of embryology at University College, London, then was director of the British Museum (Natural History) 1950–60.

In *Introduction to Experimental Embryology* 1926, De Beer observed that certain cartilage and bone cells are derived from the outer ectodermal layer of the embryo. This finally disproved the germ-layer theory, according to which these cells are formed from the mesoderm. Paedomorphism was first described in *Embryos and Ancestors* 1940, refuting the theory that the embryonic development of an organism repeats the adult stages of the organism's evolutionary ancestors.

De Beer's studies of the fossil *Archaeopteryx*, the earliest known bird, led him to propose mosaic evolution – whereby evolutionary changes occur piecemeal – to explain the presence of both reptilian and avian features.

de Broglie Maurice and Louis. French physicists; see ◊Broglie.

Debye Peter 1884–1966. Anglicized name of Petrus Josephus Wilhelmus Debije. Dutch-born US physicist. A pioneer of X-ray powder crystallography, he also worked on polar molecules, dipole moments, molecular structure, and polymers. The Debye–Hückel theory, developed with German chemist Erich Hückel, concerns the ordering of ions in solution. Nobel Prize for Chemistry 1936.

Debye was born in Maastricht and studied at the Technische Hochschule in Aachen, Germany, and at the University of Munich. He held a series of professorships in Switzerland and Germany, starting at Zürich 1910. In 1934 he went to the Max Planck Institute, Berlin. He was lecturing at Cornell University in the USA in 1940 when Germany invaded the Netherlands, so he remained at Cornell as professor 1940–52.

Debye's first major contribution was a modification of Einstein's theory of specific heats to include compressibility and expansivity.

Debye's studies of dielectric constants led to the explanation of their temperature dependence and of their importance in the interpretation of dipole moments as indicators of molecular structure. This earned him the Nobel prize, and the unit of dielectric constant is now called the ***debye***.

In crystallography he showed that the thermal motion of the atoms in a solid affects the X-ray interfaces and that randomly oriented particles can produce X-ray diffraction patterns of a characteristic kind. Powder X-ray diffraction analysis eliminated the need for large single crystals.

In the 1930s Debye showed that sound waves in a liquid can behave like a diffraction grating.

Dedekind (Julius Wilhelm) Richard 1831–1916. German mathematician who made contributions to number theory. In 1872 he introduced the ***Dedekind cut*** (which divides a line of infinite length representing all real numbers) to define irrational numbers in terms of pairs of sequences of rational numbers.

Dedekind was born in Brunswick and studied at Göttingen. He was professor at the Technische Hochschule in Brunswick 1862–94.

In 1858 he succeeded in producing a purely arithmetic definition of continuity and an exact formulation of the concept of the irrational number. This led to the Dedekind cut, explained in his book *Stetigkeit und irrationale Zahlen* 1872. In *Was sind und was sollen die Zahlen?* 1888 he devised axioms that formally and exactly represented the logical concept of whole numbers. The factorization of algebraic numbers and the theory of the ideal, which he described 1879–94, was fundamental to modern algebra; and he introduced the concept of dual groups, a precursor of lattice theory.

de Duve Christian Rene 1917– . British-born Belgian biochemist who won the Nobel Prize for Physiology or Medicine with Albert Claude and George Palade 1974 for his work on the structure and function of lysosomes (cell organelles containing enzymes that can break down the cell).

He worked with Palade using Palade's technique, differential centrifugation, to examine the function of individual inner parts of secretory cells, specifically the role of individual cell organelles in the manufacture of enzymes secreted by the cell. In these studies, de Duve discovered small, membrane-bound organelles, called lysosomes, that contain many enzymes capable of breaking down redundant structures inside and outside the cell to make way for more useful ones. Lysosomes, therefore, play an important part in cell and tissue remodelling, such as that which occurs in the nervous system and in embryonic development.

De Duve was born in Thames-Ditton in Surrey but moved to study medicine at Louvain 1947, where he was appointed professor of biochemistry 1951. He later moved to New York to take up the chair of biochemistry at the Rockefeller University 1962.

Dee John 1527–1608. English alchemist, astrologer, and mathematician who claimed to have transmuted metals into gold, although he died in poverty. He long enjoyed the favour of Elizabeth I, and was employed as a secret diplomatic agent.

De Forest Lee 1873–1961. US physicist and inventor who in 1906 invented the triode valve, which contributed to the development of radio, radar, and television.

Ambrose Fleming invented the diode valve 1904. De Forest saw that if a third electrode were added, the triode valve would serve as an amplifier as well as a rectifier, and radio communications would become a practical possibility.

De Forest was born in Council Bluffs, Iowa, and studied at Yale. Working for the Western Electric Company in Chicago, he devised ways of rapidly transmitting wireless signals, his system being used 1904 in the first wireless news report (of the Russo-Japanese War).

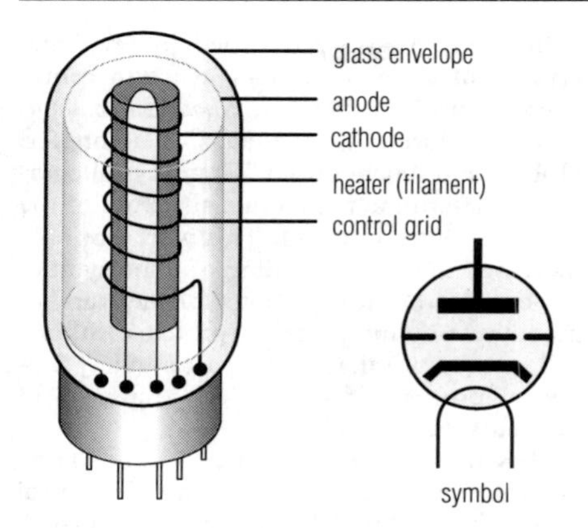

De Forest *De Forest developed the triode tube, a three-electrode vacuum tube in which a control grid allows the current flowing through the valve to be controlled by the voltage at the grid. The triode was used in amplifying circuits for radio and early computers, until largely superseded by transistors and other solid-state devices.*

De Forest set up his own wireless telegraph company, but nearly went bankrupt twice. He was prosecuted for attempting to use the US mail to defraud, by seeking to promote the 'worthless' audion tube (as he called the triode valve).

In 1912, De Forest arranged triode valves to transmit both speech and music by radio, and in 1916 he set up a radio station and began broadcasting news.

In 1923, De Forest demonstrated an early system of motion pictures carrying a soundtrack, called phonofilm. Its poor quality, and lack of interest from filmmakers, led to its demise, though the principle was later adopted.

de Gennes Pierre-Gilles 1932 French physicist who worked on liquid crystals and polymers. He showed how mathematical models, developed for studying simpler systems, are applicable to such complicated systems. Nobel Prize for Physics 1991.

It had been known for a long time that liquid crystals scatter light in an unusual way but all early explanations failed. De Gennes found the explanation in the special way that the molecules of a liquid crystal are arranged. According to de Gennes, the molecules are arranged in a similar way to the molecules of a magnet, so that they point in the same direction. De Gennes found similar analogies between the behaviour of molecules in magnetic materials and polymers. This led to the formulation of laws from which simple relations between different properties of polymers can be deduced. In this way, predictions could be made about unknown properties – predictions which have been confirmed by experiment.

De Gennes was born in Paris, France and graduated from the Ecole Normalen 1955. From 1955 to 1959 he was a research engineer at the Atomic Energy Centre at Saclay, working mainly on neutron scattering and magnetism. A brief period at Berkeley, California, and over two years in the French navy followed before he became assistant professor at Orsay. In 1971 he became professor at the College de France, Paris.

De Havilland Geoffrey 1882–1965. British aircraft designer who designed and whose company produced the Moth biplane, the Mosquito fighter-bomber of World War II, and in 1949 the Comet, the world's first jet-driven airliner to enter commercial service. Knighted 1944.

After designing a fighter and a bomber for use in World War I, he founded the De Havilland Aircraft Company 1920. This was eventually absorbed into the Hawker Siddeley conglomerate.

De Havilland was born near High Wycombe, and as a youth designed and built steam cars and motorcycles. In 1908–09, he constructed his first aeroplane. De Havilland's design had a better aerodynamic shape than earlier biplanes. However, he had never flown before (indeed, he had only ever seen one aircraft flying in the distance), and his first flight ended in the aircraft being wrecked.

In the 1920s and 1930s the De Havilland Company produced a series of light transport aircraft and the Moth series of private planes, starting with the Cirrus Moth 1925.

The all-wood Mosquito was at first rejected by the Air Ministry, but went into squadron service in Sept 1941. Faster than the Spitfire, it could out-fly virtually anything in the air.

After World War II the De Havilland company put a range of jet-powered aircraft into production, many of which used the company's own engines.

Dehmelt Hans G 1923– . German-born US physicist who contributed to the development and application of the ion trap technique, used to store single atoms long enough to make accurate spectroscopic measurements on single atoms. Nobel Prize for

Physics 1989 (shared with Norman Ramsey and Wolfgang Paul).

Dehmelt used the ion trap technique mainly for studying electrons and in 1973 he succeeded in observing a single electron in an ion trap, and holding the electron in the trap for months at a time. It proved possible to measure the magnetic moment of the trapped electron accurately to 12 digits, one of the most accurate measurements of a physical quantity ever made. Dehmelt was also able to trap and study single ions (charged atoms), pointing the way to improved atomic clocks.

Dehmelt was born and educated in Berlin, Germany. He studied physics at the University of Göttingen, attending classes held by Wolfgang Paul. After gaining his doctorate 1949, he was invited to study at the microwave laboratory at Duke University, North Carolina, USA. He later moved to the University of Washington, Washington State, USA, demonstrating 1956 the usefulness of ion trapping for high accuracy spectroscopy.

Dehn Max Wilhelm 1878–1952. German-born US mathematician who in 1907 provided one of the first systematic studies of topology. His work was mainly concerned with the geometric properties of polyhedra.

Dehn was born in Hamburg and studied at Göttingen University under David Hilbert. After World War I Dehn became professor of mathematics at Frankfurt University, but lost his position 1935 because of the Nazi anti-Semitism laws. He emigrated 1940 to the USA, and worked from 1945 at the Black Mountain College in North Carolina.

Dehn found a solution to one of Hilbert's 23 unsolved problems (concerning the existence of tetrahedra with equal bases and heights, but not equal in the sense of division and completeness).

Dehn proved an important theorem on topological manifolds 1910 that came to be known as Dehn's lemma, but was later found not to apply in all circumstances. Dehn continued to work on topological problems of transformation and isomorphism.

Deisenhofer Johann 1943– . German chemist who was the first to apply the technique of X-ray crystallography (the use of X-rays to discern atomic structure) to biological molecules. He shared the 1988 Nobel Prize for Chemistry with Robert Huber and Harmut Michel for their determination of the three-dimensional structure of a photosynthetic reaction centre from a bacterium.

In 1982, Michel succeeded in preparing crystals of a photosynthetic reaction centre from a purple bacterium, *Rhodipseudomonas viridis*. During 1982–85, Deisenhofer, Huber, and Michel used X-ray crystallography to determine the atomic structure of the reaction centre in detail. This work not only increased the understanding of photosynthesis, but also has implications for the study of membrane-bound proteins (proteins within a cell membrane) that play a part in many biological processes, such as the transport of nutrients into cells.

Deisenhofer was born in Zusamaltheim, Bavaria, and educated at the Technical University at Munich and the Max Planck Institute for Biochemistry in Munich (and later Martinsried, near Munich). Working in Huber's laboratory at Martinsried, he obtained his doctorate 1974 for the application of X-ray crystallography to biological molecules. In 1988, Deisenhofer moved to the United States and became professor of biochemistry at the University of Texas in Dallas.

De la Beche Henry Thomas, (born Beach) 1796–1855. English geologist who secured the founding of the Geological Survey 1835, a government-sponsored geological study of Britain, region by region. His main work is *The Geological Observer* 1851.

Beach was born in London, went to military school and served in the Napoleonic Wars. Gentrifying his name, he joined the Geological Society of London 1817 and travelled extensively during the 1820s through Great Britain and Europe. He persuaded the government of the need for a geological survey, and became its first director.

De la Beche wrote books of descriptive stratigraphy, above all on the Jurassic and Cretaceous rocks of the Devon and Dorset area. He also conducted important fieldwork on the Pembrokeshire coast and in Jamaica. He prided himself upon being a scrupulous fieldworker and a meticulous artist. Such works as *Sections and Views Illustrative of Geological Phenomena* 1830 and *How to Observe* 1835 insisted upon the primacy of facts and sowed distrust of theories.

de la Rue Warren 1815–1889. British astronomer and instrumentmaker who pioneereed celestial photography. Besides inventing the first photoheliographic telescope, he took the first photograph of a solar eclipse 1860 and used it to prove that the prominences observed during an eclipse are of solar rather than lunar origin.

De la Rue was born in Guernsey and joined his father in the printing business. He was one of the first printers to adopt electrotyping and in 1851 he invented the first envelope-making machine. He also invented the silver chloride battery.

De la Rue's interest in new technologies led him to apply the art of photography to astronomy. He modified his 33-cm/13-in telescope to incorporate a wet collodion plate. His first photographs were of the Moon, and their success encouraged him to build and equip a new observatory in Cranford, Middlesex.

There, de la Rue began a daily sequence of photographs of the Sun. He designed a photoheliographic telescope to take to Spain for the 1860 eclipse, after which it was set up at the Kew Observatory, London. He used it to map the surface of the Sun and study the sunspot cycle. This work led to his being able to show that sunspots are in fact depressions in the Sun's atmosphere.

Delbruck Max 1906–1981. German-born US biologist who pioneered techniques in molecular biology, studying genetic changes occurring when viruses invade bacteria. He was awarded the Nobel Prize for Physiology or Medicine 1969, which he shared with Salvador Luria and Alfred Hershey (1908–).

de Lesseps Ferdinand, Vicomte. French engineer; see ◊Lesseps, Ferdinand, Vicomte de Lesseps.

Democritus *c.* 460–*c.* 370 BC. Greek philosopher and speculative scientist who made a significant contribution to metaphysics with his atomic theory of the universe: all things originate from a vortex of tiny, indivisible particles, which he called atoms, and differ according to the shape and arrangement of their atoms.

Democritus' discussion of the constant motion of atoms to explain the origins of the universe was the most scientific theory proposed in his time. His concepts come to us through Aristotle's work in this area.

Democritus was born in Thrace and travelled widely in the East. He is reputed to have written more than 70 works, although only fragments have survived.

According to Democritus' theory, atoms cannot be destroyed (an idea similar to the modern theory of the conservation of matter) and they exist in a vacuum or void, which corresponds to the space between atoms. Atoms of a liquid are smooth and round; atoms of a solid are jagged and catch on to each other. Atoms differ only in shape, position, and arrangement.

De Morgan Augustus 1806–1871. British mathematician who initiated and developed a theory of relations in logic. He devised a symbolism that could express such notions as the contradictory, the converse, and the transitivity of a relation, as well as the union of two relations.

De Morgan was born in Madura (now Madurai), Tamil Nadu, India, and studied at Cambridge. He was the first professor of mathematics at University College, London, 1828–58.

In his books *Formal Logic* 1847 and *Syllabus of a Proposed System of Logic*, De Morgan developed a logic of noun expressions. He also extended his syllogistic vocabulary using definitions, giving rise to new kinds of inferences, both direct (involving one premise) and indirect (involving two premises). He was thus able to work out purely structural rules for transforming a premise or pair of premises into a valid conclusion.

Descartes René 1596–1650. French philosopher and mathematician. He believed that commonly accepted knowledge was doubtful because of the subjective nature of the senses, and attempted to rebuild human knowledge using as his foundation *cogito ergo sum* ('I think, therefore I am'). He also believed that the entire material universe could be explained in terms of mathematical physics, and founded coordinate geometry as a way of defining and manipulating geometrical shapes by means of algebraic expressions. Cartesian coordinates, the means by which points are represented in this system, are named after him. Descartes also established the science of optics, and helped to shape contemporary theories of astronomy and animal behaviour.

Descartes identified the 'thinking thing' (*res cogitans*) or mind with the human soul or consciousness; the body, though somehow interacting with the soul, was a physical machine, secondary to, and in principle separable from, the soul. He held that everything

Except our own thoughts, there is nothing absolutely in our power.

René Descartes

Descartes *French philosopher and mathematician René Descartes, who reformulated scientific thinking in the 17th century with his attempts to describe the whole of knowledge using mathematics. He devised the Cartesian system of coordinate geometry, which allows points to be described numerically on a set of perpendicular axes.*

has a cause; nothing can result from nothing. He believed that, although all matter is in motion, matter does not move of its own accord; the initial impulse comes from God. He also postulated two quite distinct substances: spatial substance, or matter, and thinking substance, or mind. This is called 'Cartesian dualism', and it preserved him from serious controversy with the church.

Descartes was born in La Haye (renamed Descartes in his honour), south of Tours, and studied at Poitiers. He served in the army of Prince Maurice of Orange, and in 1619, while travelling through Europe, decided to apply the methods of mathematics to metaphysics and science. He settled in the Netherlands 1628, where he was more likely to be free from interference by the Catholic church. In 1649 he visited the court of Queen Christina of Sweden, and shortly thereafter he died in Stockholm.

Descartes's great work in mathematics was *La Géométrie/Geometry* 1637. Although not the first to apply algebra to geometry, he was the first to apply geometry to algebra. He was also the first to classify curves systematically, separating 'geometric curves' (which can be precisely expressed as an equation) from 'mechanical curves' (which cannot). Other works include *Discourse on Method* 1637, *Meditations on the. First Philosophy* 1641, and *Principles of Philosophy* 1644, and numerous books on physiology, optics, and geometry.

Deslandres Henri Alexandre 1853–1948. French physicist and astronomer. His work in spectroscopy led to his construction of the spectroheliograph for his solar studies. In 1902 he predicted that the Sun would be found to be a source of radio waves, and 40 years later this was confirmed.

Deslandres was born and educated in Paris. In 1897 he moved to the observatory at Meudon, where he became director 1907. The Paris and Meudon Observatories were combined in 1926. Deslandres retired 1929.

Deslandres's early scientific work in spectroscopy led to the formulation of two simple empirical laws describing the banding patterns in molecular spectra; the laws were later found to be easily explained using quantum mechanics.

In 1891 US astronomer George Ellery Hale and Deslandres independently constructed a photographic device for studying the solar spectrum in detail. Deslandres's spectroheliograph was finished a year later than Hale's spectrograph but was particularly well adapted to studying the solar chromosphere.

Desmarest Nicolas 1725–1815. French naturalist who became a champion of volcanist geology, countering the widely held belief that all rocks were sedimentary. He wrote extensively for the *Encyclopédie.*

Desmarest, born near Troyes, moved to Paris and occupied minor government offices. Gradually he staked out for himself a scientific career, publishing charts of the English Channel and making geological tours.

Studying the large basalt deposits of central France, Desmarest traced their origin to ancient volcanic activity in the Auvergne region. In 1768 he produced a detailed study of the geology and eruptive history of the volcanoes responsible. However, he did not believe that all rocks had igneous origin, and emphasized the critical role of water in the shaping of the Earth's history.

Désormes Charles Bernard 1777–1862. French physicist and chemist who determined the ratio of the specific heats of gases 1819. He did this and almost all his scientific work in collaboration with his son-in-law Nicolas Clément (1779–1841).

Désormes was born in Dijon, Côte d'Or. He was a student at the Ecole Polytechnique in Paris from 1794, when it opened, and subsequently worked there as a demonstrator. Désormes met Clément at the Ecole Polytechnique 1801, beginning a scientific collaboration that lasted until 1824. He left the Ecole 1804 to establish an alum refinery at Berberie, Oise, with Clément and Joseph Montgolfier, who had earlier pioneered balloon flight. Désormes was elected counsellor for Oise 1830 and in 1848 to the national assembly, in which he sat with the Republicans.

Clément and Désormes correctly determined the composition of carbon disulphide (CS_2) and carbon monoxide (CO) 1801–02. In 1806 they elucidated all the chemical reactions that take place during the production of sulphuric acid by the lead chamber method, as used in industrial chemistry. In 1813 they made a study of iodine and its compounds.

De Vaucouleurs Gerard Henri 1918– . French-born US astronomer who carried out important research into extragalactic nebulae. In 1956 he suggested that there is a pattern in the location of nebulae, clusters of stars formerly thought to be randomly scattered.

De Vaucouleurs was educated at the University of Paris. He worked in Australia 1951–57 and then moved to the USA. In 1965 he became professor of astronomy at the University of Texas, Austin.

The material he published 1956 seemed to indicate that a local supercluster of nebulae exists, which includes our own Milky Way stellar system. He suggested a model in which the great Virgo cluster might be 'a dominant congregation not too far from its central region'. As evidence for its existence, De Vaucouleurs pointed out the similarity in position and extent of a broad maximum of cosmic radio noise.

De Vries Hugo Marie 1848–1935. Dutch botanist who conducted important research on osmosis in plant cells and was a pioneer in the study of plant evolution. His work led to the rediscovery of Austrian biologist Gregor Mendel's laws and the discovery of spontaneously occurring mutations.

De Vries was born in Haarlem and studied medicine at Heidelberg and Leiden. He spent most of his academic career at the University of Amsterdam, and was professor of botany there 1881–1918.

Using a plant called *Oenothera lamarckiana*, he began a programme of plant-breeding experiments 1892. In 1900, he formulated the same laws of heredity that – unknown to De Vries – Mendel had discovered 1866. De Vries further found that occasionally an entirely new variety of *Oenothera* appeared and that this variety reappeared in subsequent generations. De Vries called these new varieties mutations. He assumed that, in the course of evolution, those mutations that were favourable for the survival of the individual persisted unchanged until other, more favourable mutations occurred. He summarized this work in *Die Mutationstheorie/The Mutation Theory* 1901–03.

Dewar James 1842–1923. Scottish chemist and physicist who invented the vacuum flask (Thermos) 1872 during his research into the properties of matter at extremely low temperatures. Knighted 1904.

Working on the liquefaction of gases, Dewar found, in 1891, that both liquid oxygen and ozone are magnetic. In 1895 he became the first to produce liquid hydrogen, and in 1899 succeeded in solidifying hydrogen at a temperature of –259°C/–434°F. He also invented the explosive cordite 1889.

Dewar was born in Kincardine, Fife, and studied at Edinburgh. He became professor of chemistry at Cambridge 1875, as well as at the Royal Institute in London 1877.

Every gas had now been liquefied and solidified except one – helium. Dutch physicist Heike Kamerlingh Onnes managed to liquefy helium using Dewar's techniques in 1908, and Dewar was then able to achieve temperatures within a degree of absolute zero (–273°C/–459°F) by boiling helium at low pressure.

Dewar also carried out work in spectroscopy, particularly concerned with the absorption spectra of metals. He investigated properties such as chemical reactivity, electrical resistance, strength, and phosphorescence at low temperature. Feathers, for example, were found to be phosphorescent at these temperatures.

Dicke Robert Henry 1916– . US physicist who in 1964 proposed a version of the Big Bang theory known as the 'hot Big Bang': he suggested that the present expansion of the universe had been preceded by a collapse in which high temperatures had been generated.

Dicke was born in St Louis, Missouri, and

studied at Princeton and the University of Rochester. From 1946 he was on the staff at Princeton, becoming professor 1957.

When Arno Penzias and Robert Wilson announced they had detected an unexpected and relatively high level of radiation at a wavelength of 7 cm/2.8 in, with a temperature of about 3°5 K (–270°C/–453°F), Dicke proposed that this was cosmic black-body radiation from the hot Big Bang.

Dicke carried out experiments to verify the supposition of the general theory of relativity that a gravitational mass is equal to its inertial mass. He was able to establish the equality to an accuracy of one part in 10^{11}. In 1961, he put forward a theory (the Brans–Dicke theory) that the gravitational constant varies with time (by about 10^{-11} per year). Experiment has not supported this idea.

Dick-Read Grantly 1890–1959. British gynaecologist. In private practice in London 1923–48, he developed the concept of natural childbirth: that by the elimination of fear and tension, labour pain could be minimized and anaesthetics, which can be hazardous to both mother and child, rendered unnecessary.

Dicksee Cedric Bernard 1888–1981. British engineer who was a pioneer in developing the compression-ignition (diesel) engine into a suitable unit for road transport. This became standard in commercial vehicles.

Dicksee received his technical education at the Northampton Engineering College. He worked in the USA 1919–26 for the Westinghouse Electric and Manufacturing Company, then joined the manufacturing subsidiary of the London General Omnibus Company (LGOC) 1928. During World War II he worked on combustion-chamber design for the De Havilland series of jet engines.

Dicksee built his first engine 1930 and kept improving on it. The 1933 model became the standard; a further development of this engine used combustion chambers of a toroidal shape and was subsequently adopted for larger engines in the company's range.

Dicksee's engines ran at speeds ranging from 1,800 to 2,400 (governed) rpm, higher than comparable engines of the day. With this performance, the way was opened for the adoption of compression-ignition engines instead of petrol engines for road transport.

Dickson Leonard Eugene 1874–1954. US mathematician who gave the first extensive exposition of the theory of fields. In *History of the Theory of Numbers* 1919–23, now a standard work, he investigated abundant numbers, perfect numbers, and Pierre de Fermat's last theorem.

Dickson was born in Independence, Iowa, and studied at the universities of Texas and Chicago, spending nearly all his academic career at the latter and becoming professor 1910.

Dickson's work in mathematics spanned many topics, including the theory of finite and infinite groups, the theory of numbers, algebras and their arithmetics, and the history of mathematics. Investigating the history of the theory of numbers, Dickson studied the work of Diophantus, who lived in Alexandria in the 3rd century AD. Diophantus assumed that every positive integer is the sum of four squares, and Dickson proved an extension of this theory.

Diderot Denis 1713–1784. French philosopher. He is closely associated with the Enlightenment, the European intellectual movement for social and scientific progress, and was editor of the enormously influential *Encyclopédie* 1751–80.

An expanded and politicized version of the English encyclopedia 1728 of Ephraim Chambers (*c.* 1680–1740), this work exerted an enormous influence on contemporary social thinking with its materialism and anticlericalism. Its compilers were known as Encyclopédistes.

Diderot's materialism, most articulately expressed in *D'Alembert's Dream*, published after Diderot's death, sees the natural world as nothing more than matter and motion.

His account of the origin and development of life is purely mechanical.

Diels Otto Paul Hermann 1876–1954. German chemist. In 1950 he and his former assistant, Kurt Alder, were jointly awarded the Nobel Prize for Chemistry for their research into the synthesis of organic chemical compounds.

In 1927 Diels dehydrogenated cholesterol to produce 'Diels hydrocarbon' ($C_{18}H_{16}$), an aromatic hydrocarbon closely related to the skeletal structure of all steroids, of which cholesterol is one. In 1935 he synthesized it. This work proved to be a turning point in the understanding of the chemistry of cholesterol and other steroids.

Diels was born in Hamburg and studied at Berlin. He was director of the Chemical

Institute at the Christian Albrecht University in Kiel 1916–48.

Working with his assistant Alder, Diels developed the diene synthesis, which first achieved success in 1928, when they combined cyclopentadiene with maleic anhydride (*cis*-butenedioic anhydride) to form a complex derivative of phthalic anhydride. Generally, conjugated dienes (compounds with two double bonds separated by a single bond) react with dienophiles (compounds with one double bond activated by a neighbouring substituent such as a carbonyl or carboxyl group) to form a six-membered ring.

Diels published a textbook, *Einführung in die organische Chemie/Introduction to Organic Chemistry* 1907.

Diesel Rudolf Christian Karl 1858–1913. German engineer who patented the diesel engine. He began his career as a refrigerator engineer and, like many engineers of the period, sought to develop a better power source than the conventional steam engine. Able to operate with greater efficiency and economy, the diesel engine soon found a ready market.

Born in Paris, Diesel moved to Germany after the outbreak of the Franco-Prussian War 1870, and studied at Munich Polytechnic.

His ideas for an engine where the combustion would be carried out within the cylinder were first published 1893, one year after he had taken out his first patent. In his first engine, Diesel is thought to have used coal dust as a fuel, but he later discarded this along with several other types in favour of a form of refined mineral oil. A very high pressure must be used to compress the air before fuel injection, and he was nearly killed when a cylinder head blew off one of his prototype engines. In 1899, he founded his own manufacturing company in Augsburg.

Dillenius Johann Jacob 1687–1747. German botanist, botanical artist, and physician who studied and produced several papers on cryptogams (lower plants that do not have seeds or flowers), such as mosses and algae.

Dillenius was born in Darmstadt, Germany. He studied medicine at Giessen, where his father was the professor of medicine, and graduated 1713. While he was a doctor, he became interested in botany, producing several papers on cryptogams, such as mosses and algae. However, because he was critical of Bachmann's system of classification, he did not succeed in establishing a reputation for himself in Germany.

In 1721, he went to England where he worked with Sherard to produce an encyclopedia. In 1724, he was elected a Fellow of the Royal Society and was awarded an honorary MD by St John's College, Oxford 1734. When world renowned naturalist Carolus Linnaeus, visited Oxford 1736, it was as Dillenius's guest. In 1741, his *Historia Muscorum* was published and six years later he died of a fit of apoplexy in Oxford.

Diophantus lived AD 250. Greek mathematician in Alexandria whose *Arithmetica* is one of the first known works on problem solving by algebra, in which both words and symbols are used.

His main mathematical study was in the solution of what are now known as 'indeterminate' or 'Diophantine' equations – equations that do not contain enough facts to give a specific answer but enough to reduce

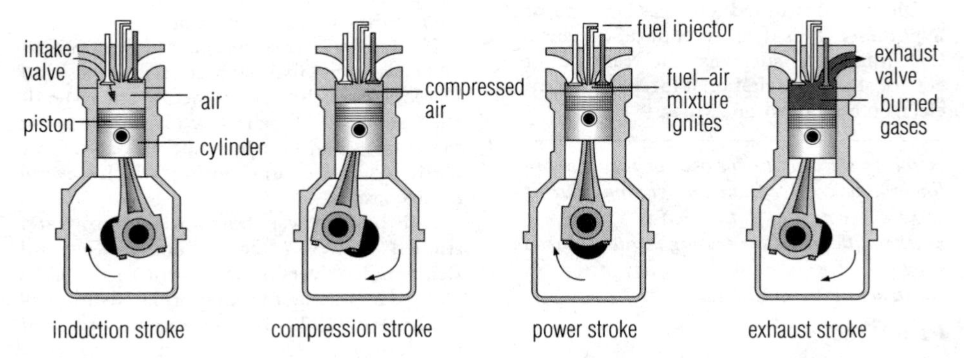

Diesel *In a diesel engine, fuel is injected on the power stroke into hot compressed air at the top of the cylinder, where it ignites spontaneously. The four stages are exactly the same as those of the four-stroke or Otto cycle.*

the answer to a definite type. These equations have led to the formulation of the theory of numbers, regarded as the purest branch of present-day mathematics.

In the solution of equations Diophantus was the first to devise a system of abbreviating the expression of his calculations by means of a symbol representing the unknown quantity. Because he invented only one symbol for the unknown, however, in equations requiring two or more variables his work can become extremely confusing in its repetition of that single symbol.

Dirac Paul Adrien Maurice 1902–1984. British physicist who worked out a version of quantum mechanics consistent with special relativity. The existence of antiparticles, such as the positron (positive electron), was one of its predictions. He shared the Nobel Prize for Physics 1933 with Austrian physicist Erwin Schrödinger.

Dirac was born and educated in Bristol and from 1923 at Cambridge, where he was professor of mathematics 1932–69. From 1971 he was professor of physics at Florida State University.

In 1928 Dirac formulated the relativistic theory of the electron. The model was able to describe many quantitative aspects of the electron, including such properties as the half-quantum spin and magnetic moment, and introduced the first antiparticle.

Dirac noticed that those particles with half-integral spins obeyed statistical rules different from the other particles. For these particles, Dirac worked out the statistics, now called ***Fermi–Dirac statistics*** because Italian physicist Enrico Fermi had done very similar work. These are used, for example, to determine the distribution of electrons at different energy levels.

Dirac also worked on the large-number hypothesis. This deals with pure, dimensionless numbers, such as the ratio of the electrical and gravitational forces between an electron and a proton, which is 10^{39}.

A theory with mathematical beauty is more likely to be correct than an ugly one that fits some experimental data. God is a mathematician of a very high order, and He used very advanced mathematics in constructing the universe.

Paul Dirac

Scientific American, May 1963

Dirichlet (Peter Gustav) Lejeune 1805–1859. German mathematician whose work in applying analytical techniques to mathematical theory resulted in the fundamental development of the theory of numbers. He was also a physicist interested in dynamics.

Dirichlet was born in Düren, near Aachen, and studied at Cologne and the Collège de France, in Paris. He became professor at the University of Berlin at the age of 23, and at Göttingen 1855.

Dirichlet's papers included studies on quadratic forms, the number theory of irrational fields (including the integral complex numbers), and the theory of units. His most important work was on the convergence of the Fourier series, which led him to the modern notion of a generalized function. In 1837 he presented his first paper on analytic number theory, proving ***Dirichlet's theorem***: in every arithmetical sequence a, $a + d$, $a + 2d$, and so on, where a and d are relatively prime (that is, have no common divisors other than 1), there is an infinite number of prime numbers.

Dirichlet applied his mathematical knowledge to various aspects of physics, such as an analysis of vibrating strings, and to astronomy in a critique of the ideas about the stability of the Solar System proposed by French mathematician Pierre Laplace.

Dirichlet's *Vorlesungen über Zahlentheorie/Lectures in Number Theory* was published posthumously 1863.

Dobzhansky Theodosius, originally Feodosy Grigorevich Dobrzhansky 1900–1975. Ukrainian-born US geneticist who established evolutionary genetics as an independent discipline. He showed that genetic variability between individuals of the same species is very high and that this diversity is vital to the process of evolution.

His book *Genetics and the Origin of Species* 1937 was the first significant synthesis of Darwinian evolutionary theory and Mendelian genetics. Dobzhansky also proved that there is a period when speciation is only partly complete and during which several races coexist.

Dobzhansky was born in Nemirov and studied at Kiev. After teaching at Kiev and Leningrad universities, he went to the USA 1927. He was at the California Institute of Technology 1929–40. He became professor of zoology at Columbia University, New York City, 1940; worked at the Rockefeller Institute (later the Rockefeller University)

1962–71; then he moved to the University of California at Davis.

His book *Mankind Evolving* 1962 had great influence among anthropologists. He wrote on the philosophical aspects of evolution in *The Biological Basis of Human Freedom* 1956 and *The Biology of Ultimate Concern* 1967.

Dodds (Edward) Charles 1899–1973. English biochemist who was largely responsible for the discovery of stilboestrol, a powerful synthetic hormone used in treating prostate conditions and also for fattening cattle. Baronet 1964.

Dodgson Charles Lutwidge, real name of writer Lewis ◊Carroll.

Doisy Edward Adelbert 1893–1986. US biochemist. In 1939 he succeeded in synthesizing vitamin K, a compound earlier discovered by Carl Dam, with whom he shared the 1943 Nobel Prize for Medicine.

Doll (William) Richard Shaboe 1912– . British physician who, working with Bradford Hill (1897–), provided the first statistical proof of the link between smoking and lung cancer in 1950. In a later study of the smoking habits of doctors, they were able to show that stopping smoking immediately reduces the risk of cancer. Knighted 1971.

Dollfus Audouin Charles 1924– . French physicist and astronomer whose preferred method of research is to use polarization of light. He was the first to detect a faint atmosphere round the planet Mercury, in 1950, and established that the Moon does not have one. In pursuit of his detailed investigations into Mars, Dollfus made the first French ascent in a stratospheric balloon.

Dollfus was born and educated in Paris. In 1946 he became an astronomer at the Meudon Observatory there.

Hoping to establish the mineral composition of the Martian deserts, Dollfus checked the polarization of light by several hundreds of different terrestrial minerals to try to find one for which the light matched that polarized by the bright Martian desert areas. He found only one, and that was pulverized limonite.

In 1966 Dollfus discovered Janus, the innermost moon of Saturn, at a time when the rings – to which it is very close – were seen from Earth edgeways on and practically invisible.

Domagk Gerhard Johannes Paul 1895–1964. German pathologist, discoverer of antibacterial sulphonamide drugs. He found in 1932 that a coal-tar dye called Prontosil red contains chemicals with powerful antibacterial properties. Sulphanilamide became the first of the sulphonamide drugs, used – before antibiotics were discovered – to treat a wide range of conditions, including pneumonia and septic wounds. Nobel prize 1939.

Domagk was born in Lagow, Brandenburg (now in Poland), and studied medicine at Kiel. From 1927, he directed research at the Laboratories for Experimental Pathology and Bacteriology of I G Farbenindustrie, Düsseldorf, a dyemaking company. But he also remained on the staff of Münster University, as professor from 1928.

In 1946, Domagk and his coworkers found two compounds (eventually produced under the names of Conteben and Tibione) which, although rather toxic, proved useful in treating tuberculosis caused by antibiotic-resistant bacteria.

Donald Ian 1910–1987. English obstetrician who introduced ultrasound (very high-frequency sound wave) scanning. He pioneered its use in obstetrics as a means of scanning the growing fetus without exposure to the danger of X-rays. Donald's experience of using radar in World War II suggested to him the use of ultrasound for medical purposes.

Donati Giovanni Battista 1826–1873. Italian astronomer who discovered six comets. He made important contributions to the early development of stellar spectroscopy and applied spectroscopic methods to the understanding of the nature of comets. He also studied cosmic meteorology.

Donati was born and educated in Pisa. From 1852 he worked at the observatory in Florence, becoming its director 1864.

During the 1850s Donati was an enthusiastic comet-seeker, and the most dramatic of his discoveries was named after him. ***Donati's comet***, first sighted 1858, had, in addition to its major tail, two narrow extra tails.

Using the new technique of stellar spectroscopy, Donati found that when a comet was still distant from the Sun, its spectrum was identical to that of the Sun. When the comet approached the Sun, it increased in magnitude (brightness) and its spectrum became completely different. Donati concluded that when the comet was still distant from the Sun, the light it emanated was simply a reflection of sunlight. As the comet approached the Sun the material in it became so heated that it emitted a light of its own, which reflected the comet's composition.

Other areas of interest which engaged Donati's attention were atmospheric phenomena and events in higher zones, such as the aurora borealis.

Donders Franciscus Cornelius 1818–1889. Dutch oculist and ophthalmologist who made important advances in the prescribing and fitting of spectacles. He wrote a classic textbook on the optics of the eye and devised sets of letters of different sizes for testing a patients visual acuity, diagnosing sight defects, and prescribing spectacles.

In his book *The Anomalies of Refraction and Accommodation* 1864, Donders separated the errors of refraction (the bending of light) from those of accommodation (the change in the shape of the lens of the eye to maintain focus). He describes the errors in refraction as constant, resulting from light being focused at a constant point either behind or in front of the retina, depending on the shape of the eye. Errors in accommodation, Donders discovered, result from defects in the eye's focusing machinery, such as weak ciliary muscles (the muscles that change the shape of the lens) or a hardening of the lenses.

Donkin Bryan 1768–1855. English engineer who made several innovations in papermaking and printing; he invented the forerunner of the rotary press. He also contributed to food preservation by taking the bottling process introduced by Nicolas Appert and modify it to use metal cans instead of glass bottles.

Donkin was born in Northumberland and apprenticed to a papermaker in Dartford, Kent. In 1803 he perfected a new type of papermaking machine, invented in France, and set up a factory in S London where he manufactured nearly 200 machines.

This success led him to investigate another recent invention from France, the preservation of food by bottling. He established his own company, but later returned to the printing and paper trade. By 1815 he had turned to civil engineering.

In printing, Donkin tackled the problem of increasing the speed of presses. The original flat-bed press, with its back-and-forth movement, was too slow. Donkin arranged four (flat) formes of type around a spindle – a rudimentary rotary press. He introduced a composition of glue and treacle for the inking rollers, an innovation which was still widely used long after his press had been superseded.

Dooley Thomas Anthony 1927–1961. US medical missionary. He founded Medico, an international welfare organization, 1957, after tending refugees in Vietnam who were streaming south after the partition of the country 1954. As well as Medico, he established medical clinics in Cambodia, Laos, and Vietnam.

Born in St Louis, USA, Dooley attended Notre Dame University, joined the navy, and received an MD degree from St Louis University 1953. His assignment in Vietnam aroused his compassion and he devoted the rest of his life to medical work in SE Asia.

Doolittle James Harold 1896–1993. US aviation pioneer who took part in the development of new aircraft designs and more efficient aircraft fuel. During World War II he saw active service and in 1942 led a daring bombing raid over Tokyo. He later participated in the invasion of North Africa and the intensive bombing of Germany.

Born in Alameda, California, USA, Doolittle attended the University of California and served as an army flying instructor during World War I. Later, in the Army Air Corps, he became an aviation specialist, earning an engineering degree from the Massachusetts Institute of Technology (MIT).

Doppler Christian Johann 1803–1853. Austrian physicist who in 1842 described the ***Doppler effect*** and derived the observed frequency mathematically in Doppler's principle.

The Doppler effect is the change in the observed frequency, or wavelength, of waves due to relative motion between the wave source and the observer; for example, the perceived change in pitch of a siren as it approaches and then recedes.

Doppler was born in Salzburg and attended the Polytechnic Institute in Vienna. In 1835 he went to Prague to teach mathematics; in 1850 he became director of the new Physical Institute at the Royal Imperial University of Vienna.

The first experimental test of Doppler's principle was made 1845 in Utrecht in the Netherlands. A locomotive was used to carry a group of trumpeters in an open carriage to and fro past some musicians able to sense the pitch of the notes being played. The variation of pitch produced by the motion of the trumpeters verified Doppler's equations.

Doppler correctly suggested that his principle would apply to any wave motion. He believed that all stars emit white light and that differences in colour are observed on Earth because the motion of stars affects the observed frequency of the light and hence its colour. In fact, stars vary in their basic colour, but it is possible to use the red shift to determine the distances of galaxies.

Dornier Claude 1884–1969. German pioneer aircraft designer who invented the seaplane and during World War II designed the Do-17 'flying pencil' bomber, the mainstay of the Luftwaffe during the air raids on Britain 1940–41.

Born in Bavaria, he founded the Dornier Metallbau works at Friedrichshafen, Lake Constance, in 1922.

Douglas David 1798–1834. Scottish botanist, explorer, and plant collector. As a botanical collector for the Royal Horticultural Society, he travelled extensively throughout North America, including the Oregon Territory and British Columbia, collecting specimens. Altogether, he described around 50 new trees and shrubs and brought back over 100 herbaceous plants to England. He also discovered several new animal species. In 1827, he travelled around the Hudson Bay, down into California, and along the Fraser River. The Douglas fir *Pseudotsuga menziesii* and the primrose genus *Douglasia* are named after him.

Douglas was born in Scone, Perthshire. As a child he disliked school work and persisted in absenting himself so that he could escape to the beautiful countryside around his home town. This prompted his father to remove him from school and place him in an apprenticeship with a gardener at Scone Palace. In 1820, he secured the post of gardener at the botanical garden at Glasgow and then became a botanical collector for the Royal Horticultural Society. During his travels, he is rumoured to have fallen in love with and possibly married an Indian princess. He died in the Sandwich Islands 12 July by being gored to death by a bull in a cattle pit.

Draper Henry 1837–1882. US astronomer who used a spectrograph of his own devising to obtain high-quality spectra of celestial objects. His work is commemorated by the *Henry Draper Catalogue* of stellar spectral types.

Draper was born in Virginia and studied medicine at the University of the City of New York. Travelling in Europe, he became interested in telescope-making and photography. In 1860, he was appointed professor of natural science at the University of the City of New York.

Draper built an observatory in Hastings-on-Hudson, New York. By 1873, he had devised a spectrograph. He was director of the photographic department of the US commission to observe the transit of Venus 1874. Later he spectrographically studied the Moon, Mars, Jupiter, the comet 1881 III, and the Orion nebula.

He also succeeded in obtaining photographs of stars that were too faint to be seen with the same telescope by using exposure times of more than 140 minutes.

The Harvard College Observatory carried out a programme 1886–97 to establish a comprehensive classification scheme for stars and a catalogue of spectra. This project was funded by Draper's legacy and the result was the *Henry Draper Catalogue*.

Drew Charles Richard 1904–1950. US surgeon who demonstrated that plasma had a longer life than whole blood and therefore could be better used for transfusion.

Drew was born in Washington DC and studied at Amherst College and McGill University Medical School. He became professor of surgery at Howard University Medical School 1942.

In 1939, he established a blood bank and was in charge of collecting blood for the British army at the beginning of World War II. In 1941, he became director of the American Red Cross Blood Bank in New York, which collected blood for the US armed forces. Drew resigned, however, when the Red Cross decided to segregate blood according to the race of the donor.

Dreyer John Louis Emil 1852–1926. Danish astronomer, in Ireland from 1874. He compiled three catalogues which together described more than 13,000 nebulae and star clusters; these achieved international recognition as standard reference material. He also wrote a biography of Danish astronomer Tycho Brahe 1890.

Dreyer was born and educated in Copenhagen. In 1874 he was appointed assistant at Lord Rosse's Observatory at Birr Castle in Parsonstown, Ireland. Four years later he took up a similar post at Dunsink Observatory at the University of Dublin, and he was director of the Armagh Observatory 1882–1916.

In 1877 and 1886, Dreyer presented data on new nebulae and corrections to the original catalogue on nebulae and star clusters compiled by English astronomer John Herschel. The Royal Astronomical Society then invited him to compile a comprehensive new catalogue, which Dreyer published 1888 with supplementary indexes 1895 and 1908. His other writings included a history of astronomy.

Driesch Hans Adolf Eduard 1867–1941. German embryologist and philosopher who was one of the last advocates of vitalism, the theory that life is directed by a vital principle and cannot be explained solely in terms of chemical and physical processes. He made several important discoveries in embryology.

Driesch was born in Bad Kreuznach, Prussia, and studied at the universities of Hamburg, Freiburg, Munich, and Jena. He travelled extensively in Europe and the Far East, working at the International Zoological Station in Naples, Italy, 1891–1900. He was professor of philosophy at Heidelberg 1911–20, moving to Cologne and Leipzig before being forced to retire 1935 by the Nazi regime.

In 1891 Driesch, experimenting with sea urchin eggs, discovered that the fate of a cell is not determined in the early developmental stages. Subsequently he produced an oversized larva by fusing two normal embryos, and in 1896 he was the first to demonstrate embryonic induction when he displaced the skeleton-forming cells of sea urchin larvae and observed that they returned to their original positions. These findings provided a great impetus to embryological research. Driesch came to believe that living activities, especially development, were controlled by an indefinable vital principle, which he called entelechy.

Druce George Claridge 1850–1932. British botanist and mayor of Oxford who helped to found the Ashmolean Natural History Society of Oxford 1880 and published *The Flora of Oxfordshire* 1886.

Druce was born in Potterspury, Northamptonshire and, although he was illegitimate, he was able to obtain an education while apprenticed to P Jeyes and Company in Northampton, a retail and manufacturing chemist. He studied for his pharmaceutical exams in the evenings and passed them 1873, enabling him to devote his free time to his childhood love of natural history.

In June 1879, he left his employer and set up as an independent chemist in Oxford. He matriculated at Magdalen College and obtained an MA in 1891. He was a Free Mason and served on the city council from 1892 until his death. He was mayor of Oxford 1900, president of the Pharmaceutical Society 1901 and 1902, and obtained a DSc from Oxford University by examination 1924.

Dubois Marie Eugène François Thomas 1858–1940. Dutch palaeontologist who in Indonesia 1891 discovered the remains of an early species of human, *Homo erectus*, known as Java man.

Dubois was born in Eijsden, on the Belgian border, and studied at the University of Amsterdam. In 1887 he joined the Army Medical Service and was posted to Java – then a Dutch possession – where he was commissioned by the Dutch government to search for fossils. He returned to Europe 1895 and became a professor at Amsterdam 1899.

Dubois set out to search for the 'missing link' in the evolutionary chain. The remains of extinct animals had been found on the banks of the Solo River in E Java, and it was there that he found teeth, a skullcap, and a femur. The teeth were intermediate between ape and human. The femur was definitely human, the ends of the bone and the straightness of the shaft suggesting that its owner had walked erect. From the brain capacity it was thought that *Pithecanthropus erectus* (as it was first called) was almost exactly halfway between ape and human on the evolutionary scale, although it is now considered to be more human and is called *Homo erectus*.

Du Bois-Reymond Emil Heinrich 1818–1896. German physiologist who showed the existence of electrical currents in nerves, correctly arguing that it would be possible to transmit nerve impulses chemically. His experimental techniques proved the basis for almost all future work in electrophysiology.

Du Bois-Reymond was born and educated in Berlin, and became professor of physiology there 1858.

Investigating the physiology of muscles and nerves, Du Bois-Reymond demonstrated the presence of electricity in animals, especially researching electric fishes. By 1849 he had evolved a delicate multiplier for measuring nerve currents, enabling him to detect an electric current in ordinary localized muscle tissues, notably contracting muscles. He observantly traced it to individual fibres, finding their interior was negative with regard to the surface.

Du Bois-Reymond denounced the vitalistic doctrines that were in vogue among German scientists and denied that nature contained mystical life forces independent of matter.

Dubos René Jules 1901–1982. French-US microbiologist who studied soil microorganisms and became interested in their antibacterial properties.

The antibacterials he discovered had limited therapeutic use since they were toxic. However, he opened up a new field of research that eventually led to the discovery of such major drugs as penicillin and streptomycin.

Duchenne Guillaume Benjamin Armand 1806–1875. French physician renowned for his description of the muscle disease *tabes dorsalis*, now known as Duchenne's muscular dystrophy, and his pioneering use of electrophysiology to study muscle function.

During eleven years work as a general practitioner, Duchenne became interested in diseases of the muscles of the body and published an account of the disease *tabes dorsalis* that was so masterly that the disease became known as Duchenne's muscular dystrophy. The disease is characterized by loss of coordination of movement, and lesions are found in the posterior columns of the spinal cord. Impaired sense of the position of ones joints worsens as the disease progresses and eventually shooting pains develop in the afflicted limbs. Muscular dystrophy develops in less than 5% of patients with untreated syphilis, and the symptoms usually become manifest 10–20 years after the primary infection.

Duchenne was born in Boulogne-sur-Mer and educated in Paris. He also pioneered the use of electrophysiology to study muscle function and was particularly interested in whether electrotherapy would rejuvenate or relieve damaged nerves and muscles. He particularly examined the changes in facial expression under emotion using electrical methodology. He published several books on this subject in which he described the effects of placing two moistened electrodes on the skin.

Dulbecco Renato 1914– . Italian biologist who shared the Nobel Prize for Physiology or Medicine 1975 with David Baltimore and Howard Temin for his work on the cancer-causing properties of the genes of papovaviruses, circular DNA viruses that integrate into the host cell's DNA, producing a cell clone in which the virus is maintained.

Dulbecco described the molecular basis of the cancer-causing (oncogenic) properties of a group of viruses, the papovaviruses, particularly polyoma and SV40. Papovaviruses induce a cancerlike state in some cell cultures, a process called cell transformation. The principal transforming factor of these viruses is the T-antigen, which alters gene expression in the transformed cell clones. If the viruses replicate and lyse (destroy by disintegrating) a cell in culture they do not integrate.

Dulbecco was born in Catanzaro in Italy and studied medicine in Turin before joining the Resistance movement during World War II. He emigrated to the USA 1947 and, after holding several academic appointments there, moved to the UK to take up the post of deputy director of research at the Imperial Cancer Research Fund Laboratories in London. He finally settled in California 1977 to work at the Salk Institute.

Dulong Pierre Louis 1785–1838. French chemist and physicist. In 1819 he discovered, together with physicist Alexis Petit, the law that now bears their names. ***Dulong and Petit's law*** states that, for many elements solid at room temperature, the product of relative atomic mass and specific heat capacity is approximately constant. He also discovered the explosive nitrogen trichloride 1811.

Dulong concluded 1829 that, under the same conditions of temperature and pressure, equal volumes of all gases evolve or absorb the same quantity of heat when they are suddenly expanded or compressed to the same fraction of their original volumes. He also deduced that the accompanying temperature changes are inversely proportional to the specific heat capacities of the gases at constant volume.

Dulong was born in Rouen and studied medicine at the Ecole Polytechnique in Paris. He was professor of chemistry at Paris 1820–30, and then director of studies at the Ecole Polytechnique. He collaborated with Petit 1815–20, and continued on his own to work on specific heat capacities, publishing his findings in 1829.

Dumas Jean Baptiste André 1800–1884. French chemist who made contributions to organic analysis and synthesis, and to the determination of atomic weights (relative atomic masses) through the measurement of vapour densities.

Dumas was born in Gard *département* and apprenticed to an apothecary. He moved to Geneva, Switzerland, and continued his studies there. In 1822 he went to Paris, where he became professor of chemistry first at the Lyceum and 1835 at the Ecole Polytechnique. After the political upheavals of 1848 Dumas abandoned much of his scientific work for politics, and held ministerial posts under Napoleon III.

Studying blood, Dumas showed that urea is present in the blood of animals from which the kidneys have been removed, proving that one of the functions of the kidneys is to remove urea from the blood, not to produce it.

In 1826, Dumas began working on atomic theory, and concluded that 'in all elastic fluids observed under the same conditions, the molecules are placed at equal distances' – that is, they are present in equal numbers.

His theory of substitution in organic compounds, which he proved by experiments, established that atoms of apparently opposite electrical charge replaced each other. This refuted the dualistic theory of chemistry proposed by Swedish chemist Jöns Berzelius.

In 1833, Dumas worked out an absolute method for the estimation of the amount of nitrogen in an organic compound, which still forms the basis of modern methods of analysis. He went on to correct the atomic masses of 30 elements – half the total number known at that time – referring to the hydrogen value as 1.

Dunlop John Boyd 1840–1921. Scottish inventor who founded the rubber company that bears his name. In 1888, to help his child win a tricycle race, he bound an inflated rubber hose to the wheels. The same year he developed commercially practical pneumatic tyres, first patented by Robert William Thomson (1822–1873) 1845 for bicycles and cars.

Thomson's invention had gone practically unnoticed, whereas Dunlop's arrived at a crucial time in the development of transport, and with the rubber industry well established.

Dunlop was born in Dreghorn, Ayrshire, and studied veterinary medicine at Edinburgh University before setting up a practice in Ireland near Belfast, in 1867. He then founded his own company for the mass production of tyres. In 1896, after trading for only about five years, Dunlop sold both his patent and his business for £3 million.

Dunlop's first simple design consisted of a rubber inner tube, covered by a jacket of linen tape with an outer tread also of rubber. The inner tube was inflated using a football pump and the tyre was attached by flaps in the jacket which were rubber-cemented to the wheel. Later, he incorporated a wire through the edge of the tyre which secured it to the rim of the wheel.

Durrell Gerald Malcolm 1925–1995. English naturalist, writer, and zoo curator. He became director of Jersey Zoological Park 1958, and wrote 37 books, including the humorous memoir *My Family and Other Animals* 1956. He was the brother of the writer Lawrence Durrell.

Anyone who has got any pleasure at all should try to put something back. Life is like a superlative meal and the world is like the maitre d'hotel. What I am doing is the equivalent of leaving a reasonable tip.

Gerald Durrell

Guardian 1971

Born in Jamshedpur, India, Durrell spent part of his childhood in Corfu, where he set up his own childhood zoo of scorpions, lizards, and eagle owls. Critical of the conditions in which most zoos kept animals, the lack of interest in breeding programmes, and the concentration on large species, such as the big cats, rhinos, and elephants, Durrell became determined to build up his own zoo, and to run it so that it could supplement conservation programmes in the wild rather than detract from them. He finally settled in Jersey, where he founded the Jersey Zoological Park of which he was Honorary Director 1958–1995. The zoo concentrated mainly on animal and plant species from oceanic islands, and established many successful breeding programmes including the golden lion tamarin and the Mauritius pink pigeon, both of which were successfully reintroduced to their habitats. Durrell also set up an international training centre near the zoo, which is now the world's foremost centre for training animal keepers, specifically those from the developing world. It has trained 700 people from over 80 countries.

Through his work in conservation – within his zoo and as chairman of the Flora and Fauna Preservation Society – and perhaps more particularly his many books and television programmes, Durrell encouraged and inspired a whole generation of naturalists, zoologists, and zoo keepers.

Du Toit Alexander Logie 1878–1948. South African geologist whose work was to form one of the foundations for the synthesis of continental drift theory and plate tectonics that created the geological revolution of the 1960s.

Du Toit was born near Cape Town and studied there, then went to the UK and studied at Glasgow and the Royal College of Science, London. He spent 1903–20 mapping for the Geological Commission of the Cape of Good Hope.

The theory of continental drift put forward by German geophysicist Alfred Wegener inspired Du Toit's book *A Geological Comparison of South America and South Africa* 1927, in which he suggested that they had probably once been joined. In *Our Wandering Continents* 1937, he maintained that the southern continents had, in earlier times, formed the supercontinent of Gondwanaland, which was distinct from the northern supercontinent of Laurasia.

Dutrochet (Rene Joachim) Henri 1776–1847. French physiologist who outlined the process of osmosis (the passive diffusion of water from high concentration to low concentration through a semi-permeable membrane, such as a cell wall) in plants and described various important parts of the plant respiratory mechanism. He was also the first to recognize the role of the pigment chlorophyll in the conversion by plants of carbon dioxide to oxygen (photosynthesis) and to identify stomata (pores) on the surface of leaves; later recognized as important in the exchange of gases between the plant and its surroundings.

By 1835, it had been recognized that plant cells and unicellular animals had a nucleus and that plant cells had cell walls. Little attention, however, had been paid to the material that lies between the nucleus and the membrane. Dutrochet realized that the viscous material, which we now know as protoplasm, that makes up the majority of the volume of most cells is important.

Dutrochet was born in Neon and graduated in medicine in Paris. He enrolled as an army physician but was forced to resign his commission after a bout of typhoid. As he was a member of a prosperous family he could afford to withdraw from public life and spent his future years studying cell respiration, particularly in plants.

Duve Christian René de 1917– . British-born Belgian biochemist; see ◊de Duve.

Du Vigneaud Vincent 1901–1978. US biochemist who was awarded the 1955 Nobel Prize for Chemistry for his investigations into biochemically important sulphur compounds and the first synthesis of a polypeptide hormone, oxytocin. Du Vigneaud also isolated vitamin H (biotin) and determined its structure.

He studied the hormones secreted by the posterior pituitary gland, especially oxytocin and vasopressin. He discovered that oxytocin was made up of eight amino acids, arranged in a particular order. In 1953 he synthesized oxytocin. This was the first protein to be synthesized. In 1956, du Vigneaud and his team synthesized vasopressin.

Du Vigneaud was born in Chicago and educated at the University of Illinois, Urbana, and the University of Rochester, New York, gaining his doctorate 1927 for work on the hormone insulin. In 1932, he became professor and head of the biochemistry department at the School of Medicine at George Washington University, Washington DC. From 1938–67, he held a similar position at Cornell University Medical College in Ithaca, New York. He was professor of chemistry at Cornell University 1967–75.

Duwez Pol 1907– . US scientist, born in Belgium, who in 1959 developed metallic glasses (alloys rapidly cooled from the melt, which combine properties of glass and metal) with his team at the California Institute of Technology.

Dyson Frank Watson 1868–1939. English astronomer especially interested in stellar motion and time determination. He initiated the public broadcasting of time signals by the British Broadcasting Corporation over the radio 1924. Knighted 1915.

Dyson was born in Ashby-de-la-Zouch, Leicestershire, and studied at Cambridge. He was Astronomer Royal for Scotland 1906–10 and for England 1910–33.

Dyson was one of a number of astronomers who confirmed the observations of Jacobus

Kapteyn on the proper motions of stars, which indicated that the stars in our Galaxy seemed to be moving in two great streams. These results were later realized to be the first evidence for the rotation of our Galaxy.

Dyson organized several expeditions to study total eclipses of the Sun. Other areas to which he made important contributions include the study of the Sun's corona and of stellar parallaxes.

Eastman George 1854–1932. US entrepreneur and inventor who founded the Eastman Kodak photographic company 1892. He patented flexible film 1884, invented the Kodak box camera 1888, and introduced daylight-loading film 1892. By 1900 his company was selling a pocket camera for as little as one dollar.

Eastman was born in Waterville, New York, and left school at 14. In 1879 he patented a photographic-emulsion coating machine, and began mass production of dry plates in Rochester, New York. In 1886 he introduced stripping film, in which paper was used only to support the emulsion and was stripped off once the negative had been transferred to glass. The image on the glass then had to be transferred to a gelatine sheet for printing. The first Kodak camera was sold loaded with a roll of the stripping film large enough for 100 exposures. After use, the camera was sent to Rochester, where the film was developed and a new film loaded.

My work is done. Why wait?

George Eastman

Suicide note

Eastman followed this up 1889 with the first commercially available transparent nitrocellulose (celluloid) roll films; the flammable nitrocellulose was later replaced by nonflammable cellulose acetate. Eastman's roll film replaced his stripping film in the Kodak camera and ushered in the era of press-button photography.

Eastwood Alice 1859–1953. US botanist who provided critical specimens for professional botanists as well as advising travellers on methods of plant collecting and arousing popular support for saving native species.

Eastwood studied plants in the Colorado Mountains while working as a schoolteacher. In 1892, she went to the California Academy of Sciences, San Francisco, where she

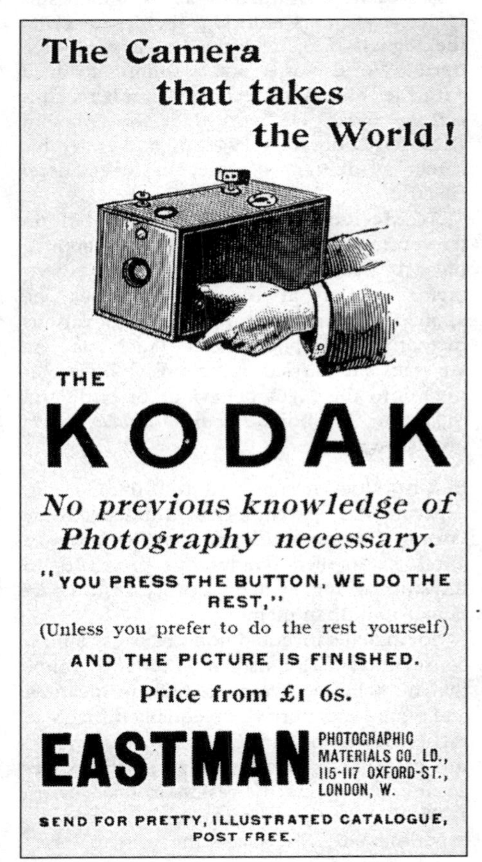

Eastman *An advertisement from 1893 for one of George Eastman's early cameras. Eastman's first camera went on sale in 1888 and was called the Detective Camera. Each roll of film took a hundred circular pictures, each about 7cm in diameter. Later that year, Eastman changed the name of the camera to the Kodak and by 1895 had sold over a hundred thousand.*

became curator of the herbarium and founded and ran the California Botanical Club. After the San Francisco earthquake of 1906, she spent years rebuilding the collections, involving field trips to the coastal ranges and the Sierra Nevada and visits to botanical gardens around the world. Between 1912 and her retirement 1949, over 340,000 specimens were added to the herbarium.

Eastwood's early fieldwork led to the publication of *A Handbook of the Trees of California* 1905.

Eastwood Eric 1910–1981. British electronics engineer who made major contributions to the development of radar for both military and civilian purposes.

Eastwood studied at Manchester University and Cambridge before entering the Signals branch of the Royal Air Force during World War II, and becoming involved with the solution of technical problems concerning radar. He worked at the Marconi Research Laboratory 1948–62 and ended his career as director of research for General Electric.

At Marconi, Eastwood concentrated on telecommunications, radar, and applied physics. With the aid of the Marconi experimental station at Bushy Hill, Essex, he applied radar methods to the study of various meteorological phenomena (such as the aurorae) and carried out extensive investigations into the flight behaviour of birds and migration; his book *Radar Ornithology* was published 1967.

Ebbinghaus Hermann 1850–1909. German experimental psychologist. Influenced by Gustav Fechner's *Elements of Psychophysics* 1860, he applied quantitative principles to the study of higher mental processes, in particular to human memory.

Ebbinghaus invented nonsensical syllables, consonant-vowel-consonant letter groups that he believed (wrongly) had no meaning and would therefore all be equally difficult to memorize. Using himself as subject, he used this material to investigate learning and forgetting, publishing the results in his *Memory* 1885. It was the first research to attempt, experimentally, to isolate the principal factors that generate learning curves. Although of great influence, Ebbinghaus's methods were later extensively criticized, notably by Frederic Charles Bartlett.

Eccles John Carew 1903– . Australian physiologist who shared (with Alan Hodgkin and Andrew Huxley) the 1963 Nobel Prize for Physiology or Medicine for work on conduction in the central nervous system. In some of his later works, he argued that the mind has an existence independent of the brain.

Eckert John Presper Jr 1919–1995. US electronics engineer and mathematician who collaborated with John Mauchly on the development of the early ENIAC (1946) and UNIVAC 1 (1951) computers.

Eckert was born in Philadelphia, Pennsylvania, and studied at the University of Pennsylvania. During World War II he worked on radar ranging systems and then turned to the design of calculating devices, building the Electronic Numerical Integrator and Calculator (ENIAC) with Mauchly. The Eckert–Mauchly Computer Corporation, formed 1947, was incorporated in Remington Rand 1950 and subsequently came under the control of the Sperry Rand Corporation.

The ENIAC weighed many tonnes and lacked a memory, but could store a limited amount of information and perform mathematical functions. It was used for calculating ballistic firing tables and for meteorological and research problems.

ENIAC was superseded by BINAC, also designed in part by Eckert, and in the early 1950s, Eckert's group began to produce computers for the commercial market with the construction of the UNIVAC 1. Its chief advance was the capacity to store programs.

Eddington Arthur Stanley 1882–1944. British astrophysicist who studied the motions, equilibrium, luminosity, and atomic structure of the stars. In 1919 his observation of stars during a solar eclipse confirmed Albert Einstein's prediction that light is bent when passing near the Sun, in accordance with the general theory of relativity. In *The Expanding Universe* 1933 Eddington expressed the theory that in the spherical universe the outer galaxies or spiral nebulae are receding from one another.

We used to think that if we knew one, we knew two, because one and one are two. We are finding that we must learn a great deal more about 'and'.

Arthur Eddington

Attributed remark

Eddington discovered the fundamental role of radiation pressure in the maintenance of stellar equilibrium, explained the method by which the energy of a star moves from its interior to its exterior, and in 1924 showed that the luminosity of a star depends almost exclusively on its mass – a discovery that caused a complete revision of contemporary ideas on stellar evolution.

Eddington was born in Kendal, Cumbria, and studied at Cambridge. Working for the Royal Observatory, Greenwich, London, he was sent to Malta 1909 to determine the longitude of the geodetic station there, and in 1912 he went to Brazil, leading an expedition to study an eclipse. In 1913, Eddington returned to Cambridge as professor, and was director of the university's observatory from 1914.

Eddington became a leading exponent of Einstein's relativity theory, and from 1930 he worked on relating the theory of relativity and quantum theory.

Eddington's first book, *Stellar Movements and the Structure of the Universe* 1914, introduced the subject of stellar dynamics. *The Internal Construction of the Stars* 1926 became one of the classics of astronomy.

We are inquiring into the deepest nature of our constitutions: How we inherit from each other. How we can change. How our minds think. How our will is related to our thoughts. How our thoughts are related to our molecules.

Gerald Edelman

Newsweek 4 July 1976

Edelman Gerald Maurice 1929– . US biochemist who worked out the sequence of 1,330 amino acids that makes up human immunoglobulin, a task completed 1969. For this work he shared the Nobel Prize for Physiology or Medicine 1972 with Rodney Porter. In 1996 he was head of the Neurosciences Institute, La Jolla, California.

Edinger Tilly (Johanna Gabrielle Ottilie) 1897–1967. German-born US palaeontologist whose work in vertebrate palaeontology laid the foundations for the study of palaeoneurology. She demonstrated that the evolution of the brain could be studied directly from fossil cranial casts.

Edinger was born in Frankfurt and studied there and at Heidelberg and Munich. With the Nazis's rise to power, she was forced to leave Germany. After a year in the UK, she went to Cambridge, Massachusetts, in 1940, to take up a job at the Museum of Comparative Zoology at Harvard.

Her research shed new light on the evolution of the brain and showed that the progression of brain structure does not proceed at a constant rate in a given family but varies over time; also that the enlarged forebrain evolved several times independently among advanced groups of mammals and there was no single evolutionary scale.

Edinger's main works are *Die fossilen Gehirne/Fossil Brains* 1929 and *The Evolution of the Horse Brain* 1948.

Genius is one per cent inspiration and ninety-nine per cent perspiration.

Thomas Edison

Life ch 24

Edison Thomas Alva 1847–1931. US scientist and inventor, with over 1,000 patents. In Menlo Park, New Jersey, 1876–87, he produced his most important inventions, including the electric light bulb 1879. He constructed a system of electric power distribution for consumers, the telephone transmitter, and the phonograph.

Edison's first invention was an automatic repeater for telegraphic messages. Later came the carbon transmitter (used as a microphone in the production of the Bell telephone), the electric filament lamp, a new type of storage battery, and the kinetoscopic camera, an early cine camera. He also anticipated the Fleming thermionic valve. He supported direct current (DC) transmission, but alternating current (AC) was eventually found to be more efficient and economical.

Edison was born in Milan, Ohio, and self-educated. As a 19-year-old telegraph operator, he took out his first patent, for an electric vote recorder. With the proceeds of an improved stock ticker (a machine that printed stock-market prices on continuous paper tape), he opened an industrial research laboratory in Newark, New Jersey, 1869, later moving to Menlo Park and then to West Orange. In 1889 he formed the Edison Light Company, which became the General Electric Company.

Edlén Bengt 1906–1993. Swedish astrophysicist who resolved the identification of certain lines in spectra of the solar corona that had misled scientists for the previous 70 years.

Edlén was born in Östergötland and educated at Uppsala University, where he was on the staff 1928–44. He was professor of physics at Lund University 1944–73.

During the eclipse of 1869, astronomers recorded unexpected spectral lines in the Sun's corona that they ascribed to the presence of a new element which they called 'coronium'. Similar lines were later discovered to originate nearer the Earth; these were attributed to 'geocoronium'.

In the early 1940s, Edlén showed that, if iron atoms are deprived of many of their electrons, they can produce spectral lines like those of 'coronium'. Similarly ionized atoms of nickel, calcium, and argon produced even more lines. It was determined that such high stages of ionization would require temperatures of about 1,000,000°C/1,800,000°F and when, in the 1950s, it was verified that such high temperatures did exist in the solar corona, it became accepted that 'coronium' did not exist.

The lines thought to be caused by 'geocoronium' were found to be produced by atomic nitrogen emitting radiation in the Earth's upper atmosphere.

Edwards George Robert 1908– . British civil and military aircraft designer, associated with the Viking, Viscount, Valiant V-bomber, VC-10, and Concorde. Knighted 1957.

Edwards Robert Geoffrey 1925– . British physiologist who with Patrick Steptoe devised a technique for fertilizing a human egg outside the body and transferring the fertilized embryo to the uterus of a woman. A child born following the use of this technique is popularly known as a 'test-tube baby'. Edwards's research has added to knowledge of the development of the human egg and young embryo.

Edwards was educated at the universities of Wales and Edinburgh. He has held academic positions in Scotland and California, and was at the National Institute of Medical Research, London, 1958–62. In 1963 he moved to the Department of Physiology at Cambridge.

In the 1950s Edwards successfully replanted mouse embryos into the uterus of a mouse and he wondered if the same process could be applied to humans. In 1965 he first attempted the *in vitro* fertilization of human eggs, not succeeding until 1967. Steptoe had just invented a new technique, laparoscopy, to view the internal organs. Edwards and Steptoe met 1968 and arranged to collaborate.

Steptoe treated volunteer patients with a fertility drug to stimulate maturation of the eggs in the ovary, while Edwards devised a simple piece of apparatus to be used with the laparoscope for collecting mature eggs from human ovaries. He then prepared them for fertilization. In 1971, once they were sure that the fertilized eggs were developing normally, Edwards and Steptoe were ready to introduce an eight-celled embryo into the uterus of a volunteer patient, but their attempts were unsuccessful until 1977, when they abandoned the use of the fertility drug.

Egas Moniz Antonio Caeteno de Abreu Freire 1874–1955. Portuguese neurosurgeon and politician who treated mental disorders with brain surgery. He was awarded the Nobel Prize for Physiology or Medicine 1949 for introducing the ***prefrontal lobotomy***.

In 1935, he performed a surgical technique to isolate the frontal lobes from the brain by injecting the connecting white matter with alcohol. This was done in patients suffering from acute emotional tension or mental disorders, such as schizophrenia, that had proved incurable. Later, he took the bold step of cutting through the connecting white matter. This operation, prefrontal lobotomy, was actually within the substance of the brain and was a very delicate procedure. It found favour at the time but has since fallen into disrepute due to its unpleasant long-term side-effects.

Egas Moniz was born in Avanca, Portugal and graduated in medicine from Lisbon University, where he later returned to become professor of neurology 1911. He rose to fame in his country from 1903–18 as a politician, reaching the position of foreign minister before returning to his work as a surgeon in Lisbon.

Eggen Olin Jenck 1919– . US astronomer whose work has included studies of high-velocity stars, red giants (using narrow- and broadband photometry), and subluminous stars.

Born in Orfordville, Wisconsin, Eggen graduated from Wisconsin University. He has spent much of his working life in senior appointments all round the world; he was director of Mount Stromlo and Siding Spring observatories, Australia, 1966–77, when he moved to the Observatory Interamericano de Cerro Tololo, Chile.

During the mid-1970s, Eggen completed a study – based on ultraviolet photometry and every available apparent motion – of all red giants brighter than apparent magnitude 5. As a result he was able to classify these stars, categorizing them as very young discs, young discs, and old discs. A few remained unclassifiable (haloes).

He also systematically investigated the efficiency of the method of stellar parallax using visual binaries originally suggested by William Herschel in 1781, and reviewed the original correspondence of English astronomers John Flamsteed and Edmond Halley.

Ehrenberg Christian Gottfried 1795–1876. German naturalist who developed one of the forerunners of the modern scheme for classification of the animal kingdom. He was the first scientist to study the fossils of microorganisms and can be regarded as the founder of micropalaeontology.

He demonstrated that fungi develop from spores and that phosphorescence (the emission of light by certain substances after they have absorbed energy) is caused by plankton. In 1838, he published *Die Infusionsthierchen als volkommene Organismen/The Infusoria as Complete Organisms*. His concept that all animals are complete organisms was refuted by Felix Dujardin.

Ehrenberg was born in Delitzsch in Saxony, Germany and attended the University of Berlin, where he graduated 1818 with an MD. In 1820, he participated in an expedition to Egypt, Libya, the Sudan, and the Red Sea. He was the only person to survive the expedition, however, and returned home 1825. He collected 34,000 animal and 46,000 plant specimens for the University of Berlin and the Prussian Academy of Sciences. In 1829, he went to Central Asia and Siberia with Alexander von Humboldt on an expedition financed by Nicholas I of Russia.

The first rule of intelligent tinkering is to save all the parts.

Paul Ehrlich

Saturday Review 5 June 1971

Ehrlich Paul 1854–1915. German bacteriologist and immunologist who produced the first cure for syphilis. He developed the arsenic compounds, in particular Salvarsan, that were used in the treatment of syphilis before the discovery of antibiotics. He shared the 1908 Nobel Prize for Physiology or Medicine with Ilya Mechnikov for his work on immunity.

Ehrlich *German bacteriologist Paul Ehrlich. Ehrlich won the Nobel Prize for Physiology or Medicine in 1908 for his work which led, in part, to the development of immunology as a discipline. His immune theory proposed that certain antibodies possessed side chains that could bind specific toxin molecules, a theory now known to be correct. Ehrlich was also an early haematologist who developed methods for distinguishing blood diseases.*

Ehrlich founded chemotherapy – the use of a chemical substance to destroy disease organisms in the body. He was also one of the earliest workers on immunology, and through his studies on blood samples the discipline of haematology was recognized.

Ehrlich was born in Strehlin, Silesia (now Strzelin, Poland). He studied at Breslau, Strasbourg, and Leipzig. In 1884 he became a professor in Berlin, but spent 1886–88 in Egypt, curing himself of tuberculosis contracted in the course of research. He set up a small private laboratory in Berlin 1889, in addition to his academic posts.

Ehrlich teamed up with bacteriologists Emil von Behring and Shibasaburō Kitasato to try to find a cure for diphtheria.

Ehrlich had studied antigen–antibody reactions using toxic plant proteins on mice, and Behring and Ehrlich were able to produce antitoxins obtained from much larger mammals which had been immunized against the diphtheria organism. These antitoxins were concentrated and purified, and successfully used on children 1894.

The search progressed for dyes that would stain only bacteria and not other cells, and from this research Ehrlich's staff continued synthesizing and testing chemical substances that could seek out and destroy the bacteria without harming the human body. Ehrlich termed these compounds 'magic bullets'. This was how the cure for syphilis was found.

Eiffel (Alexandre) Gustave 1832–1923. French engineer who constructed the ***Eiffel Tower*** for the 1889 Paris Exhibition. The tower, made of iron, is 320 m/1,050 ft high and stands in the Champ de Mars, Paris. Sightseers may ride to the top for a view.

Eiffel set up his own business in Paris 1867 and quickly established his reputation with the construction of a series of ambitious railway bridges, of which the span across the Douro at Oporto, Portugal, was the longest at 160 m/525 ft. In 1881 he provided the iron skeleton for the Statue of Liberty.

Eiffel *As well as the tower named after him, Gustave Eiffel also designed many notable bridges and viaducts and conducted pioneer researches in aerodynamics and the use of wind tunnels.*

Eiffel was born in Dijon and attended the Ecole des Arts et Manufactures in Paris. Specializing in the design of large metal structures, he was one of the first to use compresssed air for underwater foundations, for the iron railway bridge over the Garonne at Bordeaux. He also participated in the French attempt to build the Panama Canal, in the course of which he designed and partly constructed some huge locks. When the entire project collapsed 1893, Eiffel went to prison for two years. In 1900, he took up meteorology and later, using wind tunnels, carried out extensive research in aerodynamics.

Originally, the Eiffel Tower was intended to be dismantled at the conclusion of the exhibition, but it was preserved as a radio transmitting station. For some time it was by far the highest artificial structure in the world.

Eigen Manfred 1927– . German chemist who worked on extremely rapid chemical reactions (those taking less than 1 millisecond). From 1954 he developed a technique by which very short bursts of energy could be applied to solutions, disrupting their equilibrium and enabling him to investigate momentary reactions, such as the formation and dissociation of water. Nobel prize 1967.

A theory has only the alternative of being right or wrong. A model has a third possibility: it may be right, but irrelevant.

Manfred Eigen

The Physicist's Conception of Nature 1973, Ed Jagdish Mehra p 618

Eigen was born in Bochum, Ruhr, and educated at Göttingen. From 1953 he worked at the Max Planck Institute for Physical Chemistry in Göttingen, eventually becoming its director.

Eigen investigated particularly very fast biochemical reactions that take place in the body. With his colleague Ruthild Winkler he tried to relate chance and chemistry in processes that could have led to the origin of life on Earth. Eigen theorized that in the 'primeval soup' of the early Earth, cycles of chemical reactions would have occurred, one reproducing nucleic acids and one reproducing proteins, and that life arose from a combination of these.

Eijkman Christiaan 1858–1930. Dutch bacteriologist. He pioneered the recognition of vitamins as essential to health and identified vitamin B_1 deficiency as the cause of the disease beriberi. He shared the 1929 Nobel Prize for Physiology or Medicine with Frederick Gowland Hopkins.

Eilenberg Samuel 1913– . Polish-born US mathematician whose research in the field of algebraic topology led to considerable development in the theory of cohomology.

Algebraic topology, sometimes called 'combinatorial' topology, is based on homology theory – the study of closed curves, closed surfaces, and similar geometric arrangements in a given topological space. Much of Eilenberg's work was concerned with a modification of homology theory called cohomology theory. It is possible to define a 'product' of cohomology classes by means of which, together with the addition of cohomology classes, the direct sum of the cohomology classes of all dimensions becomes a ring (the cohomology ring). This is a richer structure than is available for homology groups, and allows finer results. Various very complicated algebraic operations using cohomology classes can lead to results not provable in any other way.

Eilenberg was born and educated in Warsaw. In the 1930s he emigrated to the USA. He was professor of mathematics at the University of Indiana 1946–49, and ended his career at Columbia University, New York.

Einstein Albert 1879–1955. German-born US physicist who formulated the theories of relativity, and worked on radiation physics and thermodynamics. In 1905 he published the special theory of relativity, and in 1915 issued his general theory of relativity. He received the Nobel Prize for Physics 1921. His last conception of the basic laws governing the universe was outlined in his unified field theory, made public 1953.

The theories of relativity revolutionized our understanding of matter, space, and time. The ***special theory of relativity*** started with the premises that (1) the laws of nature are the same for all observers in unaccelerated motion, and (2) the speed of light is independent of the motion of its source. Einstein postulated that the time interval between two events was longer for an observer in whose frame of reference the events occur in different places than for the observer for whom they occur at the same place. This theory contained the equation $E = mc^2$.

In the ***general theory of relativity***, the properties of space-time were to be conceived as modified locally by the presence of a body with mass; light rays should bend when they pass by a massive object. A planet's orbit around the Sun arises from its natural trajectory in modified space-time – there is no need to invoke, as Isaac Newton did, a force of gravity. General relativity theory was inspired by the simple idea that it is impossible in a small region to distinguish between acceleration and gravitation effects (as in a lift one feels heavier when it accelerates upwards), but the mathematical development was formidable.

Einstein also established that light may have a particle nature and deduced 1905 the photoelectric law that governs the production of electricity from light-sensitive metals. He

If my theory of relativity is proven correct, Germany will claim me as a German and France will declare that I am a citizen of the world. Should my theory prove untrue, France will say that I am a German and Germany will declare that I am a Jew.

Albert Einstein

Address at the Sorbonne, Paris Dec 1929

Einstein *Physicist Albert Einstein, 1944. He developed his theories by using simple 'thought experiments', but the full flowering of his ideas required very complex mathematics.*

investigated Brownian movement, also 1905, and was able to explain it so that it not only confirmed the existence of atoms but could be used to determine their dimensions. He also proposed the equivalence of mass and energy, which enabled physicists to deepen their understanding of the nature of the atom, and explained radioactivity and other nuclear processes.

Einstein was born in Ulm, Württemberg, and lived in Munich and Italy before settling in Switzerland. Disapproving of German militarism, he became a Swiss citizen and was appointed an inspector of patents in Berne. In his spare time, he took his PhD at Zürich. In 1909 he became a lecturer in theoretical physics at the university. After holding a similar post at Prague 1911, he returned to teach at Zürich 1912. In 1913 he took up a specially created post as director of the Kaiser Wilhelm Institute for Physics, Berlin. Confirmation of the general theory of relativity by the solar eclipse of 1919 made Einstein world famous. Deprived of his position at Berlin by the Nazis, he emigrated to the USA 1933 and became professor of mathematics and a permanent member of the Institute for Advanced Study at Princeton, New Jersey.

If A *is a success in life, then* A *equals* x *plus* y *plus* z. *Work is* x; y *is play; and* z *is keeping your mouth shut.*

Albert Einstein

Observer 15 Jan 1950

In 1939, Einstein drew the attention of the president of the USA to the possibility that Germany might be developing the atomic bomb. This prompted US efforts to produce the bomb, though Einstein did not take part in them. After World War II he was actively involved in the movement to abolish nuclear weapons. In 1952, the state of Israel paid him the highest honour it could by offering him the presidency, which he declined.

Einthoven Willem 1860–1927. Dutch physiologist and inventor of the electrocardiogram. He demonstrated that certain disorders of the heart alter its electrical activity in characteristic ways. He was awarded the 1924 Nobel Prize for Physiology or Medicine.

Eisenhart Luther Pfahler 1876–1965. US theoretical geometrist who formulated a unifying principle to the theory of the deformation of surfaces. In the 1920s he attempted to develop his own geometry theory from that of German mathematician Georg Riemann.

Eisenhart was born in York, Pennsylvania, and studied at Gettysburg College, Pennsylvania, and Johns Hopkins University, Baltimore. His life's work in mathematical research was spent at Princeton University 1900–45.

One of Eisenhart's major achievements was to relate his theories regarding differential geometry to studies bordering on the topological. At the age of 25 he wrote one of the first characterizations of a sphere as defined in terms of differential geometry (the paper had the somewhat daunting title 'Surfaces whose first and second forms are respectively the second and first forms of another surface').

The deformation of a surface involves the congruence of lines connecting a point and its image. Eisenhart's contribution was to realize that, in all known cases, the intersections of these surfaces with the given surface and its image form a set of curves which have special properties. He wrote his account of the theory in 1923, in *Transformations of Surface*.

Eisner Thomas 1929– . German-born US entomologist and conservation activist. He is an authority on the role of chemicals in insect behaviour. A campaigner for the preservation of biodiversity, in order to prevent the extinction of species and the loss of potentially useful chemicals he advocates 'chemical prospecting', whereby drug companies buy the rights to extract chemically rich organic matter from forests, leaving the forests themselves intact.

Thomas was born in Berlin but moved to New York with his family in 1947. His early entomological work concentrated on the bombardier beetle. He became professor of biology at Cornell University, New York, 1976, and director of the Cornell Institute for Research in Chemical Ecology. Concerned at the environmental implications of the population explosion, he is a member of Zero Population Growth.

Elders (Minnie) Joycelyn. Born Jones 1933– . US physician, surgeon general (head of the public health service) 1993–94. She was director of the Arkansas Department of Health 1987–93 and a member of Hillary Clinton's health-care task force 1993.

She was born in Schaal, Arkansas, graduated from the University of Arkansas Medical

School 1960, and became a paediatric endocrinologist, working on children's growth patterns and hormone-related illnesses. At the Arkansas Department of Health she promoted sex education, contraception, and abortion rights. As surgeon general she called for higher taxes on alcohol and tobacco and, being the second woman and first black American to be appointed to the post, was under heavy scrutiny by the media, resigning Dec 1994.

Elion Gertrude Belle 1918– . US biochemist who shared the Nobel Prize for Physiology or Medicine 1988 with her colleague George Herbert Hitchings, and Joseph Black, for her work on the development of various drugs to treat cancer, gout, malaria, and various viral infections.

She is best known for her work to develop new drugs, particularly those which inhibit the synthesis of DNA by diseased cells. Together with her colleague Hitchings she investigated the chemistry and function of two important components of DNA, pyrimidine and purine. From this they developed new drugs to treat leukaemia, malaria, kidney stones, gout, herpes, and AIDS. Elion also helped to develop early forms of immunosuppressive drugs to enable transplant patients to tolerate the presence of tissues from an unrelated donor.

Elion was born in New York City and graduated from Hunter College 1937 before obtaining her MSc from New York University. In 1944 she joined the laboratories of Burroughs Wellcome in North Carolina and has held the position of research professor of pharmacology and medicine at Duke University 1983.

Elkington George Richards 1801–1865. English inventor who pioneered the use of electroplating for finishing metal objects.

Elkington was born in Birmingham and in 1818 he became an apprentice in the local small-arms factory; in due course he became its proprietor. With his cousin Henry Elkington (1810–1852), he explored the alternatives to traditional methods of plating from about 1832. The process of plating base metals with silver and gold by electrodeposition was announced in a patent taken out by the Elkington cousins in 1840. In 1841 they established a workshop for electroplating in Birmingham, and successfully patented their ideas in France. George Elkington also established large copper-smelting works in South Wales, providing houses for his workers and schools for their children.

Ellet Charles 1810–1862. US civil engineer who designed the first wire-cable suspension bridge in the USA, in 1842. He also designed the world's first long-span wire-cable suspension bridge, crossing the Ohio River at Wheeling, West Virginia.

Ellet was born in Pennsylvania and began his career as a surveyor and assistant engineer on the Chesapeake and Ohio Canal 1828. In 1831–32 he was in Europe, enrolled at the Ecole Polytechnique in Paris and studied the various engineering works taking place in France, Germany, and Britain.

For his first wire-cable suspension bridge, over the Schuylkill River at Fairmount, Pennsylvania, Ellet introduced a technique he had learned in France of binding small wires together to make the cables.

The central span of the suspension bridge over the Ohio River was at 308 m/1,010 ft the longest ever built when completed 1849. The bridge failed under wind forces in 1854; however, Ellet's towers remained standing and the bridge was rebuilt.

Following the outbreak of the American Civil War in 1861, Ellet produced a steam-powered ship for the Union (Northern) forces to ram the Confederates on the Mississippi River. In June 1862, Ellet led a fleet of nine of these rams in the Battle of Memphis. The Union side was victorious, but in the course of the fighting Ellet was fatally wounded.

Elsasser Walter Maurice 1904–1991. German-born US geophysicist who pioneered analysis of the Earth's former magnetic fields, frozen in rocks. He also produced the dynamo model of the Earth's magnetic field.

Born in Mannheim and educated at Göttingen, Elsasser left in 1933 following Hitler's rise to power, and spent three years in Paris working on the theory of atomic nuclei. After settling in 1936 in the USA and joining the staff of the California Institute of Technology, he specialized in geophysics. Elsasser became a professor at the University of Pennsylvania 1947; in 1962 he was made professor of geophysics at Princeton.

His magnetical researches in the 1940s yielded the dynamo model of the Earth's magnetic field. The field is explained in terms of the activity of electric currents flowing in the Earth's fluid metallic outer core. The theory premises that these currents are magnified through mechanical motions, rather as currents are sustained in power-station generators.

Elton Charles Sutherland 1900–1991. British ecologist, a pioneer of the study of animal and plant forms in their natural environments, and of animal behaviour as part of the complex pattern of life. He defined the concept of food chains and was an early conservationist, instrumental in establishing the Nature Conservancy Council 1949, and much concerned with the impact of introduced species on natural systems.

Elton carried out a 20-year research project of interrelationships of animals in meadows, woods, and water near Oxford. He originated the concept of the 'pyramid of numbers' as a method of representing the structure of an ecosystem in terms of feeding relationships.

His books include *Animal Ecology and Evolution* 1930 and *The Pattern of Animal Communities* 1966.

Emeléus Harry Julius 1903– . English chemist who made wide-ranging investigations in inorganic chemistry, studying particularly nonmetallic elements and their compounds. He worked on chemical kinetics and studied the hydrides of silicon and the halogen fluorides.

Emeléus was born in London and studied at Imperial College, London, and Karlsruhe University, Germany. He was professor of inorganic chemistry at Cambridge 1945–70.

Emeléus began his researches with a study of the phosphorescence of white phosphorus. His spectrographic investigations of phosphorescence provided new information about the mechanisms of combustion reactions.

Emeléus showed that naturally occurring water exhibits a small variation in deuterium content, and that distillation, freezing, and adsorption methods can all effect some degree of separation of the two isotopic forms.

In 1945 he started studying the halogen fluorides. By 1959 he was working with the fluorides of vanadium, niobium, tantalum, and tungsten and much of his research in the 1960s concerned the fluoralkyl derivatives of metals.

Emeléus summarized much of his work in *The Chemistry of Fluorine and its Compounds* 1969.

Emin Pasha Mehmed, adopted name of Eduard Schnitzer 1840–1892. German explorer, physician, and linguist. Appointed by British general Charles Gordon chief medical officer and then governor of the Equatorial province of S Sudan, he carried out extensive research in anthropology, botany, zoology, and meteorology.

Schnitzer practised medicine in Albania. In 1876 he joined the Egyptian Service, where he was known as Emin Pasha. Isolated by his remote location in Sudan and cut off from the outside world by Arab slave traders, he was 'rescued' by an expedition led by H M Stanley in 1889. He travelled with Stanley as far as Zanzibar but returned to continue his work near Lake Victoria. Three years later he was killed while leading an expedition to the W coast of Africa.

Empedocles *c.* 493–433 BC. Greek philosopher and scientist who proposed that the universe is composed of four elements – fire, air, earth, and water – which through the action of love and discord are eternally constructed, destroyed, and constructed anew. He lived in Acragas (Agrigentum), Sicily, and according to tradition, he committed suicide by throwing himself into the crater of Mount Etna.

Encke Johann Franz 1791–1865. German astronomer whose work on star charts during the 1840s contributed to the discovery of the planet Neptune in 1846. He also worked out the path of the comet that bears his name.

Encke was born in Hamburg and studied at the University of Göttingen. He was a professor at the Academy of Sciences in Berlin and director of the Berlin Observatory 1825–65.

The new star charts took nearly 20 years to draw up, compiled from both old and new observations. They were completed 1859 but were soon improved upon by those of Friedrich Argelander.

Encke carried out continuous research on comets and the perturbations of the asteroids. What subsequently became known as Encke's comet had been reported by French astronomer Jean-Louis Pons, but little was known of its behaviour. Encke showed that the comet had an elliptical orbit with a period of just less than four years.

Enders John Franklin 1897–1985. US virologist. With Thomas Weller (1915–) and Frederick Robbins (1916–), he developed a technique for culturing virus material in sufficient quantity for experimental work. This led to the creation of effective vaccines against polio and measles. The three were awarded the Nobel Prize for Physiology or Medicine 1954.

Enders was born in West Hartford, Connecticut. He interrupted his studies at Yale to become a flying instructor during World War I and then took up a business career but left it to study English at Harvard, before changing to medicine. He remained at Harvard Medical School, becoming professor 1962.

Viruses cannot be grown, as bacteria can, in nutrient substances, and so a method had been developed for growing them in a living chick embryo. In 1948, Enders and his colleagues prepared a medium of homogenized chick embryo and blood and, adding penicillin to suppress bacteria, managed to grow a mumps virus in it.

Previously the polio virus could be grown only in living nerve tissue from primates. But using their method, Enders managed to grow the virus successfully on tissue scraps obtained from stillborn human embryos, and then on other tissue.

The general application of this technique meant that viruses could be more readily isolated and identified.

Endlicher Stephan Ladislaus 1804–1849. Austrian botanist who developed an important system of plant classification in which plants were arranged into Genera. His system was especially useful in its division of the plant kingdom into thallophytes (algae, fungi, and lichens) and cormophytes (mosses, ferns, and seed plants).

Outlined in *Genera Plantarum Secundum Ordines Naturales Disposita* 1836–40, Endlicher's system was adopted throughout Europe for half a century, despite the erroneous basis for the classification of some of the 6,835 plants that were included.

Endlicher was born in Pressburg, Hungary. He originally went to Budapest University to study theology, but later changed to medicine and natural history, which he also studied at Vienna University, gaining his MD 1840. During this time he was the curator of the Vienna Museum of Natural History and was appointed professor of botany at Vienna University 1840. However, he spent all of his financial resources on specimens and the publication of his work, and committed suicide in Vienna 28 March.

His own collection of botanical specimens contained about 30,000 plants, which he donated to the museum's herbarium on his death.

Engler (Heinrich Gustav) Adolf 1844–1930. German taxonomical botanist who wrote several important taxonomic references and developed a system of classification that was widely accepted. His *Syllabus der Pflanzennamen/Syllabus of Plant Names* 1892 is still in use. He collaborated with Alphonse de Candolle from 1878–91 on *Monographiae Phanerogamarum/Monographs of Flowering Plants* and started the *Botanical Yearbooks*.

Engler was born in Sagan, Prussia and attended the University of Breslau, graduating with a PhD 1866. Four years later, he was made a custodian at the Botanical Institute of Munich, and in 1878, moved to Kiel to take up a professorship at the university there. He returned to Breslau 1884 as the professor of botany and director of the botanical garden of the university. He moved to Berlin 1889, where he was made the director of the Berlin Botanical Garden, Dahlem. He remained in this position until 1921 taking responsibility for improving and reshaping the gardens.

He travelled throughout the world, making several trips to Africa and went on an expedition around the world 1913. He started the *Botanical Yearbooks* 1880, which he edited until his death 10 October.

Eötvös Roland, Baron von 1848–1919. Hungarian scientist who investigated problems of gravitation. He constructed the double-armed torsion balance for determining variations of gravity.

Epstein (Michael) Anthony 1921– . English microbiologist who, in collaboration with his assistant Barr, discovered 1964 the Epstein–Barr virus (EBV) that causes glandular fever in humans and has been linked to some forms of human cancer.

This was the first time a virus had been linked to the development of cancer, and it prompted many subsequent studies by other scientists into the role of viruses in the onset and progression of human tumours.

Epstein was born in London and educated at Trinity College, Cambridge before moving to the Middlesex Hospital Medical School in London and then to the University of Bristol, where he was appointed professor of pathology 1968. Upon his retirement in 1985, he moved to Oxford to continue his research at the John Radcliffe Hospital Medical School, University of Oxford.

Erasistratus *c.* 304–*c.* 250 BC. Greek physician and anatomist regarded as the founder of physiology. He came close to discovering the true function of several important systems of the body, which were not fully understood until nearly 1,000 years later.

Erasistratus was born on the Aegean island of Ceos (now Khios). He learned his skills in Athens and became court physician to Seleucus I, who governed western Asia. He then moved on to Alexandria, where he taught.

Erasistratus dissected and examined the human brain, noting the convolutions of the outer surface, and observed that the organ is divided into larger and smaller portions (the cerebrum and cerebellum). He compared the human brain with those of other animals and made the correct hypothesis that the surface area/volume complexity is directly related to the intelligence of the animal.

Erasistratus came near to discovering the principle of blood circulation, although he had it circulating in the wrong direction. Tracing the network of veins, arteries, and nerves, he postulated that the nerves carry the 'animal' spirit, the arteries the 'vital' spirit, and the veins blood. He did, however, grasp a rudimentary principle of oxygen exchange and condemned bloodletting as a form of treatment.

Eratosthenes *c.* 276–*c.* 194 BC. Greek geographer and mathematician whose map of the ancient world was the first to contain lines of latitude and longitude, and who calculated the Earth's circumference with an error of about 10%. His mathematical achievements include a method for duplicating the cube, and for finding prime numbers (Eratosthenes' sieve).

No work of Eratosthenes survives complete. The most important that remains is on geography – a word that he virtually coined as the title of his three-volume study of the Earth (as much as he knew of it) and its measurement.

Eratosthenes was born in Cyrene (now Shahhat, Libya) and educated in Athens. At the age of 30 he was invited by Ptolemy III Euergetes to become tutor to his son and to work in the library of the museum at Alexandria. In 240 BC he became the chief librarian.

Eratosthenes divided the Earth into five zones: two frigid zones around each pole; two temperate zones; and a torrid zone comprising the two areas from the equator to each tropic.

In *Chronography* and *Olympic Victors*, many of the dates he set for events (for example, the fall of Troy) have been accepted by later historians. He also wrote many books on literary criticism in a series entitled *On the Old Comedy*.

Ericsson John 1803–1889. Swedish-born US engineer who took out a patent to produce screw-propeller-powered paddle-wheel ships 1836. He built a number of such ships, including the *Monitor*, which was successfully deployed during the American Civil War.

Ericsson was born in Värmland and worked from the age of 13 doing technical drawing for the Göta Canal works. In 1826, he moved to London to seek sponsorship for a new type of heat engine he had invented, which used the expansion of superheated air as the driving force. This was unsuccessful and not until towards the end of his life did Ericsson construct small, efficient engines of this type.

In 1829 he built the *Novelty*, a steam locomotive which competed unsuccessfully against George Stephenson's *Rocket*. Ericsson turned to building ships fitted with steam engines and screw propellers, and moved to the USA 1839. In 1849, he built the *Princeton*, the first metal-hulled, screw-propelled warship and the first to have its engines below the waterline for added protection. The *Monitor* was the first warship to have revolving gun turrets.

From 1877, Ericsson also explored the possiblity of using solar energy and gravitation and tidal forces as sources of power.

Erlang Agner Krarup 1878–1929. Danish mathematician who applied the theory of probabilities to problems connected with telephone traffic, such as congestion and waiting time. The ***erlang*** is now the unit of telephone-traffic flow.

Erlang was born near Tarm in Jutland and studied at the University of Copenhagen. From 1908 he was leader of the laboratory of the Copenhagen Telephone Company.

Rules determining the amount of traffic to be carried per selector have been established by every telephone authority and company, following Erlang's formulas. Especially important are his formulas for the probability of barred access in busy-signal systems and for the probability of delay and for the mean waiting time in waiting-time systems.

Erlang also constructed a measuring

bridge to meter alternating current (the so-called Erlang complex compensator), which was a considerable improvement on earlier apparatus of similar function. Of equal significance were his investigations into telephone transformers and telephone cable theory.

Erlanger Joseph 1874–1965. US physiologist who shared the 1944 Nobel Prize for Physiology or Medicine with Herbert Gasser for their work on how nerve fibres transmit electrical impulses. They found that the smaller nerve fibres were responsible for the conduction of pain, and that the thickness of a nerve fibre dictates the speed at which a nerve can transmit electrical information.

Neurons, cells that are specialized for transmitting electrical signals from one location in the body to another, are the functional units of the nervous system. Neurons share common features. A neuron has a relatively large cell body containing the nucleus and most of the cytoplasm and organelles of the cell. They also have very long fibrelike processes, that can extend over large distances in the body and conduct messages.

Erlanger was one of the first to characterize the different types of neurons, and together with Gasser, demonstrated that the sensory fibres (dendrons) in mixed nerve trunks (nerves with both sensory neurons and motor neurons) can be arranged in decreasing orders with various physical properties. In general, they found that the smaller nerve fibres were responsible for the conduction of pain, and that the thickness of a nerve fibre dictates the speed of its transmission of electrical information.

Erlanger was born in San Francisco and graduated from Johns Hopkins Medical School. During his time as professor of physiology at the University of Washington 1910–46 he began working with the neurophysiologist Gasser, with whom he was later to share the Nobel prize.

Ernst Richard Robert 1933– . Swiss physical chemist who improved the technique of nuclear magnetic resonance (NMR) spectroscopy in the investigation of atomic nuclei by increasing the sensitivity of the NMR instrumentation and interpretation. He received the Nobel Prize for Chemistry in 1991. NMR became the ideal method for examining the structure of proteins and other biological molecules, laying the foundations for magnetic resonance imaging (MRI).

The nuclear magnetic resonance technique subjected a sample to a single high-energy radio pulse, to which many nuclei in the sample responded, emitting an incomprehensible signal. To interpret the signal, Ernst used Fourier analysis and a computer programme, thereby cutting the analysis time in half. In 1970, Ernst again made a breakthrough in the design of NMR, known as ***two-dimensional Fourier NMR***. By replacing a single radio pulse with even higher energy pulses, larger, more complex molecules could be investigated.

Ernst was born in Winterthur, Switzerland and attended the Federal Institute of Technology (ETH), Zurich where he gained a PhD in 1962. From 1963 to 1968, he worked in the laboratories of Varian Assiciates, Palo Alto, California, then he took a job with ETH becoming professor of physical chemistry in 1976.

Esaki Leo, (originally Esaki Reiona) 1925– . Japanese physicist who in 1957 noticed that electrons could sometimes 'tunnel' through the barrier formed at the junctions of certain semiconductors. The effect is now widely used in the electronics industry.

For this early discovery Esaki shared the 1973 Nobel Prize for Physics with British physicist Brian Josephson and Norwegian-born US physicist Ivar Giaever (1929–).

Esaki, born in Osaka, graduated from the University of Tokyo and worked for electronics manufacturer Sony 1956–60. He then joined IBM's research centre in Yorktown Heights, New York, but returned to Japan 1992 as president of the University of Tsukuba.

Tunnelling is a quantum-mechanical effect whereby electrons can travel through electrostatic potentials that they would be unable to overcome classically. Esaki was able to use this effect for switching and to build ultra-small and ultrafast tunnel diodes, now called Esaki diodes. He continued to research the nonlinear transport and optical properties of semiconductors, in particular multilayer superlattice structures grown by molecular-beam epitaxy techniques.

Eskola Pentti Eelis 1883–1964. Finnish geologist who was one of the first to apply physicochemical postulates on a far-reaching basis to the study of metamorphism, thereby laying the foundations of most subsequent studies in metamorphic petrology.

Eskola was born in Lellainen and educated as a chemist at the University of Helsinki before specializing in petrology. In the early 1920s he worked in Norway and in Washington DC, USA. He was professor at Helsinki 1924–53.

Throughout his life Eskola was fascinated by the study of metamorphic rocks, taking early interest in the Precambrian rocks of England. Building largely on Scandinavian studies, he was concerned to define the changing pressure and temperature conditions under which metamorphic rocks were formed. His approach enabled comparison of rocks of widely differing compositions in respect of the pressure and temperature under which they had originated.

Euclid *c.* 330–*c.* 260 BC. Greek mathematician who wrote the *Stoicheia/Elements* in 13 books, of which nine deal with plane and solid geometry and four with number theory. His great achievement lay in the systematic arrangement of previous discoveries, based on axioms, definitions, and theorems.

Euclid's works, and the style in which they were presented, formed the basis for all mathematical thought and expression for the next 2,000 years. He used two main styles of presentation: the synthetic (in which one proceeds from the known to the unknown via logical steps) and the analytical (in which one posits the unknown and works towards it from the known, again via logical steps).

Euclid went to the recently founded city of Alexandria (now in Egypt) in around 300 BC and set up his own school of mathematics there. His mathematical works survived in almost complete form because they were translated first into Arabic, then into Latin; from both of these they were then translated into other European languages.

In *Stoicheia/Elements* Euclid used the synthetic approach; he used the analytical mode of presentation in his other important mathematical work, *Treasury of Analysis*.

There is no royal road to geometry.

Euclid

[To Ptolemy I] From Proclus *Commentary on Euclid* Prologue

Eudoxus of Cnidus *c.* 400–*c.* 347 BC. Greek mathematician and astronomer. He devised the first system to account for the motions of celestial bodies, believing them to be carried around the Earth on sets of spheres. Work attributed to Eudoxus includes methods to calculate the area of a circle and to derive the volume of a pyramid or a cone.

Probably Eudoxus regarded the celestial spheres as a mathematical device for ease of computation rather than as physically real, but the idea was taken up by Aristotle and became entrenched in astronomical thought until the time of Tycho Brahe.

In mathematics Eudoxus' early success was in the removal of many of the limitations imposed by Pythagoras on the theory of proportion. Eudoxus also established a test for the equality of two ratios.

The model of planetary motion was published in a book called *On Rates*. Further astronomical observations were included in two other works, *The Mirror* and *Phaenomena*, providing the basis of the constellation system still in use today. In a series of geographical books with the overall title of *A Tour of the Earth*, Eudoxus described the political, historical, and religious customs of the countries of the E Mediterranean.

Euler Leonhard 1707–1783. Swiss mathematician. He developed the theory of differential equations and the calculus of variations, and worked in astronomy and optics. He also enlarged mathematical notation.

Euler developed spherical trigonometry and demonstrated the significance of the

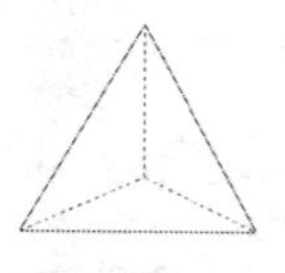
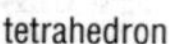

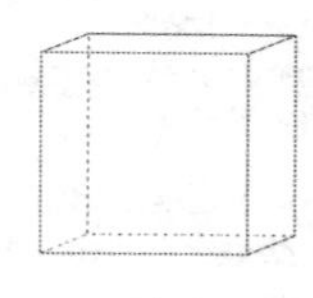

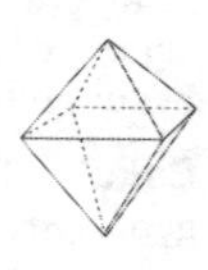

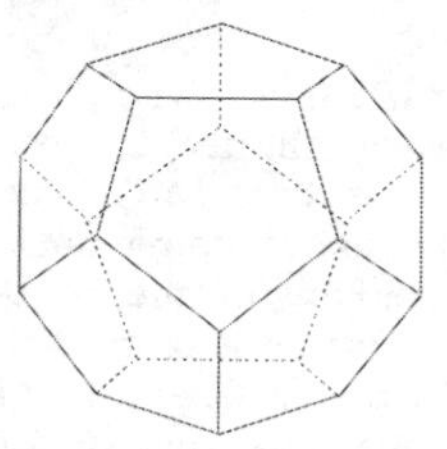

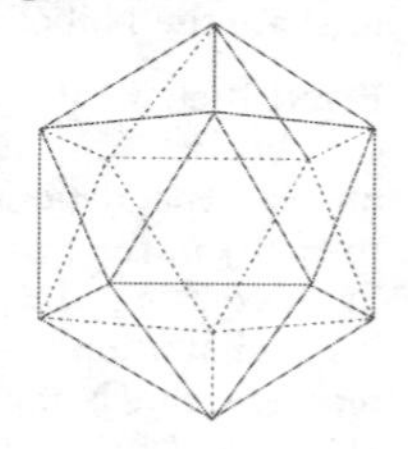

Euclid *The Platonic solids are the only regular convex polyhedra. A tetrahedron is made up of four equilateral triangular faces, a cube is six squares, an octahedron is eight equilateral triangles, a dodecahedron is twelve regular pentagons and an icosahedron is twenty equilateral triangles. They are the basis for Euclidean solid geometry.*

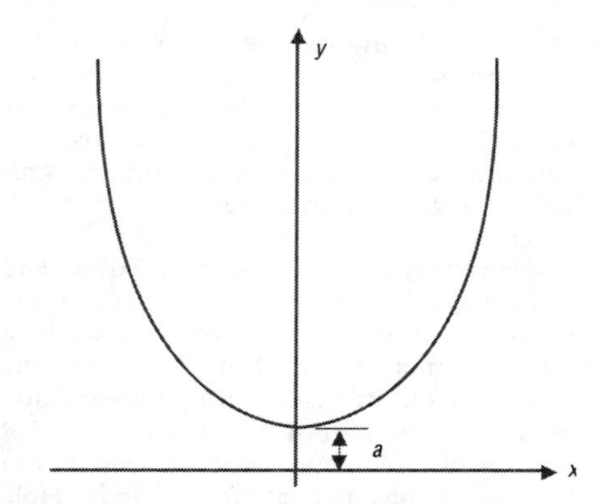

Euler *A catenary is a transcendental curve that may be represented by the equation $y = (a/2)(e^{x/a} + e^{-x/a})$, where a is a constant and e is Euler's number (2.718...).*

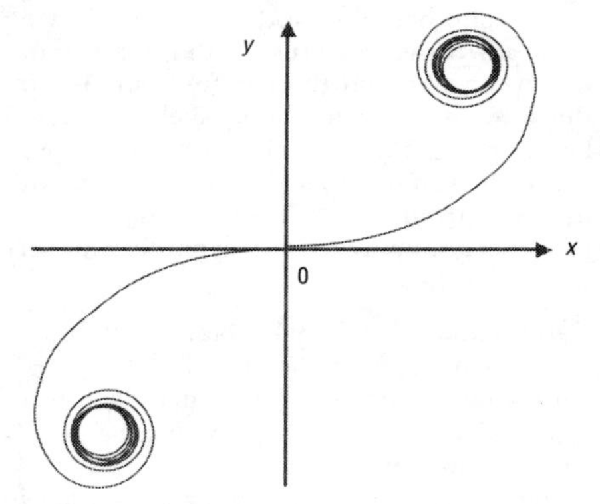

Euler *Euler's spiral*

coefficients of trigonometric expansions; ***Euler's number*** (e, as it is now called) has various useful theoretical properties and is used in the summation of particular series.

Euler was born and educated in Basel, a pupil of Johann Bernoulli. He became professor of physics at the University of St Petersburg 1730. In 1741 he was invited to Berlin by Frederick the Great, where he spent 25 years before returning to Russia.

Euler carried out research into the motion and positions of the Moon, and the gravitational relationships between the Moon, the Sun, and the Earth. His resulting work on tidal fluctuations took him into the realm of fluid mechanics.

Euler-Chelpin Hans Karl August Simon von 1873–1964. German biochemist who defined the structure of the yeast coenzyme, the nonprotein part of the yeast enzyme zymase that interlocks with the zymase and affects how it functions. Nobel Prize for Chemistry 1929.

Arthur Harden discovered in 1904 the presence of nonprotein coenzymes that interlock with an enzyme and affect how it functions. Euler-Chelpin, who studied fermentation – a process that requires the presence of a yeast enzyme called zymase – showed that vitamins A and B activate zymase. This discovery led him to define the structure of the yeast coenzyme, which he called diphosphopyridine nucleotide, today known as NAD. NAD is composed of a nucleotide similar to that found in nucleic acids.

Euler-Chelpin was born in Augsburg, Germany to Swedish parents. He received a cosmopolitan education in biochemistry, moving between the universities of Berlin, Strasbourg, and Göttingen. After studying at the Pasteur Institute in Paris, he moved to Sweden, where he lectured in physical chemistry at the University of Stockholm and later became a professor of chemistry. In 1929, he was appointed director of the Institute for the Biochemistry of Vitamins.

Eustachio Bartolommeo 1520–1574. Italian anatomist, the discoverer of the Eustachian

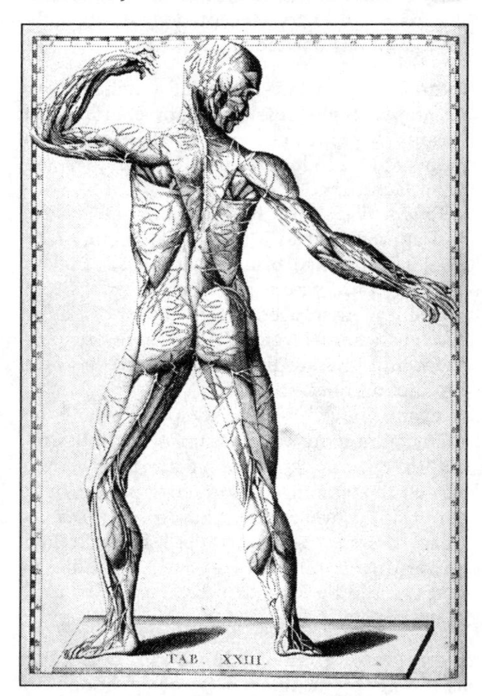

Eustachio *A sketch of human anatomy by Bartolommeo Eustachio. The drawing is taken from the second edition of a book of his work* Tabulae Anatomicae *which was published 1722, 148 years after Eustachio's death.*

tube, leading from the middle ear to the pharynx, and of the Eustachian valve in the right auricles of the heart.

Evans Alice Catherine 1881–1975. US microbiologist whose research into the bacterial contamination of milk led to the recognition of the danger of unpasteurized milk. As a result of her research the incidence of brucellosis was greatly reduced when the dairy industry accepted that all milk should be pasteurized.

Evans was born in Neath, Pennsylvania, and studied at Cornell University and the University of Wisconsin, after which she took a research post at the US Department of Agriculture studying the bacteriology of milk and cheese. In 1918 she moved to the Hygienic Laboratories of the United States Public Health Service to research into epidemic meningitis and influenza as well as milk flora.

Brucellosis in humans and cattle had been thought to be two separate diseases until Evans published her findings 1918. Her results were later confirmed by other scientists, though it was not until the 1930s that the dairy industry accepted that all milk should be pasteurized.

Evans Oliver 1755–1819. US engineer who developed high-pressure steam engines and various machines powered by them. He also pioneered production-line techniques in manufacturing.

Evans was born in Newport, Delaware, and apprenticed to a wagonmaker. In 1780, he joined his two brothers at a flour mill in Wilmington, where he helped to build machinery that used water power to drive conveyors and elevators. As a result, one person could operate the whole mill as a single production line.

Evans moved to Philadelphia, where he spent more than ten years unsuccessfully trying to develop a steam carriage. He then turned to stationary steam engines, and built about 50, as well as the *Orukter Amphibole*, a steam dredger he constructed 1804. It had power-driven rollers as well as a paddle so that it could be moved on land under its own power.

Evershed John 1864–1956. English astronomer who made solar observations. In 1909, he discovered the radial movements of gases in sunspots (the ***Evershed effect***). He also gave his name to a spectroheliograph, the Evershed spectroscope.

Ewing (William) Maurice 1906–1974. US geologist whose studies of the ocean floor provided crucial data for the plate-tectonics revolution in geology in the 1960s. He demonstrated that midocean ridges, with deep central canyons, are common to all oceans.

Ewing was born in Lockney, Texas, and studied at the Rice Institute in Houston. He developed his geological interests by working for oil companies. In 1944 he joined the Lamont–Doherty Geological Observatory, New York. From 1947 he was professor of geology at Columbia University, while also holding a position at the Woods Hole Oceanographic Institute.

Using marine sound-fixing and ranging seismic techniques and pioneering deep-ocean photography and sampling, Ewing ascertained that the crust of the Earth under the ocean is much thinner (5–8 km/3–5 mi thick) than the continental shell (about 40 km/25 mi thick). His studies of ocean sediment showed that its depth increases with distance from the midocean ridge, which gave clear support for the hypothesis of sea-floor spreading.

Eyde Samuel 1866–1940. Norwegian industrial chemist who helped to develop a commercial process for the manufacture of nitric acid that made use of comparatively cheap hydroelectricity.

Eyde was born in Arendal and studied in Germany at the Charlottenburg High School, Berlin. He worked as an engineer in various German cities before returning to Scandinavia. He obtained the hydroelectric rights on some waterfalls in Norway and became director of an electrochemical company 1904, founding a hydroelectric company 1905. He was a member of the Norwegian parliament.

In 1901, while studying the problem of the fixation of nitrogen (the conversion of atmospheric nitrogen into chemically useful compounds), he met his compatriot Christian Birkeland (1867–1917). Together they developed the Birkeland–Eyde process for the economic combination of nitrogen and oxygen (from air) in an electric arc to produce nitrogen oxides and, eventually, nitric oxide.

Eysenck Hans Jürgen 1916– . British psychologist. His work concentrates on personality theory and testing by developing behaviour therapy. He is an outspoken critic of psychoanalysis as a therapeutic method.

His theory that intelligence is almost entirely inherited and can be only slightly modified by education aroused controversy.

Eysenck was born in Berlin, but left his native Germany for the UK when the Nazis came to power in the 1930s. He studied at London University, where he became professor of psychology 1955.

That intelligence has a physical basis and is heritable is shown, he argues, by objective physiological methods, such as measures of changes in EEG (brain wave) patterns evoked by a sudden stimulus. Much of the controversy over his findings, and his view that there may be a genetic basis for racial differences in IQ, is concerned with the methods of assessing intelligence and whether or not they are free of cultural bias.

Eysenck has also made important studies of emotionality and conditioning, and social attitudes.

Fabre Jean Henri Casimir 1823–1915. French entomologist whose studies of wasps, bees, and other insects, particularly their anatomy and behaviour, have become classics.

Fabre was born in Saint-Léons, S France, and studied in Paris. In 1852 he became professor of physics and chemistry at the lycée in Avignon. He held this post for 20 years, eventually resigning because the authorities would not allow girls to attend his science classes. He then abandoned his teaching career and embarked on a serious study of entomology.

In 1878 he bought a small plot of waste land in Serignan, Provence. He built a wall around the plot and remained there for the rest of his life, treating it as an open-air laboratory. Towards the end of his life he became world famous as an authority on entomology.

In addition to numerous entomological papers, Fabre wrote the ten-volume *Souvenirs entomologiques* 1879–1907. Based almost entirely on observations Fabre made in his small plot, this work is a model of meticulous attention to detail.

History celebrates the battlefields whereon we meet our death, but scorns to speak of the ploughed fields whereby we live. It knows the names of the kings' bastards, but cannot tell us the origin of wheat.

Henri Fabre

Souvenirs entomologiques

Fabricius Geronimo 1537–1619. Latinized name of Girolamo Fabrizio. Italian anatomist and embryologist. He made a detailed study of the veins and discovered the valves that direct the blood flow towards the heart. He also studied the development of chick embryos.

Fabricius also investigated the mechanics of respiration, the action of muscles, the anatomy of the larynx (about which he was the first to give a full description) and the eye (he was the first to correctly describe the location of the lens and the first to demonstrate that the pupil changes size).

Fabricius was born in Aquapendente, near Orvieto, and studied at Padua, where he was taught by anatomist Gabriel Fallopius. In 1565 he succeeded Fallopius as professor and remained at Padua for the rest of his career. Fabricius built up an international reputation that attracted students from many countries, including William Harvey.

Fabricius publicly demonstrated the valves in the veins of the limbs 1579, and in 1603 published the first accurate description, with detailed illustrations, of these valves in *De Venarum Ostiolis/On the Valves of the Veins*. He mistakenly believed, however, that the valves' function was to retard the flow of blood to enable the tissues to absorb nutriment.

In his treatise *De formato foetu/On the Formation of the Fetus* 1600 – the first work of its kind – he compared the late fetal stages of different animals and gave the first detailed description of the placenta. In *De formatione ovi et pulli/On the Development of the Egg and the Chick* 1612, he made some erroneous assumptions; for example, that the sperm did not enter the ovum, but stimulated the generative process from a distance.

Fabricus Johann Christian 1745–1808. Danish entomologist who developed a classification system for insects. Using the mouth structure as the basis for classification, he named and described over 10,000 insects. He also wrote on evolution, long before Darwin, and was convinced that humans had evolved from the great apes.

Fabricus was born at Tünder, South Jutland, Denmark and studied at Uppsala under Linnaeus 1762–64, and then travelled throughout Europe, visiting may of the great museums. He was made professor of natural science and economics at the Universities of Copenhagen and Kiel. His publications included *Philosophia entomologica* (1778), *Betrachtungen uber die allgemeinen Einrichtugen in der natur* (1781) and *Resultate naturhistorischer Vorlesungen* (1804).

Fabry (Marie Paul Auguste) Charles 1867–1945. French physicist who specialized in optics, devising methods for the accurate measurement of interference effects. He took part in inventing a device known as the ***Fabry–Pérot interferometer***. In 1913, Fabry demonstrated that ozone is plentiful in the upper atmosphere and is responsible for filtering out ultraviolet radiation from the Sun.

Fabry was born in Marseille and studied in Paris at the Ecole Polytechnique and the Sorbonne. He became professor at the University of Marseille 1904. The Ministry of Inventions recalled him to Paris in 1914 to investigate interference phenomena in light and sound waves, and in 1921 Fabry became professor of physics at the Sorbonne.

Alfred Pérot (1863–1925) and Fabry worked together 1896–1906 on the design and uses of their invention. It consists of two parallel plates of half-silvered glass or quartz. A light source produces rays which undergo a different number of reflections before being focused. When the rays are reunited on the focal plane of the instrument, they either interfere or cohere, producing dark and light bands respectively. If the two reflecting plates are fixed in position relative to one another, the device is called a ***Fabry–Pérot etalon***.

In 1906 Fabry began to collaborate with Henri Buisson. They used the interferometer to confirm the Doppler effect for light 1914.

Fahrenheit Gabriel Daniel 1686–1736. Polish-born Dutch physicist who invented the first accurate thermometer 1724 and devised the Fahrenheit temperature scale. Using his thermometer, Fahrenheit was able to determine the boiling points of liquids and found that they vary with atmospheric pressure.

Fahrenheit was born in Danzig (Gdansk). He learned the manufacture of scientific instruments in Amsterdam from 1701, and spent ten years travelling round Europe, meeting scientists. In 1717 he became an instrumentmaker in Amsterdam, and remained in the Netherlands for the rest of his life.

Fahrenheit's first thermometers contained a column of alcohol which expanded and contracted directly, as originally devised by Danish astronomer Ole Römer in 1701. Fahrenheit substituted mercury for alcohol because its rate of expansion, although less than that of alcohol, is more constant. Furthermore, mercury could be used over a much wider temperature range than alcohol.

In order to reflect the greater sensitivity of his thermometer, Fahrenheit expanded Römer's scale so that blood heat was 90° and an ice–salt mixture was 0°; on this scale freezing point was 30°. Fahrenheit later adjusted the scale to ignore body temperature as a fixed point so that the boiling point of water came to 212° and freezing point was 32°. This is the Fahrenheit scale that is still in use today.

Fairbairn William 1789–1874. Scottish engineer who designed a riveting machine that revolutionized the making of boilers for steam engines. He also worked on many bridges, including the wrought iron box-girder construction used first on the railway bridge across the Menai Straits in North Wales. Baronet 1869.

Fairbairn was born in Kelso, Roxburghshire, and was self-educated. While apprenticed to a millwright near Newcastle-upon-Tyne, he became a friend of the engineers George and Robert Stephenson. Fairbairn used his inventive skills and engineering ability to earn a fortune by the time he was 40 years old.

In Manchester, Fairbairn set up as a manufacturer of cotton-mill machinery. In 1824 he erected two watermills in Zürich. From 1830, he concentrated on shipbuilding, first in Manchester (where he built ships in sections) and then, from 1835, on the river Thames, where his Millwall Iron Works employed some 2,000 people.

The success of Fairbairn's riveting machine led Robert Stephenson to consult him over the building of the Menai railway bridge, though they fell out 1849, a year before the bridge was completed.

Fajans Kasimir 1887–1975. Polish-born US chemist who did pioneering work on radioactivity and isotopes. He also formulated rules that help to explain valence and chemical bonding.

Fajans was born in Warsaw and educated at Leipzig, Heidelberg, Zürich, and Manchester. He worked in Germany 1911–35, becoming director of the Munich

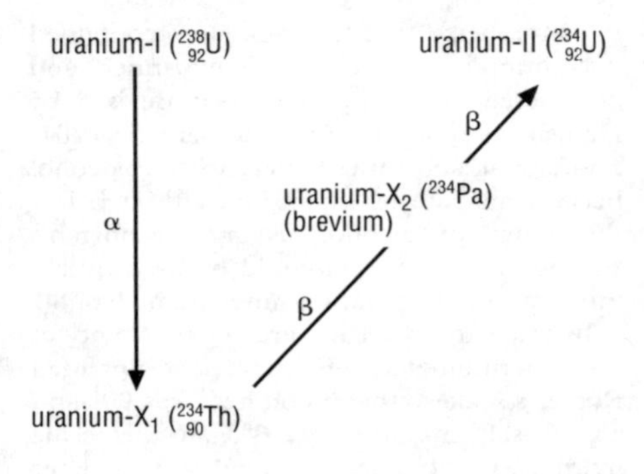

Fajans *Fajans' scheme for the decay of uranium–1 with modern symbols added. Uranium–X_1 and uranium-X_2 are actually isotopes of thorium and protactinium.*

Institute of Physical Chemistry. In 1936, he emigrated to the USA and served as a professor at the University of Michigan, Ann Arbor.

In 1913, Fajans arrived simultaneously with, but independently of, Frederick Soddy at the theory of isotopes. He used this to explain how the decay of uranium-238, by emitting first an alpha particle and then two beta particles, produces a new uranium isotope four mass units lighter: uranium-234.

In inorganic chemistry, Fajans formulated two rules to account for the diagonal similarities between elements in the periodic table in terms of the ease of formation of covalencies and electrovalencies (ionic valencies). The first rule states that covalencies are more likely to be formed as the number of electrons to be removed or donated increases, so that highly charged ions are rare or impossible. The second rule states that electrovalencies are favoured by large cations and small anions.

Fajans also used radioactivity to estimate the ages of minerals.

Falconer Hugh 1808–1865. Scottish botanist and palaeontologist who studied the vertebrae of the fossilized remains of elephants, rhinoceroses, hippopotamuses, giraffes, and crocodiles. He arranged the British Museum's collection of Indian fossils and made many expeditions into the mountains of India to collect botanical specimens. He also was responsible for commencing the cultivation of Indian tea.

Falconer was born in Forres, Scotland and attended Aberdeen University and then Edinburgh University, where he obtained an MD. In 1830, he travelled to India as a surgeon for the East India Company, and became the superintendent of the botanical gardens at Saharan Pur, near the Himalayas 1832. He returned to England between 1842 and 1848, and during his stay, arranged the British Museum's collection of Indian fossils. He was made the superintendent of the Calcutta Botanic Gardens 1848, as well as the professor of botany at Calcutta Medical College. In 1855, he was forced to retire to England due to ill health, and died in London.

Fallopius Gabriel 1523–1562. Latinized name of Gabriele Falloppio. Italian anatomist who discovered the Fallopian tubes, which he described as 'trumpets of the uterus', and named the vagina. As well as the reproductive system, he studied the anatomy of the brain and eyes, and gave the first accurate description of the inner ear.

Fallopius, born in Modena, studied at Padua under Andreas Vesalius, becoming professor of anatomy at Pisa 1548 and Padua 1551. He was the teacher of Geronimo Fabricius.

Fallopius extended Vesalius's work and corrected its details. He was the first to describe the clitoris and the tubes leading from the ovary to the uterus, which were subsequently named after him. He failed, however, to grasp the function of the Fallopian tubes. He also carried out investigations on the larynx, muscular action, and respiration.

Faraday Michael 1791–1867. English chemist and physicist. In 1821 he began experimenting with electromagnetism, and ten years later discovered the induction of electric currents and made the first dynamo. He subsequently found that a magnetic field will rotate the plane of polarization of light. Faraday produced the basic laws of electrolysis 1834.

In 1821 he devised an apparatus that demonstrated the conversion of electrical energy into motive force, for which he is usually credited with the invention of the electric motor.

Faraday's work in chemistry included the isolation of benzene from gas oils 1835. He demonstrated the use of platinum as a catalyst and showed the importance in chemical reactions of surfaces and inhibitors.

Faraday was born in Newington, Surrey,

Faraday *An artist's impression of Michael Faraday giving one of his famous lectures at the Royal Institution in the presence of Prince Albert and the Prince of Wales. The first of these immensely popular lectures took place in 1825 and they continued until 1862, long after the end of Faraday's own experimental career.*

and was apprenticed to a bookbinder; he was largely self-educated. In 1812 he began researches into electricity, and made his first electrical cell. He became a laboratory assistant to Humphry Davy at the Royal Institution 1813, and in 1833 succeeded him as professor of chemistry. Faraday delivered highly popular lectures at the Royal Institution 1825–62. He refused to take part in the preparation of poison gas for use in the Crimean War.

He experimented with high-quality steel alloys 1818. In 1820 he prepared the chlorides of carbon. In 1823 he produced liquid chlorine, and after the production of liquid carbon dioxide 1835, Faraday used this coolant to liquefy other gases.

Faraday's laws of electrolysis established the link between electricity and chemical affinity, one of the most fundamental concepts in science. It was Faraday who coined the terms 'anode', 'cathode', 'cation', 'anion', 'electrode', and 'electrolyte'.

With the device he built 1831 to produce electromagnetic induction (which, unknown to him, had been discovered 1830 by Joseph Henry), Faraday discovered the transformer.

Faraday showed in 1832 that an electrostatic charge gives rise to the same effects as current electricity. In 1837 he demonstrated that electrostatic force consists of a field of curved lines of force, and conceived of specific inductive capacity. He went on to point out that the energy of a magnet is in the field around it and not in the magnet itself, and he extended this basic conception of field theory to electrical and gravitational systems.

Finally Faraday considered the nature of light and in 1846 arrived at a form of the electromagnetic theory of light that was later developed by Scottish physicist James Clerk Maxwell.

Farman Henry 1874–1958. Anglo-French aviation pioneer. He designed a biplane 1907–08 and in 1909 flew a record distance of 160 km/100 mi.

With his brother Maurice Farman (1878–1964), he founded an aircraft works at Billancourt, Brittany, supplying the army in France and other countries. The UK also made use of Farman's inventions, for example, air-screw reduction gears, in World War II.

Fechner Gustav Theodor 1801–1887. German psychologist. He became professor of physics at Leipzig in 1834, but in 1839 turned to the study of psychophysics (the relationship between physiology and psychology). He devised ***Fechner's law***, a method for the exact measurement of sensation.

Feller William 1906–1970. Yugoslavian-born US mathematician largely responsible for making the theory of probability accessible to students of subjects other than mathematics through his textbook on the subject. In the theory of limits, he formulated the law of the iterated logarithm.

Feller was born and educated in Zagreb and also studied in Germany at Göttingen. He worked at the University of Stockholm 1933–39. On the outbreak of World War II, he emigrated to the USA. He was professor of mathematics at Cornell University, New York, 1945–50, and at Princeton from 1950.

Feller came to the conclusion, early on, that the traditional emphasis placed on averages meant that insufficient attention was paid to random fluctuations. Much of his study of probability theory focused on the nature of Markov processes (a mathematical description of random changes in a system which, for instance, can occur in either of two states). Feller demonstrated the applicability of this tool to subjects in which probability theory had not usually previously been employed; for example, in the study of genetics.

His work is set out in *Introduction to Probability Theory and its Applications* 1950–66.

Ferguson Harry George 1884–1960. Northern Irish engineer who pioneered the development of the tractor, joining forces with Henry Ford 1938 to manufacture it in the USA. He also experimented in automobile and aircraft development.

Ferguson was born near Belfast. In 1902 he joined his brother in a car- and cycle-repair business. He built his own aeroplane and flew it 1909, becoming one of the first Britons to do so. He started to import tractors, then, from 1936, designed his own. In 1946, with British government backing, the Ferguson tractor, made by the Standard Motor Company in Coventry, was launched. In the USA Ferguson and Ford fought a massive antitrust suit, largely over a similar machine produced by Ford. Ferguson set up his own US plant 1948, but sold it to Massey-Harris 1953.

For the first Ferguson tractor, he designed a plough that would not rear up and crush the driver when encountering an obstacle. But it was his system of draught control, patented 1925, that revolutionized farming methods by improving the effective traction so that expensive, heavy machines were no longer necessary.

Ferguson Margaret Clay 1863–1951. US botanist who made important contributions as a teacher and administrator. Studying plant genetics, she worked particularly on the genus *Petunia*.

Ferguson was born in Orleans, New York State, and educated at Wellesley College and Cornell University, her studies interspersed with schoolteaching. She taught at Wellesley 1893–1938, becoming professor 1906. Ferguson's department became a major centre for botanical education.

Her early work focused on the life history and reproductive physiology of a species of North American pine. Turning to *Petunia*, she analysed the inheritance of features such as petal colour, flower pattern, and pollen colour, and built up a large database of genetic information.

Fermat Pierre de 1601–1665. French mathematician who, with Blaise Pascal, founded the theory of probability and the modern theory of numbers. Fermat also made contributions to analytical geometry. In 1657, Fermat published a series of problems as challenges to other mathematicians, in the form of theorems to be proved.

Fermat's last theorem states that equations of the form $x^n + y^n = z^n$ where x, y, z, and n are all integers have no solutions if $n > 2$. Fermat scribbled the theorem in the margin of a mathematics textbook and noted that he could have shown it to be true had he enough space in which to write the proof. The theorem remained unproven for 300 years (and therefore, strictly speaking, constituted a conjecture rather than a theorem). In 1993, Andrew Wiles of Princeton University, USA, announced a proof; this turned out to be premature, but he put forward a revised proof 1994 which was accepted.

Fermat was born near Montauban and graduated in law from the University of Orléans. He became a magistrate in Toulouse, rising to the rank of King's Counsellor for the local parliament 1648–65. He refused to publish any of his achievements in mathematics, which are known only from his letters.

To divide a cube into two other cubes, a fourth power or in general any power whatever into two powers of the same denomination above the second is impossible, and I have assuredly found an admirable proof of this, but the margin is too narrow to contain it.

Pierre de Fermat

[written in the margin of his copy of Diophantus *Arithmetica*] Translated from Latin in *Source Book of Mathematics* 1929

Fermat's technique in much of his work was 'reduction analysis', a reversible process in which a particular problem is 'reduced' until it can be seen to be part of a group of problems for which solutions are already known. Analytical geometry was developed simultaneously both by Fermat (in letters written before 1636) and by René Descartes (who published his *Géométrie* 1637). There followed a protracted and bitter dispute over priority.

Fermat also derived what is now known as ***Fermat's principle***, which states that light travels by the path of least duration.

Fermi Enrico 1901–1954. Italian-born US physicist who proved the existence of new radioactive elements produced by bombardment with neutrons, and discovered nuclear reactions produced by low-energy neutrons. He took part in the Manhattan Project to construct an atom bomb. His theoretical work included study of the weak nuclear

Fermi *Italian-born US physicist Enrico Fermi pictured around 1942. Fermi won the Nobel Prize for Physics 1938 following his work with radioactive material which led to considerable insights into the structure of the nucleus.*

force, one of the fundamental forces of nature. Nobel prize 1938.

Fermi's experimental work on beta-decay in radioactive materials provided further evidence for the existence of the neutrino, as predicted by Austrian physicist Wolfgang Pauli. At the University of Chicago, Fermi built the first nuclear reactor 1942. This was the basis for studies leading to the atomic bomb and nuclear energy.

Fermi was born in Rome and studied at Pisa; Göttingen, Germany; and Leiden, the Netherlands. He was professor of theoretical physics at Rome 1926–38, when the rise of Fascism in Italy caused him to emigrate to the USA. He was professor at Columbia University, New York, 1939–42, and from 1946 at the University of Chicago.

With British physicist Paul Dirac, Fermi studied the quantum statistics of fermion particles, which are named after him.

If I could remember the names of all these particles I'd be a botanist.

Enrico Fermi

Quoted in R L Weber, *More Random Walks in Science*

Fernel Jean François 1497–1558. French physician who introduced the terms 'physiology' and 'pathology' into medicine.

Ferranti Sebastian Ziani de 1864–1930. British electrical engineer who established the principle of a national grid and an electricity-generating system based on alternating current (AC) (successfully arguing against Thomas Edison's proposal). He brought electricity to much of central London. In 1881 he made and sold his first alternator.

Ferranti also designed, constructed, and experimented with many other electrical and mechanical devices, including high-tension cables, circuit breakers, transformers, turbines, and spinning machines.

Ferranti was born in Liverpool. He started his own company at 18, in partnership with Irish physicist William Thomson (Lord Kelvin), to design and manufacture the Thomson–Ferranti alternator and install lighting systems. He was chief engineer with the London Electric Supply Company 1887–92, and worked on the design of a large power station at Deptford. He set up a company in Oldham, Lancashire, 1896, to design and build all kinds of electrical equipment, most of which was designed by Ferranti himself, and to develop high-voltage systems for long-distance transmission. He was also involved with heat engines of various kinds.

Fessenden Reginald Aubrey 1866–1932. Canadian physicist who worked in the USA. He patented the modulation of radio waves (transmission of a signal using a carrier wave), an essential technique for voice transmission. At the time of his death, he held 500 patents.

Early radio communications relied on telegraphy by using bursts of single-frequency signals in Morse code. In 1900 Fessenden devised a method of making audio-frequency speech (or music) signals modulate the amplitude of a transmitted radio-frequency carrier wave – the basis of AM radio broadcasting.

Fessenden's other major invention was the heterodyne effect. In this, the received radio wave is combined with a wave of frequency slightly different to that of the carrier wave. The resulting intermediate frequency wave is easier to amplify before being demodulated to generate the original sound wave.

Fessenden was born in East Bolton, Quebec, and educated at Bishop's University, Lennoxville, Quebec. He went to the USA to work for inventors Thomas Edison and, 1890–92, George Westinghouse. Fessenden became professor of electrical engineering at Purdue University, Lafayette, and then at the Western University of Pennsylvania (now the University of Pittsburgh). It was there that Fessenden began major work on the problems of radio communication.

In 1902 Fessenden organized the building of a 50-kHz alternator for radiotelephony by the General Electric Company. This was followed by his building a transmitting station at Brant Rock, Massachussetts. On Christmas Eve 1906, the first amplitude-modulated radio message was broadcast.

Unanimity of opinion may be fitting for a church, for the frightened or greedy victims of some (ancient or modern) myth, or for the weak and willing followers of some tyrant. Variety of opinion is necessary for objective knowledge.

Paul K Feyerabend

Against Method p 46

Feyerabend Paul K 1924–1994. Austrian-born US philosopher of science, who rejected the attempt by certain philosophers (such as Karl Popper) to find a methodology applicable to all scientific research. His works include *Against Method* 1975.

Feyerabend argues that successive theories that apparently concern the same subject (for instance the motion of the planets) cannot in principle be subjected to any comparison that would aim at finding the truer explanation. According to this notion of incommensurability, there is no neutral or objective standpoint, and therefore no rational and objective way in which one particular theory can be chosen over another. Instead, scientific progress is claimed to be the result of a range of sociological factors working to promote politically convenient notions of how nature operates. In the best-selling *Against Method*, he applied an anarchic approach to the study of knowledge and espoused practices, such as the Haitian cult of voodoo, that flew in the face of conventional scientific wisdom.

At 18, Feyerabend joined the wartime German army, and was severely wounded. After the end of World War II, he studied singing and stage management before returning to his native Vienna to take up theoretical physics, history, and philosophy, and gained a doctorate 1951. He was a British Council visiting scholar at the London School of Economics 1952–53, and was again in the UK 1957 as a lecturer at Bristol University. The following year he went to the University of California at Berkeley.

Over the next two decades he published some of his most important and controversial papers, including 'Problems of Empiricism' 1965. His philosophy has been interpreted as 'anything goes', and, while it found favour with leftist youth culture, it aroused much criticism among academic philosophers and the establishment.

Feynman Richard P(hillips) 1918–1988. US physicist whose work laid the foundations of quantum electrodynamics. For his work on the theory of radiation he shared the Nobel Prize for Physics 1965 with Julian Schwinger and Sin-Itiro Tomonaga (1906–1979). He also contributed to many aspects of particle physics, including quark theory and the nature of the weak nuclear force.

For his work on quantum electrodynamics, he developed a simple and elegant system of

Feynman *US physicist Richard Feynman, noted for the major theoretical advances he made in quantum electrodynamics. Feynman began working on the Manhattan Project while at Princeton University and then worked in Los Alamos 1943–46, on the development of the first atomic bomb.*

Feynman diagrams to represent interactions between particles and how they moved from one space-time point to another. He had rules for calculating the probability associated with each diagram.

His other major discoveries are the theory of superfluidity (frictionless flow) in liquid helium, developed in the early 1950s; his work on the weak interaction (with US physicist Murray Gell-Mann) and the strong force; and his prediction that the proton and neutron are not elementary particles. Both particles are now known to be composed of quarks.

One does not, by knowing all the physical laws as we know them today, immediately obtain an understanding of anything much.

Richard P Feynman

The Character of Physical Law

Feynman was born in New York and studied at the Massachusetts Institute of Technology and at Princeton. During World War II, he worked at Los Alamos, New Mexico, on the behaviour of neutrons in atomic explosions. Feynman was professor of theoretical physics at Caltech (California Institute of Technology) from 1950 until his death.

As a member of the committee investigating the *Challenger* space-shuttle disaster 1986, he demonstrated the faults in rubber seals on the shuttle's booster rocket.

The Feynman Lectures on Physics 1963 became a standard work. He also published two volumes of autobiography: *Surely You're Joking, Mr Feynman!* 1985 and *What Do You Care What Other People Think?* 1988.

Fibonacci Leonardo, also known as ***Leonardo of Pisa*** *c.* 1170–*c.* 1250. Italian mathematician. He published *Liber abaci/The Book of the Calculator* in Pisa 1202, which was instrumental in the introduction of Arabic notation into Europe. From 1960, interest increased in ***Fibonacci numbers***, in their simplest form a sequence in which each number is the sum of its two predecessors (1, 1, 2, 3, 5, 8, 13,...). They have unusual characteristics with possible applications in botany, psychology, and astronomy (for example, a more exact correspondence than is given by Bode's law to the distances between the planets and the Sun).

In 1220, Fibonacci published *Practica geometriae*, in which he used algebraic methods to solve many arithmetical and geometrical problems.

Fibonacci was born in Pisa. He learned mathematics in Algeria and travelled extensively in the Mediterranean region.

Returning to Pisa in about 1200, he began his mathematical writings. In 1225 he won a mathematical competition in Pisa in the presence of Holy Roman Emperor Frederick II. A marble tablet dated 1240 is thought to refer to Fibonacci as being awarded an annual pension for his accountancy services to the state.

Liber abaci was a thorough treatise on algebraic methods and problems in which he strongly advocated the introduction of the Indo-Arabic numeral system, comprising the figures 1 to 9, and the innovation of the 'zephirum' – the figure 0 (zero). Dealing with operations in whole numbers systematically, he also proposed the idea of a bar (solidus) for fractions.

Fick Adolf Eugen 1829–1901. German physiologist who worked on the physics of how the human body works, particularly the eyes, heart, and muscles. He devised techniques to quantify body functions, such as muscle contraction, impulse conduction by nerves, and blood pressure in the heart. He also described the physics of how the eye works, especially the blind spot of the retina.

In one experiment he disproved the then accepted belief that the energy of muscle contraction was derived from the oxidation of muscle cells, and that the nitrogen excreted in urine increases as a result. He and a colleague scaled the Faulhorn in Switzerland having starved themselves of nitrogenous food for the previous seventeen hours. By analyzing their own urine they found that they had done at least three times as much work as could be accounted for by the metabolism of protein.

Fick was born in Kassel, Germany and studied medicine at Marburg University before completing his doctoral studies 1851. From 1868 until his death, Fick was the professor of physiology at Wurzburg.

Field George Brooks 1929– . US theoretical astrophysicist whose main research has been into the nature and composition of intergalactic matter and the properties of residual radiation in space.

Field was born in Providence, Rhode Island, and educated at the Massachusetts Institute of Technology and Princeton

University. He became professor at the University of California at Berkeley 1965 and at Harvard University 1972; from 1973 he was also director of the Center of Astrophysics at the Harvard College Observatory and the Smithsonian Astrophysical Observatory.

One of Field's major areas of research has been to investigate why a cluster of galaxies remains a cluster rather than dispersing. They are thought to be stabilized gravitationally by intergalactic matter, mainly hydrogen and helium. By studying part of the spectrum of the radio source Cygnus A, Field found in 1958 some evidence of atomic hydrogen distributed intergalactically.

Field has also carried out research into the lines in the spectra of stars.

Finsen Niels Ryberg 1860–1904. Danish physician, the first to use ultraviolet light treatment for skin diseases. Nobel Prize for Physiology or Medicine 1903.

Fischer Edmond 1920– . US biochemist who shared the 1992 Nobel Prize for Physiology or Medicine with Edwin Krebs for isolating and describing the action of the enzymes responsible for reversible protein phosphorylation. Reversible phosphorylation is the attachment or detachment of phosphate groups to or from proteins in cells. It is at the heart of a wide range of biological processes ranging from muscle contraction to the regulation of genes.

In 1955 and 1956 Fischer and Krebs, both at the University of Washington, Seattle, USA, isolated the enzyme, called phosphorylase, involved in phosphorylation. They also discovered other enzymes, called protein kinases and phosphatase, which add and remove phosphate groups to the phosphorylase. By adding and removing phosphate groups, the enzymes change the shape of the phosphorylase molecule, switching on and off its catalytic properties.

Fischer was born in Shanghai, China, the son of Swiss parents. He was educated in chemistry at the University of Geneva, receiving his doctorate 1947. In 1953 he moved to the University of Washington, Seattle, becoming a full professor 1961.

Fischer Emil Hermann 1852–1919. German chemist who produced synthetic sugars and, from these, various enzymes. His descriptions of the chemistry of the carbohydrates and peptides laid the foundations for the science of biochemistry. Nobel prize 1902.

About 1882, Fischer began working on a group of compounds that included uric acid and caffeine. He realized that they were all related to a hitherto unknown substance, which he called purine. Over the next few years he synthesized about 130 related compounds, one of which was the first synthetic nucleotide. These studies led to the synthesis of powerful hypnotic drugs derived from barbituric acids (barbiturates).

Fischer was born near Bonn and educated there and at Strasbourg and Munich. He held professorships at Erlangen 1882–85, Würzburg 1885–92, and Berlin from 1892.

In 1884, Fischer discovered a key reaction in the study of sugars. He went on to determine the structures of glucose, fructose, mannose, and the group of sugars known collectively as hexoses.

Fischer's investigations into the chemistry of proteins began 1899. He synthesized the amino acids ornithine (1,4-diaminopentanoic acid) 1901, serine (1-hydroxy-2-aminobutanoic acid) 1902, and the sulphur-containing cystine 1908. He then combined amino acids to form polypeptides.

Fischer Ernst Otto 1918– . German inorganic chemist who showed that transition metals can bond chemically to carbon. He and English chemist Geoffrey Wilkinson shared the 1973 Nobel Prize for Chemistry for their work on organometallic compounds, which they carried out independently.

Fischer was born in Munich and educated at the Munich Technical University. He remained there, becoming professor 1959.

Investigating a synthetic compound called ferrocene, both Fischer and Wilkinson separately came to the conclusion that each molecule of ferrocene consists of a single iron atom sandwiched between two five-sided carbon rings – an organometallic compound. A combination of chemical and

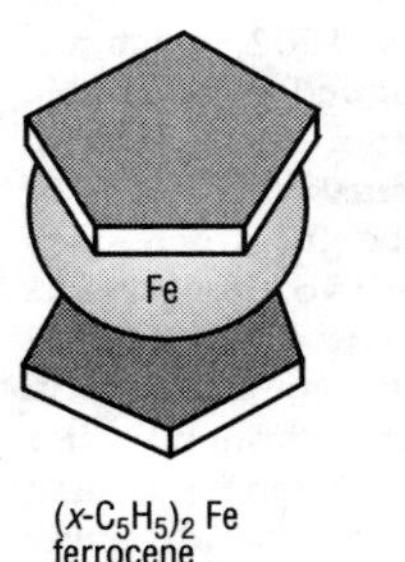

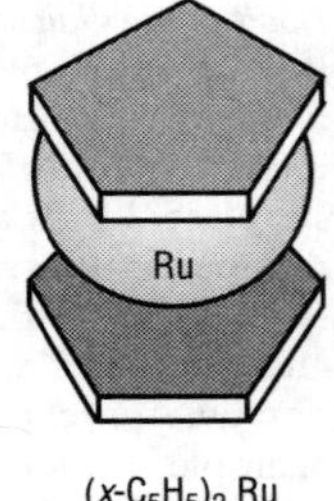

Fischer, Ernst *Structures of ferrocene and ruthenocene*

physical studies, finally confirmed by X-ray analysis, showed the compound's structure.

With this work came the general realization that transition metals can bond chemically to carbon, and other ring systems were then studied. All the elements of the first transition series have now been incorporated into molecules of this kind and all except that of manganese have the ferrocene-type structure. Only ferrocene, however, is stable in air, the others being sensitive to oxidation.

Fischer Hans 1881–1945. German chemist awarded a Nobel prize 1930 for his work on haemoglobin, the oxygen-carrying, red colouring matter in blood. He determined the molecular structures of three important biological pigments: haemoglobin, chlorophyll, and bilirubin.

Fischer was born in Höchst-am-Main, near Frankfurt, and studied at Marburg and Munich. He went to Austria as professor at Innsbruck 1915–18 and Vienna 1918–21, returning to Germany as professor at the Munich Technische Hochschule. In 1945 Fischer's laboratories were destroyed in an Allied bombing raid and in a fit of despair he committed suicide.

In 1921 Fischer began investigating haemoglobin, concentrating on haem, the iron-containing non-protein part of the molecule. By 1929 he had elucidated the complete structure and synthesized haem. Chlorophyll, he found in the 1930s, has a similar structure. He then turned to the bile pigments, particularly bilirubin (the pigment responsible for the colour of the skin of patients suffering from jaundice), and by 1944 had achieved a complete synthesis of bilirubin.

Fischer Hermann Otto Laurenz 1888–1960. German organic chemist who carried out research into the synthetic and structural chemistry of carbohydrates, glycerides, and inositols.

Fischer was born in Würzburg, Bavaria, the son of chemist Emil Fischer, and studied in the UK at Cambridge and in Germany at Berlin and Jena. In 1912, he returned to the Chemical Institute of Berlin University where he continued research with his father, until the start of World War I two years later. With the rise of Nazi leader Adolf Hitler, the Fischers left Berlin in 1932 and went to Basel, Switzerland. In 1937 Hermann moved to the Banting Institute in Toronto, Canada, where he stayed until moving, in 1948, to the University of California at Berkeley.

Between 1920 and 1932 Fischer worked out the exact structure of quinic acid and investigated the difficult chemistry of the trioses glyceraldehyde and dihydroxyacetone and the related two-, three-, and four-carbon compounds.

Fischer also worked on glyceraldehydes, extending it from 1937 to glycerides (esters of glycerol, i.e. propan-1,2,3-triol) and demonstrated the action of lipase enzymes on these biologically important substances.

Fisher Ronald Aylmer 1890–1962. English statistician and geneticist. He modernized Charles Darwin's theory of evolution, thus securing the key biological concept of genetic change by natural selection. Fisher developed several new statistical techniques and, applying his methods to genetics, published *The Genetical Theory of Natural Selection* 1930.

This classic work established that the discoveries of the geneticist Gregor Mendel could be shown to support Darwin's theory of evolution.

Fisher was born in London and studied at Cambridge. In 1919 he was appointed head of Rothampstead Experimental Station, where he made a statistical analysis of a backlog of experimental data that had built up over more than 60 years.

At Rothamstead, Fisher also bred poultry, mice, snails, and other creatures, and in his papers on genetics contributed to the contemporary understanding of genetic dominance. As a result, in 1933 he was appointed professor of eugenics at University College, London. He was professor of genetics at Cambridge 1943–57.

In statistics, Fisher evolved the rules for decision-making that are now used almost automatically, and many other methods that have since been extended to virtually every academic field in which statistical analysis can be applied.

Fitch John 1743–1798. US inventor and early experimenter with steam engines and steamships. In 1786 he designed the first steamboat to serve the Delaware River. His venture failed, so Robert Fulton is erroneously credited with the invention of the steamship.

Fitch Val Logsdon 1923– . US physicist who shared the 1980 Nobel Prize for Physics with James Cronin for their joint work in particle physics, studying the surprising way certain mesons change from matter to antimatter.

Fitch was born in Merriman, Nebraska, and educated at McGill University and Columbia University. He became professor at Princeton University 1960.

The discovery for which Fitch and Cronin received the 1980 Nobel prize was first published in 1964. They had set up an experiment with the proton accelerator at the Brookhaven Laboratory in New York to study the properties of K^0 mesons. K^0 is a mixture of two 'basic states' which have a long and a short lifetime and are therefore called $K^0{}_L$ and $K^0{}_S$ respectively. These two basic states can also mix together to form not K^0 but an antimatter particle (anti-K^0), and K^0 can oscillate from particle to antiparticle through either of its basic states. Fitch and Cronin found that decays of $K^0{}_L$ mesons sometimes violate the known rules, and so are different from all other known particle interactions.

Fitzgerald George Francis 1851–1901. Irish physicist known for his work on electromagnetics. In 1892 he explained the anomalous results of the Michelson–Morley experiment 1887 by supposing that bodies moving through the ether contracted as their velocity increased, an effect since known as the ***Fitzgerald–Lorentz contraction***.

Fitzgerald was born in Dublin and studied there at Trinity College, where he was professor of natural and experimental philosophy from 1888.

Fitzgerald predicted that a rapidly oscillating (that is, alternating) electric current should result in the radiation of electromagnetic waves – a prediction proved correct in the late 1880s by Heinrich Hertz's early experiments with radio, which Fitzgerald brought to the attention of the scientific community in Britain.

Considering the Michelson–Morley result – or lack of result – Fitzgerald worked out a simple mathematical relationship to show how velocity affects physical dimensions. The idea was independently arrived at and developed by Dutch physicist Hendrik Lorentz in 1895. In 1905, the contraction hypothesis was incorporated and given a different interpretation in Albert Einstein's general theory of relativity.

Fitzroy Robert 1805–1865. British vice admiral and meteorologist. In 1828 he succeeded to the command of HMS *Beagle*, then engaged on a survey of the Patagonian coast of South America, and in 1831 was accompanied by naturalist Charles Darwin on a five-year survey. Fitzroy was governor of New Zealand 1843–45. In 1855 the Admiralty founded the Meteorological Office, which issued weather forecasts and charts, under his charge.

Fixx James 1932–1984. US popularizer of jogging for cardiovascular fitness with his book *The Complete Book of Running* 1978. He died of a heart attack while jogging.

Fizeau Armand Hippolyte Louis 1819–1896. French physicist who in 1849 was the first to measure the speed of light on the Earth's surface. He also found that light travels faster in air than in water, which confirmed the wave theory of light, and that the motion of a star affects the position of the lines in its spectrum.

Fizeau, born in Paris, studied at the College de France and with François Arago at the Paris Observatory. Many of his discoveries were made in collaboration with Léon Foucault 1839–47.

Fizeau began to research into the new science of photography in 1839, and with Foucault developed daguerreotype photography for astronomical observations by taking the first detailed pictures of the Sun's surface 1845. They also found, in 1847, that heat rays from the Sun undergo interference and that radiant heat therefore behaves as a wave motion.

To determine the speed of light, Fizeau sent a beam through the gaps in the teeth of a rapidly rotating cog wheel to a mirror 8 km/5 mi away. On returning, the beam was brought to the edge of the wheel, the speed being adjusted so that the light was obscured. This meant that light rays which had passed through the gaps were being blocked on their return by the adjacent teeth as they moved into the position of the gaps. The time taken for the teeth to move this distance was equal to the time taken for light travel 16 km/10 mi to the mirror and back.

Flamsteed John 1646–1719. English astronomer who began systematic observations of the positions of the stars, Moon, and planets at the Royal Observatory he founded at Greenwich, London, 1676. His observations were published in *Historia Coelestis Britannica* 1725.

As the first Astronomer Royal of England, Flamsteed determined the latitude of Greenwich, the slant of the ecliptic, and the position of the equinox. He also worked out an ingenious method of observing the absolute right ascension – a coordinate of the position of a heavenly body – which removed all errors of parallax, refraction, and latitude. Having obtained the positions of 40 reference stars, he then went back and computed positions for the rest of the 3,000 stars in his catalogue.

Flamsteed was born near Derby and studied at Cambridge. He was appointed astronomer to Charles II 1675, but had to supply his own equipment.

Flamsteed began his astronomical studies at home by observing a solar eclipse 1662, about which he corresponded with other astronomers. In 1672, he determined the solar parallax from observations of Mars. His lunar calculations were urgently needed by Isaac Newton and Edmond Halley to test their theories, but Flamsteed withheld them and fell out with both in 1704.

Flavell Richard Anthony 1945– . British-born molecular biologist who is best known for his work on the nature of the genes for human globin chains (the protein components of the blood's oxygen-carrying substance haemoglobin). He showed that thalassaemia, a group of inherited anaemias, were the result of genetic defects. His work led to the development of gene therapies to correct the faulty genes and treat the condition.

He demonstrated that genetic defects in one or more of the globins cause individuals to develop thalassaemia. He went on to describe the exact nature of these genetic defects and the abnormal type of globin chain(s) they formed.

Flavell also investigated the segregation of different phenotypes (the visible traits displayed by an organism) with identical genotypes (the particular set of gene variants that produced these traits) using wheat.

Flavell studied at the University of Hull, the University of Amsterdam, and the University of Zurich. He was appointed to a staff position at the National Institutes of Medical Research, Mill Hill, London 1979 and was made president of the Biogen Corporation 1982–88. He was elected a Fellow of the Royal Society 1984. In 1988 he settled in the USA in the post of professor in both immunology and biology at Yale University Medical School.

Fleming (John) Ambrose 1849–1945. English electrical physicist and engineer who invented the thermionic valve 1904 and devised Fleming's rules. Knighted 1929.

Fleming's rules are memory aids used to recall the relative directions of the magnetic field, current, and motion in an electric generator or motor, using one's fingers. The three directions are represented by the thu*m*b (for *m*otion), *f*orefinger (*f*ield), and se*c*ond finger (*c*urrent), all held at right angles to each other. The right hand is used for generators and the left for motors.

Fleming was born in Lancaster, Lancashire, and educated at University College and South Kensington, London, and at Cambridge, where he worked in the Cavendish Laboratory and studied under Scottish physicist James Clerk Maxwell. In 1882–83, Fleming was professor at Nottingham, and from 1885 at University College, London. He was a consultant at various times to the Edison, Swan, and Ferranti electric-lighting companies and the Marconi Wireless Telegraph Company, for which he designed many parts of their early radio apparatus.

In 1904 he produced experimental proof that the known rectifying property of a thermionic valve was still operative at radio frequencies, and this discovery led to the invention and production of what was first

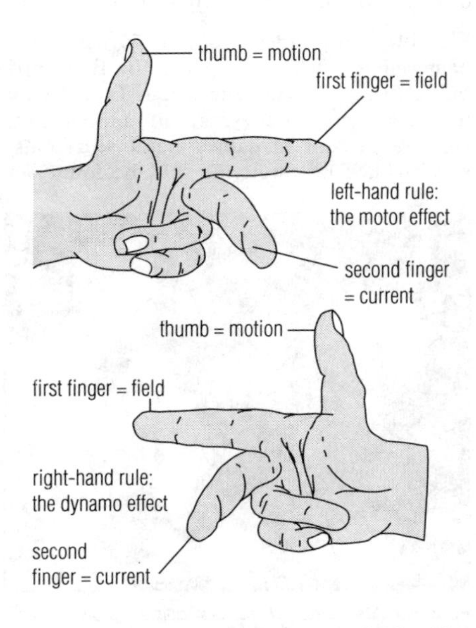

Fleming, Ambrose *Fleming's rules give the direction of the magnetic field, motion, and current in electrical machines. The left hand is used for motors, and the right hand for generators and dynamos.*

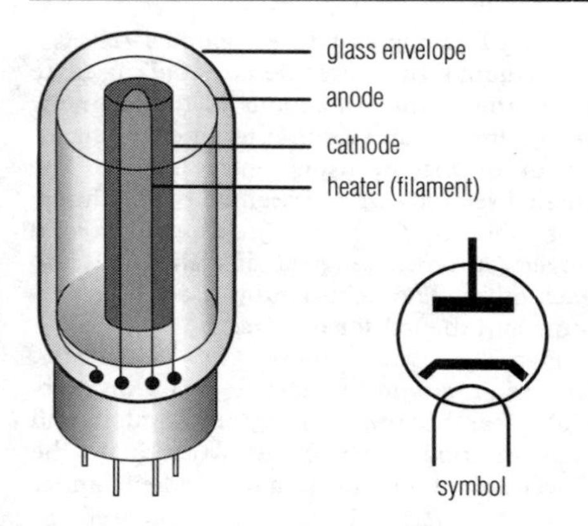

Fleming, Ambrose *Fleming's work on electric lamps led him to invent the diode valve, a two=electrode vacuum tube in which a heated filament causes thermionic emission of electrons from the cathode. The valve was used as a rectifier and a detector in radio receivers until largely superseded by the semiconductor diode.*

known as the 'Fleming valve'. He called it a valve because it allowed electrical currents to pass in only one direction. It worked by allowing one of the electrodes – the cathode – to be kept hot so that electrons could evaporate from it into the vacuum. The other electrode – the anode – was left cool enough to prevent any appreciable evaporation of electrons from it. It revolutionized the early science of radio.

Fleming Alexander 1881–1955. Scottish bacteriologist who discovered the first antibiotic drug, penicillin, in 1928. In 1922 he had discovered lysozyme, an antibacterial enzyme present in saliva, nasal secretions, and tears. While studying this, he found an unusual mould growing on a neglected culture dish, which he isolated and grew into a pure culture; this led to his discovery of penicillin. It came into use in 1941. In 1945 he won the Nobel Prize for Physiology or Medicine with Howard W Florey and Ernst B Chain, whose research had brought widespread realization of the value of penicillin.

Fleming was born in Lochfield, Ayrshire, and studied medicine at St Mary's Hospital, London, where he remained in the bacteriology department for his entire career, becoming professor 1928.

Fleming, Alexander *British biochemist Alexander Fleming who discovered penicillin 1922. The potential of the first antibiotic was realized by Howard Florey during the World War II. Fleming, Florey and Ernst Chain, who purified the drug, shared the Nobel Prize for Physiology or Medicine 1945.*

I have been trying to point out that in our lives chance may have an astonishing influence and, if I may offer advice to the young laboratory worker, it would be this – never to neglect an extraordinary appearance or happening. It may be – usually is, in fact – a false alarm that leads to nothing, but it may on the other hand be the clue provided by fate to lead you to some important advance.

Alexander Fleming

Lecture at Harvard

Fleming discovered the antibacterial properties of penicillin, but its purification and concentration was left to Florey and Chain, in Oxford. Fleming also developed methods, which are still in use, of staining spores and flagella of bacteria. He identified organisms that cause wound infections and showed how cross-infection by streptococci can occur among patients in hospital wards. He also studied the effects of different antiseptics on various kinds of bacteria and on living cells. His interest in chemotherapy led him to introduce Paul Ehrlich's Salvarsan into British medical practice.

Fleming Williamina Paton Stevens 1857–1911. Scottish-born US astronomer, assistant to Edward Pickering, with whom she compiled the first general catalogue classifying stellar spectra.

Fleming was born in Dundee and emigrated to the USA 1878. From 1879 she was employed by Pickering, director of the Harvard College Observatory, initially as a 'computer' and copy editor. In 1898 she was appointed curator of astronomical photographs.

Photographs were taken of the spectra obtained using prisms placed in front of the objectives of telescopes. In the course of her analysis of these spectra, Fleming discovered 59 nebulae, more than 300 variable stars, and 10 novae.

The spectra of the stars observed in this manner could be classified into categories. Fleming designed the system adopted in the 1890 *Draper Catalogues*, in which 10,351 stellar spectra were listed in 17 categories ('A' to 'Q'). This system was to be superseded by the work of Annie Jump Cannon at the same observatory.

Flemming Walther 1843–1905. German biologist who is best known for his work on the way cells divide; a process that he named mitosis. Using a microscope and aniline dyes Flemming could observe the duplication of chromosomes (the threadlike genetic material in the nucleus) before a cell divides.

He developed advanced techniques for imaging cells using a microscope and staining cells with aniline dyes. These techniques allowed him to describe the nuclear changes accompanying mitosis. In mitosis, the number of chromosomes in the parental cell is first doubled by the process of DNA replication and then, as the cell divides, the chromosomes split into two sets, each daughter cell inheriting an identical set that has the same number of chromosomes as the parental cell (diploid number).

Flemming also proposed an alternative form of cell division, called meiosis, the process of cell division that produces reproductive cells (gametes) with half the number of chromosomes (haploid number). During meiosis, homologous pairs of chromosomes segregate to give gametes with a single set of chromosomes. When a male gamete (sperm) and a female gamete (egg) fuse during fertilization the diploid number is regenerated.

Flemming was born in Sachsenburg and graduated in medicine before becoming professor of anatomy at Kiel University.

Florey Howard Walter, Baron Florey 1898–1968. Australian pathologist whose research into lysozyme, an antibacterial enzyme discovered by Alexander Fleming, led him to study penicillin (another of Fleming's discoveries), which he and Ernst Chain isolated and prepared for widespread use. With Fleming, they were awarded the Nobel Prize for Physiology or Medicine 1945. Knighted 1944, Baron 1965.

Florey was born in Adelaide and educated there and at Oxford University, England. He was professor of pathology at Sheffield 1932–35 and at Oxford from 1935.

Florey and his co-worker Chain found that penicillin did not behave like an antiseptic or an enzyme, but blocked the normal process of cell division. Their experiments showed that penicillin could protect against infection but that the concentration of penicillin in the human body and the length of time of treatment were vital factors.

In 1940, during the early part of World War II, a German invasion of Britain seemed imminent; Florey and his colleagues smeared spores of the *Penicillium* mould on their coat linings so that, if necessary, any one of them could continue their research elsewhere.

In 1943, Florey went to Tunisia and Sicily and used penicillin successfully on war casualties. By 1945, it was established that antibacterial activity could take place using a dilution of 1 part in 50 million and, with the war over, large-scale commercial production of penicillin began.

Florey and his co-workers resumed their researches on other antibiotics. They discovered cephalosporin C, which later became the basis of some derivatives, such as cephalothin, that can be used as an alternative antibiotic to penicillin.

As a broad principle, science has been too successful in observing human life.

Howard Florey

Flory Paul John 1910–1985. US polymer chemist who was awarded the 1974 Nobel Prize for Chemistry for his investigations of synthetic and natural macromolecules. With Wallace Carothers, he developed nylon, the first synthetic polyamide, and the synthetic rubber neoprene.

Flory was born in Sterling, Illinois, and educated at Manchester College, Indiana, and Ohio State University. He then embarked on a career as an industrial research chemist, working successively for Du Pont (with Carothers), Esso, and the Goodyear Tire and Rubber Company. He was professor of chemistry at Cornell University 1948–56 and at Stanford University from 1961.

Flory pioneered research substances made up of giant molecules, such as rubbers, plastics, fibres, films, and proteins. In addition to developing polymerization techniques, he discovered ways of analysing polymers. Many of these substances are able to increase the lengths of their component molecular chains and Flory found that one extending molecule can stop growing and pass on its growing ability to another molecule.

Flory's later researches looked for and found similarities between the elasticity of natural organic tissues – such as ligaments, muscles, and blood vessels – and synthetic and natural plastic materials.

Flourens Pierre Jean Marie 1794–1867. French physiologist who experimented widely on the effects of the removal of various parts of the central nervous system. He determined the function of different parts of the mammalian brain and the role of the semi-circular canals of the inner ear in balance.

Flourens experimented widely on the central nervous system. He removed the cerebral hemispheres in the brain of a pigeon and observed that this made the bird blind. When one cerebral hemisphere was removed, the bird lost the sight from its opposite eye. Flourens therefore demonstrated that vision depends on the integrity of the cerebral cortex. He next removed only the cerebellum and determined that while the bird could see and hear well, it stood, walked, and flew in an indecisive manner. The birds equilibrium was almost entirely abolished. Flourens later demonstrated the same results on a dog. Injury to the cerebellum therefore causes loss of coordination. Flourens introduced the idea of nervous coordination to physiology.

Flourens also did important work on the role of the semicircular canals of the inner ear in balance and demonstrated that the respiratory centre is situated in the brain stem, the medulla oblongata, the area in the brain that is responsible for the involuntary contraction of the respiratory muscles.

Fludd Robert 1574–1637. British physician and alchemist who attempted to present a comprehensive account of the universe based on Hermetic principles.

Fontana Niccolò *c.* 1499–1557. Italian mathematician and physicist known as Tartaglia.

Forbes Edward 1815–1854. British naturalist who studied molluscs and made significant contributions to oceanography. In palaeobotany, he divided British plants into five groups, and proposed that Britain had once been joined to the continent by a land bridge.

Forbes was born on the Isle of Man and studied at Edinburgh. He became palaeontologist to the Geological Society of London, then professor of natural history at Edinburgh and from 1851 at the Royal School of Mines in London.

Forbes discounted the contemporary conviction that marine life subsisted only close to the sea surface, spectacularly dredging a starfish from a depth of 400 m/1,300 ft in the Mediterranean. His *The Natural History of European Seas* 1859 was a pioneering oceanographical text. It developed his favourite idea of 'centres of creation'; that is, the notion that species had come into being at one particularly favoured location. Though not an evolutionist, Forbes' ideas could be commandeered for evolutionary purposes.

Ford Henry 1863–1947. US automobile manufacturer. He built his first car 1896 and founded the Ford Motor Company 1903. His Model T (1908–27) was the first car to

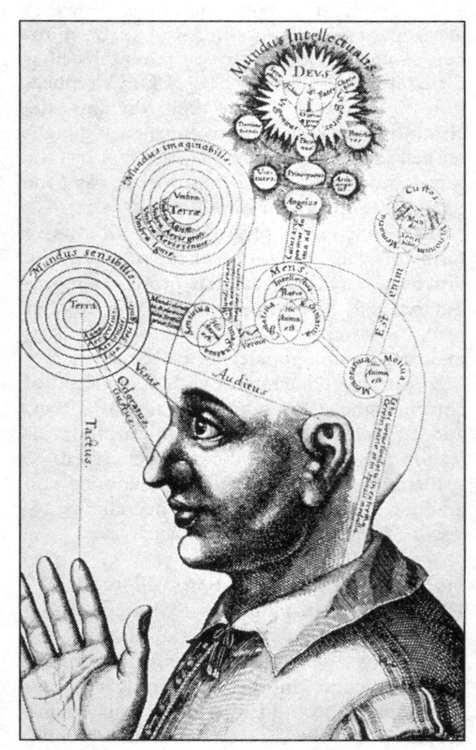

Fludd *Drawing by Robert Fludd, British physician, which attributes different functions of the mind and senses to different regions of the brain. The picture was published between 1617 and 1619. Allocation of region and function was achieved using occult science.*

Ford *US automotive engineer Henry Ford, shown in his first car. Ford's early cars were named alphabetically, albeit in a rather erratic fashion. His first eight models were the A, B, C, F, K, N, R, and S before he finally produced, in 1908, his first mass produced car, the Model T. When, after nineteen years, he ceased production of the Model T, he succeeded it not with the Model U, but another Model A.*

be constructed solely by assembly-line methods and to be mass-marketed; 15 million of these cars were sold.

Ford's innovative policies, such as a $5 daily minimum wage and a five-day working week, revolutionized employment practices, but he opposed the introduction of trade unions. In 1928 he launched the Model A, a stepped-up version of the Model T.

People can have the Model T in any colour – so long as it's black.

Henry Ford

A Nevins *Ford*

Ford was born in Dearborn, Michigan, and apprenticed to a Detroit machinist 1878. He worked for the Edison Illuminating Company 1891–99 and then for the Detroit Automobile Company before starting his own firm. Victory in a car race at Grosse Point, Michigan, 1901, brought him the publicity he sought, and in 1904 he drove his 999 to a world record of 39.4 sec for 1 mi/1.6 km over the ice on Lake St Clair.

Ford was politically active and a pacifist; he opposed US intervention in both world wars and promoted his own anti-Semitic views. In 1936 he founded, with his son Edsel Ford (1893–1943), the philanthropic Ford Foundation; he retired in 1945 from the Ford Motor Company, then valued at over $1 billion.

Forssmann Werner 1904–1979. German heart surgeon. In 1929 he originated, by experiment on himself, the technique of cardiac catheterization (passing a thin tube from an arm artery up into the heart for diagnostic purposes). He shared the 1956 Nobel Prize for Physiology or Medicine.

Forsyth Andrew Russell 1858–1942. Scottish mathematician whose *Theory of Functions* 1893 introduced the main strands of European mathematical study to British mathematicians. Bringing together the work of all the various schools in a single volume, the book completely changed the nature of mathematical thinking.

Forsyth was born in Glasgow and studied at Cambridge. He was professor at Liverpool College 1882–84, but spent most of his career at Cambridge. From 1913 to 1923 he was professor at Imperial College, London.

He formulated a theorem that generalized a large number of identities between double theta functions; because this work was also carried out independently yet simultaneously by Henry Smith (1826–1883), the theorem is now called the ***Smith–Forsyth theorem***.

Forsyth also studied languages, enabling him to translate the works of others and to introduce their ideas to the UK. His *Theory of Functions* stimulated such rapid developments in mathematics that Forsyth was soon left behind.

Fortin Jean Nicolas 1750–1831. French physicist and instrumentmaker who invented a portable mercury barometer in 1800. Any barometer in which the mercury level can be adjusted to zero is now known as a Fortin barometer.

The barometer Fortin designed incorporated a mercury-filled leather bag, a glass cylinder, and an ivory pointer for marking the mercury level. Fortin did not invent these features but he was the first to use them together in a sensitive portable barometer.

Fortin was born in Ile de France and worked in Paris at the Bureau de Longitudes, and later for the Paris Observatory, constructing instruments for astronomical studies and surveying. He also made clocks, and precision equipment for many scientists, including a balance for French chemist Antoine Lavoisier which could measure masses as little as 70 mg/0.0025 oz. In 1799 he adjusted the weight standard, the platinum kilogram, which was stored in the French National Archives.

Fossey Dian 1938–1985. US zoologist. Almost completely untrained, Fossey was sent by Louis Leakey into the African wild. From 1975, she studied mountain gorillas in Rwanda and discovered that they committed infanticide and that females were transferred to nearly established groups. Living in close proximity to them, she discovered that they led peaceful family lives. She was murdered by poachers whose snares she had cut.

The self-eulogizing attempts of expatriates to impose the notion of wildlife as a treasured legacy overlook the reality that to most of a local impoverished and inert populace wildlife is considered an obstacle...

Dian Fossey

Gorillas in the Mist

Foster Michael 1836–1907. English physiologist and founder of the School of Physiology at Cambridge University. Foster published a series of textbooks on physiology (the first was published 1876), in which he hypothesized widely about the mechanism of nerve impulse transmission and subsequent muscle contraction. He proposed that both electrical and chemical processes could be involved in transmission.

He was particularly interested in the control of the heart beat. Since a frog's heart continued to beat for some time after it was extracted from the animal, Albrecht von Haller had proposed that the cardiac muscle possesses an inherent rhythmical power that acts automatically (the myogenic theory). Foster demonstrated this theory by applying a constant current to the apex of a frog's heart devoid of nerve tissue, producing rhythmical activity.

Foster was born in Huntingdon, UK and graduated in classics from University College London. However, he had always had a keen interest in medicine and decided to study the subject, obtaining his MD 1859. Following a voyage 1860 as a ships surgeon and a short period as a general practitioner, Foster entered academic life as a lecturer in the physiology department at University College London. He went on to found the physiology department at Cambridge and to hold the first chair in physiology 1883–1903.

Foucault Jean Bernard Léon 1819–1868. French physicist who used a pendulum to demonstrate the rotation of the Earth on its axis, and invented the gyroscope 1852. In 1862 he made the first accurate determination of the velocity of light.

Foucault investigated heat and light, discovered eddy currents induced in a copper disc moving in a magnetic field, invented a polarizer, and made improvements in the electric arc. In 1860, he invented high-quality regulators for driving machinery at a constant speed; these were used in telescope motors and factory engines.

Foucault was born and educated in Paris and became a physicist at the Paris Observatory 1855. Until 1847, his scientific work was carried out in collaboration with Armand Fizeau. They took the first detailed photographs of the Sun's surface 1845. In 1847, they found that the radiant heat from the Sun undergoes interference and that it therefore behaves as a wave motion. In 1850 Foucault succeeded in showing that light travels faster in air than in water, just beating Fizeau to the same conclusion; this also supported the wave theory.

A pendulum maintains the same movement relative to the Earth's axis and the plane of vibration appears to rotate slowly as the Earth turns beneath it. Foucault made a spectacular demonstration of this by

Foucault *The demonstration of the Earth's rotation using Foucault's pendulum in the Panthéon in Paris in 1851. The swing of the pendulum remains constant but the Earth rotates so that at all positions on the Earth's surface, apart from on the equator, the table rotates in relation to the swing of the pendulum. The effect is most pronounced at the North Pole.*

suspending a pendulum from the dome of the Panthéon in Paris 1851. The invention of the gyroscope followed from this, as Foucault realized that a rotating body would behave in the same way as a pendulum.

Fourier Jean Baptiste Joseph 1768–1830. French applied mathematician whose formulation of heat flow 1807 contains the proposal that, with certain constraints, any mathematical function can be represented by trigonometrical series. This principle forms the basis of ***Fourier analysis***, used today in many different fields of physics. His idea, not immediately well received, gained currency and is embodied in his *Théorie analytique de la chaleur/The Analytical Theory of Heat* 1822.

Light, sound, and other wavelike forms of energy can be studied using Fourier's method, a developed version of which is now called harmonic analysis.

Fourier was born in Auxerre, Bourgogne. His education was interrupted by the French Revolution. He accompanied Napoleon on his Egyptian campaign 1798–1801, and on his return was appointed prefect of the *département* of Isère and later of Rhône, from which he resigned 1815 in political protest. Soon afterwards he obtained a post at the Bureau of Statistics.

Fourier laid the groundwork for the later development of dimensional analysis and linear programming. He also investigated probability theory and the theory of errors.

Fourneyron Benoit 1802–1867. French engineer who invented the first practical water turbine 1827. In 1855 he produced an improved version. He went on to build more than 1,000 hydraulic turbines of various forms and for use in different parts of the world, including Niagara Falls, USA.

Fourneyron was born in Saint-Etienne, Loir, where he studied at the New School of Mines. After graduation he worked on developing the mines at Le Creusot, oil prospecting, laying out a railway, and setting up tin plate – manufacture, which was previously an English monopoly.

Fourneyron's water turbine was an outward-flow turbine. Water passed into guide passages in the movable outer wheel. When the water impinged on these wheel vanes, its direction was changed and it escaped round the periphery of the wheel. But the outward-flow turbine was unstable and speed regulation was difficult. Fourneyron patented an improved design which incorporated a three-turbine installation 1832.

Fowler William Alfred 1911–1995. US astrophysicist. In 1983 he and Subrahmanyan Chandrasekhar were awarded the Nobel Prize for Physics for their work on the life cycle of stars and the origin of chemical elements.

Fowler was born in Pittsburgh, Pennsylvania, and obtained his bachelor's degree in physics from Ohio State University 1933. He attended the California Institute of Technology (Caltech), gained a PhD, and became a Research fellow there 1936. He spent his entire career at Caltech, rising from assistant professor to professor and, in 1970, instructor professor.

All of us, are truly and literally a little bit of stardust.

William Fowler

Fowler concentrated on research into the abundance of helium in the universe. The helium abundance was first defined as the result of the 'hot Big Bang' theory proposed by US physicist Ralph Alpher, Hans Bethe, and George Gamow 1948. In its original form, the Big Bang theory accounted only for the creation of the lightest elements, hydrogen and helium. In their classic paper 1957, Fowler, Hoyle, and the Burbages described how, in a star like the Sun, two hydrogen nuclei, or protons, combine to create the next heavier element, helium, thus generating energy. Over time, more and heavier elements are produced until, after millions of years, the star finally explodes into a supernova, scattering its material across the Universe.

Fowler and Hoyle published an even more complete exposition of stellar nuclear synthesis in 1965 and completed the work two years later with R Wagoner. Taking into account all the reactions that can occur between the light elements, and considering the build-up of heavier elements, they were able to calculate helium abundance in the universe to 1%.

Fracastoro Girolamo *c.* 1478–1553. Italian physician known for two medical books. *Syphilis sive morbus gallicus/Syphilis or the French disease* 1530 was written in verse. It was one of the earliest texts on syphilis, a disease Fracastoro named. In *De contagione/On contagion* 1546, he wrote, far ahead of his time, about 'seeds of contagion'. He was born and worked mainly in Verona.

Fraenkel Abraham Adolf 1891–1965. German-born Israeli mathematician who wrote many textbooks on set theory. He also investigated the axiomatic foundations of mathematical theories.

Fraenkel, born in Munich, studied there and at Marburg, Berlin, and Breslau (now Wroclaw, Poland). He was professor at Marburg 1922–28. In 1929 he emigrated to Israel and taught until 1959 at the Hebrew University of Jerusalem.

Fraenkel became very involved with set theory as it had been formulated in 1908, in the axiomatic system put forward by German mathematician Ernst Zermelo. He proposed in 1922 a solution for Zermelo's unexplained notion of a 'definite property', though his explanation did not become accepted. His research led him to posit an eighth axiom (to follow Zermelo's seventh), an axiom of replacement, which stated that if the domain of a single-valued function is a set, its counterdomain is also a set.

Fraenkel's works include *Einleitung in die Mengenlehre* 1919, *Abstract Set Theory* 1953, and *Foundations of Set Theory* 1958.

Fraenkel-Conrat Heinz Ludwig 1910– . German-born US biochemist who showed that the infectivity of bacteriophages (viruses that infect bacteria) is a property of their inner nucleic acid component, not the outer protein case.

Fraenkel-Conrat was born and educated in Breslau (now Wroclaw, Poland). With the Nazis' rise to power, he left Germany for the UK and the University of Edinburgh, after which he went to the USA. He became professor at the University of California 1958.

In 1955, Fraenkel-Conrat developed a technique for separating the outer protein coat from the inner nucleic acid core of bacteriophages without seriously damaging either portion. He also succeeded in reassembling the components and showed that these reformed bacteriophages are still capable of infecting bacteria. This work raised fundamental questions about the molecular basis of life. He then showed that the protein component of bacteriophages is inert and that the nucleic acid component alone has the capacity to infect bacteria. Thus, it seemed the fundamental properties of life resulted from the activity of nucleic acids.

Francis James Bicheno 1815–1892. English-born US hydraulics engineer who was active in the industrial development of New England. He made significant contributions to the understanding of fluid flow and to the development of the Francis-type water turbine.

Francis was born in Oxfordshire and became assistant to his father on canal and harbour works. In 1833 he went to the USA and worked for a locks and canals company, rising to chief engineer 1837. He advised on a number of dam projects, was a member of the Massachusetts state legislature, and was president of the Stonybrook Railroad for 20 years.

The industrialization of New England resulted initially from water power rather than steam. The leading part Francis played in the exploitation of the Merrimack River and the development of the town of Lowell, Massachusetts, was at the time more important than his work on turbines.

Francis also devised a complete system of water supply for fire protection and had it working in the Lowell district for many years before anything similar was in operation anywhere else. He designed and built hydraulic lifts for the guard gates of the Pawtucket Canal and reconstructed the Pawtucket Dam 1875–76.

His book *The Lowell Hydraulic Experiments* 1855 gives the ***Francis formula*** for the flow of fluids over weirs.

Franck James 1882–1964. German-born US physicist. He shared a Nobel prize 1925 with his co-worker Gustav Hertz (1887–1975) for their experiments of 1914 on the energy transferred by colliding electrons to mercury atoms, showing that the transfer was governed by the rules of quantum theory.

Franck was born in Hamburg and educated at Heidelberg and Berlin. In 1920 he became professor of experimental physics at Göttingen, but emigrated to the USA 1933 after publicly protesting against the Nazis' racial policies. He was a professor at the University of Chicago 1938–49. He participated in the wartime atomic-bomb project at Los Alamos but organized the 'Franck petition' 1945, which argued that the bomb should not be used against Japanese cities. After World War II he turned his research to photosynthesis.

Investigating the collisions of electrons with rare-gas atoms, Franck found that they are almost completely elastic and that no kinetic energy is lost. With Hertz, he extended this work to other atoms. This led to the discovery that there are inelastic collisions in which energy is transferred in definite amounts.

Franck also studied the formation, dissociation, vibration, and rotation of molecules, and was able to calculate the dissociation energies of molecules. Edward Condon (1902–1974) interpreted this method in terms of wave mechanics, and it has become known as the ***Franck–Condon principle***.

Frank Ilya Mikhailoivich 1908–1990. Russian physicist known for his work on radiation. In 1934 Pavel Cherenkov had noted a peculiar blue radiation sometimes emitted as electrons passed through water.

It was left to Frank and his colleague at Moscow University, Igor Tamm (1895–1971), to realize that this form of radiation was produced by charged particles travelling faster through the medium than the speed of light in the same medium.

Frank shared the 1958 Nobel Prize for Physics with Cherenkov and Tamm.

Franklin Benjamin 1706–1790. US scientist, statesman, writer, printer, and publisher. He proved that lightning is a form of electricity, distinguished between positive and negative electricity, and invented the lightning conductor. He was the first US ambassador to France 1776–85, and negotiated peace with Britain 1783. As a delegate to the Continental Congress from Pennsylvania 1785–88, he helped to draft the Declaration of Independence and the US Constitution.

A printer, Franklin wrote and published the popular *Poor Richard's Almanac* 1733–58, as well as engaging in scientific experiment and making useful inventions, including bifocal spectacles. A member of the Pennsylvania Assembly 1751–64, he was sent to Britain to lobby Parliament about tax grievances and achieved the repeal of the Stamp Act; on his return to the USA he was prominent in the deliberations leading up to independence. As ambassador in Paris he enlisted French help for the American Revolution. After independence he became president of Pennsylvania and worked hard to abolish slavery.

Franklin was born in Boston and self-educated. He became one of the most widely travelled of the leaders of the American colonies, bringing an internationalist perspective to the Constitutional Convention. He organized an effective postal system; taught himself Spanish, French, Italian, and Latin; and mapped the Gulf Stream. By flying a kite in a thunderstorm, he was able to charge up a condenser and produce sparks. He recognized the aurora borealis as an electrical phenomenon, and speculated on the existence of the ionosphere.

His autobiography first appeared 1781 (in complete form, 1868).

We must indeed all hang together, or, most assuredly, we shall all hang separately.

Benjamin Franklin

Remark to John Hancock, at Signing of the Declaration of Independence 4 July 1776

Franklin, Benjamin *US statesman and scientist Benjamin Franklin, who helped draft the Declaration of Independence and performed some of the early investigations into the nature of electricity. Franklin was also a manic inventor. He gave the world bifocals, the lightning rod, extendible grippers for taking items of high shelves, and the Franklin stove.*

Franklin Rosalind Elsie 1920–1958. English biophysicist whose research on X-ray diffraction of DNA crystals helped Francis Crick and James D Watson to deduce the chemical structure of DNA.

Fraunhofer Joseph von 1787–1826. German physicist who did important work in optics. The dark lines in the solar spectrum (***Fraunhofer lines***), which reveal the chemical composition of the Sun's atmosphere, were accurately mapped by him.

Fraunhofer determined the dispersion powers and refractive indices of different kinds of optical glass. In the process, he developed the spectroscope, and in 1821 he became the first to use a diffraction grating to produce a spectrum from white light.

Fraunhofer was born in Bavaria and started work in his father's glazing workshop at the age of ten. In 1806 he entered the optical shop of the Munich Philosophical Instrument Company, which produced scientific instruments, and by 1811 he had become a director. From 1823 he was director of the Physics Museum of the Bavarian Academy of Sciences.

In 1814, to obtain more accurate optical values, he commenced using two bright yellow lines in flame spectra as a source of monochromatic light. Comparing this with sunlight, he observed that the solar spectrum is crossed with many fine, dark lines: 574 between the red and violet ends of the spectrum.

Fraunhofer constructed both diffraction and reflection gratings. By using the wave theory of light, he was able to derive a general form of the grating equation that is still in use today.

Fraze Ermal Cleon 1913–1989. US inventor of the ring-pull on drink cans, after having had to resort to a car bumper to open a can while picnicking.

Fredholm Erik Ivar 1866–1927. Swedish mathematician and mathematical physicist who founded the modern theory of integral equations. His work provided the foundations for much of the research later carried out by German mathematician David Hilbert.

Fredholm was born in Stockholm and studied at the Polytechnic Institute there, and at the University of Uppsala. He became professor at Stockholm University 1906.

Fredholm founded much of his theory on work carried out by US astronomer George Hill (1838–1914), who used linear equations involving determinants of an infinite number of rows and columns. In Fredholm's paper 'Sur une nouvelle méthode pour la résolution du problème de Dirichlet' 1900, he first developed the essential part of the theory of what is now known as ***Fredholm's integral equation***; further, he went on to define and solve the Fredholm equation of the second type, involving a definite integral.

Fredholm discovered in 1900–03 the algebraic analogue of his own theory of integral equations. His results were used by Hilbert, who extended them in deriving his own theories, which contributed fundamentally towards quantum theory.

Frege (Friedrich Ludwig) Gottlob 1848–1925. German philosopher, the founder of modern mathematical logic. He created symbols for concepts like 'or' and 'if ... then', which are now in standard use in mathematics. His *Die Grundlagen der Arithmetik/The Foundations of Arithmetic* 1884 influenced Bertrand Russell and Ludwig Wittgenstein. Frege's chief work is *Begriffsschrift/Conceptual Notation* 1879.

Frege was born in Wismar on the Baltic coast and studied at Jena and Göttingen; he became professor at Jena 1879.

Frege incorporated improvements to the *Begriffsschrift* into his two-volume *Grundgesetze der Arithmetik/Basic Laws of Arithmetic* 1893–1903, but was devastated to receive a letter from Bertrand Russell 1902 in which Russell asked Frege how his logical system coped with a particular logical paradox in set theory. Frege's system was not able to resolve it, and he was forced to acknowledge his system to be useless.

Although at the time Frege was largely discredited, his innovations have been useful in the development of symbolic logic, and even the problem posed by Russell was resolved by later logico-mathematicians.

Fresnel Augustin Jean 1788–1827. French physicist who refined the theory of polarized light. Fresnel realized in 1821 that light waves do not vibrate like sound waves longitudinally, in the direction of their motion, but transversely, at right angles to the direction of the propagated wave.

If you cannot saw with a file or file with a saw, then you will be no good as an experimentalist.

Augustin Fresnel

In C V Boys *DSB*

Fresnel first had to confirm the wave theory of light. He demonstrated mathematically that the dimensions of light and dark bands produced by diffraction could be related to the wavelength of the light producing them if light consisted of waves. To explain double refraction, he then arrived at the theory of transverse waves.

Fresnel was born in Broglie, Normandy, and studied in Paris, becoming a civil engineer for the government. When Napoleon returned to France from Elba in 1815, Fresnel deserted his post in protest. He was placed under house arrest, taking advantage of this enforced leisure to develop his ideas on the wave nature of light into a comprehensive mathematical theory.

Napoleon's return proved to be short-lived and Fresnel was reinstated into government service.

Fresnel applied his new ideas on light to lenses for lighthouses. He produced a revolutionary design consisting of concentric rings of triangular cross-section, varying the overall curvature to produce lenses that required no reflectors to produce a bright parallel beam.

Freud Sigmund 1856–1939. Austrian physician who pioneered the study of the unconscious mind. He developed the methods of free association and interpretation of dreams that are basic techniques of psychoanalysis. The influence of unconscious forces on people's thoughts and actions was Freud's discovery, as was his controversial theory of the repression of infantile sexuality as the root of neuroses in the adult. His books include *Die Traumdeutung/The Interpretation of Dreams* 1900, *Jenseits des Lustprinzips/ Beyond the Pleasure Principle* 1920, *Das Ich und das Es/The Ego and the Id* 1923, and *Das Unbehagen in der Kultur/Civilization and its Discontents* 1930. His influence has permeated the world to such an extent that it may be discerned today in almost every branch of thought.

From 1886 to 1938 Freud had a private practice in Vienna, and his theories and writings drew largely on case studies of his own patients, who were mainly upper-middle-class, middle-aged women. The word 'psychoanalysis' was, like much of its terminology, coined by Freud, and many terms have passed into popular usage, not without distortion. His theories have changed the way people think about human nature and brought about a more open approach to sexual matters. Antisocial behaviour is now understood to result in many cases from unconscious forces, and these new concepts have led to wider expression of the human condition in art and literature. Nevertheless, Freud's theories have caused disagreement among psychologists and psychiatrists, and his methods of psychoanalysis cannot be applied in every case.

[Poets] are masters of us ordinary men, in knowledge of the mind, because they drink at streams which we have not yet made accessible to science.

Sigmund Freud

Attributed remark

Freud *Austrian physician Sigmund Freud. Freud initially investigated the relationship between the unconscious mind and the psychological health of an individual. He was a pioneer of psychoanalysis and an early investigator of human consciousness.*

Freud was born in Freiburg, Moravia (now Príbor in the Czech Republic). After first intending to study law, he studied medicine in Vienna from 1873, working under Ernst Wilhelm von Brücke. During this time Freud was a member of the research team that discovered the local anaesthetic effects of cocaine. In 1884 he became assistant physician at the General Hospital of Vienna, and was appointed lecturer in neurology 1885.

In the same year Freud began to study hypnosis as a treatment for hysteria under French physiologist Jean Charcot at the Saltpêtrière hospital, Paris. He was influenced by Charcot's belief that hysteria is of psychical origin and that ideas can produce physical changes, and in 1886 he returned to Vienna with this first inspiration that led to psychoanalysis.

Freud was also influenced by Viennese physician Josef Breuer's research into hysteria, and in 1893 he and Breuer published *Studien über Hysterie/Studies on Hysteria*, outlining the theory that hysterical cases can successfully be treated while under hypnosis by freeing the idea at the root of condition from the unconscious mind.

In about 1895 Freud abandoned hypnosis for the technique of free association, which led to an interest in the interpretation of dreams. From this point he progressed rapidly with his studies and consequent discoveries in psychoanalysis, and published successively *Die Traumdeutung/The Interpretation of Dreams* 1900, *Zur Psychopathologie des Alltagslebens/The Psychopathology of Everyday Life* 1904, and *Drei Abhandlungen zur Sexualtheorie/Three Treatises on the Sexual Theory* 1905. *Die Traumdeutung* put forward the important idea that the recollected parts of dreams are symbols of the activities of the unconscious mind during sleep when the will is ineffective and conscious self-control is suspended. Freud drew a comparison between the symbolism of dreams and of mythology and religion, stating that religion was infantile (God as the father image) and neurotic (projection of repressed wishes).

The revolutionary nature of his theories aroused great hostility, since to assert that nearly all cases of neurosis are due to the repression of sexual desires shocked the public idea of morality at the time. Informed observers, however, found much to interest them in the consequent doctrine that a disturbance in a child's sexual growth explains many cases of emotional disturbance, and that under proper direction sexual impulses may be 'sublimated' into forces which can inspire great achievements.

In 1903 he founded the Vienna Psychoanalytical Circle, and by 1906 branches were established in several other countries. By 1908 his influence had spread further, and the first International Psychoanalytical Congress was held at Salzburg, Austria. In 1909 the International Psychoanalytical Association was formed. Following the Nazi occupation of Vienna, Freud sought refuge in London 1938 and died there the following year.

Freundlich Herbert Max Finlay 1880–1941. German physical chemist who worked on the nature of colloids, particularly sols and gels. He introduced the term 'thixotropy' to describe the behaviour of gels.

Freundlich was born in Berlin and studied at Munich and Leipzig. In 1911 he became professor at the Technische Hochschule in Brunswick. He worked at the Kaiser Wilhelm Institut in Berlin 1914–33. When Adolf Hitler came to power, Freundlich emigrated first to Britain and then, in 1938, to the USA, where he became professor of colloid chemistry at the University of Minnesota.

Freundlich's research was mainly devoted to all aspects of colloid science. He investigated colloid optics, the scattering of light by dispersed particles of various shapes. He studied the electrical properties of colloids, since electrostatic charges are largely responsible for holding colloidal dispersions in place. He investigated mechanical properties such as viscosity and elasticity, and studied the behaviour of certain systems under other types of mechanical force. One application of this work has been the development of non-drip paints.

Friedel Charles 1832–1899. French organic chemist and mineralogist who with US chemist James Mason Crafts (1839–1917) discovered the ***Friedel–Crafts reaction***, which is useful in organic synthesis. Throughout his career, Friedel successfully combined his interests in chemistry and minerals.

Friedel was born in Strasbourg and studied at the Sorbonne in Paris. He qualified in both chemistry and mineralogy and in 1856 was made curator of the collection of minerals at the Ecole des Mines. In 1871 he became an instructor at the Ecole Normale and from 1876 was professor of mineralogy at the Sorbonne.

Doing research into various alcohols, Friedel synthesized glycerol (propan-1,2,3-triol) from propylene (propene) 1871.

Friedel and Crafts collaborated during the period 1874–91. In 1877 they discovered the Friedel–Crafts reaction, which uses aluminium chloride as a catalyst to facilitate the addition of an alkyl halide (halogenoalkane) to an aromatic compound. The reaction is now employed in the industrial preparation of triphenylamine dyes.

From 1879 to 1887 Friedel worked on the attempted synthesis of minerals, including diamonds, using heat and pressure. He established the similarity in properties and structure between carbon and silicon.

Friedman Aleksandr Aleksandrovich 1888–1925. Russian mathematician and cosmologist, who made fundamental contributions to the development of theories regarding the expansion of the universe.

Friedman was born and educated in St Petersburg, where he joined the mathematics faculty. From 1918 to 1920 he was professor of theoretical mechanics at Perm University, but he returned to St Petersburg in 1920 to conduct research at the Academy of Sciences.

Friedman's early research was in the fields of geomagnetism, hydromechanics. and, above all, theoretical meteorology. His work of the greatest relevance to astronomy was his independent and original approach to the solution of Albert Einstein's field equation in the general theory of relativity. Einstein had produced a static solution, which indicated a closed universe. Friedman derived several solutions, all of which suggested that space and time were isotropic (uniform at all points and in every direction), but that the mean density and radius of the universe varied with time – indicating an either expanding or contracting universe.

Friedman Jerome I 1930– . US physicist who, with Henry W Kendall and Richard Taylor, conducted pioneering investigations into high-energy electrons colliding with protons and neutrons. These were important in developing the quark model of particle physics. Nobel Prize for Physics 1990.

In 1970, Friedman, Taylor, and Kendall led a team working at the Stanford Linear Accelerator Center (SLAC) in California. Their experiments involved bombarding protons (and later on neutrons) with high-energy electrons. They knew from earlier experiments that the proton had a small but finite volume and believed that high-energy electrons would suffer only small deflections as they passed through the protons. However they found that the electrons were sometimes scattered through large angles inside the proton.

This result was interpreted by theorists James D Bjorken and Richard P Feynman. They suggested that the electrons were hitting hard pointlike objects inside the proton. These objects were soon shown to be quarks, particles whose existence had been proposed independently by Murray Gell-Mann and George Zweig 1964. Today quarks are recognized as amongst the most fundamental building blocks of matter.

Friedman was born in Chicago, Illinois, USA, and educated at the University of Chicago, where he completed his doctorate in 1956. In 1957 he joined the High Energy Physics laboratory at Stanford University, California, as a research associate and learned the techniques used in electron scattering experiments. At Stanford he began his long association with Henry Kendall and became acquainted with Richard Taylor. In 1960 he went to the physics department of the Massachusetts Insitute of Technology (MIT). Kendall joined his research group 1961. In 1963 Friedmand and Kendall began a collaboration with Taylor and others at Stanford to develop electron scattering facilities at the Stanford Linear Accelerator Center. In 1980 he became director of the laboratory for nuclear science at MIT and then served as head of the physics department 1983–88, when he returned to full-time teaching and research.

Friedman Maurice 1903–1991. US physician who in the 1930s developed the 'rabbit test' to determine if a woman was pregnant. Following injection of a woman's urine into a female rabbit, changes would occur in the animal's ovaries if the woman was pregnant. However, such changes could only be detected on dissection of the rabbit.

Fries Elias Magnus 1794–1878. Swedish botanist whose work lay the foundation for modern mycology (the study of fungi). His most prominent contribution, *Systema mycologicum* 1821, included a classification of fungi by their physical structure and stages of development. He also studied European lichens and published *Observationes Mycologicae* 1815 and *Systema orbis vegetablis* 1825.

Born in Femsjo, Sweden, Fries attended Vaxjo Gymnasium and the University of Lund, where he graduated with a degree in philosophy 1814. However, from then onwards he concentrated on studying botany and remained at the university, where his father supported him financially until he obtained a position as a university demonstrator 1828. By the time he was made professor of botany at Uppsala University 1835, he was the leading Swedish botanist.

Friese-Greene William 1855–1921. English photographer, inventor, and early experimenter in cinematography.

Friese-Greene was born in Bristol. In about 1875 he opened a portrait photography studio in Bath, moving to London 1885. Asked to produce slides for a magic lantern (forerunner of the slide projector), he became interested in moving pictures, and in 1889 patented a camera that could take ten photographs per second on a roll of sensitized paper. Using his own apparatus, he was able to project a jerky picture of people and horse-drawn vehicles moving past Hyde Park Corner – probably the first time a film of an actual event had been projected on a screen.

In 1890, he substituted celluloid film for the paper in the camera, and in the next few years he patented other inventions: a three-colour camera, moving pictures using a two-colour process, and machinery for rapid photographic processing and printing.

Although Friese-Greene's patent of 1890 was upheld in the USA in 1910, the first functional cine camera is generally credited to French physiologist Etienne-Jules Marey (1830–1904) in 1888.

Frisch Karl von 1886–1982. Austrian zoologist, founder with Konrad Lorenz of ethology, the study of animal behaviour. He specialized in bees, discovering how they communicate the location of sources of nectar by movements called 'dances'. He was awarded the Nobel Prize for Physiology or Medicine 1973 together with Lorenz and Nikolaas Tinbergen.

Frisch Otto Robert 1904–1979. Austrian-born British physicist who first described the fission of uranium nuclei under neutron bombardment, coining the term 'fission' to describe the splitting of a nucleus.

Frisch was born and educated in Vienna. Doing research at Hamburg, he fled from Nazi Germany in 1933, initially to the UK and in 1934 to the Institute of Theoretical Physics in Copenhagen. The German occupation of Denmark at the beginning of World War II forced Frisch to return to Britain. He then worked 1943–45 on the atom bomb at Los Alamos, New Mexico, USA. He was professor of natural philosophy at Cambridge University 1947–71.

Frisch worked on methods of separating the rare uranium-235 isotope that would undergo fission. He also calculated details such as the critical mass needed to produce a chain reaction and make an atomic bomb, and urged the British government to undertake nuclear research. At the first test explosion of the atomic bomb, Frisch conducted experiments from a distance of 40 km/25 mi.

He was the nephew of physicist Lise Meitner.

Frobenius Ferdinand Georg 1849–1917. German mathematician who formulated the concept of the abstract group – the first abstract structure of 'new' mathematics. His research into the theory of groups and complex number systems would prove useful to the development of quantum mechanics. He also made contributions to the theory of elliptic functions and to the solution of differential equations.

Frobenius was born in Berlin and studied at Göttingen. He was professor at the Eidgenossische Polytechnikum in Zürich, Switzerland, 1875–92, and at Berlin for the rest of his career.

Publications on abstract algebra and groups written or cowritten by Frobenius include *Über Gruppen von vertauschbaren Elementen* 1879, *Über endliche Gruppen* 1895, and *Über die reellen Darstellungen der endlichen Gruppen* 1906.

Fröhlich Herbert 1905–1991. German-born British physicist who helped lay the foundations for modern theoretical physics in the UK. He revolutionized solid-state theory by importing into it the methods of quantum field theory – the application of quantum theory to particle interactions.

In particular, he proposed a theory to explain superconductivity using the methods of quantum field theory. He made important advances in the understanding of low-temperature superconductivity. His work also led him to the idea that quantum methods might elucidate some aspects of biological systems, such as the electrical properties of cell membranes.

Fröhlich studied in Munich under Arnold Sommerfeld, one of Germany's foremost theoretical physicists. When the Nazis engineered Fröhlich's dismissal from his teaching post 1933, he went to Bristol. From 1948 he was professor of theoretical physics at Liverpool University.

Froude William 1810–1879. English engineer and hydrodynamicist who first formulated reliable laws for the resistance that water offers to ships and for predicting their stability. He also invented the hydraulic dynameter (1877) for measuring the output of high-power engines. These achievements were fundamental to marine development.

Froude was born in Devon and educated at Oxford. He remained for a time at Oxford, working on water resistance and the propulsion of ships. From 1859 he carried out tank-testing experiments at his home, first in Paignton and then in Torquay.

In 1838 Froude assisted Isambard Kingdom Brunel on the building of the Bristol and Exeter Railway. Brunel later consulted him on the behaviour of the *Great Eastern* at sea and, on his recommendation, the ship was fitted with bilge keels.

Beginning in 1867, Froude towed models in pairs, balancing one hull shape against the other. Realizing that the frictional resistance and the wave-making resistance follow different laws, he also towed submerged planks with different surface roughness. He was able to establish a formula which would predict the frictional resistance of a hull with accuracy, and formulated ***Froude's law of comparison***. This stated that the wave-making resistance of similar-shaped models varies as the cube of their dimensions if their speeds are as the square root of their dimensions. With these two analytical results, Froude had found a reliable means of estimating the power required to drive a hull at a given speed.

Froude also carried out model experiments and theoretical work on the rolling stability of ships. His general deductions are still the standard exposition of the rolling and oscillation of ships.

Fuchs Immanuel Lazarus 1833–1902. German mathematician whose work on Georg Riemann's method for the solution of differential equations led to a study of the theory of functions that was later crucial to Henri Poincaré in his investigation of function theory.

Fuchs was born in Moschin (now in Poland), and studied at Berlin, where he became professor 1866. After holding posts at Göttingen and Heidelberg, he returned to Berlin 1882.

The first proof for solutions of linear differential equations of order *n* was developed from his study of functions, as were the Fuchsian differential equations and the Fuchsian theory on solutions for singular points. His work in this field was of great importance to Poincaré's work on automorphic functions. Fuchs also carried out some research into number theory and geometry.

Fuchs Vivian Ernest 1908– . British explorer and geologist. Before World War II, he accompanied several Cambridge University expeditions to E Africa. From 1947 he worked in the Falkland Islands as director of the Scientific Bureau. In 1957–58, he led the overland Commonwealth Trans-Antarctic Expedition. He published his autobiography *A Time to Speak* in 1991. Knighted 1958.

Fukui Kenichi 1918– . Japanese industrial chemist who shared the Nobel Prize for Chemistry in 1981 with Roald Hoffman for his work on 'frontier orbital theory', predicting the change in molecular orbitals (the arrangement of electrons around the nucleus) during chemical reactions.

Fukui described a reaction involving methyl radicals as the interaction between two of the electronic orbitals of the reacting molecules. He called these frontier orbitals and discovered that the course of a chemical reaction was determined partly by the symmetry of the frontier orbitals.

The importance of his findings was initially overlooked as Fukui presented them in mathematical terms that confounded contemporaries. However, following a publication by Robert Woodward and Roald Hoffman that outlined the rules for the conservation of orbital symmetry, scientists came to realize the importance of Fukui's theories. This led to the use of frontier orbital theory in rationalizing organic reactivity and preceded sophisticated computer calculations.

Fukui was born in the district of Nara, Japan and studied industrial chemistry at Kyoto University. He spent World War II working in industry as a technician in the Army Fuel Laboratory, but in 1948 he returned to Kyoto University, where he obtained a PhD in engineering. Here he found his true vocation and advanced to become professor of hydrocarbon physical chemistry 1951, a position he held until his retirement 1982. He was president of the Japanese Chemical Society 1983–84.

Fuller Solomon Carter 1872–1953. US physician, neurologist, psychiatrist, and pathologist. He worked on degenerative brain diseases including Alzheimer's disease, which he attributed to causes other than arteriosclerosis; this was supported by medical researchers in 1953.

Fuller was born in Monrovia, Liberia, and educated at Livingstone College, Salisbury, North Carolina; Long Island College Hospital, Brooklyn, New York; and Boston University School of Medicine, where he became a faculty member 1899. He practised medicine in Boston and at his home in Framingham and taught pathology, neurology, and psychology at the University until he retired as professor emeritus 1937.

Fuller is best known for his work in neuropathology and psychiatry. His post-graduate studies included training at the Carnegie Laboratory, New York, and, 1904–05, at the University of Munich under Alois Alzheimer.

Fulton Robert 1765–1815. US engineer and inventor who designed the first successful steamships. He produced a submarine, the *Nautilus*, for Napoleon's government in France 1801, and experimented with steam navigation on the Seine, then returned to the USA. The first steam vessel of note, known as the *Clermont*, appeared on the river Hudson 1807, sailing between New York and Albany. The first steam warship was the USS *Fulton*, of 38 tonnes, built 1815.

Fulton was born in Pennsylvania and became a portrait painter in Philadelphia. He went to England 1787 to study art but was so taken with the Industrial Revolution that from 1793 he devoted himself to engineering. He designed and patented a device for hauling canal boats over difficult country, and machines for sawing marble and twisting hemp (for rope), and he built a mechanical dredger for canal construction.

In 1796, Fulton went to France, where he experimented with fitting steam engines to ships. His steamships were based on prototypes by US inventor John Fitch. The *Clermont* had paddle wheels and a 18-kW/24-hp engine and travelled at an average speed of 8 kph/5 mph. After this success, a large boatworks was built in New Jersey, and steamboats came into use along the Atlantic Coast and later in the West.

Funk Casimir 1884–1967. Polish-born US biochemist who pioneered research into vitamins. He was the first to isolate niacin (nicotinic acid, one of the vitamins of the B complex).

Funk proposed that certain diseases are caused by dietary deficiencies. In 1911 he demonstrated that rice extracts cure beriberi in pigeons. As the extract contains an amine, he mistakenly concluded that he had discovered a class of 'vital amines', a phrase soon reduced to 'vitamins'.

Funk, born in Warsaw, studied in Berne, Switzerland, and worked at research institutes in Europe before emigrating to the USA 1915. He returned to Warsaw 1923 but, because of the country's uncertain political situation, went in 1927 to Paris, where he founded a research institution, the Casa Biochemica. With the German invasion of France at the outbreak of World War II in 1939, Funk returned to the USA. In 1940, he became president of the Funk Foundation for Medical Research.

Funk failed to find the factor that prevents beriberi in human beings – thiamin, or vitamin B_1 – but once it had been isolated by Robert Williams (1886–1965), Funk determined its molecular structure 1936 and developed a method of synthesizing it.

Funk also carried our research into animal hormones, particularly male sex hormones, and into cancer, diabetes, and ulcers. He improved methods used for drug manufacture and developed several new commercial products.

Gabor Dennis 1900–1979. Hungarian-born British physicist. In 1947 he invented the holographic method of three-dimensional photography. He was awarded a Nobel prize 1971.

Born in Budapest, Gabor studied at the Budapest Technical University and then at the Technishe Hochschule in Berlin. He worked in Germany until he fled to Britain in 1933 to escape the Nazis. He was professor of applied electron physics at the Imperial College of Science and Technology, London, 1958–67.

When Gabor began work on the holograph, he considered the possibility of improving the resolving power of the electron microscope, first by using the electron beam to make a hologram of the object and then by examining this hologram with a beam of coherent light. But coherent light of sufficient intensity was not achievable until the laser was demonstrated 1960.

Gabor's other work included research on high-speed oscilloscopes, communication theory, and physical optics. In 1958 he invented a type of colour TV tube of greatly reduced depth. He took out more than 100 patents for his inventions.

Till now man has been up against Nature, from now on he will be up against his own nature.

Dennis Gabor

Inventing the Future

Gaffky Georg Theodor August 1850–1918. German bacteriologist who isolated and cultured the typhoid bacillus *Salmonella typhi*, a bacterium that inhabits the intestine, 1884. This lead to improved diagnosis and prevention of typhoid fever.

Gaffky isolated and cultured the bacillus from the spleen 1884. Until then, the study of the natural history of the bacillus had been obscured by all the other similar bacilli that make up the flora of the intestinal canal. Eventually, several different forms of the typhoid bacillus were fully characterized.

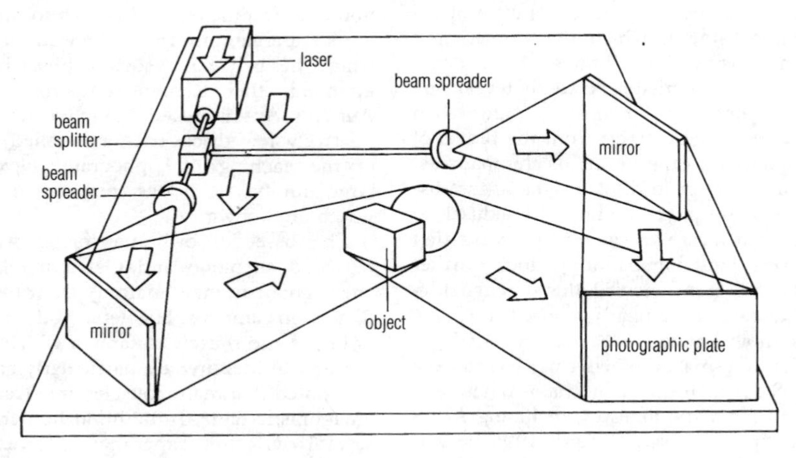

Gabor *Recording a transmission hologram.*

Gaffky was born in Hanover and graduated from the University of Berlin with an MD 1873. He was fortunate to work as a research assistant in the laboratory of the bacteriologist Robert Koch 1880–85. Gaffky witnessed Koch's discovery of the microorganisms responsible for tuberculosis and cholera, which inspired his own work on the typhoid bacillus discovered earlier by Eberth 1880. Gaffky eventually succeeded Koch as the director of the Institute for Infectious diseases in Berlin.

Gagarin Yuri (Alexeyevich) 1934–1968. Soviet cosmonaut who in 1961 became the first human in space aboard the spacecraft *Vostok 1*.

Gagarin was born in the Smolensk region. He became a pilot 1957, and on 12 April 1961 completed one orbit of the Earth, taking 108 minutes from launch to landing. He died in a plane crash while training for the *Soyuz 3* mission.

Gajdusek D(aniel) Carleton 1923– . US virologist and paediatrician who won the Nobel prize 1976 for his work on identifying and describing slow virus infections in humans. This was based on his studies of kuru, a disease of neural degeneration found in people in New Guinea.

Gajdusek was born in Yonkers, New York, and educated at the universities of Rochester and Harvard, and the California Institute of Technology. As a virologist interested in epidemiology, he travelled to New Guinea in the 1950s and made extensive studies of kuru. From 1958 he has worked at the National Institute of Neurology and Communicative Disorders and Stroke in Bethesda, Maryland.

The affected people practised a form of ritual cannibalism, in which the women and children consumed the brains of the dead. Analyses of brain tissue failed to reveal any signs of infective organisms, but when Gajdusek injected extracts from the brains of kuru victims into the brains of chimpanzees, the animals began to display signs of the disease after about a year. This led Gajdusek to propose that kuru was caused by a virus that has a very long incubation period. Further work by his team showed that Creutzfeldt–Jakob disease is similarly caused by slow viruses (now known to be caused not by a virus but by a smaller infective particle called a prion). Since then such a mechanism has been proposed for many illnesses, including AIDS and multiple sclerosis. In April 1996 he was arrested on charges of sex abuse. In February 1997 he pleaded guilty to sexual abuse of one of the 38 Micronesian children he raised and educated in the USA.

Galen *c.* 129–*c.* 200. Greek physician and anatomist whose ideas dominated Western medicine for almost 1,500 years. Central to his thinking were the theories of humours and the threefold circulation of the blood. He remained the highest medical authority until Andreas Vesalius and William Harvey exposed the fundamental errors of his system.

That physician will hardly be thought very careful of the health of others who neglects his own.

Galen

Of Protecting the Health Bk V

The humours were four kinds of fluid of which the human body was supposed to consist: phlegm, blood, choler or yellow bile, and melancholy or black bile. Physical and mental characteristics were explained by different proportions of humours in individuals.

Galen postulated a circulation system in which the liver produced the natural spirit, the heart the vital spirit, and the brain the animal spirit. He also wrote about philosophy and believed that Nature expressed a divine purpose, a belief that became increasingly popular with the rise of Christianity (Galen himself was not a Christian). This accounted for the enormous influence of his ideas.

Galen was born in Pergamum in Asia Minor and studied medicine there and at Smyrna (now Izmir), Corinth in Greece, and Alexandria in Egypt, after which he returned home to become chief physician to the gladiators at Pergamum. In 161 he went to Rome, where he became a society physician and attended the Roman emperor Marcus Aurelius Antoninus. Although Galen made relatively few discoveries and relied heavily on the teachings of Hippocrates, he wrote a large number of books, more than 100 of which are known.

The dissection of human beings was then regarded as taboo and Galen made inferences about human anatomy from his many dissections animals. His detailed descriptions of bones and muscles, many of which he was the first to identify, are particularly good; he also noted that many muscles are arranged in antagonistic pairs. In addition he performed several vivisection experiments; for example, to show that urine passes from the kidneys down the ureters to the bladder. More important, he demonstrated that arteries carry blood, not air, thus disproving Erasistratus' view, which had been taught for some 500 years.

Galileo properly Galileo Galilei 1564–1642. Italian mathematician, astronomer, and physicist. He developed the astronomical telescope and was the first to see sunspots, the four main satellites of Jupiter, and the appearance of Venus going through phases, thus proving it was orbiting the Sun. In mechanics, Galileo discovered that freely falling bodies, heavy or light, have the same, constant acceleration and that a body moving on a perfectly smooth horizontal surface would neither speed up nor slow down.

Galileo's work founded the modern scientific method of deducing laws to explain the results of observation and experiment (although the story of his dropping cannonballs from the Leaning Tower of Pisa is questionable). His observations were an unwelcome refutation of the ideas of Aristotle taught at the (church-run) universities, largely because they made plausible for the first time the heliocentric (Sun-centred) theory of Polish astronomer Nicolaus Copernicus. Galileo's persuasive *Dialogo sopra i due massimi sistemi del mondo/Dialogues on the Two Chief Systems of the World* 1632 was banned by the church authorities in Rome and he was made to recant by the Inquisition.

In questions of science the authority of a thousand is not worth the humble reasoning of a single individual.

Galileo

Arago's Eulogy of Laplace Smithsonian Report, 1874 p 164

Galileo was born and educated in Pisa, and in 1589 became professor of mathematics at the university there; in 1592 he became a professor at Padua, and in 1610 was appointed chief mathematician to the Grand Duke of Tuscany. When tried for heresy in 1633, and forced to abjure his belief that the Earth moves around the Sun, Galileo is reputed to have muttered: '*Eppur si muove*' ('Yet it does move'). He was put under house arrest for his last years.

Galileo discovered in 1583 that each oscillation of a pendulum takes the same amount of time despite the difference in amplitude.

Galileo *The Italian astronomer and mathematician Galileo revolutionized science with his experimental observations made on falling bodies and the flight of projectiles.*

He invented the thermometer and a hydrostatic balance, and discovered that the path of a projectile is a parabola.

Galileo published *De motu/On Motion* 1590. Having made his own telescopes, he published his first findings in *Sidereus nuncius/The Starry Messenger* 1610; the book was a sensation throughout Europe. He summed up his life's work in *Discorsi e dimostrazioni matematiche intorno a due nove scienze/ Discourses and Mathematical Discoveries Concerning Two New Sciences*. The manuscript of this book was smuggled out of Italy and published in Holland 1638.

Galle Johann Gottfried 1812–1910. German astronomer who located the planet Neptune 1846, close to the position predicted by French mathematician Urbain Leverrier.

Galle was born in Saxony, Prussia, and educated at Berlin. He worked at the Berlin Observatory until 1851, when he became professor of astronomy and director of the Breslau Observatory (now in Wroclaw, Poland).

In 1846, Galle received from Leverrier, director of the Paris Observatory, the latter's calculation of the position of a new planet, predicted mathematically from its apparent gravitational effect on Uranus, then the outermost known planet. Within one hour of beginning their search, Galle and his colleague Heinrich d'Arrest (1822–1875) had located Neptune, less than 1° away from the predicted position.

Galle was also the first to distinguish the Crêpe Ring, an inner ring around Saturn, in 1838. He also suggested a method of measuring the scale of the Solar System by observing the parallax of asteroids, first applying his method to the asteroid Flora in 1873. The method was employed with great success after Galle's death.

Gallo Robert Charles 1937– . US scientist credited with identifying the virus responsible for AIDS. Gallo discovered the virus, now known as human immunodeficiency virus (HIV), in 1984; the French scientist Luc Montagnier (1932–) of the Pasteur Institute, Paris, discovered the virus, independently, in 1983. The sample in which Gallo discovered the virus was supplied by Montagnier, and it has been alleged that this may have been contaminated by specimens of the virus isolated by Montagnier a few months earlier. In 1996, he headed the newly founded Institute of Virology in Baltimore.

Galois Evariste 1811–1832. French mathematician who originated the theory of groups and greatly extended the understanding of the conditions in which an algebraic equation is solvable.

Galois was born near Paris and entered the Ecole Normale Supérieure 1829. By then he had already mastered the most recent work on the theory of equations, number theory, and elliptic functions, and was in contact with mathematician Augustin Cauchy, but Galois's attempts to gain recognition for his work were thwarted by the French mathematical establishment.

In 1830, Galois joined the revolutionary movement. In the next year he was twice arrested, and was imprisoned for nine months for taking part in a republican demonstration. Shortly after his release he was killed in a duel. The night before, he had hurriedly written out his discoveries on group theory. His only published work was a paper on number theory 1830.

What has come to be known as the ***Galois theorem*** demonstrated the insolubility of higher-than-fourth-degree equations by radicals. ***Galois theory*** involved groups formed from the arrangements of the roots of equations and their subgroups, which he fitted into each other rather like Chinese boxes.

Galton Francis 1822–1911. English scientist, inventor, and explorer who studied the inheritance of physical and mental attributes with the aim of improving the human species. He was the first to use twins to try to assess the influence of environment on development, and is considered the founder of eugenics (a term he coined). Knighted 1909.

Galton believed that genius was inherited, and was principally to be found in the British; he also attempted to compile a map of human physical beauty in Britain. He invented the 'silent' dog whistle, the weather map, a teletype printer, and forensic fingerprinting, and discovered the existence of anticyclones.

Galton was born in Birmingham and studied medicine in London as well as mathematics at Cambridge. In 1850 he set out for uncharted areas of Africa, and on his return wrote two books describing his explorations.

Galton designed several instruments to plot meteorological data, and made the first serious attempt to chart the weather over large areas – described in his book *Meteorographica* 1863. He also helped to establish the Meteorological Office and the National Physical Laboratory.

In *Hereditary Genius* 1869, based on a study of mental abilities in eminent families, Galton formulated the regression law, which states that parents who deviate from the average in a positive or negative direction have children who, on average, also deviate in the same direction but to a lesser extent.

Galton invented instruments to measure mental abilities in some 9,000 subjects. In order to interpret his data, Galton devised new statistical methods of analysis, including correlational calculus, which has since become an invaluable tool in many disciplines. The results were summarized in *Inquiries into Human Faculty and its Development* 1883.

Galvani Luigi 1737–1798. Italian physiologist who discovered galvanic, or voltaic, electricity in 1762, when investigating the contractions produced in the muscles of dead frogs by contact with pairs of different metals. His work led quickly to Alessandro Volta's invention of the electrical cell, and later to an understanding of how nerves control muscles.

Galvani was born and educated in Bologna, where he taught anatomy. He was professor 1775–97, when he resigned rather than swear allegiance to Napoleon as head of the new Cisalpine Republic.

In 1786 Galvani noticed that touching a frog with a metal instrument during a thunderstorm made the frog twitch. He concluded that electricity was causing the contraction and postulated (incorrectly) that it came from the animal's muscle and nerve tissues. He summarized his findings in 1791 in a paper called 'De viribus electricitatis in motu musculari commentarius/Commentary on the Effect of Electricity on Muscular Motion', which gained general acceptance. But by 1800, Volta had proved that Galvani had been wrong and that the source of the electricity in his experiments had been two different metals and the animal's body fluids. Nevertheless, for many years current electricity was called Galvanic electricity.

Gamow George (Georgi Antonovich) 1904–1968. Russian-born US cosmologist, nuclear physicist, and popularizer of science. His work in astrophysics included a study of the structure and evolution of stars and the creation of the elements. He explained how the collision of nuclei in the solar interior could produce the nuclear reactions that power the Sun. With the 'hot Big Bang' theory, he indicated the origin of the universe.

Gamow predicted that the electromagnetic radiation left over from the universe's formation should, after having cooled down during the subsequent expansion of the universe, manifest itself as a microwave cosmic background radiation. He also made an important contribution to the understanding of protein synthesis.

Gamow was born in Odessa (now in Ukraine), and studied at Leningrad (St Petersburg) and Göttingen, Germany. He then worked at the Institute of Theoretical Physics in Copenhagen, Denmark, and at the Cavendish Laboratory, Cambridge, England. From 1931 to 1933 he was at the Academy of Science in Leningrad, and then defected to the USA, becoming professor at George Washington University in Washington DC 1934–56 and then at the University of Colorado. In the late 1940s, he worked on the hydrogen bomb at Los Alamos, New Mexico.

Gamow's model of alpha decay 1928 represented the first application of quantum mechanics to the study of nuclear structure. Later he described beta decay.

With US scientist Ralph Alpher, he investigated the possibility that heavy elements could have been produced by a sequence of neutron-capture thermonuclear reactions. They published a paper 1948, which became known as the Alpher–Bethe–Gamow (or alpha–beta–gamma) hypothesis, describing the 'hot Big Bang'.

Gamow also contributed to the solution of the genetic code. The double-helix model for the structure of DNA involves four types of nucleotides. Gamow realized that if three nucleotides were used at a time, the possible combinations could easily code for the different amino acids of which all proteins are constructed. Gamow's theory was found to be correct in 1961.

Gardner Julia (Anna) 1882–1960. US geologist and palaeontologist whose work was important for petroleum geologists establishing standard stratigraphic sections for Tertiary rocks in the southern Caribbean.

Julia Gardner was born in South Dakota and educated at Bryn Mawr College and Johns Hopkins University. She worked for the US Geological Survey 1911–54. During World War II, she joined the Military Geologic Unit where she helped to locate Japanese beaches from which incendiary bombs were being launched, by identifying shells in the sand ballast of the balloons.

Her work on the Cenozoic stratigraphic palaeontology of the Coastal Plain, Texas, and the Rio Grande Embayment in northeast Mexico led to the publication of *Correlation of the Cenozoic Formations of the Atlantic and Gulf Coastal Plain and the Caribbean Region* 1943 (with two coauthors).

Garrod Archibald Edward 1857–1936. English physician who first recognized a class of metabolic diseases, while studying the rare disease alcaptonuria, in which the patient's urine turns black on contact with air. He calculated that the cause was a failure of the body's metabolism to break down certain amino acids into harmless substances like water and carbon dioxide. KCMG 1918.

Gaskell Walter Holbrook 1847–1914. English physiologist who investigated the anatomy of the nervous system, particularly the autonomic nervous system, with John Langley. Gaskell originally worked on the nerve structures in striated muscle and the heart. He then moved on to describe the sympathetic or 'voluntary' nervous system that carries signals to the skeletal muscles in response to external stimuli.

Gassendi Pierre 1592–1655. French physicist and philosopher who played a crucial role in the revival of atomism (the theory that the world is made of small, indivisible particles), and the rejection of Aristotelianism so characteristic of the period. He was a propagandist and critic of other views rather than an original thinker.

Gasser Herbert Spencer 1888–1963. US physiologist who shared the Nobel Prize for Physiology or Medicine 1944 with Joseph Erlanger for their discoveries regarding the specialized functions of nerve fibres. Gasser was also one of the first to demonstrate the chemical transmission of nerve impulses.

Gasser and Erlanger found that the smaller nerve fibres were responsible for the conduction of pain and that the speed of electrical transmission by a nerve depends upon its diameter. Gasser also performed a great deal of experiments attempting to prove that chemical transmission occurs between nerves. He was one of the first to demonstrate that the injection of acetylcholine into bird muscles or denervated mammalian muscles results in slow contraction. Acetylcholine is now known to be a neurotransmitter, a chemical that carries nerve impulses across synapses between nerves.

Gasser was born in Plattville, USA and graduated in medicine from Johns Hopkins University 1915. He then moved to the University of Washington, St Louis, where he began working with Erlanger on the anatomy and function of the nervous system.

Gates Bill (William Henry), III 1955– . US businessman and computer scientist. He co-founded Microsoft Corporation 1975 and was responsible for supplying MS-DOS, the operating system that IBM chose to use in the IBM PC.

When the IBM deal was struck, Microsoft did not actually have an operating system, but Gates bought one from another company, renamed it MS-DOS, and modified it to suit IBM's new computer. Microsoft also retained the right to sell MS-DOS to other computer manufacturers, and because the IBM PC was not only successful but easily copied by other manufacturers, MS-DOS found its way onto the vast majority of PCs. The revenue from MS-DOS allowed Microsoft to expand into other areas of software, guided by Gates.

To many people, Gates is Microsoft: most of the company's successes have been his ideas (as have the occasional failures). His life revolves around the company and he expects similar dedication from his staff. In 1994, Gates was successful in fending off both US and European investigations into anti-competitive practices which could have seen Microsoft broken up into smaller companies.

In 1994 he invested $10 million into a biotechnology company, Darwin Molecular, with Paul Allen.

Gates Henry Louis 1950– . US academic and social activist. A scholar of African-American studies, he has republished such forgotten works as *Our Nig* by Harriet E Wilson 1859, the earliest known novel by a black American. He published *The Signifying Monkey: A Theory of African-American Literary Criticism* 1988.

Gates was born in Keyser, West Virginia. He was the first black American to get a PhD from Cambridge University, England, for a thesis on attitudes to black American and African culture in the 18th century. He received the MacArthur Foundation 'genius' award for his work on literary theory 1981. In 1991 he was made professor of humanities and chair of the department of Afro-American studies at Harvard University. His aim is to increase the number of black-studies courses in colleges in the USA in order to raise public awareness of the cultural achievements of black Americans. Other publications include *Colored People: A Memoir* 1994.

Gatling Richard Jordan 1818–1903. US inventor of a rapid-fire gun. Patented in 1862, the Gatling gun had ten barrels arranged as a cylinder rotated by a hand crank. Cartridges from an overhead hopper or drum dropped into the breech mechanism, which loaded, fired, and extracted them at a rate of 320 rounds per minute.

The Gatling gun was used in the US Civil War, in the Indian Wars that followed the settling of the American West, and in the Franco-Prussian War of 1870. By 1882 rates of fire of up to 1,200 rounds per minute were achieved, but the weapon was soon superseded by Hiram Maxim's machine gun in 1889.

Gatling was born in North Carolina; from 1870 his main factory was in Hartford, Connecticut. He manufactured other machines of his invention: for sowing, for breaking hemp, a steam plough, and a marine steam ram.

Gauquelin Michel 1928–1991. French neo-astrologist. Gauquelin trained as a psychologist and statistician, but became widely known for neo-astrology, or the scientific measurement of the correlations between the exact position of certain planets at birth and individual fame. His work attracted strong criticism as well as much interest. His book *Neo-Astrology: a Copernican Revolution* was published posthumously 1991.

Gauquelin studied the relationship between planet and personality, discovering that athletes were more likely to be born with Mars in the crucial positions, actors with Jupiter, and scientists and doctors with Saturn. Gauquelin studied thousands of eminent people to obtain his data, using thousands of non-eminent people as a control group.

Gauss Carl Friedrich 1777–1855. German mathematician who worked on the theory of numbers, non-Euclidean geometry, and the mathematical development of electric and magnetic theory. A method of neutralizing a magnetic field, used to protect ships from magnetic mines, is called 'degaussing'.

In statistics, the normal distribution curve, which he studied, is sometimes known as the Gaussian distribution.

Between 1800 and 1810 Gauss concentrated on astronomy. He developed a quick method for calculating an asteroid's orbit from only three observations and published this work – a classic in astronomy – 1809.

Gauss *German mathematician, astronomer and physicist Karl Friedrich Gauss was born in 1777 in Brunswick to poor parents. His prodigous talent for mathematics was noticed at the age of fourteen by the Duke of Brunswick, who subsequently financed the remainder of his academic schooling. Gauss went on to make discoveries in virtually every field of physics and mathematics, many of which were not discovered until after his death.*

Gauss was born in Brunswick and studied there at the Collegium Carolinum, and at Göttingen and Helmstedt. By 1799 he had already made nearly all his fundamental mathematical discoveries. He spent most of his career at Göttingen, becoming professor of mathematics and director of the observatory.

Gauss was also a pioneer in topology, and he worked besides on crystallography, optics, mechanics, and capillarity. After 1831, he collaborated with physicist Wilhelm Weber on research into electricity and magnetism, and in 1833 they invented an electromagnetic telegraph.

Disquisitiones arithmeticae 1801 summed up Gauss's work in number theory and formulated concepts and questions that are still relevant today.

Gay-Lussac Joseph Louis 1778–1850. French physicist and chemist who investigated the physical properties of gases, and discovered new methods of producing sulphuric and oxalic acids. In 1802 he discovered the approximate rule for the expansion of gases now known as Charles's law.

Gay-Lussac was born near Limoges and studied at the Ecole Polytechnique. He became assistant to chemist Claude Berthollet 1801, made balloon ascents to study the weather in 1804, and accompanied Alexander von Humboldt on an expedition 1805–06 to measure terrestrial magnetism. In 1809 Gay-Lussac became professor of chemistry at the Ecole and professor of physics at the Sorbonne. He held various government appointments, including that of superintendent of a gunpowder factory (1818) and chief assayer to the Mint (1829). He was a member of the chamber of deputies for a short time in the 1830s.

With Humboldt he accurately determined the proportions of hydrogen and oxygen in water, showing the volume ratio to be 2:1; they also established the existence of explosive limits in mixtures of the two gases.

In 1808 he formulated ***Gay-Lussac's law*** of combining volumes, which states that gases combine in simple proportions by volume and that the volumes of the products are related to the original volumes.

Geber Latinized form of Jabir ibn Hayyan *c.* 721–*c.* 776. Arabian alchemist. His influence lasted for more than 600 years, and in the late 1300s his name was adopted by a Spanish alchemist whose writings spread the knowledge and practice of alchemy throughout Europe.

The Spanish alchemist Geber probably discovered nitric and sulphuric acids, and he propounded a theory that all metals are composed of various mixtures of mercury and sulphur.

Gegenbaur Karl 1826–1903. German comparative anatomist and champion of evolutionary theory. Gegenbaur pioneered studies comparing the anatomy of different animals and argued that the similarities were evidence that mammals had evolved from a common ancestor.

By the middle of the 19th century, much of the general anatomy of both humans and lower mammalian species had been described. Indeed, descriptive anatomy was not far from where it is now. Little was known, however, about the anatomical differences between races of humans and virtually nothing was known of the features that humans share with other animals. Gegenbaur believed that common evolutionary descent is evident in the anatomical similarities between species grouped in the same taxonomic category.

For example, the same skeletal elements make up the forelimbs of cats, bats, humans, and whales. This is not because they have been engineered independently as the best way to construct the superstructure of a whale fin, human limb, and bat wing. Rather, the limbs of these different mammals are variations on a common anatomical theme that has been modified for various functions.

Gegenbaur was born in Wurzburg and graduated in medicine from the University of Wurzburg 1851. He held a teaching post at the medical schools at Jena before settling in Heidelberg 1873.

Geiger Hans (Wilhelm) 1882–1945. German physicist who produced the Geiger counter. He spent the period 1906–12 in Manchester, England, working with Ernest Rutherford on radioactivity. In 1908 they designed an instrument to detect and count alpha particles, positively charged ionizing particles produced by radioactive decay.

In 1928 Geiger and Walther Müller produced a more sensitive version of the counter, which could detect all kinds of ionizing radiation.

Geiger was born in Neustadt, Rheinland-Pfalz, and studied at Munich and Erlangen.

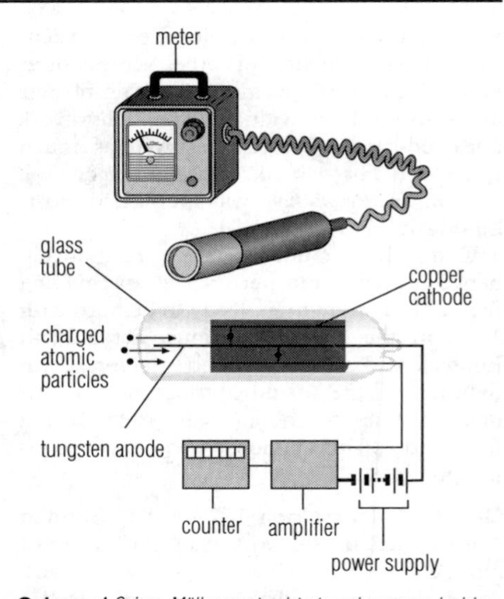

Geiger *A Geiger–Müller counter detects and measures ionizing radiation (alpha, beta, and gamma particles) emitted by radioactive materials. Any incoming radiation creates ions (charged particles) within the counter, which are accelerated by the anode and cathode to create a measurable electric current.*

On his return from Manchester 1912, Geiger became head of the Radioactivity Laboratories at the Physikalische Technische Reichsanstalt in Berlin, where he established a successful research group. He subsequently held other academic posts in Germany; from 1936 he was at the Technical University, Berlin.

Subjects that Geiger studied under Rutherford include the mathematical relationship between the amount of alpha scattering and atomic weight, the relationship between the range of an alpha particle and its velocity, the various disintegration products of uranium, and the relationship between the range of an alpha particle and the radioactive constant. From 1931 onwards Geiger mainly studied cosmic radiation.

Gell-Mann Murray 1929– . US theoretical physicist. In 1964 he formulated the theory of the quark as one of the fundamental constituents of matter. In 1969 he was awarded a Nobel prize for his work on elementary particles and their interaction.

Gell-Mann proposed in 1962 a classification system for elementary particles called the ***eightfold way***. It postulated the existence of supermultiplets, or groups of eight particles which have the same spin value but different values for charge, isotopic spin, mass, and strangeness. The model also predicted the existence of supermultiplets of different sizes.

Gell-Mann was born in New York and studied at Yale and the Massachusetts Institute of Technology. He became professor at the California Institute of Technology 1956.

Gell-Mann proposed in 1953 a new quantum number called the strangeness number, together with the law of conservation of strangeness, which states that the total strangeness must be conserved on both sides of an equation describing a strong or an electromagnetic interaction but *not* a weak interaction. This led to his theory of associated production 1955 concerning the creation of strange particles. Gell-Mann used these rules to group mesons, nucleons (neutrons and protons), and hyperons, and was thereby able to form successful predictions.

Geoffroy Saint-Hilaire Etienne 1772–1844. French zoologist who performed some of the earliest embryological experiments on chicks. He tried to halt the development of a chick embryo at the stage of a fish in order to demonstrate that evolutionary stages of development are recapitulated in the embryo.

Early on in his scientific career he worked with Georges Cuvier, but they parted company and became rivals. In 1825 Geoffroy Saint-Hilaire developed an interest in palaeontology and he identified Cuvier's fossil 'crocodile' as a *Teleosarus.*

Saint-Hilaire was born in Etampes in France. Although the youngest of fourteen children, he received special tuition from the Abbé de Tressan which ensured that he was a canon by the time he was fifteen years old. He had a promising future in the church, but when the French Revolution started he left the church in order to study first law and then medicine. In 1792 he met several of the most eminent scientists of the time at the Collége du Cardinal Lemoine in Paris. He was made a professor at the Museum d'Histoire Naturelle when he was 21. From 1798 until 1801 he served as a scientist on Napoleon Bonaparte's Egyptian campaign and wrote *Description de L'Egypte par la Commission des Sciences,* describing his discoveries. He suffered from cataracts and went blind1840.

Geoffroy Saint-Hilaire Isidore 1805–1861. French zoologist and anatomist who specialized in the study of apes and developed a system for their classification. He also studied the manner in which animals interact with their environments.

The only son of Etienne Geoffroy Saint-Hilaire, Isidore was born in Paris and, though he wanted to study mathematics, ended up working for his father in his laboratory. At the age of 25, he delivered a series of lectures at the Athénée 1837, and was appointed professor of comparative anatomy at the Faculté des Sciences, and then professor at the Museum d'Histoire Naturelle, replacing his father in both positions.

From 1844 until 1850 he was the inspector general of education before becoming the professor of zoology at the Sorbonne. His research included a system for the classification of the apes, and he identified many of their infantile characteristics. In 1856 he led a campaign to encourage the French public to eat more horse meat. In 1859 he wrote a history of the origin of species which put Darwin's proposals in the context of his French predecessors, including Buffon, Lamarck and Etienne Geoffroy Saint-Hilaire.

Germain Sophie 1776–1831. French mathematician, born in Paris. Although she was not allowed to study at the newly opened Ecole Polytechnique, she corresponded with the mathematicians Joseph Lagrange and Karl Gauss. She is remembered for work she carried out in studying Fermat's principle.

Gesell Arnold Lucius 1880–1961. US psychologist and educator. He founded the Yale Clinic of Child Development, which he directed 1911–48. Among the first to study the stages of normal development, he worked as a consultant to the Gesell Institute of Child Development, New Haven, Connecticut, which was founded 1950 to promote his educational ideas.

Gesner Konrad von 1516–1565. Swiss naturalist who produced an encyclopedia of the animal world, the five-volume *Historia animalium* 1551–58. He began a similar project on plants that was incomplete at the time of his death. He is considered the founder of zoology.

Gesner was born in Zurich, Switzerland and was the godson of the Protestant reformer Ulrich Zwingli. He studied theology at Carolinum in Zurich and then at the Fraumünster seminary. He went to Strasbourg Academy 1532 to study Hebrew and then studied medicine in Bourges, Paris, and Basel. From 1537–40, he was the professor of Greek at Lausanne Academy and obtained his doctorate 1541. After a spell in Montpellier studying botany, he returned to Zurich to become a physician. He travelled widely in Europe and worked on *Opera botanica*, which was completed and published after his death by C Schmiedel. He died in Zurich 13 March.

Giacconi Riccardo 1931– . Italian-born US physicist, the head of a team whose work has been fundamental in the development of X-ray astronomy. In 1970 they launched a satellite devoted entirely to X-ray astronomy.

Giacconi was born in Genoa and obtained his doctorate from Milan. He emigrated to the USA 1956. In 1959 he joined American Science and Engineering, Inc., rising to executive vice president 1969. In 1973 he was made professor of astronomy at Harvard University. Later he became director of the European Southern Observatory.

In 1962, a rocket sent up by Giacconi and his group to observe secondary spectral emission from the Moon detected strong X-rays from a source evidently located outside the Solar System. X-ray research has since led to the discovery of many types of stellar and interstellar material. Giacconi's team have developed a telescope capable of producing X-ray images.

Giacconi has also worked with a Cherenkov detector, by means of which it is possible to observe the existence and velocity of high-speed particles.

Giaever Ivar 1929– . Norwegian-born US physicist who worked on tunnelling (the flow of an electric current through a thin film of insulating material sandwiched between two metal plates when a voltage is applied to the plates) and superconductivity. Nobel Prize for Physics 1973 (shared with Brian Josephson and Leo Esaki).

Giaever investigated the tunelling effects previously observed by Esaki. In 1960 he found that if one of the metal plates was in a superconducting state, the behaviour of the electric current when the voltage was varied revealed much about the superconducting state. This discovery was instrumental in Josephson's discovery of the Josephson effect (the flow of electrons across an insulating film between two superconducting materials).

Giaever was born in Bergen, Norway, and studied electrical engineering at the Norwegian Institute of Technology. He emigrated to Canada 1954 and moved to the United States 1957. He joined the General Electric research laboratory in Schenectady, New York, 1958. In 1964 he gained his doctorate from the New York Rensselaer Polytechnical Institute.

Giauque William Francis 1895–1982. Canadian-born US physical chemist who specialized in chemical thermodynamics, in particular the behaviour of matter at extremely low temperatures, and received the Nobel Prize for Chemistry 1949.

Giauque used statistics to calculate the entropy (state of disorder) of different substances close to absolute zero, using the energy levels of molecules gained from spectroscopy. His research validated the third law of thermodynamics: chemical changes at absolute zero (273.15°C/459.67°F) involve no change of entropy.

He was born in Niagra Falls, Ontario. His academic career, spanning over fifty years, was spent at the University of California. It began when Giaque graduated in chemistry, obtaining his PhD 1922. He rapidly gained a reputation for hard work and was appointed professor in 1934.

Gibbon John Heysham 1903–1974. US surgeon who invented the heart–lung machine in 1953. It has become indispensable in heart surgery, maintaining the circulation while the heart is temporarily inactivated.

Gibbs Josiah Willard 1839–1903. US theoretical physicist and chemist who developed a mathematical approach to thermodynamics and established vector methods in physics. He devised the phase rule and formulated the Gibbs adsorption isotherm.

Gibbs showed how many thermodynamic laws could be interpreted in terms of the results of the movements of enormous numbers of bodies such as molecules. His ensemble method equated the behaviour of a large number of systems at once to that of a single system over a period of time.

Gibbs was born in New Haven, Connecticut, and studied at Yale and in

One of the principal objects of theoretical research in my department of knowledge is to find the point of view from which the subject appears in its greatest simplicity.

Josiah Gibbs

1881

Europe 1866–69. From 1871 he was professor of mathematical physics at Yale.

The phase rule, published 1876–78, may be stated as:

$$f = n + 2 - r$$

where f is the number of degrees of freedom, n the number of chemical components, and r the number of phases – solid, liquid or gas; degrees of freedom are quantities such as temperature and pressure that may be altered without changing the number of phases. He also described his concept of free energy, which can be used as a measure of the feasibility of a given chemical reaction. It is defined in terms of the enthalpy, or heat content, and entropy, a measure of the disorder of a chemical system. From this Gibbs developed the notion of chemical potential, which is a measure of how the free energy of a particular phase depends on changes in composition.

The Gibbs adsorption isotherm showed that changes in the concentration of a component of a solution in contact with a surface occur if there is an alteration in the surface tension.

Gibson James Jerome 1904–1979. US psychologist who did influential and highly original work on visual perception. An outspoken critic of the German physiologist Hermann Helmholtz's notion that perception involves unconscious inferences from sense data and learning-based associations, he proposed that perceptual information is gained directly from the environment, without the need for intermediate processing.

Educated at Princeton, Gibson went on to teach at Smith College, Massachusetts, 1928–49, where he was influenced by Kurt Koffka (1886–1941), and at Cornell University 1949–72.

In his experimental work, Gibson dispensed with the use of two-dimensional, static images and instead explored the perception of motion in freely moving subjects under natural conditions, publishing his results in *The Perception of the Visual World* 1950. He went on to develop what he called an ecological theory of perception in *Senses Considered as Perceptual Systems* 1966.

Giffard Henri 1825–1882. French inventor of the first passenger-carrying powered and steerable airship, called a dirigible, built 1852. The hydrogen-filled airship was 43 m/144 ft long, had a 2,200-W/3-hp steam engine that drove a three-bladed propeller, and was steered using a saillike rudder. It flew at an average speed of 5 kph/3 mph.

In the early 1850s Giffard, an engineer, began to experiment with methods for steering balloons, and then built his airship. On 24 Sept 1852 he took off from the Hippodrome in Paris and flew to Elancourt, near Trappes.

Giffard went on to build another airship 1855, and a series of large balloons. This was funded by money from other inventions, such as an injector to feed water into a steam-engine boiler to prevent it running out of steam when not in motion.

Gilbert Walter 1932– . US molecular biologist who studied genetic control, seeking the mechanisms that switch genes on and off. By 1966 he had established the existence of the lac repressor, a molecule that suppresses lactose production. Further work on the sequencing of DNA nucleotides won him a share of the 1980 Nobel Prize for Chemistry, with Frederick Sanger and Paul Berg.

Gilbert was born in Boston, Massachusetts, and educated as a physicist at Harvard and at Cambridge, England. In 1960 he changed to biology, becoming professor of biophysics at Harvard 1964, then professor of molecular biology 1969.

Gilbert began in 1965 his attempt to identify repressor substances involved in the regulation of gene activity. He devised a technique called equilibrium dialysis that enabled him to produce relatively large quantities of the repressor substance, which he then isolated and purified, and by late 1966 he had identified it as a large protein molecule. Gilbert then developed a method of determining the sequence of bases in DNA, which involved using an enzyme that breaks the DNA molecule at specific, known points.

Gilbert William 1540–1603. English scientist who studied magnetism and static electricity, deducing that the Earth's magnetic field behaves as if a bar magnet joined the North and South poles. His book on magnets, published 1600, is the first printed scientific book based wholly on experimentation and observation.

Gilbert was the first English scientist to accept Nicolas Copernicus' idea that the Earth rotates on its axis and revolves around the Sun. He also believed that the stars are at different distances from the Earth and might be orbited by habitable planets, but erroneously thought that the planets were held in their orbits by magnetic forces.

Gilbert was born in Colchester, Essex, and educated at Cambridge. In about 1573, he settled in London, where he established a medical practice. He was appointed physician to Queen Elizabeth I 1600 and later briefly to James I.

Gilbert discovered many important facts about magnetism, such as the laws of attraction and repulsion and magnetic dip. He also investigated static electricity and differentiated between magnetic attraction and electric attraction (as he called the ability of an electrostatically charged body to attract light objects). This is described in his book *De magnete, magneticisque corporibus, et de magno magnete tellure/Concerning Magnetism, Magnetic Bodies, and the Great Magnet Earth* 1600.

Gilchrist Percy Carlyle 1851–1935. British metallurgist who devised a method of producing low-phosphorus steel from high-phosphorus ores, such as those commonly occurring in the UK. This meant that steel became cheaply available to British industry.

Gilchrist was born in Lyme Regis, Dorset, and studied at the Royal School of Mines. He developed the steelmaking process 1875–77, together with his cousin Sidney Gilchrist Thomas. The product became known at first as 'Thomas steel'.

Pig iron was melted in a convector similar to that used in the Bessemer process and subjected to prolonged blowing. The oxygen in the blast of air oxidized carbon and other impurities, and the addition of lime at this stage caused the oxides to separate out as a slag on the surface of the molten metal. Continued blowing then brought about oxidation of the phosphorus.

Gill David 1843–1914. Scottish astronomer who pioneered the use of photography to catalogue stars. He also made much use of a heliometer, determining the solar parallax and measuring the distances of 20 of the brighter and nearer southern stars. KCB 1900.

Gill was born in Aberdeen and studied there at Marischal College; he then went to Switzerland to learn clockmaking. In 1872, he became director of Lord Lindsay's private observatory at Dun Echt, near Aberdeen. He was astronomer at the observatory at the Cape of Good Hope, South Africa, 1879–1906.

With Lord Lindsay, Gill went on an expedition to Mauritius 1872–78 to measure the distance of the Sun and other related constants when Venus passed across the face of the Sun in 1874.

In 1882 Gill started a huge project, in collaboration with other observatories, to chart and measure star positions using photography. This work was published as the *Cape Durchmusterung*, giving the positions and brightness of more than 450,000 southern stars. Gill also served on the council for the *International Astrographic Chart and Catalogue*, which assigned positions for all stars to a precision of the 11th magnitude. This was only completed in 1961, 61 years after all the photographs had been taken.

Gillette King Camp 1855–1932. US inventor of the Gillette safety razor.

Gilman Alfred 1941– . US pharmacologist who shared the 1994 Nobel Prize for Physiology or Medicine with Martin Rodbell for their discovery of a family of proteins that translate messages from outside a cell into action inside cells.

When an outside message – in the form of a hormone or other chemical signal – reaches a cell it enters through a specific receptor molecule on the cell surface. As it crosses the cell membrane, the message is translated, or converted into a second internal chemical signal that the cell can understand.

In the late 1970s Gilman and his colleagues isolated the molecule (dubbed a G protein) that does the translation. Since Gilman's discovery many G proteins have been identified. For example, there are specific G proteins in the rods and cones of the eye. More than 100 receptors have been identified that translate messages using G proteins.

Gilman was born in New Haven, Connecticut and educated at Yale University, New Haven, and Case Western Reserve University, receiving his doctorate in 1969. From 1969–71, he worked for the National Institute of Health, Bethesda, Maryland. He taught at the University of Virginia from 1871–81. He then became director of pharmacology at the University of Texas Southwestern Medical Center in Dallas.

Gilman Henry 1893–1986. US organic chemist who made a comprehensive study of methods of high-yield synthesis, quantitative and qualitative analysis, and uses of organometallic compounds, particularly Grignard reagents.

Gilman was born in Boston, Massachusetts, and studied at Harvard. He was professor at Iowa State University 1923–47.

Gilman investigated the organic chemistry of 26 different metals, from aluminium, arsenic, and barium to thallium, uranium, and zinc, and discovered several new types of compounds. He was the first to study organocuprates, now known as ***Gilman reagents***, and his early work with organomagnesium compounds (Grignard reagents) would later play an important part in the preparation of polythene.

Ginzburg Vitalii Lazarevich 1916– . Russian astrophysicist whose use of quantum theory in a study of Cherenkov radiation contributed to the development of nuclear physics. He was one of the first to believe that cosmic background radiation comes from beyond our own Galaxy.

Ginzburg was born and educated in Moscow and spent his career there at the Physics Institute of the Academy of Sciences.

Ginzburg also formulated the theory of a molecule containing particles with varying degrees of motion, for which he devised the first relativistically invariant wave equation. After 1950, he concentrated on problems in thermonuclear reactions. His work on cosmic rays led him to a hypothesis about their origin and to the conclusion that they can be accelerated in a supernova.

Glaser Donald Arthur 1926– . US physicist who invented the bubble chamber for detecting high-energy elementary particles in 1952, for which he received the Nobel Prize for Physics 1960.

Glaser's first bubble chamber consisted of a vessel only a few centimetres across, containing superheated liquid ether under pressure. When the pressure was released suddenly, particles traversing the chamber left tracks consisting of streams of small bubbles formed when the ether boiled locally; the tracks were photographed using a high-speed camera. In later, larger bubble chambers, liquid hydrogen was substituted for ether.

Glaser was born in Cleveland, Ohio, and educated there at the Case Institute of Technology, and at the California Institute of Technology. In 1957 he became professor at the University of Michigan, moving 1959 to the University of California. In the early 1960s he turned from physics to molecular biology.

Glashow Sheldon Lee 1932– . US particle physicist. In 1964 he proposed the existence of a fourth, 'charmed' quark, and later argued that quarks must be coloured. Insights gained from these theoretical studies enabled Glashow to consider ways in which the weak nuclear force and the electromagnetic force (two of the fundamental forces of nature) could be unified as a single force now called the electroweak force. For this work he shared the Nobel Prize for Physics 1979 with Abdus Salam and Steven Weinberg.

Glashow was born in New York and studied at Cornell and Harvard universities. He worked at the Niels Bohr Institute in Copenhagen before becoming professor at Harvard 1967.

Glauber Johann Rudolf 1604–1670. German chemist who about 1625 discovered the salt known variously as ***Glauber's salt*** and '*sal mirabile*' (sodium sulphate). He made his living selling patent medicines and used the salt to treat almost any complaint.

Glauber was born in Karlstadt, Franconia, and was self-educated. After many years of travelling in Europe, he settled in Amsterdam 1655 and built a chemical laboratory there.

Glauber investigated and developed processes that could have industrial application. He prepared nitric acid by substituting saltpetre (potassium nitrate) for salt in the reaction with sulphuric acid. He made many metal chlorides and nitrates from the mineral acids, and produced organic liquids containing such solvents as acetone (dimethylketone) and benzene – although he did not identify them – by reacting and distilling natural substances such as wood, wine, and vegetable oils.

The chemical techniques involved are described in his book *Opera omnia chymica* 1651–61. He outlined his views on a possible utopian future for Germany in *Teutschlands-Wohlfahrt/Germany's Prosperity*

Glenn John Herschel, Jr 1921– . US astronaut and politician. On 20 Feb 1962, he became the first American to orbit the Earth, doing so three times in the Mercury spacecraft *Friendship 7*, in a flight lasting 4 hr 55 min. After retiring from NASA, he was elected to the US Senate as a Democrat from Ohio 1974; re-elected 1980 and 1986. He unsuccessfully sought the Democratic presidential nomination 1984.

As a senator, he advocated nuclear-arms-production limitations and increased aid to education and job-skills programmes.

Gmelin Johan Georg 1709–1755. German botanist, natural historian, and geographer who was the first person to note that the Caspian Sea existed at a lower ground level than the Mediterranean Sea. On an expedition to Siberia 1733–41, he described the position of the Yenisei river as a boundary between Europe and Asia and participated in measuring the lowest temperature ever recorded at Yeniseysk. He described his botanical studies in Siberia in his four volume *Flora Sibirica*.

Born in Tübingen, Germany, Gmelin was the son of an apothecary. He was precocious and extremely gifted as a child. His father encouraged him to use his formidable skills as a scholar to study medicine and he graduated with a medical degree 1727 at the age of 18. He left for St Petersburg, where he became a tutor at the university 1730. After delivering a series of brilliant lectures, he was made a professor of chemistry and natural history 1731. In 1749, he returned to Tübingen to become professor of medicine, botany, and chemistry at the University and later died there in 1755.

Goddard Robert Hutchings 1882–1945. US rocket pioneer. His first liquid-fuelled rocket was launched at Auburn, Massachusetts, in 1926. By 1935 his rockets had gyroscopic control and carried cameras to record instrument readings. Two years later a Goddard rocket gained the world altitude record with an ascent of 3 km/1.9 mi.

Goddard developed the principle of combining liquid fuels in a rocket motor, the technique used subsequently in every practical space vehicle. He was the first to prove by actual test that a rocket will work in a vacuum and he was the first to fire a rocket faster than the speed of sound.

Goddard was born in Worcester, Massachusetts. At Clark University in his home town, and at Mount Wilson, California, he carried out experiments with naval signal rockets, and went on to design and build his own rocket motors. On the USA's entry into World War I, he turned his energies to investigating the military application of rockets. In 1929, instruments, and a camera to record them, were carried aloft for the first time.

The military rockets Goddard developed in World War II were more advanced than the German V2, although smaller. A few days before the end of the war he demonstrated a rocket fired from a launching tube.

Gödel Kurt 1906–1978. Austrian-born US mathematician and philosopher. He proved that a mathematical system always contains statements that can be neither proved nor disproved within the system; in other words, as a science, mathematics can never be totally consistent and totally complete. He worked on relativity, constructing a mathematical model of the universe that made travel back through time theoretically possible.

I have continued my work on the continuum problem last summer and I finally succeeded in proving the consistency of the continuum hypothesis (even the generalized form) with respect to generalized set theory. But for the time being please do not tell anyone of this.

Kurt Gödel

Letter to his teacher Karl Menger 1937

Gödel was born in Brünn, Moravia (now Brno in the Czech Republic) and educated at the University of Vienna, where he worked until 1938. When Austria was annexed by Nazi Germany, he emigrated to the USA. He settled at Princeton, where he was appointed professor 1953.

In 1930, Gödel showed that a particular logical system (predicate calculus of the first order) was such that every valid formula could be proved within the system; in other words, the system was what mathematicians call complete. He then investigated a much larger logical system – that constructed by English philosophers Bertrand Russell and Alfred Whitehead as the logical basis of mathematics. The resultant paper, 'On formally undecidable propositions of *Principia Mathematica* and related systems' 1931, is the one in which Gödel dashed the hopes of philosophers and mathematicians alike.

Godwin Harry 1901–1985. English botanist whose primary research was on the distribution of pollen deposits in the peat of the English Fens. He also made contributions to the science of carbon dating ancient material and the physiology of plants.

Godwin was born in Rotherham, Yorkshire and in 1919, won a scholarship to Clare College, Cambridge, where he graduated in natural sciences 1922. He worked as a research student with the British botanist Frederick Frost Blackman studying the mechanisms of plant respiration and obtaining a PhD 1926. In 1927, he joined the staff of the botany department at Cambridge and began a distinguished career researching the factors controlling the level of pollen in Britain using tree pollen initially as his model. He also made a significant contribution to the science of radiocarbon dating (method of dating ancient organic material using the radioactive isotope carbon-14).

His first book, *Plant Biology* 1930, was popular for its emphasis on plant physiology rather than simply the anatomy and evolution of plants. He also published *The History of the British Flora* 1956, which included most of his research on the distribution of pollen deposits in fenland peat. He was made professor of botany 1960, elected a Fellow of the Royal Society 1945 and was knighted 1970.

Goeppert-Mayer Maria, (born Goeppert) 1906–1972. German-born US physicist who studied the structure of the atomic nucleus. Her explanation of the stability of particular atoms 1948 envisaged atomic nuclei as shell-like layers of protons and neutrons, with the most stable atoms having completely filled outermost shells. She shared the 1963 Nobel Prize for Physics with Eugene Wigner and Hans Jensen.

Goeppert was born in Kattowitz, Upper Silesia (now Katowice in Poland) and studied at Göttingen. Emigrating to the USA 1930, she was professor at Chicago 1946–60 and at the University of California from 1960.

In 1945, Goeppert-Mayer developed a 'little bang' theory of cosmic origin with US physicist Edward Teller to explain element and isotope abundances in the universe. This led her to study the stability of nuclei. In 1948, she published evidence of the special stability of the following numbers of protons and neutrons: 2, 8, 20, 50, 82 and 126. These are commonly called magic numbers.

She and Jensen independently proposed a shell model, and in 1955 they wrote a book together, *Elementary Theory of Nuclear Shell Structure*.

Gold Thomas 1920– . Austrian-born US astronomer and physicist who in 1948 formulated, with Fred Hoyle and Hermann Bondi, the steady-state theory regarding the creation of the universe.

Gold was born in Vienna and studied at Cambridge. In 1956 he emigrated to the USA and became professor of astronomy at Harvard 1958 and at Cornell University from 1959. Gold has served as an adviser to NASA.

The steady-state theory assumes an expanding universe in which the density of matter remains constant because, as galaxies recede from one another, new matter is continually created (at an undetectably slow rate). The implications that follow are that galaxies are not all of the same age, and that the rate of recession is uniform.

With the discovery in the 1960s of cosmic background radiation, the steady-state hypothesis was abandoned by most cosmologists in favour of the Big Bang model.

Goldberg Leo 1913–1987. US astrophysicist who carried out research into the composition of stellar atmospheres and the dynamics of the loss of mass from cool stars. His main subject of research was the Sun.

Goldberg was born in Brooklyn, New York, and studied at Harvard, where, after 23 years at the University of Michigan, he was professor 1960–73 and director of Harvard College Observatory 1966–71. From 1971 to 1977 he was director of the Kitt Peak National Observatory in Arizona.

At Harvard, Goldberg and his colleagues designed an instrument that could function either as a spectrograph or as a spectroheliograph (a device to photograph the Sun using monochromatic light). This formed part of the equipment of *Orbital Solar Observatory IV*, launched 1967. He has also carried out research on the temperature variations and chemical composition of the Sun and of its atmosphere, in which he succeeded in detecting carbon monoxide.

Goldring Winifred 1888–1971. US palaeontologist whose research focused on Devonian fossils. She did much to popularize geology.

Goldring was born near Albany, New York State, and educated at Wellesley College. She worked 1914–54 at the New York State Museum. In 1939 she was made state palaeontologist.

Goldring began her work in palaeobotany 1916. During the late 1920s and the 1930s, as well as geologically mapping the Coxsackie and Berne quadrangles of New York, she developed and maintained the State Museum's public programme in palaeontology.

Her works include *The Devonian Crinoids of the State of New York* 1923 and *Handbook of Paleontology for Beginners and Amateurs* 1929–31.

Goldschmidt Richard Benedikt 1878–1958. German evolutionary biologist who believed that environmentally induced changes in organisms with similar genotypes could provide information about the origin of species. He worked for many years on the genetics of butterflies using the X chromosome as a model, and formulated a theory (now largely discredited) that chromosomes rather than individual genes are the units of heredity.

Goldschmidt was born in Frankfurt and appointed director of Biological Studies at the Kaiser Willhelm Institute in Berlin 1921. In 1935, just before World War II, he left Germany for the USA to become professor of zoology at the University of California at Berkeley.

Goldschmidt Victor Moritz 1888–1947. Swiss-born Norwegian chemist who did fundamental work in geochemistry, particularly on the distribution of elements in the Earth's crust. He considered the colossal chemical processes of geological time to be interpretable in terms of the laws of chemical equilibrium.

Goldschmidt was born in Zürich but moved to Norway as a child and studied at the University of Christiania (now Oslo). He was professor and director of the Mineralogical Institute 1914–29, when he moved to Göttingen, Germany. The rise of Nazism forced him to return to Norway 1935, but during World War II he had to flee again, first to Sweden and then to Britain, where he worked at Aberdeen and Rothamsted (on soil science). He returned to Norway after the end of the war.

Using X-ray crystallography, Goldschmidt was able to show that, given an electrical balance between positive and negative ions, the most important factor in crystal structure is ionic size. Exhaustive analysis of results from geochemistry, astrophysics, and nuclear physics led to his work on the cosmic abundance of the elements and the links between isotopic stability and abundance. Studies of terrestrial abundance reveal about eight predominant elements. Recalculation of atom and volume percentages lead to the remarkable notion that the Earth's crust is composed largely of oxygen anions (90% of the volume), with silicon and the common metals filling up the rest of the space.

During World War II Goldschmidt carried a cyanide capsule for suicide should the Germans have invaded Britain. When a colleague asked for one, he was told: 'Cyanide is for chemists; you, being a professor of mechanical engineering, will have to use the rope.'

Goldstein Eugen 1850–1930. German physicist who investigated electrical discharges through gases at low pressures. He discovered canal rays and gave cathode rays their name.

Goldstein was born at Gleiwitz, Upper Silesia (now Gliwice, Poland), and studied at Breslau and Berlin. He worked at the Berlin Observatory 1878–90, and was eventually appointed head of the Astrophysical Section of Potsdam Observatory.

In 1876 Goldstein showed that cathode rays can cast shadows and that the rays are emitted at right angles to the surface of the cathode. He then demonstrated the deflection of cathode rays by magnetic fields. In 1886 he performed an experiment in which he made holes in the anode and observed glowing yellow streamers coming from the holes. He called these *Kanalstrahlen* or canal rays. Goldstein later investigated the wavelengths of light emitted by metals and oxides when canal rays impinge on them, and observed that alkali metals show bright spectral lines.

In 1928 he observed that a trace of ammonia was present after the discharge in a tube containing nitrogen and hydrogen. This observation was only followed up much later to see if this phenomenon could have lead to the origination of biologically important molecules, and hence life.

Goldstein Joseph Leonard 1940– . US geneticist who shared the 1985 Nobel Prize for Physiology or Medicine with Michael Brown for their work on cholesterol metabolism. They discovered that the gene mutated in familial hypercholesterolaemia, a condition in which patients have very high blood cholesterol, is the gene for LDL-receptors involved in the removal of cholesterol from the bloodstream.

Goldstein and Brown worked out the steps involved in the uptake of cholesterol, in the form of low-density lipoprotein (LDL), in the blood plasma by the cells of the body. The LDL particles bind to their receptors and are then internalized by endocytosis (adsorption). In this manner an endocytic vesicle is formed which subsequently fuses with a lysosome (a digestive organelle inside a cell). Lysosomes contain all the degradative enzymes required to free the amino acids and cholesterol from the LDL particles. The receptors are freed to return to the plasma membrane of the cell.

They discovered that the gene mutated in familial hypercholesterolaemia is the gene for the LDL-receptor. Children who inherit the mutated gene from both parents, are therefore homozygous for the mutation and generally die of heart disease in childhood. Children who inherit one normal gene and one mutated gene, heterozygotes, have half the number of receptors that would be expected in a healthy person and have a milder disease than homozygotes for the mutation.

Goldstein was born in Sumter, South Carolina in the USA and graduated from the University of Texas in medicine 1966. While working as a junior doctor at the Massachusetts General Hospital in Boston, he met and began working with Brown; a collaboration which was to lead to the joint award of the Nobel prize

Golgi Camillo 1843–1926. Italian cell biologist who produced the first detailed knowledge of the fine structure of the nervous system. He shared the 1906 Nobel Prize for Physiology or Medicine with Santiago Ramón y Cajal, who followed up Golgi's work.

Golgi's use of silver salts in staining cells proved so effective in showing up the components and fine processes of nerve cells that even the synapses – tiny gaps between the cells – were visible. The ***Golgi apparatus***, a series of flattened membranous cavities found in the cytoplasm of cells, was first described by him 1898.

Golgi was born near Brescia and studied at Pavia, where he spent most of his academic career, becoming professor 1876. He was elected to the senate 1900.

From his examinations of different parts of the brain, Golgi put forward the theory that there are two types of nerve cells, sensory and motor cells, and that axons are concerned with the transmission of nerve impulses. He discovered tension receptors in the tendons – now called the organs of Golgi.

Between 1885 and 1893 Golgi investigated malaria. This was known to be caused by the protozoon *Plasmodium*, and Golgi showed how the fever attacks coincide with the release into the bloodstream of a new generation of the parasites. He also established a method of treatment.

Goodall Jane, Baroness van Lawick-Goodall 1934– . English primatologist and conservationist who has studied the chimpanzee community on Lake Tanganyika since 1960, and is a world authority on wild chimpanzees.

Goodall was born in London. She left school at 18 and worked as a secretary and a film production assistant, until she had an opportunity to work for anthropologist Louis Leakey in Africa. She began to study the chimpanzees at the Gombe Stream Game Reserve, on Lake Tanganyika. Goodall observed the lifestyles of chimpanzees in their natural habitats, discovering that they are omnivores, not herbivores as originally thought, and that they have highly developed and elaborate forms of social behaviour. In 1964 she married Hugo van Lawick the Dutch wildlife photographer, and he collaborated with her on several of her books and films. They divorced later. She obtained a PhD from Cambridge University 1965, despite the fact that she had never been an undergraduate. Her books include *In the Shadow of Man* 1971, *The Chimpanzees of Gombe: Patterns of Behaviour* 1986, and *Through a Window* 1990.

Often I am asked if I prefer chimpanzees to humans. The answer to that is easy – I prefer some chimpanzees to some humans, some humans to some chimpanzees!

Jane Goodall

Through a Window 1990

Goodsir John 1814–1867. Scottish anatomist who was particularly interested in the structure and function of living cells, and contributed to the development of the theory that all living things are made up of cells, proposed by Matthias Schleiden and Theodor Schwann 1839.

Goodsir made many contributions to the development and acceptance of the cell theory by confirming its fundamental tenets in his microscopic observations as an anatomist. He also advanced our knowledge of the cell as a centre of nutrition and of the secretory functions of certain cells.

Goodsir was born in Anstruther in Scotland and studied at the University of St Andrews and then the University of Edinburgh. He chose initially to specialize in surgery and joined his father's practice before being elected professor of anatomy at the University of Edinburgh in 1846. As he got older he suffered with a temporary paralysis thought to be related to depression, anxiety and stress.

Goodyear Charles 1800–1860. US inventor who developed rubber coating 1837 and vulcanized rubber 1839, a method of curing raw rubber to make it strong and elastic.

Goodyear was born in New Haven, Connecticut, and entered his father's hardware business. He began to investigate rubber in the 1830s and obtained US patents for his process 1844, but both Britain and France refused his applications because of legal technicalities. His attempts to set up companies in both countries failed, and for a while he was imprisoned for debt in Paris. When he returned to the USA, many of his patents had been pirated by associates, and he died in poverty.

Goodyear discovered the vulcanization process by accident. One day he was mixing rubber with sulphur and various other ingredients when he dropped some on top of a hot stove. The next morning the stove had cooled and the rubber had vulcanized.

Goodyear, Charles *American Charles Goodyear who invented the vulcanization process for rubber. During this process rubber is heated to high temperature in the presence of sulphur in order to increase its elasticity and strength. Goodyear went on to invent many applications for rubber including the motor-vehicle tyre.*

Gorgas William Crawford 1854–1920. US military physician who developed methods to prevent the spread of yellow fever. He eradicated yellow fever in Havana, Cuba during and after the Spanish-American war by initiating measures to improve sanitation and using quarantine to isolate carriers of the disease.

Gorgas set out to reduce the number of deaths from malaria, typhoid fever, dysentery, and yellow fever in Havana. He believed, as did many at the time, that all these diseases were 'filth' diseases and initiated measures to improve sanitary amenities. He built drains, had stable floors cemented and arranged refuse removal. The reduction in the number of cases of malaria, typhoid fever and dysentery was startling.

Gorgas noted that those who succumbed to the disease were generally Spanish soldiers, or Spanish immigrants after the war, and realized that the local population must be resistant, presumably due to infection early in life.

Born in Mobile, Alabama, Gorgas was educated at Bellevue Hospital Medical College in New York. He then joined the US Army Medical Corps as a physician. In 1898, when the fighting of the Spanish-American war was almost over, Gorgas was sent to Havana as the chief sanitary officer.

While Gorgas was in Cuba, Walter Reed, another US military physician, realized that transmission of yellow fever was via the mosquito parasite *Aedes aegypti* and that the infectious agent had to develop within the mosquito for a number of days. When Gorgas learned of this he confined victims to mosquito-proof rooms and introduced fines to penalize any house owner who was found to have mosquito larvae on his premises. By these means he eradicated yellow fever from Havana.

Gorgas's next fight against yellow fever was during the US construction of the Panama Canal, which began in 1904. While he never eradicated the disease, he greatly reduced the number of cases and was generally considered the world's greatest expert on the control of the disease until his death.

Gosse Philip Henry 1810–1888. English naturalist who built the first aquarium ever used to house marine animals long-term and wrote many books on marine zoology, including *Manual of Marine Zoology* 1855, *Actinologia Britannica* 1858, a work on sea anemones, *Introduction to Zoology* 1843, and *Evenings at the Microscope* 1859.

Gosse was born in Worcestershire and was a member of the Plymouth Brethren, an exclusive Christian group that rejected the theory of evolution completely. In 1827, he was appointed as a clerk in Carbonear,

Newfoundland, Canada, where he studied natural history in his spare time. He attempted to farm in Canada, but when this failed he taught in Alabama, America, before returning to England 1839. He retired to St Mary Church and died there 23 August.

Gosset William Sealy 1876–1937. British industrial research scientist whose work on statistical analysis of the normal distribution opened the door to developments in the analysis of variance.

Gosset was born in Canterbury and studied at Oxford and under statistician Karl Pearson at University College, London. Gosset spent his career with the Guinness brewery firm, first in Dublin, Ireland, and from 1935 in London.

When Gosset arrived in Dublin he found that there was a mass of data concerning brewing which called for sophisticated mathematical analysis. Gosset's main problem was to estimate the mean value of a characteristic on the basis of very small samples, for use by industry when large sampling was too expensive or impracticable. For any large probability, that is one of 95% or more, he was in 1908 able to compute the error e, such that it is 95% probable that:

$$(x - \mu) < e$$

where x is the value of the sample, and μ is the mean. From this was derived what came to be known as Student's t-test of statistical hypotheses (Gosset published all his papers under the pseudonym 'Student'). The test consists of rejecting a hypothesis if, and only if, the probability (derived from t) of erroneous rejection is small.

Gould John 1804–1881. English zoologist who with his wife Elizabeth (1804–1841), a natural-history artist, published a successful series of illustrated bird books. They visited Australia 1838–40 and afterwards produced *The Birds of Australia*, issued in 36 parts from 1840. *Mammals of Australia* followed 1845–63, and *Handbook to the Birds of Australia* 1865.

Gould Stephen Jay 1941– . US palaeontologist and writer. In 1972 he proposed the theory of punctuated equilibrium, suggesting that the evolution of species did not occur at a steady rate but could suddenly accelerate, with rapid change occurring over a few hundred thousand years. His books include *Ever Since Darwin* 1977, *The Panda's Thumb* 1980, *The Flamingo's Smile* 1985, and *Wonderful Life* 1990.

Gould was born in New York and studied at Antioch College, Ohio, and Columbia University. He became professor of geology at Harvard 1973 and was later also given posts in the departments of zoology and the history of science.

Gould has written extensively on several aspects of evolutionary science, in both professional and popular books. His *Ontogeny and Phylogeny* 1977 provided a detailed scholarly analysis of his work on the developmental process of recapitulation. In *Wonderful Life* he drew attention to the diversity of the fossil finds in the Burgess Shale Site in Yoho National Park, Canada, which he interprets as evidence of parallel early evolutionary trends extinguished by chance rather than natural selection.

I am, somehow, less interested in the weight and convolutions of Einstein's brain than in the near certainty that people of equal talent have lived and died in cotton fields and sweatshops.

Stephen Jay Gould

The Panda's Thumb 1980

Graaf Regnier de 1641–1673. Dutch physician and anatomist who discovered the ovarian follicles, which were later named ***Graafian follicles***. He named the ovaries and gave exact descriptions of the testicles. He was also the first to isolate and collect the secretions of the pancreas and gall bladder.

Graham Thomas 1805–1869. Scottish chemist who laid the foundations of physical chemistry (the branch of chemistry concerned with changes in energy during a chemical transformation) by his work on the diffusion of gases and liquids. ***Graham's law*** 1829 states that the diffusion rate of a gas is inversely proportional to the square root of its density.

His work on colloids (which have larger particles than true solutions) was equally fundamental; he discovered the principle of dialysis, that colloids can be separated from solutions containing smaller molecules by the differing rates at which they pass through a semipermeable membrane. The human kidney uses the same principle to extract nitrogenous waste.

Graham was born in Glasgow and studied at Glasgow and Edinburgh. In 1830, Graham became professor at Anderson's

College, Glasgow, moving to University College, London, 1837–54. In 1855, he was appointed Master of the Royal Mint.

Grandi Guido 1671–1742. Italian mathematician who worked on the definition of curves. He devised the curves now known as the 'versiera', the 'rose', and the 'cliela', and his theory of curves also comprehended the means of finding the equations of curves of known form. He was mainly responsible, in addition, for introducing calculus into Italy 1703.

Grandi was born in Cremona. He became professor of philosophy at Pisa in 1700 and of mathematics 1714.

In his fascination with the study of curves, Grandi was influenced first by English scientist Isaac Newton. In 1728 he published his complete theory in *Fleores geometrica*, an attempt (among other things) to define geometrically the curves that have the shapes of flowers, particularly multipetalled roses.

Grandi also did some work in practical mechanics and his observations regarding hydraulics were utilized by the Italian government in such public works as the drainage of the Chiana valley and the Pontine Marshes in central Italy.

Grassmann Hermann Günther 1809–1877. German mathematician and linguist who discovered a new calculus. It was one of the earliest mathematical attempts to investigate *n*-dimensional space, where *n* is greater than 3. He also studied ancient languages and comparative linguistics.

Grassmann was born in Stettin, Pomerania (now Szczecin, Poland), and studied at Berlin. He became a schoolteacher in Stettin.

Grassmann developed his method of calculus, which he called the theory of extension, from 1840. He published it 1844, but his vocabulary was so obscure that the book had virtually no impact at all until after his death. He used the method to reformulate Ampère's law and investigate the subject of algebraic curves, but again the work was ignored.

Grassmann learned and examined many ancient languages, such as Persian, Sanskrit, and Lithuanian. From his investigations he derived a theory of speech. He published a glossary to the Rig-Veda (Hindu sacred writings) 1873–75.

Gray Asa 1810–1888. US botanist and taxonomist who became America's leading expert in the field. His major publications include *Elements of Botany* 1836 and the definitive *Flora of North America* 1838, 1843. He based his revision of the Linnaean system of plant classification on fruit form rather than gross morphology.

Gray, born in Saquoit, New York, graduated from medical school but chose botany rather than medicine as his career. A friend and supporter of Charles Darwin, he was one of the founders of the American National Academy of Sciences.

His *Manual of Botany* 1850 remains the standard reference work on flora east of the Rockies.

Gray Henry *c.* 1827–1861. British anatomist who compiled a book on his subject, published 1858 with illustrations by his colleague H Vandyke Carter. Through its various editions and revisions, it has remained the definitive work on anatomy.

Gray studied at St George's Hospital, London, where he became demonstrator of anatomy and curator of the St George's Museum.

What is now known as *Gray's Anatomy* was based on his own dissections. Unlike other contemporary works on the subject, it was organized in terms of systems, rather than areas of the body. Such sections as neuroanatomy have been greatly enlarged in later editions but the section that deals with, for example, the skeletal system is almost identical to Gray's original work. It remains a standard text for students and surgeons alike.

Gray James 1891–1975. English zoologist who helped to found the study of cytology (cell structure).

Born in Wood Green, London, Gray attended Merchant Taylor's School and King's College, Cambridge. He graduated in natural sciences 1913 and then spent a year in Naples before returning to King's to take up a fellowship. He then joined the Queen's Royal West Surrey regiment and served in France and Palestine in World War I, during which time he was awarded the Military Cross and the *croix de guerre avec palme* for bravery.

He resumed his position at King's College 1919 and was appointed a lecturer 1929. He was elected Fellow of the Royal Society 1931, and in the same year his *A textbook of Experimental Cytology* was published, which helped to establish cytology as a distinct branch of zoology. However, in the early 1930s Gray began to concentrate on animal location, realizing the techniques needed to study the cell were yet to be developed. In 1937 he was made professor of zoology, having been very involved in the fund raising for the new zoology department at Cambridge, which was finally completed 1934. From 1943 to 1947 he was the Fullerian professor of physiology at the Royal Institute. He was awarded a CBE 1946 and knighted 1954.

Green George 1793–1841. English mathematician who coined the term 'potential', now a central concept in electricity, and introduced ***Green's theorem***, which is still applied in the solution of partial differential equations; for instance, in the study of relativity.

Green was born in Nottingham and studied mathematics by himself until at the age of 40 he became a student at Cambridge, and later a member of the staff.

Green's groundbreaking paper was published 1828 and in it he demonstrated the importance of 'potential function' (also known as the Green function) in both magnetism and electricity, and showed how the Green theorem enabled volume integrals to be reduced to surface integrals. It stimulated great interest, and was to influence such scientists as James Clerk Maxwell and Lord Kelvin.

Green went on to produce other important papers on fluids (1832, 1833), attraction (1833), waves in fluids (1837), sound (1837), and light (1837).

Greenstein Jesse Leonard 1909– . US astronomer who took part in the discovery of the interstellar magnetic field and the discovery and interpretation of quasars. His early work involved the spectroscopic investigation of stellar atmospheres; later work included a study of the structure and composition of white dwarf stars.

Greenstein was born in New York and studied at Harvard. In 1948, he joined the California Institute of Technology and also the staff of the Mount Wilson and Palomar Observatories. He became professor of astrophysics 1971. During the 1970s he guided both the US space agency NASA and the National Academy of Sciences in their policies.

He confirmed the hypothesis of US astronomer Maarten Schmidt (1929–) that the emission lines of quasars could be explained by a shift in wavelength. In collaboration with Schmidt, Greenstein proposed a detailed physical model of the size, mass, temperature, luminosity, magnetic field, and high-energy particle content of quasars.

By 1978 Greenstein had discovered some 500 white dwarf stars. His research enabled him to pinpoint the problems of explaining the evolutionary sequence that links red giant stars with white dwarfs. This initiated spectroscopic studies of such stars from space.

Gregg Norman McAlister 1892–1966. Australian ophthalmic surgeon who discovered 1941 that German measles in a pregnant woman could cause physical defects in her child. Knighted 1953.

Grew Nehemiah 1641–1712. English botanist and physician who made some of the early microscopical observations of plants. He studied the structure of various plants' anatomy and introduced the term 'parenchyma' to refer to the ground tissue, or unspecialized cells, of a plant. His observations were included in his book *The Anatomy of Plants* 1682.

Grew and his contemporary Marcello Malpighi were responsible for significant advances in the understanding of botanical anatomy due to their widespread use of the light microscope. He was greatly influenced by the brilliant philosopher and scientist Robert Hooke, whose treatise *Micrographia* 1665 had become an established text on the microscopic observations of plant life at that time.

Grew was born in Mancetter, Warwickshire and studied for a BA at Pembroke Hall, Cambridge. However, because he was a religious nonconformist, he had to move to Leiden in Holland to study for an MD. Following his graduation, he practised medicine in Coventry and then moved to London 1672 so that he could be more involved in the work of the Royal Society. He was secretary of the Royal Society 1677, which gave him access to the Society's microscope, a valuable asset to a would-be plant anatomist. Because he did not have any pupils, many of his unpublished results were lost after his death.

Griffin Donald Redfield 1915– . US zoologist who discovered that bats use echolocation to navigate and orientate themselves in space, that is they emit ultrasonic sounds that rebound off objects that they are then able to avoid. His later research was mainly in the areas of animal navigation, acoustic orientation and sensory biophysics, and animal consciousness.

Griffin was born in Southampton, New York and studied at Harvard University, where as an undergraduate he discovered that bats use echolocation in order to orientate themselves. He obtained a PhD from Harvard 1942, and then worked as a research assistant in the psycho-acoustic laboratory and fatigue laboratory at Harvard until 1945. In 1946 he went to Cornell University where he lectured in zoology until 1953 when he returned to Harvard. He was appointed professor at the Rockefeller University in New York 1965, also becoming a research zoologist for the New York Zoological Society in the same year.

His writing includes *Listening in the Dark* 1958, *Echoes of Bats and Men* 1959, *Animal Structure and Function* 1962, *Bird Migration* 1964, and *The Question of Animal Awareness* 1976.

Grignard (François Auguste) Victor 1871–1935. French chemist. In 1900 he discovered a series of organic compounds, the ***Grignard reagents***, that found applications as some of the most versatile reagents in organic synthesis. Members of the class contain a hydrocarbon radical, magnesium, and a halogen such as chlorine. He shared the 1912 Nobel Prize for Chemistry.

Grignard was born in Cherbourg and studied at Lyon. He became professor at Nancy 1910. During World War I he headed a department at the Sorbonne concerned with the development of chemical warfare. From 1919 he was professor at Lyon.

Grignard reagents added to formaldehyde (methanal) produce a primary alcohol; with any other aldehyde they form secondary alcohols, and added to ketones give rise to tertiary alcohols. They will also add to a carboxylic acid to produce first a ketone and ultimately a tertiary alcohol.

His multivolume *Traité de chimie organique* began publication 1935.

Grimaldi Francesco Maria 1618–1663. Italian physicist who discovered the diffraction of light. In physiology, he observed muscle action and was the first to note that minute sounds are produced by muscles during contraction.

Grimaldi was born in Bologna and was professor of mathematics at the Jesuit College there from 1648. He also helped with astronomical observations.

Grimaldi let a beam of sunlight enter a darkened room through a small circular aperture and observed that, when the beam passed through a second aperture onto a screen, the spot of light was slightly larger than the second aperture and had coloured fringes. Grimaldi concluded that the light rays had diverged slightly, becoming bent outwards, or diffracted. On placing a narrow obstruction in the light beam, he noticed bright bands at each side of its shadow on the screen. This phenomenon can be explained readily only if light is regarded as travelling in waves – contrary to the then accepted corpuscular theory of light.

A description of the experiments was published in *Physicomathesis de lumine, coloribus, et iride/Physicomathematical Thesis of Light, Colours and the Rainbow* 1665.

Grisebach August Heinrich Rudolph 1814–1879. German botanist and plant geographer who explored the Balkans and Asia Minor. In 1866, he coined the term 'geobotany' (phytogeography), the study of the geographical distribution of plants.

Grisebach was born in Hanover, Germany. His first botany instructor was his uncle, the eminent botanist Georg Friedrich Wilhelm Meyer. He studied medicine in Göttingen and Berlin, graduating 1836, and then secured a job as a Privatdocent at Göttingen. He travelled in the Balkans and Asia Minor 1839–40, an area in which the flora was virtually unknown at that time. The papers he published on his return helped him establish his reputation as a botanist. He rapidly became an expert in taxonomy and phytogeography and was made an associate professor at the University of Göttingen 1847. He died in Göttingen in May.

Grosseteste Robert *c.* 1169–1253. English scholar and bishop. His prolific writings include scientific works as well as translations of Aristotle and commentaries on the Bible. He was a forerunner of the empirical school, being one of the earliest to suggest testing ancient Greek theories by practical experiment.

He was bishop of Lincoln from 1235 to his death, attempting to reform morals and clerical discipline, and engaging in controversy with Innocent IV over the pope's finances.

Grosseteste was born in Suffolk and studied at Oxford and perhaps Paris, later becoming chancellor of Oxford University.

Guericke Otto von 1602–1686. German physicist and politician who invented the air pump and demonstrated the pressure of the atmosphere. He also constructed the first machine for generating static electricity.

Guericke was born in Magdeburg and attended several German universities, before returning to Magdeburg and becoming alderman of the city. In 1631, most of Magdeburg's population of 40,000 were butchered in the Thirty Years' War, and Guericke moved to Brunswick and Erfurt, where he worked as engineer to the Swedish government and later for that of Saxony as well. He was able to serve Magdeburg as envoy to various occupying powers – alternately French, Habsburg, and Swedish – and represented his city at various conferences. He was mayor of Magdeburg 1646–76.

Guericke constructed an air pump in an attempt to produce a vacuum, to test French mathematician René Descartes's idea that space was matter. In 1647 Guericke imploded a copper sphere from which he pumped the air via an outlet at the bottom. But when he built a stronger vessel, he succeeded in evacuating it without causing it to collapse. He also demonstrated that a candle is extinguished as the air is removed, and gradually the theory that a vacuum cannot exist was discarded.

Demonstrating the immense force that the pressure of the atmosphere exerts on an evacuated vessel, Guericke set up various public experiments. In 1657, two copper hemispheres with tight-fitting edges were placed together and the air pumped out. Two teams of horses were then harnessed to the hemispheres and proved unable to pull them apart. Once air was admitted to the hemispheres, they fell apart instantly.

Guericke went on to investigate the decrease of pressure with altitude, and the link between atmospheric pressure and the weather. In 1660 he was making weather forecasts with a water barometer and proposing a string of meteorological stations contributing data to a forecasting system.

While experimenting with a globe of sulphur constructed to simulate the magnetic properties of the Earth, Guericke discovered that it produced static electricity when rubbed; he went on to develop a primitive machine for the production of static electricity. He also demonstrated the magnetization of iron by hammering in a north–south direction.

Another first with which he is credited is the observation of coloured shadows.

Guérin Camille 1872–1961. French bacteriologist who, with Albert Calmette, developed the BCG vaccine for tuberculosis 1921.

Guettard Jean-Etienne 1715–1786. French naturalist who pioneered geological mapping. He also studied botany and medicine, as well as the origin of various types of rock.

Guettard was born in Etampes, Ile-de-France, and became keeper of the natural-history collection of the Duc d'Orléans. In 1746 he presented his first mineralogical map of France to the Académie des Sciences; and in 1766, he and chemist Antoine Lavoisier were commissioned to prepare a geological survey of France, though only a fraction was ever completed.

Research in the field suggested that the rocks of the Auvergne district of central France were volcanic in nature, and Guettard boldly identified several peaks in the area as extinct volcanoes, though he later had doubts about this hypothesis. Of basalt, he originally took the view that it was not volcanic in origin, but changed his mind after visits to Italy in the 1770s.

Guillaume Charles Edouard 1861–1938. Swiss physicist who studied measurement and alloy development.

He discovered a nickel-steel alloy, Invar, which showed negligible expansion with rising temperatures. He was awarded the Nobel Prize for Physics 1920.

As the son of a clockmaker, Guillaume came to appreciate early in life the value of precision in measurement. He spent most of his life at the International Bureau of Weights and Measures in Sèvres, France, which established the standards for the metre, litre, and kilogram.

Guillemin Roger Charles Louis 1924– . French-born US endocrinologist. He has isolated and identified various hormones, for which he received the 1977 Nobel Prize for Physiology or Medicine, together with his co-worker Andrew Schally (1926–) and US physicist Rosalyn Yalow. Guillemin also discovered endorphins.

Guillemin was born in Dijon and educated there and at Lyon. He moved to the USA 1953 and did most of his work at Baylor College of Medicine, Houston, Texas, becoming professor 1963. In 1970 he joined the Salk Institute in La Jolla, California.

Guillemin found that the brain controls the pituitary gland by means of hormones produced by central neurons – the neurosecretory cells of the hypothalamus. He worked with Lithuanian refugee Schally 1957–62, and later their investigations were parallel. Between 1968 and 1973 they isolated and synthesized three hypothalamic hormones which regulate the secretion of the anterior pituitary gland.

Gullstrand Allvar 1862–1930. Swedish physician who won the Nobel Prize for Physiology or Medicine 1911 for his work on the optics of the eye, particularly how the lens of the eye is adjusted to focus on both near and far objects. To assist his research Gullstrand invented the slit-lamp which is still used in modern instruments to illuminate the eye.

He worked extensively on the refractory system of the eye, describing the shape of the living cornea and the changes in the lens when focusing on near or far objects (accommodation).

Gullstrand was born in Landskrona, Sweden and studied medicine at the University of Uppsala, graduating 1888. He obtained his PhD in ophthalmology 1890, and became professor of ophthalmology at Uppsala.

Gunter Edmund 1581–1626. English mathematician who became professor of astronomy at Gresham College, London 1619. He is reputed to have invented a number of surveying instruments as well as the trigonometrical terms 'cosine' and 'cotangent'.

Gurdon John Bertrand 1933– . English molecular biologist who has studied nuclear transplantation and the effects of known protein fractions on gene activity.

Gurdon was born in Dippenhall, Hampshire, and educated at Oxford. In 1972 he joined the Laboratory of Molecular Biology at Cambridge.

Transplanting nuclei and nuclear constituents, such as DNA, into enucleated eggs, mainly frog's eggs, Gurdon showed how the genetic activity changes in the recipient eggs. He concluded that the changes in gene activity induced by nuclear transplantation are indistinguishable from those that occur in normal early development. He also demonstrated how nuclear transplantation and microinjection techniques can be used to elucidate the intracellular movements of proteins.

Gutenberg Johannes Gensfleisch *c.* 1398–1468. German printer, the inventor of printing from movable metal type, based on the Chinese wood-block-type method (although Laurens Janszoon Coster has a rival claim).

Gutenberg began work on the process in the 1440s and in 1450 set up a printing business in Mainz. By 1456 he had produced the first printed Bible (known as the Gutenberg Bible). It is not known what other books he printed.

He punched and engraved a steel character (letter shape) into a piece of copper to form a mould which he filled with molten metal. The letters were in the Gothic style and of equal height. By 1500, more than 180 European towns had working presses of this kind.

Gutenberg was born in Mainz and set up a printing firm in Strasbourg in the late 1430s, where he may have invented movable type. This business folded, as did the subsequent one in Mainz with Johann Fust (*c.* 1400–1466) as a backer: Fust seized the press for nonpayment of the loan. Gutenberg is believed to have gone on to set up a third press and print the Mazarin and Bamberg Bibles. In 1462, Mainz was involved in a local feud, and in the upheaval Gutenberg was expelled from the city for five years before being reinstated, offered a pension, and given tax exemption.

Gutenberg *German printer Johannes Gutenberg, who invented movable type, shown at work in his printing shop. Gutenberg produced the first major work to come off a printing press, the so-called Gutenberg Bible, of which there are 47 surviving copies.*

Haber Fritz 1868–1934. German chemist whose conversion of atmospheric nitrogen to ammonia opened the way for the synthetic fertilizer industry. His study of the combustion of hydrocarbons led to the commercial 'cracking' or fractional distillation of natural oil (petroleum) into its components (for example, diesel, petrol, and paraffin). In electrochemistry, he was the first to demonstrate that oxidation and reduction take place at the electrodes; from this he developed a general electrochemical theory.

At the outbreak of World War I in 1914, Haber was asked to devise a method of producing nitric acid for making high explosives. Later he became one of the principals in the German chemical-warfare effort, devising weapons and gas masks, which led to protests against his Nobel prize 1918.

For more than forty years I have selected my collaborators on the basis of their intelligence and their character and I am not willing for the rest of my life to change this method which I have found so good.

Fritz Haber

Letter of Resignation 30 April 1933
[Haber was unwilling to follow the Nazi requirements for racial purity.]

Haber was born in Bresslau, Silesia (now Wroclaw, Poland), and educated at Berlin, Heidelberg, and the Berlin Technische Hochschule. He was professor at Karlsruhe 1906–11, and then was made director of the newly established Kaiser Wilhelm Institute for Physical Chemistry in Berlin. When Adolf Hitler rose to power in 1933, Haber sought exile in Britain, where he worked at the Cavendish Laboratory, Cambridge.

After World War I, Haber set himself the task of extracting gold from sea water to help to pay off the reparations demanded by the Allies. Swedish scientist Svante Arrhenius had calculated that the sea contains 8,000 million tonnes of gold. The project got as far as the fitting-out of a ship and the commencement of the extraction process, but the yields were too low and the project was abandoned 1928.

Hadamard Jacques Salomon 1865–1963. French mathematician who originated functional analysis, one of the most fertile branches of modern mathematics. He also made contributions to number theory and formulated the concept of a correctly posed problem.

Hadamard was born in Versailles and studied at the Ecole Normale Supérieure in Paris. He was professor of mathematics at the Collège de France in Paris 1909–37. During the German occupation of France in World War II, he went into exile, returning 1945.

Hadamard's early work was on analytic functions; that is, functions that can be developed as power series that converge. He began to study the Riemann zeta function and in 1896 solved the problem of determining the number of prime numbers less than a given number x. Hadamard was able to demonstrate that this number was asymptotically equal to $x/\log x$, which was the most important single result ever obtained in number theory.

Hadamard became interested in the 'functions of lines' – numerical functions that depend upon a curve or an ordinary function as their variable. By extending the theory of ordinary functions to the case where the variable, or variables, would no longer be a number, or numbers, Hadamard created a new branch of mathematics. This required a redefinition, or at least a new generalization, of concepts such as continuity, derivative, and differential.

By extension from this work, Hadamard came to investigate functions of a complex variable and to define a singularity as a point at which a function ceases to be regular. He showed that the existence of a set of singular points may be compatible with the continuity of a function, and named the region formed by such a set a 'lacunary space'. The study of such spaces has occupied mathematicians ever since.

Since it is often helpful, or necessary, to find an approximate solution (in physics, for example), a correctly posed problem, according to Hadamard, is one for which a solution exists that is unique for given data, but which also depends continuously on the data. This is the case when the solution can be expressed as a set of convergent power series. The idea has been fundamental to the development of the theory of function spaces, functional analysis.

Hadfield Robert Abbott 1858–1940. English industrial chemist and metallurgist who invented stainless steel and developed various other ferrous alloys. Knighted 1908, baronet 1917.

Hadfield was born in Sheffield, Yorkshire, joined his father's steel foundry, and by the age of 24 was its manager. His research into various steel alloys began in the 1880s.

In making ordinary mild steel, pig iron is oxidized, to lower the carbon content. Hadfield carried out many experiments in which he mixed other metals to the steel. He found, for example, that a small amount of manganese gave a tough, wear-resistant steel suitable for such applications as railway track and grinding machinery. By adding nickel and chromium he produced corrosion-resistant stainless steels.

Man creates God in his own image.

Ernst Haeckel

Generelle Morphologie vol I 1866 p 174

Haeckel Ernst Heinrich Philipp August 1834–1919. German zoologist and philosopher. His theory of 'recapitulation', expressed as 'ontogeny repeats phylogeny' (or that embryonic stages represent past stages in the organism's evolution), has been superseded, but it stimulated research in embryology.

Haeckel was born in Potsdam, Prussia, and studied at Würtzburg and Berlin. He was professor at Jena 1865–1909.

In 1866, the same year that he met Charles Darwin, Haeckel introduced a method, still used today, of representing evolutionary history, or phylogeny, by means of treelike diagrams.

Haeckel tried to apply Darwin's doctrine of evolution to philosophy and religion. He denied the immortality of the soul, the freedom of the will, and the existence of a personal God. His view that the origin of life lies in the chemical and physical factors of the environment, such as sunlight, oxygen, water, and methane, has been shown to be likely. He coined the term 'ecology', and published best-selling general scientific works such as *Welträtsel/The Riddles of the Universe* 1899.

Hagenbeck Carl 1844–1913. German zoo proprietor. In 1907 he founded Hagenbeck's Zoo, near his native Hamburg. He was a pioneer in the display of animals against a natural setting, rather than in restrictive cages.

Hahn Otto 1879–1968. German physical chemist who discovered nuclear fission. In 1938 with Fritz Strassmann (1902–1980), he discovered that uranium nuclei split when bombarded with neutrons. Hahn did not participate in the resultant development of the atom bomb. Nobel Prize for Chemistry 1944.

In 1918, Hahn and Lise Meitner discovered the longest-lived isotope of a new element which they called protactinium, and in 1921 they discovered nuclear isomers – radioisotopes with nuclei containing the same subatomic particles but differing in energy content and half-life.

Hahn was born in Frankfurt-am-Main and studied at Marburg. From 1904 to 1906 he worked in London under William Ramsay, who introduced Hahn to radiochemistry, and at McGill University in Montréal, Canada, with Ernest Rutherford. Returning to Germany, Hahn was joined at Berlin in 1907 by Meitner, beginning a long collaboration. Hahn was director of the Kaiser Wilhelm Institute for Chemistry in Berlin 1928–44, and then president of the Max Planck Institute in Göttingen.

Hahnemann (Christian Friedrich) Samuel 1755–1843. German physician and founder of homoeopathy. While practising conventional medicine, he developed a theory that a disease can be cured by drugs that cause the symptoms of that disease in healthy individuals (in other words, 'like cures like'). He called this new form of treatment homoeopathy. It involved the administration of very small doses of a drug, many of which are herbal in origin, to cure or alleviate the symptoms of some diseases.

From 1798 to 1810, he attempted to practice homoeopathic medicine in various towns in Germany by administering his homoeopathic remedies free of charge, but found himself antagonizing local practitioners by his unorthodox theories and methods, to the extent that he was often prosecuted for practising dangerous, if not illegal, forms of medicine. However, he persevered in his efforts to publicize homoeopathy and by 1825, the practise had spread to the USA and various European countries.

Hahnemann was born in Meissen in Germany and studied medicine at the University of Leipzig.

I have no doubt that in reality the future will be vastly more surprising than anything I can imagine. Now my own suspicion is that the universe is not only queerer than we suppose, but queerer than we can suppose.

J B S Haldane

Possible Worlds and Other Papers 1927 p 286

Haldane J(ohn) B(urdon) S(anderson) 1892–1964. British physiologist, geneticist, and author of popular science books. In 1936 he showed the genetic link between haemophilia and colour blindness.

Haldane was born and educated at Oxford. In 1933 he became professor of genetics at University College, London. He emigrated to India 1957 in protest at the Anglo-French invasion of Suez and was appointed director of the Genetics and Biometry Laboratory in Orissa. He became a naturalized Indian citizen in 1961.

In 1924 Haldane produced the first proof that enzymes obey the laws of thermodynamics.

Haldane carried our research into how the regulation of breathing in man is affected by the level of carbon dioxide in the bloodstream. During World War II, in 1942, Haldane, who often used his own body in biochemical experiments, spent two days in a submarine to test an air-purifying system.

Haldane was convinced that natural selection and not mutation is the driving force behind evolution. In 1932, he estimated for the first time the rate of mutation of the human gene and worked out the effect of recurrent harmful mutations on a population. He is supposed to have remarked: 'I'd lay down my life for two brothers or eight cousins.'

Haldane John Scott 1860–1936. Scottish physiologist whose studies of the exchange of gases during respiration led to an interest in the health hazards of coal mining and deep-sea diving. His aim was to bridge the gap between theoretical and applied science.

Haldane was born and educated in Edinburgh. He was director of the Mining Research Laboratory (first in Doncaster, then in Birmingham) 1913–28. He also lectured at various universities in the UK, the USA, and Ireland.

Haldane devised methods for studying respiration and the blood – the Haldane gas analyser and an apparatus for determining the blood gas content. Having investigated the danger to miners of suffocation, he turned to the toxicity of carbon monoxide, which is usually present in mines after an explosion, and showed that haemoglobin in the red blood cells binds this gas in preference to oxygen.

In 1905, Haldane proposed that breathing is controlled by the concentration of carbon dioxide in arterial blood acting on the respiratory centre of the brain. In 1907, he announced the technique of decompression by stages which is still used today to allow deep-sea divers to surface safely.

He also researched the reaction of the kidneys to the water content of the blood, and the physiology of sweating.

Hale George Ellery 1868–1938. US astronomer who made pioneer studies of the Sun and founded three major observatories. In 1889 he invented the spectroheliograph, a device for photographing the Sun at particular wavelengths. In 1917 he established on Mount Wilson, California, a 2.5-m/100-in reflector, the world's largest telescope until superseded 1948 by the 5-m/200-in reflector on Mount Palomar, which Hale had planned just before he died.

In 1897 he founded the Yerkes Observatory in Wisconsin, with the largest refractor, 102 cm/40 in, ever built at that time.

Hale was born in Chicago and studied at the Massachusetts Institute of Technology. In 1892 he became professor at Chicago, moving to Mount Wilson as director 1904–23. He was elected to the governing body of Throop Polytechnic Institute in Pasadena, and through his influence this developed into the California Institute of Technology.

Hale's work on solar spectra was the stimulus for his construction of a number of specially designed telescopes. By studying the split spectral lines of sunspots, he showed the presence of very strong magnetic fields – the first discovery of a magnetic field outside the Earth. In 1919, he showed that these magnetic fields reverse polarity twice every 22–23 years.

Hales Stephen 1677–1761. English scientist who studied the role of water and air in the maintenance of life. He gave accurate accounts of water movement in plants. He demonstrated that plants absorb air, and that some part of that air is involved in their nutrition. His work laid emphasis on measurement and experimentation.

Hales's work on air revealed to him the dangers of breathing 'spent' air in enclosed places, and he invented a ventilator which improved survival rates when introduced on naval, merchant, and slave ships, in hospitals, and in prisons.

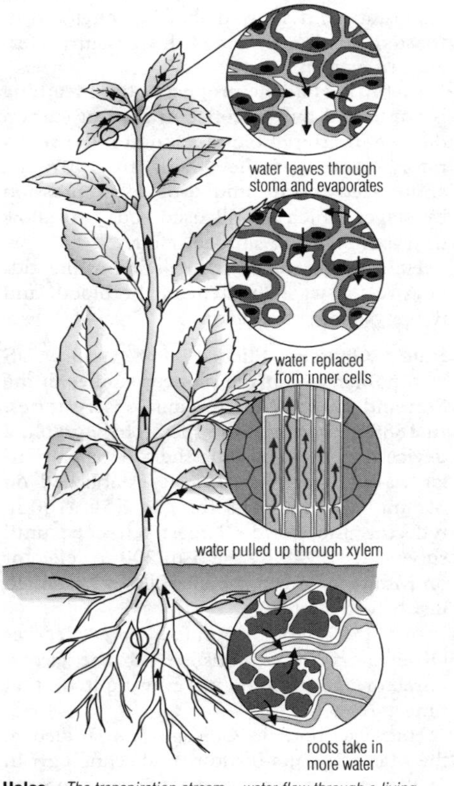

Hales was born in Kent and studied at Cambridge. A cleric, he was curate at Teddington, Middlesex, from 1709. His experiments on plants took place mainly between 1719 and 1725.

He measured the pressure of sap in growing vines, calculated its velocity, and found that the rate of flow varies in different plants. He measured plant growth and water loss, relating this to the upward movement of water from plants to leaves (transpiration). He also measured blood pressure and the rate of blood flow in animals.

Hales examined stones taken from the bladder and kidney and suggested possible chemical solvents for their nonsurgical treatment. He also invented the surgical forceps.

Hales's findings were published in his book *Vegetable Staticks* 1727, enlarged 1733 and retitled *Statical Essays, Containing Haemastaticks, etc.*

Hales *The transpiration stream – water flow through a living plant.*

Hall Asaph 1829–1907. US astronomer who discovered the two Martian satellites, Deimos and Phobos 1877. He determined the orbits of satellites of other planets and of double stars, the rotation of Saturn, and the mass of Mars.

Hall was born in Goshen, Connecticut. Apprenticed to a carpenter at 16, he later enrolled at the Central College in McGrawville, New York. In 1856, he took a job at the Harvard College Observatory in Cambridge, Massachusetts, and turned out to be an expert computer of orbits. Hall became assistant astronomer at the US Naval Observatory in Washington DC 1862, and within a year of his arrival he was made professor.

In 1875 Hall was given responsibility for a 66-cm/26-in telescope, the largest refractor in the world at the time. He noticed a white spot on Saturn which he used as a marker to ascertain the planet's rotational period. In 1884, he showed that the position of the elliptical orbit of Saturn's moon, Hyperion, was retrograding by about 20° per year.

Hall also investigated stellar parallaxes and the position of the stars in the Pleiades cluster.

Hall James 1761–1832. Scottish geologist, one of the founders of experimental geology. He provided evidence in support of the theories of Scottish naturalist James Hutton regarding the formation of the Earth's crust. Succeeded to baronetcy 1776.

Hall was born in Berwickshire (Borders region) and spent much of the 1780s travelling in Europe. He undertook extensive geological observations in the Alps and studying Mount Etna in Sicily. He was also won over to the new chemistry of Antoine Lavoisier.

Hall set out to prove his friend Hutton's 'Plutonist' geological theories (the view that heat rather than water was the chief rock-building agent and shaper of the Earth's crust). By means of furnace experiments, he showed with fair success that Hutton had been correct to maintain that igneous rocks would generate crystalline structures if cooled very slowly. Hall also demonstrated that there was a degree of interconvertibility between basaltine and granitic rocks; and that, even though subjected to immense heat, limestone would not decompose if sustained under suitable pressure.

Hall Marshall 1790–1857. English physician and physiologist who distinguished between voluntary and involuntary reflex muscle contractions, proving that the spinal cord is more than a passive nerve trunk transmitting voluntary signals from the brain and sensory signals to the brain.

Hall is best known for his work on the nervous system of frogs. He showed that if the spinal cord of a frog was severed between the front and back limbs, then the front limbs could still be moved voluntarily but the back limbs were useless. He further showed that the back legs could be stimulated to move artificially, but only once for each stimulus. These were reflex (involuntary) muscle contractions. Pain stimuli applied to the back legs were not felt by the animals. From these experiments Hall deduced that the nervous system is made up of a series of reflex arcs. In the intact spinal cord these reflex arcs are coordinated by the ascending and descending pathways in the cord to form movement patterns.

Hall also demonstrated that stimulus could not be put into the cord through a sensory nerve without it resulting in effects beyond the anatomical segment to which that nerve belongs.

Hall was born in Basford, and studied medicine at the University of Edinburgh, obtaining his PhD 1822. As a physician in Nottingham and then later in London he experimented in various areas of physiology, such as the blood circulatory system and respiration, although he is best known for his work on the nervous system of frogs.

Hall Philip 1904–1982. English mathematician who specialized in the study of group theory.

Hall was born in London and studied at Cambridge, where he spent his whole career. He was professor of pure mathematics 1953–67.

In 1928 Hall began a study of prime power groups. From this work he developed his 1933 theory of regular groups. An investigation of the conditions under which finite groups are soluble led him in 1937 to postulate a general structure theory for finite soluble groups. In 1954, he published an examination of finitely generated soluble groups in which he demonstrated that they could be divided into two classes of unequal size. At the end of the 1950s Hall turned to the subject of simple groups, and later also examined non-strictly-simple groups.

Hall Charles Martin 1863–1914. US chemist who developed a process for the commercial production of aluminium 1886. A similar process was independently but simultaneously developed in France by Paul Héroult.

He found that when aluminium was mixed with cryolite (sodium aluminium fluoride), its melting point was lowered and electrolysis became commercially viable. It had previously been as costly as gold, but by 1914 its price was 40 cents a kilogram.

Hall was born in Ohio and educated at Oberlin. He invented the aluminium process at 22 and, after initial difficulties, formed the Pittsburgh Reduction Company (later to become the Aluminum Company of America) and became a multimillionaire.

Haller Albrecht von 1708–1777. Swiss physician and scientist, founder of neurology. He studied the muscles and nerves, and concluded that nerves provide the stimulus that triggers muscle contraction. He also showed that it is the nerves, not muscle or skin, that receive sensation.

Haller was born in Berne and studied at Leiden, the Netherlands. He was professor at Göttingen, Germany, 1736–53.

Tracing the pathways of nerves, he was able to demonstrate that they always lead to the spinal cord or the brain, suggesting that these regions might be where awareness of sensation and the initiation of answering responses are located.

While carrying out his experiments, Haller discovered several processes of the human body, such as the role of bile in digesting fats. He also wrote a report on his study of embryonic development.

Haller published *De respiratione experimenta anatomica/Experiments in the Anatomy of Respiration* 1747 and *Elementa physiologiae corporis humani/The Physiological Elements of the Human Body* 1757–66.

Halley Edmond 1656–1742. English atronomer who not only identified 1705 the comet that was later to be known by his name, but also compiled a star catalogue, detected the proper motion of stars, using historical records, and began a line of research that – after his death – resulted in a reasonably accurate calculation of the astronomical unit.

Halley calculated that the cometary sightings reported in 1456, 1531, 1607, and 1682 all represented reappearances of the same comet. He reasoned that the comet would follow a parabolic path and announced 1705 that it would reappear 1758. When it did, public acclaim for the astronomer was such that his name was irrevocably attached to it.

Halley was also a pioneer geophysicist and meteorologist and worked in many other fields, including mathematics. He became the second Astronomer Royal 1720. He was a friend of Isaac Newton, whose *Principia* he financed.

Halley *English astronomer and mathematician Edmond Halley, who carried out important work on the motion and positions of planetary and cellestial bodies, and is chiefly known for the identification of the comet that now bears his name. Halley's comet approaches the Earth every 76 years and is depicted in the Bayeaux tapestry.*

Halley was born near London and studied at Oxford but left without taking a degree. He spent 1676–78 on the S Atlantic island of St Helena, charting the stars of the southern hemisphere. He became professor of geometry at Oxford 1703.

Hamilton Alice 1869–1970. US physician, social reformer, and antiwar campaigner who pioneered the study of industrial diseases and industrial toxicology.

Hamilton was born in New York State and educated at the University of Michigan, Johns Hopkins Medical School, and in Germany at Leipzig and Munich. As a member of the Illinois Commission on Occupational Diseases, she supervised in 1910 a survey of industrial poisons. She and her staff identified many hazardous procedures and consequent state legislature introduced safety measures in the workplace and medical examinations for workers at risk. The following year Hamilton was appointed special investigator for the US Bureau of Labor and rapidly became the leading authority on lead poisoning in particular and industrial diseases in general. She lectured at Harvard from 1919, almost 30 years before Harvard accepted women as medical students.

During and after World War I she attended International Congresses of Women and was a pacifist until 1940, when she urged US participation in World War II. During the 1940s and 1950s she spoke out on such subjects as contraception, civil liberties, and workers' rights. In the 1960s she was still considered worthy of attention by the Federal Bureau of Investigation when she protested against US military actions in Vietnam.

Hamilton's *Industrial Poisons in the United States* established her reputation worldwide. She also wrote the classic textbook *Industrial Toxicology* 1934 and an autobiography, *Exploring the Dangerous Trades* 1943.

Hamilton William D(onald) 1936– . British biologist. By developing the concept of inclusive fitness, he was able to solve the theoretical problem of explaining altruism in animal behaviour in terms of neo-Darwinism.

Hamilton William Rowan 1805–1865. Irish mathematician whose formulation of Isaac Newton's dynamics proved adaptable to quantum theory, and whose 'quarternion' theory was a forerunner of the branch of mathematics known as vector analysis. Knighted 1835.

Hamilton was born in Dublin and educated there at Trinity College. In 1827, while still an undergraduate, he was appointed professor of astronomy and royal astronomer of Ireland.

Hamilton showed that the sum of two complex numbers could be represented by a parallelogram and that complex numbers could be used, in general, as a useful tool in plane geometry. From couples Hamilton went on to investigate triples and this led him to his great work on quaternions. He began to lecture on them in 1848, and revealed that for quaternions the ordinary commutative principle of multiplication (that is 3 x 4 = 4 x 3) did not work, for

$$ij = -ji$$

This forced mathematicians to abandon their belief in the commutative principle as an axiom.

It was Hamilton who coined the word 'vector', and he made it possible to deal with lines in all possible positions and directions and freed them from dependence on Cartesian axes of reference.

Hammick Dalziel Llewellyn 1887–1966. English chemist whose major contributions were in the fields of theoretical and synthetic organic chemistry. He devised a rule to predict the order of substitution in benzene derivatives.

Hammick was born in London and studied at Oxford and Munich, Germany. In 1921, he returned to Oxford as a lecturer.

Hammick initially researched inorganic substances, particularly with regard to their solubilities. He studied sulphur and its compounds (such as carbon disulphide), and suggested structures for liquid sulphur and plastic sulphur (which later workers interpreted in terms of linear polymers).

In 1922, he showed that the sublimation of α-trioxymethylene results in the polymer polyoxymethylene; 40 years later this substance was to be used as a commercial polymer.

Hammick spent much time investigating organic reaction rates. He studied, for example, the decarboxylation of quinaldinic acid (the alpha-, ortho- or 2-carboxylic acid of quinoline) by aldehydes and ketones. The reaction, now known as the ***Hammick reaction***, is used in the synthesis of larger molecules.

Hanbury-Brown Robert 1916– . British radio astronomer who participated in the early development of radio-astronomy techniques and later in designing a radio interferometer that permits considerably greater resolution in the results provided by radio telescopes.

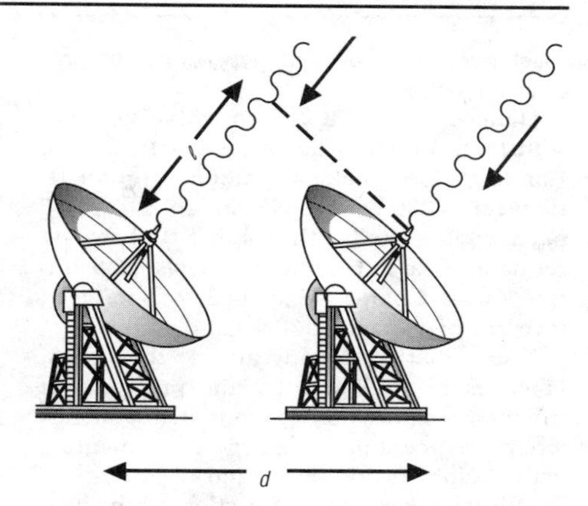

Hanbury-Brown *Hanbury-Brown developed interferometry in radio-astronomy hich uses two (or more) radio telescopes, receiving waves from the same radio source, whose path length differ by* l, *and which are out of phase, resulting in interference when they are combined. Usually one of the telescopes is mounted on rails so that it can be moved to vary the baseline* d.

Hanbury-Brown was born in Aruvankadu, India, and studied engineering at Brighton Polytechnic and the City and Guilds College before joining a radar-research team under the auspices of the Air Ministry in 1936. After World War II he joined the staff at the Jodrell Bank Observatory in Cheshire. In 1960 he was made professor at Victoria University, Manchester, moving in 1962 to the University of Sydney, Australia.

Hanbury-Brown became one of the first astronomers to construct a radio map of the sky. In 1949 he detected radio waves emanating from the Andromeda nebula at a distance of 2.2 million light years. To improve resolution, Hanbury-Brown and his colleagues devised the radio interferometer, and as a result Cygnus A became the first radio source traced to a definite optical identification, even though it had a magnitude (brightness) of only 17.9.

Hanbury-Brown developed the technique of intensity interferometry 1956, and has used the stellar interferometer at Narrabi Observatory in Australia to study the sizes of hotter stars.

Hancock Thomas 1786–1865. English inventor who developed various processes used in the rubber industry, such as the 'masticator', a machine which kneaded raw rubber to produce a solid block.

Hancock was born in Marlborough, Wiltshire. In 1820 he opened a factory in London for making rubber products. Between 1820 and 1847 he took out 17 patents connected with working rubber, and set up a research laboratory. He also collaborated with Charles Macintosh of Glasgow, inventor of a waterproof cloth.

Like Charles Goodyear in the USA, Hancock wanted to solve the problems of rubber's tackiness and inconsistency at different temperatures. After experimenting with sulphur additives and learning of Goodyear's work, Hancock adopted the heat process of vulcanization.

In 1857 he published *Personal Narrative of the Origin and Progress of the Caoutchouc or India Rubber Manufacture in England.*

Handley Page Frederick 1885–1962. English aeronautical engineer who designed the first large bomber 1915. His company produced a series of military aircraft, including the Halifax bomber in World War II. Knighted 1942.

Handley Page was born at Cheltenham, Gloucestershire. By the age of 21 he was chief designer of an electrical company, but in 1908 he set up as an aeronautical engineer and a year later, with a capital of £10,000, he established the first private British company of this kind, in Barking, Essex.

The outbreak of World War I led to his design of the first two-engined bomber, enlarged before the end of the war into a four-engined bomber with a fully laden weight of 13 tonnes.

Handley Page then turned to civil aviation, but found this not to be viable without government subsidies, and these were only to be had when the Handley Page airline merged with Imperial Airways, the forerunner of BOAC (now part of British Airways).

In 1930, Handley Page produced the first 40-seat airliner, the Hercules, a four-engined plane.

During World War II Handley Page produced the Halifax, of which 7,000 were constructed. Work on the bomber continued after the end of the war, resulting in a four-engined jet of unusual design, the Victor, which made its first flight 1952.

Hankel Hermann 1839–1873. German mathematician and mathematical historian who made significant contributions to the study of complex and hypercomplex numbers and the theory of functions. Much of his work was also in developing that of others.

Hankel was born in Halle and studied at Leipzig, Göttingen, and Berlin. He lectured at Leipzig and Tübingen.

Hankel's *Theorie der complexen Zahlensysteme* 1867 dealt with the real, complex, and hypercomplex number systems, and demonstrated that no hypercomplex number system can satisfy all the laws of ordinary arithmetic.

In *Untersuchungen über die unendlich oft oscillerenden und unstetigen Functionen*, he presented a method for constructing functions with singularities at every rational point. He also explicitly stated that functions do not possess general properties; this work was an important advance towards modern integration theory.

The ***Hankel functions*** provided a solution to the Bessel differential equation, which had originally occurred in connection with the theory of planetary motions. Today the equation is relevant in many fields.

Hankel was also the first to suggest a method for assessing the magnitude, or 'measure', of absolutely discontinuous point sets (such as the set of only irrational numbers lying between 0 and 1). The 'measure' theory of point sets has now been extensively applied to probability, cybernetics, and electronics.

Harden Arthur 1865–1940. English biochemist who investigated the mechanism of sugar fermentation and the role of enzymes in this process. For this work he shared the 1929 Nobel Prize for Chemistry. Knighted 1936.

Harden was born in Manchester and studied at Owen's College (now the University of Manchester) and at Erlangen, Germany. He worked at the British Institute of Preventative Medicine (later called the Jenner Institute) 1897–1912, when he became professor at the University of London.

Harden began work on sugar fermentation in 1898, investigating the metabolism of yeasts from 1900, three years after the discovery of enzymes. He showed that fermentation was caused by the action of the enzyme zymase on glycogen. Harden and his co-workers went on to show that zymase consists of at least two different substances: one heat-sensitive (probably a protein, the enzyme) and one heat-stable (now known as a coenzyme).

Harden then discovered the hexose sugar compound hexosediphosphate in the normal reaction mixture (he later found hexosemonophosphate as well), thus proving that phosphorylation is an intermediate step in the fermentation process. This finding stimulated great interest in intermediate metabolism.

Hardy Alister Clavering 1896–1985. English marine biologist who developed methods for ascertaining the numbers and types of minute sea organisms.

Hardy was born in Nottingham and educated at Oxford. In 1924 he joined an expedition to the Antarctic, and on his return in 1928 he was appointed professor at Hull University, where he founded the Department of Oceanography. Clavering was professor of zoology and comparative anatomy at Oxford 1946–65 and master of the Unitarian College, Manchester College, Oxford, 1958–65.

Hardy made his special study of plankton on the 1924 *Discovery* expedition. The aim of quantitative plankton studies is to estimate the numbers or weights of organisms beneath a unit area of sea surface or in a unit volume of water. He developed the Hardy plankton continuous recorder – a net that can be used behind faster-moving vessels, at a depth of 10 m/33 ft, giving a larger area in which accurate recordings can be made. Surveys using this device now annually cover many thousands of kilometres in the Atlantic, North Sea, and Icelandic waters, for the benefit of fisheries.

He also attempted to apply scientific methods to religious phenomena, conducting a number of research projects into religious experience. He concluded that such experiences were common and classified them into types. Although he held back from claiming that his results proved that religious experience was genuine and therefore that God existed, his work has been cited as evidence for the existence of God by some and of the inappropriateness of scientific methodology in religion by others.

His work is continued by the Alister Hardy Research Centre in Oxford, which is committed to scientific study of the nature and function of religious experience in the human species.

Hardy Godfrey Harold 1877–1947. English mathematician whose research was at a very advanced level in the fields of pure mathematics known as analysis and number theory. His *Course in Pure Mathematics* 1908 revolutionized the teaching of mathematics at senior school and university levels.

Hardy was born in Cranleigh, Surrey, and studied at Cambridge. He became professor at Oxford 1919, but returned in 1931 to Cambridge.

The mathematician's patterns, like the painter's or the poets, must be beautiful; the ideas, like the colours or the words, must fit together in a harmonious way. Beauty is the first test: there is no permanent place in the world for ugly mathematics.

G H Hardy

A Mathematician's Apology 1940

Hardy's researches included such topics as the evaluation of difficult integrals and the treatment of awkward series of algebraic terms. Among his successes in number theory was a new proof of the prime number theorem. Other problems on which he worked were the ways in which numbers could be partitioned into simpler numbers, 'decomposed' into squares and cubes, and so on. Hardy also worked on the question of whether every even number is the sum of two prime numbers. (This is still an unsolved problem.)

Hardy was the sole or joint author of several books, including *An Introduction to the Theory of Numbers*, *Inequalities*, and *Divergent Series*.

Hargreaves James *c.* 1720–1778. English inventor who co-invented a carding machine for combing wool 1760. About 1764 he invented his 'spinning jenny' (patented 1770), which enabled a number of threads to be spun simultaneously by one person.

Hargreaves was born near Blackburn, and was initially a weaver, making the first spinning jenny for his family's use. When he began to sell the machines, spinners with the old-fashioned wheel became alarmed by the possibility of cheaper competition and in 1768 a mob from Blackburn gutted Hargreaves's house and destroyed his equipment. Hargreaves moved to Nottingham, where he formed a partnership and built a small cotton mill in which the jenny was used.

The spinning jenny multiplied eightfold the output of the spinner and could be worked easily by children. It did not entirely supersede the spinning wheel in cotton manufacturing (and was itself overtaken by

Samuel Crompton's mule). But for woollen textiles the jenny could be used to make both the warp and the weft.

Harris Henry 1925– . Australian-born British geneticist known for his somatic cell fusion experiments and his theory that genes are activated by molecules in the cytoplasm (the part of the cell that is outside the nucleus).

Harris examined the nature of the factors controlling cell differentiation (the process by which cells become increasingly specialized for different functions) and cell activity by fusing somatic (body) cells to form hybrid cells in which both the cytoplasm and the nuclei were fused. He then showed that when two very different cells were fused, the hybrid cell displayed the characteristics of both cells, indicating that their respective genes remained active so long as activating factors from their respective cytoplasm were present.

He suggested that genes could be controlled either at the level of transcription (DNA copying into RNA) in the nucleus or at the level of translation (protein production) in the cytoplasm after the messenger RNA (mRNA) has been exported from the nucleus.

In one particular experiment he elegantly showed that once mRNA is formed and exported to the cytoplasm, the control mechanism for translation is then in the cytoplasm. If transcription is blocked by the antibiotic actinomycin D, mammalian cells continue to synthesize specific proteins for long periods in culture. Gene expression is now known to be controlled both at the level of transcription and translation.

Harris graduated from the University of Sydney before studying for his PhD at Oxford University 1954. He worked at the Sir William Dunn School of Pathology at Oxford until 1959 and then became professor of cell biology at the John Innes Institute 1960. He moved back to Oxford 1963 to take up the chair in pathology and was elected a Fellow of the Royal Society 1968 and Regius professor of medicine at Oxford 1979.

Harrison John 1693–1776. English horologist and instrumentmaker who made the first chronometers that were accurate enough to allow the precise determination of longitude at sea, and so permit reliable (and safe) navigation over long distances.

Harrison was born in Yorkshire, and learned his father's trades of carpentry and mechanics. In 1726, he made a compensated clock pendulum, which remained the same length at any temperature, making use of the different coefficients of expansion of two different metals.

In 1714, the British government's Board of Longitude announced a prize of up to £20,000 for anyone who could make an instrument to determine longitude at sea to an accuracy of 30 minutes (half a degree). Between 1735 and 1760, Harrison submitted four instruments for the award. When his fourth marine chronometer was tested at sea, it kept accurate time to within 5 seconds over the duration of two voyages to the West Indies, equivalent to just over one minute of longitude. Harrison was eventually awarded the prize money.

A unique feature that contributed to the chronometer's accuracy was a device that enabled it to be rewound without temporarily stopping the mechanism. This was subsequently incorporated into other chronometers.

Hartline Haldan Keffer 1903–1983. US physiologist who shared the Nobel Prize for Physiology or Medicine 1967 with Ragnar Granit and George Wald for his work on the analysis of vision. Hartline measured electrical activity in the eye and determined that a decrease in the neurotransmitter (chemical messenger) signals to the brain that the eye has been stimulated by light.

Hartline's first work was to measure the electrical activity generated by light in the retina (the light-sensitive inner area of the back of the eye that is connected to the brain by the optic nerve) of the vertebrate eye. He went on to investigate the more simply organized eye of the horseshoe crab *Limulus*. This eye, like the insect eye, consists of separate ommatidia ('micro-eyes') each of which has its own refracting index, photosensitive cells (receptor cells), and sensory nerve cell. Study of a single ommatidium, with its single group of fibres connecting to the central nervous system, provided Hartline with an opportunity to study electrical events following light stimulation.

Hartline's research resulted in the 'generator potential' theory in which slow, non-propagated voltage variation develops in the receptor cell. When light hits the receptor cell it decreases the permeability of the cell membrane to sodium ions. This results in the receptor cell leaking less neurotransmitter in the light than in the dark. Thus, it is actually a decrease in a receptor cell's chemical signal to neurons that serves as the messenger that photoreceptors have been stimulated by light.

Hartnett Laurence John 1898–1986. Australian engineer and company director, born in England. He was the creator of the Holden car, the first mass-produced all-Australian car. Knighted 1967.

Harvey Ethel Browne 1885–1965. US embryologist and cell biologist who discovered the mechanisms of cell division, using sea urchin eggs as her experimental model.

Ethel Browne Harvey was born in Baltimore and studied at Columbia University, New York. After her marriage she worked part-time, and visited several marine laboratories; from 1931 she was an independent research worker attached to the biology department of Princeton. She was a frequent visitor to the Stazione Zoologica in Naples.

Harvey's work concentrated on the role in cell fertilization and development of non-nuclear cell components in the cytoplasm. She undertook morphological studies and physiological experiments to examine the factors that affect the process of cell division and was able to stimulate division in fragments of sea urchin eggs that contained no nucleus. This was an important contribution to unravelling the connections between different cellular structures in controlling cell division and development.

Harvey William 1578–1657. English physician who discovered the circulation of blood. In 1628 he published his book *De motu cordis/On the Motion of the Heart and the Blood in Animals*. He also explored the development of chick and deer embryos.

Harvey's discovery marked the beginning of the end of medicine as taught by Greek physician Galen, which had been accepted for 1,400 years.

Harvey was born in Folkestone, Kent, and studied at Cambridge and at Padua, Italy, under Geronimo Fabricius. He worked at St Bartholomew's Hospital, London, and served as a professor there 1615–43. From 1618, he was court physician to James I and later to Charles I.

Both Hen and Housewife are so matched,
That her Son born is only her Son hatched,
And when her teeming hopes have prosperous been, Yet to conceive is but to lay within.

William Harvey

The Generation of Animals 1651

Harvey *English physician William Harvey, who discovered the circulation of blood after viewing it experimentally in animals. Harvey also showed that the heart acted as a pump and that blood flowed through the body via the lungs along arteries, and back to the heart along veins. His observations met with initial hostility but were generally accepted by the time of his death in 1657.*

Examining the heart and blood vessels of mammals, Harvey deduced that the blood in the veins must flow only towards the heart. He also calculated the amount of blood that left the heart at each beat, and realized that the same blood must be circulating continuously around the body. He reasoned that it passes from the right side of the heart to the left through the lungs (pulmonary circulation).

Harvey also published *Exercitationes de generatione animalium/Anatomical Exercitations concerning the Generation of Living Creatures* 1651.

Harvey William Henry 1811–1866. Irish botanist who became a leading authority on algae. He discovered several new forms of Irish flora, especially mosses, despite not having been educated in botany at university. He wrote *The Genera of South African Plants* while in South Africa and collected many botanical specimens, including algae and flowering plants. His description of algae was included in James Townsend Mackay's *Flora Hibernica* 1836.

Harvey was born in Limerick, Ireland. In 1836, he became the colonial treasurer in

Cape Town, South Africa but was forced to retire 1842 due to ill health. While in South Africa he became known as a leading specialist in algae. In 1844, he was made the keeper of the herbarium at Trinity College, Dublin. In 1846, he published *Phycologia Britannica*, an extraordinary piece of work entirely illustrated by Harvey himself. He went to America 1849, where he lectured to large and enthusiastic audiences in Boston and Washington and obtained several new specimens while he was in Florida. He also travelled to the Southern Hemisphere, writing *Phycologia Australica* 1858. He was made chair of botany at Trinity College, Dublin 1856, where he died of tuberculosis.

Hassel Odd 1897–1981. Norwegian physical chemist who established the technique of ***conformational analysis*** – the determination of the properties of a molecule by rotating it around a single bond – and received the Nobel Prize for Chemistry in 1969. Hassel described the conformations of cyclohexane (a saturated hydrocarbon in which the carbon atoms are linked in a ring).

Conformations are different spatial rearrangements of the same molecule that are not achieved by the breaking of bonds, but by the rotation of atoms around the bonds. The cyclohexane molecule exists in two main conformations: the 'boat' and 'chair', of which the chair is the most stable.

His work was brutally disrupted by World War II, when he was imprisoned by occupying Nazi troops. However, his findings came to light in the 50s and 60s and were further developed by Derek Barton with whom he shared the Nobel prize.

Hassel was born in Oslo, Norway where he went to university. He moved to Germany and studied under the guidance of Kasimir Fajans in Munich and then obtained a PhD from Berlin University in 1924. He was a deep-thinking man who felt uncomfortable with the unstable political situation in Berlin and he quickly returned to his homeland, Oslo, where he joined the staff at the university's school of physical chemistry. He spent the rest of his career there, becoming a professor and a well-respected director in 1934.

Hatch Marshall Davidson 1932– . Australian biochemist who described the mechanism by which tropical plants trap carbon dioxide in bundle sheath cells, the major sites of photosynthesis. He also published on many other areas of plant metabolism, including how plants break down starch and sugars to form ethanol and carbon dioxide.

During photosynthesis in plants, light energy is used to convert water and carbon dioxide to oxygen and energy-rich organic compounds (such as starch) in a series of enzymatic steps called the Calvin photosynthetic pathway, or C3 pathway. The rate limiting step of the Calvin pathway is a reaction catalysed by the enzyme rubisco. Since the activity of rubisco increases rapidly as the temperature rises, tropical plants might be expected to have a problem with excessively high rates of photorespiration.

In 1968, Hatch reported an alternative to the Calvin pathway, known as the C4 or Hatch–Slack–Kortschak pathway of enzymatic steps, for incorporating carbon dioxide into carbohydrates. The C4 pathway had evolved in tropical grasses and other plants experiencing climatic extremes to circumvent excessive photorespiration.

Hatch was born in Perth, Australia and studied at Sydney University and later in California. He worked for the the Colonial Sugar Refining Co in Sydney 1961–70, before being lured back to botanical research by the Commonwealth Scientific Industrial Research Organisation (CSIRO) in Canberra, where he has been since 1970.

Hauptman Herbert A 1917– . US mathematician who shared the 1985 Nobel Prize for Chemistry with Jerome Karle for discovering a general method of determining crystal structures by X-ray diffraction. Previously, crystallographers had to make assumptions or guess to determine a crystal structure using X-ray diffraction. The work of Hauptman and Karle made it possible to calculate crystal structure directly from measurements of the intensities of the diffracted X-rays.

Hauptman was born in New York City and educated in mathematics at the City College of New York and Columbia University. In 1947, he began the collaboration with Karle at the Naval Research Laboratory in Washington DC, developing mathematical methods in X-ray crystallography. At the same time Hauptman enrolled in the doctoral programme at the University of Maryland. By 1954, when he was awarded his doctorate, the two men had laid the foundations of the direct method in X-ray crystallography. In 1970, he joined the Medical Foundation of Buffalo, New York, becoming research director 1972.

Hausdorff Felix 1868–1942. German mathematician and philosopher who developed the branch of mathematics known as topology, in which he formulated the theory of point sets. He investigated general closure spaces, and formulated Hausdorff's maximal principle in general set theory.

Hausdorff was born in Breslau, Germany (now Wroclaw, Poland), and studied at Berlin, Freiburg, and Leipzig. In 1902 he was appointed professor at Leipzig, later moving to Bonn. During World War II he committed suicide rather than be sent to a concentration camp.

Hausdorff formulated a theory of topological and metric spaces, proposing that such spaces be regarded as sets of points and sets of relations among the points, and introduced the principle of duality. This principle states that an equation between sets remains valid if the sets are replaced by their complements and the symbol for union is exchanged for that of intersection.

He also created what are now called Hausdorff's neighbourhood axioms; from these four axioms he derived Hausdorff's topological spaces. A topological space is understood to be a set E of elements x and certain subsets S_x of E which are known as neighbourhoods of x.

Hausdorff also contributed extensively to several other fields of mathematics.

His major work is *Grundzuge der Mengenlehre/Basic Features of Set Theory* 1914.

Haüy René-Just 1743–1822. French mineralogist, the founder of modern crystallography. He regarded crystals as geometrically structured assemblages of units (integrant molecules), and developed a classification system on this basis.

Haüy was born in Oise *département*. He trained as a priest, then became professor of mineralogy in Paris 1802. During the Revolutionary era, he was protected from anticlerical attacks by Napoleon's patronage.

Haüy demonstrated in 1784 that the faces of a calcite crystal could be regarded in terms of the standard stacking of cleavage rhombs, provided that the rhombs were assumed to be so small that the face looked smooth. Similar principles would lead to other crystal shapes being built up from suitable simple structural units. He proposed six primary forms: parallelepiped, rhombic dodecahedron, hexagonal dipyramid, right hexagonal prism, octahedron, and tetrahedron.

His two major works are *Traité de minéralogie/Treatise of Mineralogy* 1801 and *Treatise of Crystallography* 1822.

Hawking Stephen William 1942– . English physicist whose work in general relativity – particularly gravitational field theory – led to a search for a quantum theory of gravity to explain black holes and the Big Bang, singularities that classical relativity theory does not adequately explain.

Hawking's objective of producing an overall synthesis of quantum mechanics and relativity theory began around the time of the publication of his seminal book *The Large Scale Stucture of Space-Time*, written with G F R Ellis, 1973. His most remarkable result, published in 1974, was that black holes could in fact emit particles in the form of thermal radiation – the so-called ***Hawking radiation***.

How can the complexity of the universe and all its trivial details be determined by a simple set of equations? Alternatively, can one really believe that God chose all the trivial details, like who should be on the cover of Cosmospolitan?

Stephen Hawking

Black Holes and Baby Universes

Hawking was born in Oxford, studied at Oxford and Cambridge, and became Lucasian Professor of Mathematics at Cambridge 1979.

Hawking's most fruitful work was with black holes, stars that have undergone total gravitational collapse and whose gravity is so great that nothing, not even light, can escape from them. Since 1974, he has studied the behaviour of matter in the immediate vicinity of a black hole, concluding that black holes do, contrary to expectation, emit radiation. He has proposed a physical explanation for this 'Hawking radiation' which relies on the quantum-mechanical concept of 'virtual particles' – these exist as particle–antiparticle pairs and are supposed to fill 'empty' space. Hawking suggested that, when such a particle is created near a black hole, one half of the pair might disappear into the black hole, leaving the other half, which could escape to infinity. This would be seen by a distant observer as thermal radiation.

Confined to a wheelchair because of a rare and progressive neuromotor disease, Hawking remains mentally active. His book *A Brief*

History of Time 1988 gives a popular account of cosmology and became an international bestseller. His latest book, written with Roger Penrose entitled *The Nature of Space and Time* was published in 1996.

Haworth Adrian Harvey 1766–1833. English botanist and entomologist whose *Lepidoptera Britannica* 1803 was the first complete description of several hundred British moths and butterflies. He was also an expert on succulent plants, with much of his work in this area being published in *Synopsis plantarum succulentarum* 1812. He founded the Aurelian Society 1802 and the Entomological Society of London.

Haworth was born in Hull. He was initially employed in a law office until, upon completing his articles, he became financially able to support his botanical studies and forays into the world of insects. In 1798, he moved to Chelsea in London and became a member of the Linnaean Society. He returned to his native Hull 1812 to found the Hull Botanical Gardens. Following his death from cholera, his library, herbarium, and specimen collection were turned into a museum in his former home.

Haworth (Walter) Norman 1883–1950. English organic chemist who was the first to synthesize a vitamin (ascorbic acid, vitamin C) 1933, for which he shared a Nobel prize 1937. He made significant advances in determining the structures of many carbohydrates, particularly sugars. Knighted 1947.

Haworth was born in Chorley, Lancashire, and studied at Manchester and Göttingen, Germany. He was professor at Birmingham 1925–48.

The linkages in the ring structure of carbon atoms in hexoses are now known as Haworth formulas because of his work. His team investigated polysaccharides, establishing the chain structures of cellulose, starch, and glycogen. Work on sugars led naturally to vitamin C.

In the late 1930s, Haworth studied the reactions of polysaccharides with enzymes and certain aspects of the chemistry of the hormone insulin. During World War II, his laboratory became a primary producer of purified uranium, which led to work on the preparation and properties of organic fluorine compounds.

Haworth's book *The Constitution of the Sugars* 1929 became a standard work.

Hayashi Chushiro 1920– . Japanese physicist whose research in 1950 exposed a fallacy in the 'hot Big Bang' theory proposed two years earlier by Ralph Alpher and others. Hayashi has published many papers on the origin of the chemical elements in stellar evolution and on the composition of primordial matter in an expanding universe.

Hayashi was born and educated in Kyoto, where he became professor of physics 1957.

Hayashi pointed out that in the Big Bang earlier than the first two seconds, the temperature would have been greater than 10^{10} K, which is above the threshold for the making of electron–positron pairs. This radically altered the timescale proposed in the 'hot Big Bang' theory. He also showed that the abundance of neutrons at the heart of the Big Bang did not depend on the material density but on the temperature and the properties of the weak interreactions. Provided the density is great enough for the reaction between neutrons and protons to combine at a rate faster than the expansion rate, a fixed concentration of neutrons will be incorporated into helium nuclei, however great the material density is – producing a plateau in the relationship between helium abundance and material density.

Hayem Georges 1841–1920. French physician who pioneered the study of diseases of the blood and was one of the founders of haematology. He was the first to describe in detail the different cell types present in blood. Realizing the importance of blood platelets and their role in blood clotting, Hayem was the first to perform an accurate blood-platelet count 1878, and he later published a textbook on the subject.

Hayem was born and studied medicine in Paris. After obtaining his MD 1868, he became professor of therapy and materia medica at the hospital in Tenon.

Heath Thomas Little 1861–1940. English mathematical historian who specialized in ancient Greece. His *History of Greek Mathematics* 1921 is regarded as the standard work on the subject in the English language. KCB 1909.

Heath was born in Lincolnshire, studied at Cambridge and joined the civil service. He rose through the ranks in the Treasury Office; in 1913 he was appointed joint permanent secretary to the Treasury and auditor of the Civil List, and he was comptroller general and secretary to the Commissioners for the Reduction of the National Debt 1919–26.

Between 1885 and 1912, Heath edited the works of Greek mathematicians Diophantus,

Apollonius of Perga, and Archimedes. His book *Aristarchus of Samos, the Ancient Copernicus* 1913 is a comprehensive account of Aristarchus and his work on astronomy. *Euclid's Elements* 1908 is a monumental work in three volumes. And in the great *History of Greek Mathematics*, Heath dealt with his subjects mainly according to topics and not in chronological order as others had done before him.

Heathcoat John 1783–1861. English inventor of lacemaking machinery 1807. Throughout his life, he took out patents for further inventions in textile manufacture. In 1832 he patented a steam plough to assist with agricultural improvements in Ireland.

Heathcoat was born near Derby, apprenticed in the hosiery trade, and became a master mechanic about 1803. His lace factory in Loughborough was destroyed by Luddites 1816, and Heathcoat moved to Tiverton, Devon. In 1832 he was elected to represent Tiverton in the new reformed Parliament and remained MP for the borough until 1859.

Heathcoat's lacemaking machine was acknowledged by engineer Marc Isambard Brunel, who said that Heathcoat (then 24 years old) had devised 'the most complicated machine ever invented'. The first square yard (0.83 sq m) of plain net from the machine was sold for £5; by the end of the 19th century its price had fallen to 1% of that.

Heaviside Oliver 1850–1925. English physicist. In 1902 he predicted the existence of an ionized layer of air in the upper atmosphere, which was known as the Kennelly–Heaviside layer but is now called the E layer of the ionosphere. Deflection from it makes possible the transmission of radio signals around the world, which would otherwise be lost in outer space.

His theoretical work had implications for radio transmission. His studies of electricity published in *Electrical Papers* 1892 had considerable impact on long-distance telephony, and he added the concepts of inductance, capacitance, and impedance to electrical science.

Heaviside was born in London. Because of severe hearing difficulties, he was mainly self-taught and was unemployed most of his life.

When Heaviside became involved with the passage of electricity along conductors, he modified Ohm's law to include inductance and this, together with other electrical properties, resulted in his derivation of the equation of telegraphy. On considering the problem of signal distortion in a telegraph cable, he came to the conclusion that this could be substantially reduced by the addition of small inductance coils throughout its length, and this method has since been used to great effect.

In *Electromagnetic Theory* 1893–1912, Heaviside extended Scottish physicist James Clerk Maxwell's discoveries as well as making many valuable discoveries of his own. In the third volume, he considered wireless telegraphy.

Hedwig Johannes 1730–1799. Transylvanian-born German botanist whose *Fundamentum historiae naturalis muscorum frondosorum* 1782 led to his establishment as a leading expert on the grouping and early classification of mosses. He was especially interested in the relationship between, and the reproduction of, mosses and liverworts.

Hedwig was born in Kronstadt, Transylvania and attended the University of Leipzig 1752, where he graduated in medicine 1759. He set up a practice in Chemnitz, Saxony. In his spare time he studied botany, often rising very early in the morning in order to do so.

In 1784, he was made director of the military hospital in Leipzig, where he had been living since 1781. He was then made professor of botany and director of the botanical gardens 1789, although he continued to work as a doctor. He died of a fever he caught while visiting patients in the winter.

Heine Heinrich Eduard 1812–1881. German mathematician who completed the formulation of the notion of uniform continuity. He subsequently provided a proof of the classic theorem on uniform continuity of continuous functions, which has since become known as ***Heine's theorem***.

Heine was born in Berlin and studied there and at Göttingen. In 1848, he was appointed professor at Halle University and remained there for the rest of his life.

Heine's theorem is sometimes known as the Heine–Borel theorem, because French mathematician Emile Borel formulated the covering property of uniform conformity, and proved it. Heine formulated the notion of uniform continuity and went on to prove the classic theorem of uniform continuity of continuous functions.

In all, Heine published more than 50 papers on mathematics. His speciality was spherical functions, Lamé functions, and Bessel functions. His *Handbuch der Kugelfunctionen* 1861 became the standard work on spherical functions for the next 50 years.

Heinkel Ernst Heinrich 1888–1958. German aircraft designer who pioneered jet aircraft. He founded his firm 1922 and built the first jet aircraft 1939. During World War II his company was Germany's biggest producer of warplanes, mostly propeller-driven.

Notable Heinkel aircraft of World War II include the He 111, originally designed as a civilian airliner but adapted almost immediately as a medium bomber. It was used extensively in air raids over England 1940, but was found to be vulnerable to attack and so was redeployed as a night bomber and mine-layer.

Heisenberg Werner Karl 1901–1976. German physicist who developed quantum theory and formulated the uncertainty principle, which concerns matter, radiation, and their reactions, and places absolute limits on the achievable accuracy of measurement. He was awarded a Nobel prize 1932 for work he carried out when only 24.

The ***uncertainty principle*** states that it is impossible to specify precisely both the position and the simultaneous momentum (mass multiplied by velocity) of a particle. There is always a degree of uncertainty in either, and as one is determined with greater precision, the other can only be found less exactly. The result of an action can be expressed only in terms of the probability that a certain effect will occur.

An expert is someone who knows some of the worst mistakes that can be made in his subject and how to avoid them.

Werner Heisenberg

The Part and the Whole

During World War II Heisenberg worked for the Nazis on nuclear fission, but his team were many months behind the Allied atom-bomb project. After the war he worked on superconductivity.

Heisenberg was born in Würzburg and studied at Munich. In 1923 he became assistant to Max Born at Göttingen, then worked with Danish physicist Niels Bohr in Copenhagen 1924–26. Heisenberg became professor at Leipzig 1927 at the age of only 26, and stayed until 1941. He was director of the Max Planck Institute for Physics 1942–70.

Heisenberg was concerned not to try to picture what happens inside the atom but to find a mathematical system that explained it. His starting point was the spectral lines given by hydrogen, the simplest atom. Born helped Heisenberg to develop his ideas, which he presented in 1925 as a system called matrix mechanics. This was the first precise mathematical description of the workings of the atom and with it Heisenberg is regarded as founding quantum mechanics, although matrix mechanics was soon to be replaced by wave mechanics.

Science clears the fields on which technology can build.

Werner Heisenberg

Attributed remark

In 1927 Heisenberg used the Pauli exclusion principle, which states that no two electrons can have all four quantum numbers the same, to show that ferromagnetism is caused by electrostatic intereaction between the electrons.

Helmholtz Hermann Ludwig Ferdinand von 1821–1894. German physiologist, physicist,

Helmholtz *Hermann von Helmholtz spent six years as an army surgeon before being released from military service 1848 to embark on a scientific career. Helmholtz was a remarkable scientist, publishing pioneering work in both physics and physiology. He is best known for the principle of the conservation of energy and his discoveries in the physiology of vision.*

and inventor of the ophthalmoscope for examining the inside of the eye. He was the first to explain how the cochlea of the inner ear works, and the first to measure the speed of nerve impulses. In physics he formulated the law of conservation of energy, and worked in thermodynamics.

Helmholtz's scientific work in many fields was intended to prove that living things possess no innate vital force, and that their life processes are driven by the same forces and obey the same principles as nonliving systems. He arrived at the principle of conservation of energy 1847, observing that the energy of life processes is derived entirely from oxidation of food, and that animal heat and muscle action are generated by chemical changes in the muscles.

The formation of scales and the web of harmony is a product of artistic invention, and is in no way given by the natural structure or by the natural behaviour of our hearing.

Hermann von Helmholtz

Theory of Sound, 1862

Helmholtz was born in Potsdam and studied at the Friedrich Wilhelm Institute in Berlin. He first became professor at Bonn 1855 and ended his career as director of the Physico-Technical Institute of Berlin from 1887.

Helmholtz invented the ophthalmoscope, which is used to examine the retina, 1851, and the ophthalmometer, which measures the curvature of the eye, 1855. He also revived the three-colour theory of vision first proposed 1801 by English physicist Thomas Young, by showing that a single primary colour (red, green, or violet) must also affect retinal structures sensitive to the other primary colours. This explained the colour of afterimages and the effects of colour blindness.

In acoustics, Helmholtz produced a comprehensive explanation of how the upper partials in sounds combine to give them a particular tone or timbre, and how resonance may cause this to happen.

Helmont Jean Baptiste van 1579–1644. Belgian doctor who was the first to realize that there are gases other than air, and claimed to have coined the word 'gas' (from Greek *cháos*).

Helmont identified four gases: carbon dioxide, carbon monoxide, nitrous oxide, and methane. He was the first to take the melting point of ice and the boiling point of water as standards for temperature and the first to use the term 'saturation' to signify the combination of an acid and a base. In medicine, Helmont used remedies that specifically considered the type of disease, the organ affected and the causative agent. He demonstrated acid as the digestive agent in the stomach.

Helmont was born in Brussels, studied at Louvain, and 1600–05 travelled to Spain, Italy, France, and England. From 1621 to 1642 he was persecuted by the Roman Catholic Church for his views on the cure of wounds by applying ointment to the weapon rather than the wound, which was then common. Although he did not reject the belief, he insisted that it was a natural phenomenon with no supernatural element, as set out in his treatise *De magnetica vulnerum... curatione* 1621.

Taking care to weigh the materials he used in chemistry, Helmont gained insight into the indestructibility of matter and the fact that metals dissolved in acid were recoverable, not destroyed or transmuted. He believed that all matter was composed of water and air, which he demonstrated by growing a willow tree in a measured quantity of earth, adding only water.

His works were collectively published posthumously as *Ortus medicinae* 1648.

Hench Philip Showalter 1896–1965. US physician who introduced cortisone treatment for rheumatoid arthritis, for which he shared the 1950 Nobel Prize for Physiology or Medicine with Edward Kendall and Tadeus Reichstein.

Hench noticed that arthritic patients improved greatly during pregnancy or an attack of jaundice and concluded that a hormone secreted in increased quantity during both these conditions caused the improvement. This turned out to be cortisol, a steroid converted to cortisone in the liver.

Henry Joseph 1797–1878. US physicist, inventor of the electromagnetic motor 1829 and of a telegraphic apparatus. He also discovered the principle of electromagnetic induction, roughly at the same time as Michael Faraday, and the phenomenon of self-induction. The unit of inductance, the ***henry***, is named after him.

Born in Albany, New York, Henry studied at Albany Academy, where he became professor 1826, moving 1832 to New Jersey College (later Princeton). He was the Smithsonian Institution's first director, from 1846.

By 1830, he had made powerful electromagnets by using many turns of fine insulated wire wound on iron cores. In that year he anticipated Faraday's discovery of electromagnetic induction (although Faraday published first). In 1835 he developed the relay (later to be much used in electric telegraphy).

In astronomy, Henry studied sunspots and solar radiation, and his meteorological studies at the Smithsonian led to the founding of the US Weather Bureau.

Henry William 1774–1836. English chemist and physician. In 1803 he formulated ***Henry's law***, which states that when a gas is dissolved in a liquid at a given temperature, the mass that dissolves is in direct proportion to the pressure of the gas.

Henry was born in Manchester and graduated from Edinburgh. In poor health after a childhood accident, he worked mainly for his father, an industrial chemist.

Henry worked for about 20 years on the analysis of inflammable mixtures of gases and attempted to find correlations between chemical composition and illuminative properties. He established that firedamp – the cause of many mining disasters – is methane, and confirmed the composition of methane and ethane. Like English chemist John Dalton, Henry showed that hydrogen and carbon combine in definite proportions to form a limited number of compounds.

In medicine, Henry studied contagious diseases. He believed that these were spread by chemicals which could be rendered harmless by heating; he used heat to disinfect clothing during an outbreak of cholera in 1831.

A series of chemistry lectures which he gave 1798–99 were later published as *Elements of Experimental Chemistry* and became a highly successful textbook.

Heraklides of Pontus 388–315 BC. Greek philosopher and astronomer who may have been the first to realize that the Earth turns on its axis, from west to east, once every 24 hours. He also thought that the observed motions of Mercury and Venus suggested that they orbited the Sun rather than the Earth.

Born in Heraklea, near the Black Sea, Heraklides migrated to Athens and studied at the Academy of Plato. He is said also to have attended the schools of the Pythagorean philosophers, and would thus have come into contact with Aristotle. All his writings are lost, so his astronomical theories are known only at second hand.

In proposing the doctrine of a rotating Earth (not to be accepted for another 1,800 years), Heraklides contradicted the accepted model of the universe put forward by Aristotle. Heraklides thought that the immense spheres in which the stars and planets were assumed to be fixed could not rotate so fast. He did not completely adopt the heliocentric view of the universe stated later by Aristarchus, but proposed instead that the Sun moved in a circular orbit (in its sphere) and that Mercury and Venus moved on epicycles around the Sun as centre.

Herapath John 1790–1868. English mathematician. His work on the behaviour of gases, though seriously flawed, was acknowledged by the physicist James Joule in his own more successful investigations.

Herbig George Howard 1920– . US astronomer who specialized in spectroscopic research into irregular variable stars, notably those of the T Tauri group. He also worked on binary stars.

Herbig was born in Wheeling, West Virginia, and educated at the University of California. In 1944, he became a member of the staff at the Lick Observatory, California, rising to professor 1966.

Herbig's main area of research was the nebular variables of which the prototype is T Tauri. It is believed that the members of this group are in an early stage of stellar evolution. Most of them are red and fluctuate in light intensity. In 1960, Herbig drew attention to the fact that many of them have a predominance of lithium lines, similar to the abundance of this element on Earth and in meteorites, and concluded that this might represent the original level of lithium in the Milky Way. In the Sun and other stars, lithium may have largely been lost through nuclear transformation.

Herbig has also investigated the spectra of atoms and molecules that originate in interstellar space.

Herbrand Jacques 1908–1931. French mathematical prodigy who originated some innovatory concepts in the field of mathematical logic. He formulated the Herbrand theorem, which established a link between quantification theory and sentential logic.

Herbrand was born in Paris, where he was

educated at the Ecole Normale Supérieure. At the age of 20 he published his first paper on mathematical logic for the Paris Academy of Sciences, and in the following year – 1929 – for his doctorate, he produced the paper containing the Herbrand theorem. This has since found many applications in such fields as decision and reduction problems.

Herbrand was also fascinated by modern algebra and wrote a number of papers on class-field theory.

Hermite Charles 1822–1901. French mathematician who was a principal contributor to the development of the theory of algebraic forms, the arithmetical theory of quadratic forms, and the theories of elliptic and Abelian functions. Much of his work was highly innovative, especially his solution of the quintic equation through elliptic modular functions, and his proof of the transcendence of e.

Hermite was born in Dieuze, Lorraine, and studied at the Lycée Louis le Grand and the Ecole Polytechnique, Paris. He became professor at the Ecole Normale 1869, moving to the Sorbonne 1870.

Between 1847 and 1851 he worked on the arithmetical theory of quadratic forms and the use of continuous variables. Then for he worked on the theory of invariants 1854–64.

In 1873 he worked out Hermitian forms (a complex generalization of quadratic forms) and Hermitian polynomials. In the same year, he showed that e, the base of natural logarithms, is transcendental. (Transcendental numbers are real or complex numbers that are not algebraic.)

In 1872 and 1877 Hermite solved the Lamé differential equation, and in 1878 he solved the fifth-degree (quintic) equation of elliptic functions.

Hero of Alexandria lived AD 62. Greek mathematician and engineer, the greatest experimentalist of antiquity. Among his many inventions was an automatic fountain and a kind of stationary steam engine. His books have survived mainly in Arabic.

Hero was also a teacher and in Alexandria founded a technical school with one section devoted entirely to research.

He regarded air as a substance that could be compressed and expanded, and explained the phenomenon of suction and associated apparatus, such as the pipette. His assumption that air is composed of minute particles was 1,500 years ahead of its time.

In mechanics, he devised a system of gear wheels which could lift a mass of 1,000 kg/2,200 lb by means of a mere 5 kg/11 lb. His work *Mechanics* contains the parallelogram of velocity and the laws of levers, and his construction of a variable ratio via a friction disc has been used to build a motor vehicle with a semi-automatic transmission.

Hero's book *Metrica* explains the measurement of geometrical figures. *Pneumatica* describes numerous mechanical devices operated by gas, water, steam, or atmospheric pressure, and siphons, pumps, and working automata in the likeness of animals or birds.

Herophilus of Chalcedon *c.* 330–*c.* 260 BC. Greek physician, active in Alexandria. His handbooks on anatomy make pioneering use of dissection, which, according to several ancient sources, he carried out on live criminals condemned to death.

Héroult Paul Louis Toussaint 1863–1914. French metallurgist who developed the electrolytic manufacturing process for aluminium, simultaneously with US chemist Charles Hall. Héroult also invented the electric arc furnace for the production of steels.

Héroult was born in Thury-Harcourt, Normandy, and spent a year at the Paris School of Mines. Like Hall, Héroult succeeded in producing aluminium using his electrolytic apparatus at the age of 23. He patented the system in 1886. He also met with difficulty in commercializing the process, until large-scale production was begun by a joint German–Swiss venture in Neuhausen, Switzerland. Héroult also patented a method for the production of aluminium alloys 1888.

Like Hall in the USA, Héroult used direct-current electrolysis to extract aluminium from compounds, dissolving aluminium oxide in a variety of molten fluorides to find the best combination of material, and finding cryolite (sodium aluminium flouride) the most promising. Héroult's process differed from Hall's in that the former used one large, central graphite electrode in the graphite cell holding the molten material.

Herschbach Dudley R 1932– . US chemist who shared the 1986 Nobel Prize for Chemistry with Yuan T Lee and John C Polyani for their researches into the dynamics of the processes which occur when atoms and molecules react.

In his experiments, Herschbach collided beams of molecules and atoms at one point in space. The molecules and atoms reacted to form new chemical products. The energy of the reacting molecules or atoms was carefully controlled and the properties of the products formed – their chemical composition, their angular distribution from the point of collision, their speed, and rotational and vibrational energy – measured. Using the data obtained, Herschbach was able to deduce a detailed picture of the events taking place during the chemical reaction.

Herschbach was born in San José, California, USA, and entered Stanford University, California, 1950, receiving his master's degree 1955. His education continued at Harvard University, Boston, Massachusetts, where he was awarded his doctorate in chemical physics 1958. He was appointed an assistant professor at the University of California at Berkeley 1959, and associate professor 1961. He returned to Harvard as professor of chemistry 1963.

Herschel Caroline Lucretia 1750–1848. German-born English astronomer, sister of William Herschel, and from 1772 his assistant in England. She discovered eight comets and worked on her brother's catalogue of star clusters and nebulae.

Herschel was born in Hanover. She received no formal education. In 1787, she was granted an annual salary by George III. On her brother's death 1822, she returned to Hanover but continued to work on his catalogue.

Herschel John Frederick William 1792–1871. English scientist, astronomer, and photographer who discovered thousands of close double stars, clusters, and nebulae. He coined the terms 'photography', 'negative', and 'positive', discovered sodium thiosulphite as a fixer of silver halides, and invented the cyanotype process; his inventions also include astronomical instruments. Baronet 1838.

Herschel was born in Slough, the son of William Herschel, and studied at Cambridge. He held no academic post but devoted himself to private research. He mapped the southern skies from the Cape of Good Hope Observatory in South Africa 1834–38. During the early days of photography he gave lectures on the subject and exhibited his own images. Together with James South, he won the Gold Medal of the Astronomical Society 1826 for their catalogue of double stars published 1824.

His works include *Outlines of Astronomy* 1849, which became a standard textbook; *General Catalogue of Nebulae and Clusters*, still the standard reference catalogue; and *General Catalogue of 10,300 Multiple and Double Stars*, published posthumously.

Herschel (Frederick) William 1738–1822. German-born English astronomer. He was a skilled telescope maker, and pioneered the study of binary stars and nebulae. He discovered the planet Uranus 1781 and infrared solar rays 1801. He catalogued over 800 double stars, and found over 2,500 nebulae, catalogued by his sister Caroline Herschel; this work was continued by his son John Herschel. By studying the distribution of stars, William established the basic form of our Galaxy, the Milky Way. Knighted 1816.

Herschel discovered the motion of binary stars around one another, and recorded it in his *Motion of the Solar System in Space* 1783. In 1789 he built, in Slough, Berkshire, a 1.2-m/4-ft telescope of 12 m/40 ft focal length (the largest in the world at the time), but he made most use of a more satisfactory 46-cm/18-in instrument. He discovered two satellites of Uranus and two of Saturn.

Herschel was born in Hanover and joined a regimental band at 14. He went to England 1757 and worked as a musician and composer while instructing himself in mathematics and astronomy, and constructing his own reflecting telescopes. The discovery of Uranus brought him fame and,

Herschel, William *German-born English astronomer William Herschel, who discovered the planet Uranus in 1871. Herschel named the planet, which was the first to be discovered telescopically,* Georgium Sidus *(George's Star) in honour of King George III, who appointed him his private astronomer a year later.*

in 1782, the post of private astronomer to George III.

In 1800, Herschel examined the solar spectrum using prisms and temperature-measuring equipment, and found that the hottest radiation was infrared. This was the beginning of the science of stellar photometry. He also established the motion and velocity of the Sun.

Hershey Alfred Day 1908– . US biochemist who won the Nobel Prize for Physiology or Medicine 1969 for his work using bacteriophages (viruses that infect bacteria) to demonstrate that DNA, not protein, is the genetic material. His experiments demonstrated that viral DNA is sufficient to transform bacteria.

With Martha Chase he demonstrated that DNA is the genetic material, by studying the T2 bacteriophage, a virus that infects the bacterium *E. coli.* It had been suggested by Roger Herriot 1951 that a bacteriophage acted like 'a little hypodermic needle full of transforming principles'; the virus as such never enters the cell; only the tail contacts the host and perhaps enzymatically cuts a small hole through the outer membrane and then the nucleic acid of the virus head flows into the cell.

Hershey and Chase tested this hypothesis by labelling the bacteriophage DNA with radioactive phosphorous and the protein with radioactive sulphur. A sample of *E. coli* was infected with the radiolabelled bacteriophage for a short incubation and then the two were separated by centrifugation. The bacteria were found to be full of radioactive phosphorous but practically devoid of radioactive sulphur. The bacteria were also fully competent to produce progeny virus. Hershey and Chase's conclusion was that 'a physical separation of the phage T2 into genetic and non-genetic parts is possible'.

Hershey was born in Owosso, Michigan in the USA. He graduated from Michigan State College and for the next 24 years worked at the Carnegie Institution in Washington to provide evidence for the nature of the material in genes.

Hertwig Wilhelm August Oscar 1849–1922. German zoologist who revealed that fertilization involves the fusion of the nuclei of egg and sperm. He showed for the first time that only one sperm is required to fertilize an egg.

His research was influenced by the work of Haeckel and described the mechanism of fertilization of sea urchin eggs, work which was of fundamental importance to the development of both the science of embryology and methods of in vitro fertilization over a century later. Auerbach had already demonstrated that a fertilized egg contains two nuclei, but it was Hertwig who discovered that the origins of these nuclei were from the egg and the sperm. He published his results 1875.

Hertwig was born at Friedberg, Hessen, Germany, the son of a chemist who insisted that he be educated at home. He went on to study chemistry and medicine at Jena University and was appointed assistant professor of anatomy there 1878. He was promoted to professor 1881, and finally to the founder chair of cytology and embryology when the position was created 1888.

Hertz Gustav 1887–1975. German physicist who, with US physicist James Franck, demonstrated that mercury atoms, when bombarded with electrons, absorb energy in discrete units (or quanta). Following the absorbtion of energy, the atoms return to their original state by emitting a photon of light. This was the first experimental proof that the quantum theory of atoms was correct and demonstrated the reality of atomic energy levels. Nobel Prize for Physics 1925.

Gustav Hertz was born in Hamburg, Germany, and obtained his doctorate in Berlin 1911. He became an assistant in the Berlin physics institute and began his collaboration with Franck. He was professor of experimental physics at the University of Halle 1925–1927. From 1928 to 1935 he worked at the Berlin Techniche Hochschule. Because of his Jewish descent, Hertz was forced to resign 1935. However, he remained in Germany during World War II, working as director of the Siemens Research laboratory, Berlin. After the war, he was captured by the Russians and taken to the USSR to continue his work in atomic physics. In 1955 he re-emerged as director of the Physics Institute in Leipzig, East Germany. He is the nephew of Heinrich Hertz, the discoverer of radio waves.

Hertz Heinrich Rudolf 1857–1894. German physicist who studied electromagnetic waves, showing that their behaviour resembles that of light and heat waves.

Hertz confirmed James Clerk Maxwell's theory of electromagnetic waves. In 1888, he realized that electric waves could be produced and would travel through air, and he confirmed this experimentally. He went on to determine the velocity of these waves (which were later called radio waves) and, on showing that it was the same as that of light, devised experiments to show that the waves could be reflected, refracted, and diffracted.

Hertz was born in Hamburg and studied at Munich and Berlin. He was professor at Karlsruhe 1885–89 and at Bonn from 1889.

From about 1890, Hertz gained an interest in mechanics. He developed a system with only one law of motion: that the path of a mechanical system through space is as straight as possible and is travelled with uniform motion.

The unit of frequency, the ***hertz***, is named after him.

Hertzsprung Ejnar 1873–1967. Danish astronomer and physicist who introduced the concept of the absolute magnitude (brightness) of a star, and described the relationship between the absolute magnitude and the temperature of a star, formulating his results in the form of a diagram that has become a standard reference.

Hertzsprung was born in Frederiksberg and studied chemical engineering at the Frederiksberg Polytechnic. After some years working in Russia, he learned photochemistry and became employed at observatories in Denmark and later in Germany. In 1919 he became professor of astronomy at Leiden, the Netherlands. He retired 1945 and returned to Denmark, but did not cease his astronomical research until well into the 1960s.

In 1905, Hertzsprung proposed a standard of stellar magnitude for scientific measurement, and defined this 'absolute magnitude' as the brightness of a star at the distance of 10 parsecs (32.6 light years). He also described the relationship between the absolute magnitude and the colour – the spectral class or temperature – of a star, plotting it on a graph in 1906. He did not publish

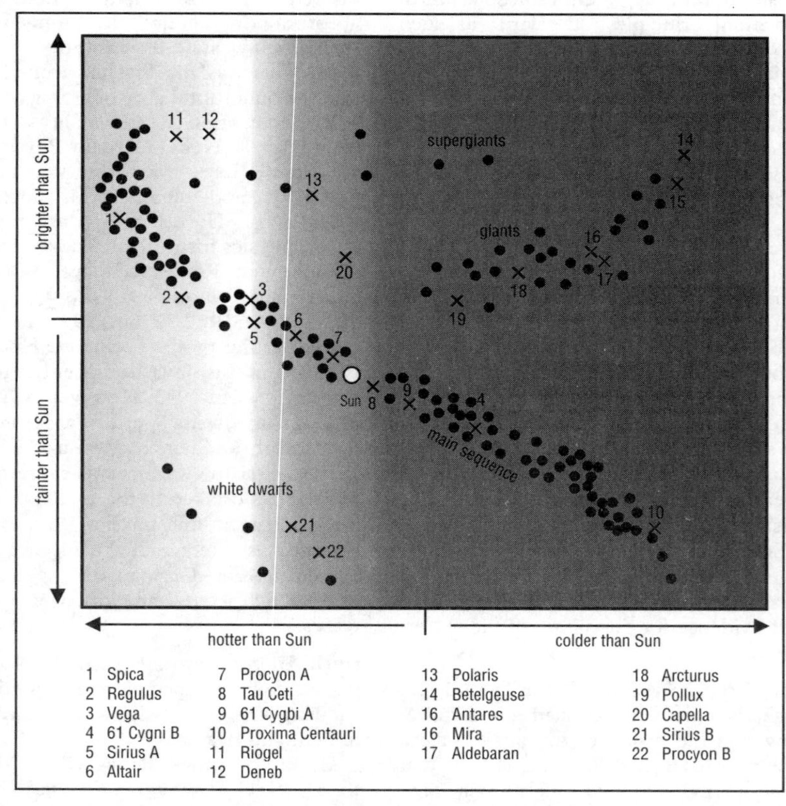

Hertzsprung *The Hertzsprung–Russell diagram relates the brightness (or luminosity) of a star to its temperature. Most stars fall within a narrow diagonal band called the main sequence. A star moves off the main sequence when it grows old. The Hertzsprung–Russell diagram is one of the most important diagrams in astrophysics.*

the diagram, which was independently arrived at by US astronomer Henry Russell in 1913.

The Hertzsprung–Russell diagram has the temperature on one axis and absolute magnitude on the other. As plotted, the stars range themselves largely along a curve running from the upper left (the blue giant stars) to the lower right (the red dwarf stars) of the graph. Approximately 90% of stars belong to this ***main sequence***; our Sun lies near the middle and is classed as a yellow dwarf.

The Hertzsprung–Russell diagram has been used in determining the distance of stars from Earth, and has also been essential to the development of modern theories of stellar evolution. As stars age and deplete their store of nuclear fuel, they are believed to leave the main sequence and become red giants. Eventually they radiate so much energy that they then cross the main sequence and collapse into blue dwarfs. Larger stars may follow a different pattern and explode into novae, or collapse to form black holes, at the end of their lifespans.

In 1922, Hertzsprung published a catalogue on the mean colour equivalents of nearly 750 stars of magnitude greater than 5.5.

Herzberg Gerhard 1904– . German-born Canadian physicist who used spectrocopy to determine the electronic structure and geometry of molecules, especially free radicals (atoms or groups of atoms that possess a free, unbonded electron). Nobel Prize for Chemistry 1971.

Herzberg was born in Hamburg and studied at the Technische Universität in Darmstadt, at Göttingen, and in the UK at Bristol. In 1935, with the rise to power of Adolf Hitler, he fled to Canada, where he became professor of physics at the University of Saskatchewan, Saskatoon, and director of the Division of Pure Physics for the National Research Council in Ottawa 1949–69. He spent 1945–49 at the Yerkes Observatory in Wisconsin, USA.

Depending on the conditions, molecules absorb or emit electromagnetic radiation of discrete wavelengths. The radiation spectrum is directly dependent on the electronic and geometric structure of an atom or molecule. Herzberg discovered new lines in the spectrum of molecular oxygen; called ***Herzberg bands***, these spectral lines have been useful in analysing the upper atmosphere. He also discovered the new molecules phosphorus nitride and phosphorus carbide; proved the existence of the methyl and methylene free radicals; and demonstrated that both neutrons and protons are part of the nucleus.

In addition, Herzberg interpreted the spectral lines of stars and comets, finding that a rare form of carbon exists in comets. He showed that hydrogen exists in the atmospheres of some planets, and identified the spectra of certain free radicals in interstellar gas.

Herzog Bertram 1929– . German-born computer scientist, one of the pioneers in the use of computer graphics in engineering design.

Herzog was born in Offenburg, near Strasbourg, but emigrated to the USA and studied at the Case Institute of Technology. He has alternated academic posts with working in industry. In 1965 he became professor of industrial engineering at the University of Michigan. Two years later he became professor of electrical engineering and computer science at the University of Colorado.

In 1963, Herzog joined the Ford Motor Company as engineering methods manager, where he extensively applied computers to tasks involved in planning and design. Herzog remained as a consultant to Ford after his return to academic life.

Hess Germain Henri 1802–1850. Swiss-born Russian chemist, a pioneer in the field of thermochemistry. The law of constant heat summation is named after him.

Hess was born in Geneva, but his family emigrated to Russia. Hess studied at the University of Dorpat (now Tartu, Estonia), and in Stockholm, Sweden, under chemist Jöns Berzelius. Returning to Russia, he took part in a geological expedition to the Urals before setting up a medical practice in Irkutsk. In 1830 he settled in St Petersburg, where he held various academic appointments, becoming professor at the Technological Institute.

Hess's law was published 1840 and states that the heat change in a given chemical reaction depends only on the initial and final states of the system and is independent of the path followed, provided that heat is the only form of energy to enter or leave the system. Every chemical change is either endothermic (absorbing heat) or exothermic (evolving heat). Hess's law is in fact an application of the law of conservation of energy, but this was not formulated until 1842.

In 1842 Hess proposed his second law, the law of thermoneutrality, which states that in exchange reactions of neutral salts in aqueous solution, no heat effect is observed.

Hess Harry Hammond 1906–1969. US geologist who in 1962 proposed the notion of seafloor spreading. This played a key part in the acceptance of plate tectonics as an explanation of how the Earth's crust is formed and moves.

Hess was born in New York and studied at Yale and Princeton, where he eventually became professor. From 1931, he carried out geophysical research into the oceans, continuing during World War II while in the navy. Later he was one of the main advocates of the Mohole project, whose aim was to drill down through the Earth's crust to gain access to the upper mantle.

Building on the recognition that certain parts of the ocean floor were anomalously young, and the discovery of the global distribution of midocean ridges and central rift valleys, Hess suggested that convection within the Earth was continually creating new ocean floor, rising at midocean ridges and then flowing horizontally to form new oceanic crust. It would follow that the further from the midocean ridge, the older would be the crust – an expectation confirmed by research 1963.

Hess envisaged that the process of seafloor spreading would continue as far as the continental margins, where the oceanic crust would slide down beneath the lighter continental crust into a subduction zone, the entire operation thus constituting a kind of terrestrial conveyor belt.

Hess Victor Francis 1883–1964. Austrian physicist who emigrated to the USA shortly after sharing a Nobel prize in 1936 for the discovery of cosmic radiation.

Hess was born in Waldstein and educated at Graz. From 1906 to 1920 he worked in Vienna, studying radioactivity and atmospheric ionization. He was professor at Graz 1920–31 and at Innsbruck 1931–38, founding a cosmic-ray observatory on the nearby Hafelekar mountain. After the Nazi annexation of Austria, he emigrated to the USA, becoming professor at Fordham University, New York.

Hess made ten balloon ascents in 1911–12 to collect data about atmospheric ionization. Ascending to altitudes of more than 5,000 m/16,000 ft, he established that the intensity of ionization decreased to a minimum at about 1,000 m/3,000 ft, then increased steadily. By making ascents at night – and one during a nearly total solar eclipse – he proved that the ionization was not caused by the Sun. He concluded that radiation enters the atmosphere from outer space.

Hevelius Johannes 1611–1687. Latinized form of Jan Hewel or Hewelcke. German astronomer who in 1647 published the first comparatively detailed map of the Moon in his *Selenographia*. He also discovered four comets and suggested that these bodies orbited in parabolic paths about the Sun.

Hevelius was born in Danzig (now Gdansk). He worked as a brewing merchant and was a city councillor, but spent his evenings on the roof of his house, where he had an observatory installed 1641. This was destroyed by fire in 1679, with some of his notes. His wife Elizabeth assisted him in his work and, after his death, she edited and published his *Prodromus astronomiae* 1690.

Between 1642 and 1645, Hevelius deduced a fairly accurate value for the period of the solar rotation and gave the first description of the bright areas in the neighbourhood of sunspots. The name he gave to them, *faculae*, is still used.

Although Hevelius used telescopes for details on the Moon and planets, he refused to apply them to his measuring apparatus, and his observations of the positions of stars were made with the naked eye, not always accurately. His *Uranographia* 1690 contains a catalogue of more than 1,500 stars and a celestial atlas with 54 plates.

Hevesy Georg Karl von 1885–1966. Hungarian-born Swedish chemist, discoverer of the element hafnium. He was the first to use a radioactive isotope to follow the steps of a biological process, for which he won the Nobel Prize for Chemistry 1943.

Hevesy was born in Budapest and educated in Germany, Switzerland, and the UK, studying at Manchester under nuclear-physics pioneer Ernest Rutherford 1911. He worked in Copenhagen at the Institute of Physics under Niels Bohr 1920–26 and 1934–43. During the German occupation of Denmark in World War II, Hevesy escaped to Sweden and became professor at Stockholm.

Hevesy first used radioactive tracers to study the chemistry of lead and bismuth salts. During the early 1930s, he began experiments with this technique on biological specimens, noting the take-up of radioactive lead by plants, and going on to use an isotope to trace the movement of phosphorus in the tissues of the human body. He used heavy water to study the mechanism of water exchange between goldfish and their surroundings and also within the human body. Using radioactive calcium to label families of mice, he showed that, of calcium atoms present at birth, about 1 in 300 are passed on to the next generation.

Hewish Antony 1924– . English radio astronomer who was awarded, with Martin Ryle, the Nobel Prize for Physics 1974 for his work on pulsars, rapidly rotating neutron stars that emit pulses of energy.

The discovery by Jocelyn Bell Burnell of a regularly fluctuating signal, which turned out to be the first pulsar, began a period of intensive research. Hewish discovered another three straight away, and more than 170 pulsars have been found since 1967.

Hewish was born in Cornwall and studied at Cambridge. He worked at the Cavendish Laboratory there, and became professor at Cambridge 1972.

Before 1950, Hewish used radio telescopes mainly to study the solar atmosphere. When new instruments became available, radio observations were extended to sources other than the Sun. Before the discovery of pulsars, Hewish examined the fluctuation in such sources of the intensity of the radiation (the scintillation) resulting from disturbances in ionized gas in the Earth's atmosphere, within the Solar System, and in interstellar space.

Hewish has patented a system of space navigation using three pulsars as reference points, which would provide coordinates in outer space accurate up to a few hundred kilometres.

Hey James Stanley 1909– . English physicist whose work in radar led to pioneering research in radioastronomy. He discovered that large sunspots were powerful ultrashortwave radio transmitters, and pinpointed a radio source in the Milky Way.

Hey was born in the Lake District and studied at Manchester. From 1940 to 1952 he was on the staff of an Army Operational Research Group. He then became a research scientist at the Royal Radar Establishment.

In Feb 1942, during World War II, the British early-warning coastal-defence radar became severely jammed. At first the jamming was attributed to enemy countermeasures, but Hey, noting that the interference began as the Sun rose and ceased as it set, concluded that the spurious radio radiation was associated with a large solar flare that had just been reported.

Intense radio sources had already been detected in the Milky Way, notably in the constellations Cygnus, Taurus, and Cassiopeia. In 1946 Hey and his colleagues announced that they had narrowed down the location of the radiation source in Cygnus to Cygnus A.

Using radio, Hey's team also noted that they could detect and follow meteors more accurately than ever before.

Heyrovský Jaroslav 1890–1967. Czech chemist who was awarded the 1959 Nobel prize for his invention and development of polarography, an electrochemical technique of chemical analysis.

Heyrovský was born in Prague and studied there at Charles University and at University College, London. From 1920 he was on the staff of the Institute of Analytical Chemistry in Prague; in 1950 he became director of the newly founded Polarographic Institute of the Czechoslovak Academy of Sciences.

Heyrovský's technique depends on detecting the discharge of ions during electrolysis of aqueous solutions. In 1925 he developed the polarograph, an instrument that traces the resulting voltage–current curve on a chart recorder. Such curves are called polarograms.

Polarography can be used to analyse for several substances at once, and is capable of detecting concentrations as low as 1 part per million. Most chemical elements can be determined by the method (as long as they form ionic species) in compounds, mixtures, or alloys. The technique has been extended to organic analysis and to the study of chemical equilibria and the rates of reactions in solutions. It can also be used for endpoint detection in titrations, a type of volumetric analysis sometimes called voltammetry.

Hilbert David 1862–1943. German mathematician, philosopher, and physicist whose work was fundamental to 20th-century mathematics. He founded the formalist school with *Grundlagen der Geometrie/Foundations of Geometry* 1899, which was based on his idea of postulates.

Hilbert attempted to put mathematics on a logical foundation through defining it in terms of a number of basic principles, which Kurt Gödel later showed to be impossible. In 1900 Hilbert proposed a set of 23 problems for future mathematicians to solve, and gave 20 axioms to provide a logical basis for Euclidean geometry.

Hilbert was born in Königsberg, Prussia (now Kaliningrad, Russia) and studied there and at Leipzig and Paris. He was professor at Königsberg 1892–95 and at Göttingen 1895–1930.

Physics is much too hard for physicists.

David Hilbert

In Constance Reid *Hilbert* 1970 p 127

Studying algebraic invariants, Hilbert had by 1892 not only solved all the known central problems of this branch of mathematics, he had introduced sweeping developments and new areas for research, particularly in algebraic topology.

From 1909 Hilbert worked on problems of physics, such as the kinetic theory of gases and the theory of relativity.

Hildegard of Bingen 1098–1179. German abbess, writer, and composer. Her encyclopedia of natural history, *Liber simplicis medicinae* 1150–60, giving both Latin and German names for the species described as well as their medicinal uses, is the earliest surviving scientific book by a woman.

Hildegard was abbess of the Benedictine convent of St Disibode, near the Rhine, from 1136.

She wrote a mystical treatise, *Liber Scivias* 1141, and collected her lyric poetry in the 1150s into one volume, providing each individual text with music. The poetry is vivid, reflecting the visions she experienced throughout her life. The melodic structure of her music is based on a small number of patterns (similar to motifs), which are repeated in different modes.

Hill Archibald Vivian 1886–1977. English physiologist who studied muscle action and especially the amount of heat produced during muscle activity. For this work he shared the 1922 Nobel Prize for Physiology or Medicine.

Hill was born in Bristol and educated at Cambridge. He was professor at Manchester 1920–23 and at the Royal Society 1923–51. He also served as scientific adviser to India 1943–44 and was a member of the War Cabinet Scientific Advisory Committee during World War II.

To record minute heat changes, Hill modified delicate thermocouples, and discovered by 1913 that contracting muscle fibres produce heat in two phases. Heat is first produced quickly as the muscle contracts. Then, after the initial contraction, further heat is evolved more slowly but often in greater amounts. Hill also showed that molecular oxygen is consumed after the work of the muscles is over but not during muscular contraction.

Hill Austin Bradford 1897–1991. English epidemiologist and statistician. He pioneered rigorous statistical study of patterns of disease and, together with Richard Doll, was the first to demonstrate the connection between cigarette smoking and lung cancer. Knighted 1961.

Hill took a degree in economics, and in 1923 began working for the Medical Research Council as a statistician. In 1933 he moved to the London School of Hygiene and Tropical Medicine, where he later became professor of medical statistics. His work on smoking and lung cancer, which involved collecting data on the smoking habits and health of over 30,000 British doctors for several years, in the precomputer age, is considered to be among the great medical achievements of the century.

Hill Robert 1899–1991. British biochemist who showed that during photosynthesis, oxygen is produced, and that this derived oxygen comes from water. This process is now known as the Hill reaction. He also demonstrated the evolution of oxygen in human blood cells by the conversion of haemoglobin to oxyhaemoglobin.

Hill was educated at Cambridge and remained researching there until 1938. From 1943 to 1966 he was a member of the scientific staff of the Agricultural Research Council.

The process of photosynthesis has been shown to occur in two separate sets of reactions, those that require sunlight (the light reactions) and those that do not (the dark reactions). Both sets of reactions are dependent on one another. In the light reactions some of the energy of sunlight is trapped within the plant and in the dark reactions this energy is used to produce potentially energy-generating chemicals, such as sugar.

Hill's experiments in 1937 confirmed that the light reactions of photosynthesis occur within the chloroplasts of leaves, as well as elucidating in part the mechanism of the light reactions. To do this, he isolated chloroplasts from leaves and then illuminated them in the presence of an artificial electron-acceptor.

Hinkler Herbert John Louis 1892–1933. Australian pilot who in 1928 made the first solo flight from England to Australia. He was killed while making another attempt to fly to Australia.

Hinshelwood Cyril Norman 1897–1967. English chemist who shared the 1956 Nobel prize for his work on chemical chain reactions. He also studied the chemistry of bacterial growth. Knighted 1948.

Hinshelwood was born in London and studied at Oxford, where he became professor 1937. During World War I he worked in the Department of Explosives at the Royal Ordnance Factory in Queensferry, Scotland.

Studying gas reactions and the decomposition of solid substances in the presence and absence of catalysts, Hinshelwood went on to demonstrate that many reactions can be explained in terms of a series – a chain – of interdependent stages. At high temperatures the chain reactions of some elements accelerate the process to explosion point. He provided experimental evidence for the role of activated molecules in initiating the chain reaction. In his bacterial-growth experiments, too, he considered that all the various chemical reactions that occurred were interconnected and mutually dependent, the product of one reaction becoming the reactant for the next.

He also investigated reaction kinetics in aqueous and nonaqueous solutions, and published *Kinetics of Chemical Change* 1926.

Hinton William Augustus 1883–1959. US bacteriologist and pathologist who worked on syphilis, in particular the development of the Hinton test.

Hinton was born in Chicago and studied at the University of Kansas and at Harvard. In 1915 he was made chief of the Wasserman Laboratory, and from 1918 he was an instructor at the Harvard Medical School. In 1949, he became the first black professor in the university's history.

Hinton developed a blood-serum test for syphilis which reduced the number of false positive diagnoses of the disease; in 1934, the US Public Health Service showed the Hinton test for syphilis to be the best. He wrote many scientific papers and his book *Syphilis and its Treatment* 1926 was the first medical textbook by a black American to be published.

Hinton was also the discoverer of the Davies–Hinton tests of blood and spinal fluid.

Hipparchus *c.* 190–*c.* 120 BC. Greek astronomer and mathematician who invented trigonometry and calculated the lengths of the solar year and the lunar month. He discovered the precession of the equinoxes, made a catalogue of 850 fixed stars, and advanced Eratosthenes' method of determining the situation of places on the Earth's surface by lines of latitude and longitude.

Hipparchus was born in Nicaea, Bithynia (now in Turkey), and lived on the island of Rhodes and in Alexandria, Egypt.

In 134 BC Hipparchus noticed a new star in the constellation Scorpio, a discovery which inspired him to put together a star catalogue – the first of its kind. He entered his observations of stellar positions using a system of celestial latitude and longitude, and taking the precaution wherever possible to state the alignments of other stars as a check on present position. He classified the stars by magnitude (brightness). His finished work, completed in 129 BC, was used by Edmond Halley some 1,800 years later.

Hippocrates *c.* 460–*c.* 377 BC. Greek physician, often called the founder of medicine. Important Hippocratic ideas include cleanliness (for patients and physicians), moderation in eating and drinking, letting nature take its course, and living where the air is good. He believed that health was the result of the 'humours' of the body being in balance; imbalance caused disease. These ideas were later adopted by Galen.

He was born and practised on the island of Kos, where he founded a medical school. He travelled throughout Greece and Asia Minor, and died in Larisa, Thessaly. He is known to have discovered aspirin in willow bark. The *Corpus Hippocraticum/Hippocratic Collection*, a group of some 70 works, is attributed to him

---

Life is short, the Art long, opportunity fleeting, experience treacherous, judgment difficult.

Hippocrates

Aphorisms I, 1

---

but was probably not written by him, although the works outline his approach to medicine. They include *Aphorisms* and the ***Hippocratic Oath***, which embodies the essence of medical ethics.

The *Corpus Hippocraticum* remains impressive for its focus on observation and the description of symptoms. Diseases are seen as being caused by an imbalance of the four basic ingredients, or humours, of the human body (blood, phlegm, yellow bile, and black bile). This remarkably influential idea perhaps gained credence by analogy with the theory of the four elements of matter of Zeno of Elea and Parmenides. Being vague, the theory of humours was also widely applicable, for instance in its link with the seasons (winter illnesses being characterized by cold and wet discharges). Where given, the treatment suggested consists of eating compensatory food such as hot and dry foods in winter. The Hippocratic tradition in medicine remained dormant until the 18th–19th century, when it was replaced by the germ theory of disease, but the image of Hippocrates as the ideal physician remains today.

Hitchings George Herbert 1905– . US pharmacologist who shared the Nobel Prize for Physiology or Medicine 1988 with his co-worker Gertrude Elion (1918–). His work was on anticancer agents, and immunosuppressive drugs and antibiotics that were of vital importance in the growing surgical field of transplantation.

Hitchings was born in Hoquiam, Washington, and educated at the University of Washington and at Harvard. In 1942 he joined the biochemical research laboratory of Burroughs Wellcome, where he was joined two years later by Elion. Hitchings rose to vice president of the company 1967.

Hitchings and Elion started to synthesize compounds that inhibited bacteria from producing DNA and therefore from multiplying, in the hope that these compounds could be used to prevent the rapid growth of cancer cells. They investigated particularly the chemistry of purines and pyrimidines, two groups of chemicals that are involved with DNA synthesis. Their work lead to the production of clinically significant anticancer drugs as well as drugs against malaria, for the treatment of gout and kidney stones, and also drugs that suppressed the normal immune reactions of the body. In the 1970s, Hitchings and Elion's research produced an antiviral compound, acyclovir, active against the herpes virus, which preceded the development by Burroughs Wellcome of AZT, the anti-AIDS compound.

Hoagland Mahlon Bush 1921– . US biochemist who was the first to isolate transfer RNA (tRNA), a nucleic acid that plays an essential part in intracellular protein synthesis.

Hoagland was born in Boston, Massachusetts. He studied at Harvard University Medical School, and worked there 1953–67. In 1967, he was appointed professor at Dartmouth Medical School and scientific director of the Worcester Foundation for Experimental Biology in Shrewsbury, Massachusetts.

In the late 1950s Hoagland isolated various types of tRNA molecules from cytoplasm and demonstrated that each type of tRNA can combine with only one specific amino acid. Each tRNA molecule has as part of its structure a characteristic triplet of nitrogenous bases that links to a complementary triplet on a messenger RNA (mRNA) molecule. A number of these reactions occur on the ribosome, building up a protein one amino acid at a time.

Hoagland has also investigated the carcinogenic effects of beryllium and the biosynthesis of coenzyme A.

Hodgkin Alan Lloyd 1914– . British physiologist engaged in research with Andrew Huxley on the mechanism of conduction in

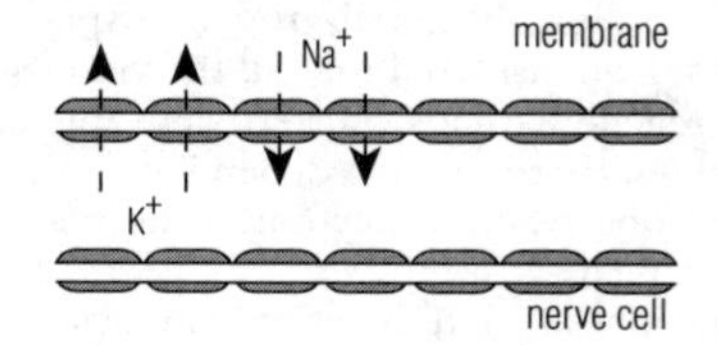

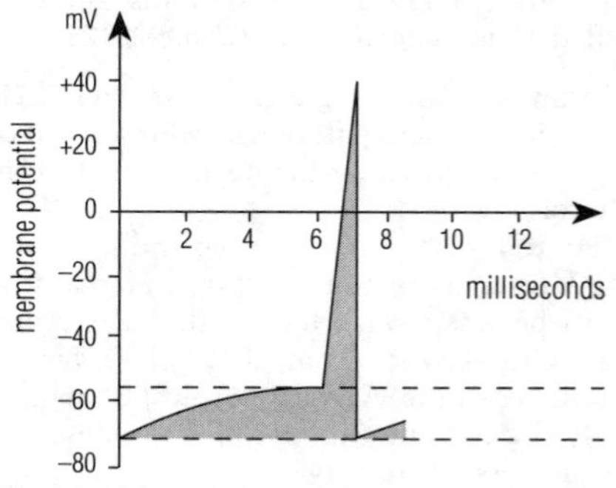

Hodgkin, Alan *The potential inside an axon changes rapidly (and reverses) with the passage of a nerve impulse, accompanied by movement of sodium and potassium ions through the cell membrane.*

peripheral nerves 1945–60. He devised techniques for measuring electric currents flowing across a cell membrane. In 1963 they shared the Nobel prize. KBE 1972.

Hodgkin was born in Banbury, Oxfordshire, and educated at Cambridge, where he spent most of his career and became professor 1952.

Hodgkin and Huxley managed for the first time to record electrical changes across the cell membrane, and Hodgkin then built on these findings working with Bernhard Katz, another cell physiologist. They proposed that during the resting phase a nerve membrane allows only potassium ions to diffuse into the cell, but when the cell is excited it allows sodium ions (which are positively charged) to enter and potassium ions to move out. The extrusion of sodium is probably dependent on the metabolic energy supplied either directly or indirectly in the form of ATP(adenosine triphosphate). The amount of sodium flowing in equals that of the potassium flowing out.

Hodgkin, Dorothy *British crystallographer Dorothy Mary Crowfoot Hodgkin, who determined the structure of penicillin and vitamin B_{12} using crystallographic methods. Hodgkin became the third woman to win a Nobel prize 1964 (for chemistry) and was admitted to the Order of Merit 1965 – the first female to do so since Florence Nightingale.*

Hodgkin Dorothy Mary Crowfoot 1910–1994. English biochemist who analysed the structure of penicillin, insulin, and vitamin B_{12}. Hodgkin was the first to use a computer to analyse the molecular structure of complex chemicals, and this enabled her to produce three-dimensional models. Nobel Prize for Chemistry 1964.

Hodgkin studied the structures of calciferol (vitamin D_2), lumisterol, and cholesterol iodide, the first complex organic molecule to be determined completely by the pioneering technique of X-ray crystallography, a physical analysis technique devised by Lawrence Bragg (1890–1971), and at the time used only to confirm formulas predicted by organic chemical techniques. She also used this technique to determine the structure of penicillin, insulin, and vitamin B_{12}.

Hodgkin was born in Cairo and educated at Somerville College, Oxford. At Cambridge 1932–34 she worked on the development of X-ray crystallography and, returning to Oxford 1934, began working on penicillin. After Howard Florey (1898–1968) and Ernst Chain (1906–1979) isolated pencillin from mould in 1939, chemists in Britain and America raced to determine its structure. Hodgkin's assertion that the core of penicillin consisted of a ring of three carbons and a nitrogen thought too unstable exist, brought from Australian chemist John Cornforth the derisive comment, 'If that's the formula of penicillin, I'll give up chemistry and grow mushrooms.' But Hodgkin was right, and she went on to determine the structure of the antibiotic cephalosporin C.

In 1948 Hodgkin began her work on vitamin B_{12}, a substance that proved to be far more complex than penicillin: the first X-ray diffraction pictures showed over a thousand atoms, compared to penicillin's 39. It took Hodgkin and co-workers eight years to solve. The structure of insulin was to take much longer still; Hodgkin first saw the diffraction pattern made by insulin in 1935, but it was to take 34 years for her to determine its structure.

In 1964 Hodgkin became the second woman to have ever received the Order of Merit (the first was Florence Nightingale) and – a committed socialist all her life – in 1987 she was awarded the Lenin Peace Prize. She became Chancellor of Bristol University 1970–88 and helped found a Hodgkin scholarship for students from the developing world.

Hodgkin Thomas 1798–1866. English physician who first recognized ***Hodgkin's***

disease, a rare type of cancer (also known as lymphoadenoma) that causes malignant inflammation of the lymph nodes. He pioneered the use of the stethoscope in the UK. He was also the first person to stress the importance of postmortem examinations.

Hodgkin was born in London and studied at Guy's Hospital, London, and at Edinburgh, and lectured at Guy's 1827–37. He was active in the Aborigines Protection Society, and died in Jaffa (now in Israel) while on a mercy mission.

His paper describing Hodgkin's disease was published 1832.

Hodgkinson Eaton 1789–1861. English civil engineer who introduced scientific methods of measuring the strength of materials. From a theoretical analysis, he devised experiments to determine the strongest iron beam, which resulted in the discovery of what is known as 'Hodgkinson's beam'.

Hodgkinson was born in Anderton, Cheshire, and became a pawnbroker. He presented a number of scientific papers to the Literary and Philosophical Society in Manchester, and in 1847–49 was a member of a Royal Commission to inquire into the application of iron to railway structures. Also in 1847 he was appointed professor of the mechanical principles of engineering at University College, London.

Hodkinson's paper 'The transverse strain and strength of materials' 1822 dealt with the 'set' or the original position of a strained body and the position it assumes when the strain is removed. He fixed the exact position of the 'neutral line' in the section of rupture or fracture and made it the basis for the computation of the strength of a beam of given dimensions.

Hodgkinson helped civil engineer Robert Stephenson in the construction of the Menai and Conway tubular bridges by fixing the best forms and dimensions of tubes.

Hoe Richard March 1812–1886. US inventor of the rotary printing press 1846, which revolutionized newspaper printing. He also improved on the cylinder press for use in lithographic and letterpress work, and introduced the web press in the USA, making it efficient enough to supersede the rotary press in the 1880s.

Hoe was born in New York and worked in his father's firm manufacturing printing presses. His inventions led to expansion in New York, Boston, and the UK; between 1865 and 1870, a large manufacturing branch was built up in London, employing 600 people. Concerned for the welfare of his employees, he ran free evening classes for apprentices.

Discarding the old flatbed printing press in the 1830s, Hoe placed the type on a revolving cylinder, which was developed into the rotary press.

The first press that would print on a continuous roll, or web, of paper had been produced 1865. In 1871, with Stephen D Tucker as a partner, Hoe designed and built the Hoe web perfecting press. The first of these to be used in the USA was installed in the *New York Tribune*. This press printed on both sides of the sheet and produced 18,000 papers per hour. Four years later, Tucker patented a rotating cylinder that folded the papers as fast as they came off the press.

Hoffman Albert Swiss-born, US physician who, in 1943, accidentally discovered lysergic acid diethylamide (LSD), the most potent psychoactive drug ever known. LSD causes hallucinatory effects, paranoia, and depression, and is illegal except for research purposes.

Hoffman, together with Swiss chemist Arthur Stoll, produced lysergic acid diethylamide (LSD) 1938 while trying to synthesize a new drug for the treatment of headaches. Since the new drug appeared to have no analgesic (pain-relieving) effect on laboratory animals, it remained untouched on a shelf for five years. Hoffman decided to perform further tests on LSD 1943, during which he accidentally ingested an unknown amount of the drug. He described his first experience of LSD intoxication as 'a kind of drunkenness which was not unpleasant and which was characterized by extreme activity of the imagination'. LSD was later shown to block or inhibit the action of the neurotransmitter seratonin in the brain.

Hoffman Roald 1937– . Polish chemist who worked on ***molecular orbital theory*** with Robert Woodward and developed the Woodward–Hoffman rules for the conservation of orbital (the arrangement of electrons around the nucleus) symmetry, which predict the conditions under which certain organic reactions can occur. He shared the Nobel Prize for Chemistry in 1981 with Woodward and Kenichi Fukui.

His findings were explained in his book *Conservation of Orbital Symmetry* 1969. From this work, chemists have a greater understanding of bonding and the prediction of chemicals synthesized by others, as well as the bonding of chemicals adsorbed on surfaces. Hoffman also experimented in various branches of inorganic chemistry.

Hoffman was born in the town of Zloczow, Poland (now Zolochev, in the Ukraine), during the start of the Nazi campaign in Poland. When he was four years old his family was detained in a labour camp under the occupation, and his father was fatally wounded by soldiers when he tried to break out. Remarkably, in 1943, Hoffman escaped undetected and was concealed by friends in a school storeroom, emerging as a refugee when the war ended.

Hoffman and his mother then travelled hundreds of miles overseas to start a new life in America, reaching New York City 1949. Hoffman gained a place at Columbia University and finished his PhD in chemical physics in 1962. He was then elected a junior fellow at Harvard University. He continued his research at Cornell University from 1965, and took the post of John A Newman professor of physical science.

An articulate, affable man with a colourful personality, Hoffman was keen to communicate the value of chemistry to the public, and to this end he wrote numerous articles in the press and presented a television series about the subject.

Hofmann August Wilhelm von 1818–1892. German chemist who studied the extraction and exploitation of coal-tar derivatives, mainly for dyes.

In 1881 he devised a process for the production of pure primary amines from amides.

Hofmann was born and educated in Giessen, Hesse. He was professor at the Royal College of Chemistry in London 1845–65 and at Berlin from 1865.

In 1858, Hofmann obtained the dye known as fuchsine or magenta by the reaction of carbon tetrachloride (tetrachloromethane) with aniline (phenylamine). Later he isolated from it a compound which he called rosaniline and used this as a starting point for other aniline dyes, including aniline blue (triphenyl rosaniline). With alkyl iodides (iodoalkanes) he obtained a series of violet dyes, which he patented 1863. These became known as 'Hofmann's violets' and were a considerable commercial and financial success.

The term 'valence' is a contraction of his notion of 'quantivalence' and he devised much of the terminology of the paraffins (alkanes) and their derivatives that was accepted at the 1892 Geneva Conference on nomenclature.

Hofmeister Wilhelm Friedrich Benedikt 1824–1877. German botanist. He studied plant development and determined how a plant embryo, lying within a seed, is itself formed out of a single fertilized egg (ovule).

Hofmeister also discovered that mosses and ferns display an alternation of generations, in which the plant has two forms, spore-forming and gamete-forming.

Hofstadter Robert 1915–1990. US nuclear physicist who made pioneering studies of nuclear structure and the elementary nuclear constituents, the proton and the neutron. He established that the proton and neutron were not pointlike, but had a definite volume and shape. He shared the 1961 Nobel Prize for Physics.

Hofstadter demonstrated that the nucleus is composed of a high-energy core and a surrounding area of decreasing density.

He helped to construct a new high-energy accelerator at Stanford University, California, with which he showed that the proton and the neutron have complex structures and cannot be considered elementary particles.

Hofstadter was born in New York and educated at City College and Princeton. From 1950 he was at Stanford, where his early work involved bouncing, or scattering, electrons from complex nuclei, such as gold. This produced accurate pictures of the charge distribution within nuclei. Gradually, smaller nuclei were studied by Hofstadter and his team, using electrons of increasing energy. By 1960, accurate data had been obtained for the proton and neutron, revealing the spatial distribution of charge and magnetization within these particles.

Hogben Lancelot Thomas 1895–1975. English zoologist and geneticist who wrote the best-selling *Mathematics for the Millions* 1933. He applied mathematical principles to genetics and was concerned with the way statistical methods were used in the biological and behavioural sciences.

Hogben was born in Southsea, Hampshire, and studied at Cambridge and London. Imprisoned as a conscientious objector in 1916 during World War I, he was released only when his health deteriorated seriously. He held various academic posts in the UK, Canada, and South Africa, becoming professor of social biology at London University 1930. During World War II he was put in charge of the medical statistics records for the British army. After the war he became professor of medical statistics at the University of Birmingham, where he remained until he retired 1961.

Hogben first began to try to apply mathematical principles to the study of genetics in the 1930s, with particular reference to his investigation of generations of the fruitfly *Drosophila* in relation to research on heredity in humans.

Hollerith Herman 1860–1929. US inventor of a mechanical tabulating machine, the first device for data processing. Hollerith's tabulator was widely publicized after being successfully used in the 1890 census. The firm he established, the Tabulating Machine Company, was later one of the founding companies of IBM.

Hollerith *The first ever mechanized data processing machine which was invented by Herman Hollerith. The data was stored on punched cards that could be read by the hinged device. The results were displayed on the panel of dials.*

Hollerith was born in Buffalo, New York, and attended the Columbia University School of Mines. From 1884 to 1896 he worked for the US Patent Office.

Working on the 1880 US census, he saw the need for an automated recording process for data, and had the idea of punching holes in cards or rolls of paper. By 1889 he had developed machines for recording, counting, and collating census data. The system was used 1891 for censuses in several countries, and was soon adapted to the needs of government departments and businesses that handled large quantities of data.

Holley Robert William 1922–1993. US biochemist who established the existence of transfer RNA (tRNA) and its function. For this work he shared the 1968 Nobel Prize for Physiology or Medicine.

Born in Urbana, Illinois, Holley studied at the University of Illinois and at Cornell. He was on the staff at Cornell from 1948, becoming professor 1962. In 1966 he moved to the Salk Institute for Biological Studies in San Diego, California.

Holley's early work at the New York State Agricultural Experimental Station at Cornell concerned plant hormones, the volatile constituents of fruits, the nitrogen metabolism of plants, and peptide synthesis. At the Salk Institute he began to study the factors that influence growth in cultured mammalian cells.

At Cornell he obtained evidence for the role of tRNAs as acceptors of activated amino acids. In 1958, he succeeded in isolating the alanine-, tyrosine- and valene-specific tRNAs from baker's yeast, and eventually Holley and his colleagues succeeded in solving the entire nucleotide sequence of this RNA.

Holmes Arthur 1890–1965. English geologist who helped develop interest in the theory of continental drift. He also pioneered the use of radioactive decay methods for rock dating, giving the first reliable estimate of the age of the Earth.

Holmes was born in Newcastle-upon-Tyne and studied at Imperial College, London. He was appointed in 1924 head of the Geology Department at Durham, moving in 1943 to Edinburgh University.

Holmes was convinced that painstaking analysis of the proportions in rock samples of elements formed by radioactive decay, combined with a knowledge of the rates of decay of their parent elements, would yield an absolute age. From 1913 he used the uranium–lead technique systematically to date fossils whose relative (stratigraphical) ages were established but not the absolute age.

In 1928, Holmes proposed that convection currents within the Earth's mantle, driven by radioactive heat, might furnish the mechanism for the continental drift theory broached a few years earlier by German geophysicist Alfred Wegener. In Holmes's view, new rocks were forming throughout the ocean ridges. Little attention was given to these ideas until the 1950s.

His books include *The Age of the Earth* 1913, *Petrographic Methods and Calculations* 1921, and *Principles of Physical Geology* 1944.

Hooke Robert 1635–1703. English scientist and inventor, originator of ***Hooke's law***, and considered the foremost mechanic of his time. His inventions included a telegraph system, the spirit level, marine barometer, and sea gauge. He coined the term 'cell' in biology.

Hooke's law, formulated 1676, states that the tension in a lightly stretched spring is proportional to its extension from its natural length.

He studied elasticity, furthered the sciences of mechanics and microscopy, invented the hairspring regulator in timepieces, perfected the air pump, and helped improve such scientific instruments as microscopes, telescopes, and barometers. His work on gravitation and in optics contributed to the achievements of his contemporary Isaac Newton.

Hooke was born in Freshwater on the Isle of Wight and educated at Oxford, where he became assistant to Irish physicist Robert Boyle. Moving to London 1663, he became curator of the newly established Royal Society, which entailed demonstrating new experiments at weekly meetings. He was also professor of geometry at Gresham College, London, from 1665.

In geology, Hooke insisted, against the prevailing, Bible-bound view, that fossils are the remains of plants and animals that existed long ago.

Hooke also designed several buildings, including the College of Physicians, London.

Hooker Joseph Dalton 1817–1911. English botanist who travelled to the Antarctic and India, and made many botanical discoveries. His works include *Flora Antarctica* 1844–47, *Genera plantarum* 1862–83, and *Flora of British India* 1875–97.

In 1865 he succeeded his father, William Jackson Hooker (1785–1865), as director of the Royal Botanic Gardens, Kew, London. Knighted 1877.

Hooker was born in Halesworth, Suffolk, and studied medicine at Glasgow. He joined an expedition to locate the magnetic South Pole 1839–43, in the course of which he visited the Falkland Islands, Tasmania, and New Zealand. From 1847 to 1850 he undertook a botanical exploration of NE India and the Himalayas, and sent back to England many previously unknown species of rhododendron. From 1855 he worked at Kew; under his directorship, the *Index Kewensis* was founded 1883; this is a list of all scientific plant names, accompanied by descriptions.

Hooker Stanley George 1907–1984. English engineer responsible for the development of aircraft engines such as the Proteus turboprop 1957, Orpheus turbojet 1958, Pegasus vectored-thrust turbofan, Olympus turbojet, and RB-211 turbofan. Knighted 1974.

Hooker was born in Sheerness, Kent, and educated at Imperial College, London, and at Oxford. He spent his career at Rolls-Royce (from 1938) and the Bristol Aeroplane Company, which merged with Rolls-Royce 1966.

Hooker developed supercharged versions of the Merlin engine, which powered many Allied aircraft in World War II. In the 1940s he was responsible for the development of the Whittle W2B turbojet and the Derwent engines, two of which powered the plane that set a world speed record 1946.

The Pegasus vectored thrust engine is the power unit for the Harrier VSTOL (vertical/short takeoff and landing) combat aircraft.

The Olympus turbojet, eventually to power the supersonic Concorde airliner, started life in 1946. The design uses two independent compressors driven by two independent turbines, giving the engine high compression (important for fuel economy) and great adaptability.

The RB-211 turbofan was on its introduction the quietest and most economical in airline service, and has a modular design so that replacing parts is easier.

Hooker William Jackson 1785–1865. English botanist and director of Kew Gardens. When he took over Kew Gardens they were only 11 acres, but he increased them to 300 acres, made them a national garden and opened them to the public for the first time. In 1804, he discovered a new moss and also illustrated botanist Dawson Turner's book *Historica fucorum*.

Hooker was born in Norwich and studied estate management at Starston Hall. In 1804, he discovered a new moss which he asked a fellow botanical enthusiast to identify for him and as a result of this meeting, he was introduced to Dawson Turner who employed Hooker to illustrate his *Historica fucorum*. He eventually married Turner's daughter Maria 1815 and bought part of the family brewery in Halesworth. However, he spent little time there, choosing to devote himself almost entirely to his botanical studies.

In 1806, he was made a fellow of the Linnaean Society. In 1809, Joseph Banks arranged for him to be the botanist on a diplomatic mission to Iceland, but all his specimens were destroyed by a fire on the journey home. He was the Regis professor of botany at Glasgow University from 1820, knighted 1836, and made director of Kew Gardens 1841. While holding the post of professor at Glasgow, he was renowned for not attending or delivering a single lecture, although his career did not seem to suffer for it. He organized several expeditions to the Western Highlands and improved the city's botanical gardens.

Hoover William Henry 1849–1932. US manufacturer who developed the vacuum cleaner. 'Hoover' soon became a generic name for vacuum cleaner.

When Hoover's business as a leather manufacturer for carriages and wagons was threatened by the advent of the automobile, he concentrated on developing a primitive existing cleaner into an effective tool for domestic use.

Hopkins Frederick Gowland 1861–1947. English biochemist whose research into diets revealed the necessity of certain trace substances, now known as vitamins, for the maintenance of health. Hopkins shared the 1929 Nobel Prize for Physiology or Medicine with Christiaan Eijkman, who had arrived at similar conclusions. Knighted 1925.

Hopkins also established that there are certain amino acids that the body cannot produce itself. Another discovery he took part in was that contracting muscle accumulates lactic acid.

Hopkins was born in Eastbourne, Sussex, and studied at the University of London and Guy's Hospital Medical School. In 1914 he was appointed professor of biochemistry at Cambridge.

Experimenting on rats fed on artificial milk, Hopkins noticed in 1906 that animals cannot survive on a diet containing only proteins, fats, and carbohydrates. When a small quantity of cow's milk was added, the rats grew. He concluded that the milk must contain accessory food factors in trace amounts, but he failed to isolate these.

Hopper Grace 1906–1992. US computer pioneer who created the first compiler and helped invent the computer language COBOL. She also coined the term 'debug'.

Hopper was educated at Vassar and Yale. She volunteered for duty in World War II with the Naval Ordinance Computation Project. This was the beginning of a long association with the Navy (she was appointed rear admiral 1983). After the war, Hopper joined a firm that eventually would become the Univac division of Sperry-Rand, to manufacture a commercial computer.

In 1945 she was ordered to Harvard University to assist Howard Aiken in building a computer. One day a breakdown of the machine was found to be due to a moth that had flown into the computer. Aiken came into the laboratory as Hopper was dealing with the insect. 'Why aren't you making numbers, Hopper?' he asked. Hopper replied: 'I am debugging the machine!'

Hopper's main contribution was to create the first computer language, together with the compiler needed to translate the instructions into a form that the computer could work with. In 1959, she was invited to join a Pentagon team attempting to create and standardize a single computer language for commercial use. This led to the development of COBOL, still one of the most widely used languages.

Horsley Victor Alexander Haden 1857–1916. English physiologist and one of the founders of neurological surgery. He was the first surgeon to successfully remove tumours from the pituitary gland (endocrine gland in the brain that secretes hormones) and the spinal cord.

He also performed research on the central nervous system, applying electrical stimulation to various points on the cerebral cortex of the brain to show that it caused muscle contraction on the opposite side of the body. These regions of the brain became known as motor regions.

Horsley also investigated ductless glands. He excised the pituitary glands from two dogs and determined that they were still well several months later. He also tested the effects of the removal of the thyroid and parathyroid glands on animals. He determined that loss of thyroid function gave rise to diseases such as myxoedema and cretinism.

Horsley was born in Kensington and graduated in medicine from University College. In 1893, he returned to the College when he was appointed professor of physiology. He was elected a Fellow of the Royal Society 1886 and knighted 1892.

Hounsfield Godfrey Newbold 1919– . English engineer, a pioneer of tomography, the application of computer techniques to X-raying the human body. He shared the Nobel Prize for Physiology or Medicine 1979. Knighted 1981.

Hounsfield was born in Newark, Nottinghamshire, and studied at the City and Guilds College, London, and the Faraday House Electrical Engineering College. He joined British electronics company EMI (now Thorn EMI) 1951 as a researcher in medical technology.

The EMI scanner he invented 1972, a computerized transverse axial tomography system for X-ray examination, enables the whole body to be screened at one time. The X-ray crystal detectors, more sensitive than film, are rotated round the body and can distinguish between, for example, tumours and healthy tissue.

Howe Elias 1819–1867. US inventor, in 1846, of a sewing machine using two threads, thus producing a lock stitch.

Howe was born in Spencer, Massachusetts, and trained as a machinist. He began work on the design of a sewing machine in about 1843, and in 1846 he was granted a US patent for a practical machine. He was the first to patent a lock-stitch mechanism, and his machine had two other important features: a curved needle with the eye (for the thread) at the point, and an under-thread shuttle (this had been invented by a Walter Hunt in 1834, probably unbeknown to Howe).

Howe went to the UK and sold the invention for £250 to a corset manufacturer named William Thomas of Cheapside, London.

Thomas secured the British patent in his own name, and although Howe worked with Thomas until 1849, his career in London was unsuccessful. Returning in poverty to the USA, he found that various people – among them Isaac Singer – were making machines similar to his own. Howe redeemed his patent, which he had pawned, and sued for infringement. The courts eventually found in his favour and Howe became a millionaire on his royalties. In his last years he manufactured the machine in Bridgeport, Connecticut.

Hoyle Fred(erick) 1915– . English astronomer, cosmologist, and writer. In 1948 he joined with Hermann Bondi and Thomas Gold in developing the steady-state theory of the universe. In 1957, with William Fowler, he showed that chemical elements heavier than hydrogen and helium may be built up by nuclear reactions inside stars. Knighted 1972.

According to Hoyle's theory on gravitation, matter is not evenly distributed throughout space, but forms self-gravitating systems. These may range in diameter from a few kilometres to a million light years. Formed from clouds of hydrogen gas, they vary greatly in density.

Hoyle has suggested that life originated in bacteria and viruses contained in the gas clouds of space and was then delivered to the Earth by passing comets. His first science-fiction novel was *The Black Cloud* 1957; he has also written many popular science books.

Hoyle was born in Bingley, Yorkshire, and studied at Cambridge. His academic career was spent at Cambridge and 1956–66 at Mount Palomar Observatory, California. On his return he became director of the Cambridge Institute of Theoretical Astronomy.

Space isn't remote at all. It's only an hour's drive away if your car could go straight upwards.

Fred Hoyle

Observer Sept 1979

Fowler and Hoyle proposed that all the elements may be synthesized from hydrogen by successive fusions. When the gas cloud reaches extremely high temperatures, the hydrogen has turned to helium and neon, whose nuclei interact, releasing particles that unite to build up nuclei of new elements.

His work on the evolution of stars was published in *Frontiers of Astronomy* 1955. His science fiction is generally set in the near future and, starting with *Fifth Planet* 1963, cowritten with his son Geoffrey Hoyle (1942–).

Hubble Edwin Powell 1889–1953. US astronomer who discovered the existence of other galaxies outside our own, and classified them according to their shape. His theory that the universe is expanding is now generally accepted.

Hubble discovered Cepheid variable stars in the Andromeda galaxy 1923, proving it to lie far beyond our own Galaxy. In 1925 he introduced the classification of galaxies as spirals, barred spirals, and ellipticals. In 1929 he announced ***Hubble's law***, which states that the galaxies are moving apart at a rate that increases with their distance.

Hubble was born in Marshfield, Missouri, and studied at Chicago and in the UK at Oxford. He briefly practised law before

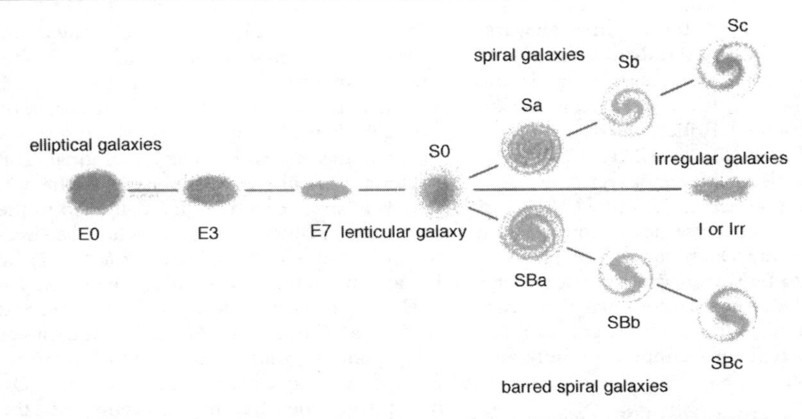

Hubble *Hubble's classification, still used today, is based on a galaxy's shape. Elliptical galaxies range from the almost spherical E0 type to the markedly elliptical E7. Type S0 – a spiral disc without arms – marks the transition from elliptical to spiral types. Types Sa, Sb and Sc are normal spirals with progressively looser arms. Barred spirals, types SBa, SBb and SBc, are also differentiated by the degree of openness of their arms.*

returning to Chicago to join Yerkes Observatory 1914. From 1919 he worked at Mount Wilson, near Pasadena, California.

His data on the speed at which galaxies were receding (based on their red shifts) were used to determine the portion of the universe that we can ever come to know, the radius of which is called the ***Hubble radius***. Beyond this limit, any matter will be travelling at the speed of light, so communication with it will never be possible. The ratio of the velocity of galactic recession to distance has been named the ***Hubble constant***.

Hubel David Hunter 1926– . US neurophysiologist who worked with Torsten Wiesel (1924–) on the physiology of vision and the way in which the higher centres of the brain process visual information. They shared the 1981 Nobel Prize for Physiology or Medicine.

Hubel was born in Ontario, Canada, and studied at McGill University in Montreal. From 1959 he worked at Harvard, becoming professor 1965.

At Harvard he met Wiesel, and they began experiments implanting electrodes into the brain of anaesthetized cats, correlating the anatomical structure of the visual cortex of the brain with the physiological responses to different types of visual stimulation. They built up a complex picture of how the brain analysed visual information by an increasingly sophisticated system of detection by the nerve cells.

Later study of the development of the visual system in young animals suggested that eye defects should be treated and corrected immediately, and the then routine ophthalmological practice of leaving a defect to correct itself was abandoned.

Huber Robert 1937– . German chemist who shared the Nobel Prize for Chemistry 1988 with Hartmut Michel and Johann Deisenhofer for his use of high resolution X-ray crystallography to successfully determine the molecular structure of the photosynthetic reaction centre in *Rhodopseudomonas viridris*. Huber pioneered the use of X-ray crystallography to study biological macromolecules, such as enzymes, and most notably, membrane-bound proteins (first crystallized by Michel) involved in photosynthesis.

Huber was born in Germany where he later attended the Technical University, obtaining his PhD in 1972. He advanced from lecturer to associate professor in 1976, and was appointed director of the Max Planck Institute for Biochemistry, Martinsried in 1971.

Hückel Erich Armand Arthur Joseph 1896–1980. German physical chemist who, with Peter Debye, developed in 1923 the modern theory that accounts for the electrochemical behaviour of strong electrolytes in solution. Hückel also made discoveries relating to the structures of benzene and similar compounds that exhibit aromaticity.

Hückel was born in Berlin and studied at Göttingen. He worked with Debye at Göttingen and the Eidgenössische Technische Hochschule, Zürich, and held various academic posts, becoming professor of theoretical physics at the University of Marburg 1937.

In 1930, Hückel began his work on aromaticity, the basis of the chemical behaviour of benzene, pyridine, and similar compounds. He developed a mathematical approximation for the evaluation of certain integrals in the calculations concerned with the exact nature of the bonding in benzene. Extending his research to other, similar chemical systems, he formulated the Hückel rule for monocyclic systems, which states that for aromaticity to occur the number of electrons contributing to the correct type of bonding (π-bonding) must be $4n + 2$, where n is a whole number. In large ring systems, however, the predicted aromaticity does not occur.

Hudson William 1734–1793. English botanist and apothecary who furthered the use and understanding of Linnaeus' system of plant classification, a taxonomic system that is based on stamens and pistils. Hudson's *Flora Anglica* was essentially a rearrangement of John Ray's *Synopsis methodica Stirpium Britanniacarum* according to Linnaeus' system of classification.

Hudson was born in Kendal in the English Lake District and apprenticed to an apothecary in London, winning the Apothecaries Company's prize for botany. In 1757, he was made sub-librarian of the British Museum, where he encountered the botanical writing of Linnaeus. He was elected a Fellow of the Royal Society 1761 and a member of the Linnaean Society 1791.

From 1765–71, he was the director of the Apothecaries Company at Chelsea Physic Garden in London. During this time Hudson also compiled an extensive insect collection and was distraught when this and his large botanical collections were destroyed in a fire at his home 1783. It is thought that he never really recovered from this event, although he worked on as an apothecary in London until his death from paralysis in 1793.

Huggins Charles Brenton 1901– . Canadian-born US physician who shared the Nobel Prize for Physiology or Medicine 1966 with Peyton Rous for his work on cancer, particularly the effects of hormones on prostate and breast cancer.

Huggins developed an interest in the hormonal treatment of cancers of the male urinogenital tract, particularly the prostate gland (the gland surrounding and opening into the urethra at the base of the bladder in males).

Huggins hypothesized that since male sex hormones largely influence the activity of the prostate gland, removing or neutralizing these hormones might affect the cancer. In 1941, he and his colleagues reported a series of experiments in which carcinoma of the prostate had been treated either by castration or by the administration of the female sex hormone oestrogen, or by both methods combined. The results were unexpectedly favourable even when extensive bony metastases were present.

The success of this treatment was so striking that very soon it was adopted by surgeons in all parts of the world. He went on to extend these studies to suggest that some forms of breast cancer in women might also respond to hormonal therapy.

Huggins was born in Halifax, Nova Scotia in Canada. In 1936, he was appointed professor of surgery at the University of Chicago.

Huggins William 1824–1910. English astronomer and pioneer of astrophysics. He revolutionized astronomy by using spectroscopy to determine the chemical make-up of stars and by using photography in stellar spectroscopy.

Huggins was born in London, where he ran the family drapery business until 1854. He then built a private observatory at Tulse Hill, London, and devoted himself entirely to science.

In 1860, with his friend W A Miller (a professor of chemistry), Huggins designed a spectroscope and attached it to the telescope. By observing the spectral lines of stars, he established that the universe was made up of well-known elements. At that time, some nebulae had been observed to be faint clusters of stars, but others could not be resolved without more powerful telescopes. Huggins realized that if they were composed of stars, they would give a characteristic stellar spectrum. However, when he turned to the unresolved nebulae in the constellation of Draco in 1864, only a single bright line was observed. Seeing this, he understood the nature of unresolved nebulae: they were clouds of luminous gas and not clusters of stars.

Hughes David 1831–1900. British-born US inventor who patented an early form of telex in 1855, a type-printing instrument for use with the telegraph. In 1857 he took the instrument to Europe, where it became widely used. *See illustration on page 246.*

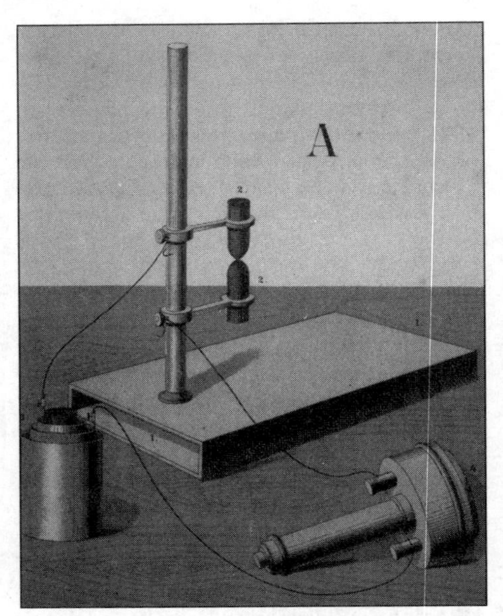

Hughes *A carbon microphone invented by David Hughes. Sound produces vibrations in a diaphragm that covers a layer of carbon granules. Compression of the carbon granules causes their resistance to vary. A current run across the carbon therefore also varies in response to the sound waves. Sound can be either amplified or transmitted using the current.*

Hulse Russell 1950– . US physicist and radio astronomer who discovered a new type of pulsar, called a binary pulsar, with Joseph H Taylor. The pulsar had an invisible companion orbiting around it and radiated gravitational waves. These waves had been predicted by Einstein's general theory of relativity but this was the first experimental confirmation of their reality. Nobel Prize for Physics 1993.

Hulse and Taylor discovered the pulsar 1974, when Hulse was a postgraduate student, supervised by Taylor, at the University of Massachusetts in Amherst, Massachusetts, USA. The two astronomers detected a small regular change in the intervals between the radio pulses emitted by a pulsar called PSR1913+16. This occurred because the pulsar was orbiting another body. Furthermore the pulses were slowing down in a regular way, losing about 75 microseconds a year. This was due to energy being lost as gravitational waves were emitted.

In 1977 Hulse moved to the Plasma Physics Laboratory at Princeton University, New Jersey, to work on computer models of how plasmas – very hot ionized gases – behave when subjected to a strong magnetic field.

Humason Milton Lasell 1891–1972. US astronomer who carried out a spectroscopic study of distant galaxies, determining their velocities from their red shift.

Humason, born in Dodge Centre, Minnesota, dropped out of school at 14 to hang around Mount Wilson Observatory, California. For a while he was a mule driver for the pack trains that carried construction materials to the observatory from the Sierra Madre. In 1917 he joined the observatory staff as a janitor, but was quickly promoted, becoming assistant astronomer 1919. From 1954 he was astronomer at the Mount Wilson and Palomar Observatories.

At Mount Wilson Observatory Humason took part in an extensive study of the properties of galaxies, initiated by Edwin Hubble 1928. The work consisted of making a series of systematic spectroscopic observations to test and extend the relationship that Hubble had found between the red shifts and the apparent magnitudes of galaxies. But because of the low surface brightness of galaxies there were severe technical difficulties. Humason developed the technique and made most of the exposures and plate measurements. The velocities of 620 galaxies were measured, and the results, published 1956, still represent the majority of known values of radial velocities for normal galaxies.

Humboldt Friedrich Wilhelm Heinrich Alexander, Baron von 1769–1859. German geophysicist, botanist, geologist, and writer who, with French botanist Aimé Bonpland (1773–1858), explored the regions of the Orinoco and Amazon rivers in South America 1800–04, and gathered 60,000 plant specimens. He was a founder of ecology.

The separate branches of natural knowledge have a real and intimate connection.

Friedrich Humboldt

Cosmos 1845

Humboldt aimed to erect a new science, a 'physics of the globe', analysing the deep physical interconnectedness of all terrestrial phenomena. He believed that geological phenomena were to be understood in terms of basic physical causes (for example, terrestrial magnetism or rotation).

Humboldt *German geologist, naturalist and explorer Alexander von Humboldt shown during his pioneering expedition across South America. Humboldt, often described as the founder of ecology, gathered a massive collection of geological, botanical, and zoological, specimens, and, on his return to Europe in 1804, aimed to create a new science – a 'physics of the globe' – based on his observations.*

One of the first popularizers of science, he gave a series of lectures later published as *Kosmos/Cosmos* 1845–62, an account of the relations between physical environment and flora and fauna.

Humboldt was born in Berlin and studied at Göttingen and the Mining Academy, Freiburg. He travelled widely in Europe before the expedition to South and Central America. Surveying, mapping, and gathering information, Humboldt covered some 9,600 km/6,000 mi. The accounts of his travels were written over the next 20 years.

In meteorology, he introduced isobars and isotherms on weather maps, made a general study of global temperature and pressure, and instituted a worldwide programme for compiling magnetic and weather observations. His studies of American volcanoes demonstrated they corresponded to underlying geological faults; on that basis he deduced that volcanic action had been pivotal in geological history and that many rocks were igneous in origin. In 1804, he discovered that the Earth's magnetic field decreased from the poles to the equator.

Hume-Rothery William 1899–1968. British metallurgist who studied the constitution of alloys. He established that the microstructure of an alloy depends on the different sizes of the component atoms, the valency electron concentration, and electrochemical differences.

Hume-Rothery was born in Worcester Park, Surrey, and educated at Oxford and the Royal School of Mines. His research was carried out at Oxford but financed by outside organizations such as the Armourers and Braziers Company, and he did not gain a formal university post until 1938. During World War II he supervised many government contracts for work on complex aluminium and magnesium alloys. He was appointed to the first chair of metallurgy 1958.

With atoms of widely different sizes, at least two types of crystal lattice may form, one rich in one metal and one rich in the other. The presence of two types of structure can increase the strength of an alloy. This is why some brasses are much stronger than their component metals zinc and copper.

If the two elements differ considerably in electronegativity, a definite chemical compound is formed. Thus steel, an 'alloy' of iron and carbon, contains various iron carbides. Hume-Rothery and his team constructed the equilibrium diagrams for a great number of alloy systems.

Hunter John 1728–1793. Scottish surgeon, pathologist, and comparative anatomist who insisted on rigorous scientific method. He was the first to understand the nature of digestion.

Hunter was born in Lanarkshire and trained in London under his elder brother William Hunter (1718–1783), anatomist and obstetrician, who became professor of anatomy in the Royal Academy 1768 and president of the Medical Society 1781. His collection of specimens and preparations is now in the Hunterian Museum of Glasgow University; John Hunter's is housed in the Royal College of Surgeons, London.

He experimented extensively on animals, and kept a number of animal specimens in his garden for dissection. He also dissected human bodies obtained from 'resurrectionists', who raided graveyards at night to sell newly buried corpses to surgeons. During the late 1760s he took up a senior surgical post at St George's Hospital, London, and was appointed physician to George III. Serving on the army surgical staff during the Seven

Years' War, he gained the knowledge for a treatise on gunshot wounds.

Hunter made studies of lymph and blood circulation, the sense of smell, the structure of teeth and bone, and various diseases. Experimenting with the transplantation of tissues, he fixed a human tooth into a cock's comb. He often carried out experiments on himself, and eventually died of syphilis with which he had injected himself in an attempt to prove it to be a type of gonorrhoea.

Hussey Obed 1792–1860. US inventor who developed one of the first successful reaping machines 1833 and various other agricultural machinery.

Hussey was born in Maine. His reaping machine used the principle of a reciprocating knife cutting against stationary guards or figures. The cutter was attached to a crank activated by gearing, connected to one of the wheels. The contraption was pulled by horses walking alongside the standing grain. During the harvest of 1834, he demonstrated the reaper to farmers and began to sell the machines. In 1851 he went to Britain and demonstrated the reaper at Hull and Barnardscastle. He was invited to show it to Prince Albert, who bought two (at £21 each).

An earlier rival design of reaper had been developed by Cyrus McCormick in 1831, although Hussey's was patented first.

Hussey also invented a steam plough, a machine for making hooks and eyes, a grinding mill for maize and a horse-powered husking machine, a sugar-cane crusher, and an ice-making machine.

Hutton James 1726–1797. Scottish geologist, known as the 'founder of geology', who formulated the concept of uniformitarianism. In 1785 he developed a theory of the igneous origin of many rocks.

His *Theory of the Earth* 1788 proposed that the Earth was incalculably old. Uniformitarianism suggests that past events could be explained in terms of processes that work today. For example, the kind of river current that produces a certain settling pattern in a bed of sand today must have been operating many millions of years ago, if that same pattern is visible in ancient sandstones.

Born in Edinburgh, Hutton studied there and at Paris and Leiden, the Netherlands. He settled in Edinburgh about 1768 and played a large role in the Scottish Enlightenment.

Hutton's theory of uniformitarianism was vehemently attacked in its day, partly on fundamentalist Christian grounds. It was later popularized by another Scottish geologist, Charles Lyell, and still forms the groundwork for much geological reasoning.

Having in the natural history of this earth, seen a succession of worlds, we may conclude that there is a system in nature. ... The result, therefore of our present enquiry is, that we find no vestige of a beginning – no prospect of an end.

James Hutton

Transactions of the Royal Society of Edinburgh 1788

Huxley Andrew Fielding 1917– . English physiologist, awarded the Nobel prize 1963 with Alan Hodgkin for work on nerve impulses, discovering how ionic mechanisms are used in nerves to transmit impulses. Knighted 1974.

Huxley was born in London, the grandson of scientist T H Huxley. He was educated at Cambridge and did military research during World War II. After the war, he returned to Cambridge. In 1960 he became professor at University College, London.

In 1945 at Cambridge, Hodgkin and Huxley began to measure the electrochemical behaviour of nerve membranes. They experimented on axons of the giant squid – each axon is about 0.7 mm/0.03 in in diameter. They inserted a glass capillary tube filled with sea water into the axon to test the composition of the ions in and surrounding the cell, which also had a microelectrode inserted into it. Stimulating the axon with a pair of outside electrodes, they showed that the inside of the cell was at first negative (the resting potential) and the outside positive, and that during the conduction of the nerve impulse the membrane potential reversed.

Huxley Hugh Esmor 1924– . English physiologist who, using the electron microscope and thin slicing techniques, established the detailed structural basis of muscle contraction.

Huxley was born in Birkenhead, Cheshire, and studied at Cambridge. He returned to Cambridge as a lecturer 1961.

Muscle fibres contain a large number of longitudinally arranged myofibrils, which, Huxley demonstrated, are composed of thick and thin filaments of the proteins myosin and actin. He showed how the filaments are attached to one another in a woven pattern, and suggested that muscle contraction is

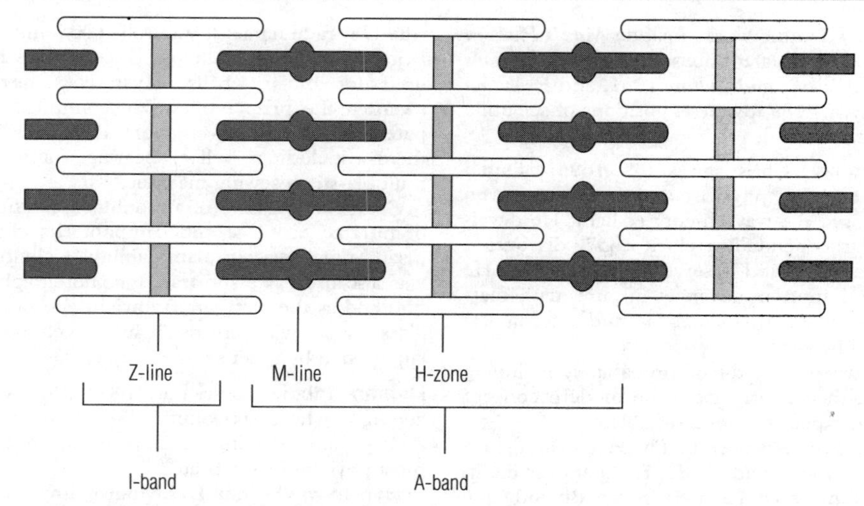

Huxley, Hugh *Banding in a fibril of striated muscle.*

brought about by sliding movements of two sets of filaments.

He discovered that the myosin filaments are able to aggregate under suitable conditions to form 'artificial' filaments of varying lengths. He proposed that the 'tails' of the myosin molecules become attached to each other to form a filament with the heads projecting from the body of the filament, and that this plays an important part in the sliding effect.

By coincidence, Andrew Huxley (no relation), working separately, came to the same conclusions at about the same time in the 1950s, although they disagree on the exact details.

Operationally, God is beginning to resemble not a ruler but the last fading smile of a cosmic Cheshire cat.

Julian Huxley

Religion without Revelation

Huxley Julian Sorell 1887–1975. English biologist, first director general of UNESCO, and a founder of the World Wildlife Fund (now the World Wide Fund for Nature). He wrote popular science books, including *Essays of a Biologist* 1923. Knighted 1958.

Huxley Thomas Henry 1825–1895. English scientist and humanist. Following the publication of Charles Darwin's *On the Origin of Species* 1859, he became known as 'Darwin's bulldog', and for many years was a prominent champion of evolution. In 1869, he coined the word 'agnostic' to express his own religious attitude, and is considered the founder of scientific humanism.

It is the customary fate of new truths to begin as heresies and to end as superstitions.

T H Huxley

Science and Culture, 'The Coming of Age of the Origin of Species'

From 1846 to 1850 Huxley was the assistant ship's surgeon on HMS *Rattlesnake* on its voyage around the South Seas. The observations he made on the voyage, especially of invertebrates, were published and made his name in the UK.

Huxley was born in London and studied medicine there at Charing Cross Hospital. In 1854 he became professor of natural history at the Royal School of Mines. His grandsons include Aldous, Andrew, and Julian Huxley.

Huxley found the system of classification introduced by French anatomist Georges Cuvier to be inadequate for the sea creatures he studied on his voyage. He reclassified the animal kingdom into Annuloida, Annulosa, Infusoria, Coelenterata, Mollusca, Molluscoida, Protozoa, and Vertebrata, and started a fundamental revision of the Mollusca. He also produced a new system of classification of birds, based mainly on the palate and other bony structures, which is the foundation of the modern system.

His scientific works include *Man's Place in Nature* 1863; later books, such as *Lay Sermons* 1870, *Science and Culture* 1881, and *Evolution and Ethics* 1893, were expositions of scientific humanism.

Huygens Christiaan 1629–1695. Dutch mathematical physicist and astronomer who proposed the wave theory of light. He developed the pendulum clock 1657, discovered polarization, and observed Saturn's rings. He made important advances in pure mathematics, applied mathematics, and mechanics, which he virtually founded.

Huygens's study of probability, including game theory, originated the modern concept of the expectation of a variable.

Huygens was born in The Hague and studied at Leiden and Breda. He spent his life in research, based 1666–81 at the Bibliothèque Royale, Paris.

Huygens also studied centrifugal force and showed, in 1659, its similarity to gravitational force; his gravitational theories successfully deal with several difficult points that his English colleague Isaac Newton carefully avoided. In the 1670s, Huygens studied motion in resisting media, becoming convinced by experiment that the resistance in such media as air is proportional to the square of the velocity.

Huygens's study of geometric optics led to the invention of a telescope eyepiece that reduced chromatic aberration. With a home-made telescope, he discovered Titan, one of Saturn's moons, 1655. *Systema Saturnium* included observations on the planets, their satellites, Saturn's rings, the Orion nebula, and the determination of the period of Mars.

Horologium oscillatorium 1673 sets out his work on the pendulum and harmonically oscillating systems. *Traité de la lumière*, containing Huygens's wave, or pulse, theory of light, was published 1678.

Hyatt John Wesley 1837–1920. US inventor who in 1869 invented celluloid, the first artificial plastic, intended as a substitute for ivory. It became popular for making a wide range of products, from shirt collars and combs to toys and babies' rattles, and is still used in the manufacture of table-tennis balls.

Hyatt was born in Starkey, New York, and worked in Illinois as a printer. In the early 1860s, the New York company of Phelan and Collender offered a prize of $10,000 for a satisfactory substitute for ivory for making billiard balls. Using pyroxylin, a partly nitrated cellulose, Hyatt developed celluloid (the US trade name: it was called Xylonite in Britain). Although celluloid did come to be used for billiard balls, Hyatt was never awarded the prize money. He continued to patent his inventions – more than 200 of them, including roller bearings and a multiple-stitch sewing machine.

Celluloid consisted of a mouldable mixture of nitrated cellulose and camphor. Its chief disadvantage was its flammability. Celluloid was also used as a substrate for photographic film and as the filling in sandwich-type safety glass for car windscreens. It has largely been superseded by other synthetic materials.

Hyman Libbie Henrietta 1888–1969. US zoologist whose six-volume *The Invertebrates* 1940–68 provided an encyclopedic account of most phyla of invertebrates.

Hyman was born in Des Moines, Iowa, and studied at the University of Chicago, where she remained as a research assistant until 1930. She then travelled to several European laboratories, working for a period at the Stazione Zoologica, Naples, Italy, before returning to New York City to begin to write a comprehensive reference book on the invertebrates, for which she was given office and laboratory space, but no salary, by the American Museum of Natural History.

Initially she worked on flatworms, but soon extended her investigations to a wide spread of invertebrates, especially their taxonomy (classification) and anatomy.

Hypatia *c.* 370–*c.* 415. Greek philosopher, born in Alexandria. She studied Neo-Platonism in Athens, and succeeded her father Theon as professor of philosophy at Alexandria. She was murdered, it is thought by Christian fanatics.

Hyrtl Joseph 1810–1894. Austrian anatomist who, in part, was responsible for the success of the New Vienna School of Medicine in the 19th century. He was the most popular and successful teacher of anatomy in Europe. His textbooks of general anatomy were widely used including his seminal work *The Handbook of Topographical Anatomy* 1845. He is also the author of scholarly works on anatomical terminology and on Hebrew and Arabic elements of anatomy.

After studying medicine in Vienna, Hyrtl was appointed professor of anatomy at the Medical School in Prague at the relatively young age of 27. He returned to a similar post in Vienna seven years later and became an influential man in the establishment of new procedures for teaching human anatomy.

I

Ingenhousz Jan 1730–1799. Dutch physician and plant physiologist who established in 1779 that in sunlight plants absorb carbon dioxide and give off oxygen. He found that plants, like animals, respire all the time and that respiration occurs in all the parts of plants.

Ingenhousz was born in Breda and studied at Louvain and Leiden and abroad at Paris and Edinburgh, after which he set up a private medical practice in Breda. In 1765, he left for England, going to work at the Foundling Hospital, London, where he successfully inoculated patients against smallpox (using the hazardous live virus). In 1768 he was sent to the Austrian court by George III, to inoculate the royal family, and became court physician there 1772–79. He then returned to England.

Apart from studying plants, Ingenhousz developed in 1776 an improved apparatus for generating large amounts of electricity; he also invented a hydrogen-fuelled lighter to replace the tinderbox, and investigated the use of an air and ether vapour mixture as a propellant for an electrically fired pistol.

Ingenhousz's work *Experiments On Vegetables, Discovering their Great Power of Purifying the Common Air in Sunshine, and of Injuring it in the Shade or at Night* 1779 laid the foundations for the study of photosynthesis.

Ingold Christopher Kelk 1893–1970. English organic chemist who specialized in the concepts, classification, and terminology of theoretical organic chemistry. He explained the mechanisms of organic reactions in terms of the behaviour of electrons in the molecules concerned. Knighted 1958.

Ingold was born in London and studied at Southampton. He became professor at Leeds 1924, moving 1930 to University College, London.

In 1926 Ingold put forward the concept of mesomerism, which allows a molecule to exist as a hybrid of a pair of equally possible structures. His ideas, first published in 1932, are still fundamental to understanding reaction mechanisms. They concerned the role of electrons in elimination and nucleophilic aliphatic substitution reactions, which he interpreted in terms of ionic organic species.

His *Structure and Mechanisms in Organic Chemistry* 1953 is a classic reference book.

Ipatieff Vladimir Nikolayevich 1867–1952. Russian-born US organic chemist who developed catalysis in organic chemistry, particularly in reactions involving hydrocarbons.

Ipatieff was born in Moscow, became an officer in the Imperial Russian Army, and studied at the Mikhail Artillery Academy in St Petersburg and in Germany at Munich, as well as a brief period in France studying explosives. He returned to Russia 1899 as professor of chemistry and explosives at the Mikhail Artillery Academy. During World War I and the Russian Revolution, he held various administrative and advisory appointments. At the age of 64, he defected to the USA. From 1931 to 1935 he was professor at Northwestern University, Illinois, and consultant to the Universal Oil Products Company, Chicago, which funded the building of the Ipatieff High Pressure Laboratory at Northwestern 1938.

In 1900 Ipatieff discovered the specific nature of catalysis in high-temperature organic gas reactions, and how using high pressures the method could be extended to liquids. He developed an autoclave called the Ipatieff bomb for heating liquid compounds to above their boiling points under high pressure. He synthesized methane and produced polyethylene by polymerizing ethylene (ethene).

In Chicago, Ipatieff began to apply his high-temperature catalysis reactions to petrol to give it a higher octane rating. Important for the production of aviation fuel during World War II, the method is still used.

Isaacs Alick 1921–1967. Scottish virologist who, with Swiss colleague Jean Lindemann, in 1957 discovered interferon, a naturally occurring antiviral substance produced by cells infected with viruses. The full implications of this discovery are still being investigated.

Isaacs was born and educated in Glasgow. From 1951 he worked at the World Influenza Centre, London, becoming its director 1961.

Isaacs began in 1947 studying different strains of the influenza virus and the body's response to them. Working with Lindemann, he eventually found that when a virus invades a cell, the cell produces interferon, which then induces uninfected cells to make a protein that prevents the virus from multiplying. Almost any cell in the body can make interferon, which seems to act as the first line of defence against viral pathogens, because it is produced very quickly (interferon production starts within hours of infection whereas antibody production takes several days) and is thought to trigger other defence mechanisms.

Issigonis Alec (Alexander Arnold Constantine) 1906–1988. Turkish-born British engineer who designed the Morris Minor 1948 and the Mini-Minor 1959 cars, comfortable yet cheaper to run than their predecessors. He is credited with adding the word 'mini' to the English language. Knighted 1969.

Overseeing the separate approaches of styling, interior packaging, body engineering, and chassis layout, Issigonis conceived the overall vehicle; specialists in his team then designed and engineered the subsystems of the car. His designs gave much greater space for the occupants together with greatly increased dynamic handling stability.

Issigonis was born in Smyrna (now Izmir), and studied engineering at Battersea Polytechnic. In 1936 he joined Morris Motors to work on suspension design. His first complete car design was the Morris Minor, which brought new standards of steering and stability to small cars. A record 1 million Morris Minors had been sold in the UK by 1961.

After working at Avis 1952–56 on an experimental car, Issigonis returned to what had now become BMC. In the wake of the Suez Crisis 1956, which threatened oil supplies, he was asked to design and produce a small and economical car, and the result was the Mini. His other major designs were the 1100 (1962), the 1800 (1964), and the Maxi (1969).

Ives Frederic Eugene 1856–1937. US inventor who developed the halftone process of printing photographs in 1878. The process uses a screen to break up light and dark areas into dots. By 1886 he had evolved the halftone process now generally in use. Among his many other inventions was a three-colour printing process (similar to the four-colour process).

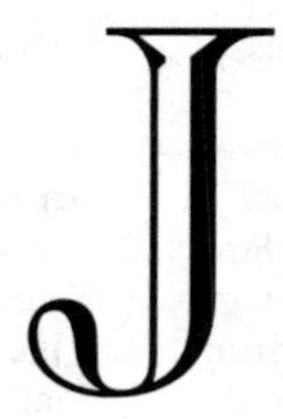

Jackson John Hughlings 1835–1911. English neurologist and neurophysiologist. As a result of his studies of epilepsy, Jackson demonstrated that specific areas of the cerebral cortex (outer mantle of the brain) control the functioning of particular organs and limbs.

He also demonstrated that Helmholtz's ophthalmoscope is a crucial diagnostic tool for disorders of the nervous system.

Jacob François 1920– . French biochemist who, with Jacques Monod and André Lwoff, pioneered research into molecular genetics and showed how the production of proteins from DNA is controlled. They shared the Nobel Prize for Physiology or Medicine 1965.

Jacob was born in Nancy and studied at the University of Paris. In 1950 he joined the Pasteur Institute in Paris as a research assistant, becoming head of the Department of Cellular Genetics 1964 and also professor of cellular genetics at the Collège de France.

Jacob began his work on the control of gene action in 1958, working with Lwoff and Monod. It was known that the types of proteins produced in an organism are controlled by DNA, and Jacob focused his research on how the amount of protein is controlled. He performed a series of experiments in which he cultured the bacterium *Escherichia coli* in various mediums to discover the effect of the medium on enzyme production. He and his team found that there were three types of gene concerned with the production of each specific protein.

Myths and science fulfil a similar function: they both provide human beings with a representation of the world and of the forces that are supposed to govern it.

François Jacob

The Possible and the Actual 1982

Jacobi Carl Gustav Jacob 1804–1851. German mathematician and mathematical physicist, much of whose work was on the theory of elliptical functions, mathematical analysis, number theory, geometry, and mechanics.

Jacobi was born in Potsdam. A child prodigy, he went to Berlin University in 1821 and graduated in the same year. In 1826 he joined the staff at Königsberg (now

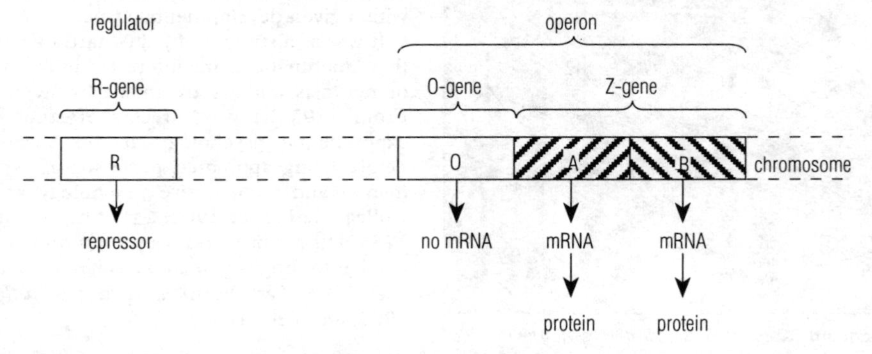

Jacob *A regulator (R-gene) produces a repressor substance that prevents an operator (O-gene) from providing messenger RNA, blocking the production of protein.*

Kaliningrad) University, becoming professor 1832.

Jacobi invented a functional determinant – now called the ***Jacobian determinant*** – which has been of considerable use in later analytical investigations, and was even supportive in the development of quantum mechanics. He advanced the theory of the configurations of rotating liquid masses by showing that the ellipsoids now known as ***Jacobi's ellipsoids*** are figures of equilibrium.

Jacobi was always trying to link together different mathematical disciplines. For instance, he introduced elliptic functions into number theory and into the theory of integration, which in turn connected with the theory of differential equations and his own principle of the last multiplier.

His book *Fundamenta nova theoriae functionum ellipticarum* 1829 introduced his own concept of hyperelliptic functions.

Jacquard Joseph Marie 1752–1834. French textile manufacturer who invented a punched-card system for programming designs on a carpetmaking loom. In 1801 he constructed looms that used a series of punched cards to control the pattern of longitudinal warp threads depressed before each sideways passage of the shuttle. On later machines the punched cards were joined to form an endless loop that represented the 'program' for the repeating pattern of a carpet.

Jacquard-style punched cards were used in the early computers of the 1940s–1960s.

Jacquard was born in Lyon and inherited a small weaving business. He invented the machine after becoming bankrupt, and worked on improving it at the Paris Conservatoire des Arts et Métiers from 1804. In Lyon and elsewhere, his machines were smashed by weavers who feared unemployment. By 1812 there were 11,000 Jacquard looms working in France, and they were introduced into many other countries.

Jacquard's attachment for pattern weaving, which was later improved by others, allowed patterns to be woven without the intervention of the weaver. Weavers had always had to plan the pattern before they began their task. This planning now became the essential feature of the weaver's job, and once the pattern had been punched onto cards, it could be used over and over again.

Jacquard *Joseph-Marie Jacquard demonstrating his loom that was able to weave complicated patterns when 'programmed' with punched cards.*

Jacuzzi Candido 1903–1986. Italian-born US engineer who invented the Jacuzzi, a pump that produces a whirlpool effect in a bathtub. The Jacuzzi was commercially launched as a health and recreational product in the mid-1950s.

Janet Pierre Marie Félix 1859–1947. French psychiatrist. He is known for his detailed work on neurosis. His early work focused on the ***psychasthenias***, a term he coined to describe anxiety, phobias, and obsessional disorders. He went on to formulate a comprehensive theory of development, in which abnormal behaviour and neuroses were seen to result from an individual's failure to fully integrate psychological functions associated with a given developmental stage.

It was as a student of Jean-Martin Charcot that Janet first became interested in the study of neurosis and its treatment by hypnosis. From 1895 he was director of studies in experimental psychology at the Sorbonne before being appointed professor of experimental and comparative psychology at the Collège de France 1902, a post he held until 1936. His major works, as yet not fully translated into English, include *Névroses et idées fixes* 1898, *Les Obsessions et la psychasthenie* 1903, and *Les Névroses* 1905.

Jansky Karl Guthe 1905–1950. US radio engineer who in 1932 discovered that the

Jansky *Radio engineer Karl Jansky pictured with an aerial array. The photograph was taken in 1928 in New Jersey. Jansky became the first radio astronomer when he detected radio waves originating in the Milky Way.*

Milky Way galaxy emanates radio waves; he did not follow up his discovery, but it marked the birth of radio astronomy.

Jansky was born in Norman, Oklahoma, and studied at the University of Wisconsin. In 1928 he joined the Bell Telephone Laboratories, New Jersey, where he investigated causes of static that created interference on radio-telephone calls.

Jansky noticed that the background hiss on a loudspeaker attached to his specially built receiver and antenna system reached a maximum intensity every 24 hours. It seemed to move steadily with the Sun but gained on the Sun by four minutes per day. This amount of time correlates with the difference of apparent motion, as seen on Earth, between the Sun and the stars, so Jansky surmised that the source must lie beyond the Solar System. By 1932, he had concluded that the source lay in the direction of Sagittarius – the centre of our Galaxy.

Janssen (Pierre) Jules César 1824–1907. French astronomer who studied the solar spectrum; he developed a spectrohelioscope 1868. In 1867 he concluded that water vapour was present in the atmosphere of Mars.

Janssen was born and educated in Paris, and built an observatory on the flat roof of his house there. He was made professor of physics at the Ecole Spéciale d'Architecture 1865, established an observatory on Mont Blanc in the Alps, and in 1875 became director of the new astronomical observatory at Meudon, near Paris. Scientific expeditions took him to many parts of the world, including Peru and Japan; during the Franco-Prussian War he travelled to Algeria by balloon to observe an eclipse.

Having constructed a special spectroscope, Janssen showed that the dark bands in the solar spectrum can be resolved into rays, and, in 1864, that the origin of the phenomenon is terrestrial. He called the rays 'telluric rays'.

In 1868, Janssen went to India to observe a total eclipse of the Sun. He was unable to correlate certain lines in the solar spectrum with wavelengths of any known elements. English scientist Norman Lockyer made the same discovery of a new, unknown element and reported it simultaneously to the French Academy of Sciences.

Atlas de photographies solaires 1904 contained Janssen's photographs of the Sun, taken from 1876 onwards.

Jeans James Hopwood 1877–1946. British mathematician and scientist. In physics he worked on the kinetic theory of gases, and on forms of energy radiation; in astronomy, his work focused on giant and dwarf stars, the nature of spiral nebulae, and the origin of the cosmos. He did much to popularize astronomy. Knighted 1928.

Jeans was born in Ormskirk, Lancashire, and studied at Cambridge. From 1905 to 1909 he was professor of applied mathematics at Princeton University in the USA, and lectured at Cambridge 1910–12. Thereafter he devoted himself to private research and writing, although he was a research associate at Mount Wilson Observatory, California, 1923–44.

Life exists in the universe only because the carbon atom possesses certain exceptional properties.

James Jeans

Mysterious Universe

In 1905 Jeans formulated the ***Rayleigh–Jeans law***, which describes the spectral distribution of black-body radiation (previously studied by English physicist Lord Rayleigh) in terms of wavelength and temperature. For some time thereafter Jeans investigated various problems in quantum theory, but in about 1912 he turned his attention to astrophysics. In 1928 he stated his belief that matter was continuously being created in the universe (a forerunner of the steady-state theory).

His *Dynamical Theory of Gases* 1904 became a standard text.

Jeffreys Alec John 1950– . British geneticist who discovered the DNA probes necessary for accurate genetic fingerprinting so that a murderer or rapist could be identified by, for example, traces of blood, tissue, or semen.

Jenner Edward 1749–1823. English physician who pioneered vaccination. In Jenner's day, smallpox was a major killer. His discovery 1796 that inoculation with cowpox gives immunity to smallpox was a great medical breakthrough.

Jenner observed that people who worked with cattle and contracted cowpox from them never subsequently caught smallpox. In 1798 he published his findings that a child inoculated with cowpox, then two months later with smallpox, did not get smallpox. He coined the word 'vaccination' from the Latin word for cowpox, *vaccinia*.

The deviation of man from the state in which he was originally placed by nature seems to have proved him to be a prolific source of diseases.

Edward Jenner

An Inquiry into the Causes and Effects of the Variolae Vaccinae, *or Cow-pox*

Jenner was born in Berkeley, Gloucestershire, and studied at St George's Hospital, London. Returning to Berkeley, he set up a medical practice there.

In 1788, an epidemic of smallpox swept Gloucestershire and inoculations with live vaccine, taken from a person with a mild attack of the disease, were used. This method had been brought to England from Turkey in 1721 by Lady Mary Wortley Montagu and had brought success to Dutch physiologist Jan Ingenhousz, but could be fatal. It was during this epidemic that Jenner made his initial observations.

Jensen (Johannes) Hans (Daniel) 1907–1973. German physicist who shared the 1963 Nobel Prize for Physics with Maria Goeppert-Mayer and Eugene Wigner for work on the detailed characteristics of atomic nuclei.

Jensen was born in Hamburg and studied there and at Freiburg. He became professor at the Institute of Technology in Hanover 1941 and at Heidelberg 1949.

Jensen proposed in 1949 the theory that a nucleus has a structure of shells or spherical layers, each filled with neutrons and protons. He explained it in *Elementary Theory of Nuclear Shell Structure* 1955, written with Goeppert-Mayer, who had developed the shell theory independently.

Jensen and his colleagues also suggested that there is a very strong interaction between the spin and the orbit of a particle and that the lower of two states is always the one with angular momentum parallel rather than antiparallel. This may account for magic numbers, certain numbers of neutrons or protons that fill the shells of particularly stable elements.

Jerne Niels Kaj 1911–1994. British-born Danish microbiologist and immunologist, who profoundly influenced the development of modern immunology by establishing its cellular basis. Nobel Prize for Physiology or Medicine 1984 (shared with Georges Köhler (1946–1995) and César Milstein(1927–)).

He developed the haemolytic plaque assay for visualizing antibody release by lymphocytes (white blood cells that are produced in the bone marrow) and proposed several important theories of how cells interact during an immune response.

Jerne's most important immunological theory is that of the idiotypic network, a regulatory interaction between lymphocytes that is produced when receptors on one lymphocyte recognize and bind the receptors on a second lymphocyte. He later described how some lymphocytes are manufactured in the thymus gland.

Jerne was born in London, and went to school in Holland, where he studied physics for two years at the University of Leiden, before moving to Copenhagen to study medicine. He was almost 40 by the time he completed his PhD in 1951, and his thesis on diphtheria antiserum was accompanied by a growing interest in antibodies and their enormous diversity within the same organism.

In 1956 he published his paper 'Natural Selection Theory of Antibody Formation' proposing that antibodies pre-exist in the body before the presence of the foreign substance, when the antibody with the best fit is selected. It was previously believed that antibodies were flexible, nonspecific molecules that adopted different shapes to wrap round foreign molecules.

I have hit the nail: others later have hit the nail on the head.

Niels Jerne

As Chief Medical Officer for WHO (1956–62), Jerne built up a network of distinguished immunologists before returning to academic life 1962 with a professorship in microbiology at Pittsburgh University. In 1969 he founded and built up the Basle Institute of Immunology, funded by Roche, of which he was also director until his retirement in 1980.

Jerne continued to theorize on the nature of antibodies and immunology, happy for others to take up his ideas and modify and develop them. 'I have hit the nail: others later have hit the nail on the head,' was how he himself summed up his contributions.

Jessop William 1745–1814. English canal engineer who built the first canal in England entirely dependent on reservoirs for its water supply (the Grantham Canal 1793–97), and designed (with Thomas Telford) the 300-m/1,000 ft long Pontcysyllte aqueduct over the river Dee. Jessop also designed the forerunner of the iron rail that later became universally adopted for railways.

Jessop was born in Devonport, Devon, and became a pupil of civil engineer John Smeaton, working on canals in England and Ireland first with him and then independently.

Jessop's first tunnel was the 2.8 km/1.7 mi long Butterley Tunnel on the Cromford Canal he built in Derbyshire, and this led to the forming of the Butterley Iron Works in 1790, making rails and bridges.

Jessop was chief engineer 1793–1805 of the Grand Union Canal, which linked London and the Midlands over a distance of 150 km/95 mi. He was also responsible for the Barnsley, Rochdale, and Trent navigation, and the Nottingham and Ellesmere canals.

Jessop was also chief engineer of the Surrey Iron Railway, built 1801–02. He worked on the construction of a large wetdock area on the Avon at Bristol, on the West India Docks and the Isle of Dogs Canal in London, on the harbours at Shoreham and Littlehampton, and on many other projects.

Jobs Steven Paul 1955– . US computer scientist. He co-founded Apple Computer Inc with Steve Wozniak 1976, and founded NeXT Technology Inc.

Jobs holds a unique position in the personal computer industry, having been responsible for the creation of three different types of computer. He produced the popular Apple II personal computer, but his greatest success came with the Apple Macintosh 1984, marketed as 'the computer for the rest of us'. A decline in Apple's fortunes led to Jobs' departure and the setting up of NeXT. The NeXT computer met limited commercial success, but its many innovative ideas, particularly in the use of object-oriented technology, have found their way into mainstream computing.

Johanssen Wilhelm Ludvig 1857–1927. Danish botanist and founder of modern genetics who coined the term ***gene*** as the unit of heredity (although it was not known at the time to consist of DNA in cells). He introduced the concept of an organism having a 'genotype' (set of variable genes) and 'phenotype' (characteristics produced by the presence of certain genes). He also studied the metabolism of germination (the initial stages of growth in a seed) in plants.

Johanssen was born in Copenhagen, Denmark. As he came from a poor family he could not afford to attend university and was apprenticed to a pharmacist 1872. He taught himself chemistry in his spare time, before working in Germany, where he developed an interest in botany. In 1879, he returned to Denmark, and, after passing his pharmacology exams 1879, became an assistant in the Carlsberg laboratory where he worked on the metabolism of germination in plants. He was made a lecturer at Copenhagen Agricultural College 1892 and professor of botany and plant physiology 1903.

Joliot-Curie Frédéric (Jean) Joliot (1900–1958) and Irène (born Curie) (1897–1956) French physicists who made the discovery of artificial radioactivity, for which they were jointly awarded the 1935 Nobel Prize for Chemistry.

Irène was the daughter of Marie and Pierre Curie and began work at her mother's Radium Institute 1921. In 1926 she married Frédéric, a pupil of her mother's, and they began a long and fruitful collaboration. In 1934 they found that certain elements exposed to radiation themselves become radioactive.

Irène Curie was born and educated in Paris, becoming professor at the Sorbonne 1937. In 1946 she became director of the Radium Institute. She died of leukaemia caused by overexposure to radioactivity.

Frédéric Joliot was born in Paris and graduated from the Ecole Supérieure de Physique et de Chimie Industrielle. He joined the Radium Institute 1925. In 1937 he became professor of nuclear physics at the Collège de France. He succeeded his wife as director of the Radium Institute 1956.

Together the Joliot-Curies worked on radioactivity and the transmutation of elements. In 1934, while bombarding light elements with alpha particles, they noticed that although proton production stopped when the alpha particle bombardment stopped, another form of radiation continued. The alpha particles had produced an isotope of phosphorus not found in nature. This isotope was radioactive and was decaying through beta-decay.

Jordan (Marie Ennemond) Camille 1838–1922. French mathematician, the greatest exponent of algebra in his day. He concentrated on research in topology, analysis, and particularly group theory, publishing *Traité des substitutions et des equations algébriques* 1870.

Jordan was born in Lyon and studied engineering at the Ecole Polytechnique in Paris, but it was as a mathematician that he joined its staff in 1873. He also gave lectures at the Collège de France.

Influenced by the work of French mathematician Evariste Galois, Jordan systematically developed the theory of finite groups and arrived at the concept of infinite groups. He also developed three theorems of finiteness.

In topology, Jordan developed an entirely new approach by investigating symmetries in polyhedra from an exclusively combinatorial viewpoint. Moreover, he formulated the proof for the 'decomposition' of a plane into two regions by a simple closed curve.

Much of Jordan's later work was concerned with the theory of functions, and he applied the theory of functions of bounded variation to the particular curve that bears his name.

Jordan's work in analysis was published in *Cours d'analyse de l'école polytechnique* 1882.

Josephson Brian David 1940– . Welsh physicist, a leading authority on superconductivity. In 1973 he shared a Nobel prize for his theoretical predictions of the properties of a supercurrent through a tunnel barrier (the ***Josephson effect***), which led to the development of the ***Josephson junction***.

The Josephson junction is a device used in superchips (large and complex integrated circuits) to speed the passage of signals by electron tunnelling. These chips respond a thousand times faster than silicon chips, but the components of the Josephson junction operate only at temperatures close to absolute zero.

Josephson was born in Cardiff and studied at Cambridge, where he has spent most of his career, becoming professor 1974.

In 1962, Josephson saw some novel connections between solid-state theory and his own experimental problems in superconductivity. He then calculated the current due to quantum mechanical tunnelling across a thin strip of insulator between two superconductors, and the current–voltage characteristics of such junctions are now known as the Josephson effect.

The Josephson effect may be used as a generator of radiation, particularly in the microwave and far infrared region, and in detecting tiny anomalies in a magnetic field.

Joule James Prescott 1818–1889. English physicist whose work on the relations between electrical, mechanical, and chemical effects led to the discovery of the first law of thermodynamics.

He determined the mechanical equivalent of heat (Joule's equivalent) 1843, and the SI unit of energy, the joule, is named after him. He also discovered Joule's law, which defines the relation between heat and electricity; and with Irish physicist Lord Kelvin in 1852 the Joule–Kelvin (or Joule–Thomson) effect.

The ***Joule–Kelvin effect*** produces cooling in a gas when the gas expands through a narrow jet. It can be felt when, for example, compressed air escapes through the valve of a bicycle tyre. The effect is caused by the conversion of heat into work done by the molecules in overcoming attractive forces between them as they move apart.

Joule was born in Salford, Lancashire, and educated by private tutors, including scientist John Dalton. Having a private income, he dedicated his life to precise scientific research. Until neighbours protested, he kept a steam engine in his house in Manchester.

Joule showed experimentally that the ratio of equivalence of the different forms of energy did not depend on how one form was converted into another or on the materials involved.

Joy Alfred Harrison 1882–1973. US astronomer who worked on stellar distances and the radial motions of stars. He observed variable stars and classified them according to their spectra; he also determined the distance and direction of the centre of the Galaxy and attempted to calculate its rotation period.

Born in Greenville, Illinois, Joy attended Greenville College.

From 1904 to 1914 he worked at the American University of Beirut, Lebanon, becoming professor of astronomy and director of the observatory. He returned to the USA 1914, and worked at the Mount Wilson Observatory 1915–52.

Together with his colleagues Walter Adams and Milton Humason at Mount Wilson, Joy ascertained the spectral type, absolute magnitude, and stellar distance of more than 5,000 stars.

Joy and his colleagues studied the Doppler displacement of the spectral lines of some stars to determine their radial velocities. They showed that many stars are spectroscopic binary stars, and deduced not only their period and orbit, but also the absolute dimensions, masses, and orbital elements of some specific stars within eclipsing binary systems.

Joy later became interested in the parts of the Galaxy where dark, absorbing clouds of gas and dust exist, and in these areas he found examples of a particular kind of variable star, called a T Tauri star. Such stars appear to be in an early stage of their evolutionary history.

Jung Carl Gustav 1875–1961. Swiss psychiatrist who collaborated with Sigmund Freud from 1907 until their disagreement in 1914 over the importance of sexuality in causing psychological problems. Jung studied myth, religion, and dream symbolism, saw the unconscious as a source of spiritual insight, and distinguished between introversion and extroversion.

Show me a sane man and I will cure him for you.

Carl Jung

Observer 19 July 1975

Jung devised the word-association test in the early 1900s as a technique for penetrating a subject's unconscious mind. He also developed his theory concerning emotional, partly repressed ideas which he termed 'complexes'. In place of Freud's emphasis on infantile sexuality, Jung introduced the idea of a 'collective unconscious' which is made up of many archetypes or 'congenital conditions of intuition'.

Jung was born near Basel and studied there and at Zürich.

Jung *Swiss psychiatrist Carl Gustav Jung. Jung was a contemporary of Sigmund Freud and they both believed that the unconscious mind plays an important role in determining the psychological health of an individual. Jung disagreed with Freud's theory that sexuality was heavily involved in this.*

He worked at the Burghölzli Psychiatric Clinic in Zürich 1902–09. In 1907 he met Freud and became his chief disciple, appointed president of the International Psychoanalytic Association on its foundation 1910. But in 1914 he resigned from the association and set up his own practice in Zürich. In 1933 he became professor at the Zürich Federal Institute of Technology.

The book that provoked his split with Freud was *Wandlungen und Symbole de Libido/The Psychology of the Unconscious* 1912. Jung introduced the concept of introverts and extroverts in *Psychologische Typen/ Psychological Types* 1921. This work also contained his theory that the mind has four basic functions: thinking, feeling, sensations, and intuition. Any particular person's personality can be ascribed to the predominance of one of these functions.

Jussieu Antoine Laurent de 1748–1836. French botanist who developed one of the first systems of classification for plants. His study of flowering plants, *Genera plantarum* 1789, became the accepted basis of classification for flowering plants. Building on the foundation laid by Linnaeus, he produced one of the first taxonomies (classifications based on the physical characteristics of plants). Many elements of Jussieu's classification remain in use today.

Jussieu was born in Lyon, the nephew of the eminent 18th century botanist Bernard Jussieu. Like his uncle he went to Paris to study medicine and graduated 1770. He then became the deputy to the professor of botany at the Jardin du Roi. In London, he studied the specimen collections and herbariums of Joseph Banks and Linnaeus. In 1793, he was made professor of botany at the Museum National d'Histoire Naturelle, which was formed from the Jardin du Roi following the French Revolution. He went on to establish the herbarium at the museum before retiring 1826.

Jussieu Bernard *c.* 1699–1777. French botanist who laid out the gardens at Trianon in Versailles for Louis XV and taught both Linnaeus and Georges Buffon. His own system of botanical classification was eventually published after his death 1789.

Jussieu was born in Lyons and came to Paris 1714 to complete his medical and botanical studies. He travelled throughout Europe with his brother Antoine, the professor of botany at the Jardin du Roi. He began work 1722 as the 'sous-démonstrateur de lextérieur des plantes' at the Jardin du Roi, where he was responsible for many improvements in the gardens. He taught and organized field trips for both Buffon and Linnaeus. Jussieu graduated with a medical degree. Apart from a collection of letters to Linnaeus, little of his work was published.

Just Ernest Everett 1833–1941. US biologist whose research focused on cell physiology and experimental embryology, particularly fertilization and experimental parthenogenesis in marine eggs.

Just was born in Charleston, South Carolina, and educated at Dartmouth College, New Hampshire. He was professor of zoology at Howard University 1912–29, and spent his summers conducting research at the Marine Biological Laboratories at Woods Hole, Massachusetts. In 1929, frustrated with the limitations imposed on him by racists, Just went to Europe to conduct research in German laboratories and at French and Italian marine stations. With the German occupation of France, Just returned to his former post at Howard in 1940.

Just became the leading authority on the embryological resources of the marine group of animals. His focus of attention was the cell and in particular the ectoplasm, which, contrary to popular belief, he stated was just as important as the nucleus, and was primarily responsible for the individuality and development of the cell.

He was a coauthor of *General Cytology* 1924 and published *Biology of the Cell Surface* 1939.

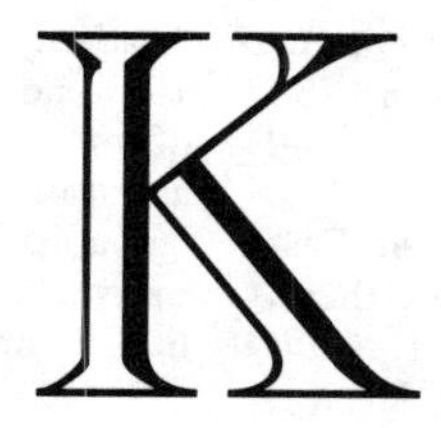

Kaiser Henry John 1882–1967. US industrialist. He developed steel and motor industries, and his shipbuilding firms became known for the mass production of vessels, including the 'Liberty ships' – cheap, quickly produced transport ships – built for the UK in World War II.

Kamerlingh Onnes Heike 1853–1926. Dutch physicist who worked mainly in the field of low-temperature physics. In 1911, he discovered the phenomenon of superconductivity (enhanced electrical conductivity at very low temperatures), for which he was awarded the 1913 Nobel prize.

Kamerlingh Onnes was born in Groningen and studied there and at Heidelberg. He was professor of experimental physics at the University of Leiden 1882–1924, and in 1894 he founded the cryogenic laboratories at Leiden, which became a world centre of low-temperature physics.

He applied the cascade method for cooling gases that had been developed by Scottish scientist James Dewar, and in 1908 succeeded in liquefying helium. In 1910, Kamerlingh Onnes managed to lower the temperature of liquid helium to 0.8 K (–272.4°C/–458.2°F). Lord Kelvin had postulated in 1902 that as the temperature approached absolute zero, electrical resistance would increase. In 1911, Kamerlingh Onnes found the reverse to be the case. He called this phenomenon supraconductivity, later renamed superconductivity.

Kamerlingh Onnes made a particular study of the effects of low temperature on the conductivity of mercury, lead, nickel, and manganese–iron alloys. He found that the imposition of a magnetic field eliminated superconductivity even at low temperatures.

Kammerer Paul 1880–1926. Austrian zoologist whose work on the genetics of the midwife toad was extremely controversial and his claims that acquired characteristics could be passed to the next generation were eventually disproved.

Kammerer's work was mainly upon amphibians, and he claimed that the male midwife toad could inherit the thick pigmented thumb pads found in other toads. His ideas met with opposition from his scientific peers and in 1923 he presented his work at Cambridge University and to the Linnaean Society in London. The toads from the controversial experiments were examined and the thumb pads were found to be stained with Indian ink. Kammerer denied any knowledge of tampering and committed suicide after the accusations were published.

Kammerer was born in Vienna and educated at the University of Vienna. He later worked in the area of genetic research.

Kant Immanuel 1724–1804. German philosopher who believed that knowledge is not merely an aggregate of sense impressions but is dependent on the conceptual apparatus of the human understanding, which is itself not derived from experience. In ethics, Kant argued that right action cannot be based on feelings or inclinations but conforms to a law given by reason, the ***categorical imperative***.

It was in his *Kritik der reinen Vernunft/Critique of Pure Reason* 1781 that Kant inaugurated a revolution in philosophy by turning attention to the mind's role in constructing our knowledge of the objective world. He also argued that God's existence could not be proved theoretically.

Who wills the end, wills also (so far as reason has a decisive influence on his actions) the means …

Immanuel Kant

The Moral Law

His other main works are *Kritik der praktischen Vernunft/Critique of Practical Reason* 1788 and *Kritik der Urteilskraft/Critique of Judgement* 1790.

Born in Königsberg (in what was then East Prussia), Kant attended the university there, and was its professor of logic and metaphysics 1770–97. His first book was *Gedanken von der wahren Schätzung der lebendigen Kräfte/Thoughts on the True Estimates of Living Forces* 1747. In *Allgemeine Naturgeschichte und Theorie des Himmels/ Universal Natural History and Theory of the Heavens* 1755, he put forward the idea that the Solar System was part of a system of stars constituting a galaxy and that there were many such galaxies making up the whole universe. Kant believed that planetary bodies in the Solar System had been formed by the condensation of nebulous, diffuse primordial matter; this later became known as the Kant–Laplacian theory (when put forward by French astronomer Pierre Laplace). Kant was also influenced by the theories of English scientist Isaac Newton.

Other works include *Prolegomena* 1783, *Metaphysik der Sitten/Metaphysic of Ethics* 1785, *Metaphysische Anfangsgründe der Naturwissenschaft/Metaphysic of Nature* 1786.

Kapitza Peter (Pyotr Leonidovich) 1894–1984. Soviet physicist who in 1978 shared a Nobel prize for his work on magnetism and low-temperature physics. He worked on the superfluidity of liquid helium and also achieved the first high-intensity magnetic fields.

The crocodile cannot turn its head. Like science, it must always go forward with all-devouring jaws.

Peter Kapitza

In A S Eve *Rutherford* 1933

Kapitza was born near St Petersburg and studied at Petrograd Polytechnical Institute, after which he went to the UK and worked at the Cavendish Laboratory, Cambridge, with nuclear physicist Ernest Rutherford. In 1930, Kapitza became director of the Mond Laboratory at Cambridge, which had been built for him. But when in 1934 he went to the USSR for a professional meeting, dictator Josef Stalin did not allow him to return. The Mond Laboratory was sold to the Soviet government at cost and transported to the Soviet Academy of Sciences for Kapitza's use. In 1946, he refused to work on the development of nuclear weapons and was put under house arrest until after Stalin's death 1953.

Kapitza was one of the first to study the unusual properties of helium II – the form of liquid helium that exists below 2.2 K (–271.0°C/–455.4°F). Helium II conducts heat far more rapidly than copper, which is the best conductor at ordinary temperatures, and Kapitza showed that this is because it has far less viscosity than any other liquid or gas. This property of helium is known as superfluidity.

In 1939, Kapitza built apparatus for producing large quantities of liquid oxygen, used in steel production. He also invented a turbine for producing liquid air cheaply in large quantities.

Kaplan Viktor 1876–1934. Austrian engineer who invented a water turbine with adjustable rotor blades. In the machine, patented 1920, the rotor was on a vertical shaft and could be adjusted to suit any rate of flow of water.

Horizontal Kaplan turbines are used at the installation on the estuary of the river Rance in France, the world's first tidal power station, which opened 1966.

Kaplan was born in Murz and educated in Vienna at the Technische Hochschule. After working in industry, he became professor at the Technische Hochschule in Brunn 1903.

Kaplan published his first paper on turbines in 1908, and set up a propeller turbine for the lowest possible fall of water. In 1913 the first prototype of the turbine was completed.

Kapteyn Jacobus Cornelius 1851–1922. Dutch astronomer who analysed the structure of the universe by studying the distribution of stars using photographic techniques. To achieve more accurate star counts he introduced the technique of statistical astronomy.

Kapteyn was born in Barneveld and studied at Utrecht. After working at the observatory at Leiden, he was professor at Groningen 1875–1921.

Kapteyn entered into an arrangement with Cape Town Observatory in South Africa whereby photographs of the stars in the southern hemisphere were analysed at Groningen and published as *Cape Photographic Durchmusterung* 1896–1900. This presented data on the brightness and positions of nearly 455,000 stars.

Studying the proper motions of stars, Kapteyn reported 1904 that these were not random, as had been believed; stars could be divided into two streams, moving in nearly opposite directions. It was later realized that Kapteyn's data had been the first evidence of the rotation of our Galaxy.

In 1906 he selected 206 specific stellar zones, aiming to ascertain the magnitudes of all the stars within these zones, as well as to collect data on their spectral type, radial velocity, proper motion, and so on. This enormous project was the first coordinated statistical analysis in astronomy and involved the cooperation of over 40 different observatories.

Karle Jerome 1918– . US chemist who, with colleague Herbert Hauptman, tested the range of available diffraction techniques, such as X-ray diffraction and the 'heavy atom' technique. The latter involved substituting an atom of a high molecular mass into the desired crystal structure, and from the change in the intensities of the diffraction patterns the phase structure of the crystalline substance could be deduced. To reduce the amount of time this took, the pair devised the 'direct method' – whereby phase structures could be inferred simply from the diffraction patterns. He received the Nobel Prize for Chemistry 1985.

Born in New York City, he was educated at the City College, New York and at the University of Michigan. He was involved in the military's Manhattan Project in Chicago in the 1940s and spent most of his working life at the Naval Research Laboratory, Washington DC.

Kármán Theodore von 1881–1963. Hungarian-born US aerodynamicist who enabled the USA to acquire a lead in rocket research. The research establishments he helped create include the Jet Propulsion Laboratory in Pasadena, California.

Kármán was born and educated in Budapest. In 1908 he went to France and witnessed an early aeroplane, which inspired him to concentrate on aeronautical engineering. Between 1913 and 1930, with the exception of the years of World War I, when he directed aeronautical research in the Austro-Hungarian army, Kármán built the new Aachen Institute, Germany, into a world-recognized research establishment. From 1928 he divided his time between the Aachen Institute and the Guggenheim Aeronautical Laboratory of the California Institute of Technology, USA.

Studying the flow of fluids round a cylinder, Kármán discovered that the wake separates into two rows of vortices which alternate like street lights. This phenomenon is called the ***Kármán vortex street***, or Kármán vortices, and it can build up destructive vibrations. The Tacoma Narrows suspension bridge was destroyed in 1940 by such vortices.

The scientist describes what is: the engineer creates what never was.

Theodor von Kármán

Biogr. Mem. FRS 1980

Kármán prompted the first research and development programme on long-range rocket-propelled missiles. He also worked on boundary layer and compressibility effects, supersonic flight, propeller design, helicopters, and gliders.

Karrer Paul 1889–1971. Russian-born Swiss organic chemist who synthesized various vitamins and determined their structural formulas, for which he shared the 1937 Nobel Prize for Chemistry. He also worked on vegetable dyes.

Karrer was born in Moscow but grew up in Switzerland. He studied at Zürich, was professor at the Zürich Chemical Institute from 1918 and its director 1919–59.

Karrer's early work concerned vitamin A and its chief precursor, carotene. Karrer worked out its correct constitutional formula in 1930, although he was not to achieve a total synthesis until 1950. He showed in 1931 that vitamin A is related to the carotenoids, substances that give a yellow, orange, or red colour to many foodstuffs. There are in fact two A vitamins. Karrer proved that there are several isomers of carotene, and that vitamin A_1 is equivalent to half a molecule of its precursor β-carotene.

In 1935, he solved the structure of vitamin B_2 (riboflavin). He also investigated vitamin E (tocopherol), which is a group of closely related compounds, and in 1938 he solved the structure of α-tocopherol, the most biologically active component.

In 1927, Karrer published *Lehrbuch der organischen Chemie/Textbook of Organic Chemistry.*

Kastler Alfred 1902–1984. French physicist who worked on double-resonance techniques in spectroscopy, in which absorbtion of optical or radio-frequency radiation is used to study the energy levels in atoms. He also developed the technique known as 'optical pumping' in which light is used to excite (energize) an atom to a higher energy level. This work led to the invention of the maser and laser. Nobel Prize for Physics 1966.

Kastler was born in Gebweiler (now Guebwiller), France, and educated at the University of Bordeaux. He became professor of physics at Bordeaux 1938. He moved to the University of Paris 1941 and remained there until he retired 1972.

Katz Bernard 1911– . British biophysicist. He shared the 1970 Nobel Prize for Physiology or Medicine for work on the biochemistry of the transmission and control of signals in the nervous system, vital in the search for remedies for nervous and mental disorders. Knighted 1969.

Katz was born in Leipzig, studied medicine at the university there, and then did postgraduate work at University College, London. Having done research in Australia 1939–42, he then served in the Royal Australian Air Force until the end of World War II, after which he returned to the UK. He spent the rest of his academic career at University College, London, becoming professor 1952.

In the 1940s, Katz joined in the Nobel-prizewinning research of Alan Hodgkin on the electrochemical behaviour of nerve membranes. During the 1950s, Katz found that minute amounts of acetylcholine were randomly released by nerve endings at the neuromuscular junction, giving rise to very small electrical potentials; he also found that the size of the potential was always a multiple of a certain minimum value. These findings led him to suggest that acetylcholine was released in discrete 'packets' (analogous to quanta) of a few thousand molecules each, and that these packets were released relatively infrequently while a nerve was at rest but very rapidly when an impulse arrived at the neuromuscular junction.

Kay John 1704–*c.* 1780. English inventor who developed the flying shuttle, a machine to speed up the work of hand-loom weaving. He patented his invention 1733.

Kay was born near Bury, Lancashire, and may have been educated in France. He invented many kinds of improved textile machinery, but was ruined by lawsuits in defence of his flying-shuttle patent. In 1753 his house in Bury was wrecked by a mob, who feared the use of machinery would cause unemployment. Kay left the country, and had some success introducing the flying shuttle to France, but is believed to have died there in poverty.

In 1730 he was granted a patent for an 'engine' for twisting and carding mohair, and for twining and dressing thread. At about the same time he improved the reeds for looms by manufacturing the 'darts' of thin polished metal instead of cane.

Up to that time the shuttle had been passed by hand from side to side through alternate warp threads. In weaving broadcloth two workers had to be employed to throw the shuttle from one end to the other. With Kay's flying shuttle, the amount of work a weaver could do was more than doubled, and the quality of the cloth was also improved.

Keeler James Edward 1857–1900. US astrophysicist who studied the rings of Saturn and the abundance and structure of nebulae. He demonstrated 1888 that nebulae resembled stars in their pattern of movement.

Keeler was born in La Salle, Illinois, and studied at Johns Hopkins University and in Germany at Heidelberg and Berlin. He was appointed astronomer at the Lick Observatory on its completion 1888; became professor and director of the Allegheny Observatory 1891; and returned to the Lick as director 1898.

In 1895, Keeler made a spectroscopic study of Saturn and its rings, in order to examine the planet's period of rotation. He found that the rings did not rotate at a uniform rate, thus proving for the first time that they could not be solid and confirming Scottish scientist James Clerk Maxwell's theory that the rings consist of meteoritic particles.

After 1898, Keeler devoted himself to a study of all the nebulae that William Herschel had catalogued a hundred years earlier. He succeeded in photographing half of them, and in the course of his work he discovered many thousands of new nebulae.

Kekulé von Stradonitz Friedrich August 1829–1896. German chemist whose theory 1858 of molecular structure revolutionized organic chemistry. He proposed two resonant forms of the benzene ring.

In 1865 Kekulé announced his theory of the structure of benzene, which he envisaged as a hexagonal ring of six carbon atoms connected by alternate single and double bonds.

... and lo, the atoms were gambolling before my eyes ... I saw frequently how two smaller atoms united to form a pair; how a larger one embraced two smaller ones; how still larger ones kept hold of three or even four of the smaller ... I saw how the longer ones formed a chain ... the cry of the conductor 'Clapham Road' awakened me from my dreaming; but I spent part of the night in putting on paper at least sketches of these dream forms.

Friedrich Kekulé

Biographical Encyclopedia of Scientists

In 1867 he proposed the tetrahedral carbon atom, which was to become the cornerstone of modern structural organic chemistry.

Kekulé was born in Darmstadt and studied at Giessen and Paris. After working in Switzerland, he went in 1854 to London, where he met many leading chemists of the day. When he returned to Germany 1855, he opened a small private laboratory in Heidelberg. In 1858 he became professor at Ghent; in 1865 at Bonn.

In 1858, Kekulé published a paper in which, after giving reasons why carbon should be regarded as a four-valent element, he set out the essential features of his theory of the linking of atoms.

Kelly William 1811–1888. US metallurgist, arguably the original inventor of the 'air-boiling process' for making steel, known universally as the Bessemer process.

Kelly was born in Pittsburgh, Pennsylvania. Having bought iron-ore lands and a furnace in Eddyville, Kentucky, he developed his steel-making process in the 1850s. Steel under the Kelly patent was first produced commercially 1864, but by then Kelly had sold his patent. He founded an axemaking business in Louisville, Kentucky, and remained in relative obscurity in spite of his invention, which became known under Henry Bessemer's name.

Experimenting, Kelly found that contrary to all ironmakers' beliefs, molten iron containing sufficient carbon became much hotter when air was blown on to it. An air blast can burn out 3–5% of carbon contained in molten cast iron. Here the carbon itself is acting as a fuel. When he heard that Bessemer had been granted a US patent on the same process, Kelly immediately applied for a patent and managed 1857 to convince the Patent Office that he was the original inventor.

After he had sold the patent, Kelly built a tilting type of converter, which produced soft steel cheaply for the first time and in large quantities. It was used for rails, bars, and structural shapes.

Kelvin William Thomson, 1st Baron Kelvin 1824–1907. Irish physicist who introduced the ***kelvin scale***, the absolute scale of temperature. His work on the conservation of energy 1851 led to the second law of thermodynamics. Knighted 1866, Baron 1892.

Kelvin's knowledge of electrical theory was largely responsible for the first successful transatlantic telegraph cable. In 1847 he concluded that electrical and magnetic fields are distributed in a manner analogous to the transfer of energy through an elastic solid. From 1849 to 1859, Kelvin also developed the work of English scientist Michael Faraday into a full theory of magnetism, arriving at an expression for the total energy of a system of magnets.

Thomson was born in Belfast and educated at Glasgow University (which he entered at the age of ten) and Cambridge. He was professor of natural philosophy at Glasgow 1846–99, and there created the first physics laboratory in a British university.

Do not imagine that mathematics is hard and crabbed, and repulsive to common sense. It is merely the etherealization of common sense.

William Kelvin

In S P Thomson *Life of Lord Kelvin* 1910

Based on the theories of French physicist Sadi Carnot on the motive power of heat, Kelvin proposed in 1848 an absolute temperature scale in which the temperature represents the total energy in a body. Kelvin differed from Carnot in seeing heat as a form of motion, an idea derived from the work of James Joule, whom Kelvin met 1847, on the determination of the mechanical equivalent of heat. In 1851, Kelvin announced that Carnot's theory and the mechanical theory of heat were compatible provided that it was accepted that heat cannot pass spontaneously from a colder to a hotter body. This is now known as the second law of thermodynamics. In 1852, Kelvin also produced the idea that mechanical energy tends to dissipate as heat.

Kelvin was very concerned with the accurate measurement of electricity, and developed an absolute electrometer 1870. He was instrumental in achieving the international adoption of many of our present-day electrical units in 1881. He also invented a tide gauge and predictor, an improved compass, and simpler methods of fixing a ship's position at sea.

Kendall Edward Calvin 1886–1972. US biochemist. In 1914 he isolated the hormone thyroxine, the active compound of the thyroid gland. He went on to work on secretions from the adrenal gland, among which he discovered the steroid cortisone. For this Kendall shared the 1950 Nobel Prize for Physiology or Medicine with Philip Hench and Tadeus Reichstein.

Kendall was born in Connecticut and studied at Columbia University, New York. From 1914 he worked at the Mayo Foundation, Minnesota, USA, becoming professor there 1921.

Hench was a physician interested in arthritis who was familiar with the experience that in some situations, such as during pregnancy, patients with arthritis improved. He and Kendall discussed whether cortisone, which Kendall's work had shown to have important metabolic effects, was involved in these temporary improvements. They discovered by giving a severely incapicitated patient cortisone that it was an effective treatment for rheumatoid arthritis.

Kendall Henry W 1926– . US physicist whose research into the collision of high-energy electrons collided with protons and neutrons was important in developing the quark model of particle physics. He collaborated with Jerome I Friedman and Richard Taylor. Nobel Prize for Physics 1990.

Kendall, Taylor, and Friedman experimented 1970 with bombarding protons (and later, neutrons) with high-energy electrons. They predicted that high-energy electrons would suffer only small deflections as they passed through the protons, but found instead that the electrons were sometimes scattered through large angles inside the proton. James Bjorken and Richard Feynman suggested that this was because the electrons were hitting hard pointlike objects inside the proton, now known to be quarks, amongst the most fundamental building blocks of matter.

Kendall was born in Boston, Massachusetts, USA, and educated at Amherst College and the Massachusetts Institute of Technology (MIT). Two years after gaining his doctorate, Kendall joined the research group of Robert Hofstadter at the Stanford Univerisity, California, which was engaged in the study of the proton and neutron structure. At Stanford Kendall met and worked with Jerome Friedman and got to know Richard Taylor. After five years at Stanford, he moved back to MIT. Friedman had moved there a year earlier and the two renewed their acquaintance. By 1964 the pair had established the collaboration with Taylor, then a research group leader at the Stanford Linear Accelerator Center, which led to the Nobel prize.

Kendrew John Cowdery 1917– . English biochemist who determined the structure of the muscle protein myoglobin. For this work Kendrew shared the 1962 Nobel Prize for Chemistry with his colleague Max Perutz. Knighted 1974.

Kendrew was born in Oxford and studied at Cambridge, where he spent most of his academic career. In 1975 he became director of the European Molecular Biology Laboratory at Heidelberg, Germany.

In 1946 at Cambridge Kendrew began working with Perutz. Their research centred on the fine structure of various protein molecules; Kendrew was assigned the task of studying myoglobin, a globular protein which occurs in muscle fibres (where it stores oxygen). He used X-ray diffraction techniques to elucidate the amino acid sequence in the peptide chains that form the myoglobin molecule. Early computers were programmed to analyse the X-ray photographs. By 1960, Kendrew had determined the spatial arrangement of all 1,200 atoms in the molecule, showing it to be a folded helical chain of amino acids with an amino ($-NH_2$) group at one end and a carboxylic ($-COOH$) group at the other. It involves an iron-containing haem group, which allows the molecule to absorb oxygen.

Kennelly Arthur Edwin 1861–1939. US engineer who gave his name to the Kennelly–Heaviside layer (now the E layer) of the ionosphere. He verified in 1902 the existence of an ionized layer in the upper atmosphere, predicted by Heaviside.

Kenyon Joseph 1885–1961. English organic chemist who studied optical activity, particularly of secondary alcohols.

Kenyon was born in Blackburn, Lancashire. Working as a laboratory assistant at Blackburn Technical College, he won a scholarship and went on to become a lecturer there. He worked on photographic developers and dyes at the British Dyestuffs Corporation, Oxford, 1916–20, and was head of the Chemistry Department at Battersea Polytechnic, London, 1920–50.

Kenyon published his research on secondary alcohols in 1911. He resolved the optically active stereoisomers of secondary octyl alcohol (octan-2-ol), and went on to obtain an optically pure series of secondary alcohols.

Kenyon put forward the 'obstacle' theory for the cause of optical activity in certain substituted diphenic acids. He synthesized and attempted to resolve some selenoxides, confirming differences between these and sulphoxides, and investigated the geometric and optical isomerism of the methylcyclohexanols.

Kepler Johannes 1571–1630. German mathematician and astronomer. He formulated what are now called ***Kepler's laws*** of planetary motion: (1) the orbit of each planet is an ellipse with the Sun at one of the foci; (2) the radius vector of each planet sweeps out equal areas in equal times; (3) the squares of the periods of the planets are proportional to the cubes of their mean distances from the Sun.

Kepler became assistant to Danish astronomer Tycho Brahe 1600, and succeeded him 1601 as imperial mathematician to Holy Roman Emperor Rudolph II. Kepler observed in 1604 a supernova, the first visible since the one discovered by Brahe 1572. Kepler completed and published the *Rudolphine Tables* 1627, the first modern astronomical tables, based on Brahe's observations. His analysis of these data led to the discovery of his three laws, the first two of which he published in *Astronomia nova* 1609 and the third in *Harmonices mundi* 1619.

Where there is matter, there is geometry.

Johannes Kepler

Attributed remark

Kepler was born in Weil der Stadt in Baden-Württemberg, and studied at Tübingen. As a Lutheran Protestant, he was expelled twice from Graz, where he had been teaching; then from Prague 1612; then from Linz, Austria, from where he moved to Ulm. His other domestic problems included the unsuccessful prosecution in Wittenberg 1618 of his mother for witchcraft.

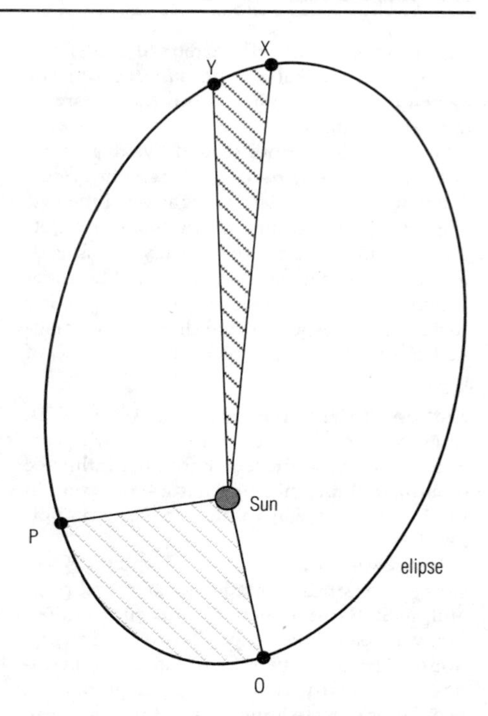

Kepler *Kepler's second law states that the dark-shaded area equals the light-shaded area if the planet moves from P to O in the same time that it moves from X to Y. The law says, in effect, that a planet moves fastest when it is closest to the Sun.*

Kepler was one of the first advocates of Sun-centred cosmology, as put forward by Copernicus. Kepler's laws are the basis of our understanding of the Solar System, and such scientists as Isaac Newton built on his ideas.

Kerr John 1824–1907. Scottish physicist who discovered the ***Kerr effect***, which produces double refraction in certain media on the application of an electric field.

Kerr was born in Ardrossan, Ayrshire, and studied at Glasgow. He was lecturer in mathematics at the Free Church Training College for Teachers, Glasgow, 1857–1901, and set up a modest laboratory there.

In 1875 Kerr demonstrated that double refraction occurs in glass and other insulators when subjected to an intense electric field. The effect was strongest, he found, when the plane of polarization was 45° to the field, and zero when perpendicular or parallel. Kerr extended the work to other materials, and constructed cells in which he could study liquids such as carbon disulphide and paraffin oil. He showed that the extent of the effect, more precisely called the ***electro-optical Kerr effect***, is proportional to the square of the field strength.

In 1876 he demonstrated what is now known as the ***magneto-optical Kerr effect***. A beam of plane-polarized light was reflected from the polished pole of an electromagnet. The beam became elliptically polarized, when the magnet was switched on. The effect depended on the position of the reflecting surface with respect to the direction of magnetization and to the plane of incidence of the light.

Kettlewell Henry Bernard Davis 1907–1979. English geneticist and lepidopterist who carried out important research into the influence of industrial melanism on natural selection in moths, showing why moths are darker in polluted areas.

Kettlewell was born in Howden, Yorkshire, and studied medicine at Cambridge and at St Bartholomew's Hospital, London. He was based at Cape Town University, South Africa, 1949–52, investigating methods of locust control and going on expeditions. Returning to the UK, he spent the rest of his career at Oxford as a genetics researcher.

Kettlewell's research into industrial melanism focused on the peppered moth *Biston betularia*. He demonstrated experimentally the efficiency of natural selection as an evolutionary force: light-coloured moths are more conspicuous than dark-coloured ones in industrial areas, where the vegetation is darkened by pollution, and are therefore easier prey for birds, but are less conspicuous in unpolluted rural areas, where the vegetation is lighter in colour, and therefore survive predation better.

Khorana Har Gobind 1922– . Indian-born US biochemist who in 1976 led the team that first synthesized a biologically active gene. In 1968 he shared the Nobel Prize for Physiology or Medicine for research on the chemistry of the genetic code and its function in protein synthesis. Khorana's work provides much of the basis for gene therapy and biotechnology.

Khorana was born in Raipur in the Punjab, now in Pakistan. He studied at Punjab University; in the UK at Liverpool; and in Switzerland at Zürich; returning to Britain 1950 to work at Cambridge. He has held academic posts in the USA and Canada, becoming professor at the University of Wisconsin 1962 and at the Massachusetts Institute of Technology 1970.

Khorana systematically synthesized every possible combination of the genetic signals from the four nucleotides known to be involved in determining the genetic code. He showed that a pattern of three nucleotides, called a triplet, specifies a particular amino acid (the building blocks of proteins). He further discovered that some of the triplets provided punctuation marks in the code, marking the beginning and end points of protein synthesis.

Khorana has also synthesized the gene for bovine rhodopsin, the pigment in the retina responsible for converting light energy into electrical energy.

Khwārizmī, al- Muhammad ibn-Mūsā *c.* 780–*c.* 850. Persian mathematician who wrote a book on algebra, from part of whose title (*al-jabr*) comes the word 'algebra', and a book in which he introduced to the West the Hindu–Arabic decimal number system.

The word 'algorithm' is a corruption of his name.

He was born in Khwarizm (now Khiva, Uzbekistan), but lived and worked in Baghdad. He compiled astronomical tables and was responsible for introducing the concept of zero into Arab mathematics.

Kimura Motō 1924– . Japanese biologist who, as a result of his work on population genetics and molecular evolution, has developed a theory of neutral evolution that opposes the conventional neo-Darwinistic theory of evolution by natural selection.

Kimura was born in Okazaki, near Nagoya, and studied at Kyoto and in the USA at Iowa State and Wisconsin universities. He spent most of his career at the National Institute of Genetics in Mishima, starting 1949 and becoming head of the Department of Population Genetics 1964.

Kimura began in 1968 the work that was to lead to the theory of neutral evolution. Comparing the amino acid compositions of the alpha and beta chains of the haemoglobin molecules in humans with those in carp, he found that the alpha chains have evolved in two distinct lineages, accumulating mutations independently and at about the same rate over a period of some 400 million years.

According to Kimura's theory, evolutionary rates are determined by the structure and function of molecules, and most variability and evolutionary change within a species is caused by the random drift of mutant genes that are all selectively equivalent and selectively neutral. Kimura's theory denies that the environment influences evolution and that mutant genes confer either advantageous or disadvantageous traits.

Kingdon Ward Francis 1885–1958. English plant collector, geographer, and author who discovered the blue poppy *Meconopsis betoniciforlia* on his travels in China. He also described many new forms of rhododendrons and primulas.

Kingdon Ward was born in Manchester, the only son of the acclaimed botanist Harry Marshall Ward. He attended St Paul's school and Christ's College, Cambridge. After studying natural science 1906, he went to teach at Shanghai Public School 1907. In 1909, he participated in an expedition to the interior of China and following his success in collecting specimens during this trip, he began to collect plant specimens as a living. His first job was for Bees of Liverpool 1911.

He travelled extensively and normally alone (until his second marriage 1947), visiting the mountains of the Chinese, Indian, and Burmese borders, and wrote extensively about his travels. His awards include the Victoria and Veitch medals from the Royal Horticultural Society, the Royal Geographical Society Founder's medal and an OBE for services to horticulture. He died in London on 8 April.

Kinsey Alfred Charles 1894–1956. US sexologist and zoologist who published controversial but ground breaking research on the subject of human sexuality, following interviews with over 18,000 men and women.

Kinsey was born in Hoboken, New Jersey and studied at Bowdoin College, Brunswick, Maine, graduating with a BSc 1916 before obtaining a PhD from Harvard University 1920. He was appointed assistant professor of zoology at Indiana University 1920 after a brief spell as a lecturer at Harvard, and was made a full professor 1929. His research during the 1930s was on gall wasps and he published *Edible Plants of Eastern North America* 1943. It is, however, for his work on human sexual behaviour that he is best known. In 1942 he founded the Institute for Sex Research at the university of Indiana and published two controversial studies: *Sexual Behaviour in the Human Male* 1948 and *Sexual Behaviour in the Human Female* 1953.

Kipping Frederic Stanley 1863–1949. English chemist who pioneered the study of the organic compounds of silicon; he invented the term 'silicone', which is now applied to the entire class of oxygen-containing polymers.

Kipping was born near Manchester and educated there. At German chemist Johann von Baeyer's laboratory in Munich, he studied under William Perkin Jr (1860–1929), with whom he was to collaborate. Kipping was professor at University College, Nottingham (later Nottingham University), 1897–1936.

Kipping's early research concentrated on the preparation and properties of optically active camphor derivatives and nitrogen compounds. In 1899 he began to look for stereoisomerism among the organic compounds of silicon, preparing them using the newly available Grignard reagents. He prepared condensation products – the first organosilicon polymers – which he called silicones. To Kipping these were chemical curiosities and it was not until World War II that they found application as substitutes for oils and greases.

Kipping and Perkin co-wrote *Organic Chemistry* 1894, a standard work for the next 50 years.

Kirchhoff Gustav Robert 1824–1887. German physicist who with R W von Bunsen developed spectroscopic analysis in the 1850s and showed that all elements, heated to incandescence, have their individual spectra.

Kirchhoff was born and educated in Königsberg (now Kaliningrad). In 1850 he was appointed professor at Breslau, where he was joined by Bunsen the following year. In 1852 Bunsen moved to Heidelberg and Kirchhoff followed him 1854. He moved to Berlin 1875.

In 1845 and 1846 he derived the laws known as ***Kirchhoff's laws*** that determine the value of the electric current and potential at any point in a network. He went on to show that electrostatic potential is identical to tension, thus unifying static and current electricity.

Observing the dark lines in the Sun's spectrum, first discovered by German physicist Joseph Fraunhofer, Kirchhoff came to the conclusion that the Fraunhofer lines are due to the absorption of light by sodium and other elements present in the Sun's atmosphere.

In 1859 Kirchhoff announced a law stating that the ratio of the emission and absorption powers of all material bodies is the same at a given temperature and a given wavelength of radiation produced. From this, Kirchhoff went on in 1862 to derive the concept of a perfect black body – one that would absorb and emit radiation at all wavelengths.

Kirkwood Daniel 1814–1895. US astronomer who identified and explained the ***Kirkwood gaps***, asteroid-free zones in the Solar System. He used the same theory to explain the nonuniform distribution of particles in the ring system of Saturn.

Kirkwood was born in Hartford County, Maryland, and became a teacher. In 1851 he became professor of mathematics at the University of Delaware; later at Indiana and at Jefferson College, Pennsylvania. From 1891 he lectured in astronomy at the University of Stanford, California.

In 1857 Kirkwood first noticed that three regions of the minor planet zone, sited at 2.5, 2.95, and 3.3 astronomical units from the Sun, lacked asteroids completely. In 1866, he proposed that the gaps arose as a consequence of perturbations caused by the planet Jupiter. The effect of Jupiter's mass would be to force any asteroid that appeared in one of the asteroid-free zones into another orbit, with the result that it would immediately leave the zone.

Kirkwood also carried out research into comets and meteors, made a fundamental critique of French astronomer Pierre Laplace's work on the evolution of the Solar System, and carried out preliminary studies on families of asteroids.

Kitasato Shibasaburō 1852–1931. Japanese bacteriologist who discovered the plague bacillus while investigating an outbreak of plague in Hong Kong. He was the first to grow the tetanus bacillus in pure culture. He and German bacteriologist Emil von Behring discovered that increasing nonlethal doses of tetanus toxin give immunity to the disease.

Kitasato was born on the island of Kyushu and studied at Kumamoto Medical School and Imperial (now Tokyo) University. In 1885 he was sent by the government to study bacteriology in Germany and went to work in the laboratory of Robert Koch in Berlin. On his return 1891, Kitasato set up a small private institute of bacteriology. When this was incorporated into Tokyo University against Kitasato's wishes 1915, he resigned and founded the Kitasato Institute, which he headed for the rest of his life.

Having made the discovery of antitoxic immunity 1890, Kitasato and von Behring rapidly developed a serum for treating anthrax. After returning to Japan, Kitasato was sent by his government to Hong Kong 1894 to investigate an epidemic of bubonic plague. France also sent a small research team, led by Swiss bacteriologist Alexandre Yersin. The two teams did not collaborate because of language difficulties and there is some doubt whether it was Kitasato or Yersin who first isolated *Pasteurella pestis*, the bacillus that causes bubonic plague, but Kitasato published his findings first. The bacillus is now called *Yersinia pestis*.

Kitasato also isolated the causative organism of dysentery in 1898 and studied the method of infection in tuberculosis.

Klaproth Martin Heinrich 1743–1817. German chemist who first identified the elements uranium and zirconium, in 1789, and was the second person to isolate titanium, chromium, and cerium. He was a pioneer of analytical chemistry.

Klaproth was born in Wernigerode, Saxony, apprenticed to an apothecary when he was 16, and in 1771 became manager of a pharmacy in Berlin. He lectured in chemistry at the Berlin School of Artillery from 1792, and when the University of Berlin was founded in 1810 he became its first professor of chemistry.

Klaproth distinguished strontia (strontium oxide) from baryta (barium oxide) and 1795 rediscovered and named titanium. He isolated chromium 1797 independently of French chemist Louis Vauquelin, but credited Franz Müller (1740–1825) with the priority for the discovery of tellurium, which Klaproth extracted 1798 and named. In 1803, Klaproth identified cerium oxide and confirmed the existence of cerium, discovered by Swedish chemist Jöns Berzelius in the same year. He also studied the rare earth minerals.

Klein (Christian) Felix 1849–1925. German mathematician and mathematical physicist who unified the various Euclidean and non-Euclidean geometries. He demonstrated that every individual geometry could be constructed purely projectively; such projective models are now called Klein models.

Klein was born in Düsseldorf and studied at Bonn, Göttingen, Berlin, and Paris. He was professor at Erlangen 1872–75, at the Technische Hochschule in Munich 1875–80, and then took up a similar post in Leipzig. From 1886 he was at Göttingen, and helped make it Germany's main centre for all the exact sciences.

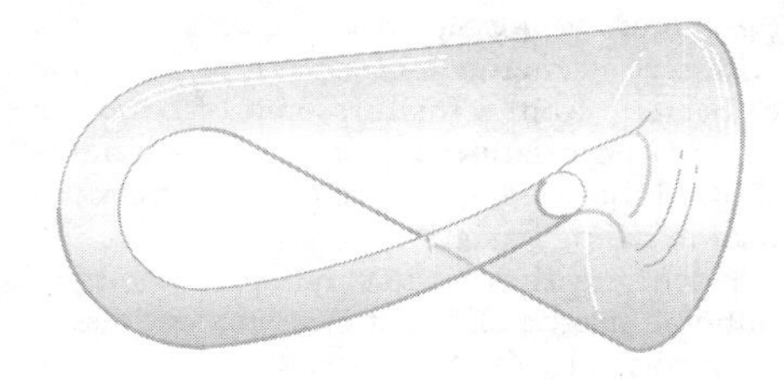

Klein *Klein bottle is a 'bottle' in name only because, although it would be possible to store liquids in it, it is in fact a one-sided surface which is closed and has no boundary.*

Klein announced 1872 what he called the Erlangen *Programm* on the unification of geometries. He showed that each of the different geometries devised during the 19th century is associated with a separate 'collection', or 'group', of tranformations. Seen in this way, the geometries could all be treated as members of one family.

In his work on number theory, group theory, and the theory of differential equations, Klein was greatly influenced by German mathematician Bernhard Riemann, and redefined a Riemann surface so that it came to be regarded as an essential part of function theory.

Klein initiated 1895 the writing of an encyclopedia of mathematics. He published books on the historical development of mathematics in the 19th century, and a textbook (with his colleague Arnold Sommerfeld) on the theory of the gyroscope.

Kliegl John H (1869–1959) and Anton T (1872–1927) German-born US technicians who in 1911 invented the brilliant carbon-arc (***klieg***) lights used in television and films. They also created scenic effects for theatre and film.

John emigrated to the USA 1888, Anton 1892, and in 1896 in New York they formed the Kliegl Brothers Universal Electric Stage Lighting Company.

Klug Aaron 1926– . South African molecular biologist who improved the quality of electron micrographs by using laser lighting. He used electron microscopy, X-ray diffraction and structural modelling to study the structures of different viruses (including the polio virus) that are too small to be visible with a light microscope or to be trapped by filters. He received the Nobel Prize for Chemistry 1982.

Klug also postulated that the protein coat of a small virus could be formed by an arrangement of quasi-equivalent (similar) protein molecules, and shed light on the formation of protein units in helical viruses. Later on in his career he applied his illuminating techniques to cell components, including chromatin and muscle filaments.

Klug was born to South African parents in Zelvas in the Baltic State of Lithuania. The family moved to South Africa when he was three years old. After studying medicine for one year at Witwatersrand University, Johannesburg, he opted to learn about science instead. In 1947, he moved to Cape Town and specialized in crystallography.

At the end of World War II, he travelled to England, working at the Cavendish Laboratory, Cambridge, and then in London at Birkbeck College as a Nuffield research fellow 1954–57. Here he collaborated with Rosalind Franklin, who worked with him during his research on viruses. In 1958, he succeeded Franklin as director of the Virus Structure Research Group, then in 1962, he returned to Cambridge to join the staff at the Medical Research Council's Laboratory of Molecular Biology. He was appointed director of the laboratory in 1986. He was elected a Fellow of the Royal Society in 1969, and received a knighthood in 1988.

Knopf Eleanora Frances, (born Bliss) 1883–1974. US geologist who studied metamorphic rocks. She introduced the technique of petrofabrics to the USA.

Bliss was born in Rosemont, Pennsylvania, and studied at Bryn Mawr College and the University of California at Berkeley. She spent most of her career working for the US Geological Survey. During the 1930s she was also a visiting lecturer at Yale and at Harvard.

In 1913 in Pennsylvania, she discovered the mineral glaucophane, previously unsighted in America east of the Pacific. In the 1920s Knopf studied the Pennsylvania and Maryland piedmont and the geologically complex mountain region along the New York–Connecticut border.

The technique of petrofabrics had been developed in Austria at Innsbruck University. Knopf applied it to the study of metamorphic rocks and wrote about it in *Structural Petrology* 1938.

Koch (Heinrich Hermann) Robert 1843–1910. German bacteriologist. Koch and his assistants devised the techniques for

Koch *German physician and pioneer bacteriologist Robert Koch, who recieved the Nobel Prize for Physiology or Medicine 1905 for his work on tuberculosis. Koch also identified and isolated the cholera germ and showed that the tsetse fly was the carrier of sleeping sickness. He also formulated a series of systematic principles, 'Koch postulates', for analysing and identifying new infectious diseases that are still used today.*

culturing bacteria outside the body, and formulated the rules for showing whether or not a bacterium is the cause of a disease. Nobel Prize for Physiology or Medicine 1905.

His techniques enabled him to identify the bacteria responsible for tuberculosis (1882), cholera (1883), and other diseases. He investigated anthrax bacteria in the 1870s and showed that they form spores which spread the infection.

Koch was a great teacher, and many of his pupils, such as Kitasato, Ehrlich, and Behring, became outstanding scientists.

Koch was born near Hanover and studied at Göttingen. He began research in the 1870s while working as a district medical officer. In 1879 he was appointed to the Imperial Health Office in Berlin to advise on hygiene and public health. He was professor at Berlin 1885–91, when he became director of the newly established Institute for Infectious Diseases, but he resigned 1904 and spent much of the rest of his life advising other countries on ways to combat various diseases.

Koch experimented with various dyes and found some that stain bacteria and make them more visible under the microscope. He also devised methods of separating one type of bacteria from a mixture, and of culturing bacteria on gelatine. Koch and his assistants showed that steam is more effective than dry heat in killing bacteria.

Koch also showed that rats are vectors of bubonic plague and that sleeping sickness is transmitted by the tsetse fly.

Köhler Georges Jean Franz 1946–1995. German immunologist who helped revolutionize medical research through the development of monoclonal antibodies, for which he shared the 1984 Nobel Prize for Physiology or Medicine.

Köhler was born in Munich and educated at Freiburg University. In 1971 he moved to the Basel Institute in Switzerland, where he gained his PhD and began to develop his interest in antibodies and their production as part of the normal immune response. In particular, he became fascinated with the enormous range of antibodies a single animal can produce.

Köhler moved to Cambridge, England 1974 where he began his collaboration with César Milstein. Within a year, the work for which they were to share the Nobel prize had been completed. Köhler revolutionized the method by which antibodies for research were produced, a process which had previously been slow and unreliable, and requiring large numbers of animals. He took lymphocytes from an immunized mouse and fused them to tumour cells, resulting in a limitless population of cloned cells that could produce a pure 'monoclonal' antibody of known specificity.

At first neither Köhler nor Milstein appreciated the enormous commercial significance of their discovery; no patents were applied for, and Köhler returned to Switzerland 1976 to pursue his original interest in normal antibody synthesis. By 1980 the monoclonal antibody technique was being used in laboratories around the world. In 1985 Köhler became Director of the Max Planck Institute for Immune Biology, in Freiburg, where he continued his research until his death.

Kolbe (Adolf Wilhelm) Hermann 1818–1884. German chemist, generally regarded as the founder of modern organic chemistry with his synthesis of acetic acid (ethanoic acid) – an organic compound – from inorganic starting materials. (Previously organic chemistry had been devoted to compounds that occur only in living organisms.)

Kolbe was born near Göttingen and educated there. He worked in the UK 1845–47, at the London School of Mines. In 1851 he was appointed professor at Marburg; by 1865 he had moved to Leipzig and embarked on establishing a laboratory bigger and better equipped than any other.

Kolbe correctly realized that organic compounds can be derived from inorganic materials by simple substitution. He introduced a modified idea of structural radicals, which contributed to the development of the structure theory, and he predicted the existence of secondary and tertiary alcohols. His most important work was on the electrolysis of the fatty (alkanoic) acids, his preparation of salicylic acid (2-hydroxybenzenecarboxylic acid) from phenol – called the Kolbe reaction, which was to lead to an easy synthesis of aspirin – and his discovery of nitromethane. But his work suffered from his refusal to abandon equivalent weights in favour of atomic weights (relative atomic masses).

Koller Carl 1857–1944. Austrian ophthalmologist who introduced local anaesthesia 1884, using cocaine.

When psychoanalyst Sigmund Freud discovered the painkilling properties of cocaine, Koller recognized its potential as a local anaesthetic. He carried out early experiments on animals and on himself, and the technique quickly became standard in ophthalmology, dentistry, and other areas in cases where general anaesthesia exposes the patient to needless risk.

Kornberg Arthur 1918– . US biochemist. In 1956 he discovered the enzyme DNA-polymerase, which enabled molecules of the genetic material DNA to be synthesized for the first time. For this work he shared the 1959 Nobel Prize for Physiology or Medicine. By 1967 he had synthesized a biologically active artificial viral DNA.

Kornberg was born in New York and studied at the University of Rochester School of Medicine. He held senior appointments at the Washington University School of Medicine (1953) and the Stamford University School of Medicine, Palo Alto, California (1959), before becoming head of the Biochemistry Department at Stamford.

In 1957, Kornberg made an artificial DNA, but this turned out to lack genetic activity. He then tried to make a simpler one, the DNA of a virus known as Phi X174, which is single-stranded and in the form of a ring; its activity (infectivity) is lost if the ring is broken. In 1966 he discovered the enzyme needed to close the ring. When the synthetic DNA was added to a culture of bacteria cells, the cells abandoned their normal activity and started to produce Phi X174 viruses.

Kornberg Hans Leo 1928– . German-born British biochemist who investigated metabolic pathways and their regulation, especially in microorganisms. He studied the way in which the cellular economy is balanced to prevent overproduction or waste, and introduced the concept of anaplerotic reactions, whereby metabolic processes are maintained by special enzymes that replenish materials syphoned off for anabolic purposes. Knighted 1978.

Kornberg was born in Herford, Westphalia, and went to Britain in 1939 as a refugee from Nazi persecution. He studied at Sheffield under biochemist Hans Krebs, and in the USA at Yale and the University of California. On his return to the UK he joined the Medical Research Council Cell Metabolism Unit headed by Krebs in Oxford. In 1960 he became professor at Leicester, and in 1975 at Cambridge.

Kornberg's research focused on the Krebs cycle, in particular the question of how an organism uses acetate to build up larger molecules while at the same time needing to break it down to provide the energy required for these anabolic functions.

Kornberg studied the regulation of enzyme activity at the genetic and cytoplasmic levels. His later research concentrated on the first step in the processing of food materials, the selective uptake of compounds across cellular membranes.

Korolev Sergei Pavlovich 1906–1966. Russian designer of the first Soviet intercontinental missile, used to launch the first Sputnik satellite 1957 and the Vostok spacecraft, also designed by Korolev, in which Yuri Gagarin made the world's first space flight 1961.

Korolev and his research team built the first Soviet liquid-fuel rocket, launched 1933. His innovations in rocket and space technology include ballistic missiles, rockets for geophysical research, launch vehicles, and crewed spacecraft. Korolev was also responsible for the *Voskhod* spaceship, from which the first space walks were made.

Korolev was born in Zhitomir, Ukraine, and trained as an aircraft designer. He was a member of the Institute for Jet Research from its foundation 1933, and worked as an engine designer 1924–46. Later he was appointed head of the large team of scientists who developed high-powered rocket systems.

Korolev published his first paper on jet propulsion in 1934. By 1939 he had designed and launched the Soviet 212 guided wing rocket. This was followed by the RIP-318–1 rocket glider, which made its first piloted flight in 1940.

The Vostok flights orbiting Earth were followed by the launching of Sun satellites and the flights of uncrewed interplanetary probes into the Solar System.

Kovalevskaia Sofya Vasilevna 1850–1891. Russian mathematician and novelist who worked on partial differential equations and Abelian integrals. In 1886 she won the Prix Bordin of the French Academy of Sciences for a paper on the rotation of a rigid body about a point, a problem the 18th-century mathematicians Euler and Lagrange had both failed to solve.

She was born in Moscow and studied in Germany at Heidelberg, Berlin, and Göttingen. Excluded from most European academic posts because of her sex, she finally obtained a lectureship in Sweden and became professor at Stockholm 1889. In addition to her mathematical work, she wrote plays and novels, including *Vera Brantzova* 1895.

Krafft-Ebing Richard, Baron von 1840–1902. German pioneer psychiatrist and neurologist. He published *Psychopathia Sexualis* 1886.

Educated in Germany, Krafft-Ebing became professor of psychiatry at Strasbourg 1872. His special study was the little-understood relationship between minor paralysis and syphilis, a sexually transmitted disease. In 1897 he performed an experiment which conclusively showed that his paralysed patients must previously have been infected with syphilis. He also carried out a far-reaching study of sexual behaviour.

Krebs Edwin Gerhard 1918– . US physician who won the Nobel Prize for Physiology or Medicine with Emil Fischer 1992 for his discovery of protein phosphorylation (the chemical bonding of a phosphate molecule to a protein) as a control mechanism in the metabolic activity of mammalian cells.

Krebs and Fischer characterized a group of enzymes, called protein kinases, that change proteins from their inactive to active form by triggering the chemical bonding of a phosphate group to the protein. This phosphorylation is the underlying switch that starts and stops a variety of cell functions, from breakdown of fats to the generation of chemical energy in response to hormonal and other signals. They determined adenosine triphosphate (ATP), a nucleotide molecule found in all cells, to be the energy-transporting compound which donated the phosphate group.

Working on muscle tissue, they also showed that protein phosphorylation was the underlying mechanism that accounted for the reversible modification of glycogen phosphorylase.

Krebs was born in Lansing, Iowa and graduated from the University of Illinois. He later joined the staff of the medical school of the University of Washington where he met and, in the 1950s, he began working with Fischer on the fundamental chemical reactions that regulate cell metabolism. In recognition of this important work, Krebs was elected a Fellow of the Royal Society 1947, and knighted 1958.

Hundreds of protein kinases have been found since their discovery and it has been estimated that perhaps 1% of genes encode one sort of protein kinase or another. Indeed it is now evident that phosphorylation controls virtually every important reaction in cells and provides the basis for understanding how integrated cellular behaviour is regulated by both intracellular control mechanisms and extracellular signals.

Krebs Hans Adolf 1900–1981. German-born British biochemist who discovered the citric acid cycle, also known as the ***Krebs cycle***, the final pathway by which food molecules are converted into energy in living tissues. For this work he shared the 1953 Nobel Prize for Physiology or Medicine. Knighted 1958.

The Krebs cycle, citric acid cycle, or tricarboxylic acid cycle, takes place in the mitochondria of the body's cells, and breaks down carbohydrate molecules in a series of small steps, producing carbon dioxide, water, and energy-rich molecules of ATP (adenosine triphosphate). The Krebs cycle is also involved in the degradation of fats and amino acids, and provides substrates for the synthesis of other compounds.

Krebs first became interested in the process by which the body degrades amino acids. He discovered that nitrogen atoms are the first to be removed (deamination) and are

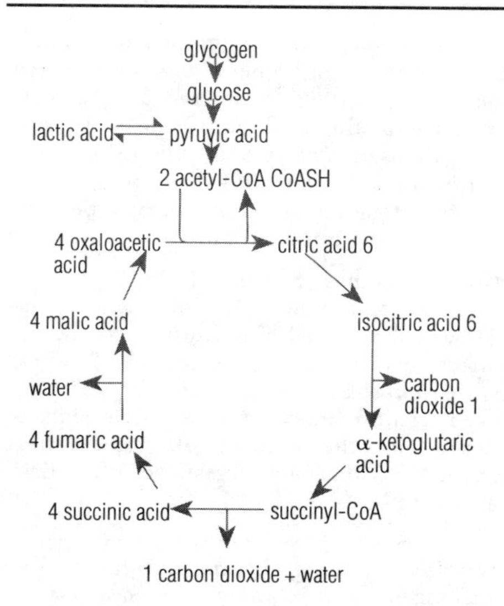

Krebs, Hans *The purpose of the Krebs (or citric acid) cycle is to complete the biochemical breakdown of food to produce energy-rich molecules, which the organism can use to fuel work. Acetyl coenzyme A (acetyl-CoA) – produced by the breakdown of sugars, fatty acids, and some amino acids – reacts with oxaloacetic acid to produce citric acid, which is then converted in a series of enzyme-catalysed steps back to oxaloacetic acid. In the process, molecules of carbon dioxide and water are given off, and the precursors of the energy-rich molecules ATP are formed. (The numbers in the diagram indicate the number of carbon atoms in the principal compounds.)*

then excreted as urea in the urine. Krebs then investigated the processes involved in the production of urea from the removed nitrogen atoms, and by 1932 he had worked out the basic steps in the urea cycle.

Krebs was born in Hildesheim and studied at the the universities of Göttingen, Freiburg, Munich, Berlin, and Hamburg. In 1933, with the rise to power of the Nazis, he moved to the UK, initially to Cambridge and in 1935 to Sheffield. He was professor at Sheffield 1945–54, and at Oxford 1954–67.

Kronecker Leopold 1823–1891. German mathematician who devised the ***Kronecker delta*** in linear algebra. He attempted to unify analysis, algebra, and elliptical functions.

Kronecker was born at Liegnitz (now Legnica, Poland), and studied at Berlin, Bonn, and Breslau. Financially independent, he devoted himself to research and lectured at Berlin from 1861. In 1883 he became professor there.

God made the integers, man made the rest.

Leopold Kronecker

Jahresberichte der deutschen Mathematiker Vereinigung Bk 2. In F Cajori *A History of Mathematics* 1919

Kronecker was obsessed with the idea that all branches of mathematics (apart from geometry and mechanics) should be treated as parts of arithmetic. He also believed that whole numbers were sufficient for the study of mathematics.

Kronecker published a system of axioms 1870, later shown to govern finite Abelian groups.

The Kronecker delta is denoted by Δ_{rs}, for which $\Delta_{rs} = 0$ when $r \neq s$ and $\Delta_{rs} = 1$ when $r = s$, where r, s are 1, 2, 3, ... He found this a useful device in the evaluation of determinants, the r and s being concerned with a row and a column of the determinant.

Krupp German family who ran the steelmaking armaments firm of the same name, founded 1811 by ***Friedrich Krupp*** (1787–1826) and developed by his son ***Alfred Krupp*** (1812–1887) by pioneering the Bessemer steelmaking process. The company developed the long-distance artillery used in World War I, and supported Hitler's regime in preparation for World War II, after which the then head of the firm, Alfred Krupp (1907–1967), was imprisoned.

Under Alfred Krupp the company became the largest armaments firm in the world, and the expansion continued under his son ***Friedrich Alfred*** (1854–1902) and his granddaughter ***Bertha*** (1886–1957). Krupp flourished as two world wars demanded iron and steel. The company developed the long-range artillery used in World War I, and supported Hitler's regime in preparation for World War II. During Wolrd War II it thrived on the benefits of the forced labour of prisoners of war and concentration camp inmates and exploiting the industrial resources of occupied territories, transporting entire plants, machinery and workers, to Germany.

With Germany's defeat 1945, Alfred Krupp was convicted of war crimes, sentenced to 12 years in prison, and ordered to sell 75% of his holdings. When no buyers could be found, Krupp was given a generous amnesty 1951 and was soon back in business. The family interest ended shortly after his death when his heir, Arndt, renounced all interest in the business and Krupp became a public corporation.

Krupp Alfred 1812–1887. German metallurgist who became known as the Cannon King because of the cast-steel guns he manufactured. He invented the first weldless steel tyre for railway vehicles. From a staff of seven, Krupp's enterprise had grown to employ 21,000 by his death.

Krupp was born in Essen and left school at 14 to join the family steel works. With the advent of the railways, the firm developed into a vast industrial empire. Despite Krupp's reputation, the output of goods – even in wartime – was predominantly for peaceful purposes.

Krupp designed and developed new machines. He invented the spoon roll for making spoons and forks, and manufactured rolling mills for use in government mints. In 1847 the firm cast its first steel cannon, but it still specialized in making fine steel suitable for dies, rolls, and machine-building. Krupp introduced the Bessemer and open-hearth steelmaking processes 1852.

Krupp created a comprehensive welfare scheme for his workers, with a sickness fund, low-cost housing, and pension schemes for all his employees. Medical care and consumer cooperatives enjoyed company backing. These institutions acted as models for the social legislation enacted in Germany in the 1880s.

Kuhn Richard 1900–1967. Austrian-born German chemist who determined the structures of vitamins A, B_2, and B_6 in the 1930s, having isolated them from cow's milk. He was awarded the 1938 Nobel Prize for Chemistry.

Kuhn was born in Vienna and educated there and at Munich, Germany. In 1929 he became professor at Heidelberg and director of the Kaiser Wilhelm (later Max Planck) Institute for Medical Research. He remained there until the late 1930s, when he was caught in a Nazi roundup of Jews and imprisoned in a concentration camp. He was unable to receive his Nobel prize until the end of World War II in 1945, when he returned to work in Heidelberg.

Kuhn's early research concerned the carotenoids, the fat-soluble yellow pigments found in plants which are precursors of vitamin A.

In the 1940s, Kuhn continued to carry out research on carbohydrates, studying alkaloid glycosides such as those that occur in tomatoes, potatoes, and other plants of the genus *Solanum*. In 1952, he returned to experiments with milk, extracting carbohydrates from thousands of litres of milk using chromatography. This work led in the 1960s to the investigation of similar sugar-type substances in the human brain.

Kuhn Thomas Samuel 1922–1996. US historian and philosopher of science, who showed that social and cultural conditions affect the directions of science. *The Structure of Scientific Revolutions* 1962 argued that even scientific knowledge is relative, dependent on the ***paradigm*** (theoretical framework) that dominates a scientific field at the time.

Nevertheless, paradigm changes do cause scientists to see the world of their research-engagement differently. In so far as their only recourse to that world is through what they see and do, we may want to say that after a revolution scientists are responding to a different world.

Thomas Kuhn

The Structure of Scientific Revolutions 1970

Such paradigms (for example, Darwinism and Newtonian theory) are so dominant that they are uncritically accepted as true, until a 'scientific revolution' creates a new orthodoxy. Kuhn's ideas have also influenced ideas in the social sciences.

Kuhne Wilhelme 1837–1900. German physiologist who coined the term enzyme (a substance produced by cells that is capable of speeding up chemical reactions) and was the first to show the reversible effect of light on the retina of the eye.

He began his research on substances responsible for the breakdown of foodstuffs in the digestive system. After working on the active substances in pancreatic juices, Kuhne coined the term enzyme, which he derived from the Greek words ***en*** which means 'in' and ***zyme*** which means 'leaven'.

A mammalian cell has thousands of different enzymes which catalyse (trigger) essential chemical reactions to meet the requirements of the cell for growth, reproduction, synthesis, and breakdown of the products of metabolism. Enzymes enhance the rate of chemical reactions frequently by as much as a million-fold and are highly specific for the reactions they catalyse.

Kuhne was born in Hamburg and as a medical student was fortunate to work with the eminent physiologists Rudolf Virchow and Claude Bernard. In 1871 he became professor of physiology at the University of Freiberg.

He also worked on the photosensitive proteins present in the retina (the photosensitive area at the back of the eye that connects to the optic nerve) of the frog's eye and was the first to show the reversible effect of light on the activity of a coloured pigment (later called rhodopsin) in the retina.

Kuiper Gerard Peter 1905–1973. Dutch-born US astronomer who made extensive studies of the Solar System. His discoveries included the atmosphere of the planet Mars and that of Titan, the largest moon of the planet Saturn.

Kuiper was adviser to many NASA exploratory missions, and pioneered the use of telescopes on high-flying aircraft. The Kuiper Airborne Observatory, one such telescope, is named after him; it was permanently grounded Oct 1995.

Kuiper was born in Harenkarspel and educated at Leiden, emigrating to the USA 1933. He joined the staff of the Yerkes Observatory (affiliated to the University of Chicago) and was its director 1947–49 and 1957–60. From 1960 he held a similar position at the Lunar and Planetary Laboratory at the University of Arizona.

In 1948, Kuiper correctly predicted that carbon dioxide was one of the chief constituents of the Martian atmosphere. He discovered the fifth moon of Uranus, which he called Miranda, also in 1948; and in 1949 he discovered the second moon of Neptune, Nereid. Kuiper's spectroscopic studies of Uranus and Neptune led to the discovery of features subsequently named ***Kuiper bands***, which indicate the presence of methane.

Kummer Ernst Eduard 1810–1893. German mathematician who introduced 'ideal numbers' in the attempt to prove Fermat's last theorem. His research into systems of rays led to the discovery of the fourth-order surface known as the ***Kummer surface***.

Kummer was born in Sorau (now Zary, Poland) and studied at Halle. He was professor at Breslau 1842–55, when he was appointed to professorships at the University of Berlin and the Berlin War College.

Ideal numbers is one of the most creative and influential ideas in the history of mathematics. With their aid Kummer was able to prove, in 1850, that the equation:

$$x^l + y^l + z^l = 0$$

was impossible in non-zero integers for all regular prime numbers (a special type of prime number related to Bernoulli numbers). He was then able to determine that the only primes less than 100 that are not regular are 37, 59, and 67. For many years Kummer continued to work on the problem and he was eventually able to prove that the equation is impossible for all primes $l < 100$.

The Kummer surface can be described as the quartic which is the singular surface of the quadratic line complex and involves the very sophisticated and complicated concept of this surface as the wave surface in space of four dimensions.

Kummer also made an important contribution to function theory by his investigations into hypergeometric series.

Kundt August Adolph Eduard Eberhard 1839–1894. German physicist who in 1866 invented ***Kundt's tube***, a simple device for measuring the velocity of sound in gases and solids. His later work entailed the demonstration of the dispersion of light in liquids, vapours, and metals.

Kundt was born in Schwerin, Mecklenburg, and educated at Leipzig and Berlin. He was professor at the Polytechnic in Zürich, Switzerland, 1868, at Würzburg from 1869, and was one of the founders of the Strasbourg Physical Institute 1872. He ended his career as director of the Berlin Physical Institute from 1888.

Kundt's tube is a glass tube containing some dry powder and closed at one end. Into the open end, a disc attached to a rod is inserted. When the rod is sounded, the vibration of the disc sets up sound waves in the air in the tube and a position is found in which standing waves occur, causing the dust to collect at the nodes of the waves. By measuring the length of the rod and the positions of the nodes in the tube, the velocity of sound in either the rod or air can be found, provided one of these quantities is known. By using rods made of various materials and different gases in the tube, the velocity of sound in a range of solids and gases can be determined.

Kusch Polykarp 1911–1933. German-born US physicist who determined the magnetic moment (or magnetic strength) of the electron. Nobel Prize for Physics 1955.

Kusch worked under Isidor Rabi on spectroscopy using atomic and molecular beams. He investigated the interactions of the constituent particles of atoms and molecules with each other and with an external magnetic field. From a study of the energy levels of certain elements, he was able to deduce a very accurate value for the magnetic moment of the electron. He found a small discrepancy between the observed magnetic moment and that predicted by theory. This discrepancy was of great significance to the theory (now known as quantum electrodynamics) describing the interactions of electrons and electromagnetic fields. His discovery was confirmed by independent experiments done by Willis Lamb, with whom he shared the Nobel prize.

Born in Blankenburg, Germany, he went to the United States in 1912. He was educated at Case Institute of Technology, Cleveland, Ohio, and the University of Illinois, where he completed his doctorate 1936. He spent his career mainly at Columbia University, New York City, rising from associate professor (1946) to dean of faculty (1969) and provost (1970). In 1972 he was appointed professor of physics at the University of Texas at Dallas.

Lacaille Nicolas Louis de 1713–1762. French astronomer who determined the positions of nearly 10,000 stars of the southern hemisphere 1750–54. He also performed a number of geodetic investigations; in particular, he made the first measurement of the arc of meridian in the southern hemisphere.

Lacaille was born in Rumigny, near Rheims, and studied theology in Paris. He began to make astronomical observations in 1737 and participated in two Academy of Sciences projects, in 1738 and 1739. At the age of 26 he became professor of mathematics at the Collège Mazarin (now the Institut de France), Paris. He was subsequently given an observatory.

Despite the lack of equipment at the Cape of Good Hope, South Africa, Lacaille catalogued a large number of seventh-magnitude stars, invisible to the naked eye. He charted and named 14 new constellations, abandoning the traditional mythological naming system and instead choosing names of contemporary scientific and astronomical instruments.

Simultaneous observations by Lacaille at the Cape and Joseph de Lalande in Berlin provided them with a baseline longer than the radius of the Earth and gave an accurate measurement of the lunar parallax.

Coelum australe stelliferum 1763 catalogued all Lacaille's data from the southern hemisphere.

Lack David 1910–1973. English ornithologist who used radar to identify groups of migrating birds. He studied and wrote about the robin, the great tit, the swift, and the finches of the Galápagos Islands.

Lack was born in London and attended Gresham's School in Norfolk and then Magdalene College, Cambridge, where he read zoology and graduated 1933. His first paper was published while he was a schoolboy and was a study of the nightjar. On leaving Cambridge he took a job as a schoolmaster at Dartington Hall where he spent his spare time studying robins, leading to the publication of *The Life of a Robin* 1943. In 1934, he travelled to Tanganyika, and, in 1935, to the USA, where he met Ernst Mayr. In 1938, he spent a year in the Galápagos Islands and wrote *Darwin's Finches*. He was the director of the Edward Grey Institute of Field Ornithology at Oxford from 1945, where he studied the great tit and swift. He was a Fellow of the Royal Society and received their Darwin medal 1972.

During World War II, he worked for the army and was one of the first people to identify migrating birds on radar. He was a devout Christian, writing *Evolutionary theory and Christian Belief* 1957.

La Condamine Charles Marie de 1701–1774. French soldier and geographer who was sent by the French Academy of Sciences to Peru 1735–43 to measure the length of an arc of the meridian. On his return journey he travelled the length of the Amazon, writing about the use of the nerve toxin curare, India rubber, and the advantages of inoculation.

Laënnec René Théophile Hyacinthe 1781–1826. French physician, inventor of the stethoscope 1816. He advanced the diagnostic technique of auscultation (listening to the internal organs) with his book *Traité de l'auscultation médiaté* 1819, which quickly became a medical classic.

Laënnec's special interest was in chest disease. Listening to the internal organs (auscultation) had been used diagnostically since the time of Hippocrates, but Laënnec introduced a wooden tube to transmit sound. He called it a stethoscope from the Greek *stethos* 'chest'.

Laënnec was born in Quimper, Brittany. As a teenager he worked in hospitals during the French Revolution; later he studied medicine at the Ecole Pratique in Paris. He was appointed personal physician to Cardinal Fesch, uncle of Napoleon I. In 1812–13, with France at war, Laënnec took charge of the wards in the Salpetrière Hospital reserved for Breton soldiers. On the restoration of the monarchy, he became physician to the Necker Hospital, retiring 1818. In 1822 he was appointed professor at the Collège de France.

Lagrange Joseph Louis 1736–1813. Italian-born French mathematician. His *Mécanique analytique* 1788 applied mathematical analysis, using principles established by Isaac Newton, to such problems as the movements of planets when affected by each other's gravitational force. He presided over the commission that introduced the metric system in 1793.

I do not know.

Joseph Lagrange

Lagrange summarizing his life's work

Lagrange was born in Turin and appointed professor of mathematics at the Royal Artillery School there when he was just 19. In 1766 Frederick the Great of Prussia invited him to become director of the Berlin Academy of Sciences, and in 1787, on the invitation of Louis XVI, Lagrange moved to Paris as a member of the French Royal Academy. He was professor at the Ecole Polytechnique from 1797.

In 1755, advancing beyond Newton's theory of sound, Lagrange settled a dispute on the nature of a vibrating string. Later he proved some of Pierre de Fermat's theorems, which had remained unproven for a century.

In *Mécanique analytique*, Lagrange succeeded in reducing the theory of solid and fluid mechanics to an analytical principle, explaining it without the aid of a single diagram or construction. His lectures at the Ecole Polytechnique were published as *Théorie des fonctions analytiques* 1797 and *Leçons sur le calcul des fonctions* 1806.

Laithwaite Eric Roberts 1921– . English electrical engineer who developed the linear motor. The idea of a linear induction motor had been suggested in 1895, but Laithwaite discovered that it is possible to arrange two linear motors back to back, so as to produce continuous oscillation without the use of any switching device.

Laithwaite was born in Yorkshire. He studied at the University of Manchester and, apart from the interruption of World War II, remained there until 1964, when he became professor of heavy electrical engineering at the Imperial College of Science and Technology at the University of London, moving to the Royal Institution 1967–76. He made many popular radio and television broadcasts.

The important feature of the linear motor is that it is a means of propulsion without the need for wheels, although often continuous motion has to be made up from reciprocating motion – achieved by the use of a ratchet, for example.

Linear motion is very common in industrial processes. In 1947 Laithwaite began research into electric linear induction motors as the shuttle drives in weaving looms, and later into the use of linear motors for conveyors and as propulsion units for high-speed railway vehicles moving on air cusions.

Lalande Joseph Jérome le Français de 1732–1807. French astronomer who observed the transit of Venus and helped to calculate the size of the Solar System. He compiled a catalogue of 47,000 stars.

Lalande was born in Bourg-en-Bresse, near Lyon, and studied in Paris. He was appointed professor at the Collège de France in 1762 and during his tenure there he published *Treatise of Astronomy* 1764. In 1795 he was made director of the Paris Observatory.

In collaboration with French astronomer Nicolas de Lacaille at the Cape of Good Hope, South Africa, in 1751, Lalande measured the lunar parallax and thus the distance from the Earth to the Moon.

Two transits of Venus, in 1761 and 1769, offered the chance to establish accurately the size of the Solar System. Such transits occur twice within a period of eight years only every 113 years. During the transit, which takes approximately five hours, Venus can be seen silhouetted across the face of the Sun; the distance of the Earth from the Sun can be deduced by measuring the different times that the planet takes to cross the face of the Sun when seen from different latitudes on Earth. Lalande was responsible for coordinating expeditions to all corners of the world and collecting the results of observations.

Lamarck Jean Baptiste de 1744–1829. French naturalist whose theory of evolution, known as ***Lamarckism***, was based on the idea that acquired characteristics (changes acquired in an individual's lifetime) are inherited by the offspring, and that organisms have an intrinsic urge to evolve into better-adapted forms. *Philosophie zoologique/ Zoological Philosophy* 1809 outlined his 'transformist' (evolutionary) ideas.

Zoological Philosophy tried to show that various parts of the body developed because they were necessary, or disappeared because of disuse when variations in the environment caused a change in habit. If these body changes were inherited over many generations, new species would eventually be produced.

Lamarck was the first to distinguish vertebrate from invertebrate animals by the presence of a bony spinal column. He was also the first to establish the crustaceans, arachnids, and annelids among the invertebrates. It was Lamarck who coined the word 'biology'.

Lamarck was born in Bazentin, Picardy. He studied medicine, meteorology, and botany, and travelled across Europe as botanist to King Louis XVI from 1781. In 1793 he was made professor of zoology at the Museum of Natural History in Paris.

So little was known about invertebrates at this time that some scientists grouped snakes and crocodiles with insects. Lamarck studied both living and fossil invertebrates, and described them in his book *Histoire naturelle des animaux sans vertèbres/Natural History of Invertebrate Animals* 1815–22.

Lamb Horace 1849–1934. English applied mathematician noted for his many books on hydrodynamics, elasticity, sound, and mechanics. His chief work is *Treatise on the Motion of Fluids* 1879, revised and updated as *Hydrodynamics* 1895–1932.

Lamb was born in Stockport, Cheshire, and studied at Cambridge. He went to Australia 1875 to take up the chair of mathematics at the University of Adelaide. In 1885 he returned to Manchester as professor at Owens College.

Lamb's contributions ranged wide over the field of applied mathematics, and he was particularly adept at applying the solution of a problem in one field to problems in another. A paper of 1882, which analysed the modes of oscillation of an elastic sphere, achieved its true recognition in 1960, when free Earth oscillations during an earthquake behaved in the way Lamb had described. A paper of 1904 gave an analytical account of propagation over the surface of an elastic solid of waves generated by given initial disturbances, and the analysis he provided is now regarded as one of the seminal contributions to theoretical seismology.

Lamb Willis Eugene 1913– . US physicist who revised the quantum theory of Paul Dirac. The hydrogen atom was thought to exist in either of two distinct states carrying equal energies. More sophisticated measurements by Lamb in 1947 demonstrated that the two energy levels were not equal. This discrepancy, since known as the ***Lamb shift***, won him the 1955 Nobel Prize for Physics.

Lanchester Frederick William 1868–1946. English engineer who began producing motorcars 1896. His work on stability was fundamental to aviation and he formulated the first comprehensive theory of lift and drag. From early on, Lanchester manufactured his cars with interchangeable parts.

Lanchester was born in London and studied at Imperial College.

He joined the Forward Gas Engine Company of Birmingham 1889, and in 1893 set up his own workshop. The Lanchester Engine Company built some 350 cars between 1900 and 1905, before going bankrupt. In 1909 he was appointed consultant to the Daimler company. He founded Lanchester's Laboratories Ltd in 1925 to provide research and development services.

Lanchester's first motorcar had a single cylinder 4-kW/5-hp engine and chain drive. His second completed a 1,600-km/1,000-mi tour in 1900.

In the early 1890s Lanchester turned his attention to the theory and practice of crewed flight. He published *Aerial Flight* 1907–08 and was invited to join Prime Minister Asquith's advisory committee for aeronautics on its formation 1909. An experimental aircraft codesigned by Lanchester did not survive its trial flight 1911, and he abandoned the practical side of aviation. However, planes that incorporated many of his ideas took to the air in the next few years.

Lanchester was also interested in radio and patented a loudspeaker and other audio equipment.

Landau Lev Davidovich 1908–1968. Russian theoretical physicist. He was awarded the 1962 Nobel Prize for Physics for his work on liquid helium.

Landé Alfred 1888–1975. German-born US quantum physicist. In 1923 he published a formula expressing a factor, now known as the Landé splitting factor, for all multiplicities as a function of the quantum numbers of the stationary state of the atom.

Landé was born in Elberfeld and studied at Marburg, Göttingen, and Munich. From 1922 he was on the staff at Tübingen, the main centre of atomic spectroscopy in Germany. He emigrated to the USA 1931, because of the rise of the Nazi regime, to settle in Columbus, Ohio, as professor of theoretical physics at Ohio State University.

In collaboration with German physicist Max Born, Landé published a paper 1918 on their conclusion that Danish physicist Niels Bohr's model of coplanar electronic orbits must be wrong, and that they must be inclined to each other. Later Landé visited Bohr in Copenhagen to discuss the Zeeman effect (the splitting of spectral lines in a magnetic field), which pointed him towards the discovery of the Landé splitting factor. This is the ratio of an elementary magnetic moment to its causative angular momentum when the angular momentum is measured in quantized units.

Landsteiner Karl 1868–1943. Austrian-born US immunologist who discovered the ABO blood group system 1900–02, and aided in the discovery of the Rhesus blood factors 1940. He also discovered the polio virus. Nobel prize 1930.

In 1927 Landsteiner found that, in addition to antigens A and B, human blood cells contain one or other or both of two heritable antigens, M and N. These are of no importance in transfusions, because human serum does not contain the corresponding antibodies, but they are of value in resolving paternity disputes.

Landsteiner was born and educated in Vienna, and also studied at other universities in Europe. He worked at the Vienna Pathology Laboratory 1898–1908, became professor at a Vienna hospital, and left Austria 1919 for the Netherlands, moving 1922 to the USA and the Rockefeller Institute for Medical Research, New York.

His book *The Specificity of Serological Reactions* 1936 helped establish the science of immunology. He also developed a test for syphilis.

Langevin Paul 1872–1946. French physicist who contributed to the studies of magnetism and X-ray emissions, especially paramagentic (weak attractive) and diamagnetic (weak repulsive) phenomena in gases. During World War I he invented an apparatus for locating enemy submarines, which is the basis of modern echolocation techniques.

Langevin was born and educated in Paris and also studied in the UK at the Cavendish Laboratories, Cambridge. He was professor at the Collège de France 1904–09 and from 1909 at the Sorbonne. In 1940, after the start of World War II and the German occupation of France, Langevin became director of the Ecole Municipale de Physique et de Chimie Industrielles, where he had been teaching since 1902, but he was soon arrested by the Nazis for his antifascist views. He escaped to Switzerland in 1944, returning after the liberation of Paris.

Langevin suggested 1905 that the alignment of molecular moments in a paramagnetic substance would be random except in the presence of an externally applied magnetic field. He extended his description of magnetism in terms of electron theory to account for diamagnetism, and showed how a magnetic field would affect the motion of electrons in the molecules to produce a moment that is opposed to the field.

He was an early supporter of Albert Einstein's theories, and the nuclear institute in Grenoble is named after him.

Langley John Newport 1852–1925. English physiologist who investigated the structure and function of the autonomic nervous system, the involuntary part of the nervous system, that controls the striated and cardiac muscles and the organs of the gastrointestinal, cardiovascular, excretory, and endocrine systems. He went on to divide up the autonomic nervous system into the sympathetic and parasympathetic branches, with specific functions being apportioned to each.

Langley did a great deal of research on the structure and function of sympathetic nerve fibres and ganglia. Ganglia are clusters of the cell bodies of sensory neurons in the peripheral nervous system which lie just outside the spinal cord. Langley blocked nervous impulses by applying various chemicals, such as nicotine, to ganglia. The cell bodies of motor neurons in the autonomic nervous system also lie in these ganglia. His findings were published in *The Autonomic Nervous System* 1921.

Langley was born in Newbury and educated at St Johns College, Cambridge. He remained in the physiology department at Cambridge and was elected professor of physiology Cambridge 1903.

Langley Samuel Pierpoint 1834–1906. US astronomer, scientist, and inventor of the bolometer, an instrument that measures radiation. His steam-driven aeroplane flew for 90 seconds in 1896 – the first flight by an engine-equipped aircraft.

He was professor of physics and astronomy at the Western University of Pennsylvania 1866–87, and studied the infrared portions of the Solar System.

From 1887 he was secretary of the Smithsonian Institution in Washington DC. He founded the Smithsonian Astrophysical Observatory in 1890 and turned to pioneering work in aerodynamics, contributing greatly to the design of early aircraft. He built and tested the first successful (but uncrewed) heavier-than-air craft (aeroplane), which he launched by catapult and which flew over the Potomac River in 1896. The subsequent catapult-launched flights of the Wright brothers at Kitty Hawk owed much to Langley's principles as well as to the more powerful engines available by the early 1900s. The Langley design was tested in later years by using a model with a modern engine; it flew successfully with a pilot aboard.

Langmuir Irving 1881–1957. US scientist who invented the mercury vapour pump for producing a high vacuum, and the atomic hydrogen welding process; he was also a pioneer of the thermionic valve. In 1932 he was awarded a Nobel prize for his work on surface chemistry.

Langmuir was born in New York and studied there at Columbia University and in Germany at Göttingen. He worked at the research laboratories of the General Electric Company 1909–50.

Langmuir's research on electric discharges in gases at very low pressures led to the discovery of the space-charge effect: the electron current between electrodes of any shape in vacuum is proportional to the 3/2 power of the potential difference between the electrodes. He also studied the mechanical and electrical properties of tungsten lamp filaments. Langmuir's introduction of nitrogen into light bulbs prevented them from blackening on the inside but increased heat loss, which was overcome by coiling the tungsten filament.

Langmuir was the first to use the terms 'electrovalency' (for ionic bonds between metals and nonmetals) and 'covalency' (for shared-electron bonds between nonmetals).

During the 1920s, Langmuir became particularly interested in the properties of liquid surfaces. He went on to propose his general adsorption theory for the effect of a solid surface during a chemical reaction.

Lankester Edwin Ray 1847–1929. English zoologist who made clear morphological distinctions between the different orders of invertebrates. He distinguished between the haemocoel (blood-containing cavity) in Mollusca and Arthropoda and the coelom (fluid-filled cavity) in worms and vertebrates for the first time, showing that whilst functionally similar they have different origins.

Lankester was born in London, son of the prominent scientific writer, Edwin Lankester, and attended Downing College, Cambridge. During his second year at university he moved to Oxford where he graduated in natural sciences from Christchurch College. In 1871 he studied marine zoology in Naples for a year before becoming a fellow and tutor at Exeter College, Oxford. He was made chair of zoology at University College, London 1874, and was elected Fellow of the Royal Society the following year (he served as vice-president 1882 and 1896, and received the Royal medal 1885).

He returned to Oxford to take up the Linacre Chair of comparative anatomy 1891, and made extensive changes to the organization of the zoology specimens in Oxford's University Museum. From 1898 he was the director of the natural history department and keeper of zoology at the British Museum. He was made KCB 1907 and awarded the Copley Medal and the Darwin Wallace Medal.

Laplace Pierre Simon. Marquis de Laplace 1749–1827. French astronomer and mathematician. In 1796, he theorized that the Solar System originated from a cloud of gas (the nebular hypothesis). He studied the motion of the Moon and planets, and published a five-volume survey of celestial mechanics, *Traité de méchanique céleste* 1799–1825. Among his mathematical achievements was the development of probability theory.

Traité de mécanique céleste contained the law of universal attraction – the law of gravity as applied to the Earth – and explanations of such phenomena as the ebb and flow of tides and the precession of the equinoxes. Marquis 1817.

Laplace was born in Normandy and studied at Caen. He became professor of mathematics at the Paris Ecole Militaire 1767. In 1799 Napoleon briefly appointed Laplace minister of the interior before elevating him to the senate. From 1814, Laplace supported the Bourbon monarchy, and in 1826 refused to sign a declaration of the French Academy supporting the freedom of the press.

In celestial mechanics, Laplace began 1784–86 by explaining the variations in the orbits of Jupiter and Saturn. In 'Théorie des attractions des sphéroïdes et de la figure des planètes' 1785 he introduced the potential function and the Laplace coefficients, both of them useful as a means of applying analysis to problems in physics.

Lapworth Arthur 1872–1941. British chemist, one of the founders of modern physical-organic chemistry. He formulated the electronic theory of organic reactions (independently of English chemist Robert Robinson).

Lapworth was one of the first to emphasize that organic compounds can ionize, and that different parts of an organic molecule behave as though they bear electrical charges, either permanently or at the moment of reaction.

With the development of theories of valency based on the electronic structure of the atom, Lapworth was able to refine some speculations about 'alternative polarities' in organic compounds into a classification of reaction centres as either 'anionoid' (nucleophilic) or 'cationoid' (electrophilic), the changes being determined by the influence of a key atom such as oxygen.

Lapworth was born in Galashiels, Scotland, and studied at Mason College, Birmingham. He became head of the chemistry department at Goldsmiths' College, London, in 1900. In 1909 he moved to Manchester, where he remained, becoming professor 1922.

Latimer Louis Howard 1848–1928. US inventor of improvements to electric lighting. He was a member of the Edison Pioneers, an organization of scientists who worked with Thomas Edison. He also supervised the installation of electric lights in New York, Philadelphia, Canada, and London.

Latimer was born in Chelsea, Massachusetts, and left school to help support the family by selling copies of the anti-slavery journal *The Liberator*. Working at a firm of patent solicitors, he learned mechanical drawing and assisted Scottish inventor Alexander Graham Bell with his application for the telephone patent. In 1880, Latimer was employed by Hiram Maxim at the US Electric Lighting Company, moving 1883 to the Edison Electric Light Company (later the General Electric Company). Between 1896 and 1911 he served as expert witness for the Board of Patent Control formed by the General Electric and Westinghouse companies to protect their patents.

After coinventing an electric lamp 1881, Latimer went on to invent a cheap method for producing long-lasting carbon light-bulb filaments 1882. Other Latimer patents included a 'Water Closet for Railroad Cars' 1874, 'Apparatus for Cooling and Disinfecting' 1886, and 'Locking Rack for Hats, Coats, and Umbrellas' 1896.

Laue Max Theodor Felix von 1879–1960. German physicist who was a pioneer in measuring the wavelength of X-rays by their diffraction through the closely spaced atoms in a crystal. His work led to the techniques of X-ray spectroscopy, used in nuclear physics, and X-ray diffraction, used to elucidate the molecular structure of complex biological materials. Nobel prize 1914.

Laue was born near Koblenz and studied at Göttingen and Berlin. He was assistant to Max Planck at the Institute of Theoretical Physics in Berlin 1905–09, and worked at the Institute of Theoretical Physics in Munich 1909–14, when he became professor at Frankfurt. After World War I he became director of the Institute of Theoretical Physics in Berlin, resigning 1943 in protest against Nazi policies. Although Laue had refused to participate in the German atomic-energy project, he was interned in Britain by the Allies after the war. He returned to Germany 1946, and in 1951 became director of the Max Planck Institute for Research in Physical Chemistry.

In 1912 Laue's idea of passing X-rays through crystal was first tested, and provided experimental demonstration of the nature of both crystal structure and X-ray radiation. He had initially considered only the interaction between the atoms in the crystal and the radiation waves, but later included a correction for the forces acting between the atoms.

Laval Carl Gustaf Patrik de 1845–1913. Swedish engineer who made a pioneering contribution to the development of high-speed steam turbines. He invented the special reduction gearing that allows a turbine rotating at high speed to drive a propeller or machine at comparatively slow speed, a principle having universal application in marine engineering.

De Laval was born in Orsa, Dalarna, and educated at the Stockholm Technical Institute and Uppsala University.

In 1887, de Laval developed a small, high-speed turbine with a speed of 42,000 revolutions per minute. He is credited with being the first to use a convergent–divergent type of nozzle in a steam turbine in order to realize the full potential energy of the expanding steam in a single-stage machine, completed 1890. He also invented various devices for the dairy industry, including a high-speed centrifugal cream separator 1878 and a vacuum milking machine, perfected 1913.

De Laval's other interests ranged from electric lighting to electrometallurgy in aerodynamics. In the 1890s he employed more than 100 engineers in developing his devices and inventions, which are exactly described in the 1,000 or more diaries he kept.

Laveran (Charles Louis) Alphonse 1845–1922. French physician who discovered that the cause of malaria is a protozoan, the first time that protozoa were shown to be a cause of disease. For this work and later discoveries of protozoan diseases he was awarded the 1907 Nobel Prize for Physiology or Medicine.

Laveran was born in Paris and studied in Strasbourg. When the Franco-Prussian war broke out he became an army surgeon, and in 1874 he was appointed professor of military medicine at the Ecole du Val-de-Grâce. Between 1878 and 1883 he was posted to Algeria. He left the army 1896 to join the Pasteur Institute in Paris, and in 1907 he used the money from his Nobel prize to open the Laboratory of Tropical Diseases at the institute.

In 1880, Laveran examined blood samples from malarial patients and discovered amoebalike organisms growing within red blood cells. They divided and formed spores, which invaded unaffected blood cells. He noted that the spores were released in each affected red cell at the same time and corresponded with a fresh attack of fever in the patient.

Laveran's studies of protozoan diseases included leishmaniasis and trypanosomiasis. His publications included *Traité des maladies et épidemies des armées/Treatise on Army Sicknesses and Epidemics* 1875 and *Trypanosomes et trypanosomiasis* 1904.

Lavoisier Antoine Laurent 1743–1794. French chemist. He proved that combustion needs only a part of the air, which he called oxygen, thereby destroying the theory of phlogiston (an imaginary 'fire element' released during combustion). With astronomer and mathematician Pierre Laplace, he showed 1783 that water is a compound of oxygen and hydrogen. In this way he established the basic rules of chemical combination.

Lavoisier established that organic compounds contain carbon, hydrogen, and oxygen. From quantitative measurements of the changes during breathing, he showed that carbon dioxide and water are normal products of respiration.

Lavoisier was born in Paris and studied at the Collège Mazarin. He worked as a tax collector and became director of the Academy of Sciences 1785. Two years later he became a member of the provincial assembly of Orléans. During the French Revolution, left-wing leader Jean-Paul Marat, whose membership of the Academy of Sciences had been blocked by Lavoisier, accused him of imprisoning Paris and preventing air circulation because of the wall he had built round the city in 1787. He fled from his home and laboratory in 1792 but was later arrested, tried, and guillotined.

When English chemist Joseph Priestley produced 'dephlogisticated air', Lavoisier, who had already been studying combustion, grasped the true explanation. He went on to burn various organic compounds in oxygen and determined their composition by weighing the carbon dioxide and water produced – the first experiments in quantitative organic analysis. He also showed by weighing that matter is conserved during fermentation as in more conventional chemical reactions.

In *Traité élémentaire de chimie* 1789, Lavoisier listed all the chemical elements then known.

Lavrentiev Mikhail 1900– . Soviet scientist who developed the Akademgorodok ('Science City') in Novosibirsk, Russia from 1957.

Lawrence Ernest O(rlando) 1901–1958. US physicist. His invention of the cyclotron particle accelerator pioneered the production of artificial radioisotopes and the synthesis of new transuranic elements. Nobel prize 1939.

During World War II, Lawrence was involved with the separation of uranium-235 and plutonium for the development of the atomic bomb, and he organized the Los Alamos Scientific Laboratories at which much of the work on this project was carried out. After the war, he continued as a believer in nuclear weapons and advocated the acceleration of their development.

Lawrence was born in South Dakota and studied there and at Minnesota, Chicago, and Yale universities. He was professor of physics at the University of California, Berkeley, from 1930 and director from 1936 of the Radiation Laboratory, which he built into a major research centre for nuclear physics.

The first cyclotrons were made in 1930 and were only a few centimetres in diameter. Each larger and improved design produced particles of higher energy than its predecessor, and a 68-cm/27-in model was used to produce artificial radioactivity. Among the results obtained from the use of the accelerated particles in nuclear transformations was the disintegration of the lithium nucleus to produce helium nuclei.

Leakey Louis Seymour Bazett 1903–1972. Kenyan archaeologist, anthropologist, and palaeontologist. With his wife Mary Leakey, he discovered extinct-animal fossils in the Olduvai Gorge in Tanzania, as well as many remains of an early human type. Leakey's conviction that human origins lie in Africa was opposed to contemporary opinion.

Leakey was born in Kabete, Kenya, and studied in the UK at Cambridge. Between 1926 and 1937 he led a series of archaeological research expeditions to E Africa. He was curator of Coryndon Museum, Nairobi, Kenya, 1945–61, and one of the founder trustees of the Kenya National Parks and Reserves. In 1961 he founded the National Museum Centre for Prehistory and Palaeontology.

Leakey began excavations at Olduvai Gorge in 1931. With Mary Leakey, he discovered a site in the Rift Valley of the Acheulian culture, which flourished between 1.5 million and 150,000 years ago. The Leakeys also found the remains of 20-million-year-old apes on an island in Lake Victoria. In 1960 they discovered the remains of *Homo habilis*, 1.7 million years old, and the skull of an Acheulian hand-axe user, *Homo erectus*, which Leakey maintained was on the direct evolutionary line of *Homo sapiens*, the modern human. In 1961 at Fort Ternan, Kenya, he found jawbone fragments of another early primate, believed to be 14 million years old.

His books for a general readership include *Stone Age Africa* 1936 and *White African* 1937.

Leakey Mary Douglas, (born Nicol) 1913–1996. British archaeologist and anthropologist. In 1948 she discovered, on Rusinga Island, Lake Victoria, E Africa, the prehistoric ape skull known as *Proconsul*, about 20 million years old; and human footprints at Laetoli, to the south, about 3.75 million years old.

Nicol was born in London and became assistant to an archaeologist. Her collaboration with Louis Leakey began when she illustrated a book he was working on. In 1936, Mary Leakey began excavating a Late Stone Age site north of Nairobi, and with her husband discovered the remains of an important Neolithic settlement.

The Leakeys together and separately carried out excavations in Kenya, and accumulated evidence that E Africa was the possible cradle of the human race. By the middle of the 1960s Mary was living almost continuously at the permanent camp they had established at the Olduvai Gorge.

She described her work in the book *Olduvai Gorge: My Search for Early Man* 1979.

At one point ... she stops, pauses, turns to the left to glance at some possible threat or irregularity, and then continues to the north. This motion, so intensely human, transcends time.

Mary Leakey

Describing footprints made by early humans 3.75 million years ago

---

Leakey Richard Erskine Frere 1944– . Kenyan palaeoanthropologist. In 1972 he discovered at Lake Turkana, Kenya, an apelike skull, estimated to be about 2.9 million years old; it had some human characteristics and a brain capacity of 800 cu cm. In 1984 his team found an almost complete skeleton of *Homo erectus* some 1.6 million years old. He is the son of Louis and Mary Leakey.

He was appointed director of the Kenyan Wildlife Service 1988, waging a successful war against poachers and the ivory trade, but was forced to resign 1994 in the face of political interference. He lost both legs in a plane crash 1993. Since resigning, he has co-founded 1995 the Kenyan political party Safina (Swahili for Noah's Ark), which aims to clean up Kenya. The party was accused of racism and colonialism by President Daniel arap Moi.

His wife Meave continues the search for fossil humans, and in 1995 announced the discovery of bones of *Australopithecus anamensis*, an upright hominid of about 4 million years ago.

Leavitt Henrietta Swan 1868–1921. US astronomer who in 1912 discovered the period–luminosity law, which links the brightness of a Cepheid variable star to its period of variation. This law allows astronomers to use Cepheid variables as 'standard candles' for measuring distances in space.

Leavitt was born in Lancaster, Massachusetts, and studied at what was to become Radcliffe College. She joined the Harvard College Observatory 1902, and was ultimately appointed head of the department of photographic stellar photometry.

Leavitt's work grew out of director Edward Pickering's research programme at the observatory, towards the establishment of a standard photographic sequence of stellar magnitudes. She discovered a total of 2,400 new variable stars and four novae.

Leavitt's period–luminosity curve for the Cepheids enabled Danish astronomer Ejnar Hertzsprung and US astronomer Harlow Shapley to read the data in terms of absolute rather than apparent magnitude. By comparing a Cepheid's apparent magnitude with its absolute magnitude, the distance of the star from the Earth could be deduced. Until then it had not been known that they are outside our Galaxy.

Lebedev Peter Nikolaievich 1866–1912. Russian physicist. He proved by experiment that light exerts a minute pressure upon a physical body, thereby confirming James Maxwell's theoretic prediction.

Lebedev was born in Moscow and studied in Germany at Strasbourg. He was professor at Moscow University 1892–1911, resigning on political grounds.

Lebedev began studying light pressure (now called radiation pressure) in the late 1890s but did not complete his investigations until 1910. Working first with solid bodies and later with gases, he not only observed the minute physical effects caused by the infinitesimal pressure exerted by light on matter but also measured this pressure using extremely lightweight apparatus in an evacuated chamber.

Lebedev also investigated the effects of electromagnetic, acoustic, and hydrodynamic waves on resonators; demonstrated the behavioural similarities between light and (as they are now known to be) other electromagnetic radiations; detected electromagnetic waves of higher frequency than had previously been studied; and researched into the Earth's magnetic field.

Lebesgue Henri Léon 1875–1941. French mathematician who developed a new theory of integration, now named after him. He also made contributions to set theory, the calculus of variations, and function theory.

Lebesgue was born in Beauvais and educated at the Ecole Normale Supérieure in Paris. He was professor at the University of Poitiers 1906–10, when he was appointed lecturer in mathematics at the Sorbonne. In 1920 he was promoted to the chair of the application of geometry to analysis, but he left the Sorbonne 1921 to take up his final academic post as professor of mathematics at the Collège de France.

Lebesgue was intrigued by problems associated with Riemannian integration. His introduction of the Lebesgue integral quickly proved to be of great importance in the development of several branches of mathematics, especially calculus, curve rectification, and the theory of trigonometric series.

Following the work of Emile Borel, Lebesgue laid the foundations of the modern theory of the functions of a real variable.

Leblanc Nicolas 1742–1806. French chemist who in the 1780s developed a process for making soda ash (sodium carbonate, Na_2CO_3) from common salt (sodium chloride, NaCl). Soda ash was widely used industrially in making glass, paper, soap, and various chemicals.

In the ***Leblanc process***, salt was first converted into sodium sulphate by the action of sulphuric acid, which was then roasted with chalk or limestone (calcium carbonate) and coal to produce a mixture of sodium carbonate and sulphide. The carbonate was leached out with water and the solution crystallized. The process was adopted throughout Europe.

Leblanc was probably born in Indre *département*. He studied medicine and became physician and assistant in 1780 to Louis Philippe Joseph (who, as duke of Orléans, was guillotined 1793). Leblanc devised his method of producing soda ash to win a prize offered 1775 by the French Academy of Sciences, but the Revolutionary government granted him only a patent (1791), which they seized along with his factory three years later. He had no money left to re-establish the process when the factory was handed back to him by Napoleon in 1802. A broken man, Leblanc committed suicide.

Lebon Phillipe 1767–1804. French engineer who in 1801 became the first person successfully to use 'artificial' gas as a means of illumination on a large scale.

Lebon was born in the charcoal-burning town of Bruchay, near Joinville on the Marne. He studied at the Ecole des Ponts et Chaussées, where he later taught mechanics. He also made some attempts at perfecting the steam engine.

In about 1797, Lebon became interested in extracting gas from wood for heating and lighting purposes. He placed some sawdust in a glass tube and held it over a flame. The gas given off caught alight as it emerged from the tube, but it smoked badly and smelled. Persevering, he patented in 1799 the Thermolampe (heat lamp). For several months in 1801, he exhibited a large version of the lamp in Paris. It attracted huge crowds, but, because he had been unable to eliminate the repulsive odour given off by the gas, the public decided that his invention was not a practical one.

It was left to William Murdoch (working independently at about the same time in Scotland) to succeed where Lebon had failed, and Murdoch has received the credit for the invention of gas lighting.

Le Châtelier Henri Louis 1850–1936. French physical chemist who formulated the principle now named after him, which states that if any constraint is applied to a system in chemical equilibrium, the system tends to adjust itself to counteract or oppose the constraint.

Le Châtelier was born in Paris and studied science and engineering at the Ecole Polytechnique. He was professor of chemistry at the Ecole des Mines 1877–98, moving first to the Collège de France and then in 1908 to the Sorbonne.

In 1887, he devised a platinum–rhodium thermocouple for measuring high temperatures by making use of the Seebeck effect. Le Châtelier also made an optical pyrometer which measures temperature by comparing the light emitted by a high-temperature object with a standard light source.

Le Châtelier's principle, formulated 1884–88, is particularly relevant in predicting the effects of changes in temperature and pressure on chemical reactions. It also agreed with the new thermodynamics being worked out in the USA by Josiah Gibbs. Le Châtelier was largely responsible for making Gibbs's researches known in Europe.

In 1895 Le Châtelier put forward the idea of the oxyacetylene torch for cutting and welding steel.

Leclanché Georges 1839–1882. French engineer. In 1866 he invented a primary electrical cell, the ***Leclanché cell***, which is still the basis of most dry batteries.

A Leclanché cell consists of a carbon rod (the anode) inserted into a mixture of powdered carbon and manganese dioxide contained in a porous pot, which sits in a glass jar containing an electrolyte (conducting medium) of ammonium chloride solution, into which a zinc cathode is inserted. The cell produces a continuous current, the carbon mixture acting as a depolarizer; that is, it prevents hydrogen bubbles from forming on the anode and increasing resistance. In a dry battery, the electrolyte is made in the form of a paste with starch.

Leclanché was born in Paris. In 1867 he gave up his job as a railway engineer to devote all his time to the improvement of the cell's design. He was successful in having it adopted by the Belgian Telegraphic Service in 1868, and the Leclanché cell rapidly came into general use wherever an intermittent supply of electricity was needed.

L'Ecluse Charles de, or ***Clusius*** 1525–1609. French botanist who published one of the earliest known books on Spanish flora. He translated several botanical works from Latin into Spanish and French and wrote *Rariorum aliquot stirpium per Hispanias observatarum historia* 1576, a description of his observations of rare flora in Spain during an expedition 1564–65.

L'Ecluse was born in Le Arras, France to a wealthy family in which it was the tradition to study law. This he dutifully did and obtained his licence to practice law 1548 from the University of Louvain. However, from childhood onwards he had expressed a great interest in plant life and, in 1551, persuaded his father to let him go to Provence in S France to collect botanical specimens for a local dignitary, Guillaume Roudelet. This led to him translating several botanical works from Latin into Spanish and French and also completing his own original works.

He was appointed professor of botany at the University of Leiden, Holland 1593. He was delighted to hold this post because the University was rapidly becoming known as a centre for science and medicine as well as theology and the classics. He continued to work in Leiden until his death at the age of 84.

Lederberg Joshua 1925– . US geneticist who showed that bacteria can reproduce sexually, combining genetic material so that offspring possess characteristics of both parent organisms. In 1958 he shared the Nobel Prize for Physiology or Medicine with George Beadle and Edward Tatum.

Lederberg is a pioneer of genetic engineering, a science that relies on the possibility of artificially shuffling genes from cell to cell. He realized 1952 that bacteriophages, viruses which invade bacteria, can transfer genes from one bacterium to another, a discovery that led to the deliberate insertion by scientists of foreign genes into bacterial cells.

Lederberg was born in New Jersey and studied at Columbia and at Yale, where he worked with Tatum. He was at the University of Wisconsin 1947–59, rising to professor, and moved 1959 to Stanford University, California, becoming director of the Kennedy Laboratories of Molecular Medicine 1962.

Lederman Leon 1921– . German-born US physicist who, with Melvin Schwartz and Jack Steinberger studied elementary particles using neutrinos, and discovered the muon neutrino. Nobel Prize for Physics 1988.

In 1961 using a huge detector weighing 10 tonnes Lederman, Steinberger, and Schwartz caught a small number of neutrinos, the world's first neutrino beam. Using this beam they found the muon neutrino, a new type of neutrino, and investigated the weak nuclear force and the quark structure of matter.

Lederman was born in New York City and educated at Columbia University, New York, receiving his doctorate in 1951, and staying on at the university for 28 years. He was appointed professor 1958. He served as director of the Fermi National Accelerator Laboratory, near Chicago 1979–89.

Lee Yuan Tseh 1936– . Taiwanese chemist who contributed much to the field of chemical reaction dynamics. He put in much of the groundwork for Dudley Herschbach's development of the molecular beam technique and shared the 1986 Nobel Prize for Chemistry with Herschbach and John Polanyi.

Born in Hsinchu, Taiwan, Lee was educated at Taiwan University and at the University of California, Berkeley, where he gained his PhD working under Herschbach. He became a professor at the University of Chicago in 1974, then accepted the post of professor of chemistry at California University.

Lee Tsung-Dao 1926– . Chinese physicist whose research centred on the physics of weak nuclear forces. In 1956 Lee proposed that weak nuclear forces between elementary particles might disobey certain key assumptions; for instance, the conservation of parity. He shared the 1957 Nobel Prize for Physics with his colleague Yang Chen Ning (1922–).

Lee trained in China; a scholarship sent him to the USA in 1946, working mostly on particle physics at the Princeton Institute of Advanced Study and at the University of California.

Leeuwenhoek Anton van 1632–1723. Dutch pioneer of microscopic research. He ground his own lenses, some of which magnified up to 300 times. With these he was able to see individual red blood cells, sperm, and bacteria, achievements not repeated for more than a century.

Leeuwenhoek was born in Delft and apprenticed to a cloth merchant. From 1660, having obtained the sinecure of chamberlain to the sheriffs of Delft, he devoted much of his time to lens grinding and microscopy. From 1672 to 1723 he described and illustrated his observations in more than 350

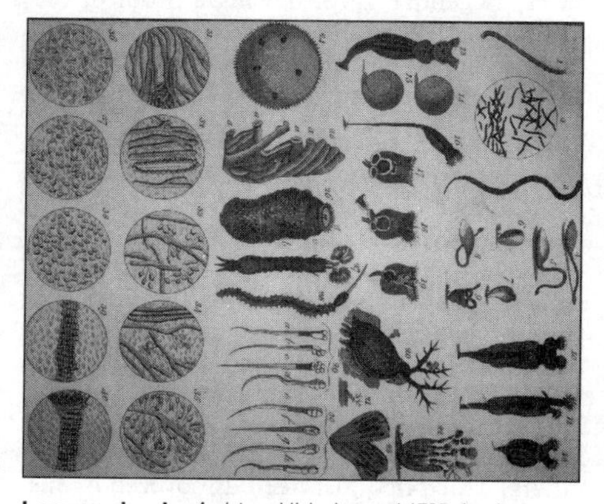

Leeuwenhoek *A plate published around 1795 showing microscopic animals, spermatozoa and plants. The diagrams were originally prepared by Anton van Leeuwenhoek, using the simple microscope that he designed and made.*

letters to the Royal Society of London. His fame was such that he was visited by several reigning monarchs, including Frederick I of Prussia and Tsar Peter the Great.

Leeuwenhoek ground more than 400 lenses, which he mounted in various ways in single-lens microscopes. Most of them were very small (some were about the size of a pinhead). In 1674 he discovered protozoa, which he called 'animalicules', and calculated their sizes. He also studied the structure of the lens in the eye, muscle striations, insects' mouthparts, the fine structure of plants, and discovered parthenogenesis in aphids.

Legendre Adrien-Marie 1752–1833. French mathematician who was particularly interested in number theory, celestial mechanics, and elliptic functions.

Legendre was born in Paris and studied there at the Collège Mazarin. During the French Revolution, he became head of the government department established to standardize French weights and measures 1794, as well as professor at the Institut de Marat. From 1813 he was chief of the Bureau de Longitudes.

In 1783–84 he introduced to celestial mechanics what are now known as Legendre polynomials. These are solutions to a second-order differential equation.

In number theory his most significant result was the law of reciprocity of quadratic residues (established more firmly by German mathematician Karl Gauss 1801) and the law of the distribution of prime numbers 1798.

In his school textbook *Eléments de géometrie* 1794, Legendre gave the single proof of the irrationality of π and the first proof of the irrationality of π^2. He published a textbook and tables on elliptical functions 1825–26.

Lehn Jean-Marie 1939– . French chemist who demonstrated for the first time how metal ions could be made to exist in a non-planar structure, tightly bound into the cavity of a crown ether molecule, explaining a possible mechanism for the transfer of metal ions across biological membranes. He was rewarded for his efforts when he received the 1987 Nobel Prize for Chemistry with Charles Pedersen and Donald Cram.

Lehn worked on the structures of crown ethers, which had just been discovered by his contemporary Pedersen and were found to have strong affinities for metal ions. Lehn demonstrated that if two nitrogen atoms replaced the oxygen atoms in the 'crown', a three-dimensional structure could be made by connecting two crowns – which Lehn called a 'cryptand'.

Lehn was then able to produce cryptands which could bind to molecules such as neurotransmitters. His research expanded the branch of organic chemistry known as host-guest chemistry and introduced the concept of supramolecules.

Lehn was born in Rosheim, France, but went to Strasbourg University. He studied the terpenes, and then at Harvard, Lehn worked on vitamin B_{12} synthesis. He moved between the two universities, gained a professorship, and finally settled at the College de France, Paris, where he held a chair in chemistry from 1979.

Leibniz Gottfried Wilhelm 1646–1716. German mathematician, philosopher, and diplomat. Independently of, but concurrently with, English scientist Isaac Newton he developed the branch of mathematics known as calculus. In his metaphysical works, such as *The Monadology* 1714, he argued that everything consisted of innumerable units, ***monads***, the individual properties of which determined each thing's past, present, and future.

The imaginary number is a fine and wonderful recourse of the divine spirit, almost an amphibian between being and not being.

Gottfried Leibniz

Attributed remark

Monads, although independent of each other, interacted predictably; this meant that Christian faith and scientific reason need not be in conflict and that 'this is the best of all possible worlds'. Leibniz's optimism is satirized in French philosopher Voltaire's novel *Candide*.

Leibniz was born in Leipzig and studied there and at Jena. From 1866 he was in the service of the archbishop and elector of Mainz, his special task being to devise plans to preserve the peace of Europe, just then emerging from the Thirty Years' War. This took him to France for three years. On the death of the elector of Mainz in 1673, Leibniz went to London, and there became acquainted with the work of Newton. The last 15 years of his life would be marred by dispute with Newton over which of them first invented the calculus. From 1676, Leibniz was librarian to the duke of Brunswick in Hanover, and often charged with diplomatic missions. From 1712 to 1714 he was an imperial privy councillor in Vienna. In 1714 the duke of Brunswick acceded to the English throne as George I, but Leibniz was denied permission to accompany him to London.

Leibniz designed a calculating machine, completed about 1672, which was able to multiply, divide, and extract roots. He worked intermittently throughout his life at devising what he called a Universal Characteristic, a language that would be accessible to everyone.

Leishman William Boog 1865–1926. Scottish army physician who discovered the protozoan parasite that causes the group of diseases now known as leishmaniasis. Knighted 1909.

Leishmaniasis encompasses various potentially fatal diseases prevalent in NE Africa and S Asia. Kala-azar, characterized by an enlarged spleen and liver, fever, and anaemia, is an example. The genus of protozoa to which the causative microorganism belongs is called *Leishmania.*

Leishman was born and educated in Glasgow, and spent his entire career in the Royal Army Medical Corps. He was posted to India 1890–97. From 1903 he was professor at the Army Medical School. In 1914 he began advising the War Office on tropical diseases, and he became the first director of pathology at the War Office in 1919. He was director-general of the Army Medical Service from 1923.

Leishman discovered the protozoan parasite that causes kala-azar in 1900, using a technique now called Leishman's stain, to examine cells from the spleen of a soldier who had died of kala-azar. He published his findings in 1903 but in the same year Charles Donovan (1863–1951) of the Indian Medical Service independently made the same discovery, as a result of which the causative protozoan came to be called *Leishmania donovani.*

Leishman also assisted his colleague Almroth Wright (1861–1947) in developing an effective antityphoid inoculation, and helped to elucidate the life cycle of the spirochaete *Spirochaeta duttoni*, which causes African tick fever.

Lejeune Jérôme Jean Louis Marie 1926–1994. French medical geneticist. Although the congenital condition known now as Down's syndrome was named after English physician J L H Down who studied it, it was Lejeune who discovered that Down's syndrome is the result of having an additional chromosome.

He believed chromosome deviations would in the future be treated by gene therapy, and vigorously opposed prenatal screening with a view to terminating Down's syndrome fetuses.

By the age of 26, Lejeune had completed his medical studies and was working with Raymond Turpin at the National Centre for Scientific Research in Paris. It was with Turpin and another colleague, Marthe Gautier, that in 1959 he published his discovery that Down's syndrome sufferers have one additional chromosome to the 46 previously determined to be the norm in humans. This was the first time that anyone had described a condition resulting from having the incorrect number of chromosomes. He went on to find that the *cri du chat* ('cat cry') syndrome is also caused by a numerical abnormality in human chromosomes where a small section of one of the longest chromosomes is missing.

In 1964 Lejeune was appointed professor of fundamental genetics in the Faculty of Medicine in Paris. He firmly believed in the 'brotherhood' of humanity; that the chromosome pattern of modern humans must have originated in a very small group of people – even one pair.

Leloir Luis Frederico 1906–1987. Argentinian chemist who studied glucose (a sugar found in the blood) metabolism and discovered an alternative mechanism for glycogen (a polysaccharide made from glucose) synthesis, involving a new enzyme. Later, he connected these reactions to glycogen synthesis in the liver and muscles. He showed that a glucose molecule is added by a process in which the reactive intermediate uridine diphosphateglucose (UDPG) transfers glucose to the growing glycogen chain. He found that galactose is broken down to yield glucose in a similar pathway. He received the Nobel Prize for Chemistry in 1970.

Gerty Cori had previously demonstrated a pathway in which glycogen is both synthesized and converted to other products. But Leloir identified the enzyme glucose1-phosphate kinase acting in a separate glycogen synthesis. The product of this reaction, glucose 1,6-bisphosphate, is a coenzyme of the glycolysis pathway enzyme, phosphoglucomutase. He went on to identify galactokinase and discovered that the product, galactose 1-phosphate, is converted into glucose 1-phosphate.

Leloir was born in Paris to Argentinian parents. He studied at Buenos Aires and spent a year at Cambridge, England under the guidance of Frederick Gowland Hopkins. He briefly returned to Argentina to work at the Institute of Physiology, however, his ideas clashed with those of President Juan Peron, and Leloir went into exile in the USA. But soon afterwards he returned to Argentinia and founded his own research institute in 1947.

Lemaître Georges Edouard 1894–1966. Belgian cosmologist who in 1933 proposed the Big Bang theory of the origin of the universe. US astronomer Edwin Hubble had shown that the universe was expanding, but it was Lemaître who suggested that the expansion had been started by an initial explosion, the Big Bang, a theory that is now generally accepted.

Lemaître was born in Charleroi, trained as a civil engineer and also ordained a priest. He studied astrophysics in the UK at Cambridge and in the USA at the Massachusetts Institute of Technology, where he became influenced by the theories of Hubble and Harvard astronomer Harlow Shapley concerning the likelihood of an expanding universe. Lemaître was made professor of astrophysics at the University of Louvain 1927.

Lemaître visualized a 'primal atom', an incredibly dense 'egg' containing all the material for the universe within a sphere about 30 times larger than the Sun. Somewhere between 20,000 and 60,000 million years ago, in his view, this atom exploded.

Lemaître's works include *Discussion on the Evolution of the Universe* 1933 and *Hypothesis of the Primal Atom* 1946.

Lenard Philipp Eduard Anton 1862–1947. Hungarian-born German physicist who investigated the photoelectric effect and cathode rays (the stream of electrons emitted from the cathode in a vacuum tube). Nobel prize 1905.

Lenard was born in Pozsony, Hungary (now Bratislava, Slovak Republic), and studied at Heidelberg and Berlin. In 1898 he became professor of experimental physics at Kiel, and held the same post at Heidelberg 1907–31. In 1924 Lenard became a Nazi. Obsessed with the idea of producing a purely 'Aryan' physics, he spent his later years reviling Albert Einstein and other Jewish physicists.

Lenard's work on cathode rays began 1892, and led him to the conclusion that an atom is mostly empty space. He also suggested that the part of the atom where the mass was concentrated consisted of neutral doublets or 'dynamids' of negative and positive electricity. This preceded by ten years the classic model of the atom proposed by Ernest Rutherford 1911.

Lenard devised the grid in the thermionic valve that controls electron flow. He showed that an electron must have a certain minimum energy before it can produce ionization in a gas. He also studied luminescent compounds and, from 1902 onwards, discovered several fundamental effects in photoelectricity.

No entry to Jews and Members of the German Physical Society

Philip Lenard

Notice on Lenard's office door

Lenoir (Jean Joseph) Etienne 1822–1900. Belgian-born French engineer and inventor who in the early 1860s produced the first practical internal-combustion engine and a car powered by it. He also developed a white enamel 1847, an electric brake 1853, and an automatic telegraph 1865.

Lenoir was born in Mussy-la-Ville.

Several people had claimed to have invented an internal-combustion engine, but not until Lenoir in 1859 did a practical model become a reality. His engine consisted of a single cylinder with a storage battery (accumulator) for the electric ignition system. Its two-stroke cycle was provided by slide valves, and it was fuelled by coal gas.

Lenoir built a small car around one of his prototypes in 1863, but it had an efficiency of less than 4% and although he claimed it was silent, this was only true when the vehicle was not under load.

The real value of his engine was for powering small items of machinery, and by 1865 more than 400 were in use, driving printing presses, lathes, and water pumps. Its use for vehicles was restricted by its size.

Lenz Heinrich Friedrich Emil 1804–1865. Russian physicist who in 1833 formulated

Lenz's law, a fundamental law of electromagnetism. He also found that the strength of a magnetic field is proportional to the strength of the magnetic induction.

Lenz's law states that the direction of an electromagnetically induced current (generated by moving a magnet near a wire or a wire in a magnetic field) will oppose the motion producing it.

Lenz was born and educated in Dorpat (now Tartu, Estonia). As geophysical scientist, he accompanied Otto von Kotzebue (1787–1846) on his third expedition around the world 1823–26. On his return, Lenz joined the St Petersburg Academy of Science, and from 1840 held posts at the University of St Petersburg.

On his voyage with Kotzebue, Lenz studied climatic conditions such as barometric pressure, and made extremely accurate measurements of the salinity, temperature, and specific gravity of sea water. On a later expedition he measured the level of the Caspian Sea.

Lenz's studies of electromagnetism date from 1831. Lenz's law is in fact a special case of the law of conservation of energy. If the induced current were to flow in the opposite direction, it would assist the motion of the magnet or wire and energy would increase without any work being done, which is impossible.

Lenz also studied the relationship between heat and current and discovered, independently of English physicist James Joule, the law now known as Joule's law.

Leonardo da Vinci 1452–1519. Italian artist, inventor, and scientist. He is regarded as one of the greatest figures of the Italian Renaissance for the universality of his genius.

Da Vinci was born on his father's estate, in Vinci, Tuscany. The illegitimate son of the Tuscan land-owner Ser Piero and a peasant girl, he was taken into his father's household in Florence, where he was the only child. He received an elementary education and in about 1467 was apprenticed to the artist Andrea del Verrocchio. There he trained in artistic as well as technical and mechanical subjects. He left the workshop in about 1477 and worked on his own until 1481. He went to Milan the following year and was employed by Ludovico Sforza, the Duke of Milan, as 'painter and engineer of the duke'. In this capacity he advised the duke on the architecture of proposed cathedrals in Milan and nearby Pavia, and was involved in hydraulic and mechanical engineering. After Milan fell to the French in 1499, he fled and began a long period of wandering.

In 1500 he visited Mantua and then Venice, where he was consulted on the reconstruction and fortification of the church San Francesco al Monte. Two years later he went into the service of Cesare Borgia as a military architect involved in the designing and development of fortifications. In 1503 he returned to Florence to investigate, on Cesare Borgia's behalf, the possibility of re-routing the river Arno so that the besieged city of Pisa would lose its access to the sea. It was at about this time that he painted his internationally renowned portrait the *Mona Lisa.* He was invited that year by the Governor of France in Milan, Charles d'Amboise, to work for the French in Milan, and in 1506 he took up the offer. There he devised plans for a castle for the governor and for the Adda Canal to connect Milan to Lake Como.

In 1513 the French were defeated and forced to leave Milan; da Vinci left with them and went to Rome to look for work. He stayed with Cardinal Giulano de' Medici, the brother of Pope Leo X, but there was little for him to do (although both Michelangelo and Raphael were working there at that time) other than to advise on the proposed reclamation of the Pontine Marshes. Three years later he left Italy for France on the invitation of King François I, and he lived in the castle of Cloux, near the king's summer residence at Amboise. Da Vinci spent the rest of his life there, sorting and editing his notes. He died on 2 May.

Da Vinci's training in Verrocchio's workshop developed his practical perception, which served him well as a technical scientist, a creative engineer and as an artist. In the years of his first visit to Milan, principally between 1490 and 1495, he produced his well-known notebooks, in mirror-writing. The illustrated treatises deal with painting, architecture, anatomy and the elementary theory of mechanics. The last was produced in the late 1490s and is now in the Biblioteca

---

The function of muscle is to pull and not to push, except in the case of the genitals and the tongue.

Leonardo da Vinci

The Notebooks of Leonardo da Vinci
vol 1, ch 3

Nacional, Madrid. In it da Vinci proposes his theory of mechanics, illustrated with sketches of machines and tools such as gears, hydraulic jacks and screw-cutting machines, with explanations of their functions and mechanical principles and of the concepts of friction and resistance.

Da Vinci's interest in mechanics developed as he realized how the laws of mechanics, motion, and force operate everywhere in the natural world. He studied the flight of birds in connection with these laws and, as a result, designed the prototypes of a parachute and of a flying machine.

During this time, he also developed his ideas about the Renaissance Church Plan, which later were considered favourably by the architect Bramante in connection with the building of the new St Peter's in Rome.

Wisdom is the daughter of Experience, Truth is only the daughter of Time.

Leonardo da Vinci

Quoted in J Huxley,
Essays in Popular Science, 1926

In about 1503, when Cesare Borgia's plan for the diversion of the river Arno failed, da Vinci also devised a project to construct a canal, wide and deep enough to carry ships, which would bypass the narrow portion of the Arno so that Florence would be linked to the sea. His hydrological studies on the properties of water were carried out at this time.

The variety of da Vinci's inventions reflect his passionate absorption in biological and mechanical details and ranged from complex cranes to pulley systems, lathes, drilling machines, a paddle-wheel boat, an underwater breathing apparatus, and a clock which registered minutes as well as hours. As a military engineer he was responsible for the construction of assault machines, pontoons, a steam cannon and a tortoise-shaped tank. For a castle in Milan he create a forced-air central heating system and also a water-pumping mechanism. His notes and diagrams established him, beyond dispute, as the greatest descriptive engineer and scientist of his age. Despite these achievements, he remains most famous as an artist, unique in the history of the world's greatest painters.

Lesseps Ferdinand Marie, Vicomte de Lesseps 1805–1894. French engineer who designed and built the Suez Canal 1859–69. He began work on the Panama Canal in 1881, but withdrew after failing to construct it without locks.

Lesseps was born in Versailles and became a diplomat. From 1825 he held posts in various capitals, including Lisbon, Tunis, and Cairo. Interested in engineering and construction, he suggested in 1854 that a passage should be cut to link the Mediterranean with the Red Sea, and was put in charge of the work. The canal was successfully completed 1869, shortening the route between Britain and India by 9,700 km/6,000 mi.

The Panama Canal project, undertaken when he was in his 70s, met with failure and bankruptcy. Lesseps was sentenced to five years' imprisonment for breach of trust, but was too ill to leave his house.

Leuckart Karl Georg Friedrich Rudolf 1822–1898. German zoologist who identified the phylum Coelenterata (now divided into separate phyla, the Cnidaria – jellyfish, sea anemones, and corals – and the Ctenophora – the comb jellies), and was responsible for establishing the first meat inspection laws in the world.

His research led to the division of the Metazoa (multicellular animals) into Coelenterata, Echinodermata (sea urchins), Annelida (segmented worms), Arthropoda (jointed limbed animals including insects, spiders, crabs, and lobsters), Mollusca (molluscs), and Vertebrata (animals with backbones, including fish, birds, and mammals), including slugs, snails, octopuses, and shellfish). He also worked on parasitology, and his work on *Trichina spiralis*, the cause of trichinosis, led to the establishment of meat inspection.

Leuckart was born in Helmstedt, Germany and educated at Helmstedt Gymnasium. As a child he was encouraged by his uncle who was the professor of zoology at Freiburg im Breisgau to collect insects. In 1842 he went to Gottingen to study medicine, becoming an assistant in Rudolf Wagner's laboratory at the Institute 1845. He was promoted to lecturer in zoology 1845 and the following year made a study of marine invertebrates along the coast of the North Sea in which he identified the phylum, Coelenterata. In 1850 he was made professor of zoology at Giessen University and, in 1869, moved to the University of Leipzig, where a new Zoological Institute was built under his guidance 1880. Between 1877 and 1878 he was rector of the university.

Leverrier Urbain Jean Joseph 1811–1877. French astronomer who predicted the existence and position of the planet Neptune, discovered in 1846.

The possibility that another planet might exist beyond Uranus, influencing its orbit, had already been suggested. Leverrier calculated the orbit and apparent diameter of the hypothetical planet, and wrote to a number of observatories, asking them to test his prediction of its position. Johann Galle at the Berlin Observatory found it immediately, within 1° of Leverrier's coordinates.

Unbeknown to Leverrier, English astronomer John Couch Adams had carried out virtually identical calculations a year earlier, but had failed to persuade anyone to act on them.

Leverrier was born in St Lô, Normandy, and studied in Paris at the Ecole Polytechnique, joining the staff 1837. He became professor 1847, and in 1849 a chair of celestial mechanics was established for him at the Sorbonne. He was politically active in the revolutions of 1848, serving as a member of the legislative assembly in 1849 and as senator in 1852. In 1854 he became director of the Paris Observatory.

After his discovery of Neptune, Leverrier compiled a comprehensive analysis of the masses and orbits of the planets of the Solar System. This was published after his death, in the *Annals* of the Paris Observatory.

Leverrier was also instrumental in the establishment of a meteorological network across continental Europe.

Levi-Civita Tullio 1873–1941. Italian mathematician who developed, in collaboration with Gregorio Ricci-Curbastro, the absolute differential calculus, published 1900. Levi-Civita also introduced the concept of parallelism in curved space 1917.

Levi-Civita was born and educated in Padua, where he was taught by Ricci-Curbastro. He was professor at the Engineering School in Padua 1897–1918, when he became professor of higher analysis at Rome. In 1938, the anti-Semitic laws promulgated by the Fascist government forced him to leave the university; he was also expelled from all Italian scientific societies.

The absolute differential calculus was a completely new calculus, applicable to both Euclidean and non-Euclidean spaces. Most significantly, it could be applied to Riemannian curved spaces, and would be fundamental to Albert Einstein's development of the general theory of relativity. Levi-Civita's idea of parallel displacement later developed into tensor calculus.

Levi-Civita also published papers on celestial mechanics and hydrodynamics. His achievements in both pure and applied mathematics established him as one of the foremost mathematicians of his age.

Levi-Montalcini Rita 1909– . Italian neurologist who discovered nerve-growth factor, a substance that controls how many cells make up the adult nervous system. She shared the 1986 Nobel Prize for Physiology or Medicine with her co-worker, US biochemist Stanley Cohen (1922–).

Levi-Montalcini was born and educated in Turin and began her research there. When the Fascist anti-Semitic laws forced her to leave the university 1939, she constructed a home laboratory. After World War II she moved to the USA and was at the Washington University in St Louis 1947–81, becoming professor 1958. In 1981 she went to Rome.

... the revelations of that day stayed permanently inscribed in my memory as marking not only the end of the long period of doubt and lack of faith in my research, but also the sealing of a lifelong alliance between me and the nervous system.

Rita Levi-Montalcini

Levi-Montalcini first discovered nerve-growth factor in the salivary glands of developing mouse embryos, and later in many tissues. She established that it was chemically a protein, and analysed the mechanism of its action. Her work has contributed to the understanding of some neurological diseases, tissue regeneration, and cancer mechanisms.

Lévi-Strauss Claude 1908– . French anthropologist who helped to formulate the principles of structuralism by stressing the interdependence of cultural systems and the way they relate to each other.

In his analyses of kinship, myth, and symbolism, Lévi-Strauss argued that, though the superficial appearance of these factors might vary between societies, their underlying structures were universal and could best be understood in terms of binary oppositions: left and right, male and female, nature and culture, the raw and the cooked, and so on.

If there is to be hope of saving mankind, mankind must first be convinced of this.

Claude Lévi-Strauss

His works include *Tristes Tropiques* 1955 – an intellectual autobiography – and *Mythologiques/Mythologies* 1964–71.

Lévi-Strauss was born in Belgium and studied philosophy at Paris. He taught in Brazil at the University of Saõ Paolo 1935–39 and later at academic institutions in New York, where he was a cultural attaché at the French embassy 1946–47. He taught at the Sorbonne 1948–58 and became professor of social anthropology at the Collège de France 1958.

His thinking was influenced by the linguistics of Ferdinand Saussure and Roman Jakobson. He also claimed to be influenced by geology, psychoanalysis, and Marxism.

Lewis Edward B 1918– . US geneticist who shared the 1995 Nobel Prize for Physiology or Medicine with Eric F Wieschaus and Christiane Nüsslein-Volhard for their discoveries concerning the genes that control early embryonic development.

Lewis at the California Institute of Technology, Pasedena, USA, began research on the development of the fruit fly in 1946. He bred generation after generation of flies, looking for rare mutations called homeotic transformations which caused one body part to substitute for another in the adult fly. For instance, one such mutation causes the body segment which usually bears tiny balancing organs called halteres to be replaced by a second pair of wings. Lewis found that many of the genes causing homeotic transformations were grouped into a single cluster on one chromosome. Within this cluster, the genes were neatly ordered: genes affecting the abdomen were at one end, and those affecting the head were at the other end. Similar principles are now known to control the development of other species, including humans.

Lewis Gilbert Newton 1875–1946. US theoretical chemist who defined a base as a substance that supplies a pair of electrons for a chemical bond, and an acid as a substance that accepts such a pair. He also set out the electronic theory of valency and in thermodynamics listed the free energies of 143 substances.

Lewis was born in Weymouth, Massachusetts, and studied at Nebraska and Harvard. He worked at the Massachusetts Institute of Technology (MIT) 1905–1912, and spent the rest of his career at the University of California, Berkeley.

In 1916, Lewis began his pioneering work on valency. He postulated that the atoms of elements whose atomic mass is higher than helium's have inner shells of electrons with the structure of the preceding rare gas. The valency electrons lie outside these shells and easily form ionic bonds, preferably covalent bonds. He also drew attention to the unusual properties of molecules that have an odd number of electrons, such as nitric oxide (nitrogen monoxide, NO).

In his later years he carried out studies on the excited electron states of organic molecules, contributing to the understanding of the colour of organic substances and the complex phenomena of phosphorescence and fluorescence.

Lewis's main works were both published 1923: *Valence and the Structure of Atoms and Molecules* and (with Merle Randall) *Thermodynamics and the Free Energy of Chemical Substances*.

Cl : Cl   { N = O   N = O }

Lewis, Gilbert *A molecule of chlorine with a covalent bond involving the sharing of a pair of electrons by the two atoms, and two hybrid forms of a molecule of nitric oxide (nitrogen monoxide), each with an odd (unpaired) electron.*

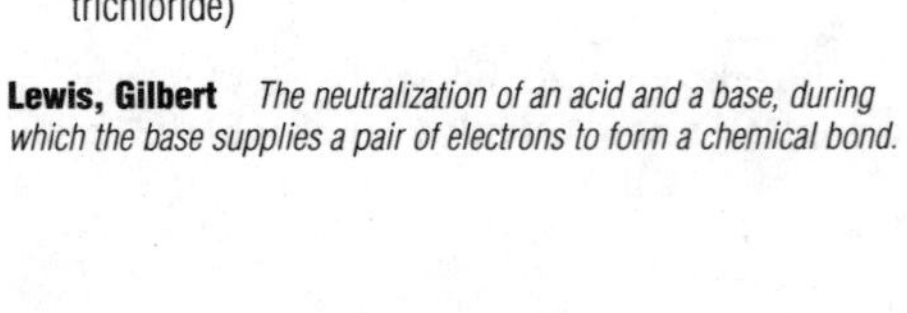

Cl–B(Cl)(Cl) + :N(H)(H)H ⟶ Cl–B(Cl)(Cl) ← N(H)(H)H

acid (boron trichloride)   base (ammonia)

Lewis, Gilbert *The neutralization of an acid and a base, during which the base supplies a pair of electrons to form a chemical bond.*

Lewis Thomas 1881–1945. Welsh cardiologist and clinical scientist who discovered that histamine, an amine compound, is released as an initial event in the inflammatory response.

His research dealt with the response of the blood vessels in the human skin to chemical, thermal, electrical and mechanical injury, around 1919. He found that the capillaries in the skin around the site of injury dilate and their endothelial linings (the smooth single layer of cells that comprises the capillary wall) acquire increased permeability to enhance the migration of white blood cells (phagocytes) to the site of injury. This inflammatory response produces the redness, heat, and swelling associated with the wound. Here the white blood cells can begin engulfing debris and any infecting microorganisms.

At the suggestion of Henry Dale, Lewis pricked histamine into the skin and found that he had produced the same immune response with histamine as an inflicted injury. He concluded that a chemical similar to histamine may have been involved in producing the inflammatory response. It is now known that histamine is released as an 'alarm' chemical signal in response to injury.

Lewis was born in Cardiff and trained as a doctor at University College, Cardiff and University College Hospital, London. He remained at the latter as a medical tutor and practitioner and was knighted 1921.

He was also involved in the determination that a small area of tissue in the right auricle of the heart functioned as the 'pace-maker', the site at which the beat of the heart is initiated.

Libby Willard Frank 1908–1980. US chemist whose development in 1947 of radiocarbon dating as a means of determining the age of organic or fossilized material won him a Nobel prize in 1960.

Libby was born in Grand Valley, Colorado, and studied at the University of California, Berkeley. During World War II he worked on the development of the atomic bomb (the Manhattan Project). In 1945 he became professor at the University of Chicago's Institute for Nuclear Studies. He was a member of the US Atomic Energy Commission 1954–59, and then became director of the Institute of Geophysics at the University of California.

Having worked on the separation of uranium isotopes for producing fissionable uranium-238 for the atomic bomb, he turned his attention to carbon-14, a radioactive isotope that occurs in the tissues of all plants and animals, decaying at a steady rate after their death. He and his co-workers accurately dated ancient Egyptian relics by measuring the amount of radiocarbon they contained, using a sensitive Geiger counter. By 1947 they had developed the technique so that it could date objects up to 50,000 years old.

Li Cho Hao 1913– . Chinese-born US biochemist who discovered in 1966 that human pituitary growth hormone (somatotropin) consists of a chain of 256 amino acids. In 1970 he succeeded in synthesizing this hormone, the largest protein molecule synthesized up to that time.

Li was born in Guangzhu and educated at the University of Nanjing. In 1935 he emigrated to the USA, where he took up postgraduate studies at the University of California at Berkeley and later joined the staff. He became professor 1950.

Li spent his entire academic career studying the pituitary-gland hormones. In collaboration with various co-workers, he isolated several protein hormones, including adreno-corticotrophic hormone (ACTH), which stimulates the adrenal cortex to increase its secretion of corticoids. In 1956, Li and his group showed that ACTH consists of 39 amino acids arranged in a specific order, and that the whole chain of the natural hormone is not necessary for its action. He isolated another pituitary hormone called melanocyte-stimulating hormone (MSH) and found that not only does this hormone produce some effects similar to those produced by ACTH, but also that part of the amino acid chain of MSH is the same as that of ACTH.

Lie (Marius) Sophus 1842–1899. Norwegian mathematician who provided the foundations for the science of topology in transformation groups known as the Lie groups. He was one of the first mathematicians to emphasize the importance of the notion of groups in geometry.

Lie was born near Bergen and studied at Christiania (now Oslo) and abroad at Berlin and Paris. He was professor at Christiania 1873–86 and 1898–99 and at Leipzig, Germany, 1886–98.

Lie's first great discovery, made in 1870, was that of his contact transformation, which mapped straight lines with spheres and principal tangent curves into curvature lines. In his theory of tangential transformations occurs the particular transformation that makes a sphere correspond to a straight line. By 1873 Lie had begun to investigate transformation groups. In this work on group theory he chose a new space element, the contact element, which is an incidence pair of point and line or of point and hyperplane. This led him to the discovery of Lie groups, one of the basic notions of which is that of infinitesimal transformation.

The Lie integration theorem, which he developed, made it possible to classify partial differential equations in such a way as to make most of the classical methods of solving such equations reducible to a single principle.

Liebig Justus, Baron von 1803–1873. German chemist, a major contributor to agricultural chemistry.

He introduced the theory of compound radicals and discovered chloroform and chloral.

Many new methods of organic analysis were introduced by Liebig, and he devised ways of determining hydrogen, carbon, and halogens in organic compounds. He demonstrated that plants absorb minerals (and water) from the soil and postulated that the carbon used by plants comes from carbon dioxide in the air rather than from the soil.

Liebig was born in Darmstadt, Hesse, and studied at Bonn (where he was arrested for his liberalist political activity), Erlangen, and Paris. At the age of 21 he became professor at Giessen, moving to Munich 1852.

In the 1820s Liebig began a long collaboration with Friedrich Wöhler. In 1832, from a study of oil of bitter almonds (benzaldehyde; phenylmethanal), they discovered the benzoyl radical (C_6H_5CO–). Based on study of related compounds, they introduced the idea of compound radicals in organic chemistry.

God has ordered all his creation by Weight and Measure.

Justus von Liebig

Notice above entrance to Liebig's laboratory

Liebig studied fermentation (but would not acknowledge that yeast is a living substance), and analysed various body fluids and urine. He calculated the calorific values of foods, emphasizing the role of fats as a source of dietary energy, and even developed a beef extract – long marketed as Liebig extract.

Lighthill (Michael) James 1924– . British mathematician who specialized in the application of mathematics to high-speed aerodynamics and jet propulsion.

Lilienthal Otto 1848–1896. German aviation pioneer who inspired US aviators Orville and Wilbur Wright. From 1891 he made and successfully flew many gliders, including two biplanes, before he was killed in a glider crash.

Lilienthal demonstrated the superiority of cambered wings over flat wings – the principle of the aerofoil. In his planes the pilot was suspended by the arms, as in a modern hang-glider. He achieved glides of more than 300 m/1,000 ft, and gliding began to catch on as a sport.

Lilienthal was born in Pomerania (then part of Prussia), and trained as an engineer. From studies of the flight of birds, he learned that curved wings allow horizontal flight without an angle of incidence to the wind, and that soaring is related to air thermals. Having shown conclusively that birds produce thrust by the action of their outer primary feathers, he built a powered machine with moving wing tips, using a carbon dioxide motor. The machine was not tested.

Lilienthal's book *Der Vogelflug als Grundlage der Fliegekunst/Bird Flight as a Basis for Aviation* 1889 greatly influenced other aviation pioneers.

Linacre Thomas *c.* 1460–1524. English humanist, physician to Henry VIII, from whom he obtained a charter in 1518 to found the Royal College of Physicians, of which he was first president.

Lindbergh Charles A(ugustus) 1902–1974. US aviator who made the first solo nonstop flight in 33.5 hours across the Atlantic (Roosevelt Field, Long Island, New York, to Le Bourget airport, Paris) 1927 in the *Spirit of St Louis*, a Ryan monoplane designed by him.

Lindbergh was born in Detroit, Michigan. He was a barnstorming pilot before attending the US Army School in Texas 1924 and becoming an officer in the Army Air Service Reserve 1925. His son Charles Jr (1930–1932) was kidnapped and killed; ensuing legislation against kidnapping was called the Lindbergh Act. Although he admired the Nazi air force and championed US neutrality in the late 1930s, he flew 50 combat missions in the Pacific theatre in

World War II. He wrote *The Spirit of St Louis* 1953 (Pulitzer Prize).

I have seen the science I worshipped, and the aircraft I loved, destroying the civilization I expected them to serve.

Charles A Lindbergh

Time 26 May 1967

Lindblad Bertil 1895–1965. Swedish astronomer who demonstrated the rotation of our Galaxy. He went on to stipulate that the speed of rotation of the stars in the Galaxy was a function of their distance from the centre (the 'differential rotation theory').

Lindblad was born in Örebro and studied at Uppsala. He was appointed director of the new Stockholm Observatory in 1927, and made professor of astronomy at the Royal Swedish Academy of Sciences.

Dutch astronomer Jacobus Kapteyn had proposed that the Solar System lay near the centre of the Galaxy, whereas US astronomer Harlow Shapley believed that the centre of the Galaxy was some 50,000 light years away in the direction of the constellation Sagittarius. Lindblad suggested that the two streams of stars Kapteyn had observed could in fact represent the rotation of all the stars in our Galaxy in the same direction, around a distant centre; he thus confirmed Shapley's hypothesis.

Lindemann (Carl Louis) Ferdinand 1852–1939. German mathematician whose discussion of the nature of π in 1882 laid to rest the old question of 'squaring the circle'.

Lindemann was born in Hanover and studied at Göttingen, Munich, and Erlangen. He was professor at Würzburg 1879–83, and at Königsberg 1883–93; from 1893 until his death, he taught at Munich.

The question whether π was a transcendental (nonalgebraic) number had never received a satisfactory answer until Lindemann proved it in his 1882 paper. He demonstrated that, except in trivial cases, every expression of the form: where A and a are algebraic numbers, must be non-zero. Therefore, since i is a root of $x^2 + 1 = 0$, and since it was known that: $1^{i\pi} + 1^0 = 0$ (that is $1^{i\pi} = -1$) then ip and therefore p (since i is algebraic) must be transcendental.

If p cannot be the root of an equation, it cannot be constructed. Therefore the 'squaring of a circle' is impossible.

Lindemann Frederick Alexander, original name of Viscount ◊Cherwell, British physicist.

Lindley John 1799–1865. English botanist and horticulturist who specialized in the study and classification of orchids.

Lindley was born in Catton, near Norwich, the son of a nurseryman. Despite not having a university education, his enthusiasm for horticulture led him to describe hundreds of species of orchid in *Genera and Species of Orchidaceous Plants* 1840. When he was nineteen he worked as Joseph Bank's assistant, helping him in the library and herbarium and went on to publish several works with the acclaimed botanist Robert Brown. In 1820, he was commissioned by the Horticultural Society to illustrate some roses, which led to his appointment as the assistant secretary of the Horticultural Society's Chiswick garden 1822. In 1828, he was elected a Fellow of the Royal Society and made professor of botany at the University of London. He continued to work for the Horticultural Society, becoming its secretary in 1858. Lindley's own herbariums were donated to Kew Gardens and Cambridge University on his death 1865, and they are still maintained in much their original state.

Linnaeus Carolus 1707–1778. Latinized form of Carl von Linné. Swedish naturalist and physician. His botanical work *Systema naturae* 1735 contained his system for classifying plants into groups depending on shared characteristics (such as the number of stamens in flowers), providing a much-needed framework for identification. He also devised the concise and precise system for naming plants and animals, using one Latin (or Latinized) word to represent the genus and a second to distinguish the species.

For example, in the Latin name of the daisy, *Bellis perennis, Bellis* is the name of the genus to which the plant belongs, and *perennis* distinguishes the species from others of the same genus. By tradition the generic name always begins with a capital letter. The author who first described a particular species is often indicated after the name, for example, *Bellis perennis* Linnaeus, showing that the author was Linnaeus.

Nature does not make jumps.

Carolus Linnaeus

Philosophia Botanica

Linnaeus *Naturalist and physician Carolus Linnaeus. During the eighteenth century Linnaeus described the classification of plants into groups with shared characteristics. This system has given a unique and concise name for every organism identified that can be recognized throughout the world.*

Linnaeus was born in Småland and studied medicine at Lund and Uppsala and in the Netherlands at Harderwijk. As a lecturer in botany, he explored Lapland 1732 for the Uppsala Academy of Sciences. He practised as a physician and was appointed professor of medicine at Uppsala 1741, but changed this position in 1742 for the chair of botany.

Linnaeus's system of nomenclature was introduced in *Species plantarum* 1753 and the fifth edition of *Genera plantarum* 1754 (first edition 1737). In 1758 he applied his binomial system to animal classification.

Linnett John Wilfrid 1913–1975. English chemist who studied molecular force fields, spectroscopy, the measurement of burning velocities in gases, the recombination of atoms at surfaces, and theories of chemical bonding.

Linnett was born in Coventry and studied at Oxford. His career was spent at Oxford and then at Cambridge, where he became professor 1965 and was vice chancellor 1973–75.

His work on explosion limits concentrated on the reaction between carbon monoxide, hydrogen, and oxygen, which led to the study of atomic reactions on surfaces of metal alloys.

In 1960, Linnett originated a modification to the octet rule concerning valency electrons. He proposed that the octet should be considered as a double quartet of electrons rather than as four pairs, and in this way he was able to explain the stability of 'odd electron' molecules such as nitric oxide.

Linnett published more than 250 scientific papers and two textbooks, in one of which (*Wave Mechanics and Valency*) he explains the processes and techniques involved in the application of wave mechanics to the electronic structures of atoms and molecules.

Liouville Joseph 1809–1882. French mathematician whose main influence was as the founder and first editor 1836–74 of the *Journal de Mathématiques Pures et Appliqués*, which became known as the *Journal de Liouville*.

Liouville was born in St Omer, Pas-de-Calais, and studied in Paris at the Ecole Polytechnique and the Ecole des Ponts et Chaussées. From 1831 for 50 years, he taught mathematics at all the leading institutions of higher learning in Paris, becoming professor 1838 at the Ecole Polytechnique. He was also for a time the director of the Bureau de Longitudes. During the revolutions of 1848, he was elected as a moderate republican to the constituent assembly, but lost his seat the following year.

The chief mathematical interest of his career was in analysis.

In that field he published more than 100 papers between 1832 and 1857. In collaboration with Charles-François Sturm (1803–1855), Liouville published papers in 1836 on vibration, which laid the foundations of the theory of linear differential equations. He also provided the first proof of the existence of transcendental functions.

Lipmann Fritz Albert 1899–1986. German-born US biochemist. He investigated the means by which the cell acquires energy and highlighted the crucial role played by the energy-rich phosphate molecule adenosine triphosphate (ATP). For this and further work on metabolism, Lipmann shared the 1953 Nobel Prize for Physiology or Medicine with Hans Krebs.

Lippershey Hans *c.* 1570–*c.* 1619. Dutch lensmaker, credited with inventing the telescope in 1608.

Lippmann Gabriel 1845–1921. French doctor. He invented the direct colour process in photography. He was awarded the Nobel Prize for Physics in 1908.

Lipschitz Rudolf Otto Sigismund 1832–1903. German mathematician who developed a hypercomplex system of number theory, which became known as Lipschitz algebra. His work in basic analysis provided a condition for the continuity of a function, now known as the Lipschitz condition, subsequently of great importance in proofs of existence and uniqueness, as well as in approximation theory and constructive function theory.

Lipschitz was born in Königsberg (now Kaliningrad) and studied there and at Berlin. From 1864 he was professor at the University of Bonn.

Lipschitz did extensive work in number theory, Fourier series, the theory of Bessel functions, differential equations, the calculus of variations, co-gradient differentiation, geometry, and mechanics. In investigating the sums of arbitrarily many squares, he derived computational rules for certain symbolic expressions from real transformations.

The investigations he began in 1869 into forms of n differentials led to his most valuable contribution to mathematics: the Cauchy–Lipschitz method of approximation of differentials.

Lipschitz's book *Grundlagen der Analysis* 1877–80 was a synthetic presentation of the foundations of mathematics and their applications. The work provided a comprehensive survey of what was then known of the theory of rational integers, differential equations, and function theory.

Lipscomb William Nunn 1919– . US chemist who studied the relationships between the geometric and electronic structures of molecules and their chemical and physical behaviour.

Lipscomb was born in Cleveland, Ohio, and studied at the University of Kentucky and the California Institute of Technology. He became professor at Harvard 1959.

Lipscomb studied the boron hydrides and their derivatives and put forward bonding theories to explain the structures of electron-deficient compounds in general. He developed low-temperature X-ray diffraction methods to study simple crystals of nitrogen, oxygen, fluorine, and other substances that are solid only below liquid nitrogen temperatures.

Lipscomb went on to investigate the carboranes and the sites of electrophilic attack on these compounds, using nuclear magnetic resonance spectroscopy (NMR). This work led to the theory of chemical shifts. The calculations provided the first accurate values for the constants that describe the behaviour of several types of molecules in magnetic or electric fields. They also gave a theoretical basis for applying quantum mechanics to complex molecules.

For me, the creative process, first of all, requires a good nine hours of sleep a night. Second, it must not be pushed by the need to produce practical applications.

William Lipscomb

New York Times 7 Dec 1977

Lissajous Jules Antoine 1822–1880. French physicist who from 1855 developed ***Lissajous figures*** as a means of visually demonstrating the vibrations that produce sound waves.

Lissajous was born in Versailles and educated at the Ecole Normale Supérieure. He became rector of the Academy of Chambéry in 1874, and then took up the same position at Besançon in 1875.

Lissajous first reflected a light beam from a mirror attached to a vibrating object such as a tuning fork to another mirror that rotated. The light was then reflected onto a screen, where the spot traced out a curve whose shape depended on the amplitude and frequency of the vibration. He then refined this

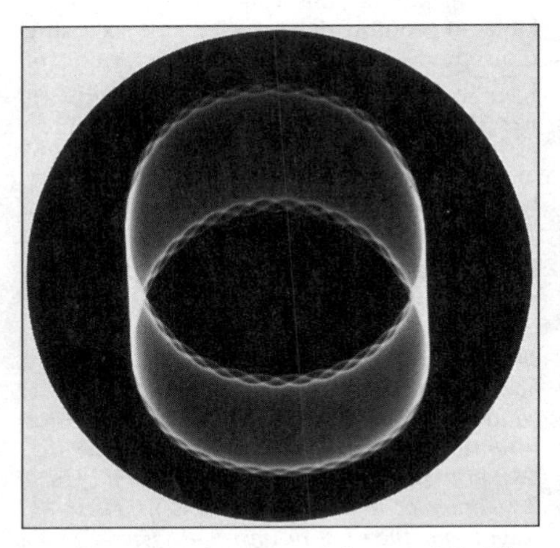

Lissajous *Lissajous figure of a radio frequency of a few MHz modulated at 400 Hz and compared with a 50 Hz reference. The radio wave and the 50 Hz modulation produce the cylinder. The 400 Hz modulation gives the frill at the top and bottom of the cylinder.*

method by using two mirrors mounted on vibrating tuning forks at right angles, and produced a wider variety of figures. By making one of the forks a standard, the acoustic characteristics of the other fork could be determined by the shape of the Lissajous figure produced.

Lissajous figures can now be demonstrated on the screen of an oscilloscope by applying alternating currents of different frequencies to the deflection plates. The curves produced depend on the ratio of the frequencies, enabling signals to be compared with each other.

Lister Joseph, 1st Baron Lister 1827–1912. English surgeon and founder of antiseptic surgery, influenced by Louis Pasteur's work on bacteria. He introduced dressings soaked in carbolic acid and strict rules of hygiene to combat wound sepsis in hospitals. Baronet 1883, Baron 1897.

The number of surgical operations greatly increased following the introduction of anaesthetics, but death rates were more than 40%. Under Lister's regime they fell dramatically.

Lister was born in Upton, Essex, and studied at University College, London. He was professor of surgery at Glasgow 1860–69, at Edinburgh 1869–77, and at King's College, London, 1877–92. In 1891 he became chair of the newly formed British Institute of Preventive Medicine (later the Lister Institute).

Sepsis was at this time thought to be a kind of combustion caused by exposing moist body tissues to oxygen. Learning of Pasteur's discovery of microorganisms, however, Lister began to use carbolic acid as a disinfectant. In 1867 he announced that his wards in the Glasgow Royal Infirmary had remained clear of sepsis for nine months. Later he adopted the method developed by Robert Koch in Germany of using steam to sterilize surgical instruments and dressings.

There are people who do not object to eating mutton chop – people who do not even object to shooting pheasant with the considerable chance that it may be only wounded and may have to die after lingering pain, unable to obtain its proper nutriment – and yet consider it something monstrous to introduce under the skin of a guinea pig a little inoculation of some microbe to ascertain its action. These seem to me the most inconsistent views.

Joseph Lister

British Medical Journal 1897

Little Clarence Cook 1888–1971. US pioneer of mouse genetics who developed the first inbred mouse strains. These mouse strains are still some of the most important in use today.

While at Harvard University, Little realized that there was a need to develop inbred mouse strains, in which the chromosome pairs in any mouse would be identical, to allow independent workers to perform experiments on the same genetic material. The original intent of the development of these strains was to demonstrate the genetic basis for various forms of cancer.

Little began a breeding programme to produce the first inbred mouse strain 1909, while he was still an undergraduate. He called this first strain DBA because it carries mutant alleles (alternative forms of the same gene at the same position on a chromosome) at three coat colour loci: dilute, brown and non-agouti. The genetic basis of breast and liver cancer was demonstrated shortly afterwards when the mouse strain C3H, which has very high susceptibility to these cancers, was compared to other inbred strains.

Little was born in Brookline, Massachusetts and studied at an agricultural college at Harvard University. His scientific career was interrupted by World War I, but he returned to his mouse breeding experiments 1918 at Cold Spring Harbor 1918–22. He was appointed president of two Universities (Maine and Michigan) before establishing the world-renowned Jackson Memorial Laboratory in Maine 1929, which he ran until 1956. Today, this laboratory stocks many of the inbred mouse strains which have been produced since Little's day and remains one of the best mouse research institutes in the world.

Llull Ramon *c.* 1232–1316. Catalan scholar and mystic. He began his career at the court of James I of Aragon (1212–1276) in Majorca. He produced treatises on theology, mysticism, and chivalry in Catalan, Latin, and Arabic. His *Ars magna* was a mechanical device, a kind of prototype computer, by which all problems could be solved by manipulating fundamental Aristotelian categories.

He also wrote the prose romance *Blanquerna* in his native Catalan, the first novel written in a Romance language.

In later life he became a Franciscan, and died a martyr at Bugia, Algeria.

Lobachevsky Nikolai Ivanovich 1792–1856. Russian mathematician who founded non-Euclidean geometry, concurrently with, but independently of, Karl Gauss in Germany and János Bolyai in Hungary. Lobachevsky published the first account of the subject in 1829, but his work went unrecognized until Georg Riemann's system was published.

In Euclid's system, two parallel lines will remain equidistant from each other, whereas in Lobachevskian geometry, the two lines will approach zero in one direction and infinity in the other. In Euclidean geometry the sum of the angles of a triangle is always equal to the sum of two right angles; in Lobachevskian geometry, the sum of the angles is always less than the sum of two right angles. In Lobachevskian space, also, two geometric figures cannot have the same shape but different sizes.

Lobachevsky was born at Nizhni-Novgorod and studied at the University of Kazan, Tatarstan. He taught there from 1814, becoming professor 1822 and rector of the university 1827–47. He also took on administrative work for the government.

Non-Euclidean geometry might find application in the intimate sphere of molecular attraction.

Nikolai Lobachevsky

Complete Geometrical Works 1883–1886

Lobachevsky developed non-Euclidean geometry between 1826 and 1856. He came to see that it was not contradictory to speak of a geometry in which all Euclid's postulates except the fifth held true. By including imaginary numbers, he made geometry more general, and Euclid's geometry took on the appearance of a special case of a wider system.

The clearest statement of Lobachevsky's geometry was made in the book *Geometrische Untersuchungen zur Theorie der Parallellinien*, published in Berlin 1840. His last work was *Pangéométrie* 1855.

Locke Joseph 1805–1860. English railway engineer, an associate of railway pioneers Isambard Kingdom Brunel and George and Robert Stephenson. He built many railway lines in the UK and France.

Lock was born near Sheffield, Yorkshire, and left school at 13. In 1823 he went to work for George Stephenson, and learned much about surveying, railway engineering, and construction. From 1847 he was Liberal member of Parliament for Honiton, Devon.

Locke's first task undertaken alone was the construction of a railway line from the Black Fell colliery to the river Tyne. He then began surveys for lines running between Leeds and Hull, Manchester and Bolton, and Canterbury and Whitstable. Locke built as straight as possible, used the terrain, and avoided the expense of tunnels whenever he could. Building the London to Southampton line, which opened 1840, he cut through the chalk Downs. He also built the Sheffield-to-Manchester route through the millstone grit of the Pennines. In 1841 he began work as chief engineer on the Paris-to-Rouen line, the first of several contracts in France.

Lockyer (Joseph) Norman 1836–1920. English scientist who studied the spectra of solar prominences and sunspots. Through his pioneering work in spectroscopy, he discovered the existence of helium. KCB 1897.

Lockyer was born in Rugby, the Midlands, and began as an amateur astronomer. In 1869 he founded the scientific journal *Nature*, which he was to edit for 50 years. He was director of the Solar Physics Observatory in South Kensington, London, 1890–1911.

In 1869 Lockyer attached a spectroscope to a 15-cm/6-in telescope and used it to observe solar prominences at times other than during a total solar eclipse. Although Lockyer had been the first to think of it, the same idea had occurred to French astronomer Pierre Janssen, then working in India, and they simultaneously notified the French Academy of Sciences of the same result. Later they worked together, Janssen providing the observations of the Sun's spectrum that led to the discovery of helium.

Lockyer also developed the theory that Stonehenge is oriented towards the direction in which the Sun rises at the time of the summer solstice. From the gradual change in position of the solstitial sunrise, he calculated that the monument must date from 1840 BC, plus or minus 200 years – later confirmed by radiocarbon dating.

Lodge Oliver Joseph 1851–1940. British physicist. He developed a system of wireless communication in 1894, and his work was instrumental in the development of radio receivers. He also proved that the ether does not exist, a discovery fundamental to the theory of relativity. After his son was killed in 1915, Lodge became interested in psychic research. Knighted 1902.

Lodge was born in Staffordshire and studied at London University. He became professor of physics at the University of Liverpool on its founding 1881; in 1900, he moved to the University of Birmingham to become its first principal.

Lodge invented a coherer, a device consisting of a container packed with metal granules whose electrical resistance varies with the passage of electromagnetic radiation. Designed to detect electromagnetic waves, this was developed into a detector of radio waves in the early investigations of radio communication, with which Lodge was closely involved.

The Michelson–Morley experiments of 1881 and 1887 had failed to detect the ether that was postulated as a medium for the propagation of light waves. This result could be explained by the ether moving with the Earth, but Lodge disproved this unlikely cause 1893 in a clever experiment in which light rays were passed between two rotating discs.

Loewi Otto 1873–1961. German physiologist whose work on the nervous system established that a chemical substance is responsible for the stimulation of one nerve cell (neuron) by another.

The substance was shown by the physiologist Henry Dale to be acetylcholine, now known to be one of the most vital neurotransmitters. For this work Loewi and Dale were jointly awarded the 1936 Nobel Prize for Physiology or Medicine.

Long Crawford William 1815–1878. US surgeon who used ether as an anaesthetic in surgical operations 1842 and is hailed as the discoverer of anaesthesia.

Long, a surgeon practising in Jefferson, Georgia, used ether as a general anaesthetic on 30 March 1842. During the next four years Long again used ether successfully a further four or five times. Since he did not publish his finding until 1849 however, his claim as the first physician to use anaesthesia has been contested. A chemistry student, William Clarke, claimed to have administered ether to a young lady before she had a tooth painlessly extracted by a dentist January 1842.

Longuet-Higgins Hugh Christopher 1923– . English theoretical chemist whose main contributions have involved the application of precise mathematical analyses, particularly statistical mechanics, to chemical problems.

Longuet-Higgins was born in Kent and studied at Oxford. He was professor of theoretical physics at King's College, London, 1952–54, and professor of theoretical chemistry at Cambridge 1954–67. He then went to Edinburgh University to study artificial intelligence and information-processing systems, which he thought had a closer bearing on true biology than purely physiochemical studies. In 1974 he moved to Sussex University, where he expanded this field into studies of the mechanisms of language and the perception of music.

Longuet-Higgins successfully predicted the structures of boron hydrides and the then unknown beryllium hydride, and the existence of the ion $(B_{12}H_{12})^{2-}$.

In 1947 Longuet-Higgins developed the orbital theory of conjugated organic molecules, deriving theoretically results that had been known experimentally for decades. He showed how the properties of conjugated systems can be derived by combining into molecular orbital theory a study of nonbonding orbitals.

He formulated a theory to describe the thermodynamic properties of mixtures, which he later extended to polymer solutions. He also investigated the optical properties of helical molecules and worked on electronic spectra.

From 1954, he used mathematical techniques to make theoretical chemical predictions. He predicted, for example, that cyclobutadiene (which had defeated all attempts to prepare it) should exist as a ligand attached to an atom of a transition metal; such a compound was successfully prepared three years later.

In a larger piece of work, he applied group theory to define the elements of symmetry of nonrigid molecules, such as hydrazine (N_2H_4), and thus was able to classify the individual quantum levels of the molecule.

Lonsdale Kathleen, (born Yardley) 1903–1971. Irish X-ray crystallographer who was among the first to determine the structures of organic molecules. She derived the structure factor formulas for all space groups. DBE 1956.

Yardley was born in Droichead Nua, County Kildare, but moved with her family to England in 1908. After graduating from Bedford College for Women in London, she joined the research team of W H Bragg at University College, London, and later at the Royal Institution. Between 1927 and 1931 she worked at Leeds and then returned to the Royal Institution. As a pacifist, she was imprisoned for a month during World War II. In 1946 she became professor of chemistry and head of the Department of Crystallography at University College, London.

At Leeds she used a grant from the Royal Society to buy an ionization spectrometer and electroscope and solved the structure of crystals of hexamethylbenzene.

Lonsdale was interested in X-ray work at various temperatures and thermal motion in crystals. She also used divergent beam X-ray photography to investigate the textures of crystals. She studied solid-state reactions, the pharmacological properties and crystal structures of methonium compounds, and the composition of bladder and kidney stones.

Lorentz Hendrik Antoon 1853–1928. Dutch physicist, winner (with his pupil Pieter Zeeman) of a Nobel prize in 1902 for his work on the Zeeman effect, in which a magnetic field splits spectral lines.

Lorentz spent most of his career trying to develop and improve Scottish scientist James Clerk Maxwell's electromagnetic theory. He also attempted to account for the anomalies of the Michelson–Morley experiment by proposing (independently of Irish physicist George Fitzgerald) that moving bodies contracted in their direction of motion. He took the matter further with his method of transforming space and time coordinates, later known as ***Lorentz transformations***, which prepared the way for Albert Einstein's theory of relativity.

Lorentz was born in Arnhem and studied at Leiden, where he became professor of theoretical physics at the age of 24. In 1912 he became director of the Teyler Institute in Haarlem.

Lorenz Konrad Zacharias 1903–1989. Austrian ethologist who studied the relationship between instinct and behaviour, particularly in birds, and described the phenomenon of ***imprinting*** 1935. Lorenz discovered that many birds do not instinctively recognize members of their own species but they do possess an innate ability to acquire this capacity. During a brief period after hatching a young bird treats the first reasonably large object it sees as representative of its species – the object becomes imprinted. In 1973 he shared the Nobel Prize for Physiology or Medicine with Nikolaas Tinbergen and Karl von Frisch.

It is a good morning exercise for a research scientist to discard a pet hypothesis every day before breakfast. It keeps him young.

Konrad Lorenz

The So-Called Evil

Lorenz was born in Vienna and studied medicine there and in the USA at Columbia University. In 1940 he was appointed professor of general psychology at the Albertus University in Königsberg, Germany. Lorenz sympathized with Nazi views on eugenics, and in 1938 applied to join the Nazi party. From 1942 to 1944 he was a physician in the German army, and then spent four years in the USSR as a prisoner of war. Returning to Austria, he successively headed various research institutes.

Together, Lorenz and Tinbergen discovered how birds of prey are recognized by other birds. All birds of prey have short necks, and the sight of any bird – or even a dummy bird – with a short neck causes other birds to fly away.

His books include *King Solomon's Ring* 1952 (on animal behaviour) and *On Aggression* 1966 (on human behaviour).

Lorenz Ludwig Valentine 1829–1891. Danish mathematician and physicist. He developed mathematical formulae to describe phenomena such as the relation between the refraction of light and the density of a pure transparent substance, and the relation between a metal's electrical and thermal conductivity and temperature.

Lorenz was born in Elsinore and studied at the Technical University in Copenhagen. He became professor at the Military Academy in Copenhagen 1876. From 1887, his research was funded by the Carlsberg Foundation.

He investigated the mathematical description for light propagation through a single homogeneous medium and described the passage of light between different media. The formula for the mathematical relationship between the refractive index and the density of a medium was published by Lorenz in

1869 and by Hendrick Lorentz (who discovered it independently) in 1870 and is therefore called the ***Lorentz–Lorenz formula***. Using his electromagnetic theory of light, Lorenz was able to derive a correct value for the velocity of light.

Lovell (Alfred Charles) Bernard 1913– . English radio astronomer, director 1951–81 of Jodrell Bank Experimental Station (now Nuffield Radio Astronomy Laboratories).

During World War II Lovell worked on developing a radar system to improve the aim of bombers in night raids. After the war he showed that radar could be a useful tool in astronomy, and lobbied for the setting-up of a radio-astronomy station. Jodrell Bank was built near Manchester 1951–57. Although its high cost was criticized, its public success after tracking the Soviet satellite *Sputnik I* 1957 assured its future. Knighted 1961.

Lovell was born in Gloucestershire and studied at Bristol. His academic career was spent at Manchester, where he became the first professor of radio astronomy 1951.

In 1950, Lovell discovered that galactic radio sources emitted at a constant wavelength and that the fluctuations ('scintillation') recorded on the Earth's surface were introduced only as the radio waves met and crossed the ionosphere.

His books include *Radio Astronomy* 1951 and *The Exploration of Outer Space* 1961.

Lovelock James Ephraim 1919– . British scientist who began the study of CFCs in the atmosphere in the 1960s (though he did not predict the damage they cause to the ozone layer) and who later elaborated the Gaia hypothesis.

Lovelock invented the electron capture detector in the 1950s, a device for measuring minute traces of atmospheric gases. In the 1970s he worked as a consultant to NASA.

Lowell Francis Cabot 1775–1817. US industrialist who imported the new technology of English textile mills to America. With the cutoff of international trade during the Anglo-American War of 1812, Lowell established the Boston Manufacturing Company, a mechanized textile mill at Waltham, Massachusetts.

Lowell was born in Newburyport, Massachusetts, and educated at Harvard. He became a successful merchant. On a trip to England 1810–12 he was impressed by the country's mechanized mills and returned to the USA to build his own, similar mills. After the war he campaigned for tariff protection for the US textile industry. In 1822 the mill town of Lowell, Massachusetts, was established and named after him.

Lowell Percival 1855–1916. US astronomer who predicted the existence of a planet beyond Neptune, starting the search that led to the discovery of Pluto 1930. In 1894 he founded the Lowell Observatory in Flagstaff, Arizona, where he reported seeing 'canals' (now known to be optical effects and natural formations) on the surface of Mars.

Lowell was born in Boston, Massachusetts, and studied mathematics at Harvard. He spent 16 years in business and diplomacy, mainly in the Far East, before taking up astronomy, becoming professor at the Massachusetts Institute of Technology 1902.

Influenced strongly by the work of Italian astronomer Giovanni Schiaparelli, Lowell set up his observatory at Flagstaff originally with the sole intention of confirming the presence of advanced life forms on Mars. He thought he could make out a complex and regular network of canals and regular seasonal variations that to him indicated agricultural activity. He led an expedition to the Chilean Andes in 1907 which produced the first high-quality photographs of the planet.

Lower Richard 1631–1691. English physician and physiologist who performed the first direct transfusion of blood 1666 and was the first to link the process of respiration with the blood. Christopher Wren, the architect and natural philosopher, was the first person to inject into the veins of animals. The medicinal fluids he injected produced effects varying from vomiting to purging to intoxication. Wren's work led directly to the first transfusion of blood from an artery of one animal to the vein of another. Lower performed this using quills to connect the two vessels but he later moved on to fine silver tubes. Jean Baptiste Denys was the first to transfuse a person. He transfused a 15 year-old boy with the blood from the artery of a lamb 1667. Lower made a direct transfusion of a man from a sheep in London five months later. Three weeks later Lower gave the same man a second transfusion with no ill effects. However, it was not until the 19th century that transfusion of human patients was performed for therapeutic purposes and blood compatabilities were determined.

Lucas Keith 1879–1916. English neurophysiologist who investigated the transmission of nerve impulses. He demonstrated that the contraction of muscle fibres follows the 'all or none' law: a certain amount of stimulus is needed in order to induce a nerve impulse and subsequent muscle contraction. Any stimulus below that threshold has no effect regardless of its duration.

Lucas showed that when two successive stimuli are given, the response to the second stimulus can not be evoked if the first nerve impulse is still in progress. He also demonstrated that following a contraction there is a period of diminished excitability during which the muscle cannot be induced to contract again. This is due to the chemical transmission of impulses over synaptic clefts (the junction between two individual neurons).

Lucas was born in Greenwich, London and graduated from the University of Cambridge in natural sciences. He joined the staff of Cambridge 1903 and began his work on structure and function of nerves and muscles.

By the outbreak of World War II, Lucas had developed highly sensitive technology for measuring the electrical currents in individual neurons. They characteristically have very long fibrelike extensions called processes which can extend over large distances in the body and conduct messages. Unfortunately, after enlisting 1914 and being trained as a pilot, he was killed in an air crash before he was able to fully apply his new experimental techniques. However, they became invaluable tools for post-war neurophysiologists.

Ludwig Carl Friedrich Wilhelm 1816–1895. German physiologist who invented graphic methods of recording events within the body. He demonstrated that the circulation of the blood is purely mechanical in nature and involves no occult vital forces.

In 1847 Ludwig invented the kymograph, a rotating drum on which a stylus charts a continuous record of blood pressure and temperature. This was a forerunner of today's monitoring systems.

Ludwig was born in Witzenhausen, Hesse, and studied at Marburg (though temporarily compelled to leave the university as a result of his political activities), Erlangen, and the surgical school in Bamberg. He was professor at Marburg 1841–49, and then held posts at Zürich, Switzerland, and in Vienna at the Austrian military medical academy. From 1865 he was professor at Leipzig.

Ludwig devised a system of measuring the level of nitrogen in urine to quantify the rate of protein metabolism in the human body. In 1859 he described his mercurial blood-gas pump, which enabled him to separate gases from a given quantity of blood. He also invented the *Stromuhr*, a flowmeter which measures the rate of the flow of blood in the veins.

Ludwig published *Das Lehrbuch der Physiologie/A Physiology Textbook* 1852–56, the first modern text on physiology.

Lumière Auguste Marie Louis Nicolas (1862–1954) and Louis Jean (1864–1948) French brothers who pioneered cinematography. In 1895 they patented their cinematograph, a combined camera and projector operating at 16 frames per second, and opened the world's first cinema in Paris to show their films.

The Lumières' first films were short static shots of everyday events such as *La Sortie des usines Lumière* 1895 about workers leaving a factory and *L'Arroseur arrosé* 1895, the world's first fiction film. Production was abandoned in 1900.

The Lumière brothers, born in Besançon, joined their father's photographic firm in Lyon. They contributed several minor improvements to the developing process, including in 1880 the invention of a better type of dry plate.

In 1894, their father purchased an Edison Kinetoscope (a peephole cineviewer). The brothers borrowed some of the ideas and developed their all-in-one machine, the cinematograph. To advertise their success they filmed delegates arriving at a French photographic congress and 48 hours later projected the developed film to a large audience.

Auguste went on to do medical research. Louis invented a photorama for panoramic shots and in 1907 a colour-printing process using dyed starch grains. Later he experimented with stereoscopy and three-dimensional films.

Lummer Otto Richard 1860–1925. German physicist who specialized in optics and thermal radiation. His investigations led directly to the radiation formula of Max Planck, which marked the beginning of quantum theory.

Lummer was born in Jena, Saxony, and attended a number of different German universities. He became an assistant to Hermann Helmholtz at Berlin 1884 and moved with him to the newly established Physikalische Technische Reichsanstalt in Berlin 1887. In 1894 Lummer was made professor there. From 1904 he was professor at Breslau (now Wroclaw, Poland).

In collaboration with Eugen Brodhun, he designed a photometer (the Lummer–Brodhun cube) and worked towards the establishment of an international standard of luminosity. Lummer and Wilhelm Wien made the first practical black-body radiator by making a small aperture in a hollow sphere. When heated to a particular temperature, it behaved like an ideal black body. Studying emission from black bodies, Lummer later confirmed Wien's displacement law but found an anomaly in Wien's radiation law.

Lummer designed a mercury vapour lamp for use when monochromatic light is required, for instance in fluorescence microscopy, and in 1902 designed a high-resolution spectroscope.

Lunardi Vincenzo 1759–1806. Italian balloonist. He came to London as secretary to the Neapolitan ambassador, and made the first balloon flight in England from Moorfields in 1784.

Luria Salvador Edward 1912–1991. Italian-born US physician who was a pioneer in molecular biology, especially the genetic structure of viruses. Luria was a pacifist and was identified with efforts to keep science humanistic. He shared the Nobel Prize for Physiology or Medicine 1969.

Luria was born in Turin. He left Fascist Italy 1938, going first to France, where he became a research fellow at the Institut du Radium in Paris, and then to the USA 1940. From 1943 he taught at a number of universities and in 1959 became a professor at the Massachusetts Institute of Technology (MIT). He founded the MIT Center for Cancer Research, which he directed 1972–85. For some time he taught a course in world literature to graduate students at MIT and at Harvard Medical School to ensure their involvement in the arts.

Lwoff André Michel 1902– . French microbiologist who proved that enzymes produced by some genes regulate the functions of other genes. He shared the 1965 Nobel Prize for Physiology or Medicine with his fellow researchers Jacques Lucien Monod and François Jacob.

Lwoff was born in Ainy-le-Château, Allier. From 1921 he worked at the Pasteur Institute. During World War II he was active in the French Resistance. He was professor at the Sorbonne 1959–68 and head of the Cancer Research Institute in Villejuif 1968–72.

In the 1920s, Lwoff demonstrated the coenzyme nature of vitamins. He also discovered the extranuclear genetic control of some characteristics of protozoa.

In the late 1940s, Lwoff worked out the mechanism of lysogeny in bacteria, in which the DNA of a virus becomes attached to the chromosome (DNA) of a bacterium, behaving almost like a bacterial gene. It is therefore replicated as part of the host's DNA and so multiplies at the same time. But certain agents (such as ultraviolet radiation) can turn the 'latent' viral DNA, called the prophage, into a vegetative form which multiplies, destroys its host, and is released to infect other bacteria.

Lyell Charles 1797–1875. Scottish geologist. In his *Principles of Geology* 1830–33, he opposed the French anatomist Georges Cuvier's theory that the features of the Earth were formed by a series of catastrophes, and expounded the Scottish geologist James Hutton's view, known as uniformitarianism, that past events were brought about by the same processes that occur today – a view that influenced Charles Darwin's theory of evolution. Knighted 1848.

Lyell suggested that the Earth was as much as 240 million years old (in contrast to the 6,000 years of prevalent contemporary theory), and provided the first detailed description of the Tertiary period, dividing it into the Eocene, Miocene, and older and younger Pliocene periods. Darwin simply applied Lyell's geological method – explaining the past through what is observable in the present – to biology.

Lyell was born in Forfarshire. He studied at Oxford, becoming a lawyer, but retired from his law practice 1827 and devoted himself full time to geology and writing. In 1831 he became professor of geology at King's College, London.

A scientific hypothesis is elegant and exciting insofar as it contradicts common sense.

Charles Lyell

Attributed by S J Gould
Ever Since Darwin 1978

Lyell arrived independently at the same theories as Hutton, and organized them into a popular and coherent form.

Lyman Theodore 1874–1954. US physicist whose work was confined to the spectroscopy of the extreme ultraviolet region.

Lyman was born in Boston and studied at Harvard, and briefly in Europe at Cambridge and Göttingen. His working life was spent at Harvard, where he became professor 1921.

When Lyman began his research, the ultraviolet end of the spectrum had been observed by enclosing the spectroscope in a vacuum with fluorite windows. Using a concave ruled grating instead of a fluorite prism, Lyman discovered false lines in the ultraviolet due to light in the visible region, and these came to be called Lyman ghosts. He established the limit of transparency of fluorite to be 1,260 Å (1 Å = 10^{-10}m) and looked at the absorbency of other suitable solids but found none better. He also examined the absorption of various gases. By 1917, Lyman had extended the spectrum to 500 Å.

A series of lines in the hydrogen spectrum discovered by Lyman 1914 was named the Lyman series. He correctly predicted that the first line would be present in the Sun's spectrum.

In the 1920s Lyman began to examine spectra in the ultraviolet region of helium, aluminium, magnesium, and neon. His last paper was published in 1935 on the transparency of air between 1,100 and 1,300 Å.

Lyman was also a traveller and naturalist. From the Altai mountains of China and Mongolia, he brought back the first specimen of a gazelle *Procapra altaica* and 13 previously unknown smaller mammalian species. A stoat became known as Lyman's stoat, *Mustela lymani.*

Lynden-Bell Donald 1935– . English astrophysicist whose theories on the structure and dynamics of galaxies predicate black holes at the centre.

Lynden-Bell was born in Dover, Kent, and studied at Cambridge.

After two years at the California Institute of Technology and Hale Observatories, he returned to Cambridge, where in 1972 he became professor and director of the Institute of Astronomy.

In 1969 Lynden-Bell proposed that quasars were powered by massive black holes. Later, continuing this line of thought, he postulated the existence of black holes of various masses in the nuclei of individual galaxies. The presence of these black holes would account for the large amounts of infrared energy that emanate from a galactic centre. Lynden-Bell further argued that in the dynamic evolution of star clusters, the core of globular star clusters evolves independently of outer parts, and that only a dissipative collapse of gas would account for that evolution.

Lynen Feodor Felix Konrad 1911–1979. German biochemist who investigated the synthesis of cholesterol in the human body and the metabolism of fatty acids. For this work he shared the 1964 Nobel Prize for Physiology or Medicine with Konrad Bloch.

Lynen was born and educated in Munich, where he spent his whole career, becoming professor 1953 and director of the Max Planck Institute for Cell Chemistry (later Biochemistry) 1954.

Lynen in Munich and Bloch in the USA corresponded and worked out the 36 steps involved in the synthesis of cholesterol. Bloch found that the basic unit is the simple acetate (ethanoate) ion, a chemical fragment containing only two carbon atoms. In 1951, Lynen found the carrier of this fragment. Bloch then found an intermediate compound, squalene – a long hydrocarbon containing 30 carbon atoms. The final stage was the transformation of the carbon chain of squalene into the four-ring molecule of cholesterol.

Lynen also worked on the biosynthesis of fatty acids, isolating from yeast an enzyme complex that acts as a catalyst in the synthesis of long-chain fatty acids from acetyl coenzyme A and malonyl coenzyme A. He also elucidated a series of energy-generating reactions that occur when fatty acids from food are respired to form carbon dioxide and water.

Lyon Mary Frances 1925– . British mouse geneticist who is best known for her theory, now known as the ***Lyon hypothesis***, that one of the X chromosomes (female sex chromosomes) is inactivated during early embryonic development of a female mammal. She also helped establish the mouse as a valid experimental model for investigating the genetic basis of inherited diseases.

The Medical Research Council's Radiobiology Unit at Harwell had been constructed at the end of World War II to define the effects of radiation on mice as a model for understanding the consequences of nuclear fallout on human beings. Lyon realized that the mutations developed in the Radiobiology Units programme could provide valuable insights into fundamental problems in mammalian genetics.

Lyon is best known for her discovery of 'transcriptional inactivation' of an X chromosome in female cells. A normal diploid cell contains two copies of each autosome (any chromosome apart from the sex chromosomes) which means it has a double dose of each gene product on these chromosomes. A female cell contains two X chromosomes, while a male cell has one X chromosome and one Y chromosome, causing an apparent divergence in the gene contribution from the X chromosome between males and females. Lyon determined that one of the X chromosomes contracts into a dense object that is transcriptionally inactive.

Lyon was born in Norwich and graduated from Girton College, Cambridge 1946. After completion of her PhD 1950, she joined the Medical Research Council's Radiobiology Unit at Harwell and became head of its Genetics Division 1962. She was elected a Fellow of the Royal Society 1973 and made deputy director of the Radiobiology Unit 1982.

Lyot Bernard Ferdinand 1897–1952. French astronomer who also designed and constructed optical instruments. He concentrated on the study of the solar corona, for which he devised the coronagraph and the photoelectric polarimeter, and he proved that some of the Fraunhofer lines in the solar spectrum represent ionized forms of known metals rather than undiscovered elements.

Lyot was born in Paris and graduated from the Ecole Supérieure d'Electricité. From 1920 he worked at the Meudon Observatory, becoming chief astronomer 1943.

Most of Lyot's research during the 1920s was devoted to the study of polarized light, reflected to the Earth from the Moon and from other planets. In addition to designing a polariscope of greatly improved sensitivity, Lyot reported 1924 that the Moon was probably covered by a layer of volcanic ash and that duststorms were a common feature of the Martian surface.

Lyot designed his coronagraph 1930, later adding improvements.

By making it possible to observe the Sun's corona in broad daylight rather than only during eclipses, the coronagraph also permitted the observation of continuous changes in the corona. This meant that the corona could be filmed, as Lyot demonstrated for the first time in 1935. He also reported the rotation of the corona in synchrony with the Sun.

Lysenko Trofim Denisovich 1898–1976. Soviet biologist who believed in the inheritance of acquired characteristics (changes acquired in an individual's lifetime) and used his position under Joseph Stalin officially to exclude Gregor Mendel's theory of inheritance. He was removed from office after the fall of Nikita Khrushchev in 1964.

The Party, the Government, and J V Stalin personally have taken an unflagging interest in the further development of the Michurian teaching.

Trofim Lysenko

Lysenko condemning Mendelism, Moscow 1948

As leader of the Soviet scientific world, Lysenko encouraged the defence of mechanistic views about the nature of heredity and speciation. This created an environment conducive to the spread of unverified facts and theories, such as the doctrine of the noncellular 'living' substance and the transformation of viruses into bacteria. Research in several areas of biology came to a halt.

Lysenko was born in Karlovka in the Russian Ukraine, and educated at the Uman School of Horticulture and the Kiev Agricultural Institute. From 1929 to 1938 he held senior positions at the Ukrainian All-Union Institute of Selection and Genetics in Odessa, becoming director of the Institute of Genetics of the USSR Academy of Sciences in 1940. In 1965 he was removed from his post and stripped of all authority.

By advocating vernalization (a method of making seeds germinate quickly in the spring), Lysenko achieved considerable increases in crop yields, and this was the basis of his political support. As his influence increased, he enlarged the scope of his theories, using his authority to remove any opposition.

Lyttleton Raymond Arthur 1911–1995. English astronomer and theoretical physicist who focused on stellar evolution and composition, as well as the nature of the Solar System.

Lyttleton was born near Birmingham and studied at Cambridge and at Princeton in the USA. He returned to Cambridge in 1937, and together with Fred Hoyle he established a research school there in theoretical astronomy. He held a number of scientific posts, including a position at the Jet Propulsion Laboratory in California 1960.

In 1939 Lyttleton and Hoyle demonstrated the presence of interstellar hydrogen on a large scale, at a time when most astronomers believed space to be devoid of interstellar gas. In the early 1940s they applied the new advances in nuclear physics to the problem of energy generation in stars.

Lyttleton published 1953 a monograph on the stability of rotating liquid masses, and later postulated that the Earth's liquid core was produced by a phase change resulting from the combined effects of intense pressure and temperature. He also stressed the hydrodynamic significance of the liquid core in the processes of precession and nutation.

In 1959 with cosmologist Hermann Bondi he proposed the electrostatic theory of the expanding universe.

McAdam John Loudon 1756–1836. Scottish engineer, inventor of the ***macadam*** road surface. It originally consisted of broken granite bound together with slag or gravel, raised for drainage. Today, it is bound with tar or asphalt.

McAdam introduced a method of road building that raised the road above the surrounding terrain, compounding a surface of small stones bound with gravel on a firm base of large stones.

A camber, making the road slightly convex in section, ensured that rainwater rapidly drained off the road and did not penetrate the foundation. By the end of the 19th century, most of the main roads in Europe were built in this way.

McAdam was born in Ayr. Emigrating to the USA at 14, he returned to the UK 1783. He was appointed paving commissioner in Bristol in 1806; ten years later he became surveyor-general of the roads in that region, and of all the roads in Britain 1827.

McAdam was also responsible for reforms in road administration, and advised many turnpike trusts. He ensured that public roads became the responsibility of the government, financed out of taxes for the benefit of everyone.

McAdam *Scottish civil engineer John McAdam. McAdam developed a procedure for the construction of roads, the principles of which are still used today. Roads are raised above the countryside and are prepared by compacting small gravel on top of larger stones to allow easy drainage.*

MacArthur Robert Helmer 1930–1972. Canadian-born US ecologist who did much to change ecology from a descriptive discipline to a quantitative, predictive science. For example, his index of vegetational complexity (foliage height diversity) 1961 made it possible to compare habitats and predict the diversity of their species in a definite equation.

MacArthur was born in Toronto and studied in the USA at Brown University and Yale. From 1965 he was professor of biology at Princeton.

He studied the relationship between five species of warbler that coexist in the New England forests; these species are now known as MacArthur's warblers.

Investigating population biology, MacArthur examined how the diversity and relative abundance of species fluctuate over time and how species evolve. In particular, he managed to quantify some of the factors involved in the ecological relationships between species.

McBain James William 1882–1953. Canadian physical chemist whose main researches were concerned with colloidal solutions, particularly soap solutions.

McBain was born in Chatham, New Brunswick, and studied at the University of

Toronto and briefly in Germany at Leipzig. From 1906 he worked in the UK at the University of Bristol, becoming professor 1919. In 1926, he went to the USA to become professor at Stanford, California.

As early as 1910 he showed that aqueous solutions of soaps such as sodium palmitate are good electrolytic conductors. He postulated the existence of a highly mobile carrier of negative electricity – the 'association ion' or 'ionic micelle'.

To determine the thermodynamic properties of soap solutions, he developed his own method based on the lowering of the dew point.

McBain proved that the surface phase in simple solutions is not just one monolayer thick: oriented underlayers exist beneath the monolayers of soap, and he devised an ingenious apparatus for determining their composition.

In the adsorption of gases and vapours by solids, various processes can take place simultaneously: physical sorption, chemisorption, and permeation of the solid by the gas or vapour. McBain introduced the generalized term 'sorption' to include all such cases. The McBain–Bakr spring balance provides a continuous record of the quantities and rate of sorption.

McClintock Barbara 1902–1992. US geneticist who discovered jumping genes (genes that can change their position on a chromosome from generation to generation). This would explain how originally identical cells take on specialized functions as skin, muscle, bone, and nerve, and also how evolution could give rise to the multiplicity of species. Nobel Prize for Physiology or Medicine 1983.

McClintock was born in Hartford, Connecticut, and studied botany at Cornell University, New York, obtaining her PhD 1927. As an undergraduate there, and from 1941 at the Carnegie Institute, New York, she studied maize chromosomes. She observed that the patterns on twin sectors of maize seedlings were the inverse of one another, and that pigmentation of certain kernels did not correspond to their genetic makeup. Realizing that as a single cell divided into sister cells, one gained what the other had lost, she deduced that not all genes behave in the same way: some genes can switch others on and off, moving from one place to another on one chromosome, or even 'jumping' from one chromosome to another. These jumping genes acted as regulators and were later discovered in bacteria and in fruit flies.

McClintock's discovery that genes are not stable overturned one of the main tenets of heredity laid down by Gregor Mendel. It had enormous implications and explained, for example, how resistance to antibiotic drugs can be transmitted between entirely different bacterial types.

She utilized X-rays to induce chromosomal aberrations and rearrangements and examined the ways in which chromosomes repair such damage. This information helped other scientists understand the problems of radiation sickness after the explosion of the atom bomb at Hiroshima, Japan.

It might seem unfair to reward a person for having so much pleasure over the years, asking the maize plant to solve specific problems and then watching its responses.

Barbara McClintock

On her lifelong research into the genetics of the maize plant. *Newsweek* 24 Oct 1983

McCollum Elmer Verner 1879–1967. US biochemist and nutritionist who originated the letter system of naming vitamins. He also researched into the role of minerals in the diet. His choice of the albino rat as a laboratory animal was to make it one of the most used animals for research.

McCollum was born in Fort Scott, Kansas, and studied at Kansas and Yale universities. He worked at the University of Wisconsin 1907–17. From 1917 he was professor at the School of Hygiene and Public Health at Johns Hopkins University.

McCollum discovered in the early 1910s that growth retardation results from a diet deficient in certain fats and that such deficiencies can be compensated for by providing a specific extract from either butter or eggs. He called this essential component 'fat-soluble A', because it dissolves in lipids. He then showed that there is another essential dietary component, which he called 'water-soluble B'. Later he found that they are not single compounds but complexes.

At Johns Hopkins, McCollum collaborated in the discovery of vitamin D.

McCormick Cyrus Hall 1809–1884. US inventor of the reaping machine 1831, which revolutionized 19th-century agriculture.

McCormick was born in Virginia. Encouraged by his father, who was an inventor, he produced in 1831 a hillside plough as well as the prototype for his reaping machine, which took nine years to perfect. He began manufacturing it in Chicago 1847. But in 1848, when his patent expired, he faced strong competition and only his good business sense kept him from being overwhelmed by other manufacturers who had been waiting to encroach on his markets. He survived and prospered, and successfully introduced his reaping machine into Europe.

It was estimated that a McCormick reaper operated by a two-person crew and drawn by a single horse could cut as much corn as could 12 to 16 people with reap hooks.

McCrea William Hunter 1904– . Irish theoretical astrophysicist and mathematician who particularly studied the evolution of galaxies and planetary systems. He also considered the formation of molecules in interstellar matter and the composition of stellar atmospheres. Knighted 1985.

McCrea was born in Dublin and studied at Cambridge and in Germany at Göttingen. He was professor of mathematics at Queen's University, Belfast, 1936–44, and at the University of London 1945–66, when he was appointed professor of theoretical astronomy at the University of Sussex.

Studying the factors that would influence the earliest stages of the evolution of stars, McCrea focused on what might happen if the condensing material encountered interstellar matter that was itself in a state of turbulence.

Together with English astrophysicist Edward Milne, McCrea found that Newtonian dynamics could be advantageously applied to the analysis of the primordial gas cloud. The model relied on the assumption that the gas cloud would be 'very large' rather than of infinite size, although for the purposes of observation it would be 'infinite'.

In physics, McCrea worked on forbidden (low-probability) transitions of electrons between energy states, analyses of penetration of potential barriers (for instance by 'tunnelling'), and relativity theory.

MacCready Paul Beattie 1925– . US designer of the *Gossamer Condor* aircraft, which made the first controlled flight by human power alone 1977. His *Solar Challenger* flew from Paris to London under solar power, and in 1985 he constructed a powered model of a giant pterosaur, an extinct flying animal.

McCulloch Warren Sturgis 1899–1969. US neurophysiologist who developed cybernetic and computational models of the brain. His papers include 'What the Frog's Eye Tells the Frog's Brain' 1959, which detailed the way in which information is transmitted from the retina to the brain to detect significant features or events in the frog's environment.

McCulloch was born in Orange, New Jersey, and educated at Yale and Columbia. His initial work, conducted at Yale in the 1930s, concentrated on primate physiology; his work on brain models was undertaken initially at the University of Illinois and later, from 1952, at the Massachusetts Institute of Technology.

Mach Ernst 1838–1916. Austrian philosopher and physicist. He was an empiricist, believing that science is a record of facts perceived by the senses, and that acceptance of a scientific law depends solely on its standing the practical test of use; he opposed concepts such as Isaac Newton's 'absolute motion'. ***Mach numbers*** are named after him. The Mach number is the ratio of the speed of a body to the speed of sound in the undisturbed medium through which the body travels. An aircraft flying at Mach 2 is flying at twice the speed of sound in air at that particular height.

Every statement in physics has to state relations between observable quantities.

Ernst Mach

Mach's Principle

Mach was born in Turas, Moravia (now in the Czech Republic), and studied at Vienna. He was professor at Graz 1864–67, Prague 1867–95, and Vienna 1895–1901. He was then appointed to the upper chamber of the Austrian parliament.

Mach studied physiological phenomena such as vision, hearing, our sense of time, and the kinesthetic sense. He investigated stimulation of the retinal field with spatial patterns and discovered a strange visual effect called ***Mach bands***. This was subsequently forgotten, and was rediscovered in the 1950s.

He investigated various aspects of optics, and wave phenomena of mechanical, electrical, and optical kinds. In 1887 he published a work on the photography of projectiles in flight, which showed the shock wave produced by the gas around the tip of the projectile. The Mach angle describes the angle between the direction of motion and the shock wave, and Mach found that it varies with the speed of the projectile. In 1929 the term Mach number came into use.

Mach postulated that all knowledge is mediated by perception, and believed that the greatest scientific advances would only arise through a deeper understanding of this process.

Mach's book *Die Mechanik* 1863 gave rise to debate on ***Mach's principle***, which states that a body could have no inertia in a universe devoid of all other mass as inertia depends on the reciprocal interaction of bodies, however distant. This principle influenced Einstein.

McIndoe Archibald Hector 1900–1960. New Zealand plastic surgeon. He became known in the UK during World War II for his remodelling of the faces of badly burned pilots. Knighted 1947.

Macintosh Charles 1766–1843. Scottish manufacturing chemist who invented a waterproof fabric, lined with rubber, that was used for raincoats – hence ***mackintosh***. Other waterproofing processes have now largely superseded this method.

Mackenzie James 1848–1924. Scottish physician and cardiologist who was a pioneer of modern cardiac medicine. He was first to identify a large number of irregularities in the heart's beat and establish which were caused by serious disease and which were of no consequence.

In his work as a general practitioner he became interested in how a person's pulse is generated and invented an ink polygraph which enabled him to compare venous pulse, arterial pulse and the beat of the heart itself. He published *The Study of the Pulse* 1902 and *Diseases of the Heart* 1908. He realized that an important principal in some heart disease is to diagnose whether a heart is able to compensate for a disordered valve rather than to assume that all heart murmurs are totally debilitating.

Mackenzie was born in Scone in Scotland and studied medicine at the University of Edinburgh. In recognition of his cardiovascular research, he was elected a Fellow of the Royal Society and knighted 1915.

Mackenzie was particularly interested in the total irregularity of the heart beat which occurs in the later stages of the disease of the mitral valve, probably the commonest of all chronic heart conditions. He found that the characteristic wave caused by contraction of the auricles in the jugular vein was absent in mitral valve disease. This is because the muscle fibres are contracting independently of each other and at a rate that is far higher than that which can be conducted to the ventricles. Hence only some pulses get through at irregular intervals. This condition is now treated with drugs, or in the most severe of cases, by mitral valve transplant.

Maclaurin Colin 1698–1746. Scottish mathematician who played a leading part in establishing the hegemony of Isaac Newton's calculus in the UK. Maclaurin was the first to present the correct theory for distinguishing between the maximum and minimum values of a function.

Maclaurin was born in Argyllshire and studied at Glasgow. At the age of 19, he was appointed professor at the Marischal College of Aberdeen. On a visit to London in 1719 he first met Isaac Newton. Maclaurin left Aberdeen 1722 to become travelling tutor to the son of an English diplomat, returning to Scotland 1724 to become professor at Edinburgh. He won the admiration of Edinburgh society for his public lectures and demonstrations in experimental physics and astronomy. In 1745, during the Jacobite rebellion, he organized the defence of Edinburgh.

Maclaurin's *Treatise of Fluxions* 1742 was an attempt to prove Newton's doctrine of prime and ultimate ratios and to provide a geometrical framework to support Newton's fluxional calculus. So influential was the treatise that it contributed to the ascendancy of Newtonian mathematics which cut off Britain from developments in the rest of the world.

Macleod John James Rickard 1876–1935. Scottish physiologist who shared the Nobel Prize for Physiology or Medicine with Frederick Banting 1923 for their part in the discovery of insulin, the hormone in the pancreas that reduces blood glucose (sugar) levels. Since its discovery, insulin has been used extensively as the main treatment for diabetes.

Macleod was born in Cluny, Scotland and studied medicine at the University of Aberdeen 1898. He went on to study biochemistry at the University of Leipzig and published various papers on biochemical aspects of respiration and carbohydrate metabolism. While professor of physiology in Toronto (1918 onwards) he invited Banting to work with him and his colleague Charles Best in their search for the hormone in the pancreas that reduces blood glucose levels. The hormone responsible for this, insulin, was first isolated in Macleod's laboratory by Banting and Best 1921 while Macleod was away on holiday.

However, when Banting and Best wanted to present their work to the American Physiological Society in December of 1921 they were unable to, because neither of them were members of the Society. Macleod, who was a member, attached his name and the resulting communication was published under the three names 1922. In 1923, the Nobel Prize for Medicine was awarded to Macleod and Banting 'for their discovery of insulin'. Banting expressed his disapproval by sharing half of his prize with Best, and Macleod thereupon gave half of his prize to Collip, who characterized insulin after its discovery.

McMillan Edwin Mattison 1907–1991. US physicist. In 1940 he discovered neptunium, the first transuranic element, by bombarding uranium with neutrons. He shared a Nobel prize with Glenn Seaborg 1951.

In 1943 McMillan developed a method of overcoming the limitations of the cyclotron, the first accelerator, for which he shared, 20 years later, an Atoms for Peace award with I Veksler, director of the Soviet Joint Institute for Nuclear Research, who had come to the same discovery independently.

McMillan was born in Los Angeles and studied at the California Institute of Technology and Princeton. From 1932 he was on the staff at the University of California, as professor 1946–73, except during World War II when he worked on radar and Seaborg took up his work at Berkeley. The discovery of plutonium facilitated the construction of the first atom bomb, in which McMillan also took part.

MacMillan Kirkpatrick 1813–1878. Scottish blacksmith who invented the bicycle 1839. His invention consisted of a 'hobby-horse' that was fitted with treadles and propelled by pedalling.

McNaught William 1813–1881. Scottish mechanical engineer who invented the compound steam engine 1845. This type of engine extracts the maximum energy from the hot steam by effectively using it twice – once in a high-pressure cylinder (or cylinders) and then, when exhausted from this, in a second, low-pressure cylinder.

McNaught was born in Paisley and joined his father's steam-engine firm. In 1849 he moved to Manchester. The firm of J and W McNaught was established 1860, manufacturing steam engines until 1914.

By 1845, the cotton and wool mill owners in the north of England thought that they would have to replace their steam engines with bigger ones to keep up with the demand for more power. McNaught's conversion to compound action saved them this financial outlay. His method became known as 'McNaughting', and was to be a valuable energy-saving system for many years to come.

Three types of new engine were to emerge using McNaught's principle. The first used cylinders mounted side by side, the second had cylinders in a line (a tandem compound), and the third and rarest type had the high-pressure cylinder enclosed by the low-pressure one.

Magendie François 1783–1855. French physician, a pioneer of modern experimental pharmacology. He helped to introduce into medicine the range of plant-derived compounds known as alkaloids as well as strychnine, morphine and codeine, and quinine. With Jean Pierre Flourens, he introduced the idea that the nervous system co-ordinates the functions of different parts of the body.

Using extensive vivisection and a certain amount of self-experimentation, Magendie conducted trials on plant poisons, deploying animals to track precise physiological effects. He demonstrated that the stomach's role in vomiting is essentially passive, and analysed emetics. He investigated the role of proteins in human diet; he was interested in olfaction; and he studied the white blood cells.

Magendie also had a great influence on the search for the active principles of drugs. He published *Formulary for the preparation and use of several new drugs, such as nux vomica, morphine, prussic acid, strychnine, veratrine, the cinchona alkaloids, emetine, iodine* 1821, a pocket formulatory for practising physicians. He was the first to use alkaloids in the treatment of disease, and his book deals almost entirely with the clinical use of the new remedies.

Magendie worked protractedly on the nerves of the spine and the skull and showed that there are two roots to the spinal canal: the anterior root is motor, while the posterior root is sensory. A canal leading from the fourth ventricle (cavity) of the brain is now known as the 'foramen of Magendie'. He also confirmed the work of Flourens, demonstrating that if the cerebellum is removed from a mammal, physical equilibrium is almost entirely abolished resulting in lack of coordination.

Magendie was born in Bordeaux and studied at Paris, where he became physician to the Hôtel Dieu. Elected a member of the Académie des Sciences 1821, he became its president 1837. In 1831, he was appointed professor of anatomy at the Collège de France.

Maiman Theodore Harold 1927– . US physicist who in 1960 constructed the first working laser.

Maiman was born in Los Angeles and studied at Columbia and Stanford universities. From 1955 to 1961 he worked at the Hughes Research Laboratories. In 1962 he founded the Korad Corporation to manufacture lasers; in 1968 he founded Maiman Associates, a laser and optics consultancy; he cofounded the Laser Video Corporation 1972. In 1975 he joined the TRW Electronics Company, Los Angeles.

In 1955, Maiman began improving the maser (microwave amplifier), first designed in 1953 by US physicist Charles Townes. Townes had also demonstrated the theoretical possibility of constructing an optical maser, or laser, but Maiman was the first to build one. His laser consisted of a cylindrical, synthetic ruby crystal with parallel, mirror-coated ends, the coating at one end being semitransparent to allow the emission of the laser beam. A burst of intense white light stimulated the chromium atoms in the ruby to emit noncoherent red light. This red light was then reflected back and forth by the mirrored ends until eventually some of the light emerged as an intense beam of coherent red light – laser light. Maiman's apparatus produced pulses; the first continuous-beam laser was made in 1961 at the Bell Telephone Laboratories.

Malpighi Marcello 1628–1694. Italian physiologist who made many anatomical discoveries in his pioneering microscope studies of animal and plant tissues. For example, he discovered blood capillaries and indentified the sensory receptors (papillae) of the tongue, which he thought could be nerve endings.

Studying the lungs of a frog, Malpighi found them to consist of thin membranes containing fine blood vessels covering vast numbers of small air sacs. This discovery made it easier to explain how air (oxygen) seeps from the lungs to the blood vessels and is carried around the body. He also investigated the anatomy of insects and found the tracheae, the branching tubes that open to the outside in the abdomen and supply the insect with oxygen for respiration.

Malpighi was born in Crevalcore, near Bologna, and studied at Bologna. He first lectured in logic at Bologna, and although he was professor of theoretical medicine at Pisa 1656–59 and at Messina 1662–66, he returned to Bologna in between. In 1667, the Royal Society in England made him an honorary member and supervised the printing of his later works. In 1691, Malpighi moved to Rome and retired there as chief physician to Pope Innocent XII.

Malthus Thomas Robert 1766–1834. English economist. His *Essay on the Principle of Population* 1798 (revised 1803) argued for population control, since populations increase in geometric ratio and food supply only in arithmetic ratio, and influenced Charles Darwin's thinking on natural selection as the driving force of evolution.

Malthus saw war, famine, and disease as necessary checks on population growth. Later editions of his work suggested that 'moral restraint' (delaying marriage, with sexual abstinence before it) could also keep numbers from increasing too quickly, a statement seized on by later birth-control pioneers (the 'neo-Malthusians').

Population, when unchecked increases in a geometrical ratio. Subsistence only increases in an arithmetical ratio.

Thomas Malthus

Essay on the Principle of Population 1798

Malthus was born near Dorking, Surrey, studied at Cambridge and became a cleric. In 1805 he became professor of history and political economy at Haileybury College.

Now it is accepted that Malthus was probably correct up to a point. Populations throughout the plant and animal kingdoms also tend to increase faster than the resources

Malthus *English cleric and economist Thomas Robert Malthus, whose work on the principles of population had a major effect on the formulation of the theory of evolution. Malthus was an advocate of population control and argued that population increased faster than the supply of food and natural resources needed to sustain it.*

to support them until some check builds up sufficiently. After reading Malthus's work, Darwin postulated that the transformation or extinction of species depends on their response (a function of variability) to changing environmental factors.

Malthus's main books on economics are *An Inquiry into the Nature and Progress of Rent* 1815 and *Principles of Political Economy* 1820.

Malus Etienne Louis 1775–1812. French physicist who discovered the polarization of light by reflection from a surface. He also found the law of polarization that relates the intensity of the polarized beam to the angle of reflection.

Malus was born in Paris and studied there at the Ecole Polytechnique. From 1796 he was in the army, taking part in Napoleon's campaign in Egypt and Syria in 1798 and eventually rising to major, but he was also an examiner for the Ecole Polytechnique.

Malus began doing experiments on double refraction in 1807. This phenomenon causes a light beam to split in two on passing through Iceland spar and certain other crystals. An empirical description had been given by Dutch physicist Christiaan Huygens, based on the assumption that light is wavelike in character, and Malus's results confirmed Huygens's laws.

In 1808, Malus held a piece of Iceland spar up to some light reflecting off a window. To his surprise, the light beam emanating from the crystal was single, not double. He then noted that of the two beams that normally emerge from the crystal, only one was reflected from a water surface if the crystal was held at a certain angle. The other passed into the water and was refracted. If the crystal was turned perpendicular, the second beam was reflected and the first refracted. He described the light as being 'polarized'.

Mandelbrot Benoit B 1924– . Polish-born French mathematician who coined the term ***fractal*** to describe geometrical figures in which an identical motif repeats itself on an ever-diminishing scale. The concept is associated with chaos theory.

Another way of describing a fractal is as a curve or surface generated by the repeated subdivision of a mathematical pattern.

Mandelbrot was born in Warsaw and studied at the Ecole Polytechnique and the Sorbonne, Paris, and the California Institute of Technology. His academic career was divided mainly between France and the USA. In 1958 he began an association with IBM's research laboratories in New York. In 1987, he became professor at Yale.

Mandelbrot's research has provided mathematical theories for erratic chance phenomena and self-similarity methods in probability. He has also carried out research on sporadic processes, thermodynamics, natural languages, astronomy, geomorphology, computer art and graphics, and the fractal geometry of nature.

His books include *Logique, Langage et Théorie de l'Information* (with L Apostel and A Morf) 1957, *Fractals: Form, Chance and Dimension* 1977, and *The Fractal Geometry of Nature* 1982.

Mannesmann Reinhard 1856–1922. German ironfounder who invented a method of making seamless steel tubes, patented 1891.

Mannesmann was born in Remscheid, Westphalia, and joined the family ironworks.

The idea behind the seamless tube had been conceived by his father in 1860. One of the first Mannesmann plants was installed in Swansea, South Wales, 1887. In 1893 the Mannesmannröhrenwerke Aktiengesellschaft was formed. Mannesmann also formed a company 1908 to exploit mineral resources in Morocco, but was forced by the French to withdraw.

The Mannesmann process involved the passing of a furnace-heated bar between two rotating rolls. Because of their geometrical configuration, the rolls drew the bar forward and at the same time produced tensions in the hot metal that caused it to tear apart at the centre. A stationary, pointed mandrel caused the ingot of metal to open out and form a tube.

Manson Patrick 1844–1922. Scottish physician who showed that insects are responsible for the spread of diseases like elephantiasis and malaria. KCMG 1903.

Manson was born in Oldmeldrum, Aberdeenshire, and studied at Aberdeen. He spent 23 years in practice in the Far East. Returning to the UK 1892, he founded the School of Tropical Medicine in London 1899, and taught there until 1914.

In 1876 Manson began studying filariasis infection in humans. Having gained a clear idea of the life history of the invading parasite, he correctly conjectured that the disease was transmitted by an insect, a common brown mosquito. He went on to study other parasitic infections; for instance, the fluke parasite, ringworms, and guinea worm. He developed the thesis that malaria was also spread by a mosquito 1894; the work that proved this was carried out by Manson and British physician Ronald Ross.

Manton Sidnie Milana (married name Harding) 1902–1979. English embryologist who specialized in the arthropods (jointed limbed animals including insects, spiders, crabs, and lobsters), concentrating mainly on their embryology and functional morphology in relation to evolution. She summarized her findings in her book *The Arthropoda: Habits, Functional Morphology and Evolution* 1977.

Manton born in London, the daughter of George Manton, an eminent dental surgeon. She attended the Froebel Educational Institute and Girton College, Cambridge. She graduated in natural sciences 1925, coming top of the year in her zoology finals paper. After a year working at Imperial College, she returned to Cambridge as a university demonstrator in comparative anatomy during which time she obtained a PhD (1928). From 1928 until 1948 she was a fellow at Girton and, became a reader at King's College, London 1949, and made an honorary associate of the British Museum. Both she and her sister, Irene Manton, were Fellows of the Royal Society and were awarded the Linnaean Society's Gold medal. She married John Philip Harding 1937, (later the keeper of zoology at the British Museum).

Marconi Guglielmo 1874–1937. Italian electrical engineer and pioneer in the invention and development of radio. In 1895 he achieved radio communication over more than a mile, and in England 1896 he conducted successful experiments that led to the formation of the company that became Marconi's Wireless Telegraph Company Ltd. He shared the Nobel Prize for Physics 1909.

After reading about radio waves, Marconi built a device to convert them into electrical signals. He then tried to transmit and receive radio waves over increasing distances. In 1898 he successfully transmitted signals across the English Channel, and in 1901 established communication with St John's, Newfoundland, from Poldhu in Cornwall, and in 1918 with Australia.

Marconi was an Italian delegate to the Versailles peace conference 1919 after World War I.

Marconi was born in Bologna. He studied under a number of Italian professors but never enrolled for a university course.

Having obtained a patent in 1896 for wireless telegraphy, he enabled Queen Victoria to send a message to the Prince of Wales aboard the royal yacht, and by 1897 he had established a commercial enterprise in London. From 1921 he lived aboard his yacht *Elettra*, which served as a home, laboratory, and receiving station.

Marconi based his apparatus on that used by German physicist Heinrich Hertz, but used a coherer to detect the waves. (The coherer was invented by Edouard Branly in France and was designed to convert the radio waves into electric current.) Marconi improved Hertz's design by earthing the transmitter and receiver, and found that an insulated aerial enabled him to increase the distance of transmission.

Marconi's later inventions included the

Marconi *The Italian inventor Guglielmo Marconi on board his yacht* Elettra, *which was equipped as a floating laboratory. The equipment seen here was used to investigate new methods of maritime communication and navigation by wireless.*

magnetic detector 1902, horizontal direction telegraphy 1905, and the continuous wave system 1912. During World War I he worked on the development of very short wavelength beams.

Marcus Rudolph Arthur 1923– . Canadian chemist who advanced the theory of electron-transfer reactions (involving soluble molecules and/or ions) which drive many biological processes. By focusing on the solvent molecules in the reaction, Marcus realized that electron-transfer reaction rate coefficients could be determined that have an explanatory and predictive value. The theory has also been applied to proton-transfer reactions. He received the Nobel Prize for Chemistry in 1992.

Born in Montréal, Canada, Marcus attended the city's McGill University and studied chemistry. In 1946, he obtained his PhD and moved to Brooklyn, New York, where he taught at the Polytechnic Institute of Brooklyn. From 1968 to 1978, he taught at the University of Illinois, Urbana, and became professor of chemistry at Caltech in California. In 1987, he was elected a foreign member of the Royal Society and became an honorary fellow of the Royal Society of Chemistry in 1991.

Mariotte Edme 1620–1684. French physicist and priest known for his recognition in 1676 of Boyle's law about the inverse relationship of volume and pressure in gases, formulated by Irish physicist Robert Boyle 1672. He had earlier, in 1660, discovered the eye's blind spot.

Markham Beryl 1903–1986. British aviator who made the first solo flight from east to west across the Atlantic 1936.

Markov Andrei Andreyevich 1856–1922. Russian mathematician, formulator of the Markov chain, an example of a stochastic (random) process.

Markov was born in Ryazan, near Moscow, and studied at St Petersburg, where he became professor 1893. At the same time he became involved in liberal political movements. He refused to accept tsarist decorations and in 1907 renounced his membership of the electorate when the government dissolved the fledgling representative *duma*, or parliament.

Markov's early work was devoted primarily to number theory – continued fractions, approximation theory, differential equations, integration in elementary equations – and to the problem of moments and probability theory. Throughout he used the method of continued fractions.

A Markov chain may be described as a chance process that possesses a special property, so that its future may be predicted from the present state of affairs just as accurately as if the whole of its past history were known. Markov believed that the only real examples of his chains were to be found in literary texts, and he illustrated his discovery by calculating the alteration of vowels and consonants in Pushkin's *Eugene Onegin*. Markov chains are now used in the social sciences, atomic physics, quantum theory, and genetics.

Marr David Courtenay 1945–1980. English psychologist who developed computer-based models of the visual system. Drawing on neurophysiology and the psychology of vision, he applied his models to a number of issues, notably the problem of how objects in the perceptual field are represented within the brain. His findings are summarized in *Vision* 1982, published posthumously.

Marr was born in Essex and studied at Cambridge. From 1975 he worked in the USA at the artificial-intelligence laboratory of the Massachusetts Institute of Technology.

Marsh Othniel Charles 1831–1899. US palaeontologist. As official palaeontologist for the US Geological Survey from 1882, he identified many previously unknown fossil species and was an early devotee of Charles Darwin. He wrote *Odontornithes* 1880,

Dinocerata 1884, and *Dinosaurs of North America* 1896.

Marsh was born in Lockport, New York, and educated at Yale University and in Germany. As first professor of palaeontology in the USA at the Yale faculty, he mounted fossil-hunting expeditions to the West from 1870. He served as president of the American Academy of Sciences 1883–95.

Martin Archer John Porter 1910– . British biochemist who received the 1952 Nobel Prize for Chemistry for work with Richard Synge on paper chromatography in 1944.

Martin was born in London and studied at Cambridge. He has held both commercial and academic research posts; he worked at the Wool Industries Research Association in Leeds 1938–46, at the National Institute for Medical Research 1952–59, was director of the Abbotsbury Laboratory 1959–70, and taught at the University of Sussex 1973–78.

Martin and Synge began the development of partition chromatography 1941 for separating the components of complex mixtures. A drop of the solution to be analysed is placed at one end of a strip of filter paper and allowed to dry. That end of the strip is then immersed in a solvent, which deposits the various components of the mixture as it permeates the strip of paper. The dried strip is sprayed with a reagent that produces a colour change with the components; Martin and Synge used ninhydrin to reveal the positions of amino acids. The developed strip is called a chromatogram.

In 1953, Martin began working on gas chromatography, which separates chemical vapours by differential adsorption on a porous solid.

Martin James 1893–1981. Northern Irish aeronautical engineer who designed and manufactured ejection seats. At the time of his death about 35,000 ejection seats were in service with the air forces and navies of 50 countries. Knighted 1965.

Martin was born in Crossgar, County Down. In his early 20s he designed a three-wheeled car, and then set up a business in London. In the early 1930s he built a two-seater monoplane made of round-section, thin-gauge steel tubing. The design of this machine, Martin-Baker MB-1, marked the start of a partnership between Martin and Captain Valentine Henry Baker, the company's chief test pilot. During the next ten years to 1944, Martin designed three fighter aircraft that performed well and could be produced cheaply, but he received no orders.

Before World War II, Martin designed a barrage-balloon cable cutter, which was employed by most RAF Bomber Command aircraft.

The ejection seat was invented during World War II to improve the Spitfire pilot's chances of escape by parachute. Martin-Baker seats were fitted in new British military jet aircraft from 1947. Martin continued to develop the ejection seat for use at higher speeds, greater altitudes, vertical takeoff, multiple crew escape, and underwater ejection.

Maskelyne Nevil 1732–1811. English astronomer who made observations to investigate the reliability of the lunar distance method for determining longitude at sea. In 1774 he estimated the mass of the Earth by noting the deflection of a plumb line near Mount Schiehallion in Scotland.

Maskelyne, the fifth Astronomer Royal 1765–1811, began publication 1766 of the *Nautical Almanac*. This contained astronomical tables and navigational aids, and was probably his most enduring contribution to astronomy.

Maskelyne was born in London and studied at Cambridge. He was ordained a cleric but began working at the Greenwich Observatory in London. His first big project in observational astronomy was an expedition to St Helena, under the auspices of the Royal Society, in order to study the solar parallax during the 1761 transit of Venus. At the appropriate moment, however, the weather turned bad and, in any case, he had lost confidence in the instruments he had brought with him. It was not until 1772 that Maskelyne perfected his technique for observing transits, by which time the 1769 transit of Venus had already occurred (and the next would not take place in his lifetime).

Mästlin Michael 1550–1631. German astronomer and mathematician who was one of the first scholars to accept and teach Polish astronomer Copernicus's observation that the Earth orbits the Sun. One of Mästlin's pupils was German mathematician Johannes Kepler.

Mästlin was born in Göppingen, Baden-Württemberg, and studied at Tübingen. In 1580 he became professor of mathematics at Heidelberg and in 1584 at Tübingen, where he taught for 47 years.

In 1573, Mästlin published an essay concerning the nova that had appeared the previous year. Its location in relation to known stars convinced him that the nova was a new star – which implied, contrary to traditional belief, that things could come into being in the spheres beyond the Moon.

Observation of the comets of 1577 and 1580 convinced Mästlin that they were also located beyond the Moon. Together with other observations, this led him explicitly to argue against the traditional cosmology of Aristotle.

However, Mästlin's *Epitome of Astronomy* 1582, a popular introduction to the subject, propounded a traditional cosmology because this was easier to teach.

Masursky Harold 1923– . US geologist who has conducted research into the surface of the Moon and the planets. Working for the US space agency NASA from 1962, he participated in the Mariner, Apollo, Viking, Pioneer, and Voyager programmes.

Masursky was born in Fort Wayne, Indiana, and studied at Yale.

After 11 years with the US Geological Survey, he transferred to the branch for astrogeological studies and began work at the National Aeronautics and Space Administration (NASA).

Masursky was responsible for the surveying of lunar and planetary surfaces, particularly in regard to the choice of landing sites. He was a member of the working groups that monitored and guided the Moon landing 1969 and analysed the data afterwards. He led the team that monitored observations of Mars made by *Mariner 9* 1971, then selected landing sites on Mars for the Viking probes 1975.

In 1978, Masursky joined the Venus Orbiter Imaging Radar Science Working Group. The surface of Venus, hidden from visual or televisual observation by its thick layer of cloud, was mapped on the basis of radar readings taken from the orbiting Pioneer probe.

Mauchly John William 1907–1980. US physicist and engineer who, in 1946, constructed the first general-purpose computer, the ENIAC, in collaboration with John Eckert. Their company was bought by Remington Rand 1950, and they built the UNIVAC 1 computer 1951 for the US census.

The work on ENIAC was carried out by the two during World War II, and was commissioned to automate the calculation of artillery firing tables for the US Army. In 1949 Mauchly and Eckert designed a small-scale binary computer, BINAC, which was faster and cheaper to use. Punched cards were replaced with magnetic tape, and the computer stored programs internally.

Mauchly was born in Cincinnati, Ohio, and studied at Johns Hopkins University, becoming professor of physics at Ursinus College in Collegeville, Pennsylvania. In 1941 he moved to the Moore School of Electrical Engineering of the University of Pennsylvania, and became principal consultant on the ENIAC project. A dispute over patent policy with the Moore School caused Mauchly and Eckert to leave and set up a partnership 1948. Mauchly was a consultant to Remington Rand (later Sperry Rand) 1950–59 and again from 1973, after setting up his own consulting company 1959.

Maudslay Henry 1771–1831. English engineer and toolmaker who improved the metalworking lathe so that it could be employed for precise screw cutting. He also designed a bench micrometer, the forerunner of the modern instrument.

Maudslay was born in London and went to work at the Woolwich Arsenal at the age of 12, apprenticed to the metalworking shop. For a time he worked for inventor Joseph Bramah, and then started his own business.

Maudslay's new screw-cutting lathe gave such precision as to allow previously unknown interchangeability of nuts and bolts and standardization of screw threads. He was also able to produce sets of taps and dies. In 1801–08, in conjunction with engineer Marc Brunel, he constructed machines for making wooden pulley blocks at Portsmouth dockyard.

Maudslay's firm went on to produce marine steam engines.

Maury Antonia Caetana de Paiva Pereira 1866–1952. US expert in stellar spectroscopy who specialized in the detection of binary stars. She also formulated a classification system to categorize the appearance of spectral lines.

Maury was born in Cold Spring-on-Hudson, New York, and educated at Vassar. She became an assistant at the Harvard College Observatory, under the direction of Edward Pickering, even before she graduated in 1887. Later she lectured in astronomy in various US cities, and taught

Maury *US naval officer and oceanographer Matthew Maury. He wrote* Navigation after a Voyage Around the World *1830, and* The Physical Geography of the Sea *1855.*

privately. In 1908 she returned to Harvard to study the complex spectrum of Beta Lyrae.

Having assisted Pickering in establishing that the star Mizar (Zeta Ursae Majoris) is a binary star, with two distinct spectra, Maury was the first to calculate the 104-day period of this star. In 1889, she discovered a second such star, Beta Aurigae, and established that it has a period of only four days.

Studying the spectra of bright stars, Maury found three major divisions among spectra, depending upon the width and distinctness of the spectral lines. In 1896 she published her new classification scheme for spectral lines, based on the examination of nearly 5,000 photographs and covering nearly 700 bright stars in the northern sky.

Maury Matthew Fontaine 1806–1873. US naval officer, founder of the US Naval Oceanographic Office. His system of recording oceanographic data is still used today.

Maxim Hiram Stevens 1840–1916. US-born British inventor of the first fully automatic machine gun, in 1884. Its efficiency was further improved by Maxim's development of a cordlike propellant explosive, cordite. Knighted 1901.

Maxim was born in Sangerville, Maine, and spent his early life in various apprenticeships. While working as chief engineer for the US Electric Lighting Company, he came up with a way of manufacturing carbon-coated filaments for the early light bulbs that ensured that each filament was evenly coated. Deciding to concentrate on arms manufacture, he settled in Britain and set up a small laboratory in London. The company created to produce the Maxim gun soon became absorbed into Vickers Limited. By 1889, the British Army had adopted the gun for use.

The Maxim gun of 1884 used the recoil from the shots to extract, eject, load, and fire cartridges. With a water-cooled barrel, it could fire ten rounds per second.

Maxim's first patent, when he was 26, was for a curling iron. Later inventions ranged from mousetraps to gas-powered engines. He was particularly interested in powered flight and the relative efficiencies of aerofoils and airscrews. By 1894 he had produced a steam-driven aircraft launched on rails. After three trials the aircraft succeeded in leaving the tracks, but it was clear that the amount of water needed to generate power made it too heavy.

Maxwell James Clerk 1831–1879. Scottish physicist. His main achievement was in the understanding of electromagnetic waves: ***Maxwell's equations*** bring together electricity, magnetism, and light in one set of relations. He studied gases, optics, and the sensation of colour, and his theoretical work in magnetism prepared the way for wireless telegraphy and telephony.

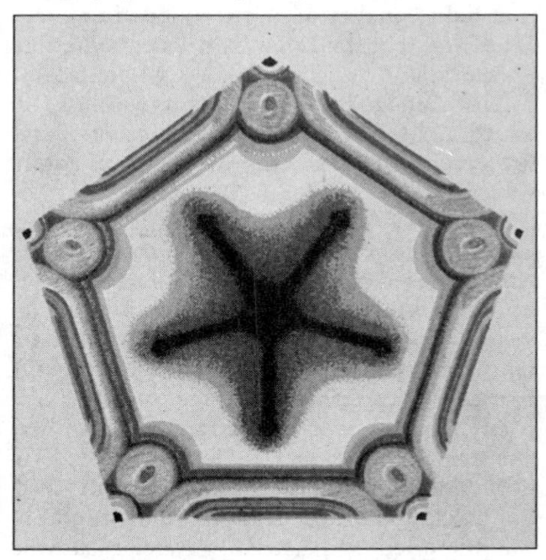

Maxwell *James Clerk Maxwell's painting of strain in a pentagon of unannealed glass visualized following the polarization of light by refraction. Maxwell originally proposed the existence of many forms of electromagnetic radiation*

In developing the kinetic theory of gases, Maxwell gave the final proof that heat resides in the motion of molecules.

Studying colour vision, Maxwell explained how all colours could be built up from mixtures of the primary colours red, green, and blue. Maxwell confirmed English physicist Thomas Young's theory that the eye has three kinds of receptors sensitive to the primary colours, and showed that colour blindness is due to defects in the receptors. In 1861 he produced the first colour photograph to use a three-colour process.

Maxwell was born in Edinburgh and educated there and at Cambridge. He was professor of natural philosophy at Aberdeen 1856–60, and of natural philosophy and astronomy at London 1860–65. From 1871 he was professor of experimental physics at Cambridge, where he set up the Cavendish Laboratory 1874.

The only laws of matter are those which our minds must fabricate, and the only laws of mind are fabricated by matter.

James Maxwell

Attributed

Maxwell studied Saturn's rings 1856–59 and established that they must be composed of many small bodies in orbit.

Having explained all known effects of electromagnetism, Maxwell went on to infer that light consists of electromagnetic waves. He also established that light has a radiation pressure, and suggested that a whole family of electromagnetic radiations must exist, of which light was only one. (This was confirmed in 1888 with the discovery of radio waves.)

Maxwell's mathematical basis for the kinetic theory of gases dates from 1860, when he used a statistical treatment to arrive at a formula to express the distribution of velocity in gas molecules, and related it to temperature. In 1865, Maxwell and his wife carried out experiments to measure the viscosity of gases over a wide range of pressure and temperature, and their findings led to new equations. Maxwell's kinetic theory still did not fully explain heat conduction, and it was modified by Austrian physicist Ludwig Boltzmann in 1868, resulting in the Maxwell–Boltzmann distribution law. Both men thereafter contributed to successive refinements of the kinetic theory and it proved fully applicable to all properties of gases.

Maxwell's works include *Perception of Colour* 1860, *Theory of Heat* 1871, *Treatise on Electricity and Magnetism* 1873, and *Matter and Motion* 1876.

Maybach Wilhelm 1846–1929. German engineer and inventor who worked with Gottlieb Daimler on the development of early motorcars.

Maybach invented the float-feed carburettor, which allowed petrol to be used as a fuel for internal-combustion engines, most of which up to that time had been fuelled by gas.

Maybach and Daimler went into partnership 1882 in Stuttgart, where they produced one of the first petrol engines. In 1895, Maybach became technical director of the Daimler Motor Company. He left Daimler in 1907 to set up his own factory for making engines for Zeppelin's airships.

In 1901, Maybach designed the first Mercedes car. He invented the spray-nozzle or float-feed carburettor in 1893. It made the fuel enter through a jet as a fine spray. The vaporized fuel mixed with air to produce a combustible mixture for the engine's cylinders.

Maybach's other inventions include the honeycomb radiator, an internal expanding brake 1901, and an axle-locating system for use with independent suspensions.

Mayer Christian 1719–1783. Austrian astronomer, mathematician, and physicist. He was the first to investigate and catalogue double stars, though his equipment was unable to distinguish true binary stars (in orbit round each other) from separate stars seen together only by the coincidence of Earth's viewpoint.

Mayer was born in Moravia. He became a Jesuit priest and in 1752 professor of mathematics and physics at Heidelberg. When the elector palatine Karl Theodor built an observatory at Schwetzingen, and then a larger one at Mannheim, Mayer was appointed court astronomer. He lost this post with the pope's dissolution of the Jesuit order in 1773, although he managed to continue his astronomical studies.

Mayer measured the degree of the meridian, based on work conducted in Paris and in the Rhineland Palatinate, and observed the transits of Venus in 1761 and 1769. The latter observation was conducted in Russia at the invitation of Catherine II.

Mayer Johann Tobias 1723–1762. German cartographer, astronomer, and physicist who improved standards of observation and navigation. He produced a map of the Moon's surface and concluded that it had no atmosphere.

Mayer was born in Marbach, near Stuttgart. He learned architectural drawing and surveying and taught himself mathematics, French, Italian, and English. He published his first book, on the application of analytical methods to the solution of geometrical problems, at the age of 18. In 1746 he began work for the Homann Cartographic Bureau in Nuremberg, and he ended his career as professor at the Georg August Academy in Göttingen.

At the Homann Cartographic Bureau, Mayer drew up some 30 maps of Germany. These established exacting new standards for using geographical data in conjunction with astronomical details to determine latitudes and longitudes on Earth. To obtain some of the astronomical details, he observed lunar oscillations and eclipses using a telescope of his own design.

Mayer's Lunar Tables 1753 were correct to one minute of arc.

Mayer also invented a simple and accurate method for calculating solar eclipses, compiled a catalogue of zodiacal stars, and studied stellar proper motion.

Mayer Julius Robert von 1814–1878. German physicist who in 1842 anticipated James Joule in deriving the mechanical equivalent of heat, and Hermann von Helmholtz in the principle of conservation of energy.

In 1845, Mayer extended the principle to show that living things are powered solely by physical processes utilizing solar energy and not by any kind of innate vital force. He described the energy conversions that take place in living organisms, realizing that plants convert the Sun's energy into food that is consumed by animals to provide a source of energy to power their muscles and provide body heat.

Mayer was born in Heilbronn and studied at Tübingen. In 1840, he took a position as a ship's physician and sailed to the East Indies for a year. He then settled in his native city and built up a medical practice. Despairing because others were making the same discoveries independently and gaining priority, Mayer tried to kill himself in 1850, but achieved recognition before his death.

During his 1840 voyage, Mayer found that the venous blood of European sailors was much redder in the tropics than at home. He put this down to a greater concentration of oxygen in the blood caused by the body using less oxygen, since less heat was required in the tropics to keep the body warm. From this, Mayer made a conceptual leap to the idea that work such as muscular force, heat such as body heat, and other forms of energy such as chemical energy produced by the oxidation of food in the body are all interconvertible. The amount of work or heat produced by the body must be balanced by the oxidation of a certain amount of food, and therefore work or energy is not created but only transformed from one form to another.

In 1842, Mayer stated the equivalence of heat and work more definitely. He published a theoretical attempt to determine the mechanical equivalent of heat from the heat produced when air is compressed. His result was inaccurate but the principle of the conservation of energy was demonstrated for the first time.

Maynard Smith John 1920– . British biologist. He applied game theory to animal behaviour and developed the concept of the evolutionary stable strategy as a mathematical technique for studying the evolution of behaviour.

His books include *The Theory of Evolution* 1958 and *Evolution and the Theory of Games* 1982.

In recent years there have been claims – in the daily press, on television, and by retired cosmologists – that Darwin may have been wrong. ... However, to see Darwinism as being under serious threat would, I think, be a false perception.

John Maynard Smith

Evolution Now

Mayo William James 1861–1939. US surgeon, founder, with his brother Charles Horace Mayo (1865–1939), of the ***Mayo Clinic*** for medical treatment 1889 in Rochester, Minnesota.

Mayr Ernst Walter 1904– . German-born US zoologist who was influential in the development of modern evolutionary theories. He led a two-year expedition to New Guinea and the Solomon Islands where he studied the effects of founder populations and speciation on the evolution of the indigenous birds and animals. This research caused him to support neo-Darwinism a synthesis of the ideas of Darwin and Mendel, being developed at that time.

Mayr was born in Kempten, Germany. He studyied medicine at Greifswald University 1923–25 and graduated with a PhD from the University of Berlin 1926, where he continued to work until 1930. He carried out taxonomic research at the American Museum of Natural History in New York 1931–53 where he was responsible for avian taxonomy. He published more than 100 papers in this area, including *Birds of the Southwest Pacific* 1945. In 1937 he contributed to Dobzhansky's *Genetics and the Origin of Species*, which outlined the neo-Darwininist synthesis of evolution and Mendelian genetics and was crucial in the widespread acceptance of the theory of evolution.

In 1950 he proposed an alternative classification for fossils, which included the hominid fossils and was widely accepted. He was appointed Alexander Agassiz professor of zoology at Harvard University 1953, and from 1961 until his retirement 1970, held the post of director of the Museum of Comparative Zoology. He has written and edited a number of books, including several upon the development of evolutionary thought, which are standard texts on university courses to this day.

Mechnikov Ilya Ilich 1845–1916. Russian-born French zoologist who discovered the function of white blood cells and phagocytes (amoebalike blood cells that engulf foreign bodies). He also described how these 'scavenger cells' can attack the body itself (autoimmune disease). He shared the Nobel Prize for Physiology or Medicine 1908.

Mechnikov was born in Ivanovka, near Kharkov, and studied at Kharkov University and afterwards in Germany. He was professor of zoology and anatomy at Odessa 1867–82, when he moved to Messina in Italy to continue his research. He briefly returned to Odessa, but left Russia 1888 to join the Pasteur Institute in Paris, becoming director on Louis Pasteur's death 1895.

While studying the transparent larvae of starfish, Mechnikov observed that certain cells surrounded and engulfed foreign particles that entered the bodies of the larvae. Later he demonstrated that phagocytes exist in higher animals, and form the first line of defence against acute infections.

Mechnikov spent the last decade of his life trying to demonstrate that lactic acid-producing bacteria in the intestine increase a person's lifespan.

Medawar Peter Brian 1915–1987. British immunologist who, with Macfarlane Burnet, discovered that the body's resistance to grafted tissue is undeveloped in the newborn child, and studied the way it is acquired. They shared a Nobel prize 1960. Knighted 1965.

Medawar's work has been vital in understanding the phenomenon of tissue rejection following transplantation.

Medawar was born in Rio de Janeiro and studied in the UK at Oxford. He was professor of zoology at Birmingham University 1947–51 and at University College, London, 1951–62; he was director of the National Institute for Medical Research 1962–75. In 1977 he was appointed professor of experimental medicine at the Royal Institution.

The human mind treats a new idea in the same way the body treats a strange protein: it rejects it.

Peter Medawar

Acting on Burnet's hypothesis that an animal's ability to produce a specific antibody is not inherited, Medawar inoculated mouse embryos of one strain with cells from mice of another strain, and found that the embryos did not produce antibodies against the cells of the other strain.

Medawar wrote essays for the general reader, collected in, for example, *The Hope of Progress* 1972, *The Art of the Soluble* 1967, and *The Limits of Science* 1985.

Mee Margaret Ursula 1909–1988. English botanical artist. In the 1950s she went to Brazil, where she accurately and comprehensively painted many plant species of the Amazon basin.

She is thought to have painted more species than any other botanical artist.

Meitner Lise 1878–1968. Austrian-born Swedish physicist who worked with German radiochemist Otto Hahn and was the first to realize that they had inadvertently achieved the fission of uranium. They also discovered protactinium 1918. She refused to work on the atom bomb.

Life need not be easy, provided only that it is not empty.

Lise Meitner

Meitner was born in Vienna and studied there and at Berlin. She joined Hahn to work with him on radioactivity at the Kaiser Wilhelm Institute for Chemistry, but was not given permission to work in the laboratory by their supervisor Emil Fischer, because she was a woman. Instead they had to set up a small laboratory in a carpenter's workroom. Nonetheless, Meitner was made joint director of the institute with Hahn 1917 and was also appointed head of the Physics Department. In 1912, Meitner had also become an assistant to Max Planck at the Berlin Institute of Theoretical Physics, and she was made professor at Berlin 1926. But in 1938 she was forced to leave Nazi Germany, and soon found a post at the Nobel Physical Institute in Stockholm. In 1947, a laboratory was established for her by the Swedish Atomic Energy Commission, and she later worked on an experimental nuclear reactor.

During the 1920s, Meitner studied the relationship between beta and gamma irradiation. She was the first to describe the emission of Auger electrons, which occurs when an electron rather than a photon is emitted after one electron drops from a higher to a lower electron shell in the atom.

In 1934 Meitner began to study the effects of neutron bombardment on uranium with Hahn. It was not found until after Meitner had fled from Germany that the neutron bombardment had produced not transuranic elements, as they expected, but three isotopes of barium. Meitner and her nephew Otto Frisch realized that the uranium nucleus had been split; they called it fission. A paper describing their analysis appeared 1939.

Meitner continued to study the nature of fission products. Her later research concerned the production of new radioactive species using the cyclotron, and also the development of the shell model of the nucleus.

Mellanby Edward 1884–1955. English pharmacologist who discovered that rickets (disease characterized by the softening of the bones) is the result of a vitamin deficiency.

In 1918, Mellanby produced puppies with rickets by feeding them a diet which was deficient in a compound found in animal fats. While it had been commonly held until 1900 that rickets was a chronic infectious disease, others had proposed before Mellanby that it was caused by poor living conditions (lack of fresh air, sunlight, and exercise).

Mellanby demonstrated conclusively by his experiments that rickets was the result of a vitamin deficiency and could be cured with cod-liver oil. In 1919, Kurt Huldschinsky in Berlin, demonstrated that rickets could also be cured by artificial sunlight.

Mellanby was born in Hartlepool and educated at Emmanuel College, Cambridge. He studied medicine at St Thomas's Hospital in London before taking the chair of physiology at King's College for Women (now Queen Elizabeth College), London 1913–20. He then moved to the chair of pharmacology at the University of Sheffield.

Mellanby Kenneth 1908– . British ecologist and entomologist who in the 1960s drew attention to the environmental effects of pollution, particularly by pesticides. He advocated the use of biological control methods, such as introducing animals that feed on pests.

Mellanby was born in West Hartlepool, County Durham, and studied at London University. On the staff of the London School of Hygiene and Tropical Medicine 1930–45, he went to E Africa to study the tsetse fly; while doing his World War II military service, he investigated scrub typhus in Burma (now Myanmar) and New Guinea. In 1947 he became the principal of University College, Ibadan, Nigeria's first university, which he played a part in creating. He was head of the Entomology Department at Rothamsted Experimental Station 1955–61, when he founded and became director of the Monks Wood Research Station (now called the Institute of Terrestrial Ecology) in Huntingdon, Cambridgeshire.

Mellanby Lady May, (born Tweedy) 1882–1978. British nutritional scientist who discovered that the onset of caries, a form of tooth decay, is determined partly by the deficiency of certain substances in the diet, particularly salts of calcium and phosphorus. She also determined that the onset of caries is dependent on the structure of the teeth, which is determined largely by the diet during development.

Mellanby grew up in pre-Soviet Russia and was educated at Girton College, Cambridge where she was awarded the equivalent of a degree 1906 (women were not allowed to formerly graduate from Cambridge at that time).

Mendel Gregor Johann 1822–1884. Austrian biologist, founder of genetics. His experiments with successive generations of peas gave the basis for his theory of particulate inheritance rather than blending, involving dominant and recessive characters. His results, published 1865–69, remained unrecognized until the early 20th century.

From his findings Mendel formulated his law of segregation and law of independent assortment of characters, which are now recognized as two of the fundamental laws of heredity.

Mendel was born in Heinzendorf (now Hyncice in the Czech Republic), and entered the Augustinian monastery in Brünn, Moravia (now Brno, Czech Republic) 1843. Later he studied at Vienna. In 1868 he became abbot of the monastery.

Much of his work was performed on the edible pea *Pisum*, which he grew in the monastery garden. He carefully self-pollinated and wrapped (to prevent accidental pollination by insects) each individual plant, collected the seeds produced by the plants, and studied the offspring of these seeds. Seeing that some plants bred true and others not, he worked out the pattern of inheritance of various traits.

Mendel *Austrian monk and botanist Gregor Johann Mendel, who, by characteristically breeding pea plants in a monastery garden, discovered the basic laws of heredity. Mendel published his findings in 1865, but the importance of his results were not recognized at the time. It was not until 1900, sixteen years after his death, that his work was rediscovered and became the basis of modern genetics.*

Mendeleyev Dmitri Ivanovich 1834–1907. Russian chemist who framed the periodic law in chemistry 1869, which states that the chemical properties of the elements depend on their relative atomic masses. This law is the basis of the periodic table of the elements, in which the elements are arranged by atomic number and organized by their related groups.

There will come a time, when the world will be filled with one science, one truth, one industry, one brotherhood, one friendship with nature ... this is my belief, it progresses, it grows stronger, this is worth living for, this is worth waiting for.

Dmitri Mendeleyev

In Yu A Urmantsev *The Symmetry of Nature and the Nature of Symmetry* 1974 p 49

Mendeleyev was the first chemist to understand that all elements are related members of a single ordered system. From his table he predicted the properties of elements then unknown, of which three (gallium, scandium, and germanium) were discovered in his lifetime. Meanwhile Lothar Meyer in Germany presented a similar but independent classification of the elements.

Mendeleyev was born in Tobol'sk, Siberia, and studied at St Petersburg and in Germany at Heidelberg. He became professor at the Technical Institute in St Petersburg 1864. But in 1890, for supporting a student rebellion, he was retired from the university and became controller of the Bureau for Weights and Measures.

Mendeleyev was convinced that the future held great possibilities for human flight, and in 1887 he made an ascent in a balloon to observe an eclipse of the Sun.

His textbook *Principles of Chemistry* 1868–70 was widely adopted.

Menzel Donald Howard 1901–1976. US physicist and astronomer whose work on the spectrum of the solar chromosphere revolutionized much of solar astronomy. He was one of the first scientists to combine

			Ti = 50	Zr = 90	? = 180
			V = 51	Nb = 94	Ta = 182
			Cr = 52	Mo = 96	W = 186
			Mn = 55	Rh = 104.4	Pt = 197.4
			Fe = 56	Ru = 104.4	Ir = 198
			Ni = Co	Pd = 106.6	Os = 198
			= 50		
H = 1			Cu = 63.4	Ag = 108	Hg = 200
	Be = 9.4	Mg = 24	Zn = 65.2	Cd = 112	
	B = 11	Al = 27.4	? = 68	Ur = 116	Au = 200
	C = 12	Si = 28	? = 70	Sn = 118	
	N = 14	P = 31	As = 75	Sb = 122	Bi = 197?
	O = 16	S = 32	Se = 79.4	Te = 128?	
	F = 19	Cl = 35.5	Br = 80	I = 127	
Li = 7	Na = 23	K = 39	Rb = 85.4	Cs = 133	Te = 204
		Ca = 40	Sr = 87.6	Ba = 137	Pb = 207
		? = 45	Ce = 92		
		?Er = 56	La = 94?		
		?Yt = 60	Di = 95		
		?In = 75.6	Th = 118?		

Mendeleyev *Mendeleyev's original (1869) periodic table, with similar elements on the same horizontal line.*

astronomy with atomic physics.

Menzel was born in Florence, Colorado, and studied at the University of Denver. At Princeton University as a graduate assistant, he learned astrophysics from Henry Russell, the co-originator of the Hertzsprung–Russell diagram. In 1932, Menzel joined Harvard University Observatory, where he was to become director some 30 years later. The coronagraph he constructed there was the beginning of High Altitude Observatory for solar physics research. During his career Menzel took part in the setting-up or development of several observatories in the USA. He retired from Harvard 1971 to become scientific director of a company manufacturing antennae for communications and radio astronomy.

In addition to this work, Menzel devised a technique for computing the temperature of planets from measurements of water cell transmissions and he made important contributions to atmospheric geophysics, radio propagation, and even lunar nomenclature. He also held patents on the use of gallium in liquid ball bearings and on heat transfer in atomic plants.

Mercator Gerardus 1512–1594. Latinized form of Gerhard Kremer. Flemish mapmaker who devised the first modern atlas, showing ***Mercator's projection*** in which the parallels and meridians on maps are drawn uniformly at 90°. It is often used for navigational charts, because compass courses can be drawn as straight lines, but the true area of countries is increasingly distorted the further north or south they are from the equator.

Mergenthaler Ottmar 1854–1899. German-born US inventor of the Linotype machine 1884–86. Casting hot-metal type in complete lines, this greatly speeded typesetting and revolutionized printing and publishing.

The machine was rather like a large typewriter, with a store of letter matrices (moulds) at the top. The keyboard operator caused the matrices to drop into position in a line. As each line was completed, a cast was made to form a metal 'slug' with the letters in relief on one side. The slug fitted into a page of type for printing, while the matrices were returned for reuse. A person operating one of Mergenthaler's keyboards could set type up to three or four times faster than by hand. Books became cheaper and newspapers more up to date, for the Linotype made it possible to change and reset copy within minutes of going to press.

Mergenthaler was born in Württemberg and apprenticed to a watchmaker. He emigrated to the USA in 1872 and settled in Baltimore 1876, working for James O Clephane. They tried to develop a writing machine but abandoned the idea and in 1884 produced a prototype of the Linotype machine. By 1886, the first machines were in use and a company was formed for their production. Over the years, Mergenthaler contributed as many as 50 modifications to the original design.

Merriam Clinton Hart 1885–1942. US ornithologist and physician who directed the Bureau of Biological Survey and helped to found the National Geographic Society.

Merriam was born in New York, and even as a child collected birds. When he was 16 years old he participated in Hayden's geological and geographical survey of the territories to collect bird skins and eggs. He studied medicine at Yale, graduating 1879. He returned to Locust Grove, where he had grown up, to practise medicine.

He gave up medicine 1905 to become an ornithologist at the Bureau of Biological Survey. He continued to direct the Bureau until 1910 when he received funding from the Harriman fund to research Californian Indians. By this time, he had published many papers and built up a large collection of mammals at the Biological Survey. He was extremely interested in conservation and was one of the founders of the National Geographic Society.

Merrifield R Bruce 1921– . US chemist who was awarded the 1984 Nobel Prize for Chemistry for his development of a method for synthesizing large organic molecules using a solid support or matrix. Merrifield and his co-workers later reached another milestone when they synthesized the enzyme ribonuclease, which contains 124 amino acids.

Merrifield worked on the synthesis of peptides, important biological molecules consisting of chains of amino acids. He first attached an amino acid to an insoluble polymer, a plastic material in the form of small spheres. Then other amino acids were added one after another, until the peptide chain was built up. The chain was then released from the polymer. Thousands of different peptides of various sizes, as well as proteins, hormones, and other organic molecules have now been synthesized using this method.

Merrifield was born in Fort Worth, Texas and educated at the University of California at Los Angeles, gaining his doctorate 1949. In the same year he joined the Rockefeller Institute for Medical Research (now Rockefeller University), New York City, and worked on the synthesis of peptides, which are important biological molecules consisting of chains of amino acids. In 1984, he was appointed John D Rockefeller Jr professor of the Rockefeller University.

Mersenne Marin 1588–1648. French mathematician and philosopher who, from his base in Paris, did much to disseminate throughout Europe the main advances of French science. In mathematics he defined a particular form of prime number, since referred to as a ***Mersenne prime***.

Meselson Matthew Stanley 1930– . US molecular biologist who, with Franklin Stahl, confirmed that replication of the genetic material DNA is semiconservative (that is, the daughter cells each receive one strand of DNA from the original parent cell and one newly replicated strand).

Meselson was born in Denver, Colorado, and studied physical chemistry at the California Institute of Technology (Caltech). He remained at Caltech, rising to professor of biology, until 1976, when he moved to Harvard. In 1963 he became a consultant to the US Arms Control and Disarmament Agency.

Meselson and Stahl used the bacterium *Escherichia coli* for their work on DNA. They grew the bacteria in a culture medium containing nitrogen-15 (a heavy isotope of nitrogen), and then transferred the bacteria to a medium containing nitrogen-14 (the normal nitrogen isotope). Later, when they extracted the DNA, they obtained three different types: one containing only nitrogen-14, one containing only nitrogen-15, and a hybrid containing both nitrogen isotopes. On heating, the hybrid separated into two halves, one from the parental DNA and one that had been newly synthesized. These findings demonstrated that the double helix of DNA splits into two strands when the DNA replicates, with each of the single strands acting as a template for the synthesis of a complementary strand.

Meselson has also investigated the molecular biology of nucleic acids, the mechanisms of DNA recombination and repair, and the processes of gene control and evolution.

Mesmer Friedrich Anton (or Franz) 1734–1815. Austrian physician, an early experimenter in hypnosis, which was formerly (and popularly) called 'mesmerism' after him.

He claimed to reduce people to trance state by consciously exerted 'animal magnetism', their willpower being entirely subordinated to his. Expelled by the police from Vienna, he created a sensation in Paris in 1778, but was denounced as a charlatan in 1785.

Messerschmitt Willy (Wilhelm Emil) 1898–1978. German aeroplane designer whose Me-109 was a standard Luftwaffe fighter in World War II, and whose Me-262 (1944) was the first mass-produced jet fighter.

Messerschmitt aeroplanes were characterized by simple concept, minimum weight and aerodynamic drag, and the possibility of continued development. He designed cantilever monoplanes from the early 1920s when the market still looked for biplanes with visible struts and bracing wires.

The Me-109 held the world speed record of 610 kph/379 mph from 1937.

Messerschmitt was born in Frankfurt-am-Main and studied at the Technische Hochschule in Munich. A glider he designed with gliding pioneer Friedrich Harth achieved an unofficial world duration record in 1921. The following year they set up a flying school and in 1923, while still a student, Messerschmitt formed his own company in Bamberg. Its first product was the S-14 cantilever monoplane glider. He produced his first powered aircraft in 1925, the ultralight sports two-seater Me-17. In 1926 came the Me-18 small transport. The Me-37 1934 was the archetypal low-wing four-seater cabin monoplane, with retractable landing gear and flaps.

By 1938, Messerschmitt was appointed chair and general director of the company manufacturing his designs, Bayerische Flugzeugwerke, renamed Messerschmitt Aktiengesellschaft. He and his company went on to produce numerous designs for fighter, bomber, and transport aircraft. From the end of World War II Messerschmitt was held prisoner for two years by the Allies, and was then banned from manufacturing aircraft in Germany. Instead, he designed and produced a two-seater bubble car.

Messerschmitt took up aircraft design again in 1952 under contract with Spanish manufacturer Hispano. Between 1956 and 1964 he worked in association with the German Bolkow and Heinkel companies and developed the VJ-101 supersonic V/STOL (vertical/short takeoff and landing) combat aircraft. His company merged with Bolkow in 1963 and they were later joined by Hamburger Flugzeugbau. In 1969, Messerschmitt became chair of the resulting Messerschmitt Bolkow Blohm (MBB) group.

Messier Charles 1730–1817. French astronomer who discovered 15 comets and in 1784 published a list of 103 star clusters and nebulae. Objects on this list are given M (for Messier) numbers, which astronomers still use today, such as M1 (the Crab nebula) and M31 (the Andromeda galaxy).

Messier was born in Badonviller, Lorraine, and joined the Paris Observatory. Watching for the predicted return of Halley's comet, he was one of the first people to spot it, an experience that inspired him with the desire to go on discovering new comets for the rest of his life. Louis XV nicknamed him the 'Comet Ferret'.

But Messier's search was continually hampered by rather obscure forms which he came to recognize as nebulae. During the period 1760–84, therefore, he compiled a list of these nebulae and star clusters, so that he and other astronomers would not confuse them with possible new comets.

Metchnikoff Elie 1845–1916. Russian immunologist who was a pioneer of cellular immunology and shared the Nobel Prize for Physiology or Medicine 1908 with Paul Ehrlich for his discovery of the innate immune response.

While at a teaching post in Italy, he started working on the immune cells of the starfish and discovered ***innate immunity***; the process by which mobile white blood cells (phagocytes) engulf and digest potentially harmful microorganisms. Innate immunity is the inborn, non-specific defence against infection, since prior exposure to the pathogen is not required. Alternatively, ***adaptive*** or ***acquired immunity*** refers to the production of a specific antibody against a particular germ and is the basis of vaccination. Innate immunity is important to the fight against any infection, since it provides defence against pathogens during the three to five days it takes to elicit the acquired immune response.

Metchnikoff was born in the Ukraine and studied at the University of Karkov before completing his PhD at the University of Petersburg 1867. He taught zoology and anatomy at Odessa and St Petersburg and was professor of zoology and comparative anatomy at Odessa 1873–82 before accepting a teaching post in Italy, where he did most of his research on the immune system.

In 1886, he returned to Russia to head the Institute of Bacteriology in Odessa but was forced to leave a year later after a number of criticisms were made by senior clinicians of a non-clinician being appointed to the directorship of this new, prestigious Institute. He left for Paris upon the invitation of Louis Pasteur to start a new laboratory at the Pasteur Institute.

Meyer Alfred 1895–1990. German-born British neuropathologist. His most significant work was on the anatomical aspects of frontal leucotomy, and the nature of the structural abnormalities in the brain associated with temporal-lobe epilepsy.

Meyer was born in Krefeld. His successful early career was threatened by the rise of the Nazis, and he escaped to the UK in 1933, beginning work at the Maudsley Hospital, London. His study of frontal leucotomy led to a classic book on the subject with Elizabeth Beck, published 1954; and he undertook pioneering work on the pathology of epilepsy while professor of neuropathology at the Institute of Psychiatry.

Meyer (Julius) Lothar 1830–1895. German chemist who, independently of his Russian contemporary Dmitri Mendeleyev, produced a periodic law describing the properties of the chemical elements.

Meyer was born in Varel, Lower Saxony, and studied at Zürich, Switzerland, and at several German universities. He was professor of chemistry at Karlsruhe Polytechnic 1868–76, and at Tübingen University for the rest of his life.

In his book *Die modernen Theorien der Chemie/Modern Chemical Theory* 1864, Meyer drew up a table presenting all the elements according to their atomic weights (relative atomic masses), relating the weights to chemical properties. In 1870 he published a graph of atomic volume (atomic weight divided by density) against atomic weight, which demonstrated the periodicity in the variation of the elements' properties. He showed that each element will not combine with the same numbers of hydrogen or chlorine atoms,

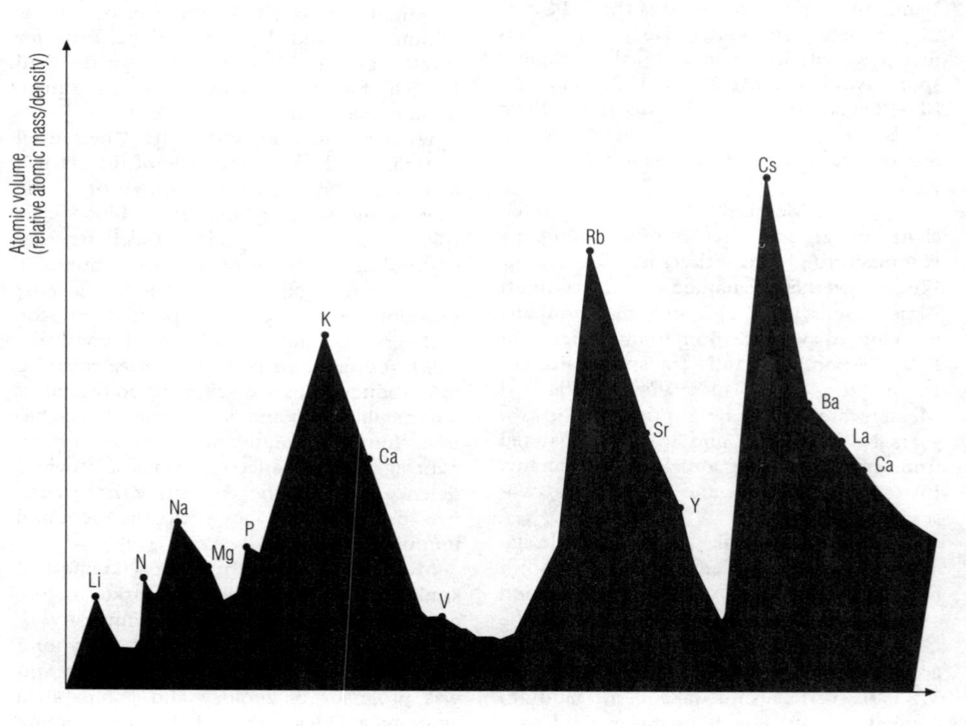

Meyer, Lothar *Part of the Lothar Meyer curve (with modern additions).*

establishing the concept of valency and grouping elements as univalent, bivalent, trivalent, and so on.

Meyer never claimed priority for his findings and, unlike Mendeleyev, he made no predictions about the composition and properties of any elements still to be discovered.

Meyer Viktor 1848–1897. German organic chemist who invented an apparatus for determining vapour densities (and hence molecular weights), now named after him. He was also the discoverer of the heterocyclic compound thiophene.

Meyer was born in Berlin and studied there and at Heidelberg and Württemberg. At the age of 22 he was appointed professor at Stuttgart Polytechnic, in 1885 at Göttingen, and in 1889 at Heidelberg.

In 1871, Meyer experimentally proved Avogadro's hypothesis by measuring the vapour densities of volatile substances (molecular weight, or relative molecular mass, is twice the vapour density). He went on to determine the vapour densities of inorganic substances at high temperatures.

From benzene obtained from petroleum, Meyer in 1883 isolated thiophene, a heterocyclic compound containing sulphur, which much later was to become an important component of various synthetic drugs.

Meyer published *Textbook of Organic Chemistry* 1883–96 and, with his brother Karl, *Pyrotechnical Research* 1885.

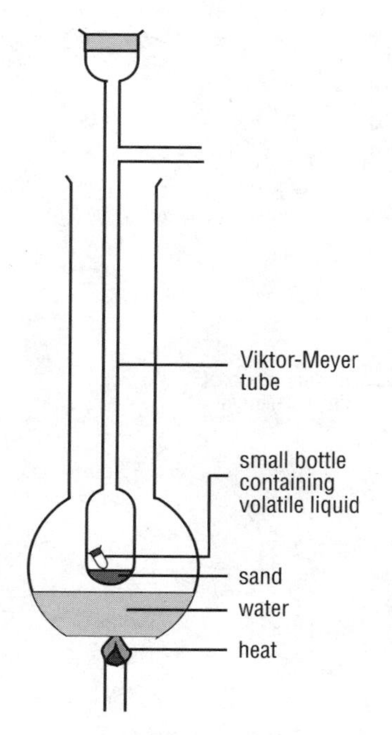

Meyer, Viktor *Viktor Meyer's apparatus for measuring vapour density.*

Meyerhof Otto 1884–1951. German-born US biochemist who carried out research into the metabolic processes involved in the action of muscles. For this work he shared the 1922 Nobel Prize for Physiology or Medicine.

Meyerhof was born in Hanover and studied at a number of German universities. From 1912 he worked at the University of Kiel, becoming professor 1918. He headed a department specially created for him at the Kaiser Wilhelm Institute for Biology in Berlin 1924–29, when he moved to Heidelberg. As a result of Adolf Hitler's rise to power in the 1930s, Meyerhof left Germany 1938 and went to Paris, where he became director of research at the Institut de Biologie Physiochimique. In 1940, when France fell to Germany in the early part of World War II, he fled to the USA, and was given a professorship at the University of Pennsylvania.

In 1920 Meyerhof showed that, in anaerobic conditions, the amounts of glycogen metabolized and of lactic acid produced in a contracting muscle are proportional to the tension in the muscle. He also demonstrated that 20–25% the lactic acid is oxidized during the muscle's recovery period and that energy produced by this oxidation is used to convert the remainder of the lactic acid back to glycogen. Meyerhof introduced the term glycolysis to describe the anaerobic degradation of glycogen to lactic acid, and showed the cyclic nature of energy transformations in living cells. The complete metabolic pathway of glycolysis is known as the ***Embden–Meyerhof pathway*** after Meyerhof and Gustav George Embden (1874–1933).

Michaelis Leonor 1875–1949. German-born, US biochemist who derived a mathematical model to describe the kinetics of how enzymes catalyse (trigger) reactions. The work of Michaelis and German scientist Maude Menton enabled several subsequent generations of biochemists to correctly assess the nature and efficiency of the key, enzyme-driven steps in cell metabolism.

Although Michaelis published many papers on the measurement of pH using various indicators, including early versions of the pH electrode, he is best known for his work on the physical properties and activities of enzymes. He worked for many years on pepsin, a digestive enzyme normally produced by the stomach. His joint work with Menton lead to the formulation of a general, mathematical model to describe the kinetics of how enzymes catalyse reactions, called the Michaelis–Menton equation, 1913.

In their mathematical calculations they correctly assumed that an enzyme works by rapidly and reversibly binding to a specific molecule (called the substrate) to form an enzyme–substrate complex, triggering a reaction that generates a product molecule. Once the reaction has occurred the enzyme is released unchanged. Michaelis and Menton then correlated the speed of the enzymatic reaction with the concentrations of both the enzyme and the substrate.

Michaelis was born in Berlin, where he later became a professor 1908. He took up a staff positions in Japan 1922, and then at Johns Hopkins University 1926–29 and Rockefeller University 1929–40 in the USA.

Michaux André 1746–1802. French botanist and explorer. As manager of a royal farm, he was sent by the French government 1782 and 1785 on expeditions to collect plants and select timber for shipbuilding. Together with his son François (1770–1852), he travelled to Carolina, Florida, Georgia, and, in 1792, to Hudson's Bay. On his return to France, he compiled the first guide to the flora of E America.

François Michaux wrote the first book on American forest trees 1810–13.

Michel Hartmut 1948– . German biochemist who shared the Nobel Prize for Chemistry in 1988 with Robert Huber and Johann Diesenhofer for his role in determining the molecular structure of photosynthetic reaction centres. Michel crystallized the protein from a reaction centre membrane of the bacterium *Rhodopseudomonas viridis*, allowing Huber to analyze the protein using the technique of X-ray crystallography to establish the protein's overall structure.

Photosynthesis occurs within reaction

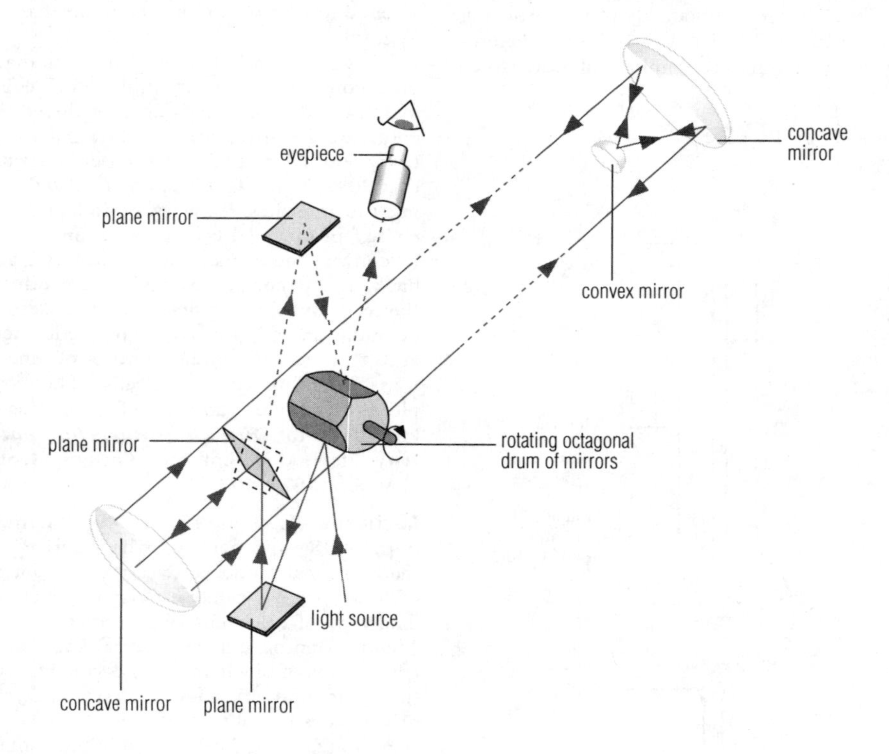

Michelson *Michelson determined the velocity of light by using a rotating drum faced with eight or more mirrors to time a light beam over a distance of more than 70 km/43 mi.*

centres embedded in membrane vesicles. Michel was convinced that membrane-bound proteins in reaction centres held the key to photosynthesis. Michel crystallized the protein, which was only partially water soluble, by using a molecule that was hydrophilic (water-loving) at one end and hydrophobic (water-hating) at the other. The hydrophobic ends of the protein membranes could be bound up, while the hydrophilic ends were free to bind with water molecules, so the protein became soluble.

Born in Ludwigsburg, W Germany, he was educated at the University of Warburg and the University of Munich. He worked at the Max Planck Institute of Biochemistry and was appointed director of the Max Planck Institute of Biophysics in 1987. In 1990, he published his findings in *Crystallization of Membrane Proteins*.

Michelson Albert Abraham 1852–1931. German-born US physicist. With his colleague Edward Morley, he performed in 1887 the ***Michelson–Morley experiment*** to detect the motion of the Earth through the postulated ether (a medium believed to be necessary for the propagation of light). The failure of the experiment indicated the nonexistence of the ether, and led Albert Einstein to his theory of relativity. Michelson was the first American to be awarded a Nobel prize, in 1907.

He invented the ***Michelson interferometer*** to detect any difference in the velocity of light in two directions at right angles. In the interferometer, a light beam is split into two by a semi-silvered mirror. The two beams are then reflected off fully silvered mirrors and recombined to form an interference pattern of dark and light bands.

The negative result of the Michelson–Morley experiment demonstrated that the velocity of light is constant whatever the motion of the observer. Michelson also made a precise measurement of the speed of light.

Michelson was born in Strelno (now Strzelno, Poland); his family emigrated to the USA and Michelson attended the US Naval Academy at Annapolis. Later he studied in Europe at Berlin and Paris. He was professor at the Case School of Applied Science in Cleveland, Ohio, 1882–89, and at Chicago 1892–1929.

Michelson first experimented with his interferometer in 1881. The presence of ether would have caused a change in the interference pattern, but he could detect none. In collaboration with Morley, he constructed a much more sensitive interferometer, but again found no change.

Michelson developed his interferometer into a precision instrument for measuring the diameters of heavenly bodies and in 1920 announced the size of the giant star Betelgeuse, the first star to be measured.

Michotte Albert 1881–1965. Belgian experimental psychologist. He is known for his investigations of perceptual causality. By means of ingenious and careful experimentation, he studied the dynamic organization of the perceptual world and was particularly concerned with the role of language in the analysis of perceptual phenomena. His book *La Perception de la causalité/The Perception of Causality* 1946 has become a classic.

In his experiments, subjects looking through a slit saw what appeared to them as two small rectangular spots in motion. Alternatively, they looked at a screen on which small moving shapes were projected. When one object A was seen to bump into another B, A appeared to give B a push or set it in motion, which Michotte termed the launching effect. If object A on reaching object B was seen to move with it and at the same speed, A appeared to carry B, which he termed the entraining effect. Michotte observed these phenomena, and others, under various experimental conditions. His work is important not least because it is a scientific investigation of a topic that has mainly been the province of philosophers.

Midgley Thomas 1889–1944. US industrial chemist and engineer whose two main discoveries, universally adopted, were later criticized as damaging to the environment. He found in 1921 that tetraethyl lead is an efficient antiknock additive to petrol (preventing pre-ignition in car engines), and in 1930 introduced Freons (a group of chlorofluorocarbons) as the working gases in refrigerators, freezers, and air-conditioning units.

Midgley was born in Beaver Falls, Pennsylvania, and studied mechanical engineering at Cornell. From 1916 he worked for the Dayton Engineering Laboratories Company, Ohio, later taken over by General Motors. He became a vice president of the Ethyl Corporation 1923 and ten years later a director of the Ethyl-Dow Chemical Company.

Midgley discovered empirically that ethyl iodide (iodo-ethane) prevents pre-ignition in car engines using low-octane fuel. He spent several years teaching himself the relevant chemistry and looking for a less expensive additive, and came up with tetraethyl lead 1921. It became a standard additive, but because of the hazardous effects of airborne lead compounds emitted in exhaust fumes, from the 1980s unleaded petrol became increasingly common.

In 1930 Midgley introduced Freon (CF_2Cl_2) as a nonflammable, nontoxic refrigerant. Freon compounds also became extensively used as propellants in aerosol containers. Because chlorofluorocarbons contribute to the destruction of the Earth's ozone layer, their use was phased out from the 1970s as propellants, and in the 1990s practical substitutes were found for use in refrigeration.

Miller Stanley Lloyd 1930– . US chemist. In the early 1950s, under laboratory conditions, he tried to recreate the formation of life on Earth. To water under a gas mixture of methane, ammonia, and hydrogen, he added an electrical discharge. After a week he found that amino acids, the ingredients of protein, had been formed.

Miller was born in Oakland, California, and studied at the universities of California and Chicago. From 1960 he held appointments at the University of California in San Diego, rising to professor of chemistry.

Miller made his experiment while working for his PhD under Harold Urey, using the components that had been proposed for the Earth's primitive atmosphere by Urey and Russian biochemist Alexandr Oparin. The electrical discharge simulated the likely type of energy source.

Miller William 1801–1880. Welsh crystallographer, developer of the ***Miller indices***, a coordinate system of mapping the shapes and surfaces of crystals.

Millikan Robert Andrews 1868–1953. US physicist, awarded a Nobel prize 1923 for his determination of Planck's constant (a fundamental unit of quantum theory) 1916 and the electric charge on an electron 1913.

His experiment to determine the electronic charge, which took five years to perfect, involved observing oil droplets, charged by external radiation, falling under gravity between two horizontal metal plates connected to a high-voltage supply. By varying the voltage, he was able to make the electrostatic field between the plates balance the gravitational field so that some droplets became stationary and floated. If a droplet of weight W is held stationary between plates separated by a distance d and carrying a potential difference V, the charge, e, on the drop is equal to Wd/V.

Millikan was born in Illinois and studied at Oberlin College and Columbia University, and in Germany with Max Planck at Berlin and Hermann Nernst at Göttingen. He worked at the University of Chicago 1896–21, becoming professor 1910, and at the California Institute of Technology 1921–45.

Millikan also carried out research into cosmic rays, a term that he coined in 1925, when he proved that the rays do come from space.

Mills William Hobson 1873–1959. English organic chemist who worked mainly on stereochemistry and the synthesis of cyanine dyes.

Mills was born in London and studied at Cambridge, where he was professor from 1912. He also collected 2,200 specimens of the British bramble plant *Rubus fructiosus* for the university.

Mills and his co-workers investigated cyanine dyestuffs for preparing photographic emulsions, mainly for use by the military in World War I. Other early research concerned stereochemistry, particularly optical isomerism – the phenomenon in which pairs of (usually organic) compounds differ only in the arrangements of their atoms in space. In the 1920s he investigated substituted derivatives of naphthalene, quinolene, and benzene.

Milne Edward Arthur 1896–1950. English astrophysicist and mathematician who formulated a theory of relativity which he called kinematic relativity, parallel to Albert Einstein's general theory.

Milne was born in Hull, Yorkshire, and studied at Cambridge, where he spent his academic career until 1925. He then moved to Manchester as professor of mathematics. From 1929 he was professor at Oxford.

Studying stellar structure, Milne suggested that a decrease in luminosity might cause the collapse of a star, and that this would be associated with nova formation. Some of his cosmological theories were developed with Irish astrophysicist William McCrea.

In 1932, Milne began his attempt to explain the properties of the universe by kinematics (the movement of bodies). Basing his theory on Euclidean space and on Einstein's special theory of relativity alone, Milne was able to formulate systems of theoretical cosmology, dynamics, and electrodynamics. He also gave a more acceptable estimate for the overall age of the universe (10,000 million years) than that provided by the general theory of relativity.

Milstein César 1927– . Argentine-born British molecular biologist who developed monoclonal antibodies, giving immunity against specific diseases. He shared the Nobel Prize for Physiology or Medicine 1984 with Georges Köhler and Niels Jerne.

Monoclonal antibodies are cloned cells that can be duplicated in limitless quantities and, when introduced into the body, can be targeted to seek out sites of disease. Milstein and his colleagues had thus devised a means of accessing the immune system for purposes of research, diagnosis, and treatment.

Milstein was born in Bahia Blanca and studied at Buenos Aires.

From 1963 he worked in the UK at the Laboratory of Molecular Biology, Cambridge, and later became joint head of its Protein Chemistry Division.

Milstein and his colleagues were among the first to determine the complete sequence of the short, low-molecular-weight part of the immunoglobulin molecule (known as the light chain). He then determined the nucleotide sequence of a large portion of the messenger RNA for the light chain. His findings led him to the technique for preparing monoclonal antibodies.

Minkowski Hermann 1864–1909. Russian-born German mathematician whose introduction of the concept of space-time was essential to the genesis of the general theory of relativity.

Minkowski was born near Kaunas and studied at Königsberg, Germany (now Kaliningrad, Russia). In 1896 he became professor at the Federal Institute of Technology in Zurich, Switzerland, returning to Germany 1902 as professor at Göttingen.

Minkowski's concept of the geometry of numbers constituted an important addition to number theory, and his research into that topic led him to consider certain geometric properties in a space of *n* dimensions and so to hit upon his notion of the space-time continuum. The principle of relativity, already put forward by Jules Poincaré and Albert Einstein, led Minkowski to the view that space and time were interlinked. In *Raum und Zeit/Space and Time* 1909, he proposed a four-dimensional manifold in which space and time became inseparable. The central idea was, as Einstein allowed, necessary for the working-out of the general theory of relativity.

Space by itself and time by itself must sink into the shadows, while only a union of the two preserves indpendence.

Hermann Minkowski

Minkowski Rudolph Leo 1895–1976. German-born US astrophysicist, responsible for the compilation of a set of photographs now found in every astronomical library, the National Geographic Society Palomar Observatory Sky Survey. A leading authority on novae and planetary nebulae, he was a pioneer in the science of radio astronomy.

Minkowski was born in Strasbourg (then in Germany, now in France) and studied at Breslau, Silesia (now Wroclaw, Poland).

He became professor of physics at Hamburg 1922, but the rise of Nazism in Germany caused him to emigrate to the USA 1935. He worked at the Mount Wilson and Palomar observatories in California until 1959, and at the University of California at Berkeley 1960–65.

Minkowski divided supernovae into two principal types, and more than doubled the number of known planetary nebulae. In collaboration with Walter Baade, who had also been his colleague at Hamburg, Minkowski identified a discrete radio source, Cygnus A, in 1951. This was the first time an extragalactic radio source was optically identified.

Minkowski determined the optical red shift of the radio source 3C 295, which was then the farthest point on the velocity–distance diagram of cosmology, in his last observing run at the Palomar 500-cm/200-in telescope.

Minot George Richards 1885–1950. US physician who shared the Nobel Prize for Physiology or Medicine with William Murphy and George Whipple 1934 for his research into the cause of pernicious anaemia, a disease which reduces the oxygen-carrying capacity of the blood.

Whipple had previously demonstrated that anaemia might be the result of a vitamin deficiency. Working with dogs that he made anaemic by regularly drawing their blood, he discovered that more blood had to be drawn to keep the animals anaemic if they were fed liver, beef, or spinach than if they were fed salmon and bread. Minot, therefore, began advising patients with pernicious anaemia to eat beef and liver. This diet did not result in much improvement in their condition.

Then in 1925, he and Murphy performed a clinical trial in which they fed patients half a pound of lightly cooked beef liver every day. This resulted in an astonishing regeneration of the red blood cells within a single week and the continued good health of the patients while they remained on the diet of liver.

Minot was born in Boston and studied medicine at Harvard University. He worked for three years at Johns Hopkins University and then returned to a staff position at Harvard, where he began working on pernicious anaemia. Until the work of Minot 1925 anaemia was invariably fatal.

Mises Richard von 1883–1953. Austrian mathematician and aerodynamicist who made valuable contributions to statistics and the theory of probability, in which he emphasized the idea of random distribution.

Von Mises was born in Lemberg (now Lvov, Ukraine), and educated at Vienna. He was professor at the University of Strassburg (now Strasbourg, France) 1909–18. In 1920 he was appointed director of the Institute for Applied Mathematics at Berlin, but with the coming to power of Adolf Hitler in 1933, von Mises emigrated to Turkey and taught at the University of Istanbul. In 1939 he went to the USA to join the faculty of Harvard, where he was professor from 1944.

Von Mises's first interest was fluid mechanics, especially in relation to aerodynamics and aeronautics. He learned to fly and in 1913 gave the first university course in the mechanics of powered flight. He made significant improvements in boundary-layer-flow theory and aerofoil design and, in 1915, built an aeroplane for the Austrian military. During World War I he served as a pilot.

Von Mises was drawn into the field of probability theory and statistics by his association (from 1907 until the 1920s) with the Viennese school of logical positivism. He came to the conclusion that a probability cannot be simply the limiting value of a relative frequency, and added the proviso that any event should be irregularly or randomly distributed in the series of occasions in which its probability is measured.

Von Mises's ideas were contained in two papers which he published in 1919. Little noticed at the time, they have come to influence all modern statisticians.

Mitchell Peter Chalmers 1864–1945. Scottish zoologist, journalist, and secretary of London Zoo.

Mitchell was born in Dunfermline and graduated from King's College, Aberdeen with an MA 1884. He then went to Christ Church College, Oxford where he gained a first in natural sciences and became a university demonstrator and assistant to the Linacre professor of comparative anatomy. Following two years employment with Oxford County Council, he became a lecturer at London Hospital Medical college 1894. He was the secretary of the Zoological Society of London 1903–35, and it was during this time that the animal houses in Regent's Park were rebuilt and Whipsnade Zoo was opened.

Mitchell's career as a journalist stretched from his time in Oxford until after his retirement to Malaga. He was scientific correspondent for the *Times* and the biological editor of *Encyclopaedia Britannica* (11th edition). He was knighted 1929. He was well known for his radical left-wing political views, describing himself as an anarchist.

Mitchell Peter Dennis 1920–1992. English chemist. He received a Nobel prize in 1978 for work on the conservation of energy by plants during respiration and photosynthesis. He showed that the transfer of energy during life processes is not random but directed.

Mitchell was born in Mitcham, Surrey, and studied at Cambridge. He worked at the Chemical Biology Unit in the Department of Zoology at Edinburgh 1955–63, and then established the privately run Glynn Research Institute at Bodmin, Cornwall.

It had been believed that the energy absorbed by animals from food and by plants from sunlight was utilized in cells by purely chemical means. The cell was seen as a bag of enzymes in which random and directionless processes took place. Mitchell proved that currents of protons pass through cell walls, which, instead of being simple partitions between cells, are, in fact, full of directional pathways. This discovery demonstrated the existence of a reverse kind of electricity (Mitchell called it 'proticity'), which he successfully used to run an engine.

Mitchell R(eginald) J(oseph) 1895–1937. English aircraft designer whose Spitfire fighter was a major factor in winning the Battle of Britain during World War II.

Mitchell, born near Stoke-on-Trent, became apprenticed to a firm of locomotive builders and continued his education in the evenings at technical colleges. From 1916 he worked for the Supermarine Aviation Company, which specialized in flying boats.

His first aeroplane, the Sea Lion 1919, was adapted from a standard design; faster models of Sea Lion followed. In 1924, Mitchell was allowed to design his own aeroplane, the Supermarine S-4. It was a monoplane and the whole wing section was made in one piece. He won speed records with several of the S-model planes in the 1920s.

The single-engined Spitfire prototype was produced 1936. More than 19,000 Spitfires were eventually built. The manoeuvrability and adaptability of the Spitfire accounted for its success.

Mitscherlich Eilhard 1794–1863. German chemist who discovered isomorphism (the phenomenon in which substances of analogous chemical composition crystallize in the same crystal form). He also synthesized many organic compounds for the first time.

Mitscherlich was born in Jever, Lower Saxony, and entered Heidelberg University to study Oriental languages, later continuing this study at Paris, but having to abandon it with the fall of Napoleon. He instead studied science at Göttingen and then worked with Swedish chemist Jöns Berzelius in Stockholm for two years. Mitscherlich became professor at Berlin 1825.

Mitscherlich began studying crystals in 1818. Observing that crystals of potassium phosphate and potassium arsenate appear to be nearly identical in form, he learned exact crystallographic methods and then applied spherical trigonometry to the data he obtained. He extended his researches to phosphates, arsenates, and carbonates, publishing the results 1822 and introducing the term isomorphism.

In 1834, Mitscherlich synthesized benzene, which he termed *Benzin*. He showed that yeast (which in 1842 he identified as a microorganism) can invert sugar in solution.

Mitscherlich published his influential *Lehrbuch der Chemie/Textbook of Chemistry* 1829.

Möbius August Ferdinand 1790–1868. German mathematician and theoretical astronomer, discoverer of the Möbius strip and considered one of the founders of topology.

Möbius formulated his ***barycentric calculus*** in 1818, a mathematical system in which numerical coefficients were assigned to points. The position of any point in the system could be expressed by varying the numerical coefficients of any four or more noncoplanar points.

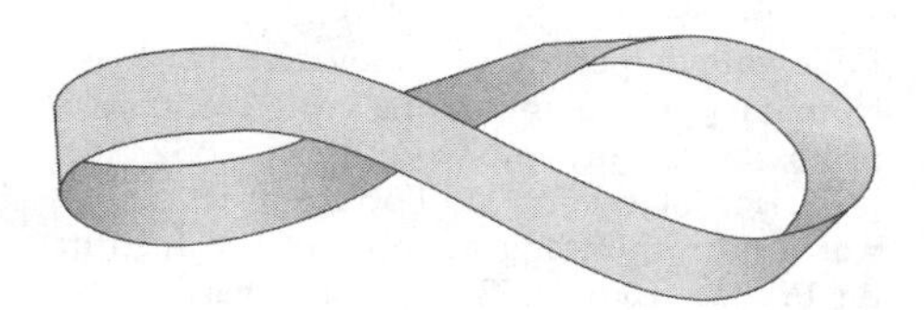

Möbius *The Möbius strip has only one side and one edge. It consists of a strip of paper connected at its ends with a half-twist in the middle.*

Möbius was born near Naumburg and studied at Leipzig and Göttingen. From 1815 he was on the staff at Leipzig, where he became professor of astronomy and higher mechanics 1844 and director of the observatory 1848.

In addition to the Möbius strip, he discovered the Möbius net, later of value in the development of projective geometry; the Möbius tetrahedra, two tetrahedra that mutually circumscribe and inscribe each other, which he described 1828; and the Möbius function in number theory, published 1832.

During a lecture of 1840, Möbius set the problem to find the least number of colours required on a plane map to distinguish political regions, given that each boundary line should separate two differently coloured regions. It has now been proved by computer analysis that four colours will always suffice.

Mohl Hugo von 1805–1872. German botanist who coined the term protoplasm (the living fluid material inside cells), describing it for the first time (although the Czech cell physiologist Johannes Purkinje had already used this term 1829, he had not applied it to plant cells). Mohl was also the first to describe the cell membrane, nucleus, utricle, and the relevance of osmosis with regard to plant cell function. He constructed a prototype of a light microscope, which he used to study and report on the structure of plant cells, and wrote a manual 1846 on its construction and applications.

Mohl was born in Stuttgart and received a classical education, although his personal preference was always for the botanical sciences. He studied medicine at Tübingen, but went on to study botany. He obtained his doctorate 1828 with a thesis on the constitution of the stomata (pores) of plants.

Following a period of innovative work on stomatal development, he was appointed professor of physiology in Bern 1832 and professor of botany in Tübingen 1835. He was a corresponding member of the Institut de France from 1838 and was awarded the Order of the Crown of Wür Hemberg 1843.

Mohs Friedrich 1773–1839. German mineralogist who 1812 devised ***Mohs' scale*** of minerals, classified in order of relative hardness.

Mohs was born in the Saxon Hartz Mountains. He studied at Halle and at the Freiberg Mining Academy, and was appointed professor of mineralogy at Graz 1812, and later at Freiberg. He ended his career in Austria as professor at Vienna from 1826.

Mohs achieved eminence for his system of the classification of minerals, dividing these into genera and species in the manner of Swedish botanist Linnaeus. Although recognized as useful, the strategy was widely criticized for failing to take sufficient account of chemical composition, and it is no longer used. He also classified crystals on the basis of the orientation of their axes.

Mohs' scale

number	*defining mineral*	*other substances compared*
1	talc	
2	gypsum	$2\frac{1}{2}$ fingernail
3	calcite	$3\frac{1}{2}$ copper coin
4	fluorite	
5	apatite	$5\frac{1}{2}$ steel blade
6	orthoclase	$5\frac{3}{4}$ glass
7	quartz	7 steel file
8	topaz	
9	corundum	
10	diamond	

note *that the scale is not regular; diamond, at number 10 the hardest natural substance, is 90 times harder in absolute terms than corundum, number 9*

Moissan (Ferdinand Frédéric) Henri 1852–1907. French chemist. For his preparation of pure fluorine 1886, Moissan was awarded the 1906 Nobel Prize for Chemistry. He also attempted to create artificial diamonds by rapidly cooling carbon heated to high temperatures. His claims of success were treated with suspicion.

Moivre Abraham De 1667–1754. French mathematician who pioneered the development of analytical trigonometry, for which he formulated his theorem regarding complex numbers. He also devised a means of research into the theory of probability.

De Moivre was born in Vitry-le-François, Champagne, and studied in Paris. With the revocation of the Edict of Nantes 1685, he was imprisoned as a Protestant for 12 months; on his release he went immediately to England. In London he became a close friend of Isaac Newton and Edmund Halley, but he never obtained a permanent position; he eked out a precarious living by tutoring and acting as a consultant for gambling syndicates and insurance companies.

His *The Doctrine of Chances* was published first in Latin and then in expanded English versions 1718, 1738, and 1756. It was one of the first books on probability, and made an approximation to the normal or Gaussian distribution, which was incorporated into statistical studies for the next 200 years. De Moivre was the first to derive an exact formulation of how 'chances' and stable frequency are related.

Analysing mortality statistics, De Moivre laid the mathematical foundations of the theory of annuities, for which he devised formulae based on a postulated law of mortality and constant rates of interest on money. First published 1725, his work became standard in textbooks of all subsequent commercial applications.

In analytical trigonometry he discovered an equation that is now named after him:

$$(\cos z + \mathrm{i} \sin z)^n = \cos nz + \mathrm{i} \sin nz$$

First stated 1722, it had been anticipated by related forms 1707. Although it was to become one of the most useful steps in the early development of complex number theory, much of De Moivre's work was appreciated only after his death.

Molina Mario 1943– . Mexican chemist who shared the 1995 Nobel Prize for Chemistry with Paul Crutzen and F Sherwood Rowland for their work in atmospheric chemistry, particularly concerning the formation and decomposition of ozone. They explained the chemical reactions which are destroying the ozone layer.

The power of nitrogen oxides to decompose ozone was pointed out in 1970 by Crutzen. In 1974 Rowland and Molina published a widely read article on the threat to the ozone layer posed by chlorofluorocarbons (CFCs) used in refrigerators and aerosol cans. They pointed out that CFCs could gradually be carried up into the ozone layer.

Here, under the influence of intense ultraviolet light, the CFCs would decompose into their constituents, notably chlorine atoms which decompose ozone in similar ways as do nitrogen oxides. They calculated that if the use of CFCs continued at an unaltered rate the ozone layer would be seriously depleted after a few decades. Molina's and Rowland's work led to restrictions on CFC use in the late 1970s and early 1980s.

Molina was born in Mexico City and educated in physical chemistry at the University of California, Berkeley, USA. The first Mexican to receive a Nobel prize for science, Molina is currently at the Massachusetts Institute of Technology, Cambridge, Massachusetts, USA.

Mond Ludwig 1839–1909. German-born British chemist who invented a process for recovering sulphur during the manufacture of alkali. He gave his name to a method of extracting nickel from nickel carbonyl, one of its volatile organic compounds.

His son Alfred Mond, 1st Baron Melchett (1868–1930), was a founder of Imperial Chemical Industries (ICI).

Mond was born in Kassel and studied chemistry at Marburg and Heidelberg. In 1859, working in a small soda works near Kassel, he initiated the new process for the recovery of sulphur. This gained him an invitation from a Lancashire industrial chemist, and Mond moved to the UK 1862. In 1873, he helped to found the firm of Brunner, Mond, and Company, which pioneered the British chemical industry.

In 1879, Mond became interested in the production of ammonia. One outcome was the development of the Mond producer gas process, in which carbon monoxide and hydrogen are produced by alternately passing air and steam over heated coal or coke (and the hydrogen used to convert nitrogen into ammonia). By the early 1900s, Mond's Dudley Port Plant in Staffordshire was using 3 million tonnes of coal each year to make producer gas (Mond gas).

Monge Gaspard 1746–1818. French mathematician and chemist, the founder of descriptive geometry. His application of analysis to infinitesimal geometry also paved the way for later developments.

Monge was born in Beaune, Burgundy, and studied at the Collège de la Trinité in Lyon. In 1771 he was drawn into the scientific circle attached to the Academy of Sciences in Paris, and in 1780 he was given official duties as assistant geometer to the Academy. In 1785 he was appointed examiner of naval cadets by the French government. By the time the French Revolution broke out in 1789, Monge was an active supporter of the radicals. He was appointed minister of the navy 1792, but was forced to resign the following year. As a member of the Committee on Arms 1793–94, he supervised the Paris armaments workshops and helped to develop military balloons. He also served on the commission to standardize weights and measures. He was instrumental in establishing the institution that became the Ecole Polytechnique, was briefly its director, and taught there until 1809.

In 1796, Monge's friendship with Napoleon began. He was sent to the newly conquered Italy as a member of various commissions. In 1798 he assisted in the preparation for the Egyptian campaign; he then accompanied Napoleon and was appointed president of the Institut d'Egypte established in Cairo. Monge also went with Napoleon on expeditions to the Suez region and Syria in 1799. Subsequently Napoleon appointed him a senator. When Napoleon was overthrown in 1815, Monge was discredited.

Independently of French chemist Antoine Lavoisier, Monge synthesized water 1783, but he also worked with Lavoisier in 1785 on the synthesis and analysis of water. Monge quickly embraced Lavoisier's new chemistry and played an energetic part in getting it accepted.

Monge asserted the autonomy of analytical geometry as a separate branch of mathematics. In particular, he devoted himself to the families of surfaces as defined by their mode of generation, and the properties of surfaces and space curves. He also established the distinction between ruled surfaces and developable surfaces.

In a paper on infinitesimal geometry 1776 Monge introduced lines of curvature and the congruences of straight lines. His main work is *Géométrie Descriptive/Descriptive Geometry* 1799.

Moniz Antonio Egas 1874–1955. Portuguese neurologist, pioneer of prefrontal leucotomy (surgical separation of white fibres in the prefrontal lobe of the brain) to treat schizophrenia and paranoia; the treatment is today considered questionable. He shared the 1949 Nobel Prize for Physiology or Medicine.

Monod Jacques Lucien 1910–1976. French biochemist who shared the 1965 Nobel Prize for Physiology or Medicine with his co-workers André Lwoff and François Jacob for research in genetics and microbiology.

Monod was born and educated in Paris. From 1945 he worked at the Pasteur Institute, where he collaborated with Lwoff and Jacob. In 1953, Monod became director of the Department of Cellular Biochemistry at the Pasteur Institute and also a professor at the University of Paris. In 1971, he was appointed director of the entire Pasteur Institute.

There are living systems; there is no 'living matter'.

Jacques Monod

Lecture Nov 1967

Working on the way in which genes control intracellular metabolism in microorganisms, Monod and his colleagues postulated the existence of a class of genes (which they called operons) that regulate the activities of the genes that actually control the synthesis of enzymes within the cell. They further hypothesized that the operons suppress the activities of the enzyme-synthesizing genes by affecting the synthesis of messenger RNA.

In his book *Chance and Necessity* 1971, Monod summoned contemporary biochemical discoveries to support the idea that all forms of life result from random mutation (chance) and Darwinian selection (necessity).

Monro Alexander 1697–1767. Also known as Monro primus. English-born anatomist who helped to found the Edinburgh Royal Infirmary and to make Edinburgh one of the key centres for medical teaching in Europe.

Monro primus was born in London, the son of a prominent surgeon, and studied medicine in London, Paris, and Leiden before settling in Edinburgh first as a lecturer in anatomy, then as the professor of anatomy 1725–59. He was an enthusiastic, well-organized, and immensely popular teacher who, at the start of his career, had only 57 students in his anatomy class but several hundred by the time of his retirement.

He trained his son Monro secundus, who succeed Monro primus at the age of 21, thereby continuing a period of 120 years in which Munros ran anatomy teaching at Edinburgh Medical School.

Monro Alexander 1733–1817. Also known as Monro secundus. Scottish anatomist who determined the role of the lymphatic system, distinguishing it from the circulatory system, and wrote one of the first works of comparative anatomy.

Monro secundus, in addition to being a very famous teacher of anatomy, was also a distinguished physician and researcher. He spent a short time working under the distinguished anatomist William Hunter in London, on a study of the lymphatic system. He then claimed in a subsequent publication, *De venis lymphaticis* 1757, that he had discovered the anatomy and role of the lymph vessels, distinguishing the lymph system from the circulatory system. This publication prompted a bitter, public dispute between him and Hunter, the latter claiming that he himself had made these discoveries earlier. Monro secundus went on to publish early accounts of the anatomy of the nervous system, eyes and ears. In 1785 he wrote *Structure and Physiology of Fishes*, one of the first works of comparative anatomy.

Monro secundus was born in Edinburgh and was trained in medicine largely by his father, Monro primus, who was professor of anatomy at the University of Edinburgh. Monro secundus succeeded his father 1754 at the young age of 21. Monro secundus handed the chair in anatomy to his own son, Monro tertius (1773–1859) 1798. The third Monro did not maintain the teaching successes of his two predecessors.

Montagnier Luc 1932– . French molecular biologist who first identified the human immunodefiecency virus (HIV) 1983, the single-stranded RNA retrovirus that causes AIDS. Patients with the disease die from rare infections because their immune systems are crippled by the virus.

Montagnier was educated in Paris and held a number of research posts before being appointed to run the Viral Oncology Unit 1985 and the Department of AIDS and Retroviruses 1990 at the Pasteur Institute in Paris.

The US virologist Robert Gallo, claimed to have been the first to identify the virus and the work of the Montagnier was largely discredited, until a few years ago when the the original work of Gallo's team was shown to have been incorrect and possibly even fraudulent.

Montgolfier Joseph Michel (1740–1810) and Jacques Etienne (1745–1799) French brothers whose hot-air balloon was used for the first successful human flight 21 Nov 1783.

On 5 June 1783 they first sent up a balloon filled with hot air. After further experiments with wood-fuelled fabric-and-paper balloons, and one crewed ascent in a tethered ballon, they sent up two people who travelled for 20 minutes above Paris, a journey of 9 km/6 mi. The Montgolfier experiments greatly stimulated scientific interest in aviation.

The Montgolfier brothers were papermakers of Annonay, near Lyon. Jacques invented vellum paper.

The first hot-air balloon was launched in the marketplace of Annonay. Next, they took their invention to the palace of Versailles, where, before a large audience (which included Louis XVI and Marie Antionette), their balloon ascended, carrying a sheep, a cock, and a duck, and made an eight-minute flight of approximately 3 km/2 mi.

Joseph Montgolfier later developed a type of parachute, a calorimeter, and a hydraulic ram and press.

Moon William 1818–1894. English inventor of the ***Moon alphabet*** for the blind. Devised in 1847, it uses only nine symbols in different orientations. From 1983 it has been possible to write it with a miniature typewriter.

Moore Patrick Alfred Caldwell 1923– . British broadcaster, writer, and popularizer of astronomy. He began presenting the BBC television series *The Sky at Night* 1968.

Moore served with the Royal Air Force during World War II, and for seven years after the war he assisted at a training school for pilots. From 1965 to 1968 he was director of the Armagh Planetarium, Northern Ireland. He has never worked as an astronomer.

His many books include *Atlas of the Universe* 1970, *Guide to the Planets* 1976, *Guide to the Stars* 1977, and *Can You Speak Venusian?* 1977.

Moore Stanford 1913–1982. US biochemist who shared the Nobel Prize for Chemistry 1972 with his colleague William Stein for their determination of the base sequence of RNA in 1958, and the probable location of the active site on the molecule, and for the development of the column chromatographic method for separating and identifying amino acids.

The chromatographic column apparatus consisted of a 1.5-m/5-ft column filled with resin. The sample was washed through the resin with solutions of various acidity, and the amino acids separated out according to their affinity for the resin and the solution.

By careful manipulation, the amino acids could be tapped from the base of the column, one at a time. They were detected using the reagent ninhydrin, which turns blue in the presence of amino acids and heat. A continuous plot of the intensity of the blue colour shows a series of peaks, each corresponding to a certain amino acid. By measuring the area under the peak, the amount of each amino acid can be found.

Moore and Stein also recorded the first example of convergent evolution, involving a bacterial protease (an enzyme capable of breaking down proteins) and a plant protease, which have similar enzymatic activity but different molecular structures.

Born in Chicago, Illinois, Moore was educated at Vanderbilt University and the University of Wisconsin. He spent all of his working life at the Rockefeller Institute, where, from 1952, he was professor of biochemistry.

Morgagni Giovanni Battista 1682–1771. Italian anatomist who developed the view that disease was not an imbalance of the body's humours but a result of alterations in the organs. His work *De sedibus et causis morborum per anatomen indagatis/On the Seats and Causes of Diseases as Investigated by Anatomy* 1761 formed the basis of pathology.

Morgagni was born in Forli and studied at Bologna. As professor of anatomy at Padua, he carried out more than 400 autopsies. He did not use a microscope and he regarded each organ of the body as a composite of minute mechanisms.

Morgagni was the first to delineate syphilitic tumours of the brain and tuberculosis of the kidney. He grasped that where only one side of the body is stricken with paralysis, the lesion lies on the opposite side of the brain. His explorations of the female genitals, of the glands of the trachea, and of the male urethra also broke new ground.

Morgan Ann Haven 1882–1966. US zoologist who promoted the study of ecology and conservation. She particularly studied the zoology of aquatic insects and the comparative physiology of hibernation.

Morgan was born in Waterford, Connecticut, and studied at Wellesley College and Cornell University. From 1912 to 1947 she taught at Mount Holyoke College, becoming professor 1918. She spent her summers at a variety of research laboratories, including the Marine Biological Laboratory at Woods Hole in Massachusetts, and also worked at the Tropical Research Laboratory of British Guiana (now Guyana). She was a member of the National Committee on Policies in Conservation Education.

Her *Field Book of Ponds and Streams: An Introduction to the Life of Fresh Water* 1930 attracted amateur naturalists as well as providing an authoritative taxonomic guide for professionals.

Morgan Conwy Lloyd 1852–1936. English psychologist. Of immense influence in the field of contemporary comparative psychology, he is renowned for his observational studies of animals in natural settings. In 1894 he was the first to describe ***trial-and-error learning*** in animals.

In trial-and-error learning an animal learns a trick, for example, how to lift a latch to open a gate, by making several attempts without any insights into what the trick involves. Eventually it lifts the latch, an act that is immediately reinforced by the freedom that follows.

Morgan was appointed professor of geology and zoology at the University College of Bristol (later the University of Bristol) 1884. In 1899, following the publication of two influential books *Animal Life and Intelligence* 1890 and *An Introduction to Comparative Psychology* 1894, he was elected a Fellow of the Royal Society, the first to be elected for research in psychology. Recognizing the tendency of previous researchers, notably G J Romanes (1848–1894), to attribute human characteristics to animals, he developed rigorous procedures of control and analysis in order to avoid such prejudice.

Morgan Garrett A 1875–1963. US inventor who patented the gas mask 1914 and the automatic three-way traffic signal 1923.

Garrett Morgan was born in Paris, Kentucky. He received only elementary-school education and at the age of 14 he left home to work in Ohio. From 1895 he lived in Cleveland, doing repairs, then opening a tailoring shop 1909, and went on to establish the G A Morgan Hair Refining Company in 1913 as a result of discovering a human hair-straightening process. In 1916 he set up a company to manufacture and sell the safety hood, as he called the gas mask.

During World War I the design of the gas mask was improved and it became part of the standard field equipment of US soldiers.

In 1920 Morgan started his own newspaper, the *Cleveland Call*, later the *Call and Post*. He sold the right to the traffic signal to the General Electric Corporation for $40,000. He also invented a friction-drive clutch.

Morgan served as treasurer of the Cleveland Association of Colored Men from 1914 until it merged with the National Association for the Advancement of Colored People, of which he remained an active member all his life.

Morgan Thomas Hunt 1866–1945. US geneticist who helped establish that the genes are located on the chromosomes, discovered sex chromosomes, and invented the techniques of genetic mapping. He was the first to work on the fruit fly *Drosophila*, which has since become a major subject of genetic studies. Nobel Prize for Physiology or Medicine 1933.

Morgan was born in Lexington, Kentucky, and studied at Johns Hopkins University. He was professor of experimental zoology at Columbia University 1904–28, when he was appointed director of the Laboratory of Biological Sciences at the California Institute of Technology.

Following the rediscovery of Austrian scientist Gregor Mendel's work, Morgan's interest turned from embryology to the mechanisms involved in heredity, and in 1908 he began his research on the genetics of *Drosophila*. From his findings he postulated that certain characteristics are sex-linked, that the X-chromosome carries several discrete hereditary units (genes), and that the genes are linearly arranged on chromosomes. He also demonstrated that sex-linked characters are not invariably inherited together, from which he developed the concept of crossing-over and the associated idea that the extent of crossing-over is a measure of the spatial separation of genes on chromosomes.

Morgan published a summary of his work in *The Mechanism of Mendelian Heredity* 1915.

Morison Robert 1620–1683. Scottish botanist who was the first professor of botany at Oxford University 1669 and was physician and botanist to England's King Charles II.

Morison was born in Dundee in Scotland and attended Aberdeen University, graduating with an MA 1638. He taught at Aberdeen University until the outbreak of the civil war, and in 1644, he fled to France, because he was a Royalist. He studied medicine in Angers and Paris before becoming a gardener to Gaston D'Orleans at Blois in order to further his interest in botany. He travelled throughout Europe to collect botanical specimens. He returned to England 1660 after the Restoration to become Charles II's physician and botanist.

In 1669, he published *Praeludia botanica*, which included a classification of the plants at Blois, although he was unable to produce a succinct classification system for these observations.

Morley Edward Williams 1838–1923. US physicist who collaborated with Albert Michelson on the ***Michelson–Morley experiment*** 1887. In 1895 he established precise and accurate measurements of the densities of oxygen and hydrogen.

Morley was born in Newark, New Jersey, and studied at Williams College. He spent his career at Adelbert College, Cleveland, Ohio, becoming professor 1869. From 1873 to 1888, he was simultaneously professor at Cleveland Medical College.

All Morley's research involved the use of precision instruments. Using a eudiometer he had made accurate to within 0.0025%, he was able to confirm on the basis of the oxygen content of the air and meteorological data that, under certain conditions, cold air derives from the downward movement of air from high altitudes rather than from the southward movement of northerly cold air.

Morris Desmond John 1928– . British zoologist, a writer and broadcaster on animal and human behaviour. His book *The Naked Ape* 1967 was a best seller.

Morris studied at Birmingham and at Oxford, working on animal behaviour under Nikolaas Tinbergen. He became head of the Granada Television and Film Unit at the Zoological Society in London in 1956. Three years later he was appointed Curator of Mammals at London Zoo and from 1967 to 1968 served as director of the Institute of Contemporary Arts in London.

In his book *The Human Zoo* 1969, Morris scrutinizes the society that the naked ape has created for itself. He compares civilized humans with their captive animal counterparts and shows how confined animals seem to demonstrate the same neurotic behaviour patterns as human beings often do in crowded cities. He believes the urban environment of the cities to be the human zoo.

He is proud that he has the biggest brain of all the primates, but attempts to conceal the fact that he also has the biggest penis.

Desmond Morris

The Naked Ape Introduction

What hath God wrought?

Samuel Morse

First message sent on his electric telegraph 24 May 1844

Morse Samuel Finley Breese 1791–1872. US inventor. In 1835 he produced the first adequate electric telegraph, and in 1843 was granted $30,000 by Congress for an experimental line between Washington DC and Baltimore. With his assistant Alexander Bain (1810–1877) he invented the Morse code.

Morse was born in Charlestown, Massachusetts. After graduating from Yale, he went to the UK and studied art at the Royal Academy in London. He became professor of the art of design at New York University.

Between 1832 and 1836 he developed his idea that an electric current could be made to convey messages. The signal current would be sent in an intermittent coded pattern and

Morse *US artist and inventor Samuel Morse, shown next to a printing telegraph. Backed by $30,000 from Congress, Morse erected the first telegraph line between Baltimore and Washington and sent the first message on 11 May 1844. His invention was such a success that within four years of Morse's first public demonstration, America had five thousand miles of telegraph wire.*

would cause an elctromagnet to attract intermittently to the same pattern on a piece of soft iron to which a pencil or pen would be attached and which in turn would make marks on a moving strip of paper.

Morton William Thomas Green 1819–1868. US dentist who in 1846, with Charles Thomas Jackson (1805–1880), a chemist and physician, introduced ether as an anaesthetic. They were not the first to use it but they patented the process and successfully publicized it.

Morton was born in Massachusetts. He set up his own dental practice in Boston in 1844 and began investigating ways to deaden pain during dental surgery. Jackson advised him to try ether. Later, Morton attempted to claim sole credit as its discoverer and spent the rest of his life in costly litigation with Jackson.

In 1846, Morton extracted a tooth from a patient under ether and later in the same year staged a public demonstration of ether anaesthesia in an operation at the Massachusetts General Hospital to remove a facial tumour.

Morton's contribution to medicine lay in making the value of anaesthesia generally known and appreciated. Crawford Long (1815–1878) had in 1842 successfully used ether anaesthesia during an operation, although he did not publish this work until 1849.

Moseley Henry Gwyn Jeffreys 1887–1915. English physicist. From 1913 to 1914 he devised the series of atomic numbers (reflecting the charges of the nuclei of different elements) that led to the revision of Russian chemist Dmitri Mendeleyev's periodic table of the elements.

Moseley was born in Weymouth, Dorset, and studied at Oxford. He worked in the Manchester laboratory of Ernest Rutherford, the pioneer of atomic science, 1910–13, and then at Oxford. Moseley was killed during the Gallipoli campaign of World War I.

In 1913 Moseley introduced X-ray spectroscopy and found that the X-ray spectra of the elements were similar but with a deviation that changed regularly through the series. A graph of the square root of the frequency of each radiation against the number representing the element's position in the periodic table gave a straight line. He called this number the atomic number of the element; the equation is known as ***Moseley's law***. When the elements are arranged by atomic number instead of atomic mass, problems appearing in the Mendeleyev version are resolved. The numbering system also enabled Moseley to predict correctly that several more elements would be discovered.

Mössbauer Rudolf (Ludwig) 1929– . German physicist who discovered in 1958 that under certain conditions an atomic nucleus can be stimulated to emit very sharply defined beams of gamma rays – a phenomenon that became known as the Mössbauer effect. For this work he shared the 1961 Nobel Prize for Physics.

Mössbauer was born in Munich, studied at the Munich Institute of Technology, and did postgraduate research in Heidelberg at the Max Planck Institute. In 1960 he went to the USA and a year later became professor of physics at the California Institute of Technology, Pasadena. He remained in this position while simultaneously holding a professorship at Munich.

Mössbauer began research into the effects of gamma rays on matter in 1953. The absorption of a gamma ray by an atomic nucleus usually causes it to recoil, so affecting the wavelength of the re-emitted ray. Mössbauer found that at low temperatures crystals will absorb gamma rays of a specific wavelength and resonate, so that the crystal as a whole recoils while the nuclei do not because they are tightly bound in the crystal lattice. This recoilless nuclear resonance absorption became known as the Mössbauer effect. The effect is exploited in ***Mössbauer spectroscopy***, a useful tool in the study of the structure of solids.

Mott Nevill Francis 1905–1996. English physicist who researched the electronic properties of metals, semiconductors, and noncrystal-line materials. He shared the Nobel Prize for Physics 1977. Knighted 1962.

Mott was born in Leeds and studied at Cambridge. He was at Bristol 1933–54, first as professor of theoretical physics and then as director of the Henry Herbert Wills Physical Laboratories. From 1954 to 1971, he was professor at Cambridge.

Mott initially studied dislocations and other defects in crystalline structure. He was the first to put forward a comprehensive theory of the process involved when a photographic film is exposed to light.

Mott and his colleagues discovered special electrical characteristics in glassy semiconductors and laid down fundamental laws of behaviour for their materials. As a result of this work, more efficient photovoltaic cells can now be produced, and the memory capacity of computers increased.

Mottelson Ben(jamin Roy) 1926– . US-born Danish physicist who with Aage Bohr and James Rainwater shared the 1975 Nobel Prize for Physics for their work on the structure of the atomic nucleus.

Mottelson was born in Chicago, Illinois, and educated at Purdue University. Based in Copenhagen from 1950, he was at the Institute of Theoretical Physics to 1953, then held a position at CERN (European Centre for Nuclear Research), and became professor at Nordita 1957.

In the early 1950s, Mottelson and Bohr together confirmed experimentally the theory worked out by Rainwater about the structure of the nucleus. Mottelson published several books and many scientific papers jointly with Bohr.

Muir John 1838–1914. Scottish-born US conservationist. From 1880 he headed a campaign that led to the establishment of Yosemite National Park. He was named adviser to the National Forestry Commission 1896 and continued to campaign for the preservation of wilderness areas for the rest of his life.

Born in Scotland, Muir emigrated to the USA with his family 1849. After attending the University of Wisconsin, he travelled widely and compiled detailed nature journals of his trips. He moved to California 1868 and later explored Glacier Bay in Alaska and mounted other expeditions to Australia and South America.

Muir Thomas 1844–1934. Scottish mathematician whose five-volume treatise on the history of determinants 1906–30 made the work of other mathematicians accessible to scholars. Knighted 1915.

Muir was born in Stonebrye, Lanarkshire, and studied at Glasgow and in Germany at Berlin. In 1874 he resigned a position as assistant professor at Glasgow to become head of the Mathematics and Science Department at Glasgow High School.

From 1892 he was superintendent-general of education in South Africa and vice chancellor of the University of Cape Town.

Although not himself a creative mathematician, Muir published 307 papers, most of them on determinants and allied subjects. His books include *A Treatise on the Theory of Determinants* 1882 and *The Theory of Determinants in its Historical Order of Development* 1890.

Muller Hermann Joseph 1890–1967. US geneticist who discovered 1926 that mutations can be artificially induced by X-rays. This showed that mutations are nothing more than chemical changes. He was awarded the Nobel Prize for Physiology or Medicine 1946.

Muller campaigned against the needless use of X-rays in diagnosis and treatment, and pressed for safety regulations to ensure that people who were regularly exposed to X-rays were adequately protected. He also opposed nuclear-bomb tests.

Muller was born in New York and studied at Columbia. In 1920 he joined the University of Texas, Austin, later becoming professor of zoology. The constraints on his freedom to express his socialist political views caused him to leave the USA 1932, and from 1933 he worked at the Institute of Genetics in the USSR. But the false ideas of Trofim Lysenko began to dominate Soviet biological research; openly critical of Lysenkoism, Muller was forced to leave in 1937. After serving in the Spanish Civil War, he worked at the Institute of Animal Genetics in Edinburgh. In 1940 he returned to the USA, becoming professor at Indiana University 1945.

In 1919 Muller found that the mutation rate was increased by heat, and that heat did not always affect both of the chromosomes in a chromosome pair. From this he concluded that mutations involved changes at the molecular or submolecular level. Next he experimented with X-rays as a means of inducing mutations, and by 1926 he had proved the method successful.

Muller's research convinced him that almost all mutations are deleterious. In the normal course of evolution, deleterious mutants die out and the few advantageous ones survive, but he believed that if the mutation rate is too high, the number of imperfect individuals may become too large for the species as a whole to survive.

Müller Johannes German astronomer, see ◊Regiomontanus.

Müller Johannes Peter 1801–1858. German comparative anatomist whose studies of nerves and sense organs opened a new chapter in physiology by demonstrating the physical nature of sensory perception. His name is associated with a number of discoveries, including the ***Müllerian ducts*** in the mammalian fetus and the lymph heart in frogs.

Müller K Alexander 1927– . Swiss physicist who worked on high-temperature superconductivity in ceramic materials. The discovery of these materials was a significant step towards the use of superconductors in computers, magnetic levitation trains, and the more efficient generation and distribution of electricity. Nobel Prize for Physics 1987 (shared wth Georg Bednorz).

Superconductivity is the resistance-free flow of electrical current which occurs in many metals and metallic compounds at very low temperatures, within a few degrees of absolute zero (0 K/–273.16°C/–459.67°F). In 1986 Müller and Bednorz showed that a ceramic oxide of lanthanum, barium and copper became superconducting at temperatures above 30 K, much hotter than for any previously known superconductor.

Müller was born in Basle, Switzerland, and studied physics and mathematics at the Swiss Federal Institute of Technology in Zürich. After graduating 1958, he joined the Battelle Memorial Institute in Geneva. While in Geneva he was appointed lecturer (becoming a professor 1970) at the University of Zürich. In 1963 he joined the IBM Zürich Research Laboratories at Rüschlikon.

Müller Paul Herman 1899–1965. Swiss chemist who discovered the first synthetic contact insecticide, DDT, 1939. For this he was awarded a Nobel prize 1948.

DDT (dichloro-diphenyl-trichloroethane) was to have a profound effect on the health of the world, both by killing insect vectors such as the mosquitoes that spread malaria and yellow fever and by combating insect pests that feed on food crops. Gradually the uses of DDT in public hygiene and in agriculture became limited by increasing DDT-resistance in insect species.

DDT is highly toxic and persists in the environment and in living tissue, disrupting food chains and presenting a hazard to animal life. Its use is now banned in most countries.

Müller was born in Olten, Solothurn, and studied at Basel. He went to work for the chemical firm of J R Geigy, researching principally into dyestuffs and tanning agents; he subsequently joined the staff of Basel University.

In 1935, Müller started the search for a substance that would kill insects quickly but have little or no poisonous effect on plants and animals, unlike the arsenical compounds then in use. He concentrated his search on chlorine compounds and in 1939 synthesized DDT.

The Swiss government successfully tested DDT against the Colorado potato beetle in 1939 and by 1942 it was in commercial production. Its first important use was in Naples, Italy, where a typhus epidemic 1943–44 was ended when the population was sprayed with DDT to kill the body lice that are the carriers of typhus.

Mulliken Robert Sanderson 1896–1986. US chemist and physicist who received the 1966 Nobel Prize for Chemistry for his development of the molecular orbital theory.

Mullis Kary Banks 1944– . US molecular biologist who developed the polymerase chain reaction (PCR) technique, which allows specific regions of DNA to be copied many times from a tiny sample, thus amplifying it to a large enough quantity to be analysed. This technique signalled an end to the laborious method of producing DNA fragments in vivo. He shared the Nobel Prize for Chemistry 1993 with Michael Smith.

In PCR, a small sample of DNA (for example, from a blood sample) is cut into segments and denatured (broken down) into single strands. Short oligonucleotide probes complementary to the DNA are prepared, and added in large amounts to the denatured DNA, and then incubated at 50–60°C. The probe joins to its correct site on the DNA, and acts as a starting point for synthesis of a new DNA chain. After the synthesis has finished the mixture is heated to 95°C to melt the newly formed DNA duplexes. The cycle can be repeated in order to amplify the desired sequence, and in just a few hours more than 100 billion copies can be made. These can then be used for analysis. The technique is now commonly used worldwide and is particularly important for disease diagnosis, for example, in the test for HIV.

Born in Lenoir, North Carolina, Mullis was educated at the Georgia Institute of Technology and the University of California, Berkeley, where he obtained his PhD in 1973. After conducting research at the University of Kansas Medical School and at the San Fransisco campus of the University of California, Mullis joined the Cetus Corporation of Emeryville, California in 1979.

Murchison Roderick Impey 1792–1871. Scottish geologist responsible for naming the

Silurian period (in his book *The Silurian System* 1839). Expeditions to Russia 1840–45 led him to define another worldwide system, the Permian, named after the strata of the Perm region. Knighted 1846, baronet 1866.

Murchison was born in Ross-shire. He entered the army at 15 and fought in the Peninsular War. Often accompanied by geologists Adam Sedgwick or Charles Lyell, Murchison made field explorations in Scotland, France, and the Alps. In 1855 he became director-general of the UK Geological Survey. An ardent imperialist, for many years he was also president of the Royal Geographical Society, encouraging African exploration and annexation.

Murchison believed in a universal order of the deposition of strata, indicated by fossils rather than solely by lithological features. Fossils showed a clear progression in complexity. The Silurian system contained, in his view, remains of the earliest life forms. He based it on studies of slate rocks in South Wales.

With Sedgwick's cooperation, Murchison also established the Devonian system in SW England.

Murdock William 1754–1839. Scottish inventor and technician. Employed by James Watt and Matthew Boulton to build steam engines, he was the first to develop gas lighting on a commercial scale, holding the gas in gasometers, from the 1790s.

Murdock was born in Auchinleck, Ayrshire. Between 1777 and 1830 he worked for Boulton and Watt, mainly in Birmingham and Cornwall.

In 1792, Murdock lighted his house in Redruth, Cornwall, by gases produced by distilling coal or wood, but it was not until about 1799 that he perfected methods for making, storing, and purifying gas. Watt and Boulton's Birmingham factory was lit by gas from 1802–03, and the manufacture of gasmaking plant seems to have begun about this period, probably in connection with apparatus for producing oxygen and hydrogen for medical purposes. A paper Murdock read before the Royal Society 1808 is the earliest practical essay on the subject.

Murdock made improvements to the steam engine, though he failed to persuade Boulton and Watt that a steam carriage was a practical idea. He was the first to devise an oscillating engine, about 1784. He also experimented with compressed air, and in 1803 constructed a steam gun.

Murphy William Parry 1892–1987. US physician who shared the Nobel Prize for Physiology or Medicine 1934 with George Minot and George Whipple for his discovery that liver cured patients suffering from pernicious anaemia, a disease that reduces the number of red blood cells and lessens the blood's oxygen-carrying capacity.

Murphy and Minot worked to find an effective treatment for pernicious anaemia, a disease that reduces the number of red blood cells in a patient's blood. It affects people in middle to old age and was invariably fatal.

Whipple, while working on dogs that he made anaemic by regularly drawing their blood, demonstrated that more blood had to be drawn to keep the animals anaemic if they were fed liver, beef, or spinach than if they were fed salmon and bread. In 1925, Minot and Murphy performed a clinical trial in which they fed patients half a pound of lightly cooked beef liver every day, which resulted in an astonishing regeneration of the red blood cells within a single week and the continued good health of the patients while they remained on the diet of liver.

Murphy was born in Stoughton, Wisconsin in the USA and studied medicine at the University of Oregon and later at Harvard Medical School. In 1925, Murphy went into private practice as a physician while working part-time in a research programme with Minot at Harvard.

Murray Joseph Edward 1919– . US surgeon whose work in the field of controlling rejection of organ transplants earned him a shared Nobel Prize for Physiology or Medicine 1990.

Muybridge *Series of photographs from Eadweard Muybridge's book* Animal Locomotion *(1877). Using a complex array of trip wires and cameras, Muybridge captured a wide variety of animal motion. He was also a pioneer of cinematography: he invented the zoopraxiscope, which used still images to give the impression of motion.*

Muybridge Eadweard, adopted name of Edward James Muggeridge 1830–1904. English-born US photographer. He made a series of animal locomotion photographs in the USA in the 1870s and proved that, when a horse trots, there are times when all its feet are off the ground. He also explored motion in birds and humans.

Nageli Karl Wilhelm von 1817–1891. Swiss botanist and early microscopist. He accurately described cell division, identifying chromosomes as 'transitory cytoblasts'. He was also the first to describe the antheridia (male reproductive organs) and spermatozoids (male gametes) of the fern family. He and Hugo von Mohl were responsible for distinguishing the protoplasm from the cell wall in plants.

Nageli was responsible for several important discoveries in plant biology, however, he also held fast to several erroneous beliefs and he rejected Gregor Mendel's paper that outlined the Mendelian laws of inheritance. His own work included the development of the idea of a meristem, a group of formative cells that are always capable of further division, but he wrongly believed that the apical cells (the growth cells at the tip of a plant's root) were the most important meristematic zone in all plants.

Nageli was born in Kilchberg, Switzerland and studied under Lorenz Oken and Augustin Pyrame de Candolle at the University of Geneva, and Matthias Jakob Schleiden at the University of Jena. He was made a professor at the Universities of Zurich, Freiburg, and Munich early in his career.

Nagell Trygve 1895– . Norwegian mathematician whose most important work was in the fields of abstract algebra and number theory.

Nagell was born and educated in Oslo, and remained at Oslo university until 1931, when he became professor at Uppsala, Sweden, retiring 1962.

Nagell first made his name with a series of papers in the early 1920s on indeterminate equations, investigations which led to the publication of a treatise on indeterminate analysis in 1929. He also published papers and books, in Swedish and English, on number theory. From the late 1920s onwards his chief interest was the study of algebraic numbers, and his 1931 study of algebraic rings was perhaps his most important contribution to abstract algebra.

Nansen Fridtjof 1861–1930. Norwegian explorer and scientist. In 1893, he sailed to the Arctic in the *Fram*, which was deliberately allowed to drift north with an iceflow. Nansen, accompanied by F Hjalmar Johansen (1867–1923), continued north on foot and reached 86° 14′ N, the highest latitude then attained. After World War I, Nansen became League of Nations high commissioner for refugees. Nobel Peace Prize 1923.

The ***Nansen passport*** issued to stateless persons is named after him.

Nansen was born in Store-Froen in Norway and studied at Christiana University (now Oslo University) and Naples before leaving to explore arctic regions aboard the vessel *Viking* 1882. Upon his return, he was made keeper of the museum of natural history at Bergen, but could not resist the invitation 1888 to join an expedition to cross the whole of Greenland. He was appointed professor of zoology 1897 and professor of oceanography at Christiania University 1908.

Following his academic career, Nansen became a statesman and was involved in the peaceful negotiations that led to the separation of Norway from Sweden. He became Norway's first ambassador to Great Britain 1906. Following World War I he was involved in the repatriation of prisoners of war in Europe, and from 1921 to 1923, Nansen was in charge of the Red Cross relief of Volga and South Ukraine in Russia.

Napier John, 8th Laird of Merchiston 1550–1617. Scottish mathematician who invented logarithms 1614 and 'Napier's bones', an early mechanical calculating device for multiplication and division.

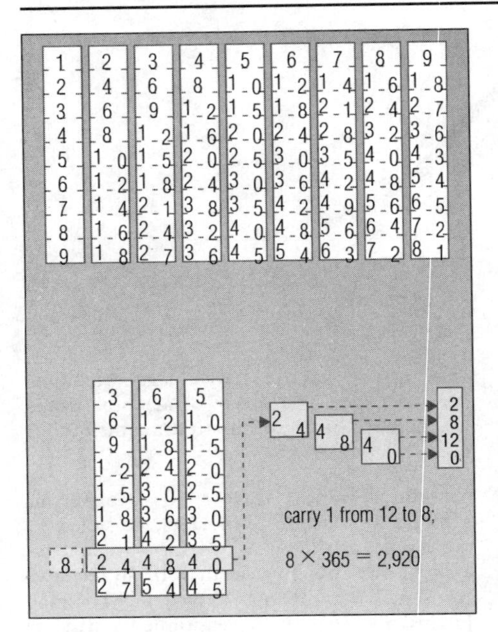

Napier *In 1617 Scottish mathematician John Napier published his description of what was arguably the first mechanical calculator – a set of numbered rods, usually made of bone or ivory and therefore known as Napier's bones. Using them, multiplication became merely a process of reading off the appropriate figures and making simple additions.*

It was Napier who first used and then popularized the decimal point to separate the whole number part from the fractional part of a number.

Napier was born in Merchiston Castle, near Edinburgh, and studied at St Andrews. He never occupied any professional post.

English mathematician Henry Briggs went to Edinburgh in 1616 and later to discuss the logarithmic tables with Napier. Together they worked out improvements, such as the idea of using the base ten.

Napier also made advances in scientific farming, especially by the use of salt as a fertilizer. In 1597 he patented a hydraulic screw by means of which water could be removed from flooded coal pits.

Napier published a denunciation of the Roman Catholic Church, *A Plaine Discovery of the Whole Revelation of St John* 1593, as well as *Mirifici logarithmorum canonis descriptio/Description of the Marvellous Canon of Logarithms* 1614 and *Mirifici logarithmorum canonis constructio* 1619. In *Rabdologiae* ('numeration by little rods') 1617 he explained his mechanical calculating system and showed how square roots could be extracted by the manipulation of counters on a chessboard.

Nasmyth James 1808–1890. Scottish engineer and machine-tool manufacturer whose many inventions included the steam hammer 1839 for making large steel forgings.

Nasmyth was born in Edinburgh and left school at the age of 12. As assistant to English toolmaker Henry Maudslay, he devised a flexible shaft of coiled spring steel for drilling holes in awkward places. After the death of Maudslay 1831, Nasmyth set up his own workshop, first in Edinburgh, then in Manchester, manufacturing machine tools, locomotives, and other machinery. His steam hammer was first used to make the propeller shaft for Isambard Kingdom Brunel's steamship *Great Britain*.

Nasmyth devised many other tools, including a vertical cylinder-boring machine which speeded up the production of steam engines, and all manner of lateral, transverse, and rotating cutting machines.

Nathans Daniel 1928– . US microbiologist who shared the 1978 Nobel Prize for Physiology or Medicine with his colleague Hamilton Smith for their work on restriction enzymes, special enzymes that can cleave genes into fragments.

Born in Wilmington, Delaware, Nathans studied at the University of Delaware and at Washington University, St Louis, Missouri. From 1962 he worked at Johns Hopkins University, Baltimore, becoming director of the Department of Microbiology 1972.

In addition to the work done with Smith, Nathans also performed much original research of his own in this field. Using the carcinogenic SV40 virus, he showed in 1971 that it could be cleaved into 11 specific fragments, and in the following year he determined the order of these fragments.

Natta Giulio 1903–1979. Italian chemist who worked on the production of polymers. He shared a Nobel prize 1963 with German chemist Karl Ziegler. Natta's early work on heterogeneous catalysts formed the basis for many important industrial syntheses.

Natta was born in Imperia, near Monaco, and studied at the Polytechnic Institute in Milan. After holding professorships at Pavia, Rome, and Turin, he returned to Milan Polytechnic 1938 as director of the Industrial Chemistry Research Institute, to work on artificial rubber. In 1953 he became a consultant to Montecatini, a company that had a licence arrangement with Ziegler.

Natta used Ziegler's catalysts to polymerize propylene (propene, $CH_3CH = CH_2$). In 1954 he found that part of the polymer is highly crystalline and coined the term 'isotactic' to describe the polymer's symmetrical structure.

The isotactic polymers Natta discovered after 1954 showed properties of commercial importance, such as high melting point, high strength, and an ability to form films and fibres. It was realized that a new type of polymerization, called coordination polymerization, was involved.

Chinese civilization has the overpowering beauty of the wholly other, and only the wholly other can inspire the deepest love and the profoundest desire to learn.

Joseph Needham

The Grand Titration

Needham Joseph 1900–1995. English biochemist and sinologist, historian of Chinese science. He worked first on problems in embryology. In the 1930s he learned Chinese and began to collect material. The first volume of his *Science and Civilization in China* was published 1954 and by 1990 sixteen volumes had appeared.

Needham was born in London and studied at Cambridge, where he spent his academic career. The arrival of some Chinese biochemists 1936 prompted him to learn their language, and in 1942–46 he travelled through China as head of the British Scientific Mission. From 1946 to 1948 he was head of the Division of Natural Sciences at the United Nations, after which he returned to Cambridge.

In *Chemical Embryology* 1931, Needham concluded that embryonic development is controlled chemically. The discovery of morphogenetic hormones and later of the genetic material DNA confirmed this view.

Needham became increasingly interested in the history of science, particularly of Chinese science, and he progressively reduced his biochemical investigations. *Science and Civilization in China* is a huge synthesis of history, science, and culture in China.

The place where we do our scientific work is a place of prayer.

Joseph Needham

The Harvest of a Quiet Eye

Néel Louis Eugène Félix 1904– . French physicist who worked on the magnetic properties of solids and predicted the existence of antiferromagnetism, a form of magnetism. Nobel Prize for Physics 1970.

In antiferromagnetic material, the molecular magnets are arranged in alternate directions and exhibit a very low magnetic susceptibility that increases with temperature. Above a certain temperature, now called the ***Néel temperature***, the susceptibility falls and the material becomes paramagnetic (weakly magnetic). This behaviour was experimentally confirmed 1938.

Néel also pointed out in 1947 that materials could exist in which the molecular magnets were of unequal strength. This phenomena is called ferrimagnetism and occurs in lodestone and some ferrites. He also explained the weak magnetism of certain rocks that made it possible to study the past history of the Earth's magnetic field.

Néel was born in Lyons, France, and gained his doctorate from the Ecole Normale Supérieure, Strasbourg 1932. He became professor of physics at the University of Strasbourg 1937. Néel later moved to the University of Grenoble where he was professor of physics until 1976. He became director (1954) and president (1971) of the Grenoble Polytechnic Institute, and director of the Grenoble Institute of Nuclear Studies 1956.

Neher Erwin 1944– . German cell physiologist who shared the Nobel Prize for Physiology or Medicine 1991 with Bert Sakmann for his studies on ion channels and beta-endorphin (a messenger hormone secreted by the pituitary gland).

Neher and Sakmann developed the patch-clamp technique 1976 to measure the electrical activity of very small portions of cell membranes. This technique revolutionized the study of ion channels. To perform the technique a glass pipette with a tip diameter of about one micrometer is pressed against a cell and slight suction is then applied to seal the cell membrane against the pipette. The technique allows the flow of ions through a single channel and transitions between different states of a channel to be monitored with a time resolution of microseconds.

Using this method, Neher and Sakmann investigated the effect of beta-endorphin on the membrane of cells. Beta-endorphin is a neurohormone (a messenger chemical that is made by nerve cells) secreted by the pituitary gland and an opiate that has been found to play a clinical role in the perception of pain, behavioural patterns, obesity, diabetes, and psychiatric disorders. Neher and Sakmann demonstrated that beta-endorphin acts not only on nerves in the brain to regulate their secretion of neurotransmitters but also, via calcium channels, acts on the walls of arteries in the brain.

Neher was born in Landsberg in Germany and trained originally as a physicist in Munich and at the University of Wisconsin. While working at the Max Planck Institute of Psychiatry in Munich, he took a year-long sabbatical to work with the physiologist Sakmann at Yale University.

Nernst (Walther) Hermann 1864–1941. German physical chemist who won a Nobel prize 1920 for work on heat changes in chemical reactions. He proposed in 1906 the principle known as the ***Nernst heat theorem*** or the third law of thermodynamics: chemical changes at the temperature of absolute zero involve no change of entropy (disorder).

Knowledge is the death of research.

Hermann Nernst

[On examinations] In *The Dictionary of Scientific Biography* ed C G Gillespie 1981

Nernst was born in Briesen, Prussia (now Wabreźno, Poland), and studied at Graz and Würzburg. He became professor of chemistry at Göttingen 1894, moving to Berlin 1905. During World War I, he was the first scientist to propose using chemical agents as a weapon.

In solution chemistry, every pH measurement depends on theories Nernst presented in the 1880s, as does the use and theory of indicators and buffer solutions.

In 1911, with British physicist Frederick Lindemann (later Lord Cherwell), Nernst constructed a special calorimeter for measuring specific heats at low temperatures.

With German chemist Fritz Haber he studied equilibria in commercially important gas reactions, such as the reversible reaction between hydrogen and carbon dioxide to form water and carbon monoxide. In 1918, Nernst investigated reactions that are initiated by light.

Having invented a substitute for the carbon filament in an electric lamp 1897, Nernst used the money from his patent to become a pioneer motorist. Many early automobiles had difficulty climbing hills, but Nernst devised a method of injecting nitrous oxide (dinitrogen monoxide) into the cylinders when the engine got into difficulties. In the 1920s he invented a 'Neo-Bechstein' piano which amplified sounds produced at low amplitudes.

His *Theoretische Chemie/Theoretical Chemistry* 1895 became a standard textbook.

Neugebauer Gerald 1932– . German-born US astronomer whose work has been crucial in establishing infrared astronomy. He has been closely involved with the space agency NASA's interplanetary missions and the design of new infrared telescopes.

Neugebauer was born in Göttingen, Germany, and studied at Cornell University and at the California Institute of Technology, where he spent his academic career from 1962. He became professor 1970 and director of the Palomar Observatory 1981. His involvement with NASA began 1969 and included work on the infrared radiometers carried aboard the Mariner missions to Mars. In 1976 he became the US principal scientist on the Infrared Astronomical Satellite.

During the mid-1960s Neugebauer and his colleagues began to establish the first infrared map of the sky. Some 20,000 new infrared sources were detected and most of these did not coincide with known optical sources. Among the brightest and strangest of these sources is in the Orion nebula, and is known as the Becklin–Neugebauer object after its discoverers. Carbon monoxide is blowing outwards from it at a high velocity. The object is thought to be a very young star.

Newcomb Simon 1835–1909. Canadian-born US mathematician and astronomer who compiled charts and tables of astronomical data with phenomenal accuracy. His calculations of the motions of the bodies in the Solar System were in use as daily reference all over the world for more than 50 years, and the system of astronomical constants for which he was most responsible is still the standard.

Newcomb was born in Wallace, Nova Scotia, and had little or no formal education.

In his teens he ran away to the USA, and eventually enrolled at Harvard. In 1861 he joined the navy, where he was assigned to the US Naval Observatory at Washington DC, and in 1877 put in charge of the American Nautical Almanac office. From 1884 he was also professor of mathematics and astronomy at Johns Hopkins University. He retired with the rank of rear admiral.

At the Nautical Almanac office, Newcomb started the great work that was to occupy the rest of his life: the calculation of the motions of the bodies in the Solar System. The results were published in *Astronomical Papers Prepared for the Use of the American Ephemeris and Nautical Almanac*, a series that he founded 1879.

With his British counterpart Arthur Matthew Weld Downing (1850–1917), Newcomb established a universal standard system of astronomical constants. This was adopted at an international conference 1896, and again 1950.

Newcomen Thomas 1663–1729. English inventor of an early steam engine. His 'fire engine' 1712 was used for pumping water from mines until James Watt invented one with a separate condenser.

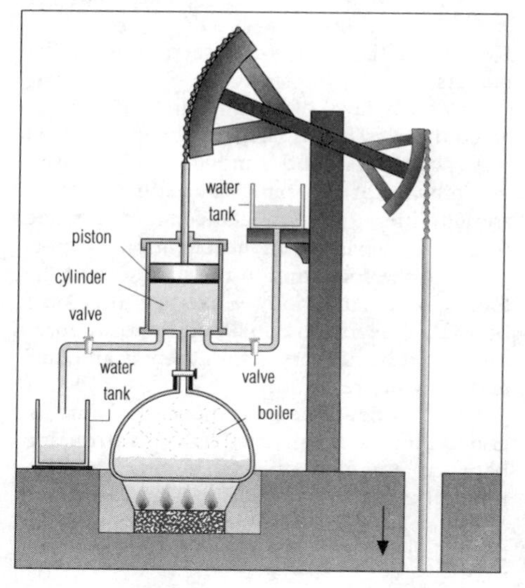

Newcomen *Thomas Newcomen's steam engine, invented 1712, was the first practical steam engine and was used to power pumps in the tin mines of Cornwall and the coal mines of N England. Steam from the boiler entered the cylinder as the piston moved up (pulled by the weight of a wooden beam). Water from a tank was then sprayed into the cylinder, condensing the steam and creating a vacuum so that air pressure forced down the piston and activated the pump.*

Newcomen was born in Dartmouth, Devon, and set up a blacksmith's shop there, assisted by a plumber called John Calley (died 1717). The first authenticated Newcomen engine was erected in 1712 near Dudley Castle, Wolverhampton, but a number of earlier machines must have been operated to develop the engine to this point. The whole situation is confused by a patent granted to Thomas Savery to 'raise water by the force of fire'; in later years Newcomen paid royalties to Savery.

Newcomen's engine consumed an enormous amount of coal because fresh hot steam had to be raised for each piston stroke. The early engines were very expensive because the cylinder was made of brass; later, iron cylinders were produced, but they were thick-walled and consequently even less efficient in terms of coal consumed. However, they were mostly used in coal mines.

It was with the Newcomen engine that the age of steam began.

Newlands John Alexander Reina 1837–1898. English chemist who worked as an industrial chemist; he prepared in 1863 the first periodic table of the elements arranged in order of relative atomic masses, and pointed out 1865 the 'law of octaves' whereby every eighth element has similar properties. He was ridiculed at the time, but five years later Russian chemist Dmitri Mendeleyev published a more developed form of the table, also based on atomic masses, which forms the basis of the one used today (arranged by atomic number).

Newlands was born in London and studied there at the Royal College of Chemistry. In 1860 he served as a volunteer with Giuseppe Garibaldi in his campaign to unify Italy (Newlands was of Italian descent on his mother's side). He set up in practice as an analytical chemist 1864, and in 1868 became chief chemist in a sugar refinery, where he introduced a number of improvements in processing. Later he left the refinery and again set up as an analyst.

Like many of his contemporaries, Newlands first used the terms 'equivalent weight' and 'atomic weight' without any distinction in meaning, and in his first paper 1863 he used the values accepted by his predecessors. The incompleteness of a table he drew up 1864 he attributed to the possible existence of additional, undiscovered elements; for example, he predicted the existence of germanium.

Newton Alfred 1829–1907. English ornithologist with a special interest in the great auk and the ornithology of Iceland.

Newton was born in Geneva, the son of William Newton, MP for Ipswich. He went to Magdalene College, Cambridge 1848 and graduated 1853. Despite the fact that he suffered from diseased hip joints and walked with the aid of two sticks, he travelled throughout Lapland, Iceland, the West Indies, and North America 1854–63. During these expeditions he studied ornithology and became particularly interested in the great auk. Upon his return he published the *Zoology of Ancient Europe* and *Ornithology of Iceland.*

In 1866 he was made the first professor of zoology and comparative anatomy at Cambridge, a position which he held until his death. He published a *Dictionary of Birds* 1893–36 and was elected Fellow of the Royal Society 1870. In 1900 he was awarded both the Royal medal and the Linnaean Society Gold medal.

Newton Isaac 1642–1727. English physicist and mathematician who laid the foundations of physics as a modern discipline. During 1665–66, he discovered the binomial theorem, differential and integral calculus, and that white light is composed of many colours. He developed the three standard laws of motion and the universal law of gravitation, set out in *Philosophiae naturalis principia mathematica* 1687 (usually referred to as the *Principia*). Knighted 1705.

Newton's greatest achievement was to demonstrate that scientific principles are of universal application. He clearly defined the nature of mass, weight, force, inertia, and acceleration.

In 1679 Newton calculated the Moon's motion on the basis of his theory of gravity and also found that his theory explained the laws of planetary motion that had been derived by German astronomer Johannes Kepler on the basis of observations of the planets.

I do not know what I may appear to the world, but to myself I seem to have been only a boy playing on the sea-shore, and diverting myself in now and then finding a smoother pebble or a prettier shell than ordinary, whilst the great ocean of truth lay all undiscovered before me.

Isaac Newton

Quoted in L T More *Isaac Newton*

Newton, Isaac *A reconstruction of Isaac Newton investigating the properties of light refracted through a prism. He found that a prism dispersed white light into a spectrum of colours and that each colour had a different refrangibility. As a result, in 1668, after concluding that it was impossible to generate a distinct image with a single lens because the constituent coloured rays would be brought to focus at slightly different points, Newton constructed the first reflecting telescope using curved mirrors.*

Newton's laws of motion are: (1) Unless acted upon by an unbalanced force, a body at rest stays at rest, and a moving body continues moving at the same speed in the same straight line. (2) An unbalanced force applied to a body gives it an acceleration proportional to the force and in the direction of the force. (3) When a body A exerts a force on a body B, B exerts an equal and opposite force on A; that is, to every action there is an equal and opposite reaction.

Newton developed his general theory of gravitation as a universal law of attraction between any two objects, stating that the force of gravity is proportional to the masses of the objects and decreases in proportion to the square of the distance between the two bodies.

Newton was born at Woolsthorpe Manor, Lincolnshire, and studied at Cambridge, where he became professor at the age of 26. He resisted James II's attacks on the liberties of the universities, and sat in the parliaments of 1689 and 1701–02 as a Whig. Appointed warden of the Royal Mint in 1696, and master in 1699, he carried through a reform of

If I have seen farther it is by standing on the shoulders of giants.

Isaac Newton

Letter to Robert Hooke Feb 1675

the coinage. Most of the last 30 years of his life were taken up by studies of theology and chronology, and experiments in alchemy.

Newton and German mathematician Gottfried Leibniz worked independently on the development of a differential calculus, both making significant advances, but Newton claimed to be its sole inventor. When Leibniz appealed to the Royal Society for a fair hearing, Newton appointed a committee of his own supporters and even wrote their report himself. The result of this controversy was to isolate English mathematics and to set it back many years, for it was Leibniz's terminology that came to be used. A similar dispute arose between Newton and English scientist Robert Hooke, who claimed prior discovery of the inverse square law of gravitation.

Newton began to investigate the phenomenon of gravitation in 1665, inspired, legend has it, by seeing an apple fall from a tree. But he was also active in algebra and number theory, classical and analytical geometry, computation, approximation, and even probability.

A by-product of his experiments with light and prisms was the development of the reflecting telescope. Newton investigated many other optical phenomena, including thin film interference effects, one of which, 'Newton's rings', is named after him.

De motu corporum in gyrum/On the Motion of Bodies in Orbit was written in 1684. The publication of the *Principia* was financed by his friend Edmond Halley. In 1704, Newton summed up his life's work on light in *Opticks*.

Nice Margaret, (born Morse) 1883–1974. US ornithologist who made an extensive study of the life history of the sparrow. She also campaigned against the indiscriminate use of pesticides.

Morse was born in Amherst, Massachusetts, and studied at Mount Holyoke College and Clark University in Worcester, Massachusetts, graduating in child psychology. She never had an academic appointment.

Her first ornithological research was a detailed study of the birds of Oklahoma. In 1927 she moved to Ohio, where she carried out the study of sparrows that established her as one of the leading ornithologists in the world, recording the behaviour of individual birds over a long period of time. A family move to Chicago provided fewer opportunities for Nice to study living birds, so she spent more of her time writing, and became involved in conservation issues.

Nicol William 1768–1851. Scottish physicist and geologist who invented the first device for obtaining plane-polarized light – the Nicol prism – in 1828.

Nicol was born in Edinburgh and lectured at the university there. He did not publish any of his research findings until 1826.

Nicol made his prism by bisecting a parallelepiped of Iceland spar (a naturally occurring, transparent crystalline form of calcium carbonate) along its shortest diagonal, then cementing the two halves together with Canada balsam. Light entering the prism is refracted into two rays, one of which emerges as plane-polarized light. Nicol prisms greatly facilitated the study of refraction and polarization, and were later used to investigate molecular structures and optical activity of organic compounds.

In 1815, Nicol developed a method of preparing extremely thin sections of crystals and rocks for microscopical study. His technique (which involved cementing the specimen to a glass slide and then carefully grinding until it was extremely thin) made it possible to view mineral samples by transmitted rather than reflected light and therefore enabled the minerals' internal structures to be seen.

Nicolle Charles Jules Henri 1866–1936. French bacteriologist whose discovery in 1909 that typhus is transmitted by the body louse made the armies of World War I introduce delousing as a compulsory part of the military routine. Nobel Prize for Physiology or Medicine 1928.

His original observation was that typhus victims, once admitted to hospitals, did not infect the staff; he speculated that transmission must be via the skin or clothes, which were washed as standard procedure for new admissions. The experimental evidence was provided by infecting a healthy monkey using a louse recently fed on an infected chimpanzee.

Nicolle was born in Rouen and studied in Paris. He became director of the Pasteur Institute in Tunis 1902. From 1932 he was professor at the Collège de France. He was also a novelist.

Niepce Joseph Nicéphore 1765–1833. French pioneer of photography. Niepce invented heliography, a precursor of photography that fixed images onto pewter plates coated with pitch and required eight-hour exposures. He produced the world's first photograph from nature 1826 and later collaborated with Daguerre on the faster daguerreotype process.

Niepce was born in Chalon-sur-Saône and became administrator of Nice 1795. From 1801 he devoted himself to research.

The first photograph was a positive image, a view from Niepce's attic bedroom. The image was captured, after an eight-hour exposure, in a camera obscura on a metal plate coated with light-sensitive bitumen. The plate has survived and is to be found in the University of Texas.

Nirenberg Marshall Warren 1927– . US biochemist who shared the 1968 Nobel Prize for Physiology or Medicine for his work in deciphering the chemistry of the genetic code.

Nirenberg was born in New York and studied at the universities of Florida and Michigan. From 1957 he was at the National Institute of Health, later moving to the Laboratory of Biochemical Genetics at the National Heart, Lung, and Blood Institute in Bethesda, Maryland.

Nirenberg was interested in the way in which the nitrogen bases – adenine (A), cytosine (C), guanine (G), and thymine (T) – specify a particular amino acid. To simplify the task of identifying the RNA triplet (codon) responsible for each amino acid, he used a simple synthetic RNA polymer. He found that certain amino acids could be specified by more than one codon, and that some triplets did not specify an amino acid at all. These 'nonsense' triplets signified the beginning or the end of a sequence. He then worked on finding the orders of the letters in the triplets, and obtained unambiguous results for 60 of the possible codons.

Second to agriculture, humbug is the biggest industry of our age.

Alfred Nobel

Attributed remark

Nobel Alfred Bernhard 1833–1896. Swedish chemist and engineer. He invented dynamite in 1867, gelignite 1875, and ballistite, a smokeless gunpowder, in 1887. Having amassed a large fortune from the manufacture of explosives and the exploitation of the Baku oilfields in Azerbaijan, near the Caspian Sea, he left this in trust for the endowment of five Nobel prizes.

Nobel *Swedish chemist and industrialist Alfred Nobel spent most of his working life developing explosives, particularly dynamite and gelignite. On his death in 1896, Nobel left all his considerable fortune to a foundation that funds the annual awards that bear his name. The first Nobel prizes – for physics, physiology or medicine, chemistry, literature, and contributions to peace – were awarded 1901.*

Nobel was born in Stockholm and studied in Europe and North America. He worked in his father's company in St Petersburg during the Crimean War 1853–56, producing munitions. After the war his father went bankrupt, and in 1859 the family returned to Sweden. During the next few years Nobel developed several new explosives and factories for making them. In 1864, a nitroglycerine factory blew up, killing Nobel's younger brother and four other people.

In 1863 Nobel invented a mercury fulminate detonator for use with nitroglycerine. Dynamite was invented to make the handling of nitroglycerine safer, by mixing it with kieselguhr, a porous diatomite mineral. Gelignite, a colloidal solution of nitrocellulose (gun cotton) in nitroglycerine, was safer still: less sensitive to shock and strongly resistant to moisture.

Nobel also worked in electrochemistry, optics, biology, and physiology, and helped to solve many problems in the manufacture of artificial silk, leather, and rubber.

Noether (Amalie) Emmy 1882–1935. German mathematician who became one of the leading figures in abstract algebra. Modern work in the field of a general theory of ideals dates from her papers of the early 1920s.

Noether was born in Erlangen, the daughter of mathematician Max Noether. Despite a rule barring women from university study, she was awarded a doctorate from Erlangen in 1907 for a thesis on algebraic invariants. But as a woman she could not hold a post in the university faculty. She persisted with her research independently and at the request of mathematician David Hilbert was invited to lecture at Göttingen 1915. There she worked with Hilbert on problems arising from Albert Einstein's theory of relativity, and in 1922 became associate professor. She remained at Göttingen until the Nazi purge of Jewish university staff in 1933. The rest of her life was spent as professor of mathematics at Bryn Mawr College in Pennsylvania, USA.

Noether first made her mark as a mathematician with a paper 1920 on noncommutative fields (where the *order* in which the elements are combined affects the result). For the next few years she worked on the establishment and systematization of a theory of ideals, and introduced the concept of primary ideals. After 1927 she returned to the subject of noncommutative algebras, her chief investigations being conducted into linear transformations of noncommutative algebras and their structure.

Noether Max 1844–1921. German mathematician who contributed to the development of algebraic geometry and the theory of algebraic functions.

Noether was born in Mannheim and studied at Heidelberg. He spent his career there and at Erlangen, where he became professor 1888. His daughter Emmy Noether became a notable mathematician.

In 1873 he published his one outstanding result, the theorem concerning algebraic curves which contains the 'Noether conditions'. Given two algebraic curves, $\Phi(x,y) = 0$ and $\Psi(x,y) = 0$, which intersect at a finite number of isolated points, the equation of an algebraic curve that passes through all the points of intersection may be expressed as:

$$A\Phi + B\Psi = 0$$

where A and B are polynomials in x and y if, and only if, certain conditions (the 'Noether conditions') are satisfied.

Noguchi Hideyo, (born Noguchi Seisaku) 1876–1928. Japanese bacteriologist who studied syphilitic diseases, snake venoms, trachoma, and poliomyelitis. He discovered the parasite of yellow fever, a disease from which he died while working in British W Africa.

Norrish Ronald George Wreyford 1897–1978. English physical chemist who studied fast chemical reactions, particularly those initiated by light. He shared the 1967 Nobel Prize for Chemistry with his co-worker George Porter. Norrish was largely responsible for the advance of reaction kinetics to a distinct discipline within physical chemistry.

Norrish was born and educated in Cambridge and spent his academic career there, becoming professor 1937.

Norrish began working in photochemistry in 1923. His interest in using intense flashes of light to initiate photochemical reactions seems to have been stimulated by his work during World War II with his student George Porter, investigating methods of suppressing the flash from guns and developing incendiary materials. By varying the time delay between two flashes, Norrish was able to study the kinetics of the formation and decay of very short-lived radicals or ions.

Norrish went on to apply these techniques to the study of chain reactions. He also made pioneering studies of the kinetics of polymerization. He and his co-workers discovered the gel effect, which occurs in the later stages of free-radical polymerization.

Northrop John Howard 1891–1987. US chemist. In the 1930s he crystallized a number of enzymes, including pepsin and trypsin, showing conclusively that they were proteins. He shared the 1946 Nobel Prize for Chemistry with Wendell Stanley and James Sumner.

Noyce Robert Norton 1927–1990. US scientist and inventor, with Jack Kilby, of the integrated circuit (chip), which revolutionized the computer and electronics industries in the 1970s and 1980s. In 1968 he and six colleagues founded Intel Corporation, which became one of the USA's leading semiconductor manufacturers.

Noyce was awarded a patent for the integrated circuit 1959. In 1961 he founded his first company, Fairchild Camera and Instruments Corporation, around which Silicon Valley was to grow. The company was the first in the world to understand and exploit the commercial potential of the integrated circuit. It quickly became the basis for such products as the personal computer, the pocket calculator, and the programmable microwave oven. At the time of his death, he was president of Sematech Incorporated, a government–industry research consortium created to help US firms regain a lead in semiconductor technology that they had lost to Japanese manufacturers.

Nurse Paul Maxime 1949– . English microbiologist who has contributed much to our knowledge of the molecular mechanisms of cell growth. He worked on the genetic and enzymatic control of the cell cycle in yeast, a microorganism that he used as a model system for mammalian cells.

Nurse showed that genes for protein kinases (a family of enzymes that chemically bond a phosphate molecule to other cellular proteins) were crucial in controlled cell growth. His work is significant for cancer research as it contributes to the understanding of why cancer cells might undergo uncontrolled cell growth and how this might be prevented by new, highly specific drugs.

Nurse was born in Norwich and studied biology at the University of Birmingham. He obtained his PhD at the University of East Anglia before working in London, Bern, Edinburgh, and Sussex. He was elected Iveagh professor of microbiology 1991, and director of research at the Imperial Cancer Research Laboratories in London 1993, then director general 1996. He was elected a Fellow of the Royal Society 1989 in recognition of his pioneering work on the genetic and enzymatic control of the cell cycle.

Nusslein-Volhard Christine 1942– . German geneticist who shared the Nobel Prize for Physiology or Medicine 1995 with Edward Lewis and Eric Wieschaus for her work on the genes controlling early embryonic development of the fruit fly *Drosophila melanogaster*. She examined 40,000 random gene mutations for their effect on the fly's development and identified 150 genes. She has since cloned several of those genes and worked out their interactions.

Nusslein-Volhard performed her experiments to identify all of the genes involved in the development of the fruit fly at the European Laboratory for Molecular Biology at Heidelberg. The mutant strains of flies she created have since been worked on by many other developmental biologists.

She was inspired in the late 1970s by the pioneering work of Edward Lewis, who had identified the transformations in the fruit fly that cause substitution of one segment of the body for another. These transformations were found to be the result of mutations in a gene family called the bithorax complex. Genes at the beginning of the complex were found to control anterior body segments, while genes further down the genetic map controlled more posterior body segments. This work was shown to be of particular importance when it was demonstrated that the gene ordering of this complex is conserved in humans. Since 1986 she has been director of the genetics division of the Max Planck Institute for Developmental Biology, Tübingen.

Nuttall Thomas 1786–1859. English-born US naturalist who explored the Arkansas, Red, and Columbia rivers and wrote books ranging from ornithology to botany.

Nuttall was born in Settle, Yorkshire, but in 1808 emigrated to Philadelphia. He was a botanist and ornithologist, but he also had an interest in geology. Between 1809 and 1811 he participated in an expedition up the Missouri River. In 1818–20, he explored the Arkansas and Red rivers, and the mouth of the Columbia River 1834–35. In 1820, he studied the geology of the Mississippi Valley. He was made curator of the botanical gardens in Harvard 1822, a position which he held for ten years.

He published *Genera of North American Plants* 1818 and *A Manual of Ornithology* 1832. He returned to England where he died near Liverpool.

Nyholm Ronald Sydney 1917–1971. Australian inorganic chemist who worked on the coordination compounds (complexes) of the transition metals. Knighted 1967.

Nyholm was born in Broken Hill, New South Wales, and studied at Sydney and, after World War II, at University College, London, where he became professor 1955. As chair of the Chemistry Consultative Committee, he was largely responsible for the Nuffield chemistry course taught in British schools and for changes to the examination syllabuses. He advocated an integrated approach to the teaching of chemistry.

Nyholm was able to prepare stable compounds of transition metals in valence states that previously had been thought to be unstable. For example, he prepared an octahedral complex of nickel(III), in which the nickel has a coordination number of six. He also made the diarsine complexes of the tetrachloride and tetrabromide of titanium(IV), the first example of an 8-coordination compound of a first-row transition metal.

Nyholm systematically exploited physical methods to study the structures and properties of coordination compounds. He employed X-ray crystallography and nuclear magnetic resonance spectroscopy, and found that magnetic moment seemed to give the closest connection between electronic structure, chemical structure, and stereochemistry.

Ochoa Severo 1905–1993. Spanish-born US biochemist who discovered an enzyme able to assemble units of the nucleic acid RNA 1955. For his work towards the synthesis of RNA, Ochoa shared the 1959 Nobel Prize for Physiology or Medicine.

Ochoa was one of the pioneers in molecular biology and genetic engineering. His early work concerned biochemical pathways in the human body, especially those involving carbon dioxide, but his main research was into nucleic acids and how their nucleotide units are linked, either singly (as in RNA) or to form two helically wound strands (as in DNA). In 1955 Ochoa obtained an enzyme from bacteria that was capable of joining together similar nucleotide units to form a nucleic acid, a type of artificial RNA. Nucleic acids containing exactly similar nucleotide units do not occur naturally, but the method of synthesis used by Ochoa was the same as that employed by a living cell.

He was born in Luarca, Asturias region, the youngest son of a lawyer. He studied at Málaga and Madrid, and obtained a degree in medicine from the University of Madrid 1929. He lectured at Madrid 1931–35, until the threat of the Spanish Civil War forced him to flee Spain for Germany, where he worked at the University of Heidelberg.

With the rise of Hitler, he fled again 1937, this time to Britain, where he worked at Oxford University. He moved to the USA 1940, and worked at Washington University 1941–42, before moving to New York University, first as a research associate in the college of medicine, and then 1954–75 as a professor in the department of biochemistry. He joined the Roche Institute of Molecular Biology 1975, and became a US citizen 1956.

Ochoa, who counted Dali and Lorca among his friends, was a very dedicated world citizen and a music lover. He was president of the International Union of Biochemistry, and a member of the Soviet Academy of Sciences. He returned to Spain permanently 1985, where his fame is attested by the fact that most cities have a street named after him.

Oersted Hans Christian 1777–1851. Danish physicist who founded the science of electromagnetism. In 1820 he discovered the magnetic field associated with an electric current.

Oersted was born at Rudkøbing, Langeland, and studied at Copenhagen. He worked as a pharmacist before making a tour of Europe 1801–03 to complete his studies in science. On his return, Oersted gave public lectures with great success, and was professor of physics at Copenhagen 1806–29, when he became director of the Polytechnic Institute in Copenhagen.

Oersted *Danish physicist Hans Christian Oersted shown discovering the effect of an electric current on a magnetic needle during a lecture at the University of Copenhagen 1820. As well as his pioneering work on the electromagnetic effect, Oersted was also the first scientist to isolate the element aluminium.*

Believing that all forces must be interconvertible, Oersted had predicted in 1813 that an electric current would produce magnetism when it flowed through a wire, just as it produced heat and light. His 1820 experiment involved a compass needle placed beneath a wire connected to a battery. He found that a circular magnetic field is produced around a wire carrying a current.

In 1822, Oersted turned to the compressibility of gases and liquids, devising a useful apparatus to determine compressibility. He also investigated thermoelectricity, in 1823.

Ohm Georg Simon 1789–1854. German physicist who studied electricity and discovered the fundamental law that bears his name. The SI unit of electrical resistance, the ***ohm***, is named after him, and the unit of conductance (the inverse of resistance) was formerly called the ***mho***, which is 'ohm' spelled backwards.

Ohm's law states that the steady electrical current in a metallic circuit is directly proportional to the constant total electromotive force in the circuit.

Ohm was born and educated in Erlangen, Bavaria. He worked as a schoolteacher until 1833, when he became professor of physics at the Polytechnic Institute, Nuremberg, moving to Munich University 1849.

Ohm began the work that led him to his law of electricity in 1825. He investigated the amount of electromagnetic force produced in a wire carrying a current, expecting it to decrease with the length of the wire in the circuit. Using a thermocouple because it produced a constant electric current, he employed an electroscope to measure how the tension varied at different points along a conductor to verify his law, and presented his arguments in mathematical form in his great work *Die Galvanische Kette* 1827.

Olah George Andrew 1927– . Hungarian-born US chemist who was awarded the 1992 Nobel Prize for Chemistry for his isolation of carbocations, electrically charged fragments of hydrocarbon molecules (molecules containing only hydrogen and carbon). His work launched a new branch of organic chemistry and led to the development of new carbon-based fuels.

In the early 1960s the chemical reactions of the hydrocarbons were poorly understood. It was assumed that hydrocarbons reacted by forming unstable, short-lived intermediate compounds, but these intermediates had never been observed. In 1962 Olah discovered that these intermediates, now known as carbocations, could be created and kept stable in solutions of very strong acids, billions of times stronger than automobile battery acid. This discovery not only provided proof that the intermediates existed but also allowed chemists to study their properties.

Olah was born in Budapest and educated at the University of Budapest, Hungary, receiving his doctorate 1949. He taught at the university until 1954, when he moved to the Hungarian Academy of Sciences. In 1956, after the Soviet invasion of Hungary, he and his wife fled to Canada and then the United States, where he worked for the Dow Chemical Company. In 1965 he was appointed professor at Case Western Reserve University, Cleveland, Ohio. He moved to the University of Southern California at Los Angeles 1977, becoming director of the Loker Hydrocarbon Research Institute in 1980.

Olbers Heinrich Wilhelm Matthäus 1758–1840. German astronomer, a founder member of the ***Celestial Police***, a group of astronomers who attempted to locate a supposed missing planet between Mars and Jupiter. During his search he discovered two asteroids, Pallas 1802 and Vesta 1807. Also credited to Olbers are a number of comet discoveries, a new method of calculating cometary orbits, and the stating of Olbers' paradox.

Olbers' paradox is the question, put forward 1826: If the universe is infinite in extent and filled with stars, why is the sky dark at night? Olbers explained the darkness of the night sky by assuming that space is not absolutely transparent and that some interstellar matter absorbs a very minute percentage of starlight. This effect is sufficient to dim the light of the stars, so that they are seen as points against the dark sky. In fact, darkness is now generally accepted as a by-product of the red shift caused by stellar recession.

Olbers was born near Bremen and studied medicine at Göttingen. He practised as a physician in Bremen 1781–1823, but was a keen amateur astronomer, with an observatory at the top of his house.

Olbers' main interest was the search for comets, and his efforts were rewarded with the discovery of four more, including one, discovered in 1815, which has an orbit of 72 years, similar to Halley's. Olbers calculated the orbits of 18 other comets. Noticing that comets consist of a starlike nucleus and a parabolic cloud of matter, he suggested that this matter was expelled by the nucleus and repelled by the Sun.

Oldenburg Henry 1615–1677. German official, residing in London from 1652, who founded and edited in 1665 the first-ever scientific periodical, *Philosophical Transactions*. He was secretary to the Royal Society 1663–77 and through his extensive correspondence acted as a clearing house for the science of the day.

Olds Ransom Eli 1864–1950. US car manufacturer. In 1895 he produced a gas-powered car and in the following year founded the Olds Motor Vehicle Company. Reorganizing the operation as the Olds Motor Works, he produced his popular Oldsmobiles from 1899 in Detroit. He pioneered the assembly-line method of car production that would later be refined by Henry Ford.

Olds was born in Geneva, Ohio, and raised in Michigan. He experimented with steam-powered prototype automobiles from 1886. After selling the Olds Motor Works 1904, Olds established the Reo Motor Car Company, serving as its president 1904–24 and chair of the board 1924–36.

Omar Khayyám *c.* 1050–*c.* 1123. Persian astronomer, mathematician, and poet. In the West, he is chiefly known as a poet through Edward Fitzgerald's version of 'The Rubaiyat of Omar Khayyám' 1859.

Khayyám was born in Nishapur. He founded a school of astronomical research and assisted in reforming the calendar. The result of his observations was the *Jalālīa* era, begun 1079. He wrote a study of algebra, which was known in Europe as well as in the East.

Onsager Lars 1903–1976. Norwegian-born US physical chemist. He worked on the application of the laws of thermodynamics to systems not in equilibrium, and received the 1968 Nobel Prize for Chemistry.

Onsager was born in Christiania (now Oslo) and studied at Norges Tekniske Høgskole in Trondheim. After working in Zürich, Switzerland, as research assistant to Dutch chemist Peter Debye, Onsager emigrated to the USA in 1928. As a lecturer first at Brown University and from 1933 at Yale, he was a failure: the students named his courses 'Sadistical Mechanics' and 'Advanced Norwegian I and II'.

At Brown University Onsager submitted a PhD thesis on what is now a classic work on reversible processes, but the authorities turned it down. It was published in 1931 but ignored until the late 1940s; in 1968 it earned Onsager the Nobel Prize. At Yale his paper called 'Solutions to the Mathieu equation of period 4π and certain related functions' was passed in incomprehension among the chemistry, physics, and mathematics departments before Onsager got his PhD.

In Zürich Onsager put forward a modification to the Debye–Hückel ionization theory. Now known as the Onsager limiting law, this gave better agreement between calculated and actual conductivities.

Investigating the connection between microscopic reversibility and transport processes, Onsager found that the key to the problem is the distribution of molecules and energy caused by random thermal motion. Ludwig Boltzmann had shown that the nature of thermal equilibrium is statistical and that the statistics of the spontaneous deviation is determined by the entropy. Using this principle Onsager derived a set of equations known as Onsager's law of reciprocal relations, sometimes called the fourth law of thermodynamics.

In 1949, he established a firm statistical basis for the theory of liquid crystals.

Oort Jan Hendrik 1900–1992. Dutch astronomer. In 1927, he calculated the mass and size of our Galaxy, the Milky Way, and the Sun's distance from its centre, from the observed movements of stars around the Galaxy's centre. In 1950 Oort proposed that comets exist in a vast swarm, now called the ***Oort cloud***, at the edge of the Solar System.

In 1944 Oort's student Hendrik van de Hulst (1918–) calculated that hydrogen in space would emit radio waves at 21 cm/8.3 in wavelength, and in the 1950s Oort's team mapped the spiral structure of the Milky Way from the radio waves given out by interstellar hydrogen.

Oort was born in Franeker, Friesland, and studied at Groningen. He spent most of his career at Leiden, becoming professor 1935 and director of the observatory 1945.

Oort confirmed the calculations of astronomers Bertil Lindblad and Harlow Shapley and went on to show that the stars in the Milky Way were arranged like planets revolving round a sun, in that the stars nearer the centre of the Galaxy revolved faster round the centre than those farther out. He established radio observatories at Dwingeloo and Westerbork, which put the Netherlands in the forefront of radio astronomy.

Oparin Alexandr Ivanovich 1894–1980. Russian biochemist who in the 1920s developed one of the first of the modern theories about the origin of life on Earth, postulating a primeval soup of biomolecules.

Oparin was born near Moscow and studied plant physiology at Moscow State University. In 1929 he became professor of plant biochemistry at Moscow State University. He was a cofounder of the Bakh Institute of Biochemistry in Moscow 1935, and its director from 1946.

Oparin's ideas about the origin of life contained three basic premises: that the first organisms arose in the ancient seas, which contained many already formed organic compounds that the organisms used as nutriment; that there was a constant, virtually limitless supply of external energy in the form of sunlight; and that true life was characterized by a high degree of structural and functional organization, contrary to the prevailing view that life was basically molecular. Oparin's theory, first published 1924, stimulated much research into the origin of life, notably US chemist Stanley Miller's attempt in 1953 to reproduce primordial conditions in the laboratory.

Oparin also researched into enzymology and did much to provide a technical basis for industrial biochemistry in the USSR.

His works include *The Origin of Life on Earth* 1936.

Öpik Ernst Julius 1893–1985. Estonian astronomer whose work on the nature of meteors and comets was instrumental in the development of heat-deflective surfaces for spacecraft on their re-entry into the Earth's atmosphere.

Öpik was born near Rakvere and studied at Tartu, where he spent most of his academic career 1921–44. He then moved to Germany, becoming professor at the Baltic University in 1945. Three years later, Öpik moved to Northern Ireland, where he eventually became director of the Armagh Observatory. From 1956 onward he held a concurrent post at the University of Maryland, USA.

Öpik was the originator of a method for counting meteors that requires two astronomers to scan simultaneously. His theories on surface events in meteors upon entering the Earth's atmosphere at high speed (the ablation, or progressive erosion, of the outer layers) proved to be extremely important in the development of heat shields and other protective devices to enable a spacecraft to withstand the friction and the resulting intense heat upon re-entry.

Much of Öpik's other work was directed at the analysis of comets that orbit our Sun. He postulated that the orbit of some of these comets may take them as far away as 1 light year.

Oppenheimer J(ulius) Robert 1904–1967. US physicist. As director of the Los Alamos Science Laboratory 1943–45, he was in charge of the development of the atom bomb (the Manhattan Project). When later he realized the dangers of radioactivity, he objected to the development of the hydrogen bomb, and was alleged to be a security risk 1953 by the US Atomic Energy Commission (AEC).

Investigating the equations describing the energy states of the atom, Oppenheimer showed in 1930 that a positively charged particle with the mass of an electron could exist. This particle was detected in 1932 and called the positron.

Oppenheimer was born in New York and studied at Harvard, going on to postgraduate work with physicists Ernest Rutherford at Cambridge, England, and Max Born at Göttingen, Germany. Between 1929 and

Oppenheimer *US physicist J Robert Oppenheimer led the Manhattan Project, which produced the atomic bomb. When later he opposed the the construction of the hydrogen bomb and advocated the international control of atomic energy, he was accused of communist sympathies. A man of wide learning, he wrote several non-technical books, including* Science and the Common Understanding 1954.

1942 he was on the staff of both the University of California, Berkeley, and the California Institute of Technology. After World War II he returned briefly to California and then in 1947 was made director of the Institute of Advanced Study at Princeton University. Oppenheimer also served as chair of the General Advisory Committee to the AEC 1946–52.

During World War II he reported to the Federal Bureau of Investigation friends and acquaintances who he thought might be communist agents; physicist David Bohm was one such.

The atomic bomb ... made the prospect of future war unendurable. It has led us up those last few steps to the mountain pass; and beyond there is different country.

Robert Oppenheimer

Quoted in R Rhodes *The Making of the Atomic Bomb* 1987

Oppolzer Theodor Egon Ritter von 1841–1886. Austrian astronomer and mathematician whose interest in asteroids, comets, and eclipses led to his compiling meticulous lists of such bodies and events for the use of other astronomers.

Oppolzer was born in Prague (now in the Czech Republic) and studied medicine, but had a private observatory. In 1866 he became lecturer in astronomy at the University of Vienna, and professor 1875. He was made director of the Austrian Geodetic Survey in 1873.

Oppolzer sought, by observation and calculation, to establish the orbits of asteroids. He was the originator of a novel technique for correcting orbits he found to be inaccurate.

In 1868, Oppolzer participated in an expedition to study a total eclipse of the Sun. Afterwards, he decided to calculate the time and path of every eclipse of the Sun and every eclipse of the Moon for as long a period as possible. The resulting *Canon der Finsternisse* 1887 covered the period 1207 BC–AD 2163.

Ore Oystein 1899–1968. Norwegian mathematician whose work concentrated on the fields of abstract algebra, number theory, and the theory of graphs.

Ore was born in Christiania (now Oslo), and studied at Oslo University, briefly visiting the University of Göttingen, Germany, where he was influenced by mathematician Emmy Noether. In 1926 he became professor at Oslo, but moved a year later to Yale in the USA. In 1945 he returned to Norway.

Ore investigated linear equations in non-commutative fields, summarizing his work in a book on abstract algebra 1936. He then turned to an examination of number theory, and in particular of algebraic numbers.

Ore also wrote a book (1967) on the four-colour problem, the theory that maps require no more than four colours for each region of the map to be coloured but with no zone sharing a common border with another zone of the same colour. German mathematician August Möbius had raised this problem 1840.

Osborn Henry Fairfield 1857–1935. US palaeontologist who did much to promote the acceptance of evolutionary theory in the USA. He emphasiszed that evolution was the result of pressures from four main directions: external environment, internal environment, heredity, and selection.

Osborn was born in Fairfield, Connecticut, and studied at Princeton University. He made his first fossil-hunting expedition to Colorado and Wyoming 1877. In 1891 he became professor of biology at Columbia. He was staff palaeontologist with the US Geological Survey 1900–24 and president of the American Museum of Natural History 1908–33.

Osborn's evolutionary studies focused on the problem of the adaptive diversification of life. He was particularly concerned with the parallel but independent evolution of related lines of descent, and with the explanation of the gradual appearance of new structural units of adaptive value.

Osborn wrote an influential textbook, *The Age of Mammals* 1910.

Ostwald (Friedrich) Wilhelm 1853–1932. Latvian-born German chemist who devised the Ostwald process (the oxidation of ammonia over a platinum catalyst to give nitric acid). His work on catalysts laid the foundations of the petrochemical industry. Nobel Prize for Chemistry 1909.

Ostwald was born in Riga and studied at the University of Dorpat (Tartu) in Estonia. He was professor at Riga 1881–87 and at Leipzig 1887–1906, and was from 1898 the first director of Leipzig's Physicochemical Institute.

In 1888, he proposed the ***Ostwald dilution law***, which relates the degree of dissociation of an electrolyte, α, to its total concentration c expressed in moles per litre (dm^3). It states that:

$$k = \alpha^2 c/(1 - \alpha)$$

The constant, k, neglects the activity coefficient and is therefore not a true thermodynamic constant K. The equation is important historically because it was the form in which the law of mass action was first applied to solutions of weak organic acids and bases. Ostwald then worked on the theory of acid–base indicators.

From 1909 Ostwald became interested in the methodology and organizational aspects of science, in a world language, internationalism, and pacifism. He also built a laboratory for colour research.

Other peoples still live under the regime of individualism, whereas we [Germans] live under the regime of organization.

Wilhelm Ostwald

In J Labadie (ed) *L'Allemagne, a-t-elle le Secret de L'organisation?* 1916

Otis Elisha Graves 1811–1861. US engineer who developed a lift that incorporated a safety device, making it acceptable for passenger use in the first skyscrapers. The device, invented 1852, consisted of vertical ratchets on the sides of the lift shaft into which spring-loaded catches would engage and lock the lift in position in the event of cable failure.

Otis was born in Halifax, Vermont, and became a builder and mechanic. During the construction of a factory in Yonkers, New York, he had to make a hoist and invented his safety device to prevent accidents to the workforce.

Otis patented and began manufacturing his invention. At the Crystal Palace Exposition in New York 1854, he demonstrated it by letting himself be hoisted into the air, and then a mechanic cut the hoisting rope. This was a grand advertisement and the orders started to come in. In 1857, the first public passenger lift was installed in New York. Generally the lifts were powered by steam engines and in 1860 Otis patented and improved the double oscillatory machine specially designed for his lifts. Also from the workshops of his company, Otis invented and patented railway trucks and brakes, a steam plough, and a baking oven.

Otto Nikolaus August 1832–1891. German engineer who in 1876 patented an effective internal-combustion engine.

The four-stroke cycle, also known as the Otto cycle, is now used in most petrol and diesel engines. The stroke is an upward or downward movement of a piston in a cylinder. In a petrol engine the cycle begins with the induction of a fuel mixture as the piston goes down on its first stroke. Going up, the piston compresses the mixture in the top of the cylinder. An electric spark ignites the mixture, and the gases produced force the piston down on its third, power stroke. On the fourth stroke the piston expels the burned gases from the cylinder into the exhaust. In a diesel engine, the first two strokes are slightly different.

Otto was born in Holzhausen, Nassau. In 1861 he built a small experimental gas engine, and three years later, with two others, formed a company to market such engines. At the Paris Exhibition of 1867 the firm's product won a gold medal in competition

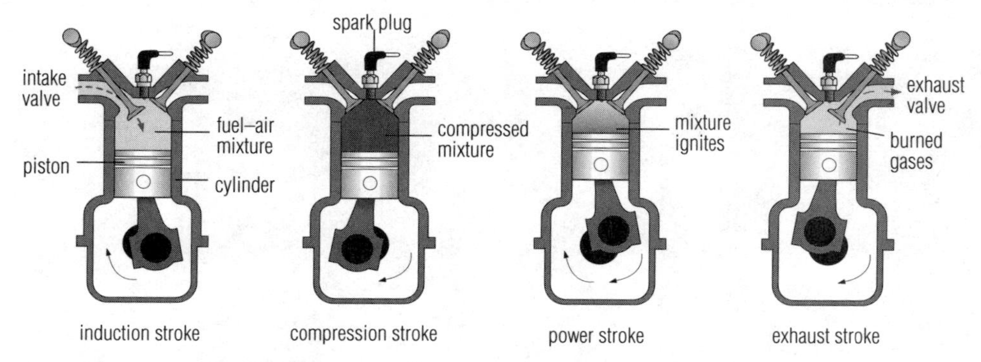

Otto *The four-stroke cycle of a modern petrol engine. The cycle is called the Otto cycle after German engineer Nikolaus Otto, who introduced it in 1876. It improved on earlier engine cycles by compressing the fuel mixture before it was ignited.*

with 14 other gas engines. A new factory, the Gasmotorenfabrik, was built at Deutz near Cologne in 1869. Otto concentrated on the administrative side of the business, and in 1872 Gottlieb Daimler and Wilhelm Maybach joined on the engineering side.

Otto first designed a successful vertical atmospheric gas engine in 1867. In 1876, he described the four-stroke engine. His patent was invalidated 1886 when his competitors discovered that Alphonse Beau de Rochas (1815–1893) had described the principle of the four-stroke cycle in an obscure pamphlet, but Otto is believed to have reached his results independently of Rochas.

Oughtred William 1575–1660. English mathematician, credited as the inventor of the slide rule 1622. His major work *Clavis mathematicae/The Key to Mathematics* 1631 was a survey of the entire body of mathematical knowledge of his day. It introduced the '×' symbol for multiplication, as well as the abbreviations 'sin' for sine and 'cos' for cosine.

Owen Richard 1804–1892. British anatomist and palaeontologist. He attacked the theory of natural selection and in 1860 published an anonymous and damaging review of Charles Darwin's work. As director of the Natural History Museum, London, he was responsible for the first public exhibition of dinosaurs. KCB 1884.

Owen was born in Lancaster and studied medicine at Edinburgh University and St Bartholomew's Hospital, London. He became professor at the Royal College of Surgeons and 1858–62 at the Royal Institution. In 1856, he was made the first superintendent of the Natural History Departments of the British Museum, and was promoted to director when the collections were moved to South Kensington.

Owen published more than 360 monographs on recent and fossil invertebrates and vertebrates, notably the pearly nautilus, the moa and other birds of New Zealand, the dodo from Mauritius, and the *Archaeopteryx* – his reconstruction of that extinct bird on comparative anatomical principles is regarded as a classic. Other works include *History of British Fossil Reptiles* 1849–84 and a popular textbook, *Palaeontology* 1860.

Page Frederick Handley 1885–1962. British aircraft engineer, founder 1909 of one of the earliest aircraft-manufacturing companies and designer of long-range civil aeroplanes and multi-engined bombers in both world wars; for example, the Halifax, flown in World War II. Knighted 1942.

Paget James 1814–1899. English surgeon, one of the founders of pathology. He described two conditions now named after him: Paget's disease of the nipple and Paget's disease of the bone. Baronet 1871.

Paget was born in Great Yarmouth, Norfolk, and studied at St Bartholomew's Hospital in London. He was one of the original 300 fellows of the Royal College of Surgeons of England in 1843, and was professor there 1847–52. Having tended the Princess of Wales 1878, he was appointed surgeon extraordinary to Queen Victoria.

Paget's disease of the nipple was described 1874 and is an eczematous skin eruption that indicates an underlying carcinoma of the breast, although the eruption is not simply an extension of the cancer cells inside the breast.

When Paget described the disease of the bone in 1877, he referred to it as osteitis deformans. This implies an inflammation of the bone, which is not accurate, and it is now called osteodystrophia deformans. This condition can affect the elderly. The bones soften, giving rise to deformity of the limbs, which may also fracture easily. If the skull is affected, bony changes cause enlargement of the head, and pressure on the VIIIth cranial nerve can cause deafness.

Palade George 1912– . Romanian cell biologist who shared the Nobel Prize for Physiology or Medicine 1974 with Albert Claude and Christian René de Duve for his work on the parts of the cell involved in protein secretion, particularly his discovery of ribosomes, which he showed to be rich in RNA, or ribonucleic acid (nucleic acid involved in the process of translating genetic material into proteins).

While he was working at the Rockefeller Institute, Palade developed a method called cell fractionation, in which elements of the cell are labelled with radioactive molecules and then segregated by breaking the cell apart and spinning down the organelles to isolate them from one another. He used this technique, along with electron microscopy, to look inside whole preserved cells and describe the organelles involved in the secretion of proteins, especially in cells that are specialized for the secretion of enzymes.

The cytoplasm (the part of the cell outside the nucleus) of these cells is packed with endoplasmic reticulum studded with ribosomes which Palade correctly determined were involved in protein secretion. He elucidated that the protein is made on the ribosomes, moved into the interior of the endoplasmic reticulum and from there to a series of flattened sacs in the cytoplasm, called the Golgi apparatus. He observed that the protein was then pushed into small vesicles and that these were then moved to the cell surface to secrete their protein contents.

Palade was born in Iassy in Romania and trained as a doctor in Bucharest. He worked as professor of anatomy in Bucharest before emigrating to the USA 1946 to take up a staff position at the Rockefeller Institute in New York. He was elected to the chair of cell biology at the medical school of Yale University 1972 and then the chair of cellular and molecular biology at the University of California 1990.

Pallas Peter Simon 1741–1811. German naturalist who classified corals and sponges and whose work in comparative anatomy, including *Zoographia Rosso-Asiatica*, established him as a predecessor of Georges Cuvier.

Pallas was born in Berlin. His father was the professor of the Berlin Medical-Surgical Academy, where Pallas studied medicine 1745–59. He also studied at the universities in Halle, Leiden, and Göttingen. A contemporary of Georges Buffon and Linnaeus, he obtained his doctorate from Leiden University with a thesis that refuted Linnaeus's classification of worms. His *Elenchus zoophytorum* 1766 was a classification of corals and sponges. From 1761–66, he continued with this research in Holland and England. He was elected a Fellow of the Royal Society 1763. In 1767, he went to St Petersburg Academy of Sciences and conducted several Russian expeditions. In 1795, he moved to Simferpol in the Crimea, where he remained until returning to Berlin 1810.

Paneth Friedrich Adolf 1887–1958. Austrian chemist who contributed to the development of radioactive tracer techniques. He worked on unstable metal hydrides and developed sensitive methods for determining trace amounts of helium. From 1929 to the end of his life, meteorites dominated his interests.

Paneth was born in Vienna. He studied and worked at a number of European institutions, including the university of Glasgow, the Vienna Institute for Radium Research, and the Prague Institute of Technology. In 1929 he became professor at Königsberg, but left Germany for the UK in 1933 because of the rise of the Nazis. He was professor at Durham 1939–53, and during World War II he was head of the chemical division of the Joint British and Canadian Atomic Energy Team in Montreal. In 1953 Paneth returned to Germany to become director of the Max Planck Institute for Chemistry in Mainz.

Paneth worked out that radium D and thorium B are isotopes of lead and that radium E and thorium C are isotopes of bismuth. He prepared a new tin hydride, SnH_4, and investigated its properties.

In the late 1930s Paneth succeeded in obtaining measurable amounts of helium by the neutron bombardment of boron: he had induced an artificial transmutation. He then began to investigate the trace elements in the stratosphere, and determined the helium, ozone, and nitrogen dioxide content of the atmosphere.

Papanicolaou Georges Nicholas 1883–1962. Greek-born US physiologist and microscopic anatomist who derived the vaginal smear test and associated the changes of the vaginal lining with the phases of the ovarian cycle.

While at Cornell, Papanicolaou became interested in reproductive physiology and, in 1917, he and Stockard demonstrated that the vaginal epithelium (lining) of certain mammals, including guinea pigs and humans, undergoes characteristic changes during each phase of the ovarian cycle. These changes were shown later to be influenced by the steroid hormones, oestrogen and progesterone, made by the ovaries.

He went on to show that abnormal cells in the cervix could be identified if a small scrape of the wall of the cervix was taken and observed under the microscope. This work formed the basis of cervical smear testing for cervical cancer.

Papanicolaou was born in Kimi in Greece and studied medicine at the University of Athens, obtaining his MD 1904. After completing a PhD in Munich 1910, he decided to emigrate to the USA and took teaching posts in anatomy in New York and at Cornell Medical School 1914. He was appointed professor of clinical anatomy at Cornell 1924 and then director of the Miami Cancer Institute.

Papin Denis 1647–*c.* 1712. French physicist and technologist who in 1679 invented a vessel that was the forerunner of the pressure cooker and the autoclave, together with a safety valve.

Papin was born in Blois, on the Loire, and studied medicine at Angers. His first job was as assistant to Dutch physicist Christiaan Huygens in Paris, and in 1675 he went to London as secretary to scientist and inventor Robert Hooke. Papin spent 1681–84 in Venice, returned to London and a job at the Royal Society, then became professor of mathematics at Marburg 1687–96. He returned to London in 1707.

Papin worked with Huygens and Irish physicist Robert Boyle on an air pump and invented the condensing pump. It was in London with Boyle that he invented the 'steam digester' – a vessel with a tightly fitting lid that prevented steam from escaping. The high pressure generated caused the boiling point of the water to rise considerably.

In 1690, Papin suggested a cylinder-and-piston steam engine, but his scheme was unworkable, because he proposed to use one vessel as both boiler and cylinder. He proposed the first steam-driven boat in 1690 and in 1707 he built a paddle boat, but the paddles were turned by human power and not by steam.

Pappus of Alexandria *c.* 300–*c.* 350. Greek mathematician, astronomer, and geographer whose book *Synagogue/Collection* deals with nearly the whole body of Greek geometry, mostly in the form of commentaries on texts the reader is assumed to have to hand.

Nothing is known of his life and many of his writings survive only in translations from the original Greek.

Without the *Collection*, much of the geometrical achievement of his predecessors would have been lost for ever.

It reproduces known solutions to problems in geometry, and also frequently gives Pappus' own solutions, or improvements and extensions to existing solutions. For example, he handles the problem of inscribing five regular solids in a sphere in a way quite different from Euclid.

Among Pappus' other works are a commentary on Ptolemy's *Almagest* and a commentary on Euclid's *Elements*. Pappus is also believed to be the author of the *Description of the World*, a geographical treatise that has come down to us only in Armenian and bearing the name of Moses of Khoren as its author.

Paracelsus adopted name of Theophrastus Bombastus von Hohenheim 1493–1541. Swiss physician, alchemist, and scientist who developed the idea that minerals and chemicals might have medical uses (iatrochemistry). He introduced the use of laudanum (which he named) for pain-killing purposes. His rejection of traditional lore and insistence on the value of observation and experimentation make him a leading figure in early science.

Overturning the contemporary view of illness as an imbalance of the four humours, Paracelsus sought an external agency as the source of disease. This encouraged new modes of treatment, supplanting, for example, bloodletting, and opened the way for new ideas on the source of infection.

Paracelsus was extremely successful as a doctor. His descriptions of miners' diseases first identified silicosis and tuberculosis as occupational hazards. He recognized goitre as endemic and related to minerals in drinking water, and originated a medical account of chorea, rather than believing this nervous disease to be caused by possession by spirits. Paracelsus was the first to distinguish the congenital from the infectious form of syphilis, and showed that it could be treated with carefully controlled doses of a mercury compound.

Paracelsus was born in Einsiedeln, Schwyz canton. Like many of his contemporaries, he became a wandering scholar, studying at Vienna, Basel, and several universities in Italy. He was a military surgeon in Venice and the Netherlands and is said to have visited England, Scotland, Russia, Egypt, and Constantinople. Having practised as a physician in Austria, he became professor of medicine at Basel 1527, but scandalized other academics by lecturing in German rather than Latin and by his savage attacks on the classical medical texts – he burned the works of Galen and Avicenna in public – and was forced to leave Basel 1528. In 1541 he was appointed physician to Duke Ernst of Bavaria.

What is accomplished with fire is alchemy, whether in the furnace or the kitchen stove.

Paracelsus

In J Bronowski *The Ascent of Man* 1975

Paracelsus was the disseminator in Europe of the medieval Islamic alchemists' theory that matter is composed of only three elements: salt, sulphur, and mercury. His study of alchemy helped to develop it into chemistry and produced new, nontoxic compounds for medicinal use; he discovered new substances arising from the reaction of metals and described various organic compounds, including ether. He was the first to devise such advanced laboratory techniques as the concentration of alcohol by freezing. Paracelsus also devised a specific nomenclature for substances already known but not precisely defined, and his attempt to construct a system of grouping chemicals according to their susceptibility to similar processes was the first of its kind.

Pardee Arthur Beck 1921– . US biochemist who discovered that enzymes can be inhibited by the final product of the reaction they catalyse (trigger); a regulatory strategy used by cells, called feedback inhibition.

Pardee worked on many aspects of how cells control their own processes, especially those involving enzymes. For example, Pardee worked with John Gerhart on the enzyme aspartate transcarbamoylase, an enzyme that catalyses the formation of

N-carbamoylaspartate from aspartate and carbamoyl phosphate. This reaction is the first step in the biosynthesis of the pyrimidines cytidine and thymidine, which are necessary substrates for the synthesis of DNA.

Pardee and Gerhardt discovered that the aspartate transcarbamoylase enzyme, which they purified from the bacterium *Escherichia coli*, is inhibited by the end product of the biosynthetic pathway, known as cytidine triphosphate. This process is in effect a control mechanism that prevents N-carbamoylaspartate and subsequent intermediates in the pathway from being produced when the final product is abundant.

Pardee was born in Chicago in the USA and was educated at the University of California, Berkeley, where he worked until 1961. He was then appointed to a staff position at Princeton University and subsequently to professor of pharmacology at Harvard Medical School in Boston 1975.

Pardee and Gerhart also discovered that the binding of the aspartate and carbamoyl phosphate to the aspartate transcarbamoylase enzyme is cooperative and that the enzyme is activated by adenosine triphosphate (ATP). ATP is a source of energy within cells and activation of the enzyme by this molecule serves to signal that there is sufficient energy available for DNA replication.

Paré Ambroise *c.* 1509–1590. French surgeon who introduced modern principles to the treatment of wounds. As a military surgeon, Paré developed new ways of treating wounds and amputations, which greatly reduced the death rate among the wounded. He abandoned the practice of cauterization (sealing with heat), using balms and soothing lotions instead, and used ligatures to tie off blood vessels.

Paré eventually became chief surgeon to Charles IX. He also made important contributions to dentistry and childbirth, and invented an artificial hand.

Paré was born in Mayenne *département* and trained in Paris. His book *La Méthode de traicter les playes faites par les arquebuses et aultres bastons à feu/Method of Treating Wounds Inflicted by Arquebuses and Other Guns* 1545 became a standard work in European armies, and was followed by a number of works on anatomy.

Parkinson James 1755–1824. British neurologist who first described Parkinson's disease.

Parsons Charles Algernon 1854–1931. English engineer who invented the Parsons steam turbine 1884, a landmark in marine engineering and later universally used in electricity generation to drive an alternator. KCB 1911.

Parsons developed more efficient screw propellers for ships and suitable gearing to widen the turbine's usefulness, both on land and sea. He also designed searchlights and optical instruments, and developed methods for the production of optical glass.

Parsons was born in London and studied at Trinity College, Dublin, and at Cambridge. He worked for various engineering firms in NE England until 1889, when he set up his first company near Newcastle-upon-Tyne. He developed turbogenerators of various kinds and increasing capacities, which formed the basic machinery for national (and much of international) electricity production.

With new propulsion machinery devised by Parsons, his steamship *Turbinia* reached a record-breaking speed of 34.5 knots in 1897. Parsons turbines fitted to the liners *Lusitania* and *Mauritania* gave them high speed with less vibration, and developed some 70,000 hp/52,000 kW.

Parsons William, 3rd Earl of Rosse 1800–1867. Irish astronomer, engineer, and politician who built the largest telescope then in use. He found 15 spiral nebulae and named the Crab nebula. He was among the first to take photographs of the Moon. Earl 1841.

Parsons was born in York and studied at Oxford. As the eldest son of a titled landowner, he was elected to Parliament while still an undergraduate to represent King's County, a seat he then held for 13 years. In 1831 he became Lord Lieutenant of County Offaly, and in 1841, on the death of his father, he entered the House of Lords. During and after the potato famine of 1846, Parsons worked to alleviate the living conditions of his tenants.

Determined to construct a large telescope, Parsons learned to cast and grind mirrors. Fourteen years after his experiments began, he was able to make a 92-cm/36-in solid mirror, and in 1842 he cast the 'Leviathan of Parsonstown', a disc 1.8 m/72 in in diameter which weighed nearly 4 tonnes and was incorporated into a telescope with a focal length of 16.2 m/54 ft. It took three years to put together. He also invented a clockwork drive for the large equatorial mounting of an observatory.

Pascal Blaise 1623–1662. French philosopher and mathematician. He contributed to the development of hydraulics, the calculus, and the mathematical theory of probability.

In mathematics, Pascal is known for his work on conic sections and, with Pierre de Fermat, on the probability theory. In physics, Pascal's chief work concerned fluid pressure and hydraulics. ***Pascal's principle*** states that the pressure everywhere in a fluid is the same, so that pressure applied at one point is transmitted equally to all parts of the container. This is the principle of the hydraulic press and jack.

The eternal silence of these infinite spaces terrifies me.

Blaise Pascal

Pensées

Pascal's triangle is a triangular array of numbers in which each number is the sum of the pair of numbers above it. In general the nth ($n = 0, 1, 2, \ldots$) row of the triangle gives the binomial coefficients ${}^{n}C_{r}$, with $r = 0, 1, \ldots, n$.

His *Pensées* 1670 was part of an unfinished defence of the Christian religion.

Pascal was born in Clermont-Ferrand. In Paris in his teens he met mathematicians Descartes and Fermat. From 1654 Pascal was closely involved with the Jansenist monastery of Port Royal. He defended a prominent Jansenist, Antoine Arnauld (1612–1694), against the Jesuits in his *Lettres provinciales/Provincial Letters* 1656. His last project was to design a public transport system for Paris, which was inaugurated 1662.

1
1 1
1 2 1
1 3 3 1
1 4 6 4 1
1 5 10 10 5 1
1 6 15 20 15 6 1
1 7 21 35 35 21 7 1

Pascal *In Pascal's triangle, each number is the sum of the two numbers immediately above it, left and right – for example, 2 is the sum of 1 and 1, and 4 is the sum of 3 and 1. Furthermore, the sum of each row equals a power of 2 – for example, the sum of the 3rd row is $4 = 2^2$; the sum of the 4th row is $8 = 2^3$.*

Between 1642 and 1645, Pascal constructed a machine to carry out the processes of addition and subtraction, and then organized the manufacture and sale of these first calculating machines. (At least seven of these 'computers' still exist. One was presented to Queen Christina of Sweden in 1652.)

Pascal's work in hydrostatics involved repeating the experiment by Italian physicist Evangelista Torricelli to prove that air pressure supports a column of mercury. This led rapidly to investigations of the use of the mercury barometer in weather forecasting.

Pasteur Louis 1822–1895. French chemist and microbiologist who discovered that fermentation is caused by microorganisms and developed the germ theory of disease. He also created a vaccine for rabies, which led to the foundation of the Pasteur Institute in Paris 1888.

Pasteur saved the French silkworm industry by identifying two microbial diseases that were decimating the worms. He discovered the pathogens responsible for anthrax and chicken cholera, and developed vaccines for these diseases. He inspired his pupil Joseph Lister's work in antiseptic surgery. ***Pasteurization*** to make dairy products free from the tuberculosis bacteria is based on his discoveries.

When meditating over a disease, I never think of finding a remedy for it, but, instead, a means of preventing it.

Louis Pasteur

Address to the Fraternal Association of Former Students of the Ecole Centrale des Arts et Manufactures, Paris, 15 May 1884

Pasteur was born in Dôle in E France and studied in Paris at the Ecole Normale Supérieure. He was professor at Strasbourg 1849–63, moving to the Ecole Normale Supérieure to institute a teaching programme that related chemistry, physics, and geology to the fine arts. Also in 1863 he became dean of the new science faculty at Lille University, where he initiated the novel concept of evening classes for workers. In 1867 a laboratory was established for him with public funds, and from 1888 to his death he headed the Pasteur Institute.

A query from an industrialist about wine- and beermaking prompted Pasteur's research into fermentation. Using a microscope he

Pasteur *An obituary issue of the popular French publication* Le Petit Journal *mourning the death of Louis Pasteur 1895. In 1888 the Pasteur Institute was created in Paris in recognition of his work in developing a vaccine for rabies. He headed the institute until his death seven years later.*

found that properly aged wine contains small spherical globules of yeast cells whereas sour wine contains elongated yeast cells. He proved that fermentation does not require oxygen, yet it involves living microorganisms, and that to produce the correct type of fermentation (alcohol-producing rather than lactic acid-producing) it is necessary to use the correct type of yeast. Pasteur also realized that after wine has formed, it must be gently heated to about 50°C/122°F – pasteurized – to kill the yeast and thereby prevent souring during the ageing process.

Paul Wolfgang 1913–1993. German nuclear physicist who made fundamental contributions to molecular beam spectroscopy, mass spectrometry, and electron acceleration technology. He developed the ion trap, or 'Paul trap', used to store single atoms long enough to make accurate spectroscopic measurements on single atoms. Nobel Prize for Physics 1989 (together with US scientists Norman Ramsey and Hans Dehmelt).

In 1957 he helped found the DESY accelerator laboratory in Hamburg which has made important contributions to particle physics. From 1964–67, Paul was director of the nuclear physics laboratory of CERN, the joint European laboratory in Particle physics in Geneva.

Paul was trained at the Munich and Berlin Institutes of Technology and obtained his MSc 1937, examined by Hans Geiger (who invented the Geiger counter). In 1939 Paul obtained his doctorate in Kiel. At this time he had a close association with the distinguished spectroscopist Hans Kopfermann and the theoretical physicist Richard Becker. These men were both non-Nazis and, under their influence, Paul took no part in projects, such as the German atom-bomb research, which had military applications. As a result, Paul was one of the few German physicists of his generation untainted by the war years.

In 1952 Paul became professor and director of the Physics Institute at the University of Bonn, Germany, a post that he held until he retired in 1981. At Bonn, he developed the sextupole focusing of molecular beams, the radio frequency quadrupole mass spectrometer and the ion trap. With collegues he built two electron accelerators.

Pauli Wolfgang 1900–1958. Austrian-born Swiss physicist who originated the ***exclusion principle***: in a given system no two fermions (electrons, protons, neutrons, or other elementary particles of half-integral spin) can be characterized by the same set of quantum numbers. He also predicted the existence of neutrinos. Nobel prize 1945.

The exclusion principle, announced 1925, involved adding a fourth quantum number to the three already used (*n*, *l*, and *m*). This number, *s*, would represent the spin of the electron and would have two possible values. The principle also gave a means of determining the arrangement of electrons into shells around the nucleus, which explained the classification of elements into related groups by their atomic number.

I don't mind your thinking slowly: I mind your publishing faster than you think.

Wolfgang Pauli

Attributed (from H Coblaus)

Pauli was born in Vienna and studied in Germany at Munich. He then went to Göttingen as an assistant to German physicist Max Born, moving on to Copenhagen to study with Danish physicist Niels Bohr. From 1928 Pauli was professor of Experimental Physics at the Eidgenössische Technische Hochschule, Zürich, though he spent World War II in the USA at the Institute for Advanced Study, Princeton.

The neutrino was proposed in 1930 to explain the production of beta radiation in a continuous spectrum; it was eventually detected 1956.

Pauling Linus Carl 1901–1994. US theoretical chemist and biologist whose ideas on chemical bonding are fundamental to modern theories of molecular structure. He also investigated the properties and uses of vitamin C as related to human health. He won the Nobel Prize for Chemistry 1954 and the Nobel Peace Prize 1962, having campaigned for a nuclear test-ban treaty.

Pauling was born in Portland, Oregon, and studied at Oregon State Agricultural College, getting his PhD from the California Institute of Technology (Caltech). In Europe 1925–27, he met the chief atomic scientists of the day. He became professor at Caltech 1931, and was director of the Gates and Crellin Laboratories 1936–58 and of the Linus Pauling Institute of Science and Medicine in Menlo Park, California 1973–75.

Pauling's work on the nature of the chemical bond included much new information about interatomic distances. Applying his knowledge of molecular structure to proteins in blood, he discovered that many proteins have structures held together with hydrogen bonds, giving them helical shapes.

He was a pioneer in the application of quantum mechanical principles to the structures of molecules, relating them to interatomic distances and bond angles by X-ray and electron diffraction, magnetic effects, and thermochemical techniques. In 1928, Pauling introduced the concept of hybridization of bonds. This provided a clear basic insight into the framework structure of all carbon compounds, that is, of the whole of organic chemistry. He also studied electronegativity of atoms and polarization (movement of electrons) in chemical bonds. Electronegativity values can be used to show why certain substances, such as hydrochloric acid, are acid, whereas others, such as sodium hydroxide, are alkaline. Much of this work was consolidated in his book *The Nature of the Chemical Bond* 1939.

In his researches on blood in the 1940s, Pauling investigated immunology and sickle-cell anaemia. Later work confirmed his conviction that the disease is genetic and that normal haemoglobin and the haemoglobin in sickle cells differ in electrical charge. Pauling's work provided a powerful impetus to Crick and Watson in their search for the structure of DNA.

Pauling was coauthor of *Introduction to Quantum Mechanics* 1935; he published two textbooks, *General Chemistry* 1948 and *College Chemistry* 1950.

During the 1950s he became politically active, his especial concern being the long-term genetic damage resulting from atmospheric nuclear bomb tests. In this, he conflicted with the US establishment and with several of his science colleagues. He was denounced as a pacifist, and a communist, his passport was withdrawn 1952–54, and he was obliged to appear before the US Senate Internal Security Committee. One item in his sustained wide-ranging campaign was his book *No More War!* 1958. He presented to the UN a petition signed by 11,021 scientists from 49 countries urging

Pauling *Double Nobel prize-winning American chemist Linus Pauling. Pauling won the prize for Chemistry in 1954 following his work investigating the nature of the chemical bond. In 1962 he also won the Peace prize following his campaign for a nuclear test-ban treaty.*

an end to nuclear weapons testing, and during the 1960s spent several years on a study of the problems of war and peace at the Center for the Study of Democratic Institutions in Santa Barbara, California.

Pavlov Ivan Petrovich 1849–1936. Russian physiologist who studied conditioned reflexes in animals. His work had a great impact on behavioural theory and learning theory. Nobel Prize for Physiology or Medicine 1904.

While investigating the secretory mechanisms of digestion 1890–1900, Pavlov made the discovery of the conditioned reflex. He confined a dog in a room with no distractions and established that if the sound of a bell was presented simultaneously with the sight of food and the combination repeated often enough, the sound of the bell alone caused salivation. Many inborn reflexes may be conditioned by Pavlov's method, including responses of the skeletal muscles (knee-jerking and blinking) and of the smooth muscles and glands.

Pavlov was born in Ryazan and studied in St Petersburg at the university and the Imperial Medical Academy, where he became professor 1890.

Studying the physiology of the circulatory system and the regulation of blood pressure, Pavlov devised animal experiments such as the dissection of the cardiac nerves of a living dog to show how the nerves that leave the cardiac plexus control heartbeat strength.

Pavlov's work relating to human behaviour and the nervous system also emphasized the importance of conditioning. He deduced that the inhibitive behaviour of a psychotic person is a means of self-protection. The person shuts out the world and, with it, all damaging stimuli. Following this theory, the treatment of psychiatric patients in Russia involved placing a sick person in completely calm and quiet surroundings.

Pavlov summarized his Nobel prizewinning work in *Die Arbeit der Verdauungsdrüsen/ Lectures on the Work of the Principal Digestive Gland* 1897.

For the kind of social experiment you are making, I would not sacrifice a frog's hind legs!

Ivan Pavlov

In opposition to Soviet science

Pavlov *Russian physiologist Ivan Pavlov, who developed the idea of the 'conditional reflex', after showing that if a bell was sounded whenever a dog was presented with food, the dog would eventually begin to salivate whenever the bell was rung, regardless of whether food was presented.*

Payne-Gaposchkin Cecilia Helena, (born Payne) 1900–1979. English-born US astronomer who studied stellar evolution and galactic structure. Her investigation of stellar atmospheres during the 1920s gave some of the first indications of the overwhelming abundance of the lightest elements (hydrogen and helium) in the Galaxy.

Payne was born in Wendover, Buckinghamshire, and studied at Cambridge and at Harvard College Observatory in Cambridge, Massachusetts, under US astronomer Harlow Shapley. In 1927 she was appointed an astronomer at the observatory, and in 1956 she became the first woman professor at Harvard.

Payne-Gaposchkin employed a variety of spectroscopic techniques in the investigation of stellar properties and composition, especially variable stars. Her studies of the Large and Small Magellanic Clouds were carried out in collaboration with her husband Sergei I Gaposchkin.

Other areas of her interest included the devising of methods to determine stellar magnitudes, the position of variable stars on the Hertzsprung–Russell diagram, and novae.

Peano Giuseppe 1858–1932. Italian mathematician who was a pioneer in symbolic logic. His concise logical definitions of natural numbers were devised in order to derive a complete system of notation for logic. He also discovered a curve that fills topological space.

Peano was born near Cuneo, Piedmont, and studied at Turin. On graduating, he joined the staff of the university and remained there for the rest of his life, first becoming a professor there in 1890. He was also professor at Turin Military Academy 1886–1901.

Peano's first work in logic, published in 1888, contained his rigorously axiomatically derived postulates for natural numbers. He acknowledged his debt for some of the work to German mathematician Richard Dedekind. Some of Peano's work was used by English philosopher Bertrand Russell.

Peano also applied the axiomatic method to other fields, such as geometry, first in 1889 and again in 1894. A treatise on this work contained the beginnings of geometrical calculus. Peano provided new definitions of the length of an arc of a curve and of the area of a surface.

Formulario mathematico 1895–1908, comprising his work and that of collaborators, contains 4,200 theorems.

Pearson Karl 1857–1936. British statistician who followed Francis Galton in introducing statistics and probability into genetics and who developed the concept of eugenics (improving the human race by selective breeding). He introduced the term standard deviation into statistics.

Pearson introduced in 1900 the χ^2 (chi-squared) test to determine whether a set of observed data deviates significantly from what would have been predicted by a 'null hypothesis' (that is, totally at random). He demonstrated that it could be applied to examine whether two hereditary characteristics (such as height and hair colour) were inherited independently.

Pearson was born in London and studied at Cambridge, where he persuaded the authorities to abolish the mandatory classes in Christianity for undergraduates. In 1884 he became professor of mathematics at University College, London; from 1911 he was professor of eugenics at London University. In order to publish work on statistics as applied to biological subjects, he founded 1901 the journal *Biometrika*, which he edited until his death.

Pearson's discoveries included the Pearson coefficient of correlation (1892), the theory of multiple and partial correlation (1896), the coefficient of variation (1898), work on errors of judgement (1902), and the theory of random walk (1905).

Pearson's *Biometrika* for 1901 is a book of tables of the ordinates, integrals, and other properties of Pearson's curves.

Pecquet Jean 1622–1674. French anatomist who discovered the thoracic duct, one of two ducts through which fluid collected by the lymphatic system (the network that transports fluids to and from the tissues in the body) empties into the blood stream.

Pecquet discovered the thoracic duct during animal dissection 1647. The thoracic duct is connected to the lymphatic system, the network of vessels and organs that transports nutrients, oxygen, and white blood cells to the tissues in the body and carries waste matter and carbon dioxide away from the tissues. Lymphatic vessels have very thin walls and the fluid passes into them more easily than back into the blood capillaries. The thoracic duct functions to carry lymph (waste fluid) and chyle (the fluid containing all the nutrients derived from digestion) from the intestinal lymphatic vessels into the cardiovascular system.

Pecquet was born in Dieppe and trained as a doctor first at Paris and then at Montpelier from where he obtained his MD 1652. He reported his findings in *Experimenta nova anatomica* 1651 and his work prompted a considerable period of research and debate on the structure and function of the lymphatic system in the 17th century.

Pedersen Charles 1904–1990. US organic chemist who shared the Nobel Prize for Chemistry in 1987 with Jean Lehn and Donald Cram for his discovery of 'crown ether', a cyclic polyether – a molecule with 12 carbon atoms and six oxygen atoms arranged in a crown-like structure. Its discovery opened up the field of guest-host chemistry.

Pedersen found that crown ether, now part of a class of crown ethers, has unusual properties, such as being able to dissolve sodium hydroxide because it binds alkali metal ions very strongly, forming a complex. This discovery enabled scientists to study metal ion transport across cell membranes.

Pedersen was born in Pusan, Korea. His mother was Japanese and his father was a Norwegian mining engineer, and in the 1920s, the family settled in America. After studying chemical engineering at the University of Dayton, Ohio, he completed a master's degree in organic chemistry at the Massachusetts Institute of Technology. He spent most of his working life as part of the research team for Du Pont de Nemours, until his retirement in 1969. His discovery came by accident. Whilst working on synthetic rubber, he spotted that one of his preparations had been contaminated with an impurity.

Peierls Rudolf Ernst 1907–1995. German-born British physicist who contributed to the early theory of the neutron–proton system. He helped to develop the atomic bomb 1940–46. Knighted 1968.

Peierls was born in Berlin and studied at universities in Germany and Switzerland under leading atomic physicists. From 1933 he worked in the UK, and was professor at Birmingham 1937–63 and at Oxford 1963–74.

I hope I have not only added a few small bricks to the growing edifice of science, but also contributed a little to the fight against its misuse.

Rudolf Peierls

Bird of Passage

In 1940, at Birmingham, Austrian physicist Otto Frisch and Peierls made an estimate of the energy released in a nuclear chain reaction, which indicated that a fission bomb would make a weapon of terrifying power. They drew the attention of the British government to this in 1940, and Peierls was placed in charge of a small group concerned with evaluating the chain reaction and its efficiency. In 1943, when Britain decided not to continue its work on nuclear energy, Peierls moved to the USA to help in the work of the Manhattan Project, first in New York and then at Los Alamos.

Pelletier Pierre-Joseph 1788–1842. French chemist whose extractions of a range of biologically active compounds from plants founded the chemistry of the alkaloids. The most important of his discoveries was quinine, used against malaria.

Pelletier was born in Paris and qualified as a pharmacist. He was professor at the Ecole de Pharmacie 1825–40.

Pelletier began with the analysis of gum resins and the colour pigments in plants. In 1817, together with chemist Joseph Caventou (1795–1877), he isolated the green pigment in leaves, which they named chlorophyll. In 1818 they turned to plant alkaloids: strychnine 1818, brucine and veratrine 1819, and quinine 1820. Their powerful effects made it possible to specify chemical compounds in pharmacology instead of the imprecise plant extracts and mixtures used previously.

Working with chemist Jean Baptiste Dumas, Pelletier obtained firm evidence for the presence of nitrogen in alkaloids 1823. He later carried out researches on strychnine and developed procedures for its extraction.

In 1832 Pelletier discovered a new opium alkaloid, narceine; he also claimed to have been the first to isolate thebaine (which he called paramorphine). In a study (1837–38) of an oily by-product of pine resin he discovered toluene (now methylbenzene).

Pelton Lester Allen 1829–1918. US engineer who developed a highly efficient water turbine used to drive both mechanical devices and hydroelectric power turbines using large heads of water. The Pelton wheel remains the only hydraulic turbine of the impulse type in common use today.

From Ohio, Pelton joined the California gold rush at the age of 20. He observed the water wheels used at the mines to power machinery, and came up with improvements. By 1879 he had tested a prototype at the University of California. A patent was granted 1889, and he later sold the rights to the Pelton Water Wheel Company of San Francisco.

The energy to drive these wheels was supplied by powerful jets of water which struck the base of the wheel on hemispherical cups. Pelton's discovery was that the wheel rotated more rapidly, and hence developed more power, with the jet striking at the inside edge of the cups, rather than the centre; he built a wheel with split cups.

By the time of his death, Pelton wheels developing thousands of horsepower in hydroelectric schemes at efficiencies of more than 90% were in operation.

Penney William George 1909–1991. English scientist who worked at Los Alamos, New Mexico, 1944–45, developing the US atomic bomb. He also headed the team that constructed Britain's first atomic bomb and directed its testing programme at the Monte Bello Islands off Western Australia 1952. He subsequently directed the UK hydrogen bomb project, tested on Christmas Island 1957, and developed the advanced gas-cooled nuclear reactor used in some UK power stations. KBE 1952, Baron 1967.

Penney was born in Sheerness, Kent, and studied at London and Cambridge, and in the USA at Wisconsin. A mathematician by training, he became an explosives expert. He was director of the Atomic Weapons Research Establishment 1953–59 and chair of the UK Atomic Energy Authority 1964–67.

Pennington Mary Engle 1872–1952. US chemist who set standards for food refrigeration. She also designed both industrial and household refrigerators.

Pennington was born in Nashville, Tennesse, and studied at the University of Pennsylvania but was initially denied a degree on account of her sex. In 1898 she opened her own Philadelphia Clinical Laboratory. She was head of the Food Research Laboratory of the Department of Agriculture 1908–19. From 1922 she worked as an independent consultant on the storage, handling, and transportation of perishable goods.

In Philadelphia, from research into the preservation of dairy products, Pennington developed standards of milk inspection that were later used by health boards across the USA. During World War I she conducted experiments into railway refrigeration cars and recommended the standards that remained in use into the 1940s. As a consultant, she turned her interest to frozen food.

Pennycuick Colin James 1933– . English zoologist who has studied the various processes involved in flight. He discovered that many migratory birds have minimal energy reserves and must stop to feed at regular intervals. The destruction of the intermediate feeding places of these birds could lead to their extinction, even if their summer and winter quarters are conserved.

Pennycuick was born in Virginia Water, Surrey, and studied at Oxford and Cambridge. In 1964 he became a lecturer at Bristol University, with a break 1971–73 spent researching at Nairobi University in Kenya.

Pennycuick's research is unusual in that it interrelates an extremely large number of factors and therefore gives a very detailed account of flight. In flying vertebrates, for example, he has investigated the mechanics of flapping; the aerodynamic effects of the feet and tail; the physiology of gaseous exchange; heat disposal; the relationship between the size and anatomy of a flying creature and the power it develops; and the frequency of wing beats.

Pennycuick has also hypothesized that migratory birds navigate using the Sun's altitude and its changing position.

Penrose Lionel Sharples 1898–1972. English physician and geneticist who carried out pioneering work on mental retardation and Down's syndrome. He was the first to demonstrate the significance of the mother's age.

Penrose was born in London and educated at Cambridge. He was director of psychiatric research for Ontario, Canada, 1939–45. In 1945 he became professor of eugenics at London University.

While working as research medical officer at the Royal Eastern Counties Institution, Colchester, 1930–39, Penrose produced an influential survey of patients and their families (*A Clinical and Genetic Study of 1,280 Cases of Mental Defect* 1938), showing that there were many different types and causes of mental defect and that normality and subnormality were on a continuum.

Early in his career, Penrose advanced the study of schizophrenia and developed a test for its diagnosis. His subsequent work concentrated on the causes of Down's syndrome.

Penrose Roger 1931– . English mathematician who formulated some of the fundamental theorems that describe black holes, including the singularity theorems, developed jointly with English physicist Stephen Hawking, which state that once the gravitational collapse of a star has proceeded to a certain degree, singularities (which form the centre of black holes) are inevitable. Penrose has also proposed a new model of the universe.

Penrose was born in Colchester, Essex, and studied at University College, London.

While he worked for his doctorate at Cambridge in 1957, Penrose and his father were devising geometrical figures, the construction of which is three-dimensionally impossible. (They became well known when incorporated by Dutch artist M C Escher into a couple of his disturbing lithographs.) Penrose was professor at Birkbeck College, London, 1966–73, and then moved to Oxford University.

The existence of a trapped surface within an 'event horizon' (the interface between a black hole and space-time), from which little or no radiation or information can escape, implies that some events remain hidden to observers outside the black hole. Penrose has put forward the hypothesis of 'cosmic censorship' – that all singularities are so hidden – which is now widely accepted.

Calculations in the world of ordinary objects (including Einstein's general theory of relativity) use real numbers, whereas the world of quantum theory often requires a system using complex numbers, containing imaginary components that are multiples of the square root of –1. Penrose holds that all calculations about both the macroscopic and microscopic worlds should use complex numbers, requiring reformulation of the major laws of physics and of space-time. He has proposed a model of the universe whose basic building blocks are what he calls 'twistors'.

His works include *The Emperor's New Mind: Concerning Computers, Minds, and the Laws of Physics* 1989 and *Shadows of the Mind: A Search for the Missing Science of Consciousness* 1994. His latest book is *The Nature of Space and Time* 1996, written with Stephen Hawking.

Penston Michael 1943–1990. British astronomer at the Royal Greenwich Observatory 1965–90. From observations made with the Ultraviolet Explorer Satellite of hot gas circulating around the core of the galaxy NGC 4151, he and his colleagues concluded that a black hole of immense mass lay at the galaxy's centre.

NGC 4151 is the brightest of a class of objects known as Seyfert galaxies, which are like scaled-down versions of quasars. Supermassive black holes had long been suspected to lie at the centres of such objects, because of their strange behaviour, but Penston's result was the first direct observational evidence in favour of this theory.

Penzias Arno Allan 1933– . German-born US radio engineer who in 1964, with radio astronomer Robert Wilson, was the first to detect cosmic background radiation. This radiation had been predicted on the basis of the 'hot Big Bang' model of the origin of the universe. Penzias and Wilson shared the 1978 Nobel Prize for Physics.

Penzias was born in Munich. His parents left Nazi Germany for the USA, and Penzias studied at the City College of New York and Columbia University. In 1961 he joined the staff of the Radio Research Laboratory of the Bell Telephone Company, becoming its director 1976 and vice president of research 1981. Concurrently he has held a series of academic positions at Princeton, Harvard, and from 1975 as professor at the State University of New York at Stony Brook.

In 1963, Penzias and Wilson were assigned by Bell to the tracing of radio noise that was interfering with the development of a communications programme involving satellites. By May 1964 they had detected a surprisingly high level of microwave radiation which had no apparent source (that is, it was uniform in all directions). The temperature of this background radiation was 3.5 K (–269.7°C/–453.4°F), later revised to 3.1 K (–270°C/–454.1°F).

They took this enigmatic result to physicist Robert Dicke at Princeton, who had predicted that this sort of radiation should be present in the universe as a residual relic of the intense heat associated with the birth of the universe following the Big Bang. His department was in the process of constructing a radio telescope designed to detect precisely this radiation when Penzias and Wilson presented their data.

Penzias's later work has been concerned with developments in radio astronomy, instrumentation, satellite communications, atmospheric physics, and related matters.

Peregrinus Petrus, adopted name of Peregrinus de Maricourt *c.* 1220–*c.* 1270/90. French scientist and scholar who published *Epistola de magnete* 1269. In it he described a simple compass (a piece of magnetized iron on a wooden disc floating in water) and outlined the laws of magnetic attraction and repulsion.

Peregrinus was an engineer in the French army under Louis IX, and was active in Paris in the middle of the 13th century. There he advised English scientist Roger Bacon. Peregrinus took part in the siege of Lucera in Italy in 1269.

His ideas on magnetism, based largely on experiment, were taken up 250 years later by English physicist William Gilbert.

Perey Marguérite (Catherine) 1909–1975. French nuclear chemist who discovered the radioactive element francium in 1939. Her career, which began as an assistant to Marie Curie 1929, culminated with her appointment as professor of nuclear chemistry at the University of Strasbourg 1949 and director of its Centre for Nuclear Research 1958.

Perkin William Henry 1838–1907. British chemist. In 1856 he discovered mauve, the dye that originated the aniline-dye industry and the British synthetic-dyestuffs industry generally. Knighted 1906.

Perkin was born in London and studied at the Royal College of Chemistry. He was only 18 when he discovered and patented the process for mauve. With the help of his father he set up a factory to manufacture the dye. Perkin's factory introduced new dyes based on the alkylation of magenta, and in 1868 he established a new route for the synthesis of alizarin. By 1871 Perkin's company was producing one tonne of alizarin every day. He sold the factory and retired from industry in 1874 to continue his academic research.

In one of his early home experiments Perkin produced the first example of the group of azo dyes produced from naphthalene. The aniline dyes – later named mauve by French textile manufacturers – are based on coal tar. The starting point for alizarin, a orange-red colour, was anthracene, another coal-tar derivative.

In the late 1860s, Perkin prepared unsaturated acids by the action of acetic anhydride on aromatic aldehydes, a method known as the Perkin synthesis. In 1868 he synthesized coumarin, the first preparation of a synthetic perfume.

Perl Martin L 1927– . US physicist who discovered the tau particle, one of the elementary particles which make up all matter. Nobel Prize for Physics 1995 (shared with Frederick Reines).

Perl and his colleagues discovered, through a series of experiments 1974–77, at the Stanford Linear Accelerator Center (SLAC), California, that the electron has a relative, some 3,500 times heavier, which is now called the tau. This was the first sign that a third 'family' of elementary particles existed. Without the existence of this third family, the standard model of elementary particles would be incomplete and unable to explain the properties of the elementary particles.

Perl was born in New York City in 1927 and educated in chemical engineering at the Polytechnic Institute of Brooklyn. His studies were interrupted by World War II; he received his Bachelor's degree 1948. From 1948 to 1950 he worked as a chemical engineer with the General Electric Company in Schenectady, New York. He entered his physics doctoral programme at Columbia University, New York 1950 and was awarded his doctorate in 1955. He moved to the University of Michigan 1955 as an instructor, later becoming assistant professor and associate professor of physics. In 1963 he was appointed professor of physics and group leader of the High Energy Physics Faculty, at the Stanford Linear Accelerator Center. In 1991 he became chairman of faculty.

Perrin Jean Baptiste 1870–1942. French physicist who produced the crucial evidence that finally established the atomic nature of matter. Assuming the atomic hypothesis, Perrin demonstrated how the phenomenon of Brownian motion could be used to derive precise values for Avogadro's number. Nobel prize 1926.

Perrin also contributed to the discovery that cathode rays are electrons. His experiments included imposing a negative electric charge on a fluorescent screen onto which various rays were focused. As the negative charge was increased, the intensity of fluorescence fell.

Perrin was born in Lille and studied in Paris at the Ecole Normale Supérieure. He worked at the Sorbonne from 1897, becoming professor 1910, but in 1940, during World War II, his outspoken antifascism caused him to flee the German occupation. He went to New York.

His book *Les Atomes/Atoms* 1913 describes his Nobel prizewinning study of Brownian motion.

Persoon Christiaan Hendrik 1761–1836. South African botanist who was one of the first to fully describe fungi. His *Synopsis fungorum* 1801 was a founding text of modern mycology. Although his observations of microscopic plant life were extremely influential in his day, Persoon held the erroneous view that some fungi grew from spores, while others formed by spontaneous generation.

Persoon was born at the Cape of Good Hope. He went to Europe to be educated 1775. He attended the gymnasium at Lingen, studied theology at Halle, and began his medical studies at Leiden. In 1787, he went to Güttingen, where he studied natural sciences as well as medicine. He was awarded a PhD from the University of Erlangen 1799.

In 1802 he moved to Paris, but he did not ever have a paid job and was never a wealthy man. In 1828 he gave his botanical collections to the Rijksherbarium in Leiden in exchange for a pension to live on in his old age.

Perutz Max Ferdinand 1914– . Austrian-born British biochemist who shared the 1962 Nobel Prize for Chemistry with his co-worker John Kendrew for work on the structure of the haemoglobin molecule.

True science thrives best in glass houses, where everyone can look in. When the windows are blacked out, as in war, the weeds take over; when secrecy muffles criticism, charlatans and cranks flourish.

Max Perutz

Is Science Necessary?

Perutz, born and educated in Vienna, moved to Britain in 1936 to work on X-ray crystallography at the Cavendish Laboratory, Cambridge. After internment in Canada as an enemy alien during World War II, he returned to Cambridge and in 1947 was appointed head of the new Molecular Biology Unit of the Medical Research Council (MRC). From 1962 he was chair of the MRC's new Laboratory of Molecular Biology.

Perutz first applied the methods of X-ray diffraction to haemoglobin in 1937, but it was not until 1953 that he discovered that if he added a single atom of a heavy metal such as gold or mercury to each molecule of protein, the diffraction pattern was altered slightly. By 1960 he had worked out the precise structure of haemoglobin.

Later, Perutz tried to interpret the mechanism by which the haemoglobin molecule transports oxygen in the blood, realizing that an inherited disorder such as sickle-cell anaemia could be caused by a mutation of this molecule.

Petit Alexis Thérèse 1791–1820. French physicist, co-discoverer with Pierre Dulong of ***Dulong and Petit's law***, which states that, for a solid element, the product of relative atomic mass and specific heat capacity is approximately constant.

Petit was born in Vesoul, Haute-Saône, and studied in Paris at the Ecole Polytechnique, becoming professor there 1815.

Petit's early research was conducted in collaboration with French scientist Dominique Arago. They examined the effect of temperature on the refractive index of gases. Their results led Petit to become an early supporter of the wave theory of light.

Petit and Dulong began their collaboration 1815, and in 1819 announced their law of atomic heats. Chemists who at that time were having difficulty determining atomic weights (and distinguishing them from equivalent weights) now had a method of estimating the approximate weight merely by measuring the specific heat of a sample of the element concerned.

Pfeffer Wilhelm Friedrich Philipp 1845–1920. German physiological botanist who was the first to measure osmotic pressure, in 1877. He also showed that osmotic pressure varies according to the temperature and concentration of the solute.

Pfeffer was born in Grebenstein, near Kassel, and studied at Göttingen. His first professorship was at Bonn 1873, and from 1887 he was at Leipzig.

Pfeffer made the first ever quantitative determinations of osmotic pressure, using a

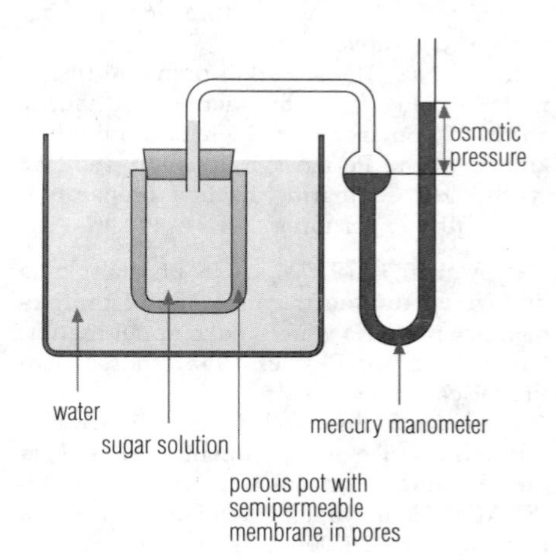

Pfeffer *Pfeffer's apparatus for measuring osmotic pressure.*

semipermeable container of sugar solution immersed in a vessel of water. He connected a mercury-filled manometer to the top of the semipermeable container. Pfeffer's work on osmosis was of fundamental importance in the study of cells, because semipermeable membranes surround all cells and play a large part in controlling their internal environment.

Pfeffer also studied respiration, photosynthesis, protein metabolism, and transport in plants. His *Handbuch der Pflanzenphysiologie/Physiology of Plants* 1881 was an important text for many years.

Pfeiffer Richard Friedrich Johannes 1858–1945. German bacteriologist who determined how to distinguish the organism that causes cholera and was the first to observe the body's immune response to an invading microbe.

Pfeiffer found that if a guinea pig is immunized against cholera, cholera vibrio (cholera bacteria) can be injected into its peritoneal (abdominal) cavity without any ill effects. He then withdrew some of the peritoneal fluid and examined the cholera germs microscopically. He observed that they swelled up and disintegrated. They had been lysed (destroyed) by some substance in the peritoneal fluid. Other organisms that resembled the cholera vibrio were not affected in this manner; the reaction was very specific. Pfeiffer managed to produce the reaction in vitro (in an artificial environment). He then demonstrated that the cholera vibrae were not lysed if the peritoneal fluid was heated to 60°C.

Pfeiffer was educated in Berlin as a military surgeon. He worked with the great bacteriologist Robert Koch at the Institute for Hygiene in Berlin and then became the professor of hygiene at Koenigsberg. In 1909, he was elected professor of hygiene at Breslau.

Philips Anton 1874–1951. Dutch industrialist and founder of an electronics firm. The Philips Bulb and Radio Works 1891 was founded with his brother Gerard, at Eindhoven. Anton served as chair of the company 1921–51, during which time the firm became the largest producer of electrical goods outside the USA.

Piaget Jean 1896–1980. Swiss psychologist whose studies of the development of thought processes in children have been influential in early-childhood research and on school curricula and teaching methods.

The subjects of Piaget's studies of intellectual development were his own children. He postulated four main stages in the development of mental processes: sensorimeter (birth to the age of two), preoperational (two to seven), concrete operational (seven to twelve), and formal operational, characterized by the development of logical thought.

Piaget was born in Neuchâtel. He was a child prodigy in zoology, and by the age of 15 had gained an international reputation for his work on molluscs. Subsequently he studied at Neuchâtel, Zürich, and Paris, where he researched into the reasons why children fail intelligence tests. This gained him the directorship of the Institut J J Rousseau in Geneva 1921. During his subsequent career Piaget held many academic positions, some of them concurrent. In 1955 he founded the International Centre of Genetic Epistemology at Geneva University. He also held several positions with UNESCO at various times.

Piaget's works include *La Naissance de l'intelligence chez l'enfant/The Origins of Intelligence in the Child* 1936 and *The Child's Construction of Reality* 1936.

I noticed with amazement that the simplest reasoning task ... presented for normal children up to the age of eleven or twelve difficulties unsuspected by the adult.

Jean Piaget

Piazzi Giuseppe 1746–1826. Italian astronomer who in 1801 identified the first asteroid, which he named Ceres.

Piazzi was born in Ponte di Valtellina (now in Switzerland). He studied in various Italian cities and in 1764 entered the Theatine Order as a monk. In 1779 he became professor of mathematics at the Palermo Academy in Sicily. When it was decided to establish observatories in Palermo and Naples, Piazzi travelled to observatories in England and France to obtain advice and equipment. He was director of Palermo Observatory from 1790 and from 1817 also of Naples Observatory. Additional responsibilities included the reformation of the Sicilian system of weights and measures 1812.

In 1803, Piazzi published his first catalogue of fixed stars, which located 6,748 stars with unprecedented accuracy.

Picard (Charles) Emile 1856–1941. French mathematician whose work was mainly in the fields of mathematical analysis and algebraic geometry. He formulated two theorems on integral functions. He applied mathematical principles as much as possible to other branches of science, particularly physics and engineering.

Picard was born in Paris and studied at the Ecole Normale Supérieure. At the age of 23 he was appointed professor in Toulouse, but returned to Paris two years later and became professor at the Sorbonne 1885.

Picard's 'little theorem' states that an integral function of the complex variable takes every finite value, with one possible exception. In 1879 he expressed it in this way:

Let $f(z)$ be an entire function. If there exist two values of A for which the equation $f(z) = A$ does not have a finite root, then $f(z)$ is a constant. From this it follows that if $f(z)$ is an entire function that is not a constant, there cannot be more than one value of A for which $f(z) = A$ has no solution.

This was followed 1880 by Picard's 'big theorem':

Let $f(z)$ be a function, analytic everywhere except at a where it has an essential isolated singularity; the equation $f(z) = A$ has in general an infinity of roots in any neighbourhood of a. Although the equation can fail for certain exceptional values of the constant A, there cannot be more than two such values.

Picard's work on the integrals attached to algebraic surfaces, together with the associated topological questions, developed into an area of algebraic geometry that had applications in topology and function theory.

Much of Picard's work was recorded in the three-volume *Traité d'analyse*.

Piccard Auguste Antoine 1884–1962. Swiss scientist. In 1931–32, he and his twin brother, ***Jean Félix*** (1884–1963), made ascents to 17,000 m/55,000 ft in a balloon of his own design, resulting in useful discoveries concerning stratospheric phenomena such as cosmic radiation. He also built and used, with his son ***Jacques Ernest*** (1922–), bathyscaphs for research under the sea.

Auguste Piccard was born and educated in Basel, and joined the balloon section of the Swiss army 1915. From 1922 he was professor of physics at Brussels, Belgium.

Jacques Piccard twice, in 1953 and 1960, set a world record with the depth he reached in a bathyscaph. The second time he descended to 10,917 m/35,820 ft in the Mariana Trench near Guam in the Pacific Ocean.

Pickering Edward Charles 1846–1919. US astronomer who was a pioneer in three practical areas of astronomical research: visual photometry, stellar spectroscopy, and stellar photography. He established an international astronomical colour index: a measure of the apparent colour of a star and thus of its temperature.

Pickering was born in Boston, Massachusetts. He was director of the Harvard College Observatory from 1876. Unusually for his generation, he encouraged women to take up astronomy as a career.

As a basis for the more than 1.5 million photometric readings he carried out, Pickering made two critical decisions. First, he adopted a scale on which a change of one magnitude represented a change of a factor of 2.512 in brightness. Second, choosing the Pole Star (Polaris), then thought to be of constant brightness, as the standard magnitude and arbitrarily assigning a value of 2.1 to it, he redesigned the photometer to reflect a number of stars round the meridian at the same time so that comparisons were immediately visible. In 1908 Harvard published a catalogue with the magnitudes of more than 45,000 stars.

The *Henry Draper Catalogue* 1918 contained the spectra of no fewer than 225,000 stars, work begun by Pickering and classified according to the new system devised by Annie Jump Cannon.

The first *Photographic Map of the Entire Sky*, published 1903, contained photographs taken at Harvard and at its sister station in the southern hemisphere, at Arequipa in Peru, where Pickering's brother ***William Pickering*** (1858–1938) was director. In addition, Pickering built up a 300,000-plate Harvard photographic library.

Pike Magnus Alfred 1908–1992. British food scientist and broadcaster. He enjoyed an extraordinary period of celebrity following his retirement 1973, co-presenting *Don't Ask Me*, which became the most popular science series on British television.

Pike was author of the standard work *The Manual of Nutrition* 1945. He ran Genochil Research Station, Scotland 1955–73, after which he became Secretary of the British Association for the Advancement of Science.

Pike had a mission to explain, which he achieved brilliantly. Much in demand as a lecturer, his arms flailed as he sought the clearest answers to the questions his audience asked.

He referred to his period of fame as a television scientist as his 'sixth life'. His 'fifth life' was spent working for Distillers Company in Scotland and bringing up his family; the fourth was marked by the publication of *The Manual of Nutrition*. His 'third life' was passed in Canada, working as a farm labourer in the summer, and studying at McGill University in the winter; his second was his education at St Paul's School, London; while the first encompassed his early childhood in west London. His 'seventh life' began when *Don't Just Sit There* – a further highly successful television series – ended, and he retired again, in 1980.

Pilcher Percy Sinclair 1867–1899. English aviator who was the first Briton to make a successful flight in a heavier-than-air craft, called the *Bat*, 1895. Like Otto Lilienthal, Pilcher made flights only downhill from gliders, using craft resembling the modern hang glider. Pilcher's next successful aircraft was the *Hawk*, launched 1896 at Eynsford, Kent, by a tow line. He was killed 1899 flying the *Hawk* near Rugby in the Midlands.

Pincus Gregory Goodwin 1903–1967. US biologist who, together with Min Chueh Chang (1908–) and John Rock (1890–1984), developed the contraceptive pill in the 1950s.

As a result of studying the physiology of reproduction, Pincus conceived the idea of using synthetic hormones to mimic the condition of pregnancy in women. This effectively prevents impregnation.

Pincus was born in Woodbine, New Jersey, and studied at Cornell and Harvard. He joined the staff of Harvard 1930, and in 1944 cofounded the Worcester Foundation for Experimental Biology in Shrewsbury, Massachusetts. It was there he began his research on steroid hormones, which was encouraged by birth-control pioneer Margaret Sanger. The pill was first marketed 1960.

Pippard (Alfred) Brian 1920– . English physicist who applied microwaves to the study of superconductivity. The research he initiated has transformed understanding of the dynamical laws governing the motion of electrons in metals. Knighted 1975.

Pippard was born in London and studied at Cambridge, returning there after World War II. In 1960 he became professor.

Pippard worked on the way in which electric currents flow without resistance in a thin layer at the surface of the metal. He measured the thickness (about 1000 Å/10^{-7} m) of this penetration layer and examined variations with temperature and purity. When he tried to change the properties at one point by applying a disturbance, he influenced the metal over a distance greater than the penetration layer thickness. Because of this, he said that the electrons of superconductors possess a property which he called 'coherence', and that impurities in the metal could shorten the coherence length. From this starting point, he worked out an equation relating current to magnetic field.

Pirquet Clemens von 1874–1929. Austrian paediatrician and pioneer in the study of allergy.

Pixii Hippolyte 1808–1835. French inventor who in 1832 made the first practical electricity generator. It could produce both direct current and alternating current.

Pixii was an instrumentmaker, trained by his father. Learning of Michael Faraday's electromagnetic induction and his suggestions for making a simple dynamo, Pixii constructed a device that consisted of a permanent horseshoe magnet, rotated by means of a treadle, and a coil of copper wire above each of the magnet's poles. The two coils were linked and the free ends of the wires connected to terminals, from which a small alternating current was obtained when the magnet rotated. This device was first exhibited at the French Academy of Sciences in Paris 1832. Later, at the suggestion of physicist André Marie Ampère, a commutator (a simple switching device for reversing the connections to the terminals as the magnet is rotated) was fitted so that Pixii's generator could produce direct-current electricity. This revised generator was taken to Britain in 1833 and exhibited in London.

We have no right to assume that any physical laws exist, or if they have existed up to now, that they will continue to exist in a similar manner in the future.

Max Planck

The Universe in the Light of Modern Physics

Planck Max Karl Ernst 1858–1947. German physicist who framed the quantum theory 1900. His research into the manner in which heated bodies radiate energy led him to report that energy is emitted only in indivisible amounts, called 'quanta', the magnitudes of which are proportional to the frequency of the radiation. His discovery ran counter to classical physics and is held to have marked the commencement of the modern science. Nobel Prize for Physics 1918.

Planck was born in Kiel and studied at Munich. He became professor at Kiel 1885, but moved 1888 to Berlin as director of the newly founded Institute for Theoretical Physics. He was also professor of physics at Berlin 1892–1926. Appointed president of the Kaiser Wilhelm Institute 1930, he resigned 1937 in protest at the Nazis' treatment of Jewish scientists. In 1945, after World War II, the institute was renamed the Max Planck Institute and moved to Göttingen. Planck was reappointed its president.

An important scientific innovation rarely makes its way by gradually winning over and converting its opponents: it rarely happens that Saul becomes Paul. What does happen is that its opponents gradually die out, and that the growing generation is familiarized with the ideas from the beginning.

Max Planck

Scientific Autobiography 1949

Planck's idea that energy must consist of indivisible particles, not waves, was revolutionary. But an explanation for photoelectricity was provided by Albert Einstein in 1905 using Planck's quantum theory, and in 1913 Danish physicist Niels Bohr successfully applied the quantum theory to the atom.

Planck's constant, a fundamental constant (symbol *h*), is the energy of one quantum of electromagnetic radiation divided by the frequency of its radiation.

Plaskett John Stanley 1865–1941. Canadian astronomer and engineer who discovered many new binary stars, including Plaskett's Twins, previously thought to be a single, massive star. He also carried out research into stellar radial velocities.

Plaskett was born near Woodstock, Ontario, and trained as a mechanic before studying at Toronto. From 1903 he was in charge of astrophysical work at the new Dominion Observatory in Ottawa. He designed a 1.8-m/72-in reflector telescope for the Dominion Astrophysical Observatory in Victoria, and was its director 1917–35. He then supervised the construction of a 205-cm/82-in mirror for the MacDonald Observatory at the University of Texas.

Plaskett's work on the radial velocities of galactic stars enabled him to confirm the contemporary discovery of the rotation of the Galaxy and to indicate the most probable location of its gravitational centre. This led to a study of the motion and distribution of galactic interstellar matter, particularly calcium.

Plato *c.* 427–347 BC. Greek philosopher. He was a pupil of Socrates, teacher of Aristotle, and founder of the Academy school of philosophy. He was the author of philosophical dialogues on such topics as metaphysics, ethics, and politics. Central to his teachings is the notion of Forms, which are located outside the everyday world – timeless, motionless, and absolutely real.

Plato's philosophy has influenced Christianity and European culture, directly and through Augustine, the Florentine Platonists during the Renaissance, and countless others.

Of his work, some 30 dialogues survive, intended for performance either to his pupils or to the public. The principal figure in these ethical and philosophical debates is Socrates and the early ones employ the Socratic method, in which he asks questions and traps the students into contradicting themselves; for example, *Iron*, on poetry. Other dialogues include the *Symposium*, on love, *Phaedo*, on immortality, and *Apology and Crito*, on Socrates' trial and death. It is impossible to say whether Plato's Socrates is a faithful representative of the real man or an articulation of Plato's own thought.

Plato's philosophy rejects scientific rationalism (establishing facts through experiment) in favour of arguments, because mind, not matter, is fundamental, and material objects are merely imperfect copies of abstract and eternal 'ideas'. His political philosophy is expounded in two treatises, *The Republic* and *The Laws*, both of which

There is only one good, namely knowledge, and only one evil, namely ignorance.

Plato

Dialogues

describe ideal states. Platonic love is inspired by a person's best qualities and seeks their development.

Born of a noble family, Plato entered politics on the aristocratic side, and in philosophy became a follower of Socrates. He travelled widely after Socrates' death, and founded an educational establishment, the Academy, in order to train a new ruling class.

In science, geometry – with its premise of symmetry and the irrefutable logic of its axioms – had the most appeal to Plato. The universe had to be spherical because the sphere was the perfect volume; for the same reason the movements of the heavenly bodies had to be circular and uniform. Moreover, the Earth, which lay at the exact centre of the cosmos, was a sphere and was surrounded by a band of crystalline spheres which held in place the Sun, the Moon, and the planets.

Pliny the Elder (Gaius Plinius Secundus) c. AD 23–79. Roman scientific encyclopedist and historian. Many of his works have been lost, but in *Historia naturalis/Natural History*, probably completed AD 77, Pliny surveys all the known sciences of his day, notably astronomy, meteorology, geography, mineralogy, zoology, and botany.

Pliny states that he has covered 20,000 subjects of importance drawn from 100 selected writers, to whose observations he has added many of his own. Botany, agriculture, and horticulture appear to interest him most. To Pliny the world consisted of four elements: earth, air, fire, and water. The light substances were prevented from rising by the weight of the heavy ones, and vice versa. This is the earliest theory of gravity.

There is always something new out of Africa.

Pliny the Elder

Historia Naturalis II. 8

Pliny was born in Como, completed his studies in Rome and took up a military career in Germany, where he became a cavalry commander and friend of Vespasian. He kept out of harm's way while Nero was on the throne, but when in AD 69 Vespasian was made emperor, Pliny returned to Rome and took up various public offices. In AD 79 he was in command of a fleet in the bay of Naples when the volcano Vesuvius erupted. Going to take a closer look, he was killed by poisonous fumes.

According to Pliny, the Earth was surrounded by seven stars: the Sun, the Moon, Mercury, Venus, Mars, Jupiter, and Saturn. Pliny took the Moon to be larger than the Earth, since it obscured the Sun during an eclipse.

Plücker Julius 1801–1868. German mathematician and physicist. He made fundamental contributions to the field of analytical geometry and was a pioneer in the investigations of cathode rays that led eventually to the discovery of the electron.

Plücker was born in Elberfeld, Wuppertal, and studied at Bonn, Heidelberg, Berlin, and Paris. He was professor of mathematics at Bonn 1828–33 and again 1836–47, when he became professor of physics there.

Plücker introduced six equations of higher plane curves which have been named Plücker's coordinates. His work led to the foundation of line geometry.

He was one of the first to recognize the potential of gas spectroscopy in analysis, and found the first three hydrogen lines.

Experimenting with electrical discharge in gases at high pressures, Plücker found in 1858 that the discharge caused a fluorescent glow to form on the glass walls of the vacuum tube, and that the glow could be made to shift by applying an electromagnet to the tube, thus creating a magnetic field. It was left to one of his students to show that the glow was produced by cathode rays.

Poincaré Jules Henri 1854–1912. French mathematician who developed the theory of differential equations and was a pioneer in relativity theory. He suggested that Isaac Newton's laws for the behaviour of the universe could be the exception rather than the rule. However, the calculation was so complex and time-consuming that he never managed to realize its full implication.

Poincaré wrote on the philosophy of science. He believed that some mathematical ideas precede logic, and stressed the role played by convention in scientific method.

He also published the first paper devoted entirely to topology (the branch of geometry that deals with the unchanged properties of figures).

Poincaré was born in Nancy and studied in Paris at the Ecole Polytechnique and the Ecole des Mines. He was professor at the Sorbonne from 1881.

Thought is only a flash between two long nights, but the flash is everything.

Jules Poincaré

Attributed

In his 1906 paper on the dynamics of the electron, Poincaré obtained many of the results of the theory of relativity later credited to Albert Einstein.

In the field of celestial mechanics, Poincaré studied the mutual gravitational and other effects of three bodies, or *n* bodies, close together in space. He developed new mathematical techniques, including the theories of asymptotic expansions and integral invariants. From his theory of periodic orbits he developed the new subject of topological dynamics.

Poiseuille Jean Léonard Marie 1799–1869. French physiologist who made a key contribution to our knowledge of the circulation of blood in the arteries. He also studied the flow of liquids in artificial capillaries.

Poiseuille was born in Paris and studied at the Ecole Polytechnique. In 1842 he was elected to the Académie de Médicine in Paris.

Poiseuille improved on earlier measurements of blood pressure by using a mercury manometer and filling the connection to the artery with potassium carbonate to prevent coagulation. He used this instrument, known as a hemodynamometer, to show that blood pressure rises during expiration (breathing out) and falls during inspiration (breathing in). He also discovered that the dilation of an artery fell to less than $\frac{1}{120}$ of its normal value during a heartbeat.

Poisson Siméon Denis 1781–1840. French applied mathematician and physicist. In probability theory he formulated the ***Poisson distribution***. ***Poisson's ratio*** in elasticity is the ratio of the lateral contraction of a body to its longitudinal extension. The ratio is constant for a given material.

Much of Poisson's work involved applying mathematical principles in theoretical terms to contemporary and prior experiments in physics, particularly with reference to electricity, magnetism, heat, and sound. Poisson was also responsible for a formulation of the 'law of large numbers', which he introduced in his work on probability theory, *Recherches sur la probabilité des jugements/Researches on the Probability of Opinions* 1837.

Poisson was born in Pithiviers, Loirel, and studied in Paris at the Ecole Polytechnique, where he became professor 1806. In 1808 he was appointed astronomer at the Bureau des Longitudes, and the following year he was appointed professor of mechanics at the Faculty of Sciences. From 1820 he was an administrator at the highest level in France's educational system.

Poisson's works include *Treatise on Mechanics* 1833, *Mathematical Theory of Heat* 1835, and *Researches on the Movement of Projectiles in Air* 1835, the first account of the effects of the Earth's rotation on motion.

Polanyi John Charles 1929– . German physical chemist whose research on infrared light given off during chemical reactions (infrared chemical luminescence) laid the foundations for the development of chemical lasers. He analyzed the chain reaction between hydrogen and chlorine, and showed that two distinct states of hydrochloric acid are formed. He shared the Nobel Prize for Chemistry in 1986 with Dudley Herschbach and Yuan Tseh Lee.

Polanyi studied radiation from strongly exothermic (heat-emitting) reactions to see how the released energy was distributed in the reaction products. In the hydrogen–chlorine reaction, he found one state of hydrochloric acid with high vibrational and rotational excitation, but low translational energy; and another with low vibrational and rotational energy but high translational energy.

Polanyi carried on his research into reaction dynamics, and has published numerous papers on science policy and on control of armaments. His work on infrared light emittence during chemical reactions (infrared chemical luminescence) laid the foundations for the development of chemical lasers by Pimental and Kaspar 1960.

Born in Berlin, Germany, the son of the Hungarian–British physical chemist and philosopher Michael Polanyi, he followed in his father's footsteps and studied chemistry at Manchester University. He then spent some time at the National Research Council in Ottawa and at Princeton. In 1952, he obtained his PhD and moved to Toronto University, where he became a professor of chemistry in 1962. Polanyi received many accolades and, in 1971, he was elected a Fellow of the Royal Society, receiving its Royal Medal. He became an Honorary Fellow of the Royal Society of Chemistry in 1991.

Polanyi Michael 1891–1976. Hungarian chemist, social scientist, and philosopher. As a scientist, he worked on thermodynamics, X-ray crystallography, and physical adsorption. As a philosopher and social scientist, he was concerned about the conflicts between personal freedom and central planning, and the impact of the conflict upon scientists.

Polanyi, born in Budapest, studied medicine there and later turned to chemistry. In 1933 he resigned from the Kaiser Wilhelm Institute in Berlin over the dismissal of Jewish scientists, and moved to Manchester University, England, as professor of physical chemistry. He was professor of social studies at Manchester 1948–58, and then moved to Oxford.

Polanyi introduced the idea of the existence of an attractive force between a solid surface and the atoms or molecules of a gas; he also suggested that the adsorbed surface is a multilayer and not subject to simple valency interactions.

He advocated that scientific research need not necessarily have a pre-stated function and expressed the belief that a commitment to the discovery of truth is the prime reason for being a scientist. He also analysed the nature of knowledge, skills, and discovery.

Polanyi's works include *Personal Knowledge* 1958, *Knowing and Being* 1969, and *Scientific Thought and Social Reality* 1974.

Pólya George 1887–1985. Hungarian mathematician who worked on function theory, probability, and applied mathematics. ***Pólya's theorem*** 1920 is a solution of a problem in combinatorial analysis theory and method.

Pólya was born in Budapest and studied at the Eötvös Lorand University. From 1914 he lectured at the Swiss Federal Institute of Technology in Zürich, becoming professor 1928. In 1940 he emigrated to the USA, ending his career at Stanford University, Palo Alto, California, 1946–53.

Pólya published studies on analytical functions in 1924 and on algebraic functions in 1927. He also worked on linear homogeneous differential equations (1924) and transcendental equations (1930). One of his studies in mathematical physics was an investigation into heat propagation 1931.

Other subjects he examined included the study of complex variables, polynomials, and number theory.

Poncelet Jean-Victor 1788–1867. French mathematician and military engineer who advanced projective geometry. His book *Traité des propriétés projectives des figures* 1822 deals with the properties of plane figures that remain unchanged when projected.

Poncelet was born in Metz. He took part in Napoleon's campaign against Russia, and was captured. During his two years as a prisoner of war, he began his work on geometry. On his release 1814, he returned to Metz and was engaged on projects in military engineering there until 1824, when he became professor of mechanics at a local military school. In 1830 he was elected a member of Metz Municipal Council and secretary of the Conseil-Général of the Moselle. He moved to the University of Paris 1838 and ended his career as commandant of the Ecole Polytechnique with the rank of general.

Poncelet had been a pupil of Gaspard Monge, the originator of modern synthetic geometry, but Poncelet also used analytical geometry and contributed greatly to the development of the relatively new synthetic (projective) geometry. He became the centre of controversy over the principle of continuity, and developed the circular points at infinity.

Poncelet developed a new model of a variable counterweight drawbridge, which he described and publicized in 1822. His most important technical contributions were concerned with hydraulic engines, such as Poncelet's water wheel, with regulations and with dynamometers, as well as in devising various improvements to his own fortification techniques.

Pond John 1767–1836. English astronomer who as Astronomer Royal 1811–35 reorganized and modernized Greenwich Observatory. Instituting new methods of observation, he went on to produce a catalogue of more than 1,000 stars in 1833.

Pond was born in London and studied at Cambridge. He travelled in several Mediterranean and Middle Eastern countries, making astronomical observations wherever possible. When he returned to England in 1798, he established a small private observatory near Bristol. The observations he published led to his appointment as Astronomer Royal.

At the age of 15, Pond noticed errors in the observations being made at the Greenwich Observatory and began a thorough investigation of the declination of a number of fixed stars. By 1806 he had publicly demonstrated that the quadrant at Greenwich had become deformed with age and needed replacing. It was this in particular that prompted his programme to modernize the whole observatory.

Pons Jean-Louis 1761–1831. French astronomer who discovered 37 comets. He was the first to recognize the return of Encke's comet.

Pons was born in Peyre, near Dauphine. At the age of 28 he became a porter and doorkeeper at the Marseille Observatory. Noting his interest in astronomy, the directors of the observatory gave him instruction, and he turned out to be good at practical observation. Pons was made assistant astronomer 1813 and assistant director 1818. In 1819 he became director of a new observatory at Lucca, N Italy, moving to the Florence Observatory 1822.

In 1818 Pons discovered three small, tailless comets, among which was one that he claimed had first been seen in 1805 by Johann Encke of the Berlin Observatory. Alerted to this possibility, Encke carried out further observations and calculations, and finally ascribed to it a period of 1,208 days – which meant that it would return in 1822. Its return was duly observed, in Australia, only the second instance ever of the known return of an identified comet. Encke wanted the comet to be named after Pons, but it continued to be called after its discoverer.

Popov Alexander Stepanovich 1859–1905. Russian physicist who devised the first aerial, in advance of Marconi (although he did not use it for radio communication). He also invented a detector for radio waves.

Popper Karl Raimund 1902–1994. Austrian-born British philosopher of science. His theory of falsificationism says that although scientific generalizations cannot be conclusively verified, they can be conclusively falsified by a counterinstance; therefore, science is not certain knowledge but a series of 'conjectures and refutations', approaching, though never reaching, a definitive truth. For Popper, psychoanalysis and Marxism are falsifiable and therefore unscientific. Knighted 1965.

One of the most widely read philosophers of the century, Popper's book *The Open Society and its Enemies* 1945 became a modern classic. In it he investigated the long history of attempts to formulate a theory of the state. Animated by a dislike of the views of Freud and Marx, Popper believed he could show that their hypotheses about hidden social and psychological processes were falsifiable.

His major work on the philosophy of science is *The Logic of Scientific Discovery* 1935. Other works include *The Poverty of Historicism* 1957 (about the philosophy of social science), *Conjectures and Refutations* 1963, and *Objective Knowledge* 1972.

Born and educated in Vienna, Popper served for a while as an assistant to the psychologist Alfred Adler before emigrating to New Zealand in 1937. Returning to Europe, he became a naturalized British subject 1945 and was professor of logic and scientific method at the London School of Economics 1949–69. He opposed Wittgenstein's view that philosophical problems are merely pseudoproblems. Popper's view of scientific practice has been criticized by T S Kuhn and other writers.

Porritt Jonathon 1950– . British environmental campaigner, director of Friends of the Earth 1984–90. He has stood in both British and European elections as a Green (formerly Ecology) Party candidate.

Green consumerism is a target for exploitation. There's a lot of green froth on top, but murkiness lurks underneath.

Jonathon Porritt

Speech at a Friends of the Earth Conference 1989

Porsche Ferdinand 1875–1951. German car designer and engineer who designed the Volkswagen Beetle, first mass produced 1945. By 1972 more than 15 million had been sold, making it the world's most popular model. Porsche sports cars were developed by his son Ferry Porsche from 1948.

Ferdinand Porsche designed his first racing

car in the mid-1930s, which was successfully developed by Auto-Union for their racing team. Ferry's Porsche Company produced Grand Prix cars, sports cars, and prototypes. Their Formula One racing car was not successful and it was at sports-car and Can-Am racing that they proved to be more dominant.

Ferdinand Porsche was born in Bohemia. He was technical director with Daimler-Benz 1923–29 and formed his own company 1931. In 1932 he devised the first torsion-bar suspension system, which was incorporated in the Volkswagen prototype he began working on 1934. In 1936, he received a contract from the German government to develop the Volkswagen and plan the factory where it would be built. World War II halted this development, so Porsche designed the Leopard and Tiger tanks used by German Panzer regiments and helped to develop the V1 flying bomb.

In the 1930s Porsche also designed light tractors, and worked on aviation engines and plans and designs for wind-driven power plants – large windmills with automatic sail adjustment.

Porter George 1920– . English chemist. From 1947 he and Ronald Norrish developed a technique by which flashes of high energy are used to bring about extremely fast chemical reactions. They shared a Nobel prize 1967. Knighted 1972.

Porter was born in Stainforth, Yorkshire, and studied at Leeds. After World War II he carried out research at Cambridge under Norrish. Porter was professor at Sheffield 1955–66, director of the Royal Institution 1966–85, and president of the Royal Society 1985–90. In the 1960s he made many appearances on British television.

Porter began using quick flashes of light to study transient species in chemical reactions, particularly free radicals and excited states of molecules. In 1950 he could detect entities that exist for less than a microsecond; by 1975, using laser beams, he had reduced the time limit to a picosecond (10^{-12} sec). His early work dealt with reactions involving gases (mainly chain reactions and combustion reactions), but he later extended the technique to solutions and also studied the processes that occur in the first nanosecond of photosynthesis in plants. He developed a method of stabilizing free radicals by trapping them in the structure of a supercooled liquid (a glass), a technique called matrix isolation.

Porter Rodney Robert 1917–1985. English biochemist. In 1962 he proposed a structure for human immunoglobulin (antibody) in which the molecule was seen as consisting of four chains. He was awarded the 1972 Nobel Prize for Physiology or Medicine.

Porter was born in Liverpool and studied there and at Cambridge. He became professor of immunology at St Mary's Hospital Medical School, London, 1960; from 1967 he was professor of biochemistry at Oxford.

Basing his research on the work of US immunologist Karl Landsteiner, Porter studied the structural basis of the biological activities of antibodies, proposing in 1962 a structure for gamma-globulin. He also worked on the structure, assembly, and activation mechanisms of the components of a substance known as complement. This is a protein that is normally present in the blood, but disappears from the serum during most antigen–antibody reactions. In addition, Porter investigated the way in which immunoglobulins interact with complement components and with cell surfaces.

Poulsen Valdemar 1869–1942. Danish engineer who in 1900 was the first to demonstrate that sound could be recorded magnetically – originally on a moving steel wire or tape; his device was the forerunner of the tape recorder.

Powell Cecil Frank 1903–1969. English physicist who investigated the charged subatomic particles in cosmic radiation by using photographic emulsions carried in weather balloons. This led to his discovery of the pion (π meson) 1947, a particle whose existence had been predicted by Japanese physicist Hideki Yukawa 1935. Powell was awarded a Nobel prize 1950.

Powell was born in Tonbridge, Kent, and studied at Cambridge, where he carried out research at the Cavendish Laboratory under Ernest Rutherford and C T R Wilson, taking photographs of particle tracks in a cloud chamber. From 1928 Powell worked at the Wills Physics Laboratory at Bristol University, becoming professor 1948.

In 1938, instead of photographing the cloud-chamber tracks, Powell made the ionizing particles trace paths in the emulsions of a stack of photographic plates. The technique received a boost with the development of more sensitive emulsions during

World War II and Powell used it in his discovery of the pion. He collaborated with Italian physicist Giuseppe Occhialini (1907–), and together they published *Nuclear Physics in Photographs* 1947, which became a standard text on the subject.

Powell John Wesley 1834–1902. US geologist whose enormous and original studies produced lasting insights into erosion by rivers, volcanism, and mountain formation. His greatness as a geologist and geomorphologist stemmed from his capacity to grasp the interconnections of geological and climatic causes.

Powell was born in New York State and self-educated. In the 1850s he became secretary of the Illinois Society of Natural History. Fighting in the Civil War, he had his right arm shot off, but continued in the service, rising to the rank of colonel. After the end of the war, Powell became professor of geology in Illinois, while continuing with intrepid fieldwork (he was one of the first to steer a way down the Grand Canyon).

In 1870, Congress appointed Powell to lead an official survey of the natural resources of the Utah, Colorado, and Arizona area.

Powell was appointed director of the US Geological Survey 1881. He drew attention to the aridity of the American southwest, and campaigned for irrigation projects and dams, for the geological surveys necessary to implement adequate water strategies, and for changes in land policy and farming techniques. Failing to win political support on such matters, he resigned in 1894 from the Geological Survey.

Poynting John Henry 1852–1914. English physicist, mathematician, and inventor. He devised an equation by which the rate of flow of electromagnetic energy (now called the ***Poynting vector***) can be determined.

In 1891 he made an accurate measurement of Isaac Newton's gravitational constant.

Poynting was born near Manchester and studied there at Owens College, and at Cambridge. From 1880 he was professor of physics at Mason College, Birmingham (which became Birmingham University in 1900).

In *On the Transfer of Energy in the Electromagnetic Field* 1884, Poynting published the equation by which the magnitude and direction of the flow of electromagnetic energy can be determined. This equation is usually expressed as:

$$S = (1/\mu)EB \sin \theta$$

where S is the Poynting vector, μ is the permeability of the medium, E is the electric field strength, B is the magnetic field strength, and θ is the angle between the vectors representing the electric and magnetic fields.

In 1903, he suggested the existence of an effect of the Sun's radiation that causes small particles orbiting the Sun to gradually approach it and eventually plunge in. This idea was later developed by US physicist Howard Percy Robertson (1903–1961) and is now known as the Poynting–Robertson effect. Poynting also devised a method for measuring the radiation pressure from a body; his method can be used to determine the absolute temperature of celestial objects.

Poynting's other work included a statistical analysis of changes in commodity prices on the stock exchange 1884.

Prain David 1857–1944. Scottish botanist and director of the Royal Botanical Garden in Calcutta and the botanical gardens at Kew in London.

Prain was born in Fellencairn, Aberdeenshire and sent to study at the University of Aberdeen. He won a bursary and taught during the university vacations in order to finance his studies for his MA in natural sciences, which he gained 1878. In 1880 he began to study medicine and on completion of these studies in 1883, he went to India with the Indian Medical Service.

In 1885 he was curator of the Royal Botanical Garden at Calcutta, and he gave up the practice of medicine 1887 in order to work full time there. He was appointed professor of botany at the Medical College in Calcutta 1895 and became superintendent of the Royal Botanical Garden when George King retired 1898. In 1905, he was made the director of the botanical gardens at Kew, where he switched his attentions to British and African plants. He retired 1922 and was knighted the same year.

Prandtl Ludwig 1875–1953. German physicist who put fluid mechanics on a sound theoretical basis and originated the boundary-layer theory. His work in aerodynamics resulted in major changes in wing design and streamlining of aircraft.

Prandtl was born in Freising, Bavaria, and studied at the Technische Hochschule in Munich. In 1901 he became professor at the Technische Hochschule in Hanover; from 1904 he was professor of applied mechanics at Göttingen. There he constructed the first German wind tunnel in 1909 and built up a centre for aerodynamics.

In a paper of 1904, Prandtl proposed that no matter how small the viscosity of a fluid, it is always stationary at the walls of the pipe. This thin static region or boundary layer has a profound influence on the flow of the fluid, and an understanding of the effects of boundary layers was developed to explain the action of lift and drag on aerofoils during the following half century. In 1907, Prandtl investigated supersonic flow.

In 1926, Prandtl developed the concept of mixing length – the average distance that a swirling fluid element travels before it dissipates its motion – and produced a theory of turbulence.

Pregl Fritz 1869–1930. Austrian chemist who, during his research on bile acids, devised new techniques for microanalysis (the analysis of very small quantities). He scaled down his analytic equipment and designed a new balance that could weigh to an accuracy of 0.001 mg. This breakthrough in organic chemistry paved the way for modern biochemistry. Nobel Prize for Chemistry in 1923.

Pregl was born on 3 September 1869 in Laibach (now Ljubljana), Slovenia. He read medicine at Graz University, qualifying in 1893, and took up the post of assistant in physiological chemistry in 1899. In 1910 he became head of the chemistry department at Innsbruck, but he returned to Graz in 1913 when he was appointed professor of medical chemistry.

Prelog Vladimir 1906– . Bosnian-born Swiss organic chemist who studied alkaloids and antibiotics. The comprehensive molecular topology that evolved from his work on stereochemistry is gradually replacing classical stereochemistry. He shared a Nobel prize 1975.

Prelog was born in Sarajevo and studied in Czechoslovakia at the Institute of Technology in Prague. In 1935 he went back to Yugoslavia to lecture at Zagreb, but in 1941, after the German occupation at the beginning of World War II, Prelog moved to the Federal Institute of Technology, Zürich, where he became professor 1957.

Alkaloids were the subject of Prelog's early research, and he derived the structures of quinine, strychnine, and steroid alkaloids from plants of the genera *Solanum* and *Veratrum*. His studies of lipoid extracts from animal organs also resulted in the discovery of various steroids and the elucidation of their structures. He investigated metabolic products of microorganisms and with a number of other researchers isolated various new complex natural products that have interesting biological properties. These include antibiotics and bacterial growth factors.

Prévost Pierre 1751–1839. Swiss physicist who first showed, in 1791, that all bodies radiate heat, no matter how hot or cold they are. In challenging the notion then prevalent that cold was produced by the entry of cold into an object rather than by an outflow of heat, Prévost made a basic advance in our knowledge of energy.

Prévost was born in Geneva and studied and travelled widely. For a year he was professor of literature at Geneva, then worked in Paris on the translation of a Greek drama. Returning to Geneva 1786, he became active in politics as well as carrying out research into magnetism and heat. He was professor of philosophy and general physics at Geneva 1793–1823. In his later years, Prévost chose to study the human ageing process. He used himself for his observations, noting down in detail every sign of advancement that his mind, body, and mirror showed.

Prévost found by experiment that dark, rough-textured objects give out and absorb more radiation than smooth, light-coloured bodies, given that both are at the same temperature. He conceived of heat as being a fluid composed of particles and this led to Prévost's theory of heat exchanges. If several objects at different temperatures are placed together, they exchange heat by radiation until all achieve the same temperature. They then remain at this temperature if they are receiving as much heat from their surroundings as they radiate away.

Priestley Joseph 1733–1804. English chemist and Unitarian minister. He identified oxygen 1774 and several other gases. Dissolving carbon dioxide under pressure in water, he began a European craze for soda water.

Swedish chemist Karl Scheele independently prepared oxygen in 1772, but his tardiness in publication resulted in Priestley being credited with the discovery.

Priestley discovered nitric oxide (nitrogen monoxide, NO) 1772 and reduced it to nitrous oxide (dinitrogen monoxide, N_2O). In the same year he became the first person to isolate gaseous ammonia by collecting it over mercury (previously ammonia was known only in aqueous solution). In 1774 he found a method for producing sulphur dioxide (SO_2).

Priestley was born near Leeds and became a cleric; as a Dissenter, he was barred from English universities. A meeting with US polymath Benjamin Franklin 1766 aroused his interest in science. As librarian and literary companion to Lord Shelburne 1773–80, Priestley accompanied him to France in 1774 and there met chemist Antoine Lavoisier. Priestley moved to Birmingham 1780 and joined the Lunar Society, an association of inventors and scientists that included James Watt, Matthew Boulton, Josiah Wedgwood, and Erasmus Darwin. In 1791 Priestley's chapel and house were sacked by a mob because of his support for the French Revolution. He fled to London, then emigrated to the USA 1794, settling in Pennsylvania.

Priestley's early work was in physics, particularly electricity and optics. He established that electrostatic charge is concentrated on the outer surface of a charged body and that there is no internal force. From this observation he proposed an inverse square law for charges, by analogy with gravitation.

Prigogine Ilya, Viscount Prigogine 1917– . Russian-born Belgian chemist who, as a highly original theoretician, has made major contributions to the field of thermodynamics. Earlier theories had considered systems at or about equilibrium; Prigogine began to study 'dissipative' or nonequilibrium structures frequently found in biological and chemical reactions. Nobel Prize for Physics 1977. Viscount 1989.

Prigogine was born in Moscow. He studied at Brussels and became professor there 1951, and in 1959 director of the Instituts Internationaux de Physique et de Chemie. He was professor at the Enrico Fermi Institute at the University of Chicago 1961–66, and from 1967 director of the Center for Statistical Mechanics and Thermodynamics at the University of Texas in Austin, concurrently with his professorship in Brussels.

When Prigogine began studying dissipative systems in the 1940s, it was not understood how a more orderly system, such as a living creature, could arise spontaneously and maintain itself despite the universal tendency towards disorder. It is now known that order can be created and preserved by processes that flow 'uphill' in the thermodynamic sense, compensated by 'downhill' events. Dissipative systems can exist only in harmony with their surroundings. Close to equilibrium, their order tends to be destroyed.

These ideas have been applied to examine how life originated on Earth, to ecosystems, to the preservation of world resources, and even to the prevention of traffic jams.

Pringsheim Ernst 1859–1917. German physicist whose experimental work on the nature of thermal radiation led directly to the quantum theory. In 1881 he developed a spectrometer that made the first accurate measurements of wavelengths in the infrared region.

Pringsheim was born in Breslau (now Wroclaw, Poland) and studied at several German universities. He was professor at Berlin 1896–1905 and at Breslau from 1905.

Pringsheim began in 1896 to collaborate with Otto Lummer on a study of black-body (a hypothetical object that absorbs heat) radiation. This led to a verification of the Stefan–Boltzmann law that relates the energy radiated by a body to its absolute temperature, but in 1899 they found anomalies in laws that had been devised to express the energy of the radiation in terms of its frequency and temperature. The results encouraged Max Planck to find a new radiation law that would account for the experimental results and in 1900, Planck arrived at such a law by assuming that the energy of the radiation consists of indivisible units that he called quanta. This marked the founding of the quantum theory.

Pringsheim Nathaniel 1823–1894. German botanist who showed that lower cryptogams (spore-producing plants) and algae reproduce by sexual union, confirming what Wilhelm Hofmeister had previously shown to occur in higher cryptogams. Pringsheim also worked on fungi and unsuccessfully on chlorophyll. Like his contemporaries Ferdinand Cohn, Hugo von Mohl, and Hofmeister, he concentrated his study more upon the physiology and dynamics of cell development and life history than upon the traditional classification and collection of specimens.

Pringsheim was born in Wziesko, Silesia and educated first at home and then at the Oppeln Gymnasium and the Breslau Gymnasium. In 1843, he went to the University in Breslau to study medicine, and moved 1844 to Leipzig University and then to Berlin University, where he studied botany. He obtained his PhD 1848 and worked in Paris and London before becoming a Privatdocent in Berlin.

In 1864 he was made professor of botany at Jena University, where he masterminded the construction of a new botanical institute. In 1868, he resigned from the university and continued with his research in private laboratories.

Prokhorov Aleksandr Mikhailovich 1916– . Russian physicist whose fundamental work on microwaves with Nikolai Basov led to the construction of the first practical maser (the microwave equivalent of the laser). They shared the 1964 Nobel Prize for Physics with Charles Townes.

Proust Joseph Louis 1754–1826. French chemist. He was the first to state the principle of constant composition of compounds – that compounds consist of the same proportions of elements wherever found.

Proust was born in Angers and trained as an apothecary. In the 1780s he went to Spain and spent the next 20 years in Madrid. He taught at various academies and carried out his research in a laboratory provided by his patron, King Charles IV of Spain.

In 1808, Napoleon invaded Spain and French soldiers wrecked Proust's laboratory. He returned to France a poor man.

French chemist Claude Berthollet. had stated that the composition of compounds could vary, depending on the proportions of reactants used to produce them. In 1799 Proust prepared and analysed copper carbonate produced in various ways and compared the results with those obtained by analysing mineral deposits of the same substance; he found that they all had the same composition. Similar results with other compounds led Proust to propose the law of constant composition. After a long controversy Berthollet conceded that Proust was right.

Prout William 1785–1850. British physician and chemist. In 1815 Prout published his hypothesis that the relative atomic mass of every atom is an exact and integral multiple of the mass of the hydrogen atom. The discovery of isotopes (atoms of the same element that have different masses) in the 20th century bore out his idea.

In 1827, Prout became the first scientist to classify the components of food into the three major divisions of carbohydrates, fats, and proteins.

Prout was born in Gloucestershire and studied medicine at Edinburgh. He set up a medical practice in London and established a private chemical laboratory. From 1813 he wrote about and gave lectures in 'animal chemistry'.

Studying various natural secretions and products, Prout became convinced that they derive from the chemical breakdown of body tissues. In 1818, he isolated urea and uric acid for the first time, and six years later he found hydrochloric acid in digestive juices from the stomach.

In his anonymous paper of 1815, Prout concluded, from the determinations of atomic weights that had been made, that hydrogen was the basic building block of matter. In 1920 Ernest Rutherford named the proton – the hydrogen nucleus, which is a constituent of every atomic nucleus – after Prout.

Prout also studied the gases of the atmosphere and in 1832 made accurate measurements of the density of air. The Royal Society adopted his design for a barometer as the national standard.

Ptashne Mark 1940– . US molecular biologist who has contributed much to our understanding of the mechanisms of transcription (formation of messenger RNA from DNA) in prokaryotic cells, such as bacteria or cyanobacteria. He has worked for many years on the mechanisms that switch genes on and off.

In genetic studies using the bacterium *Escherichia coli*, he postulated repressor proteins might be produced within a cell to bind to specific sites on or close to the gene to inactivate it. In one set of elegant studies, he showed that the gene for the enzyme b-galactosidase is turned off by a repressor molecule when lactose is not present. Ptashne showed that the repressor prevents transcription by binding to a specific sequence of the DNA called the operator, thus preventing the activity of the enzymes necessary for the transcription of neighbouring DNA sequences. This process can be reversed by the presence of lactose in the cell.

Ptashne was born in Chicago and obtained his BA and PhD at Reed College and

Harvard University respectively. He joined the staff of the biochemistry department at Harvard 1968 and was elected professor of biochemistry 1971.

Ptashne's work acted as the first model for the reversible activity of repressor proteins in the control of transcription and was the forerunner for subsequent studies showing the involvement of many other such molecules in gene expression.

Ptolemy (Claudius Ptolemaeus) c. AD 100–c. AD 170. Egyptian astronomer and geographer. His *Almagest* developed the theory that Earth is the centre of the universe, with the Sun, Moon, and stars revolving around it. In 1543 the Polish astronomer Copernicus proposed an alternative to the ***Ptolemaic system***. Ptolemy's *Geography* was a standard source of information until the 16th century.

Ptolemy produced vivid maps of Asia and large areas of Africa.

He may have been born in the town of Ptolemais Hermii, on the banks of the Nile. He worked in Alexandria and had an observatory on the top floor of a temple.

The *Almagest* (he called it *Syntaxis*) contains all his works on astronomical themes. Probably inspired by Plato, Ptolemy began with the premise that the Earth was a perfect sphere. All planetry orbits were circular, but those of Mercury and Venus, and possibly Mars (Ptolemy was not sure), were epicyclic (the planets orbited a point that itself was orbiting the Earth). The sphere of the stars formed a dome with points of light attached or pricked through.

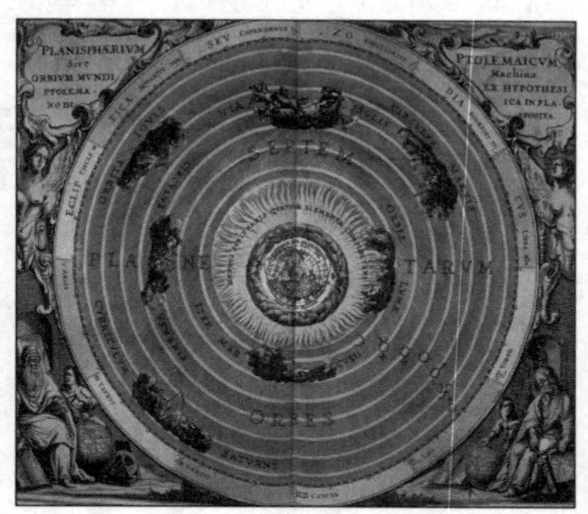

Ptolemy *The Ptolemaic system of the universe, showing the Earth at the centre surrounded by air, fire and water. Ptolemy, an Egyptian astronomer, developed the theory that the Earth is at the centre of the universe, with the Sun, Moon, planets, and stars revolving around it.*

In his thesis on astrology, *Tetrabiblios*, Ptolemy suggests that some force from the stars may influence the lives and events in the human experience.

Pullman George Mortimer 1831–1897. US engineer and entrepreneur who developed the Pullman railway car from 1864. It was not the first sleeping carriage but it was the most luxurious. Pullman cars were initially staffed entirely by black porters whose only income was from tips. In 1937 the Brotherhood of Sleeping Car Porters became the first black union recognized by a US corporation.

Pullman left school at 14 to become a cabinetmaker. In an attempt to improve the standard of comfort of rail travel, he built his first Pioneer Sleeping Car 1864. He formed the Pullman Palace Car Company 1867 and in 1881 the town of Pullman, Illinois, was built for his workers. One of the most famous strikes in US labour history took place at Pullman in 1894.

Purcell Edward Mills 1912– . US physicist who, with Felix Bloch, developed the nuclear magnetic resonance (NMR) method of measuring the magnetism of atomic nuclei in solids and liquids. The method hinges on the fact that any magnetic nuclei will absorb radio-frequency radiation by a resonance effect when in a magnetic field. Measurement of the absorbtion provides information on the properties of the nuclei and their molecular environment. The technique is now used in chemistry as a powerful analytical tool. Nobel Prize for Physics 1952.

In astronomy, Purcell was the first to detect the microwave emission from neutral hydrogen in interstellar space at the wavelength of 21 cm/8 in. This discovery made possible the mapping of a large part of our galaxy and the calculation of the temperature and motion of gas in interstellar space.

Purcell graduated in electrical engineering from Purdue University, Illinois, USA, 1933 and gained his doctorate from Harvard 1938, having also spent a year studying physics in Germany at the Technische Hochschule in Karlsruhe. He became a full professor of physics at Harvard University, Cambridge, Massachusetts, USA, 1949. During the war years, 1941–45, he worked on the development of radar at the Massachusetts Institute of Technology Radiation Laboratories.

Purkinje Jan Evangelista 1787–1869. Czech physiologist who made pioneering studies of vision, the functioning of the brain and heart, pharmacology, embryology, and cells and tissue. Purkinje described 1819 the visual phenomenon in which different-coloured objects of equal brightness in certain circumstances appear to the eye to be unequally bright; this is now called the ***Purkinje effect***.

Purkinje was born in Libochovice, Bohemia (now in the Czech Republic), and studied at Prague. In 1823 he was appointed professor at Breslau (now Wroclaw in Poland) – perhaps through the influence of German poet Wolfgang von Goethe, who had befriended him. In addition to his scientific work, Purkinje also translated the poetry of Goethe and Friedrich von Schiller. At Breslau, Purkinje founded the world's first official physiological institute. In 1850, he returned to Prague University.

In 1832, he was the first to describe what are now known as Purkinje's images: a threefold image of a single object seen by one person reflected in the eye of another person. This effect is caused by the object being reflected by the surface of the cornea and by the anterior and posterior surfaces of the eye lens.

Purkinje cells are large nerve cells with numerous dendrites found in the cortex of the cerebellum; he discovered these 1837, and the Purkinje fibres in the ventricles of the heart 1839. Also in 1839, in describing the contents of animal embryos, Purkinje was the first to use the term 'protoplasm' in the scientific sense.

In 1823 he recognized that fingerprints can be used as a means of identification. He discovered the sweat glands in skin, described ciliary motion, and observed that pancreatic extracts can digest protein. In 1837 he outlined the principal features of the cell theory.

Pye John David 1932– . English zoologist who has studied the way bats use echolocation, and also the use of ultrasound in other animals.

Pye was born in Mansfield, Nottinghamshire, and studied at the University College of Wales at Aberystwyth and London University. In 1973 he became professor at Queen Mary College, London.

A surprisingly large number of animals use ultrasound (which has a frequency above about 20 kHz and is inaudible to humans) – bats, whales, porpoises, dolphins, and many insects, for example. Because of the lack of detection devices, the phenomenon was not discovered until 1935.

In 1971 Pye calculated the resonant frequencies of the drops of water in fog and found that these frequencies coincided with the spectrum of frequencies used by bats for echolocation. In other words, bats cannot navigate in fog. Pye also found that ultrasound seems to be important in the social behaviour of rodents and insects.

Pyman Frank Lee 1882–1944. English organic chemist who worked on pharmaceuticals, particularly studying the properties of glyoxalines (glyoxal is ethanedial).

Pyman was born in Malvern, Worcestershire, and studied at Manchester and in Switzerland at Zürich Polytechnic. On his return to Britain, Pyman took a job in the Experimental Department of the Wellcome Chemical Works in Dartford, Kent. He was professor at Manchester and head of the Department of Applied Chemistry at the College of Technology 1919–27 and then became director of research at the Boots Pure Drug Company's laboratories at Nottingham.

The glyoxalines are cyclic amidines with therapeutic properties, and Pyman studied especially the relationship between their chemical constitution and their physiological action. Study of the constitution of the anhydro-bases made from glycoside revealed the existence of a substance whose molecules contained a ten-membered heterocyclic ring. He also examined alkaloids, and became the first person to isolate a natural substance containing an asymmetric nitrogen atom.

During World War I he worked on the preparation of drugs needed to treat British troops overseas.

There is geometry in the humming of the strings. There is music in the spacings of the spheres.

Pythagoras

Quoted in Aristotle *Metaphysics*

Pythagoras *c.* 580–500 BC. Greek mathematician and philosopher who formulated Pythagoras' theorem.

Much of Pythagoras' work concerned numbers, to which he assigned mystical properties. For example, he classified numbers into triangular ones (1, 3, 6, 10, ...), which can be represented as a triangular array, and square ones (1, 4, 9, 16, ...), which form squares. He also observed that

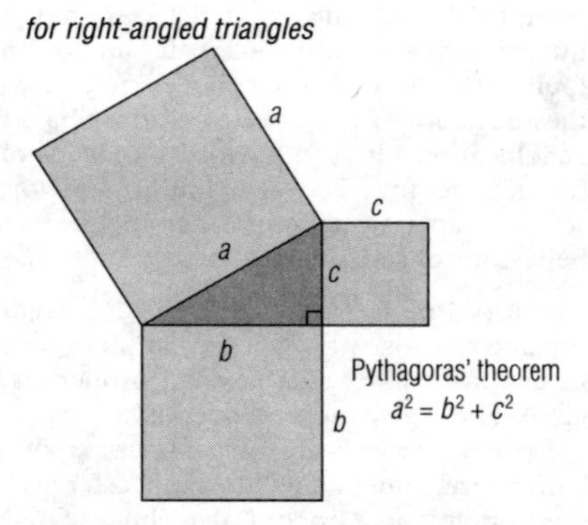

Pythagoras *Pythagoras' theorem for right-angled triangles is likely to have been known long before the time of Pythagoras. It was probably used by the ancient Egyptians to lay out the pyramids.*

any two adjacent triangular numbers add to a square number (for example, 1 + 3 = 4; 3 + 6 = 9; 6 + 10 = 16).

Pythagoras' theorem states that in a right-angled triangle, the area of the square on the hypothenuse (the longest side) is equal to the sum of the areas of the squares drawn on the other two sides.

Pythagoras was born on the island of Samos and may have been obliged to flee the despotism of its ruler. He went on to found a school and religious brotherhood in Croton, S Italy. Its tenets included the immortality and transmigration of the soul. As a mathematical and philosophical community the Pythagorean brotherhood extended science; politically its influence reached the western Greek colonies. This caused conflict which saw Pythagoras exiled to Metapontum, on the Gulf of Taranto, until he died. The school continued for 50 or 60 years before being totally suppressed.

Using geometrical principles, the Pythagoreans were able to prove that the sum of the angles of any regular-sided triangle is equal to that of two right angles (using the theory of parallels), and to solve any algebraic quadratic equations having real roots. They formulated the theory of proportion (ratio), which enhanced their knowledge of fractions, and used it in their study of harmonics upon their stringed instruments: the harmonic of the octave was made by touching the string at $\frac{1}{2}$ its length, of a fifth at $\frac{2}{3}$ its length, and so on. Pythagoras himself is said to have made this the basis of a complete system of musical scales and chords.

Quetelet Lambert Adolphe Jacques 1796–1874. Belgian statistician. He developed tests for the validity of statistical information, and gathered and analysed statistical data of many kinds. From his work on sociological data came the concept of the 'average person'.

Rabi Isidor Isaac 1898–1988. Russian-born US physicist who developed techniques to measure accurately the strength of the weak magnetic fields generated when charged elementary particles, such as the electron, spin about their axes. The work won him the 1944 Nobel Prize for Physics.

Rainwater (Leo) James 1917– . US physicist who with the Danes Aage Bohr and Ben Mottelson shared the 1975 Nobel Prize for Physics for their work on the structure of the atomic nucleus.

Rainwater was born in Council, Idaho, and studied at the California Institute of Technology and at Columbia University, where he became professor of physics 1952. During 1942–46, he worked on the Manhattan Project to construct the atom bomb.

In 1950, Rainwater wrote a paper in which he observed that most of the particles in the nucleus of an atom form an inner nucleus, while the other particles form an outer nucleus. Each set of particles is in constant motion at very high velocity and the shape of each set affects the other set. He postulated that if some of the outer particles moved in similar orbits, this would create unequal centrifugal forces of enormous power, which could be strong enough permanently to deform an ideally symmetrical nucleus. This was confirmed experimentally by Bohr and Mottelson, and paved the way for nuclear fusion.

Raman Chandrasekhara Venkata 1888–1970. Indian physicist who in 1928 discovered what became known as the ***Raman effect***: the scattering of monochromatic (single-wavelength) light when passed through a transparent substance. The Raman spectra produced are used to obtain information on the structure of molecules. Knighted 1929. Nobel prize 1930.

Raman was born in Trichinopoly, Madras, and studied at Madras. He joined the civil service as an accountant in Calcutta but pursued his studies privately. His work on vibration in sound and the theory of musical instruments and on diffraction led to his becoming professor of physics at the University of Calcutta 1917–33. From 1948 he was director of the Raman Research Institute, built for him by the government in Bangalore.

Raman showed 1921 that the blue colour of the sea is produced by the scattering of light by water molecules. Continuing to work on the scattering of light, he arrived at the Raman effect. It is caused by the internal motion of the molecules encountered, which may impart energy to the light photons or absorb energy in the resulting collisions. Raman scattering therefore gives precise information on the motion and shape of molecules.

Raman's other research included the effects of sound waves on the scattering of light in 1935 and 1936, the vibration of atoms in crystals in the 1940s, the optics of gemstones, particularly diamonds, and of minerals in the 1950s, and the physiology of human colour vision in the 1960s.

Ramanujan Srinavasa Ayengar 1887–1920. Indian mathematician who did original work especially in function theory and number theory.

Ramanujan was born near Kumbakonam, Madras (now Tamil Nadu), and taught himself mathematics from just one textbook, *A Synopsis of Elementary Results in Pure and Applied Mathematics* by G S Carr, published 1880. In 1914 he won a scholarship to Cambridge, England, but tuberculosis forced him to return to India 1919.

Carr's textbook, and particularly the section on pure mathematics, was the basis of all Ramanujan's work. From the knowledge he gained he was able to proceed beyond the material published and develop his own results in many fields, but also kept 'discovering' already well-known theorems.

With his mentor, Cambridge mathematician G H Hardy, Ramanujan published a theory on the methods for partitioning an integer into a sum of smaller integers, called summands. In function theory he found accurate approximations to π, and worked on modular, elliptic, and other functions.

His collected papers were published by Hardy 1927.

Ramón y Cajal Santiago 1852–1934. Spanish cell biologist and anatomist whose research revealed that the nervous system is based on units of nerve cells (neurons). He shared the 1906 Nobel Prize for Physiology or Medicine.

Ramón y Cajal was born in Petilla de Aragon, studied at Zaragoza, and then joined the army medical service. He was professor at Valencia 1884–87, at Barcelona 1887–92, and at Madrid 1892–1921. In 1900 he became director of the new Instituto Nacional de Higiene, and in 1921 of the Cajal Institute in Madrid, founded in his honour.

Ramón y Cajal demonstrated that the axons of neurons end in the grey matter of the central nervous system and never join the endings of other axons or the cell bodies of other nerve cells – findings indicating that the nervous system is not a network. In 1897 he investigated the human cerebral cortex, described several types of neurons, and demonstrated that structure might be related to the localization of a particular function to a specific area. Within the cell body he found neurofibrils 1903, and recognized that the cell body itself was concerned with conduction.

His books include *Structure of the Nervous System of Man and other Vertebrates* 1904 and *The Degeneration and Regeneration of the Nervous System* 1913–14.

Ramsay William 1852–1916. Scottish chemist who, with Lord Rayleigh, discovered argon 1894. In 1895 Ramsay produced helium and in 1898, in cooperation with Morris Travers, identified neon, krypton, and xenon. In 1903, with Frederick Soddy, he noted the transmutation of radium into helium, which led to the discovery of the density and relative atomic mass of radium. KCB 1902. Nobel prize 1904.

In his book *The Gases of the Atmosphere* 1896, Ramsay repeated a suspicion he had stated 1892 that there was an eighth group of new elements at the end of the periodic table. During the next decade Ramsay and Travers sought the remaining rare gases by the fractional distillation of liquid air.

Ramsay *Cartoon of Scottish chemist William Ramsay published in* Vanity Fair *in December 1908. Ramsey won the Nobel Prize for Chemistry in 1904 following his involement in the discovery of five of the inert gases.*

Ramsay was born in Glasgow and studied there and in Germany at Tübingen. In 1880 he was appointed professor at the newly created University College of Bristol (later Bristol University) and a year later became principal of the College. From 1887 he was professor at University College, London.

Helium was known from spectrographic evidence to be present on the Sun but yet to be found on Earth. Certain uranium minerals were known to produce an unidentified inert gas on heating, and Ramsay obtained sufficient of the gas to send a sample to English scientist William Crookes for spectrographic analysis. Crookes confirmed that it was helium.

Ramsey Norman F 1915– . US physicist who invented a method of storing atoms and observing them for long periods of time, and applied this in the hydrogen maser and atomic clock. One important application of Ramsey's work is the caesium atomic clock which has been the basis for the definition of the unit of time (the second) since 1967. Nobel Prize for Physics 1989 (shared with Hans Dehmelt and Wolfgang Paul).

Ramsey's invention is called the method of separated oscillatory fields. In this method, a beam of atoms passes through an oscillating (rapidly varying) magnetic field at one end of a storage box. Inside the box, the atoms bounce around for a period of time before emerging through a second oscillating field at the opposite end of the box.

Ramsey was born in Washington, DC, USA, and educated at Columbia University, New York, and Cambridge University, England. After receiving from Cambridge his second bachelors degree, he returned to Columbia to study for his doctorate. During World War II he worked on radar and at Los Alamos, New Mexico, on the atomic bomb project. At the end of the war he returned to Columbia as professor. In 1946 he became the first head of the physics department at the newly established Brookhaven National Laboratory, Long Island, New York. In 1947 he moved to Harvard where he taught for 40 years, retiring 1986.

Rankine William John Macquorn 1820–1872. Scottish engineer and physicist who was one of the founders of the science of thermodynamics, especially in reference to the theory of steam engines.

Rankine was born in Edinburgh and trained as a civil engineer.

From 1855 he was professor at Glasgow.

In 1849 he delivered two papers on the subject of heat, and in 1849 he showed the further modifications required to French physicist Sadi Carnot's theory of thermodynamics.

In *A Manual of the Steam Engine and other Prime Movers* 1859, Rankine described a thermodynamic cycle of events (the ***Rankine cycle***), which came to be used as a standard for the performance of steam-power installations where a considerable vapour provides the working fluid. Rankine here explained how a liquid in the boiler vaporized by the addition of heat converts part of this energy into mechanical energy when the vapour expands in an engine. As the exhaust vapour is condensed by a cooling medium such as water, heat is lost from the cycle. The condensed liquid is pumped back into the boiler.

Raoult François Marie 1830–1901. French chemist. In 1882, while working at the University of Grenoble, Raoult formulated one of the basic laws of chemistry. ***Raoult's law*** enables the relative molecular mass of a substance to be determined by noting how much of it is required to depress the freezing point of a solvent by a certain amount.

Ray John 1627–1705. English naturalist who devised a classification system accounting for some 18,000 plant species. It was the first system to divide flowering plants into monocotyledons and dicotyledons, with additional divisions made on the basis of leaf and flower characters and fruit types.

In *Methodus plantarum nova* 1682, Ray first set out his system. He also established the species as the fundamental unit of classification.

Ray believed that fossils are the petrified remains of dead animals and plants. This concept, which appeared in his theological writings, did not gain general acceptance until the late 18th century.

Ray was born near Braintree, Essex, and studied at Cambridge, where he became a lecturer and, in 1660, was ordained a priest. In 1662 he lost his livelihood because of the Act of Uniformity, which required from all clerics a declaration that he refused to sign. With naturalist Francis Willughby (1635–1672), Ray toured Europe 1663–66, and on their return to England, Ray lived at Willughby's home, where they collaborated on publishing the results of their studies. After Willughby's death, Ray remained in the Willughby household until 1678, and then returned to the village where he was born.

In 1670 Ray, with Willughby's help, published *Catalogus plantarum Angliae/Catalogue of English Plants*. Ray and Willughby then began working on a definitive catalogue and classification of all known plants and animals. *Historia generalis plantarum* 1686–1704 covered about 18,600 species (most of which were European) and contained much information on the morphology, distribution, habitats, and pharmacological uses of individual species as well as general aspects of plant life, such as diseases and seed germination.

Ray also wrote several books on zoology, giving details of individual species in addition to classification: *Synopsis methodica animalium quadrupedum et serpentini generis/Synopsis of Quadrupeds* 1693, *Historia insectorum/History of Insects* 1710, and *Synopsis methodica avium et piscium/Synopsis of Birds and Fish* 1713.

Rayet George Antoine Pons 1839–1906. French astronomer who, in collaboration with Charles Wolf (1827–1918), detected a new class of peculiar white or yellowish stars whose spectra contain broad hydrogen and helium emission lines. These stars are now called ***Wolf–Rayet stars***.

Rayet was born near Bordeaux and studied at the Ecole Normale Supérieure. He worked in the new weather forecasting service created by astronomer Urbain Leverrier at the Paris Observatory until dismissed over a disagreement about the practical forecasting of storms. From 1876 Rayet was professor of astronomy at Bordeaux, and from 1879 also director of the new observatory at nearby Floirac.

At the Paris Observatory, Rayet collaborated with Charles Wolf and in 1865 they photographed the penumbra of the Moon during an eclipse. In 1866 a nova appeared, and after its brilliance had significantly diminished, Rayet and Wolf detected bright bands in its spectrum. The bands were the result of a phase that can occur in the later stages of evolution of a nova. The two astronomers went on to investigate whether permanently bright stars exhibit this phenomenon and in 1867 they discovered three such stars in the constellation of Cygnus. Wolf–Rayet stars are now known to be relatively rare.

Some proofs command assent. Others woo and charm the intellect. They evoke delight and an overpowering desire to say 'Amen, Amen'.

John Strutt, 3rd Baron Rayleigh

H E Hunter *The Divine Proportion* 1970

Rayleigh John William Strutt, 3rd Baron Rayleigh 1842–1919. English physicist who wrote the standard treatise *The Theory of Sound* (1877–78), experimented in optics and microscopy, and, with William Ramsay, discovered argon. Baron 1873. Nobel prize 1904.

Rayleigh *Cartoon of English physicist John William Strutt, third Baron Rayleigh published in* Vanity Fair *in December 1899. Rayleigh discovered argon in collaboration with William Ramsay. He was elected as President of the Royal Society, a post he held 1905–08.*

Rayleigh was born in Essex and studied at Cambridge. He set up a laboratory at his home and was professor of experimental physics at Cambridge 1879–84, making the Cavendish Laboratory an important research centre.

In 1871, Rayleigh explained that the blue colour of the sky arises from the scattering of light by dust particles in the air, and was able to relate the degree of scattering to the wavelength of the light. He also made the first accurate definition of the resolving power of diffraction gratings, which led to improvements in the spectroscope. He completed in 1884 the standardization of the three basic electrical units: the ohm, ampere, and volt. His insistence on accuracy prompted the designing of more precise electrical instruments.

After leaving Cambridge, Rayleigh continued to do research in a broad range of subjects including light and sound radiation, thermodynamics, electromagnetism, and mechanics.

An inconsistency in the Rayleigh–Jeans equation, published by Rayleigh 1900 (amended 1905 by James Jeans), which described the distribution of wavelengths in black-body radiation, led to the formulation shortly after of the quantum theory by German physicist Max Planck.

Réaumur Réné Antoine Ferchault de 1683–1757. French scientist. His work on metallurgy *L'Arte de convertir le fer forge en acier* 1722 described how to convert iron into steel and stimulated the development of the French steel industry. He produced a six-volume work 1734–42 on entomology, *L'Histoire des insectes/History of Insects*, which threw much new light on the social insects. He also contributed to other areas of science.

Reber Grote 1911– . US radio engineer who pioneered radio astronomy. He attempted to map all the extraterrestrial sources of radio emission that could be traced.

Reber was born in Wheaton, Illinois, and studied at the Illinois Institute of Technology. He built his own apparatus for studying cosmic radio waves, and held posts at several US institutions. From 1954 he worked mainly in Australia at the Commonwealth Scientific and Industrial Research Organization in Tasmania, though he spent 1957–61 at the National Radio Astronomy Observatory at Green Back, West Virginia.

Reber's first instrument was a bowl-shaped reflector 9 m/30 ft in diameter, with an antenna at its focus, built in the back garden of his Illinois home in 1957. For a number of years, Reber's was probably the only radio telescope in existence. With it, he could identify only a general direction from which radio waves were coming. The most intense radiation he recorded emanated from the direction of Sagittarius, near the centre of the Galaxy.

In Hawaii a new radio telescope was constructed, sensitive to lower frequencies, and he worked there 1951–54. His last project, in Tasmania, was to complete a map of radio sources emitting waves around 144 m/473 ft in length.

Redi Francesco 1626–1697. Italian physician and poet who disputed the spontaneous-generation theory, the erroneous belief that animals can be generated spontaneously from decaying matter, and performed one of the first controlled biological experiments.

He was the first to provide experimental evidence to dispute the theory that fair-sized animals can be generated spontaneously from decaying matter; a well-established and accepted view of the time.

He showed that maggots do not breed spontaneously in decaying flesh, in a series of experiments using meat samples, half of which were exposed to air, while the other half were sealed. Only the exposed meat developed maggots, proving that maggots are, in fact, derived from eggs laid in meat by flies. Despite this advance, however, Redi continued to believe that spontaneous generation could occur in some instances and cited the example of grubs developing in oak trees.

Redi was born in Arezzo in Italy and studied medicine in Pisa and Florence before working as a physician.

Redman Roderick Oliver 1905–1975. English astronomer who was chiefly interested in stellar spectroscopy and the development of spectroscopic techniques, and in solar physics.

Redman was born in Gloucestershire and studied at Cambridge under English astrophysicist Arthur Eddington. Redman worked mainly at the Cambridge observatories and organized their re-equipping after World War II. He became professor 1947 and in 1972 was made director of the amalgamated observatories and Institute of Theoretical Physics. He also established a solar observatory in Malta.

Redman applied the method of photographic photometry to the study of elliptical galaxies and in the 1940s, in Pretoria, South Africa, also to the study of bright stars, for which he developed the narrow-band technique, which was of great value in stellar photometry.

He went all over the world in order to observe total eclipses of the Sun, during which he was able to identify thousands of the emission lines in the chromospheric spectrum and to investigate the chromospheric temperature.

Redman's final contribution to astronomy was his initiation of a large stellar photometry programme.

Reed Walter 1851–1902. US physician and medical researcher who isolated the aedes mosquito *Aedes aegypti* as the sole carrier of yellow fever. This led to the local eradication of the deadly virus disease by destruction of mosquito breeding grounds. His breakthrough work was carried out during a yellow-fever epidemic 1900–01 in Cuba.

Born in Belroi, Virginia, Reed received an MD degree from the University of Virginia 1869. He joined the Army Medical Corps 1875, served as an army surgeon in Arizona

1876–89 and Baltimore 1890–93, and was professor at the Army Medical College 1893–1902.

His 1898 research into the causes and transmission of typhoid fever brought about significant control of the disease in army camps. He also worked on malaria and diphtheria.

Regiomontanus adopted name of Johannes Müller, 1436–1476. German astronomer who compiled astronomical tables, translated Ptolemy's *Almagest* from Greek into Latin, and assisted in the reform of the Julian calendar.

Johannes Müller adopted the name Regiomontanus as a Latinized form of his birthplace Königsberg while studying at Vienna. At the age of 15, he was appointed to the Faculty of Astronomy at Vienna. In 1471 he moved to Nuremberg, where he had a printing press in his house and so became one of the first publishers of astronomical and scientific literature. He went to Rome in 1475, invited by the pope to assist in amending the notoriously incorrect ecclesiastical calendar.

In 1467, Regiomontanus started compiling trigonometric and astronomical tables, but these too were not published until more after his death. Regiomontanus's *Ephemerides* 1474 was the first publication of its kind to be printed (by himself); it gave the positions of the heavenly bodies for every day from the year 1475 to 1506.

After Regiomontanus's death, the statement 'the motion of the stars must vary a tiny bit on account of the motion of the Earth' was found in his handwriting. This has led some people to believe that Regiomontanus gave Copernicus the idea that the Earth moves round the Sun.

Regnault Henri Victor 1810–1878. German-born French physical chemist who showed that Boyle's law applies only to ideal gases. He also invented an air thermometer and a hygrometer, and discovered carbon tetrachloride (tetrachloromethane).

Regnault was born in Aachen and studied in Paris at the Ecole Polytechnique and the Ecole des Mines as well as in various parts of Europe. He returned to the Ecole Polytechnique in 1836 as an assistant to Joseph Gay-Lussac and in 1840 succeeded him as professor of chemistry. He became professor of physics at the Collège de France in 1841. From 1854 Regnault was director of the porcelain factory at Sèvres. During the Franco-Prussian war, his son was killed, and Prussian soldiers destroyed his books and instruments. He never recovered from these losses.

In chemistry, Regnault studied the action of chlorine on ethers, leading to the discovery of vinyl chloride (monochloroethene), dichloroethylene (dichloroethene), trichloroethylene (trichloroethene), and carbon tetrachloride (tetrachloromethane).

In 1842 Regnault was commissioned to redetermine all the physical constants involved in the design and operation of steam engines. This led him to study the thermal properties of gases. He measured the coefficients of expansion of various gases and by 1852 had shown how real gases depart from the behaviour required by Boyle's law.

Regnault also calculated that absolute zero is at –273°C/–459°F. He redetermined the composition of air, and performed experiments on respiration in animals.

Reichstein Tadeus 1897–1996. Swiss biochemist who investigated the chemical activity of the adrenal glands. By 1946 Reichstein had identified a large number of steroids secreted by the adrenal cortex, some of which would later be used in the treatment of Addison's disease. Reichstein shared the 1950 Nobel Prize for Physiology or Medicine with Edward Kendall and Philip Hench.

Reines Frederick 1918– . US physicist who experimentally detected the neutrino, one of the fundamental particles which make up all matter. Nobel Prize for Physics 1995 (shared with Martin Perl).

The existence of the neutrino was postulated 1930 by Austrian physicist Wolfgang Pauli to explain the energy losses that occur during beta decay, a type of radioactivity. However, the experimental detection of the neutrino proved difficult since neutrinos pass easily through solid material.

In the late 1950s US physicists Frederick Reines and Clyde Cowan detected neutrinos produced by a nuclear reactor at the Savannah River nuclear plant, in the SE USA. In the experiment neutrinos produced by the reactor passed through a tank of water. Some of the neutrinos reacted with protons in the water, producing flashes which could be detected by a photomultiplier or 'electric eye'. Nowdays, neutrinos from space are routinely detected by using underground 'telescopes' which are similar to the Reines and Cowan set-up.

Reines was born in Paterson, New Jersey, USA, and gained his doctorate in physics from New York University 1944. He is professor of physics at the University of California.

Remak Robert 1815–1865. Polish physician and neuroanatomist who pioneered the use of electrotherapy for nervous disorders. He discovered that axons of nerves were continuous with nerve cell bodies in the spinal cord. In 1838, he was the first to describe the presence of the myelin sheath, the insulating layer surrounding nerve fibres, that acts to speed up the passage of nerve impulses. He also microscopically explored the six cortical cell layers of the cerebrum and was a pioneer of the embryology of the nervous system.

Remak was born in Poznan but lived and worked in Germany and studied at Berlin University. Although he was not allowed to hold a senior teaching position in a German university or medical school because he was Jewish, he became a respected neuroanatomist.

Remington Eliphalet 1793–1861. US inventor, gunsmith, and arms manufacturer and founder (with his father) of the Remington firm. He supplied the US army with rifles in the Mexican War 1846–48, then in 1856 the firm expanded into the manufacture of agricultural implements. His son Philo continued the expansion.

In 1816, Eliphalet Remington made a flintlock rifle at his father's forge in Utica, New York, with an accuracy that soon attracted great attention. This led to the setting-up of E Remington & Son in Ilion, beside the Erie Canal.

Remington Philo 1816–1889. US inventor of the breech-loading rifle that bears his name. He began manufacturing typewriters 1873, using the patent of Christopher Sholes, and made improvements that resulted five years later in the first machine with a shift key, thus providing lower-case letters as well as capital letters.

The Remington rifle and carbine, which had a falling block breech and a tubular magazine, were developed in collaboration with his father Eliphalet Remington.

Remington was born in Litchfield, New York State, and entered the family arms-manufacturing business.

US humorist Mark Twain bought one of the earliest Remington typewriters, becoming the first author to provide his publisher with a typescript.

Philo and Eliphalet Remington made many improvements to guns and their manufacture; for example, a special lathe for the cutting of gunstocks, a method of producing extremely straight gun barrels, and the first US drilled rifle barrel from cast steel.

Rennie John 1761–1821. Scottish engineer who built three bridges over the river Thames in London, later demolished: Waterloo Bridge 1811–17, Southwark Bridge 1814–19, and London Bridge 1824–34; he also built bridges, canals, dams (Rudyard dam, Staffordshire, 1800), and harbours.

Rennie was born near Haddington, Lothian, studied at Edinburgh University and then worked for James Watt from 1784. He started his own engineering business about 1791, and built the London and East India docks, as well as construction work on harbours in the UK, Malta, and Bermuda.

Waterloo Bridge was demolished 1934–36 and replaced 1944; London Bridge was demolished 1968, bought by a US oil company and reassembled at Lake Havana, Arizona.

Reynolds Osborne 1842–1912. Irish physicist and engineer who studied fluid flow and devised the ***Reynolds number***, which gives a numerical criterion for determining whether fluid flow under specified conditions will be smooth or turbulent.

Reynolds was born in Belfast and studied at Cambridge. From 1868 he was professor at Owens College (now Manchester University).

Reynolds showed, using dye, that the flow of a liquid will be turbulent unless it has a high viscosity, low density, and an open surface. He applied much of what he had learned about turbulent flow to the behaviour of the water in river channels and estuaries. For one study he made an accurate model of the mouth of the river Mersey, and pioneered the use of such models in marine and civil engineering projects. He also worked on multistage steam turbines, and, using the experimental steam engine which he designed for his engineering department, he determined the mechanical equivalent of heat so accurately that it has remained one of the classical determinations of a physical constant.

Ricardo Harry Ralph 1885–1974. English engineer who played a leading role in the development of the internal-combustion engine. During World War I and World War II, his work enabled British forces to fight with the advantage of technically superior engines. His work on combustion and detonation led to the octane-rating system for classifying fuels for petrol engines. Knighted 1948.

Ricardo was born in London. At the age of 12 he built a steam engine, and as a student at Cambridge he designed and built a motorcycle, with its power unit. In 1905 he designed and built a two-cylinder, two-stroke engine and a four-cylinder version to power his uncle's large car. During World War I, Ricardo worked on aircraft engines and designed the engine for the Mark V tank. In 1917 he set up a research and consultancy company, which worked on engine development and categorization of fuels according to their ease of detonation.

Ricardo designed an effective combustion chamber, which was also used in his aircraft engines during World War II.

Ricci-Curbastro Gregorio 1853–1925. Italian mathematician whose systematization of absolute differential calculus (the ***Ricci calculus***) enabled Albert Einstein to derive the theory of relativity.

Ricci-Curbastro was born in Lugo, Romagna, and studied at several Italian universities and in Germany at Munich. From 1880 he was professor of mathematical physics at Padua. He was also a magistrate and local councillor.

By introducing an invariant element – an element that can also be used in other systems – Ricci-Curbastro was able to modify differential calculus so that the formulas and results retained the same form regardless of the system of variables used. In 1896, he applied the absolute calculus to the congruencies of lines on an arbitrary Riemann variety; later, he used the Riemann symbols to find the contract tensor, now known as the ***Ricci tensor*** (which plays a fundamental role in the theory of relativity). He also discovered the invariants that occur in the theory of the curvature of varieties.

With a former pupil, Tullio Levi-Civita, he published *Méthodes de calcul différentiel absolu et leurs applications*, describing the use of intrinsic geometry as an instrument of computation dealing with normal congruencies, geodetic laws, and isothermal families of surfaces. The work also shows the possibilities of the analytical, geometric, mechanical, and physical applications of the new calculus.

Rice Peter 1935–1992. British structural engineer. He was responsible for some of the most exciting buildings of the 1970s and 1980s: the Sydney Opera House, the Pompidou Centre in Paris, and the Lloyd's building in London.

Born in Ulster, Rice studied engineering at Queen's University, Belfast and Imperial College, London. His first employment was with Ove Arup; at the age of 23 he was assigned to raise the sail-like roofs of the Sydney Opera House, designed by Finnish architect Joern Utzon. There followed the Centre Pompidou, the Menil Art Collection Museum in Houston, the San Nicola World Cup Stadium in Bari, the Kansai International Airport in Japan, and the Pavilion of the Future at the Seville Expo. In Britain, he is known for the terminal at Stansted Airport as well as the Lloyd's building. The architects with whom he collaborated – notably Richard Rogers, Renzo Piano and Norman Foster – held him in high regard. Piano, the Italian architect with whom he built the Centre Pompidou and the Menil museum, said he designed structures 'like a pianist who can play with his eyes shut; he understands the basic nature of structures so well that he can afford to think in the darkness about what might be possible beyond the obvious'.

In July 1992 Rice received the UK's Royal Gold Medal for Architecture, an honour rarely awarded to engineers. In his acceptance speech, he remarked that structural engineers were often seen as reducing every soaring idea an architect might have by insisting on rationality in design. The structural engineer's true role, Rice said, was not to reduce and restrict, but to explore materials and structures as had the great Victorian engineers and medieval cathedral builders; rigour and imagination could go hand in hand.

Richard of Wallingford *c.* 1292–1335. English mathematician, abbot of St Albans from 1326. He was a pioneer of trigonometry, and designed measuring instruments.

Richards Dickinson Woodruff 1895–1973. US physician who succeeded in performing catheterization of the heart and shared the Nobel Prize for Physiology or Medicine 1956 with Werner Forssman and André Cournand for his work in cardiology.

In 1929, the German surgeon Forssman placed a hollow tube into a vein in his arm and advanced until its tip entered the right atrium of his heart. This procedure, called cardiac catheterization, was developed further by Richards and Cournand so that the tube could be inserted into the right side of the heart and the pulmonary artery in order to study congenital heart disease. They also succeeded in the catheterization of the left side of the heart by passing the tube through the septum which separates the two sides of the heart.

This technique made it possible to measure blood pressure in the various chambers of the heart and proved enormously valuable for the development of noninvasive surgery for some forms of congenital heart disease and narrowed arteries.

Richards was born in New Jersey and studied medicine at Yale University before joining the teaching staff of Columbia University, where he specialized in cardiology.

Richards Theodore William 1868–1928. US chemist who determined as accurately as possible the relative atomic masses of a large number of elements. He also investigated the physical properties of the elements, such as atomic volumes and the compressibilities of nonmetallic solid elements. Nobel prize 1914.

Richards was born in Germantown, Pennsylvania, and studied at Harvard, where he was professor from 1901.

Introducing various new analytical techniques, Richards made accurate atomic weight measurements for 25 elements; his co-workers determined 40 more. In 1913, he detected differences in the atomic weights of ordinary lead and samples extracted from uranium minerals (which had arisen by radioactive decay) – one of the first convincing demonstrations of the uranium decay series and confirming English chemist Frederick Soddy's prediction of the existence of isotopes.

Richardson Owen Willans 1879–1959. British physicist. He studied the emission of electricity from hot bodies, giving the name thermionics to the subject. At Cambridge University, he worked under J J Thomson in the Cavendish Laboratory. Nobel prize 1928. Knighted 1939.

Richet Charles Robert 1950–1935. French physiologist who discovered chemical hypersensitiveness, or anaphylaxis, and was awarded the Nobel Prize for Physiology or Medicine 1913.

Richet discovered 1902 that certain chemicals produced hypersensitivity when injected into dogs. Richet and colleagues determined the smallest lethal dose of Portuguese man-of-war extract able to kill a dog in two or three days. They then injected other dogs with a smaller dose, which caused transient symptoms. When these dogs were reinjected with a similar dose after an interval of several weeks, a violent reaction occurred which resulted in death.

The first dose had sensitized the animal to the second dose; a phenomenon he called anaphylaxis. Richet extended these studies by making healthy animals anaphylactic by injecting them with sera from dogs already with the condition, thereby indicating that substances from the blood were involved in anaphylaxis.

Richet was born and educated in Paris, graduating in medicine from the University of Paris 1877 and later becoming professor of physiology in Paris.

Richter Burton 1931– . US particle physicist. In the 1960s he designed the Stanford Positron–Electron Accelerating Ring (SPEAR), a machine designed to collide positrons and electrons at high energies. In 1974 Richter and his team used SPEAR to produce a new subatomic particle, the Ψ meson. This was the first example of a particle formed from a charmed quark, the quark whose existence had been postulated by Sheldon Glashow ten years earlier. Richter shared the 1976 Nobel Prize for Physics with Samuel Ting, who had discovered the particle independently.

Richter Charles Francis 1900–1985. US seismologist, deviser of the Richter scale used to measure the strength of the waves from earthquakes. Each point on the scale

Richter scale

magnitude	*relative amount of energy released*	*examples*
1		
2		
3		
4	1	Carlisle, England, 1979
5	30	San Francisco and New England, USA, 1979 Wrexham, Wales, 1990
6	1,000	San Fernando, California, 1971
7	30,000	Santa Cruz, California, 1989 Armenia, USSR, 1988 Kobe, Japan, 1995
8	1,000,000	Tangshan, China, 1976 San Francisco, 1906 Lisbon, Portugal, 1755 Alaska, 1964 Gansu, China, 1920

represents a thirtyfold increase in energy over the previous point.

Ricketts Howard Taylor 1871–1910. US pathologist who discovered the *Rickettsia* (named after him), a group of unusual microorganisms that have both viral and bacterial characteristics. The ten known species in the *Rickettsia* genus are all pathogenic in human beings, causing such diseases as Rocky Mountain spotted fever and typhus.

Ricketts was born in Findlay, Ohio, and educated at the University of Nebraska and Northwestern University, Chicago. From 1902 he worked at Chicago University. In 1909 he went to Mexico City to investigate typhus, and while there he became fatally infected with the disease.

Ricketts began studying Rocky Mountain spotted fever in 1906 and discovered that the disease is transmitted to human beings by the bite of a particular type of tick that inhabits the skins of animals. In 1908, he found the causative microorganisms in the blood of infected animals and in the bodies and eggs of ticks. In his studies of typhus in Mexico, Ricketts demonstrated that the microorganisms are transmitted to humans by the body louse. Before he died from the disease, he also showed that typhus can be transmitted to monkeys, and that, after recovery, they are immune to further attacks.

Rickover Hyman George 1900–1986. Russian-born US naval officer. During World War II, he worked on the atomic-bomb project, headed the navy's nuclear reactor division, and served on the Atomic Energy Commission. He was responsible for the development of the first nuclear submarine, the *Nautilus*, 1954. After retiring 1982, he became an outspoken critic of the dangers of nuclear research and development.

Rickover emigrated to the USA with his family 1906 and graduated from the US Naval Academy 1922. After further studies in engineering, he became a specialist in the electrical division of the Bureau of Ships. He was promoted to the rank of admiral 1973.

Riemann Georg Friedrich Bernhard 1826–1866. German mathematician whose system of non-Euclidean geometry, thought at the time to be a mere mathematical curiosity, was used by Albert Einstein to develop his general theory of relativity. Riemann made a breakthrough in conceptual understanding within several other areas of mathematics: the theory of functions, vector analysis, projective and differential geometry, and topology.

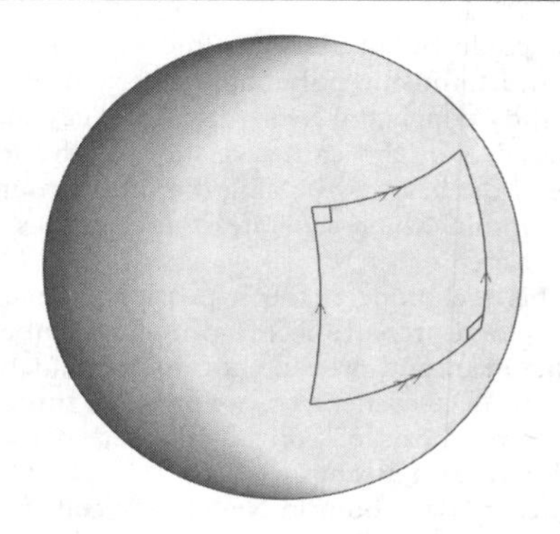

Riemann *Georg Riemann showed that in spherical geometry (a form of non-Euclidean geometry), although a parallelogram may have two opposite angles of 90°, it does not necessarily follow that the other two angles are also 90°.*

Riemann took into account the possible interaction between space and the bodies placed in it; until then, space had been treated as an entity in itself, and this new point of view was to become a central concept of 20th-century physics.

Riemann was born in Hanover and studied at Göttingen and Berlin. He spent his academic career at Göttingen, becoming professor 1859.

Riemann's paper on the fundamental postulates of Euclidean geometry, written in the early 1850s but not published until 1867, was to open up the whole field of non-Euclidean geometry and become a classic in the history of mathematics.

He also published a paper on hypergeometric series, invented 'spherical' geometry as an extension of hyperbolic geometry, and in 1855–56 lectured on his theory of Abelian functions, one of his fundamental developments in mathematics.

The Riemann–Christoffel symbols concern the theory of invariants.

He developed the ***Riemann surfaces*** to study complex function behaviour. These multiconnected, many-sheeted surfaces can be dissected by cross-cuts into a singly connected surface. By means of these surfaces he introduced topological considerations into the theory of functions of a complex variable, and into general analysis. He showed, for example, that all curves of the same class have the same Riemann surface.

Ritter Johann Wilhelm 1776–1810. German physicist who carried out early work on electrolytic cells and discovered ultraviolet radiation.

Ritter was born in Samnitz, Silesia (now in Poland), and studied medicine at Jena. Until 1804 he also taught at Jena and at Gotha, before moving to Munich as a member of the Bavarian Academy of Science.

In 1800, Ritter electrolysed water to produce hydrogen and oxygen and two years later developed a dry battery, both of which phenomena convinced him that electrical forces were involved in chemical bonding. He also compiled an electrochemical series. At about the same time he was studying the effect of light on chemical reactions, and from the darkening of silver chloride in light he discovered ultraviolet radiation.

Roberts Richard 1789–1864. Welsh engineer and inventor of such machinery as a screw-cutting lathe and a metal-planing machine. In 1845 he invented an electromagnet. He also designed a steam brake and a system of standard gauges to which all his work was constructed.

Roberts was born in Montgomeryshire and became a toolmaker. Moving to London, he worked as an apprentice to English engineer Henry Maudslay. In 1814 Roberts set up in business in Manchester. From 1828 to 1842 he was a partner in the firm of Sharpe, Thomas and Company, which manufactured machines to his design. When the Liverpool and Manchester Railway opened 1830, the firm began building locomotives. Their products were bought by railway companies throughout Europe. When the firm split up, Roberts retained the part of the company known as the Globe Works.

In 1824, at the request of some manufacturers, Roberts built a self-acting spinning mule which was a vast improvement on that devised by Samuel Crompton in 1779. In 1848 he invented a machine for punching holes in steel plates. Incorporating the Jacquard method, he devised a machine for punching holes of any pitch or pattern in bridge plates and boiler plates.

Roberts Richard John 1943– . British molecular biologist who shared the 1993 Nobel Prize for Physiology or Medicine with Phillip Sharp for the discovery of split genes (genes interrupted by nonsense segments of DNA).

In the early 1970s most information about the structure of genes was derived from experiments on simple bacteria. These experiments indicated that a gene was a continuous sequence of nucleotides, the basic building blocks of DNA, holding the genetic information. Molecular biologists assumed that a similar situation would be found in higher organisms. However Roberts and Sharp, working independently, discovered that genes in higher organisms contained long stretches of DNA that contained no genetic information. Roberts and Sharp published their results within weeks of each other 1977.

The fragmented nature of genes in higher organisms has far reaching consequences. One consequence is that the 'nonsense segments' of a gene must be removed by a process called gene splicing. This process can introduce errors in the gene that lead to inherited diseases or cancer.

Roberts was born in Derby, England, and educated at the University of Sheffield, receiving his doctorate 1968. He moved to the United States 1969 and did research at Harvard University, Cambridge, Massachusetts. In 1972 he joined Cold Spring Habor Laboratory in New York. He moved to New England Biolabs in Beverly, Massachusetts, 1992 to become research director.

Robertson Robert 1869–1949. Scottish chemist who worked on explosives for military use, such as TNT (trinitrotoluene, 2,4,6-trinitromethylbenzene), and made improvements to cordite. KBE 1918.

Robertson was born in Cupar, Fife, and studied at St Andrews. Working at the Royal Gunpowder Factory in Essex, he was put in charge of the main laboratory in 1900 and transferred to the Research Department at Woolwich Arsenal, London, 1907. In 1921 Robertson became the government chemist. He left government service in 1936 for the Royal Institution, but returned to Woolwich for the duration of World War II.

Robertson's appointment to Woolwich in 1907 coincided with the analysis of defects in British ammunition that had been revealed during the South African War. The new explosives tetryl (trinitrophenylmethylnitramine) and amatol were developed.

His investigations as government chemist included the carriage of dangerous goods by sea, the determination of sulphur dioxide and nitrous gases in the atmosphere, the elimination of sulphur dioxide from the emissions at power stations, the possible effects on health of tetraethyl lead additives to petrol, and the preservation of photographic reproductions of valuable documents.

Robinson Robert 1886–1975. English chemist, Nobel prizewinner 1947 for his research in organic chemistry on the structure of many natural products, including flower pigments and alkaloids. He formulated the electronic theory now used in organic chemistry. Knighted 1939.

Robinson's studies of the sex hormones, bile acids, and sterols were fundamental to the methods now used to investigate steroid compounds. His discovery that certain synthetic steroids could produce the same biological effects as the natural oestrogenic sex hormones paved the way for the contraceptive pill.

Robinson was born near Chesterfield, Derbyshire, and studied at Manchester. He first became professor at Sydney, Australia, 1912, returning to the UK 1915 and ending his career at Oxford 1929–55.

Robinson studied the composition and synthesis of anthocyanins (red and blue plant pigments) and anthoxanthins (yellow and brown pigments), and related their structure to their colour.

In his research on alkaloids he worked out the structure of morphine in 1925 and by 1946 he had devised methods of synthesizing strychnine and brucine, which influenced all structural studies of natural compounds that contain nitrogen.

During World War II, Robinson investigated the properties of the antibiotic penicillin and elucidated its structure. His methods were later applied to structural investigations of other antibiotics.

Rodbell Martin 1925– . US molecular biochemist who shared the 1994 Nobel Prize for Physiology or Medicine with Alfred Gilman for their discovery of a family of proteins that translate messages from outside a cell into action inside cells.

When an outside message – in the form of a hormone or other chemical signal – reaches a cell it enters through a specific receptor molecule on the cell surface. As it crosses the cell membrane, the message is translated, or converted into a second internal chemical signal that the cell can understand.

In the late 1960s Rodbell showed that the cell needed a separate molecular component to carry out the translation process. The molecule was identified in the late 1970s by Gilman and his colleagues at the University of Virginia in Charlottesville and dubbed a G protein. Since Gilman's discovery many G proteins have been identified. For example, there are specific G proteins in the rods and cones of the eye. More than 100 receptors have been identified that translate messages using G proteins.

Rodbell was born in Baltimore, Maryland, and educated at Johns Hopkins University, Baltimore, and the University of Washington, receiving his doctorate 1954. He worked at the National Institutes of Health in Bethesda, Maryland until 1985 when he joined the National Institute of Environmental Health Sciences, near Durham, North Carolina. He retired in 1994.

Roe (Edwin) Alliott Verdon 1877–1958. English aircraft designer, the first Briton to construct and fly an aeroplane, in 1908. He designed the Avro series of aircraft from 1912.

Roe was born in Patricroft, near Manchester, was apprenticed to a railway company, then entered the motor industry. He became interested in aircraft design, and his biplane flew a distance of 23 m/75 ft nearly a year before the first officially recognized flight in England by John Moore-Brabazon (1884–1964). In 1910 Roe founded the firm of A V Roe and Company, manufacturing aircraft, but in 1928 he left the firm and turned his attention to the design of flying boats. He founded the Saunders–Roe Company based on the Isle of Wight.

The first aircraft from the Manchester works was the Avro 500, one of the first machines to be ordered for use by the British army. Two of these formed the strength of the Central Flying School of the Royal Flying Corps.

In 1913 the company produced its first seaplane, a large biplane known as the Avro 503. The Avro 504, also 1913, was considerably in advance of its contemporaries, and would be used for decades of safe flying instruction. Although not basically a military aircraft, it was used extensively in World War I. Many modifications followed; the 504H was the first aircraft to be successfully launched by catapult.

Rohrer Heinrich 1933– . Swiss physicist involved in the invention of the scanning tunnelling electron microscope (STM), an ultra-powerful microscope capable of imaging individual atoms. Nobel Prize for Physics

1986 (shared with Ernst Ruska and Gerd Binnig).

The scanning tunnelling electron microscope produces a magnified image by using a tiny tungsten probe, with a tip so fine that it may consist of a single atom, which moves across a specimen. The probe tip moves so close to the specimen surface that electrons jump (or tunnel) across the gap between the tip and the surface. The magnitude of the electron flow (current) depends upon the distance from the tip to the surface, and so by measuring the current the contours of the surface can be determined. The contours can be used to form an image on a computer screen of the surface.

Rohrer was born in Buchs, St Gallen, Switzerland, and educated in Zürich. After two postgraduate years at Rutgers University, New Jersey, USA, he joined the IBM Zürich Research Laboratories at Rüschlikon, Switzerland 1963. The first STM was built at IBM with Gerd Binnig 1981.

Rolls Charles Stewart 1877–1910. British engineer who joined with Henry Royce in 1905 to design and produce cars. In 1906 a light model 20, driven by Rolls, won the Tourist Trophy and also broke the Monte Carlo-to-London record.

Rolls trained as a mechanical engineer at Cambridge, where he developed a passion for engines of all kinds. After working at the railway works in Crewe, he set up a business in 1902 as a motor dealer. Rolls was the first to fly nonstop across the English Channel and back 1910. Before the business could flourish, he died in a flying accident.

Romer Alfred Sherwood 1894–1973. US palaeontologist and comparative anatomist who made influential studies of vertebrate evolution. His *The Vertebrate Body* 1949 is still a standard textbook today.

Romer was born in White Plains, New York, and studied at Amherst College and Columbia University. From 1934 he was professor of biology at Harvard; he also became director of the Museum of Comparative Zoology 1946.

Romer spent almost all his career investigating vertebrate evolution. Using evidence from palaeontology, comparative anatomy, and embryology, he traced the basic structural and functional changes that took place during the evolution of fishes to primitive terrestrial vertebrates and from these to modern vertebrates. In these studies he emphasized the evolutionary significance of the relationship between the form and function of animals and the environment.

Römer Ole (or Olaus) Christensen 1644–1710. Danish astronomer who first calculated the speed of light, in 1679. This was all the more remarkable in that most scientists of his time considered light to be instantaneous in propagation.

Römer was born in Århus, Jutland, and studied at Copenhagen. In 1671 Jean Picard (1620–1682), who had been sent by the French Academy to verify the exact position of Tycho Brahe's observatory, was impressed by Römer's work and invited him back to Paris with him. In Paris, Römer was made a member of the Academy and tutor to the crown prince. He conducted observations, designed and improved scientific instruments, and submitted various papers to the Academy. He returned to Denmark in 1681 to take up the dual post of Astronomer Royal to Christian V and director of the Royal Observatory in Copenhagen. He also accepted a number of civic duties.

It was through the precision of both his observations and his calculations that Römer not only demonstrated that light travels at a finite speed but also put a rate to it. Noticing that the length of time between eclipses of the satellite Io by Jupiter was not constant, he realized that it depended on the varying distance between the Earth and Jupiter. He was able to announce in Sept 1679 that the eclipse of Io by Jupiter predicted for 9 Nov would occur ten minutes later than expected. Römer's prediction was borne out; his interpretation of the delay provoked a sensation. He said that the delay was caused by the time it took for the light to traverse the extra distance across the Earth's orbit.

Röntgen (or ***Roentgen***) Wilhelm Konrad, 1845–1923. German physicist who discovered X-rays 1895. While investigating the passage of electricity through gases, he noticed the fluorescence of a barium platinocyanide screen. This radiation passed through some substances opaque to light, and affected photographic plates. Developments from this discovery revolutionized medical diagnosis. Nobel prize 1901.

Röntgen was born in Lennep, Prussia, and studied in Switzerland at the Zürich Polytechnic. He was professor at Giessen 1879–88, then director of the Physical Institute at Würzburg, and ended his career in the equivalent position at Munich 1900–20.

Röntgen *German physicist Wilhelm Röntgen. Röntgen won the Nobel Prize for Physics in 1901 following his discovery of X-rays. The roentgen, named after him, is a unit of radiation exposure.*

It was at Würzburg Röntgen conducted the experiments that resulted in the discovery of the rays formerly named after him; they are now called X-rays. Today, the unit of radiation exposure is called the ***roentgen***, or röntgen (symbol R). He refused to make any financial gain out of his findings, believing that the products of scientific research should be made freely available to all.

Röntgen worked on such diverse topics as elasticity, heat conduction in crystals, specific heat capacities of gases, and the rotation of plane-polarized light. In 1888 he made an important contribution to electricity when he confirmed that magnetic effects are produced by the motion of electrostatic charges.

Ross Ronald 1857–1932. British physician and bacteriologist, born in India. From 1881 to 1899 he served in the Indian Medical Service, and during 1895–98 identified mosquitoes of the genus *Anopheles* as being responsible for the spread of malaria. Nobel prize 1902. KCB 1911.

Ross studied at St Bartholomew's Hospital in London. On retiring from the Indian Medical Service in 1899, he returned to Britain, eventually becoming professor of tropical medicine at Liverpool. During World War I he was consultant on malaria to the War Office, and when the Ross Institute of Tropical Diseases was opened 1926, he became its first director.

While on leave in England in 1894, Ross became acquainted with Scottish physician Patrick Manson, who suggested that malaria was spread by a mosquito. Returning to India, Ross collected mosquitoes, identifying the various species and dissecting their internal organs. In 1897 he discovered in an *Anopheles* mosquito a cyst containing the parasites that had been found in the blood of malarial patients.

Later, using caged birds with bird malaria, Ross was able to study the entire life history of the parasite inside a mosquito, and the mode of transmission to the victim.

Rothschild Nathaniel Mayer Victor, 3rd Baron 1910–1990. English scientist and public servant. After working in military intelligence during World War II he joined the zoology department at Cambridge University 1950–70, at the same time serving as chair of the Agricultural Research Council 1948–58 and Shell Research 1963–70. In 1971 he was asked by prime minister Edward Heath to head his new think tank, the Central Policy Review Staff, a post he held until 1974.

Rous (Francis) Peyton 1879–1970. US pathologist who received the Nobel Prize for

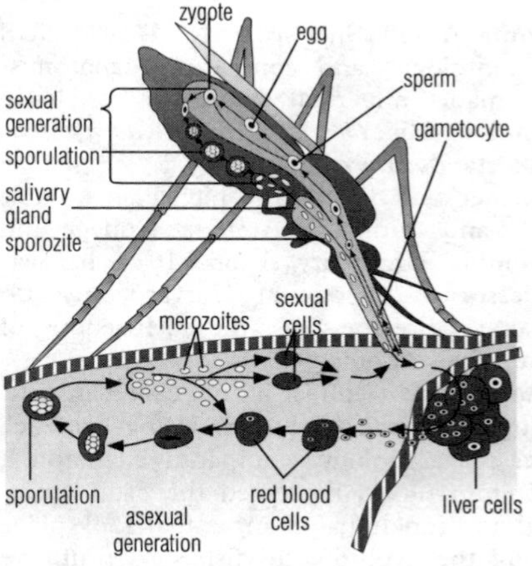

Ross *British physician Ronald Ross showed that malaria is transmitted to human beings by the bite of the* Anopheles *mosquito.*

Physiology or Medicine 1966 for his discovery that certain viruses could cause cancer in chickens.

Working at the Rockefeller Institute in New York, Rous first identified cancer-causing viruses. In 1909, a poultry farmer took a chicken that had a tumour to Rous, who then prepared a cell-free filtrate from the tumour and injected it into normal chickens. He discovered that the normal chickens developed highly malignant tumours. Rous also isolated the tumour-causing agent in the filtrate: a virus now known as the ***Rous sarcoma virus*** or avian sarcoma virus. This virus subsequently has been shown to transform some cells into cancerlike cells using a protein produced by an oncogene called the *src* gene.

Roux Wilhelm 1850–1924. German anatomist and zoologist who carried out research into developmental mechanics in embryology. He also investigated the mechanisms of functional adaptations, examining the physical stresses that cause bones, cartilage, and tendons to adapt to malformations and diseases.

Roux was born in Jena and studied there and at Berlin and Strasbourg. At Breslau (now Wroclaw in Poland) from 1879, he eventually became director of his own Institute of Embryology.

He was professor at Innsbruck, Austria, 1889–95, when he became director of the Anatomical Institute at Halle.

Roux's embryological investigations were performed mainly on frogs' eggs. Puncturing the eggs at the two-cell stage of development (a technique Roux pioneered), he found that they grew into half-embryos: the fate of the parts had already been determined. He also researched into the earliest structures in amphibian development.

Rowland F Sherwood 1927– . US chemist who shared the 1995 Nobel Prize for Chemistry with Mario Molina and Paul Crutzen for their work in atmospheric chemistry, particularly concerning the formation and decomposition of ozone. They explained the chemical reactions which are destroying the ozone layer.

The power of nitrogen oxides to decompose ozone was pointed out in 1970 by Crutzen. In 1974 Rowland and Molina published a widely read article on the threat to the ozone layer posed by chlorofluorocarbons (CFCs) used in refrigerators and aerosol cans. They pointed out that CFCs could gradually be carried up into the ozone layer. Here, under the influence of intense ultraviolet light, the CFCs would decompose into their constituents, notably chlorine atoms which decompose ozone in similar ways to nitrogen oxides. They calculated that if the use of CFCs continued at an unaltered rate the ozone layer would be seriously depleted after a few decades. Molina's and Rowland's work led to restrictions on CFC use in the late 1970s and early 1980s.

Rowland was born in Delaware, Ohio, USA and educated in chemistry at the University of Chicago, USA, receiving his doctorate in 1952. He is currently at the University of California at Irvine, California, USA.

Rowland Henry Augustus 1848–1901. US physicist who developed the concave diffraction grating 1882, which made the analysis of spectra much faster and more accurate. He also carried out the precise determination of certain physical constants.

Rowland was born in Honesdale, Pennsylvania, and studied at the Rensselaer Polytechnic Institute at Troy, New York, later joining its staff. From 1876 he was professor at Johns Hopkins in Baltimore, and made the physics laboratories at the newly established university among the best equipped in the world.

Rowland provided in 1875 the first demonstration that an electric current could be regarded as a sequence of electric charges in motion, by showing that a rapidly rotating charged body was able to deflect a magnet.

Rowland was able to produce greatly improved diffraction gratings and went on to introduce a concave metal or glass grating. This was self-focusing and thus eliminated the need for lenses, which absorbed some wavelengths of the spectrum. He put his invention to use 1886–95 by remapping the solar spectrum, publishing the wavelengths for 14,000 lines with an accuracy ten times better than his predecessors had managed.

Roxburgh William 1751–1815. Scottish botanist who enlarged the Royal Botanic Gardens in Calcutta. Until he took over the care of these gardens they had served a simply commercial purpose, the East India Company had established them in order to acclimatize plants that they wished to introduce to the Indian subcontinent. Roxburgh brought specimens from all over India and developed a large herbarium.

Roxburgh was born in Craigie in Ayrshire and was educated at Edinburgh University. He found his first employment as a surgeon's mate with the East India Company 1780. While stationed at Samulcotta, he became a keen botanist and collected a large number of botanical specimens, although many of these were unfortunately lost in a flood of the area 1787. However, he was fortunate to have previously used an Indian artist to sketch many of them and published a series of illustrations of these and other local plants in a three-volume series *Plants of the Coast of Coromandel from 1795–1819*. In recognition of his studies, he was elected superintendent of the Royal Botanic Gardens in Calcutta, India 1793.

Royce (Frederick) Henry 1863–1933. British engineer, who so impressed Charles Rolls by the car he built 1904 that Rolls-Royce Ltd was formed 1906 to produce automobiles and aeroengines. Baronet 1930.

Royce was born in Huntingdonshire and became an apprentice engineer with the Great Northern Railway. He worked on the pioneer scheme 1882–83 to light London's streets with electricity, and as chief electrical engineer on the project to light the streets of Liverpool. In 1884 he founded the firm of F H Royce and Company, manufacturing electric cranes and dynamos. He built three cars of his first design 1904 and soon afterwards teamed up with Rolls.

Rubbia Carlo 1934– . Italian physicist and 1989–93 director-general of CERN, the European nuclear research organization. In 1983 he led the team that discovered the weakons (W and Z particles), the agents responsible for transferring the weak nuclear force. Rubbia shared the Nobel Prize for Physics 1984 with his colleague Simon van der Meer (1925–).

Rubbia was born in Gorizia and studied at Pisa, Rome, and in the USA at Columbia. He worked at CERN from 1960 and concurrently served as professor of physics at Harvard 1972–88.

Rubik Ernö 1944– . Hungarian architect who invented the ***Rubik cube***, a multicoloured puzzle that can be manipulated and rearranged in only one correct way, but about 43 trillion wrong ones. Intended to help his students understand three-dimensional design, it became a fad that swept around the world.

Rudbeck Olof 1630–1702. Swedish physiologist who discovered that the thoracic duct is connected to the intestinal lymphatics (the lymph vessels that carry chyle, the fluid containing all the nutrients derived from digestion).

Jean Pecquet claimed that he had discovered the thoracic duct during his animal dissections 1647. However, Rudbeck, working independently of Pecquet, discovered that the thoracic duct is connected to the lymphatic system, in particular the intestinal lymph vessels, which carry chyle, the fluid containing all the nutrients derived from digestion. The thoracic duct functions to carry lymph and chyle into the cardiovascular system. The lymphatic system recovers interstitial fluid from the loose connective tissues throughout the body. Lymphatic vessels have very thin walls and the fluid passes into them more easily than back into the blood capillaries.

Rudbeck studied medicine at the University of Uppsala and, in Leiden, embarked on detailed studies of the lymphatic system of the body; work which led to his appointment as professor of medicine at Uppsala.

Ruffini Paolo 1765–1822. Italian mathematician, philosopher, and physician. He published a theorem stating that it was impossible to give a general solution to equations of greater than the fourth degree using only radicals (such as square roots, cube roots, and so on). This became known as the ***Abel–Ruffini theorem*** when endorsed by Norwegian mathematician Niels Abel.

Ruffini was born in Valentano, Viterbo, and studied at Modena. From 1788 he was a professor there, though when Napoleon entered Modena in 1796, Ruffini found himself appointed an official of the French republic. Two years later he refused to swear the oath of allegiance to the republic and was for a time barred from teaching.

Ruffini made a substantial contribution to the theory of equations, developing the so-called theory of substitutions, which was the forerunner of modern group theory. His work became incorporated into the general theory of the solubility of algebraic equations developed by the ill-starred French mathematician Evariste Galois.

In addition, Ruffini considered the possibility that living organisms had come into existence as the result of chance, thus anticipating more modern work on probability.

Ruhmkorff Heinrich Daniel 1803–1877. German-born French instrumentmaker who invented the Ruhmkorff induction coil 1851, a type of transformer for direct current that outputs a high voltage from a low-voltage input.

Ruhmkorff was born in Hanover and went to Paris in 1819. Working as a porter in the laboratory of physicist Charles Chevalier (1804–1850), he became interested in electrical equipment and soon began to manufacture scientific instruments. He opened his own workshop in 1840, and eventually became famous throughout Europe for his scientific apparatus.

Ruhmkorff's first notable invention was a thermoelectric battery 1844. The principles of the induction coil had been worked out by English scientist Michael Faraday in 1831. Ruhmkorff's induction coil consisted of a central cylinder of soft iron on which were wound two insulated coils: an inner primary coil comprising only a few turns of relatively thick copper wire, and an outer secondary coil with a large number of turns of thinner copper wire. An interrupter automatically made and broke the current in the primary coil, thereby inducing an intermittent high voltage in the secondary coil.

Rumford Benjamin Thompson, Count von Rumford 1753–1814. American-born British physicist and inventor. In 1798, impressed by the seemingly inexhaustible amounts of heat generated in the boring of a cannon, he published his theory that heat is a mode of vibratory motion, not a substance.

Rumford spied for the British in the American Revolution, and was forced to flee from America to England 1776. He travelled in Europe, and was knighted and created a count of the Holy Roman Empire for services to the elector of Bavaria 1784.

Rumford devised the domestic range – the 'fire in a box' – and fireplaces incorporating all the features now considered essential in open fires and chimneys, such as the smoke shelf and damper.

Rumford was born in Massachusetts and was self-educated. In Bavaria he became war and police minister as well as grand chamberlain to the elector. He cofounded the Royal Institution in London 1799, and two years later moved to France.

In Bavaria, Rumford employed beggars from the streets to manufacture military uniforms, and took responsibility for feeding them. A study of nutrition led him to devise many recipes, emphasizing vegetable soup and potatoes. Meanwhile soldiers were being employed in gardening to produce the vegetables. His search for an alternative to alcoholic drinks led to the promotion of coffee and the design of the first percolator.

He invented the Rumford shadow photometer and established the standard candle, which was the international unit of luminous intensity until 1940. He also devised a calorimeter to compare the heats of combustion of various fuels.

Rumford even planned the large park in Munich called the Englischer Garten.

It frequently happens that in the ordinary affairs ... of life opportunities present themselves of contemplating the most curious operations of nature.

Benjamin Thompson, Count Rumford

Addressing the Royal Society 1798

Ruska Ernst August Friedrich 1906–1988. German physicist who worked in opto-electronics and on the construction of the first electron microscope (called the transmission electron microscope). Nobel Prize for Physics 1986 (shared with Gerd Binnig and Heinrich Rohrer).

By the 1920s it was known that electrons could, in appropriate circumstances, behave like waves. In 1926 Hans Busch in Berlin showed how a magnetic coil could be used to focus electron waves. In 1931 Max Knoll and Ernst Ruska, also in Berlin, combined these phenomena to make the first electron microscope, which was capable of magnifying 17 times. Knoll left the team soon after and Ruska, working alone as a doctoral student, produced the first electron microscope which surpassed the optical microscope. By 1933 he had made an instrument which magnified 12,000 times (the best optical microscopes magnify around 2,000 times).

Ruska was born in Heidelberg and studied electrical engineering in Munich and Berlin. He worked on the development of commercial electron microscopes from December 1933. In 1938 he completed the first prototype of a commercial electron microscope for the German firm of Siemens. During World War II, he worked in a disused bakery until it was destroyed by Soviet troops. After the war, he continued development work with Seimens. From 1949 to 1971 he was a professor at the Free University in Berlin.

Russell Bertrand Arthur William, 3rd Earl Russell 1872–1970. British philosopher and mathematician who contributed to the development of modern mathematical logic

Russell *English philosopher and mathematician Bertrand Russell (right) meeting United Nations secretary general U Thant in London 1962. During a long, wide-ranging intellectual career Russell wrote numerous books on philosophy, education, morals, and religion. A life-long campaigner on a range of social issues, he was imprisoned 1961 at the age of 89 for taking part in a nuclear disarmament demonstration in London.*

and wrote about social issues. His works include *Principia Mathematica* 1910–13 (with A N Whitehead), in which he attempted to show that mathematics could be reduced to a branch of logic; *The Problems of Philosophy* 1912; and *A History of Western Philosophy* 1946. He was an outspoken liberal pacifist. Earl 1931. Nobel Prize for Literature 1950.

Most people would sooner die than think: in fact they do so.

Bertrand Russell

Quoted in *Observer* 12 July 1925

Russell was born in Monmouthshire, the grandson of Prime Minister John Russell. He studied mathematics and philosophy at Trinity College, Cambridge, where he became a lecturer 1910. His pacifist attitude in World War I lost him the lectureship, and he was imprisoned for six months for an article he wrote in a pacifist journal. His *Introduction to Mathematical Philosophy* 1919 was written in prison. He and his wife ran a progressive school 1927–32. After visits to the USSR and China, he went to the USA 1938 and taught at many universities. In 1940, a US court disqualified him from teaching at City College of New York because of his liberal moral views. He later returned to England and resumed his fellowship at Trinity College.

Russell was a life-long pacifist except during World War II. From 1949 he advocated nuclear disarmament and until 1963 was on the Committee of 100, an offshoot of the Campaign for Nuclear Disarmament.

Among his other works are *Principles of Mathematics* 1903, *Principles of Social Reconstruction* 1917, *Marriage and Morals* 1929, *An Enquiry into Meaning and Truth* 1940, *New Hopes for a Changing World* 1951, and *Autobiography* 1967–69.

The true spirit of delight, the exaltation, the sense of being more than Man, which is the touchstone of the highest excellence, is to be found in mathematics as surely as in poetry.

Bertrand Russell

Russell Frederick Stratten 1897–1984. English marine biologist who studied the life histories and distribution of plankton. He also discovered a means of distinguishing between different species of fish shortly after they have hatched, when they are almost identical in appearance. Knighted 1965.

Russell was born in Bridport, Dorset, and studied at Cambridge. From 1924 he worked for the Marine Biological Association in Plymouth, becoming its director 1945.

Having investigated the different types of behaviour of individual species of fish at various times of the year, and the distribution of two kinds of plankton, Russell was able to establish certain types of plankton as indicators of different types of water in the English Channel and the North Sea. He also offered a partial explanation for the difference in abundance of herring in different areas. Russell's studies of plankton and of water movements provided valuable information on which to base fishing quotas, the accuracy of which is essential to prevent overfishing and the depletion of fish stocks.

Russell also elucidated the life histories of several species of medusa by rearing the hydroids from parent medusae, and he published *The Medusae of the British Isles* 1953–70.

Russell Henry Norris 1877–1957. US astronomer who was the first to put in graphic form what became known as the Hertzsprung–Russell diagram 1913.

Russell was born in Oyster Bay, New York, and studied at Princeton, where in 1905 he was made professor and director of the observatory. In 1921 he moved to the Mount Wilson Observatory, California.

Like Ejnar Hertzsprung, Russell concluded that stars could be grouped in two main classes, one much brighter than the other. He used Annie Cannon's system of spectral classification, which also indicated surface temperature. Most of the stars were grouped together in what became known as the 'main sequence', but there was a group of very bright stars outside the main sequence. Russell put forward the theory that all stars progress at one time or another either up or down the main sequence, depending on whether they are contracting (and therefore becoming hotter) or expanding (thus cooling), but the progression he proposed was discredited within a decade.

Russell's lifelong study of binary stars resulted in a method for calculating the mass of each star from a study of its orbital behaviour. He pioneered a system using both orbits and masses in order to compute distance from Earth.

All science is either physics or stamp collecting.

Ernest Rutherford

Quoted in J B Birks *Rutherford at Manchester*

Rutherford Ernest, 1st Baron Rutherford of Nelson 1871–1937. New Zealand–born British physicist, a pioneer of modern atomic science. His main research was in the field of radioactivity, and he discovered alpha, beta, and gamma rays. He was in 1911 the first to recognize the nuclear nature of the atom. Nobel prize 1908. Knighted 1914, Baron 1931.

Rutherford produced the first artificial transformation, changing one element to another, in 1919, bombarding nitrogen with alpha particles and getting hydrogen and oxygen. After further research he announced that the nucleus of any atom must be composed of hydrogen nuclei; at Rutherford's suggestion, the name 'proton' was given to the hydrogen nucleus in 1920. He speculated that uncharged particles (neutrons) must also exist in the nucleus.

Rutherford *New Zealand-born physicist Ernest Rutherford. Rutherford described atomic structure. The vast majority of an atom's mass is made up of neutrons and protons which exist in an infinitesimally small nucleus. The nucleus is surrounded by space in which electrons orbit.*

In 1934, using heavy water, Rutherford and his co-workers bombarded deuterium with deuterons and produced tritium. This may be considered the first nuclear fusion reaction.

Rutherford was born near Nelson on South Island and studied at Christchurch. In 1895 he went to Britain and became the first research student to work under English physicist J J Thomson at the Cavendish Laboratory, Cambridge. Rutherford obtained 1898 his first academic position with a professorship at McGill University, Montréal, Canada, which then boasted the best-equipped laboratory in the world. He returned to the UK 1907, to Manchester University. From 1919 he was director of the Cavendish Laboratory, where he directed the construction of a particle accelerator, and was also professor of natural philosophy at the Royal Institution from 1921.

Rutherford began investigating radioactivity 1897 and had by 1900 found three kinds of radioactivity with different penetrating power: alpha, beta, and gamma rays. When he moved to Montréal, he began to use thorium as a source of radioactivity instead of uranium. English chemist Frederick Soddy helped Rutherford identify its decay products, and in 1903 they were able to to explain that radioactivity is caused by the breakdown of the atoms to produce a new element. In 1904 Rutherford worked out the series of transformations that radioactive elements undergo and showed that they end as lead.

In 1914, Rutherford found that positive rays consist of hydrogen nuclei and that gamma rays are waves that lie beyond X-rays in the electromagnetic spectrum.

Anyone who expects a source of power from the transformation of these atoms is talking moonshine.

Ernest Rutherford

Rutherford *Physics Today* Oct 1970

Rutherfurd Lewis Morris 1816–1892. US spectroscopist and astronomical photographer. He produced a classification scheme of stars based on their spectra that turned out to be similar to that of Italian astronomer Angelo Secchi.

Rutherfurd was born in Morrisania, New York, and studied at Williams College, Massachusetts. In 1856 he had his own observatory built and spent the rest of his life working there.

From 1858 Rutherfurd produced many photographs that were widely admired of the Moon, Jupiter, Saturn, the Sun, and stars down to the fifth magnitude. He went on to map the heavens by photographing star clusters, and devised a new micrometer that could measure the distances between stars more accurately.

In 1862 he began to make spectroscopic studies of the Sun, Moon, Jupiter, Mars, and 16 fixed stars. To help this work, Rutherfurd devised highly sophisticated diffraction gratings.

Ruzicka Leopold Stephen 1887–1976. Swiss chemist who began research on natural compounds such as musk and civet secretions. In the 1930s he investigated sex hormones, and in 1934 succeeded in extracting the male hormone androsterone from 31,815 litres/7,000 gallons of urine and synthesizing it. Born in Croatia, Ruzicka settled in Switzerland in 1929. Ruzicka shared the 1939 Nobel Prize for Chemistry with Adolf Butenandt.

Rydberg Johannes Robert 1854–1919. Swedish physicist who discovered a mathematical expression that gives the frequencies of spectral lines for elements. It includes a constant named the ***Rydberg constant*** after him.

Rydberg was born in Halmstad and studied at Lund, where he spent his whole career, becoming professor 1897.

Rydberg began by classifying spectral lines into three types: principal (strong, persistent lines), sharp (weaker but well-defined lines), and diffuse (broader lines). Each spectrum of an element consists of several series of these lines superimposed on each other. He then sought to find a mathematical relationship that would relate the frequencies of the lines in a particular series. In 1890, he achieved this by using a quantity called the wave number, which is the reciprocal of the wavelength. The formula expresses the wave number in terms of a constant common to all series (the Rydberg constant), two constants that are characteristic of the particular series, and an integer. The lines are then given by changing the value of the integer.

Rydberg then went on to produce another formula, which would express the frequency of every line in every series of an element.

Ryle Martin 1918–1984. English radio-astronomer. At the Mullard Radio Astronomy Observatory, Cambridge, he developed the technique of sky-mapping using 'aperture synthesis', combining smaller dish aerials to give the characteristics of one large one. His work on the distribution of radio sources in the universe brought confirmation of the Big Bang theory. He was knighted 1966, and won, with his co-worker Antony Hewish, the Nobel Prize for Physics 1974.

Ryle was born in Brighton, Sussex, and studied at Oxford.

During World War II he was involved in the development of radar. After the war he joined the Cavendish Laboratory in Cambridge, and in 1959 he became the first Cambridge professor of radio astronomy, responsible for most of the radiotelescope developments. He was Astronomer Royal 1972–82.

Larger and larger radio telescopes were built at the Cambridge sites, resulting in the Cambridge Catalogue Surveys, numbered 1C–5C, giving better and better maps of radio sources in the northern sky. The 3C survey, published 1959, is used as reference by all radio astronomers. The 4C survey catalogued 5,000 sources.

The first 'supersynthesis' telescope, in which a fixed aerial maps a band of the sky using solely the rotation of the Earth, and another aerial maps successive rings out from it concentrically, was built in 1963, and a 5-km/3-mi instrument was completed in 1971. The programmes for which it is in use includes the mapping of extragalactic sources and the study of supernovae and newly born stars. It can provide as sharp a picture as the best ground-based optical telescopes.

Sabatier Paul 1854–1941. French chemist. He found in 1897 that if a mixture of ethylene and hydrogen was passed over a column of heated nickel, the ethylene changed into ethane. Further work revealed that nickel could be used to catalyse numerous chemical reactions. Nobel prize 1912.

Sabatier was born in Carcassone, Aude, and studied at the Ecole Normale Supérieure in Paris. From 1884 he was professor at Toulouse.

With his assistant Abbé Jean-Baptiste Senderens (1856–1936), Sabatier extended the nickel-catalyst method to the hydrogenation of other unsaturated and aromatic compounds, and synthesized methane by the hydrogenation of carbon monoxide. He later showed that at higher temperatures the same catalysts can be used for dehydrogenation, enabling him to prepare aldehydes from primary alcohols and ketones from secondary alcohols.

Sabatier later explored the use of oxide catalysts, such as manganese oxide, silica, and alumina. Different catalysts often gave different products from the same starting material.

Alumina, for example, produced olefins (alkenes) with primary alcohols, which yielded aldehydes with a copper catalyst.

Sabin Albert Bruce 1906–1993. Russian-born US microbiologist who developed a highly effective, live vaccine against polio. The earlier vaccine, developed by physicist Jonas Salk, was based on heat-killed viruses. Sabin was convinced that a live form would be longer-lasting and more effective, and in 1957 he succeeded in weakening the virus so that it lost its virulence. The vaccine can be given by mouth.

Sabin was born in Bialystok, Russia (now in Poland). Emigrating with his parents to America 1921, he was educated at New York University. After a period at the Rockefeller Institute of Medical Research, he was appointed professor at the University of Cincinnati College of Medicine 1946–60. He became interested in polio research while working at the Rockefeller Institute. In 1936, he and a co-worker were able to make polio viruses from monkeys grow in tissue cultures from the brain cells of a human embryo. He concentrated on developing a live-virus vaccine because it would not, like the Salk vaccine, have to be injected. Sabin succeeded in finding virus strains of all three types of polio, each producing its own variety of antibody. The single-dose vaccine worked by inducing a harmless infection of the intestinal tract, causing rapid antibody formation and thereby providing lasting immunity.

Sabin was unable to test his new vaccine in America because, at an earlier stage of the Salk vaccine's development in 1954, a faulty batch caused paralytic polio in some children. However, Sabin managed to interest the Russians in his vaccine, and subsequently was able to report 1959 that 4.5 million vaccinations had been successfully carried out. The vaccine was commercially available by 1961.

Sabin Florence Rena 1871–1953. US medical researcher who studied the development of the lymphatic system and tuberculosis. She also campaigned to modernize US public health laws. Sabin was the first woman to be elected to the National Academy of Sciences.

Sabine Edward 1788–1883. Irish geophysicist who made intensive studies of terrestrial magnetism. He was able to link the incidence of magnetic storms with the sunspot cycle. KCB 1869.

Sabine was born in Dublin and educated at the Royal Military Academy, Woolwich, London. He served in the Royal Artillery, rising to the rank of major general in 1859.

In 1818, Sabine was official astronomer on an expedition to explore the Northwest Passage. The following year he went to the Arctic, and 1821–22 to the southern hemisphere. Sabine collaborated with English mathematician Charles Babbage from 1826 on a survey of magnetism in Britain, a project that was repeated by Sabine himself in the late 1850s.

At Sabine's urging, an expedition to establish observatories in the southern hemisphere was sent out in 1839 and with the data thus accumulated, Sabine in 1851 discovered a 10–11-year periodic fluctuation in the number of magnetic storms. He then correlated this magnetic cycle with data German astronomer Samuel Schwabe had collected on a similar variation in solar activity.

Sachs Julius von 1832–1897. German botanist and plant physiologist who developed several important experimental techniques and showed that photosynthesis occurs in the chloroplasts (the structure in a plant cell containing the green pigment chlorophyll) and produces oxygen. He was especially gifted in his experimental approach; some of his techniques are still in use today, such as the simple iodine test, which he used to show the existence of starch in a whole leaf.

Sachs was born in Breslau, Silesia. He was educated at Breslau Gymnasium, but had to leave school early because of the death of his parents. He managed to find a job as an assistant to the physiologist Johannes Purkinje in Prague 1850 and was later able to complete his schooling. He attended Prague University, graduating with a PhD 1856.

In 1857 he was appointed as a lecturer in plant physiology, the first of his kind in Germany. In 1859, he moved to be an assistant in plant physiology at the Agricultural and Forestry College in Tharandt. In 1861, he went to work at the Agricultural College in Poppelsdorf before becoming a professor at Freiburg in Breisgau 1867. In 1868, he was appointed professor of botany at Würzburg University and was a gifted and influential teacher with students coming from across Europe to study with him. He was made rector of the university 1871 and a privy counsellor 1877.

Sagan Carl Edward 1934–1996. US physicist and astronomer who has popularized astronomy through writings and broadcasts. His main research has been on planetary atmospheres. His books include *Broca's Brain: Reflections on the Romance of Science* 1979 and *Cosmos* 1980, based on his television series of that name.

Sagan has also done research into the origin of life on Earth, the probable climatic effects of nuclear war, and the possibility of life on other planets.

Sagan was born in New York and studied at Chicago. He became professor of astronomy and space science at Cornell University 1970, and has provided data for several NASA space-probe missions, especially *Mariner 9* to Mars.

Our loyalties are to the species and the planet. We speak for Earth. Our obligation to survive is owed not just to ourselves but also to that Cosmos, ancient and vast, from which we spring.

Carl Sagan

Cosmos 1980

In the early 1960s Sagan determined the surface temperature of Venus. He then turned his attention to the early planetary atmosphere of the Earth, and, like US chemist Stanley Miller before him, was able to produce amino acids by irradiating a mixture of methane, ammonia, water, and hydrogen sulphide. In addition, his experiment produced glucose, fructose, nucleic acids, and traces of adenosine triphosphate (ATP, used by living cells to store energy).

Other works by Sagan are *Cosmic Connection: An Extraterrestrial Perspective* 1973 and the science-fiction novel *Contact* 1985.

Saint-Claire Deville Henri Etienne 1818–1881. French inorganic chemist who worked on high-temperature reactions and was the first to extract metallic aluminium in any quantity.

Saint-Claire Deville was born on the island of St Thomas, Virgin Islands (then Danish territory), the son of the French consul there. He studied science and medicine in France and in 1845 became professor at Besançon. In 1851 he moved to the Ecole Normale in Paris and from 1859 he was at the Sorbonne.

In 1827, German chemist Friedrich Wöhler had isolated small quantities of impure aluminium from its compounds by heating them with metallic potassium. Saint-Claire Deville substituted the safer sodium. He first had to prepare sufficient sodium metal, but by 1855 he had obtained enough aluminium to cast a block weighing 7 kg/ 15 lb. The process was put into commercial production and within four years the price of aluminium had fallen to one-hundredth of its former level.

Sakharov Andrei Dmitrievich 1921–1989. Soviet physicist, an outspoken human-rights campaigner who with Igor Tamm (1895–1971) developed the hydrogen bomb. He later protested against Soviet nuclear tests and was a founder of the Soviet Human Rights Committee 1970, winning the Nobel Peace Prize 1975. For criticizing Soviet action in Afghanistan, he was in internal exile 1980–86.

Sakharov was elected to the Congress of the USSR People's Deputies 1989, where he emerged as leader of its radical reform grouping before his death later the same year.

Every day I saw the huge material, intellectual and nervous resources of thousands of people being poured into the creation of a means of total destruction, something capable of annihilating all human civilisation. I noticed that the control levers were in the hands of people who, though talented in their own ways, were cynical.

Andrei Sakharov

Sakharov Speaks 1974

Sakharov was born and educated in Moscow and did all his research at the P N Lebedev Institute of Physics. In 1948, Sakharov and Tamm outlined a principle for the magnetic isolation of high-temperature plasma, and their subsequent work led directly to the explosion of the first Soviet hydrogen bomb in 1953. But by 1950 they had also formulated the theoretical basis for controlled thermonuclear fusion – which could be used for the generation of electricity and other peaceful ends.

In the early 1960s, Sakharov was instrumental in breaking biologist Trofim Lysenko's hold over Soviet science and in giving science some political immunity. Sakharov's scientific papers in the 1960s concerned the structure of the universe. He also began publicly to argue for a reduction of nuclear arms by all nuclear powers, an increase in international cooperation, and the establishment of civil liberties in Russia. Such books as *Sakharov Speaks* 1974, *My Country and the World* 1975, and *Alarm and Hope* 1979 made him an international figure but also brought harassment from the Soviet authorities.

Sakmann Bert 1942– . German cell physiologist who shared the Nobel Prize for Physiology or Medicine with Erwin Neher 1991 for their studies of the electrical activity of nerve çell membranes. They also determined the role of the neurohormone beta-endorphin.

In 1976 Sakmann worked with Neher to develop a technique called the patch-clamp technique, which greatly enhanced the ability of researchers to measure the electrical activity of nerves and revolutionized the study of ion channels in membranes.

Using the patch-clamp technique Neher and Sakmann also investigated the role of beta-endorphin. Beta-endorphin is a neurohormone which is secreted by the pituitary gland and reaches all body tissues carried in the blood. It is a peptide opiate that has been found to play a clinical role in the perception of pain, behavioural patterns, obesity and diabetes, and psychiatric disorders. They demonstrated that beta-endorphin acts not only to regulate release of neurotransmitter substances by nerves in the brain, but also, via calcium channels, on the walls of the arteries of the brain.

Sakmann was born in Stuttgart and trained as a doctor at the University of Stuttgart. He currently heads the prestigious Max Planck Institute for Medical Research in Heidelberg.

Salam Abdus 1926–1996. Pakistani physicist who proposed a theory linking the electromagnetic and weak nuclear forces. In 1979 he became the first person from his country to receive a Nobel prize.

Salam shared the Nobel prize with US physicists Sheldon Glashow and Steven Weinberg for unifying the theories of electromagnetism and the weak force, the force responsible for a neutron transforming into a proton, an electron, and a neutrino during radioactive decay. Building on Glashow's

The whole history of particle physics, or of physics, is one of getting down the number of concepts to as few as possible.

Abdus Salam

L Wolpert and A Richards, *A Passion for Science*, 1988

Salam *Theoretical physicist Abdus Salam who was instrumental in setting up the International Centre for Theoretical Physics in Trieste, Italy, to stimulate science and technology in developing countries.*

work, Salam and Weinberg independently arrived at the same theory 1967.

Salam was born in Jhang near Faisalabad, in what was then part of British India. He attended Government College in Lahore before going to Cambridge University in England. From 1957 he was professor at Imperial College, London, and he was chief scientific adviser to the president of Pakistan 1961–74.

The theory actually involves two new particles (the W^0 and B^0), which combine in different ways to form either the photon or the Z^0. This was verified experimentally at CERN, the European particle-physics laboratory near Geneva, in 1973, though the W and Z particles were not detected until 1983. Weinberg and Salam also predicted that the electroweak interaction should violate left–right symmetry and this was confirmed by experiments at Stanford University in California.

Salisbury Edward James 1886–1978. English botanist and economist. His research was primarily on the woodland ecology of Hertfordshire and the reproductive capacity of plants (especially their seed production). His most famous work *The Living Garden* 1935 was enormously popular.

Salisbury was born in Harpenden, Hertfordshire and obtained a second class degree in botany from University College, London 1905. In 1913, he gained a DSc with a thesis on fossilized seed. In 1914, he was appointed a senior lecturer at East London College (now Queen Mary's College), but returned as a senior lecturer to University College, where he was made a reader in plant ecology 1924, and Quain professor of botany 1929.

In 1943 he became the director of Kew Gardens, the first Director to be appointed without the traditional background in taxonomic botany. He was responsible for the post-war restoration of the gardens before his retirement 1956. He retired to Bognor Regis but continued to publish until his death. He was a Fellow of the Royal Society and won their Royal Medal. He was also awarded the Veitch medal by the Royal Horticultural Society 1936. In 1939 he received the CBE and was knighted 1946.

It is courage based on confidence, not daring, and it is confidence based on experience.

Jonas Salk

On administering the then-experimental polio vaccine to himself and his family 1955

Salk Jonas Edward 1914–1995. US physician and microbiologist. In 1954 he developed the original vaccine that led to virtual eradication of paralytic polio in industrialized countries.

Salk was born and educated in New York. He began working on virus epidemics in the 1940s. He was director of the Salk Institute for Biological Studies, University of California, San Diego, 1963–75.

Salk set about finding a way of treating the polio virus so that it was unable to cause the disease but was still able to produce an antibody reaction in the human body. He collected samples of spinal cord from many polio victims and grew the virus in a live-cell culture medium, and used formaldehyde (methanol) to render the virus inactive. By 1952 he had produced a vaccine effective against the three strains of polio virus common in the USA; he tested it on monkeys, and later on his own children. In 1955, in a big publicity campaign, some vaccine was prepared without adequate precautions and about 200 cases of polio, with 11 deaths, resulted from the clinical trials. More stringent control prevented further disasters.

The people – could you patent the Sun?

Jonas Salk

On being asked who owned the patent on his polio vaccine

Samuelsson Bengt Ingemar 1934– . Swedish biochemist who shared the Nobel Prize for Physiology or Medicine with Sune Bergström and John Vane 1982 for the purification of prostaglandins, chemical messengers produced by the prostate gland.

Ulf Muller originally discovered that human semen and extracts of sheep seminal vesicular glands had peculiar properties. Both substances caused contraction of smooth muscle in vitro (in an artificial environment, such as a test-tube) and sharp decreases in the blood pressure in experimental animals. Muller called the active agents in these substances prostaglandins, because they were primarily made in the prostate gland.

The purification of the prostaglandins was complicated by the very low amounts present in seminal fluid and their extremely short half lives. In 1957, Bergstrom and Samuelsson managed to obtain crystals from two prostaglandins, alprostadil (PGE1) and PGF1a, which cause the contraction of smooth muscle. They reported the chemical characterization of these two prostaglandins 1962.

Samuelsson was born in Halmstad, Sweden and, after graduating in medicine from the University of Lund, went to work with the eminent biochemist Bergstrom at the Karolinska Institute in Stockholm. Together they isolated and characterized a family of chemical messengers called prostaglandins. When released by cells in the body, these compounds either act locally to alter the activity of neighbouring cells or act on cells some distance away by entering the bloodstream.

Samuelsson received a doctor's degree in biochemistry 1960 and in medicine 1961. He remained at the Karolinska Institute and became dean of its medical faculty.

Sanctorius Sanctorius 1561–1636. Italian physiologist who pioneered the study of metabolism and invented the clinical thermometer and a device for measuring pulse rate.

Sanctorius introduced quantitative methods into medicine. For 30 years he weighed both himself and his food, drink, and waste products. He determined that over half of normal weight loss is due to 'insensible perspiration'.

Sanger Frederick 1918– . English biochemist, the first person to win a Nobel Prize for Chemistry twice: the first in 1958 for determining the structure of insulin, and the second in 1980 for work on the chemical structure of genes.

Sanger's second Nobel prize was shared with two US scientists, Paul Berg and Walter Gilbert, for establishing methods of determining the sequence of nucleotides strung together along strands of RNA and DNA. He also worked out the structures of various enzymes and other proteins.

Sanger was born in Gloucestershire and studied at Cambridge, where he spent his whole career. In 1961 he became head of the Protein Chemistry Division of the Medical Research Council's Molecular Biology Laboratory.

Between 1943 and 1953, Sanger and his co-workers determined the sequence of 51 amino acids in the insulin molecule. By 1945 he had discovered a compound, ***Sanger's reagent*** (2,4-dinitrofluorobenzene), which attaches itself to amino acids, and this enabled him to break the protein chain into smaller pieces and analyse them using paper chromatography.

From the late 1950s, Sanger worked on genetic material, and in 1977 he and his co-workers announced that they had established the sequence of the more than 5,000 nucleotides along a strand of RNA from a bacterial virus called R17. They later worked out the order for mitochondrial DNA, which has approximately 17,000 nucleotides.

Sanger Margaret Louise, (born Higgins) 1883–1966. US health reformer and crusader for birth control. In 1914 she founded the National Birth Control League. She founded and presided over the American Birth Control League 1921–28, the organization that later became the Planned Parenthood Federation of America, and the International Planned Parenthood Federation 1952.

Sanger was born in Corning, New York;

she received nursing degrees from White Plains Hospital and the Manhattan Eye and Ear Clinic. As a nurse, she saw the deaths and deformity caused by self-induced abortions and became committed to providing health and birth-control education to the poor. In 1917 she was briefly sent to prison for opening a public birth-control clinic in Brooklyn 1916.

Her *Autobiography* appeared 1938.

Saunders Cicely Mary Strode 1918– . English philanthropist, founder of the hospice movement, which aims to provide a caring and comfortable environment in which people with terminal illnesses can die. DBE 1980.

She was the medical director of St Christopher's Hospice in Sydenham, S London, 1967–85, and later became its chair. She wrote *Care of the Dying* 1960.

Saussure Horace Bénédict de 1740–1799. Swiss geologist who made the earliest detailed and first-hand study of the Alps. He was a physicist at the University of Geneva. The results of his Alpine survey appeared in his classic work *Voyages des Alpes/Travels in the Alps* 1779–86.

Saussure Nicholas Théodore de 1767–1845. Swiss botanist, chemist, and plant physiologist who established the discipline of phytochemistry (the study of the chemistry of plants) and showed that plants gain weight during photosynthesis. He also studied the formation of carbonic acid in plants.

His most important research was in the area of photosynthesis. In 1804, he demonstrated that plants gain weight by converting carbon dioxide to oxygen. Originally, he concluded correctly that this reaction was dependent upon light and incorrectly that carbon and oxygen were the products formed from the carbon dioxide. However, he later realized that more weight was gained than was due to the carbon, and he deduced that water must also be incorporated into the plant's dry weight.

Saussure was born in Geneva, the son of a scientist who chose to educate his son himself. They made several scientific expeditions together, ascending Mont Blanc 1787 and conducted experiments that confirmed the work of Edmé Mariotte on the weight of air at various altitudes. Saussure developed an interest in botany, especially plant physiology, during these alpine expeditions. In 1797, he published several papers on the formation of carbonic acid in plants.

In 1802, he was made professor of mineralogy and geology at Geneva. Despite holding this position until 1835, he never delivered any lectures as he had wanted a position in plant chemistry. He continued with his botanical research, writing *Recherches chimiques sur la vegetation* 1804, which established the discipline of phytochemistry.

Savery Thomas *c.* 1650–1715. British engineer who invented the steam-driven water pump 1696. It was the world's first working steam engine, though the boiler was heated by an open fire.

The pump used a boiler to raise steam, which was condensed (in a separate condenser) by an external spray of cold water. The partial vacuum created sucked water up a pipe; steam pressure was then used to force the water away, after which the cycle was repeated. Savery patented his invention 1698, but it appears that poor-quality work and materials made his engines impractical.

Savery was born in Devon. His first patent was in 1696 for a machine for cutting, grinding, and polishing mirror glass. He also invented a mechanism for measuring the distance sailed by a ship. From 1705 to 1714 he was treasurer for Sick and Wounded Seamen. In 1714 he was appointed surveyor of the waterworks at Hampton Court and he designed a pumping system, driven by a water wheel, for supplying the fountains.

His pump was called the Miner's Friend and was intended to raise water from mines, but there are no records of any engines being

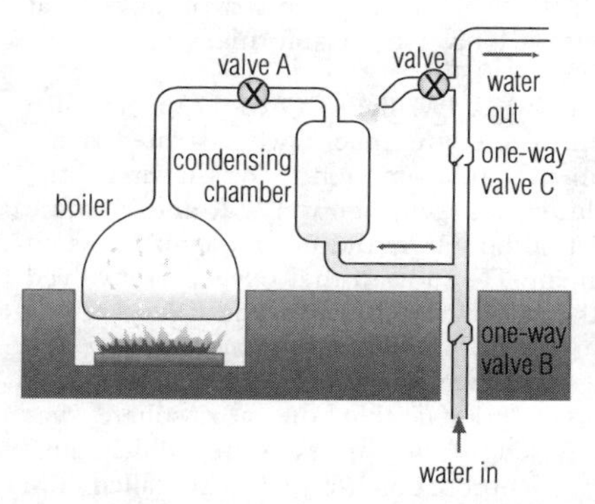

Savery *Thomas Savery's steam pump, the 'Miners' Friend', has been described as the precursor of the steam engine. However, it achieved only limited success and was not adopted widely, probably because of faulty materials and poor workmanship.*

installed in mines. An engine built at York Buildings waterworks had continuous problems with blowing steam joints.

Schaudinn Fritz Richard 1871–1906. German zoologist and microbiologist who, together with Paul Hoffmann (1868–1959), discovered that *Spirochaeta pallida* causes syphilis. He also determined the life cycle of the Coccidiae (scale insects).

Schaudinn was born in Roseningken in East Prussia and initially studied Germanic philosophy at Berlin University, but developed an interest in zoology and graduated in this 1894. After working as an assistant in the Berlin Institute of Zoology, he became a zoology lecturer at Berlin University 1898 and was then put in charge of the laboratory of protozoology in the Kaiserliche Gesundheitsamt, Berlin, and made a professor in the Institute for Tropical Diseases in Hamburg.

Schawlow Arthur Leonard 1921– . US physicist who worked in laser spectroscopy and is generally considered, with Charles Townes, to be co-inventor of the laser. Schawlow used the laser as a tool to study atomic spectra and their associated energy levels. He derived improved values for atomic constants such as the Rydberg constant. Nobel Prize for Physics 1981 (shared with Nicolaas Bloembergen and Kai Siegbahn).

Schawlow was born in Mount Vernon, New York, USA, and educated at the University of Toronto, Canada, receiving his doctorate 1949. He joined the Bell Telephone Laboratories, New Jersey, USA, 1951. There Schawlow worked with Charles Townes on the early work on the maser. In 1961 he became a professor of physics at Stanford University, California.

Scheele Karl Wilhelm 1742–1786. Swedish chemist and pharmacist who isolated many elements and compounds for the first time, including oxygen, about 1772, and chlorine 1774, although he did not recognize it as an element. He showed that oxygen is involved in the respiration of plants and fish.

In the book *Abhandlung von der Luft und dem Feuer/Experiments on Air and Fire* 1777, Scheele argued that the atmosphere was composed of two gases. One, which supported combustion (oxygen), he called 'fire air', and the other, which inhibited combustion (nitrogen), he called 'vitiated air'. He thus anticipated Joseph Priestley's discovery of oxygen by two years.

Scheele *Swedish chemist Karl Wilhelm Scheele, who discovered and isolated many compounds and elements, most notably oxygen and chlorine. Scheele released only one major publication during his life and, as a result, many of his discoveries have been attributed to other scientists. He suffered from extremely bad health, perhaps as a result of his work with toxic chemicals, and died aged only 43 in 1786.*

Scheele was born in Stralsund, Pomerania (now in Germany). At 14 he became an apothecary's apprentice in Göteborg. Practising as an apothecary, he moved via Malmö to Stockholm and finally Uppsala in 1770, where his talents as a chemist were recognized. Although offered academic positions in Germany and England, from 1775 he ran a pharmacy in the small town of Köping on Lake Mälaren in Västmanland.

Scheele's discoveries include arsenic acid, benzoic acid, calcium tungstate (scheelite), citric acid, copper arsenite (Scheele's green), glycerol, hydrogen cyanide and hydrocyanic acid, hydrogen fluoride, hydrogen sulphide, lactic acid, malic acid, manganese, nitrogen, oxalic acid, permanganates, and uric acid. He also discovered that the action of light modifies certain silver salts (50 years before they were first used in photographic emulsions).

Scheiner Christoph 1573–1650. German astronomer who carried out one of the earliest studies of sunspots and made significant improvements to the helioscope and the telescope. In about 1605 he invented the pantograph, an instrument used for copying plans and drawings to any scale.

Scheiner was born near Mindelheim, Bavaria, and studied at Ingolstadt, where he became professor of mathematics and Hebrew 1610. There he began to make astronomical observations and organized public debates on current issues in astronomy. In 1616 Scheiner was invited to the court in Innsbruck, Austria, and the following year he was ordained to the priesthood. From 1633 he lived in Vienna and from 1639 in Neisse (now Nysa in Poland).

Scheiner built his first telescope in 1611, one of the first properly mounted telescopes. He projected the image of the Sun onto a white screen so that it would not damage his eyes. When he detected spots on the Sun, he believed they were small planets circling the Sun. His Jesuit superiors did not wish him to publish his observations in case he might discredit their order, so he communicated his discovery to a friend who under a pseudonym passed it on to astronomers Galileo and Kepler. Galileo nonetheless identified Scheiner and claimed priority for the discovery of sunspots, hinting that Scheiner was guilty of plagiarism. Scheiner also concluded that Venus and Mercury revolve around the Sun, but because of his religious beliefs, he did not extend this observation to the Earth.

In his *Sol ellipticus* 1615 and *Refractiones caelestes* 1617, Scheiner drew attention to the elliptical form of the Sun near the horizon, which he explained as being due to the effects of refraction. In his major work, *Rosa ursina sive sol* 1626–30, he accurately described the inclination of the axis of rotation of the sunspots to the plane of the ecliptic.

Schiaparelli Giovanni Virginio 1835–1910. Italian astronomer who drew attention to linear markings on Mars, which gave rise to popular belief that they were canals. These markings were soon shown by French astronomer Eugène Antoniadi to be optical effects and not real lines. Schiaparelli also gave observational evidence for the theory that all meteor showers are fragments of disintegrating comets.

During his mapping of Mars, beginning 1877, Schiaparelli noted what his colleague in Rome, Pietro Secchi, had called 'channels' (*canali*). Schiaparelli adopted this term and also wrote of 'seas' and 'continents', but he made it quite clear that he did not mean the words to be taken literally. Nevertheless, fanciful stories of advanced life on Mars proliferated on the basis of the 'canals'.

Schiaparelli was born in Savigliano, Piedmont, and studied at Turin. From 1860 he was astronomer at the Brera Observatory in Milan. He discovered asteroid 69 (Hesperia) 1861, but mainly studied comets until more sophisticated instruments became available at Milan, and then turned his attention to the planets. He also studied ancient and medieval astronomy.

Schiaparelli concluded that Mercury and Venus revolved in such a way as always to present the same side to the Sun. Other observations included a study of binary stars in order to deduce their orbital systems.

Schick Bela 1877–1967. Hungarian-born US paediatrician who developed a test for diphtheria in infants. His work to diagnose and effectively treat diphtheria saved many children in Europe in the first part of the 20th century from this major killer of newborn infants.

The ***Schick test*** was developed by Schick when he was professor of paediatrics in Vienna 1913. He showed that susceptibility to diphtheria could be detected by the skin reaction after injection into the skin of minute doses of diphtheria toxin. He found that newborn infants are seldom susceptible to the toxin and that their susceptibility increases up to one year of age then gradually decreases.

In 1923 Schick emigrated to New York. He held the position of professor of diseases of children at Columbia University 1936–42.

Schimper Andreas Franz Wilhelm 1856–1901. German botanist and plant geographer who classified the plant life in Africa according to the terrain and climate of the natural habitat, and in 1880, he showed for the first time that starch is an important form of stored energy in plants.

Schimper was born in Strasbourg (then in Germany) and educated at the University of Strasbourg, where he was influenced by Heinrich de Bary. He obtained a doctorate in natural philosophy 1878 and replaced his father as director of the Museum of Natural History 1880. He also went to Wurzburg to work with Julius Sachs and travelled throughout America in the early 1880s. He was a fellow at Johns Hopkins University in Baltimore. In 1882, he was made a lecturer and then a professor in physiological botany at the University of Bonn, where he worked with Edvard Strasburger.

He travelled to the West Indies, South America, Ceylon, Java, and Africa, where he contracted malaria, which was to plague him for the rest of his life. He worked intensively to divide plant life according to the terrain and climate of the natural habitat. In 1899, his contribution to botanical research was recognized; he was made professor of botany at Basel University. He remained in Basel till his death.

Schimper Karl Friedrich 1803–1867. German naturalist and poet who proposed many of his scientific ideas in poetic form, including botany, geology, and the formation of the Alps during the Ice Age.

Schimper was born in Mannheim, Germany. His parents soon divorced and he had an unhappy childhood. He became interested in botany while he was still at school, collaborating with his teacher F W L Succow on *Flora Manhemiensis*. In 1822, he went to the University of Heidelberg and won a scholarship to study theology. In 1826, he began to study medicine. He went to Munich 1827, where he completed his studies 1829.

He returned to Mannheim 1841, because he had been unable to get an academic appointment in Munich. By 1849, he had still not managed to find regular employment and moved to Schwetzingen, where he died of dropsy. He never married despite two engagements. He was a lifelong friend of both Alexander Braun and Louis Agassiz.

Schimper Wilhelm Philipp 1808–1880. German botanist who was the director of the Natural History Museum in Strasbourg and professor of natural history and geology at Strasbourg University. He studied the anatomy of moss and travelled throughout Europe to collect botanical specimens.

Schimper was born in Dossenheim in Alsace, France, and attended Strasbourg University 1826–33, studying philosophy, philology, mathematics, and theology. In 1834, he took part in a botanical expedition to the Alps and returned to Strasbourg 1835 to become an assistant in the Natural History Museum. In 1836, he and Philipp Bruch published *Bryologica Europea*. He graduated with a scientific degree 1845 and obtained a doctorate 1848 with a thesis on the anatomy of moss. He remained in Strasbourg when it was conquered by the Germans in the Franco-Prussian war.

Schleiden Matthias Jakob 1804–1881. German botanist who identified the fundamental units of living organisms when, in 1838, he announced that the various parts of plants consist of cells or derivatives of cells. This was extended to animals by Theodor Schwann the following year.

Schleiden was born in Hamburg and studied at a number of German universities. He was professor at Jena 1831–62 and at Dorpat, Estonia, 1862–64, after which he returned to Germany.

The existence of cells had been discovered by British physicist Robert Hooke 1665, but Schleiden was the first to recognize their importance. He also noted the role of the nucleus in cell division, and the active movement of intracellular material in plant tissues.

Schmidt Bernhard Voldemar 1879–1935. Estonian lens- and mirrormaker who devised a special lens to work in conjunction with a spherical mirror in a reflecting telescope. The effect of this was to nullify 'coma', the optical distortion of focus away from the centre of the image, and thus to bring the entire image into a single focus.

Schmidt was born on the island of Naissaar. He lost most of his right arm in a childhood experiment with gunpowder. He studied engineering in Göteborg, Sweden, and at Mittweida in Germany. He stayed in Mittweida making lenses and mirrors for astronomers; in 1905 he made a 40-cm/27-in mirror for the Potsdam Astrophysical

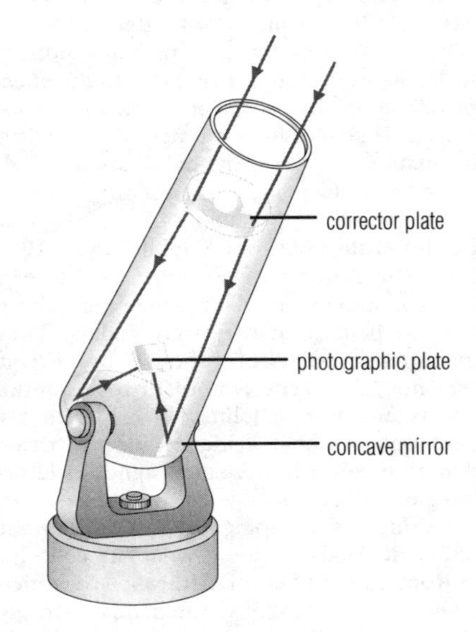

Schmidt *In a Schmidt camera, a corrector plate allows a distortion-free image to be formed on a curved photographic plate or film.*

Observatory. From 1926 he was attached to the Hamburg Observatory. He worked on the mountings and drives of the telescopes, as well as on their optics. It was in Hamburg that he perfected his lens and built it into the observatory telescope, specifically for use in photography.

By replacing the parabolic mirror of a telescope with a spherical one plus his correcting lens, Schmidt could produce an image that was sharply focused at every point (generally on a curved photographic plate, although on later models he used a second lens to compensate for the use of a flat photographic plate).

Schoenheimer Rudolf 1898–1941. German-born US biochemist who introduced the use of isotopes as tracers to study biochemical processes in 1935.

Schoenheimer was born and educated in Berlin. In 1933 he emigrated to the USA, working at the University of Columbia. He committed suicide.

Schoenheimer used deuterium to replace some of the hydrogen atoms in molecules of fat, which he fed to laboratory animals.

On analysing the body fat of rats four days later, he found that about half of the labelled fat was being stored. This meant that, contrary to previous belief, there was a constant changeover in the body between stored fat and fat that was used.

Schoenheimer used the isotope nitrogen-15 (prepared by US chemist Harold Urey, also at Columbia) to label amino acids, the basic building blocks of proteins, and again found that component molecules of the body are continually being broken down and built up.

He summarized his findings in his book *The Dynamical State of Bodily Constituents* 1942.

Schrieffer John Robert 1931– . US physicist who, with John Bardeen and Leon Cooper, developed the first satisfactory theory of superconductivity (the resistance-free flow of electrical current which occurs in many metals and metallic compounds at very low temperatures). He has also worked on ferromagnetism, surface physics and dilute alloys. Nobel Prize for Physics 1972.

In 1956 Cooper showed that, at low temperatures, electrons in a metal can weakly attract one another by distorting the lattice of metal atoms, forming a bound pair. In 1957 Bardeen, Cooper, and Schrieffer showed these bound pairs of electrons can move through the metal without resistance. This theory of superconductivity (called the BCS theory) is amazingly complete and explains all known phenomena associated with superconductivitity in metals and alloys (except the high-temperature superconducting ceramics discovered in the late 1980s).

Schrieffer was born at Oak Park, Illinois, USA. He studied engineering and physics at the Massachusetts Institute of Technology (MIT), and the University of Illinois, where he worked for his doctorate (awarded 1957) under Bardeen on superconductivity. Schrieffer became professor of physics at the University of Pennsylvania 1964.

Schrödinger Erwin 1887–1961. Austrian physicist who advanced the study of wave mechanics to describe the behaviour of electrons in atoms. He produced in 1926 a solid mathematical explanation of the quantum theory and the structure of the atom. Nobel prize 1933.

Schrödinger's mathematical description of electron waves superseded matrix mechanics, developed 1925 by Max Born and Werner Heisenberg, which also described the structure of the atom mathematically but, unlike wave mechanics, gave no picture of the atom. It was later shown that wave mechanics is equivalent to matrix mechanics.

The task is ... to think what nobody has yet thought, about that which everybody sees.

Erwin Schrödinger

L Bertalanffy *Problems of Life* 1952

Schrödinger was born and educated in Vienna. He was professor at Zürich, Switzerland, 1921–33. With the rise of the Nazis in Germany, Schrödinger went to Oxford, England, 1933. Homesick, he returned to Austria in 1936 to take up a post at Graz, but the Nazi takeover of Austria in 1938 forced him into exile, and he worked at the Institute for Advanced Studies in Dublin, Ireland, 1939–56. He spent his last years at the University of Vienna.

French physicist Louis de Broglie had in 1924, using ideas from Albert Einstein's special theory of relativity, shown that an electron or any other particle has a wave associated with it. In 1926 both Schrödinger and de Broglie published the same wave equation, which Schrödinger later

Schrödinger *Nobel laureate Austrian physicist Erwin Schrödinger. Schrödinger's equation describes the wave function associated with an electron. He shared the Nobel Prize for Physics with Paul Dirac and Werner Heisenberg in 1933.*

formulated in terms of the energies of the electron and the field in which it was situated. He solved the equation for the hydrogen atom and found that it fitted with energy levels proposed by Danish physicist Niels Bohr.

In the hydrogen atom, the wave function describes where we can expect to find the electron. Although it is most likely to be where Bohr predicted it to be, it does not follow a circular orbit but is described by the more complicated notion of an orbital, a region in space where the electron can be found with varying degrees of probability. Atoms other than hydrogen, and also molecules and ions, can be described by Schrödinger's wave equation, but such cases are very difficult to solve.

Schultze Max Johann Sigismund 1825–1874. German zoologist who adopted the term protoplasm to refer to the contents of cells. His work emphasized its role in both plant and animal cells. He also worked on the retina, correctly arguing that the cones were responsible for colour vision and the rods for the perception of light

Schultze was born in Freiburg im Breisgam, Germany and was educated at the Greifswald Gymnasium 1845. He went to the University of Greifswald to study medicine and graduated 1849. He went to Halle as assistant professor of anatomy 1854 and, in 1859, was appointed professor of anatomy at Bonn.

Schumacher Fritz (Ernst Friedrich) 1911–1977. German economist who believed that the increasing size of institutions, coupled with unchecked economic growth, creates a range of social and environmental problems. He argued his case in books like *Small is Beautiful* 1973, and established the Intermediate Technology Development Group.

Schumacher studied at Oxford and held academic posts there and in the USA at Columbia in the 1930s and 1940s. After World War II he was economic adviser to the British Control Commission in Germany 1946–50 and to the UK National Coal Board 1950–70. He also served as president of the Soil Association and as director of the Scott-Bader Company, which manufactures polymers and is based on common ownership. He advised many governments on problems of rural development.

His book *A Guide for the Perplexed* 1977 deals with philosophy.

Schwabe Samuel Heinrich 1789–1875. German astronomer who was the first to measure the periodicity of the sunspot cycle. This may be considered as marking the beginning of solar physics.

Schwabe was born in Dessau, studied at Berlin, and worked as a pharmacist. In 1829 he sold his pharmacy and became an astronomer. He published 109 scientific papers and left 31 volumes of astronomical data to the Royal Astronomical Society.

Schwabe began to watch the Sun in 1825 with a 5-cm/2-in telescope and noticed sunspots, making daily counts of them for most of the rest of his life. In 1843 he was able to announce a periodicity: he declared that the sunspots waxed and waned in number according to a ten-year cycle.

In 1827 Schwabe rediscovered the eccentricity of Saturn's rings, and in 1831 he drew a picture of the planet Jupiter on which the Great Red Spot was shown for the first time.

Schwann Theodor 1810–1882. German physiologist who, with Matthias Schleiden, is credited with formulating the cell theory, one of the most fundamental concepts in biology. Schwann also did important work on digestion, fermentation, and the study of tissues.

Schwann was born in Neuss and studied at Bonn, Würzburg, and Berlin. He spent 1834–38 working as an assistant to German physiologist Johannes Peter Müller at the Museum of Anatomy in Berlin. In 1839, however, Schwann's work on fermentation attracted so much adverse criticism that he left Germany for Belgium, where he was professor at Louvain 1839–48 and then at Liège.

In 1836, Schwann isolated from the lining of the stomach a chemical responsible for protein digestion, which he called pepsin. This was the first enzyme to be isolated from animal tissue.

Schwann showed 1836–37 that the fermentation of sugar is a result of the life processes of living yeast cells (he later coined the term 'metabolism' to denote the chemical changes that occur in living tissue).

In *Mikroskopische Untersuchungen über die übereinstimmung in der Struktur und dem Wachstum der Tiere und Pflanzen/Microscopical Researches on the Similarity in the Structure and Growth of Animals and Plants* 1839, he concluded that all organisms consist entirely of cells or of products of cells and that the life of each individual cell is subordinated to that of the whole organism. In addition, he noted that an egg is a single cell that eventually develops into a complex organism.

Schwartz Melvin 1932– . US physicist who, with Jack Steinberger and Leon Lederman, used neutrinos to study elementary particles and discovered the muon neutrino. Nobel Prize for Physics 1988.

Using a large proton accelerator as a source of neutrinos, Schwartz, Steinberger, and Lederman produced the world's first beam of neutrinos 1961. Using this they investigated the weak nuclear force and the quark structure of matter. They also found the muon neutrino, a new type of neutrino, which is paired with and can be transformed into the muon particle. The other, previously recognized, neutrino is similarly paired with the electron and is called the electron neutrino.

Schwartz was educated at the Bronx High School of Science in New York City and at Columbia University, New York. He became assistant professor at Columbia 1958 and a full professor 1963. In 1966, Schwartz moved to Stanford University, California and, during the 1970s, he became chief executive of a commercial firm in California.

Schwarzschild Karl 1873–1916. German astronomer and theoreticical physicist who was the first to substitute a photographic plate at the telescope in place of the eye and then measure densities with a photometer. He designed and constructed some of his own instruments.

Schwarzschild was born in Frankfurt and studied at Strasbourg and Munich. In 1902 he was appointed professor at Göttingen and director of the observatory. From 1909 he was director of the Astrophysical Observatory at Potsdam. He was the father of Martin Schwarzschild.

In 1900, he suggested that the geometry of space was possibly not in conformity with Euclidean principles. (This was 16 years before the publication of Albert Einstein's general theory of relativity.) He later gave the first exact solution of Einstein's field equations.

Schwarzschild introduced the concept of radiative equilibrium in astrophysics and was probably the first to see how radiative processes were important in conveying heat in stellar atmospheres. In 1906, he published work on the transfer of energy at and near the surface of the Sun.

He devised a multi-slit interferometer and used it to measure the separation of close double stars. During a total solar eclipse 1905, he obtained spectrograms that gave information on the chemical composition of regions at various heights on the Sun. He later designed a spectrographic objective that provided a quick, reliable way to determine the radial velocities of stars.

Schwarzschild Martin 1912–1997 . German-born US astronomer whose most important work was in the field of stellar structure and evolution. He greatly narrowed the estimated range of mass that stars can have.

Schwarzschild was born in Potsdam, the son of astronomer Karl Schwarzschild. After studying at Göttingen, he emigrated to the USA 1935. He was professor at Columbia 1947–51 and at Princeton from 1951.

Schwarzschild worked out a quantity (Z_{He}) for the total mass density of the elements heavier than helium, using the density of hydrogen as one unit. The values of Z_{He} are smallest for old stars (0.003) and largest for young stars (0.04), implying that the most recently formed stellar objects were formed out of a medium of interstellar gas and dust that was already enriched with heavy elements. These elements were probably produced in stellar interiors and expelled by the oldest stars.

In 1938, Schwarzschild suggested that the star's deepest interior pulsates, but that in the outermost regions the elements of gas do not all vibrate in unison, causing a lag in the light curve by the observed amount.

Reverence for Life.

Albert Schweitzer

My Life and Thought ch 13

Schweitzer Albert 1875–1965. French Protestant theologian, organist, and missionary surgeon. He founded the hospital at Lambaréné in Gabon in 1913, giving organ recitals to support his work there. He wrote a life of German composer J S Bach and *Von reimarus zu Wrede/The Quest for the Historical Jesus* 1906. He was awarded the Nobel Peace Prize in 1952 for his teaching of 'reverence for life'.

Schwinger Julian Seymour 1918–1994. US quantum physicist. His research concerned the behaviour of charged particles in electrical fields. This work, expressed entirely through mathematics, combines elements from quantum theory and relativity theory into a new theory called ***quantum electrodynamics***, the most accurate physical theory of all time. Schwinger shared the Nobel Prize for Physics 1963 with Richard Feynman and Sin-Itiro Tomonaga (1906–1979).

Described as the 'physicist in knee pants', he entered college in New York at the age of 15, transferred to Columbia University and graduated at 17. At the age of 29 he became Harvard University's youngest full professor.

He went to work on nuclear physics problems at Berkeley (in association with J Robert Oppenheimer) and at Purdue University. From 1943 to 1945 he worked on problems relating to radar at the Massachusetts Institute of Technology and, after the war, moved to Harvard, where he developed his version of quantum electrodynamics. He calculated the anomalous magnetic moment of the electron soon after its discovery. In 1957, Schwinger anticipated the existence of two different neutrinos associated with the electron and the muon (heavy electron), which was confirmed experimentally in 1963. He also speculated that weak nuclear forces are carried by massive, charged particles. This was confirmed in 1983 at CERN (the European Laboratory for Particle Physics) in Geneva. In 1972 Schwinger became Professor of Physics at the University of California, Los Angeles.

Scott Dukinfield Henry 1854–1934. English botanist who studied the anatomy of plants and, with William Crawford Williamson, described the evolutionary links between ferns and cycads, research that led to the development of phylogenetic theories of plants. His best known studies were in the field of palaeobotany, including an excellent account of the fruiting bodies of fossil plants 1904.

Scott was born in London and educated mainly at home by tutors. His mother instilled his interest in field botany before his 14th birthday. He was competent in the use of a microscope and studied plant structure. However, he read classics at Christchurch, Oxford, and completed his training as an engineer 1872–76. In 1879, he went to the University of Wurzburg, Germany to study botany under the inspiring and eminent scientist Julius Sachs. In 1882, he returned to England to take up a position at University College, London, and in 1885, he went to work under Thomas Henry Huxley at the Normal School of Science (Imperial College).

In 1892, he became the honorary director of the Jodrell Laboratory and was elected to the Royal Society 1894, receiving both the Royal medal and the Darwin medal. He was elected president of the Linnaean Society from 1908–12 and received their Gold medal 1921.

Scott Peter Markham 1909–1989. British naturalist, artist, and explorer, founder of the Wildfowl Trust at Slimbridge, Gloucestershire, England, 1946, and a founder of the World Wildlife Fund (now World Wide Fund for Nature). He was knighted in 1973.

He was the son of Antarctic explorer R F Scott; he studied at Cambridge, in Germany, and at the Royal Academy School, London. In 1936 he represented Britain in the Olympic Games, gaining a bronze medal for the single-handed sailing event. During World War II he served with the Royal Navy.

In 1949 he led his first expedition, which was to explore the uncharted Perry River area in the Canadian Arctic. Scott also led ornithological expeditions to Iceland, Australasia, the Galápagos Islands, the Seychelles, and the Antarctic. He was the first president of the World Wildlife Fund 1961–67.

Scott's paintings were usually either portraits or bird studies.

He published many books on birds, including *Key to the Wild Fowl of the World* 1949 and *Wild Geese and Eskimos* 1951, and an autobiography 1961, *The Eye of the Wind.*

Seaborg Glenn Theodore 1912– . US nuclear chemist. For his discovery of plutonium and research on the transuranic elements, he shared a Nobel prize 1951 with his co-worker Edwin McMillan.

Seaborg was born in Michigan and studied at the University of California. During part of World War II he was at the metallurgical laboratory at Chicago University, where much of the early work on the atomic bomb was carried out. He was professor at the University of California at Berkeley 1945–61, and chair of the Atomic Energy Commission 1961–71, encouraging the rapid growth of the US nuclear-power industry. He returned to Berkeley 1971.

People must understand that science is inherently neither a potential for good nor for evil. It is a potential to be harnessed by man to do his bidding.

Glenn T Seaborg

Associated Press interview with Alton Blakeslee, 29 Sept 1964

Transuranic elements are all radioactive and none occurs to any appreciable extent in nature; they are synthesized by transmutation reactions. Seaborg was involved in the identification of plutonium (atomic number 94) 1940, americium (95) 1944–45, curium (96) 1944, berkelium (97) 1949, californium (98) 1950, einsteinium (99) 1952, fermium (100) 1953, mendelevium (101) 1955, and nobelium (102) 1957.

Secchi Pietro Angelo 1818–1878. Italian astronomer and astrophysicist who classified stellar spectra into four classes based on their colour and spectral characteristics. He was the first to classify solar prominences, huge jets of gas projecting from the Sun's surface.

Secchi was born in Reggio nell'Emilia and became a Jesuit priest, lecturing in physics and mathematics at the Collegio Romano from 1839. In 1848 he was driven into exile for being a Jesuit and went first to Stonyhurst College, England, then to Georgetown University in Washington DC. He returned to Italy 1849 as director of the observatory at the Collegio Romano and professor of astronomy.

With English astronomer William Huggins, Secchi was the first person to adapt spectroscopy to astronomy in a systematic manner and he made the first spectroscopic survey of the heavens. He proposed that the differences in stellar spectra reflected differences in chemical composition. His classification system of 1867 is the basis of the modern system.

Secchi was among the first to use the new technique of photography for astronomical purposes. By 1859 he had a complete set of photographs of the Moon.

Sedgwick Adam 1785–1873. English geologist who contributed greatly to understanding the stratigraphy of the British Isles, using fossils as an index of relative time. Together with Scottish geologist Roderick Murchison, he identified the Devonian system in SW England.

Sedgwick was born in Dent, Yorkshire, and studied mathematics at Cambridge, where he became professor of geology 1818.

An energetic champion of field work, Sedgwick explored such diverse districts as the Isle of Wight, Devon and Cornwall, the Lake District, and NE England. In the 1830s, he unravelled the stratigraphic sequence of fossil-bearing rocks in North Wales, naming the oldest of them the Cambrian period (now dated at 500–570 million years ago). In South Wales, his companion Murchison had concurrently developed the Silurian system. The question of where the boundary lay between the older Cambrian and the younger Silurian sparked a dispute that was not resolved until 1879, when Charles Lapworth (1842–1920) coined the term Ordovician for the middle ground.

Segrè Emilio Gino 1905–1989. Italian-born US physicist who in 1955 discovered the antiproton, a new form of antimatter. He shared the 1959 Nobel Prize for Physics with his co-worker Owen Chamberlain. Segrè discovered the first synthetic element, technetium (atomic number 43), in 1937.

Segrè was born near Rome and studied

there, working with Enrico Fermi. Segrè became professor at Palermo 1936 but was forced into exile by the Fascist government and, apart from wartime research at Los Alamos, New Mexico, worked from 1938 at the University of California at Berkeley, where he became professor 1947.

In 1940 Segrè discovered another new element, now called astatine (atomic number 85). He again met up with Fermi, now at Columbia University, to discuss using plutonium-239 instead of uranium-235 in atomic bombs. Segrè began working on the production of plutonium at Berkeley and then moved to Los Alamos to study the spontaneous fission of uranium and plutonium isotopes.

In 1947, Segrè started work on proton–proton and proton–neutron interaction, using a cyclotron accelerator at Berkeley. This was how he detected the antiproton, which confirmed the relativistic quantum theory of English physicist Paul Dirac.

Seguin Marc 1786–1875. French engineer who, in 1825, built the first successful suspension bridge in Europe using cables of iron wire. He also invented the multitubular boiler.

Seguin was born in Annonay, Ardèche, and was self-taught in engineering science. Seguin's first suspension bridge was built in Geneva in association with Swiss engineer Henri Dufour (1786–1875). Over the next 20 years, Seguin and his brothers erected cable suspension bridges in France, beginning with one over the river Rhône at Tournon in 1827. Seguin established France's first modern railway between Lyon and St Etienne, completed 1832.

In physics, Seguin argued that matter consisted of small, dense molecules constantly on the move in miniature solar systems and that magnetic, electrical, and thermal phenomena were the result of their particular velocities and orbits. He identified heat as molecular velocity and explained that its conversion to a mechanical effect occurs when the molecules transmit their velocities to external objects. In *De l'Influence des chemins de fer* 1839 he tried unsuccessfully to determine the numerical relationship between heat and mechanical power.

Seki Kowa, (also called Takakazu) *c.* 1642–1708. Japanese mathematician who created a new mathematical notation system and used it to discover many of the theorems and theories that were being – or were shortly to be – discovered in the West.

Seki was born in Fujioka in Gunma prefecture. Much of his reputation stems from the social reform he introduced in order to develop the study of mathematics in Japan and make it widely accessible.

He introduced Chinese ideograms to represent unknowns and variables in equations, and although he was obliged to confine his work to equations up to the fifth degree – his algebraic alphabet (*endan-jutsu*) was not suitable for general equations of the *n*th degree – he was able to create equations with literal coefficients of any degree and with several variables, and to solve simultaneous equations. In this way he was able to derive the equivalent of $f(x)$, and thereby to arrive at the notion of a discriminant – a special function of the root of an equation expressible in terms of the coefficients.

Another of Seki's contributions was the rectification of the circle; he obtained a value for π that was correct to the 18th decimal place.

Seki is also credited with major discoveries in calculus.

Semenov Nikolai Nikolaevich 1896–1986. Russian physical chemist who studied chemical chain reactions, particularly branched-chain reactions, which can accelerate with explosive velocity. For his work in this area, in 1956 he became the first Russian to gain the Nobel Prize for Chemistry.

Semenov was born in Saratov and studied at Petrograd (now St Petersburg). He became professor at the Physical-Technical Institute there 1928. He was director of the Institute for Chemical Physics at the Soviet Academy of Sciences 1931–44, and then moved to Moscow State University.

Semenov did his Nobel prizewinning work in the 1920s and summarized his results in his book *Chemical Kinetics and Chain Reactions* 1934.

Semenov also played an important part in resisting narrow interpretations of Marxism–Leninism in its application to chemistry.

Semmelweis Ignaz Philipp 1818–1865. Hungarian obstetrician who unsuccessfully pioneered asepsis (better medical hygiene), later popularized by the British surgeon Joseph Lister.

Semmelweis was an obstetric assistant at the General Hospital in Vienna at a time when 10% of women were dying of puerperal (childbed) fever. He realized that the cause was infectious matter carried on the hands of doctors treating the women after handling corpses in the postmortem room. He introduced aseptic methods (hand-washing in chlorinated lime), and mortality fell to almost zero. Semmelweis was dismissed for his efforts, which were not widely adopted at the time.

Senebier Jean 1742–1809. Swiss botanist, plant physiologist, and pastor, whose research on photosynthesis (the process by which green plants use light energy to make carbohydrates) showed that 'fixed air' (now known to be carbon dioxide) was converted to 'pure air' (oxygen) in a light-dependent process.

His work on photosynthesis was a forerunner of Claude Bernard's work in the early 1800s. Senebier showed that it was the light and not the warmth of sunlight that was necessary for photosynthesis to occur, and that photosynthesis does not occur in boiled water from which the gases have been excluded. His *Action de la lumière sur la végétation* 1779 is an important paper on photosynthesis.

Senebier was born in Geneva. Although he developed an interest in botany from an early age, he was ordained as a pastor in the Protestant church of Geneva 1765 in order to please his family. He was appointed pastor of Chancy near Geneva 1769. Prior to taking up this appointment, he had travelled to Paris, where he conducted several experiments in plant physiology. In 1769, he was made librarian for the Republic of Geneva. He was able to continue with his research and developed a reputation as a plant physiologist. He taught Pierre Huber, Augustin Pyrame de Candolle, and Nicholas Theodore de Saussure.

I will burn, but this is a mere incident. We will continue our discussion in eternity.

Michael Servetus

To his judges

Servetus Michael (Miguel Serveto) 1511–1553. Spanish Christian Anabaptist theologian and physician. He was a pioneer in the study of the circulation of the blood and found that it circulates to the lungs from the right chamber of the heart. He was burned alive by the church reformer Calvin in Geneva, Switzerland, for publishing attacks on the doctrine of the Trinity.

Seward Albert Charles 1863–1941. English palaeobotanist whose work on Palaeozoic and Mesozoic plants established the new field of palaeobotany. His *Fossil Plants as Tests of Climate* 1892 was one of the earliest works of biogeochemistry and along with *Jurassic Flora* 1900–03 was widely acclaimed.

Seward was born in Lancaster and educated at St John's College, Cambridge, where he gained a first in natural sciences 1886. He spent the next year studying palaeobotany at Manchester under William Crawford Williamson and then made a trip to Europe to study the collections of fossil plants in the continent's major museums. In 1890, he became a lecturer at Cambridge, establishing an international reputation as a palaeobotanist.

He was elected a Fellow of the Royal Society 1898. In 1906, he took up the professorship of the Cambridge School of Botany, building up the department's popularity to the extent that it had to be extended in 1934. From 1915–36, he was the master of Downing College and was made vice-chancellor of Cambridge University 1924. He was awarded the Royal Medal 1925, the Darwin Medal 1934, and was knighted 1936.

Seyfert Carl Keenan 1911–1960. US astronomer and astrophysicist who studied the spectra of stars and galaxies, and identified and classified the type of galaxy that now bears his name.

A ***Seyfert galaxy*** has a small, bright centre caused by hot gas moving at high speed around a massive central object, possibly a black hole. Almost all Seyferts are spiral galaxies. They seem to be closely related to quasars, but are about 100 times fainter.

Seyfert was born in Cleveland, Ohio, and studied at Harvard. He was director of Barnard Observatory 1946–51, and from 1951 professor at Vanderbilt University and director of the observatory there. He was a member of the National Defence Research Committee.

In 1943, Seyfert was studying 12 active spiral galaxies with bright nuclei. His investigations showed that these galaxies contain hydrogen as well as ionized oxygen, nitrogen, and neon. On the basis of their spectra, Seyfert divided the galaxies into two types, I and II. Seyfert galaxies emit radio waves, infrared energy, X-rays, and nonthermal radiation. The gases at their centres are subject to explosions which cause them to move violently, with speeds of many thousands of kilometres per second relative to the centre of the galaxy in the case of type I, and of several hundreds of kilometres per second in the case of type II.

In 1951, Seyfert began a study of the objects now known as ***Seyfert's Sextet***: a group of diverse extragalactic objects, of which five are spiral nebulae and one an irregular cloud. One member of the group is moving away from the others at a velocity nearly five times that at which the others are receding from each other.

Shannon Claude Elwood 1916– . US mathematician who founded the science of information theory. He argued that entropy is equivalent to a shortage of information content (a degree of uncertainty), and obtained a quantitative measure of the amount of information in a given message.

Shannon reduced the notion of information to a series of yes/no choices, which could be presented by a binary code. Each choice, or piece of information, he called a 'bit'. In this way, complex information could be organized according to strict mathematical principles. He also wrote the first effective program for a chess-playing computer.

His book *The Mathematical Theory of Communication* 1949 was written with W Weaver (1894–1978).

Shannon was born in Gaylord, Michigan, and studied at the University of Michigan and the Massachusetts Institute of Technology (MIT). From 1941 he worked at the Bell Telephone Laboratories, but he also held academic positions at MIT from 1956, and in 1958 he left Bell to become professor of science.

As early as 1938 Shannon began examining the question of a mathematical approach to language. His methods, although devised in the context of engineering and technology, were soon seen to have applications not only to computer design but to virtually every subject in which language was important, such as linguistics, psychology, cryptography, and phonetics; further applications were possible in any area where the transmission of information in any form was important.

Shapley Harlow 1885–1972. US astronomer who established that the Galaxy was much larger than previously thought. The Sun was not at the centre of the Galaxy as then assumed, but two-thirds of the way out to the rim; globular clusters were arranged in a halo around the Galaxy.

Shapley was born in Missouri and studied at Laws Observatory. He worked at Mount Wilson Observatory, California, 1914–21, and was then appointed director of the Harvard College Observatory. Alleged by Senator Joseph McCarthy in 1950 to be a communist, he was interrogated by the House of Representatives Committee on Un-American Activities.

Shapley obtained nearly 10,000 measurements of the sizes of stars in order to analyse some 90 eclipsing binaries. He also showed that Cepheid variable stars were pulsating single stars, not double stars. He discovered many previously unknown Cepheid variables and devised a statistical procedure to establish the distance and luminosity of a Cepheid variable. Shapley's surveys recorded tens of thousands of galaxies in both hemispheres.

Sharman Helen 1963– . The first Briton to fly in space, chosen from 13,000 applicants for a 1991 joint UK–Soviet space flight. Sharman, a research chemist, was launched on 18 May 1991 in *Soyuz TM-12* and spent six days with Soviet cosmonauts aboard the *Mir* space station.

Sharp Phillip Allen 1944– . English-born US molecular biologist who shared the Nobel Prize for Physiology or Medicine 1993 with Richard Roberts for their discovery of split genes. Using his technique for measuring nucleic acid fragments on genes, he found that genes consist of regions of DNA (genetic material) separated by regions that do not contain genetic information, called introns.

Sharp developed S1 nuclease mapping, a technique to measure the size of nucleic acid fragments in genes. He then used this technique to determine that genes are split into several regions.

In 1977, he and Roberts independently reported that a single adenovirus messenger RNA molecule corresponded to four distinct regions of DNA, and that different regions of DNA are separated by introns that do not carry genetic information.

Up until that meeting it had generally been thought that genes were continuous stretches of DNA which served as direct templates for messenger RNA (mRNA) molecules, and that these mRNA molecules were themselves templates for protein synthesis. However, before the protein can be translated from the messenger RNA, these introns have to be spliced out. Most eukaryotic genes are now known to have introns and failure to be able to remove an intron is known to cause genetic diseases, for example some forms of thalassaemia.

Sharp was born in Kentucky and educated at Union College and the University of Illinois. He worked at the Cold Spring Harbor laboratories in New York and then at the Massachusetts Institute of Technology in Boston.

Sharpey-Schafer Edward Albert, (born Schäfer) 1850–1935. English physiologist, one of the founders of endocrinology. He made important discoveries relating to the hormone adrenaline, and to the pituitary and other endocrine, or ductless, glands. Knighted 1913.

He also introduced the supine position for artificial respiration, which improved on existing techniques.

Schäfer was born in London and studied at University College; he later took the name Sharpey-Schafer to honour one of his professors, and himself became professor there 1883. From 1899 he was professor at Edinburgh.

In 1894 Sharpey-Schafer and his co-worker George Oliver (1841–1915) discovered that an extract from the central part of an adrenal gland injected into the bloodstream of an animal caused a rise in blood pressure by vasoconstriction. They also noted that the smooth muscles of the animal's bronchi relaxed. These effects were caused by the action of the hormone adrenaline.

Sharpey-Schafer also suspected that another hormone was produced by the islets of Langerhans in the pancreas. He adopted for it the name 'insulin' (from the Latin for 'island').

Shaw (William) Napier 1854–1945. English meteorologist who introduced the millibar as the meteorological unit of atmospheric pressure (in 1909, but not used internationally until 1929). He also invented the tephigram, a thermodynamic diagram widely used in meteorology, in about 1915. Knighted 1915.

Shaw was born in Birmingham and studied at Cambridge, where he worked at the Cavendish Laboratory for experimental physics 1877–1900. He was director of the Meteorological Office 1905–20, and professor of meteorology at the Royal College of Science (part of London University) 1920–24.

Shaw pioneered the study of the upper atmosphere by using instruments carried by kites and high-altitude balloons. His work on pressure fronts formed the basis of a great deal of later work in the field.

His books include *Life History of Surface Air Currents* 1906 and *The Air and its Ways* 1923 (both with his colleague R Lempfert) and (with J. S. Owens) *The Smoke Problem of Great Cities* 1925, an early work on atmospheric pollution. Shaw's *Manual of Meteorology* 1926–31 is still a standard reference work.

Shepard Alan Bartlett 1923– . US astronaut, the fifth person to walk on the Moon. He was the first American in space, as pilot of the suborbital *Mercury-Redstone 3* mission on board the *Freedom 7* capsule May 1961, and commanded the *Apollo 14* lunar landing mission 1971.

Shepard was born in New Hampshire and studied at the US Naval Academy in Annapolis. After working as a naval pilot and test pilot, he was selected to be a NASA astronaut 1959; he resigned 1974.

Sherrington Charles Scott 1857–1952. English neurophysiologist who studied the structure and function of the nervous system. *The Integrative Action of the Nervous System* 1906 formulated the principles of reflex action. Nobel Prize for Physiology or Medicine 1932. GBE 1922.

He showed that when one set of antagonistic muscles is activated, the opposing set is inhibited. This theory of reciprocal innervation is known as ***Sherrington's law***.

Sherrington also identified the regions of the brain that govern movement and sensation in particular parts of the body.

Sherrington was born in London and studied there at St Thomas's Hospital and at Cambridge. He became professor at London University's veterinary institute 1891, at Liverpool 1895, and was professor of physiology at Oxford 1913–35. During World War I, for three months he worked incognito as a labourer in a munitions factory, and the observations he made there did much to improve safety for factory workers.

One of Sherrington's findings, published 1894, was that the nerve supply to muscles contains 25–50% sensory fibres, as well as motor fibres concerned with stimulating muscle contraction. The sensory fibres carry sensation to the brain so that it can determine, for example, the degree of tension in the muscles. Sherrington divided the sense organs into three groups: interoceptive, characterized by taste receptors; exteroceptive, such as receptors that detect sound, smell, light, and touch; and proprioceptive, which involve the function of the synapse (Sherrington's word) and respond to events inside the body.

In 1906 Sherrington investigated the scratch reflex of a dog, using an electric 'flea', and found that the reflex stimulated 19 muscles to beat rhythmically five times a second, and brought into action a further 17 muscles which kept the dog upright. The exteroceptive sensors initiated the order to scratch, and the proprioceptors initiated the muscles to keep the animal upright.

Sherrington also carried out significant work in the development of antitoxins, particularly those for cholera and diphtheria.

Shockley William Bradford 1910–1989. US physicist and amateur geneticist who worked with John Bardeen and Walter Brattain on the invention of the transistor. They were jointly awarded a Nobel prize 1956. During the 1970s Shockley was criticized for his claim that blacks were genetically inferior to whites in terms of intelligence.

He donated his sperm to the bank in S California established by the plastic-lens millionaire Robert Graham for the passing-on of the genetic code of geniuses.

Shrapnel Henry 1761–1842. British army officer who invented shells containing bullets, to increase the spread of casualties, first used 1804; hence the word ***shrapnel*** to describe shell fragments.

Shrapnel was born in Bradford-on-Avon, Wiltshire. He received a commission in the Royal Artillery in 1779, and in the following year he went to Newfoundland, returning to England 1783. He served in the Duke of York's unsuccessful campaign against France 1793, being wounded in the siege of Dunkirk. In 1804, he was appointed inspector of artillery at the Royal Arsenal in Woolwich, London. He retired with the rank of lieutenant general. Shrapnel had spent several thousand pounds of his own money in perfecting his inventions. The Treasury eventually granted him a pension of £1,200 a year for life.

Shrapnel's shell was fused and filled with musket balls, plus a small charge of black powder to explode the container after a predetermined period of time. The first shells were round; later they were of an elongated form with added velocity. Shrapnel's shells continued to be used until World War I.

Shull Clifford G 1915– . US physicist who, with Bertram Brockhouse, developed neutron diffraction techniques used for studying the structure and properties of matter. Shull used neutron scattering techniques to answer questions that the similar technique of X-ray diffraction had failed to answer, such as where the atoms of the light element hydrogen are located in an ice crystal. He also showed how neutrons can reveal the magnetic properties of metals and alloys. Nobel Prize for Physics 1994.

One of Shull's discoveries was that the diffraction pattern of a proton is different from that of a deuteron – a proton and a neutron combined. He produced new insights into the hydrogen bond in water by comparing the diffraction patterns produced by water and deuterium oxide. He also showed that magnetic interactions between neutrons and atoms also produced diffraction patterns. He established the existence of antiferromagnetic materials, in which one half of the atoms align magnetically in one direction and half in the other.

Shull was born in Pittsburgh, Pennsylvania, USA. He studied physics at the Carnegie Institute of Technology (now Carnegie-Mellon University), Pittsburgh, and moved to New York University for graduate studies, completing his doctorate in 1941. He became familiar with diffraction and crystallography when working in the research laboratory of The Texas Company in Beacon, New York, 1941–46. In 1946 he moved to the Clinton Laboratory (now Oak Ridge National Laboratory) in Tennessee where he worked on neutron diffraction. In 1955 he moved to the Massachusetts Institue of Technology (MIT), in Cambridge, Massachusetts. He retired from MIT in 1986.

Sidgwick Nevil Vincent 1873–1952. English theoretical chemist who made contributions to the theory of valency and chemical bonding.

Sidgwick was born in Oxford and studied there and in Germany. He spent his entire career at Oxford, becoming professor 1935.

Sidgwick became absorbed by the study of atomic structure and its importance in chemical bonding. He explained the bonding in coordination compounds (complexes), with a convincing account of the significance of the dative bond. Together with his students he demonstrated the existence and wide-ranging importance of the hydrogen bond.

His works include *The Organic Chemistry of Nitrogen* 1910, *The Electronic Theory of*

Valency 1927, *Some Physical Properties of the Covalent Link in Chemistry* 1933, and the definitive *The Chemical Elements and their Compounds* 1950.

Siegbahn Kai 1918– . Swedish physicist who developed the use of electron spectroscopy for chemical analysis. He studied the electrons emitted from substances irradiated by X-rays by focussing a narrow beam of X-rays of a single wavelength onto a specimen and measuring the energy spectrum of the ejected electrons. The spectrum showed peaks and troughs which were characteristic of the atoms in the specimen. He shared the 1981 Nobel Prize for Physics with Nicolaas Bloembergen and Arthur Schawlow.

The position of the peaks in the spectrum were shifted by chemical effects and depended upon the way in which the atoms were linked together to form molecules. By analysing these 'chemical shifts', Siegbahn was able to identify the constituents of the specimen. He also worked on a related technique using ultraviolet light.

Kai Siegbahn was the son of Karl Manne Siegbahn who won the Nobel Prize for Physics 1924.

Siegbahn Karl Manne Georg 1886–1978. Swedish physicist who worked in the field of X-ray spectroscopy – the study of the X-rays produced by bombardment of a sample with electrons or photons. Each element emits X-rays with characteristic wavelengths and these can be used to identify the sample. The X unit or Siegbahn unit, used for expressing wavelengths of X-rays and gamma rays, is named for him. One Siegbahn unit equals 1.00202×10^{-13} m. Nobel Prize for Physics 1924.

He began his studies in X-ray spectroscopy 1914. By gradually refining his equipment and technique, Siegbahn was able to measure the wavelengths of X-rays with unprecedented accuracy. This enabled him to make corrections to the theory of X-ray diffraction and discover several new series of X-ray spectral lines. He also developed the use of the diffraction grating for studying longer wavelengths. In 1925 Siegbahn and his collegues proved that X-rays are bent or refracted as they passed through prisms, in the same way as light, and thus were of the same nature as light.

Siegbahn was educated at the University of Lund, Sweden, obtaining his doctorate 1911.

Siemens Ernst Werner von 1816–1892. German electrical engineer and inventor who discovered the dynamo principle 1867. He organized in 1870 the construction of the Indo-European telegraph system between London and Calcutta via Berlin, Odessa, and Tehran.

Siemens was born near Hanover and studied at a military academy in Berlin. In 1847 he founded, with scientific instrumentmaker Johan Halske (1814–1890), the firm of Siemens-Halske to manufacture and construct telegraph systems.

In addition, he cooperated with his brothers who founded Siemens Brothers in the UK. As scientific consultant to the British government, Siemens helped to design the first cable-laying ship. He also helped to establish scientific standards of measurement.

Siemens's inventions include a process for gold- and silver-plating and a method for providing the wire in a telegraph system with seamless insulation, using gutta-percha.

Siemens introduced the double-T armature and succeeded in connecting the armature, the electromagnetic field, and the external load of an electric generator in a single current. His companies became pioneers in the development of electric traction – making trams, for example – and also electricity-generating stations.

Sikorsky Igor Ivan 1889–1972. Ukrainian-born US engineer who built the first successful helicopter 1939 (commercially produced from 1943). His first biplane flew 1910, and in 1929 he began to construct multi-engined flying boats.

The first helicopter was followed by a whole series of production designs using one, then two, piston engines. During the late 1950s piston engines were replaced by the newly developed gas-turbine engines.

Sikorsky was born in Kiev and was inspired to make a helicopter by the notebooks of Italian Renaissance inventor Leonardo da Vinci. In 1908 he met US aviation pioneer Wilbur Wright in France, and in 1909 Sikorsky began to construct his first helicopter, but had to abandon his attempts until better materials and engines became available.

Instead, he built fixed-wing aeroplanes and began a lifelong practice of taking the controls on the first flight. In 1911 his S-5 aeroplane flew for more than an hour and achieved altitudes of 450 m/1,480 ft. His aeroplanes *Le Grand* and the even larger *Ilia Mourometz* had four engines, an enclosed cabin for crew and passengers, and even a toilet. They became the basis for the four-engined bomber that Russia used during World War I.

Following the revolution, he emigrated to the USA 1918. He founded the Sikorsky Aero Engineering Corporation, which was taken over by the United Aircraft Corporation. Sikorsky continued to work as a designer and engineering manager until 1957. His S-40 American Clipper flying boat 1931 allowed Pan American Airways to develop routes in the Caribbean and South America.

Simon Franz Eugen 1893–1956. German-born British physicist who developed methods of achieving extremely low temperatures (nearly as low as one millionth of a degree above absolute zero). He experimentally established the validity of the third law of thermodynamics. Knighted 1954.

Simon was born in Berlin and studied at several German universities. He became professor at the Technical University in Breslau 1931, but, with the rise to power of the Nazis, emigrated to the UK 1933 and spent the rest of his career at Oxford, as professor from 1945. During World War II, Simon worked on the atomic bomb.

Simon solidified gases by the use of high pressure, and showed that helium could be solidified at a temperature ten times as high as its liquid/gas critical point. He worked on the properties of fluids at high pressure and low temperature, and in 1932 he worked out a cheap and simple method for generating liquid helium.

In the 1930s, Simon developed magnetic cooling to investigate properties of substances below 1K. He then went on to investigate nuclear cooling, showing that the cooling effect is limited by interaction energies.

Simon John 1816–1904. English surgeon and public health reformer who cleaned up the City of London in the 19th century. The eight annual reports that Simon presented to the Corporation of London are the most famous health reports ever written. They embody an incredible record of success, and years later legislation was based on them, culminating in the great Public Health Act of 1875.

Simon was the first sanitary inspector appointed by the Corporation of London 1848. During the eight years that Simon was responsible for the City of London, he completely transformed the general amenities of the area. Cesspools were entirely abolished for rich and poor in the square mile of London, although for a long period afterwards they remained common in the residential areas of the rich. Simon had a great influence on the construction of the sewage system of London. He also arranged with the General Registrar to visit the houses of all those who had died in order to take measures to contain infectious diseases.

Simpson (Cedric) Keith 1907–1985. British forensic scientist, head of department at Guy's Hospital, London, 1962–72. His evidence sent John Haig (an acid-bath murderer) and Neville Heath to the gallows. In 1965 he identified the first 'battered baby' murder in England.

Simpson George Clark 1878–1965. English meteorologist who studied atmospheric electricity, ionization and radioactivity in the atmosphere, and the effect of radiation on the polar ice. KCB 1935.

Simpson was born in Derby and studied at Owens College (later Manchester University). In 1905 he became the first lecturer in meteorology at a British university (Manchester), and began working at the Meteorological Office; he was its director 1920–38. With the outbreak of World War II, he returned from retirement to take charge of Kew Observatory, continuing research into the electrical structure of thunderstorms until 1947.

Simpson travelled widely. In 1902 he visited Lapland to investigate atmospheric electricity. He spent a period inspecting meteorological stations throughout India and Burma; travelled to the Antarctic in 1910 on Captain Scott's last expedition; and visited Mesopotamia during World War I as a meteorological adviser to the British Expeditionary Force. Later he was a member of the Egyptian government's Nile Project Commission.

Simpson's revised form of the Beaufort scale of wind speed was in international use 1926–39.

Simpson concluded that excessive solar radiation would increase the amount of cloud and that the resultant increase in precipitation would lead to enlargement of the polar ice caps, which might explain the ice ages.

Simpson George Gaylord 1902–1984. US palaeontologist who studied the evolution of mammals. He applied population genetics to the subject and to analyse the migrations of animals between continents.

Simpson was born in Chicago; he attended

the University of Colorado and Yale. From 1927 he worked at the American Museum of Natural History, becoming curator 1942. He was professor at Columbia 1945–59, at Harvard 1959–70, and at Arizona from 1967.

Simpson's work in the 1930s concerned early mammals of the Mesozoic era and the Palaeocene and Eocene epochs, which entailed many extensive field trips throughout the Americas and to Asia to study fossil remains. In the 1940s he began applying genetics to mammalian evolution and classification.

He wrote several textbooks, including *The Meaning of Evolution* 1949, *The Major Features of Evolution* 1953, and *The Principles of Animal Taxonomy* 1961, which were influential in establishing the neo-Darwinian theory of evolution.

Simpson James Young 1811–1870. Scottish physician, the first to use ether as an anaesthetic in childbirth 1847, and the discoverer, later the same year, of the anaesthetic properties of chloroform, which he tested by experiments on himself. Baronet 1866.

Simpson was born near Linlithgow and studied at Edinburgh, where he became professor of midwifery 1840. From 1847 he was requested to attend Queen Victoria during her stays in Scotland. By this time he had a thriving private practice and was making pioneering advances in gynaecology; he was eventually appointed physician to Queen Victoria.

Simpson's *Account of a New Anaesthetic Agent* 1847 aroused opposition from Calvinists, who regarded labour pains as God-given. It was Queen Victoria's endorsement of Simpson's use of chloroform during the birth of her seventh child 1853 that made his techniques universally adopted.

Simpson Thomas 1710–1761. English mathematician and writer who devised ***Simpson's rule***, which simplifies the calculation of areas under graphic curves. He also worked out a formula that can be used to find the volume of any solid bounded by a ruled surface and two parallel planes.

Simpson was born in Market Bosworth, Leicestershire, and self-educated. After an eclipse of the Sun, he took up astrology and gained a reputation in the locality for divination. But after he had apparently frightened a girl into having fits by 'raising a devil' from her, he was obliged to flee with his wife to Derby. In 1735 or 1736 he moved to London and worked as a weaver at Spitalfields, teaching mathematics in his spare time. It was there that he published his first mathematical works, which won some acclaim. Soon after 1740 he was elected to the Royal Academy of Stockholm, and in 1743 he was appointed professor of mathematics at the Royal Academy in Woolwich, London.

Simpson's first mathematical work, in 1737, was a treatise on 'fluxions' (calculus). This was followed by *The Nature and Laws of Chance* 1740, *The Doctrine of Annuities and Reversions* 1742, *Mathematical Dissertation on a Variety of Physical and Analytical Subjects* 1743, *A Treatise of Algebra* 1745, *Elements of Geometry* 1747, *Trigonometry, Plane and Spherical* 1748, *Select Exercises in Mathematics* 1752, and *Miscellaneous Tracts on Some Curious Subjects in Mechanics, Physical Astronomy and Special Mathematics* 1757.

I am not a management type. I am an inventor. I am awful at managing established businesses.

Clive Sinclair

Observer June 1985

Sinclair Clive Marles 1940– . British electronics engineer who produced the first widely available pocket calculator, pocket and wristwatch televisions, a series of home computers, and the innovative but commercially disastrous C5 personal transport (a low cyclelike three-wheeled vehicle powered by a washing-machine motor).

Singer Isaac Merrit 1811–1875. US inventor of domestic and industrial sewing machines. Within a few years of opening his first factory 1851, he became the world's largest sewing-machine manufacturer (despite infringing the patent of Elias Howe), and by the late 1860s more than 100,000 Singer sewing machines were in use in the USA alone.

To make his machines available to the widest market, Singer became the first manufacturer to offer hire-purchase terms.

Singer's machines were very reliable and long-lived. So, in order to reduce the supply of second-hand machines, he would break up any old machines taken in part exchange.

Singer was born in Pittstown, New York, and became a machinist. During his early working life he patented a rock-drilling machine and later a metal- and wood-carving one. One day he was asked to carry out some repairs to a Lerow and Blodgett sewing machine. Singer decided he could add many improvements to the design. Eleven days later he produced a new model, which he patented.

Singer used the best of Howe's design and altered some of the other features. The basic mechanism was the same: as the handle turned, the needle paused at a certain point in its stroke so that the shuttle could pass through the loop formed in the cotton. When the needle continued, the threads were tightened, forming a secure stitch.

Sitter Willem de 1872–1934. Dutch astronomer, mathematician, and physicist, who contributed to the birth of modern cosmology. He was influential in English-speaking countries in bringing the relevance of the general theory of relativity to the attention of astronomers.

De Sitter was born in Sneek, Friesland, and studied at the University of Groningen and the Royal Observatory, Cape Town, South Africa. He was professor of theoretical astronomy at the University of Leiden from 1908 as well as director of its observatory from 1919.

In 1911 de Sitter outlined how the motion of the constituent bodies of our Solar System might be expected to deviate from predictions based on Newtonian dynamics if Albert Einstein's special relativity theory were valid. After the publication of Einstein's general theory of relativity 1915, de Sitter expanded his ideas and introduced the 'de Sitter universe' (as distinct from the 'Einstein universe'). His model later formed an element in the theoretical basis for the steady-state hypothesis regarding the creation of the universe. He presented further models of a nonstatic universe: he described both an expanding universe and an oscillating one.

Education is what survives when what has been learnt has been forgotten.

B F Skinner

New Scientist 21 May 1964

Skinner B(urrhus) F(rederic) 1904–1990. US psychologist, a radical behaviourist who rejected mental concepts, seeing the organism as a 'black box' where internal processes are not significant in predicting behaviour. He studied operant conditioning (influencing behaviour patterns by reward or punishment) and held that behaviour is shaped and maintained by its consequences.

He invented the 'Skinner box', an enclosed environment in which the process of learned behaviour can be observed. In it, a rat presses a lever, and learns to repeat the behaviour because it is rewarded by food. Skinner also designed a 'baby box', a controlled, soundproof environment for infants. His own daughter was partially reared in such a box until the age of two.

The real problem is not whether machines think but whether men do.

B F Skinner

Contingencies of Reinforcement, 1969

His radical approach rejected almost all previous psychology; his text *Science and Human Behavior* 1953 contains no references and no bibliography. His other works include *Walden Two* 1948 and *Beyond Freedom and Dignity* 1971. Both these books argue that an ideal society can be attained and maintained only if human behaviour is modified – by means of such techniques as conditioning – to fit society instead of society adapting to the needs of individuals.

Skinner was born in Susquehanna, Pennsylvania, and studied at Harvard. He was professor at Indiana 1945–48 and at Harvard 1948–74.

Skinner attempted to explain even complex human behaviour as a series of conditioned responses to outside stimuli. He opposed the use of punishment, arguing that it did not effectively control behaviour and had unfavourable side effects.

After the Skinner box, he developed a teaching machine. This presents information to a student at a pace determined by the student, and then tests the student on the material previously presented; correct answers are rewarded, thereby reinforcing learning.

Skolem Thoralf Albert 1887–1963. Norwegian mathematician who did important work on Diophantine equations and who helped to provide the axiomatic foundations for set theory in logic.

Skolem was born at Sandsvaer and educated at Oslo, where he became professor 1938. He wrote 182 scientific papers, but they remained largely unread, partly because they were written in Norwegian.

Skolem's main work was in the field of formal mathematical logic. From papers published in the 1920s emerged what is now known as the Löwenheim–Skolem theorem, one consequence of which is ***Skolem's paradox***: if an axiomatic system (such as axiomatic set theory) is consistent (that is, satisfiable), then it must be satisfiable within a countable domain; but Georg Cantor had shown the existence of a neverending sequence of transfinite powers in mathematics (that is, uncountability). Skolem's answer was that there *is* no complete axiomatization of mathematics.

Before such subjects as model theory, recursive function theory, and axiomatic set theory had become separate branches of mathematics, he introduced a number of the fundamental notions that gave rise to them.

Slater Samuel 1768–1835. British-born US industrialist whose knowledge of industrial technology and business acumen as a mill owner and banker made him a central figure in the New England textile industry. At first working for American firms, he established his own manufacturing company 1798.

Slater was born in England. He trained in machinery manufacture and worked as a mechanical supervisor in a textile mill. Emigrating to the USA 1789, he was quickly hired by several American machine manufacturing firms. In 1791 he supervised the construction of a mill, based on the design of the English model, in Providence, Rhode Island.

Slayton Deke (Donald Kent) 1924–1993. US astronaut, one of the original seven chosen by NASA in 1959 for the Mercury series of flights. Grounded for health reasons, he became director of NASA's flight-crew operations for ten years. In 1972 he was returned to flight status, and in 1975 he finally flew in space, then the oldest person to have done so at the age of 51.

Although chosen for the second orbital mission, he was grounded because of a minor heart irregularity. As NASA's flight-crew director, he was responsible for astronaut training and the selection of crews for various flights, including the Apollo Moon missions. He finally flew himself on the Apollo–Soyuz joint mission, in which a US Apollo docked in orbit with a Soviet Soyuz.

He subsequently left NASA to join a private company, Space Services Incorporated, which unsuccessfully attempted to develop rockets for a commercial launch service.

Slipher Vesto Melvin 1875–1969. US astronomer who established that spiral nebulae lie beyond our Galaxy. He also discovered the existence of particles of matter in interstellar space. His work in spectroscopy increased our knowledge of planetary and nebular rotation, planetary and stellar atmospheres, and diffuse and spiral nebulae.

Slipher was born in Mulberry, Indiana, attended Indiana University, and in 1902 joined the Lowell Observatory in Arizona. He was director of the observatory 1926–52.

Slipher measured the period of rotation for Venus, Mars, Jupiter, Saturn, and Uranus. His work on Jupiter first showed the existence of bands in the planet's spectrum, and he and his colleagues were able to identify the bands as belonging to metallic elements, including iron and copper. He also showed that the diffuse nebula of the Pleiades had a spectrum similar to that of the stars surrounding it and concluded that the nebula's brightness was the result of light reflected from the stars.

Slipher's measurements of the radial velocities of spiral nebulae 1912–25 suggested that they must be external to our Galaxy. This paved the way for an understanding of the motion of galaxies and for cosmological theories that explained the expansion of the universe.

Sloane Hans 1660–1753. British physician, born in County Down, Ireland. He settled in London, and in 1721 founded the Chelsea Physic Garden. He was president of the Royal College of Physicians 1719–35, and in 1727 succeeded the physicist Isaac Newton as president of the Royal Society. His library, which he bequeathed to the nation, formed the nucleus of the British Museum. 1st baronet 1716.

Smalley Richard E 1943– . US chemist who, with colleagues Robert Curl and Harold Kroto, discovered buckminsterfullerene (carbon 60) 1985. Smalley also pioneered the technique used to discover buckminsterfullerene, supersonic jet laser-beam spectroscopy. Smalley, Curl and Kroto shared the Nobel prize for Chemistry in 1996.

Smeaton John 1724–1792. England's first civil engineer. He rebuilt the Eddystone lighthouse in the English Channel 1756–59, having rediscovered high-quality cement, unknown since Roman times.

Smeaton adopted the term 'civil engineer' in contradistinction to the fast-growing number of engineers graduating from military colleges. He was also a consultant in the field of structural engineering, and from 1757 onwards he was responsible for projects including bridges, power stations operated by water or wind, steam engines, and river and harbour facilities.

Smeaton was born near Leeds and qualified as a lawyer, but then became a maker of scientific instruments.

Smeaton's research led to the abandonment of the established undershot water wheel (which operates through the action of the flow of water against blades in the wheel) in favour of the overshot wheel (which is operated by water moving the wheel by the force of its weight). Experimenting with models, Smeaton showed that overshot wheels were twice as efficient as undershot ones.

Smeaton performed extensive tests on the experimental steam engine of Thomas Newcomen, which led to improvements in its design and efficiency.

Smith Francis Graham 1923– . English radio astronomer who with his colleague Martin Ryle mapped the radio sources in the sky in the 1950s. Smith discovered the strongly polarized nature of radiation from pulsars 1968, and estimated the strength of the magnetic field in interstellar space. He was Astronomer Royal 1982–90. Knighted 1986.

Smith was born in Roehampton, Surrey, and studied at Cambridge. During World War II he was assigned to the Telecommunications Research Establishment, and when he returned to Cambridge, he joined the radio research department at the Cavendish Laboratory. He was appointed professor of astronomy at Manchester 1964 and worked at Jodrell Bank until 1974. He was director of the Royal Greenwich Observatory 1976–81. In 1981, he moved back to Jodrell Bank to become director there.

In 1948, Smith and Ryle, investigating a source of radio waves in the constellation of Cygnus, detected a second source in the constellation Cassiopeia. Smith spent the following years trying to determine the precise location of both sources. Finally, astronomers at Mount Palomar, California, were able to pinpoint optical counterparts. Cassiopeia A was shown to derive from a supernova explosion within our Galaxy; Cygnus A is a double radio galaxy.

Smith and Ryle were the first to publish (in 1957) a paper on the possibility of devising an accurate navigational system that depended on the use of radio signals from an orbiting satellite.

In 1962 Smith installed a radio receiver in *Aeriel II*, one of a series of joint US–UK satellites, enabling it to make the first investigation of radio noise above the ionosphere.

Smith Grafton Elliot 1871–1937. Australian-born anatomist and anthropologist who combined his interests in human anatomy and anthropology in order to conduct research into the mummification of the ancient Egyptians.

Smith was born in Grafton, New South Wales and began his training as a doctor at Sydney University Medical School 1888. He graduated 1892 and following a brief period of clinical practice began to concentrate upon his research 1894, completing his MD thesis on the brains of non-placental mammals (monotremes) 1895. He then came to England to study under Alexander MacAlister at Cambridge, and in 1900 he was appointed professor of anatomy in the New Medical School in Cairo.

While he was in Egypt he began to undertake an archaeological survey of the Nubia combining his interests in anatomy and anthropology by studying mummification and palaeopathology. In 1909 he became professor of anatomy at Manchester University and was appointed to a similar post in University College London 1919. In 1933 he became the Fullerian professor of physiology at the Royal Institute and despite suffering a stroke 1932 he remained in this position until his retirement 1936. He was knighted 1934.

Smith Hamilton Othanel 1931– . US microbiologist who shared the 1978 Nobel Prize for Physiology or Medicine with his colleague Daniel Nathans for their work on restriction enzymes, special enzymes that can cleave genes into fragments.

Smith was born in New York and studied at the University of California, Berkeley, and at Johns Hopkins University. In 1964 he returned to Johns Hopkins, becoming professor 1973.

Werner Arber (1929–), a Swiss microbiologist, discovered restriction enzymes in the 1960s. Smith, working independently of Arber, verified Arber's findings and was also able to identify the gene fragments. Smith collaborated with Nathans on some of this work.

As a result of the work of Nathans, Smith, and Arber (who also shared the Nobel prize), it is now possible to determine the chemical formulas of the genes in animal viruses, to map these genes, and to study the organization and expression of genes in higher animals.

Smith James Edward 1759–1828. English botanist and owner of Linnaeus's library and collection of specimens who founded the Linnaean Society. He studied Linnaeus's collections and translated some of his work.

Born in Norwich, Smith's father encouraged him to study medicine, despite his preference for botany. In 1781, he went to Edinburgh University, and in 1783, he went to London to study anatomy. In London he met Joseph Banks and through him was able to purchase Linnaeus's specimens, herbarium, library, and manuscripts for £1,000. He studied the collection, reorganized it, and translated some of Linnaeus's work, including *Reflections on the Study of Nature* 1785, *Dissertation on the Sexes of Plants* 1786, and *Flora Laponica*.

In 1821, he published a collection of Linnaeus's correspondence and was made a Fellow of the Royal Society 1785 and knighted 1818. He instructed the queen and princesses in botany. In 1788, he founded the Linnaean Society. Smith was the first president of the Society until his death 1828, when the society was annoyed to have to pay £3,000 for the Linnaean Collection.

Smith John Maynard, British biologist, see ◊Maynard Smith.

Smith Keith Macpherson 1890–1955 and Ross Macpherson Smith 1892–1922. Australian aviators and brothers who made the first England–Australia flight 1919. Both KBE 1919.

Smith Michael 1932– . British molecular biologist who received the Nobel Prize for Chemistry in 1993 for his technique ***site-specific mutagenesis***, a technique that replaced the way scientists established the function of a particular protein or gene, by using single strands of viral DNA to mutate the genetic code at precise locations.

Before Smith's technique, it was difficult to pinpoint the effects of a single mutant gene, because a mutation had to be induced in the gene, and available mutagens (substances that cause mutations) – chemicals or radiation – could not be controlled. Random and multiple mutations were inevitable. Site-specific mutagenesis uses viral DNA as the mutagen.

Smith synthesized a segment of complementary DNA that differed at a single site, and allowed it to bind to the original viral DNA. In the virus, the mutated genes produced mutated proteins. When these were compared with normal proteins in the virus, the impact of the mutation, and therefore the role of the normal gene, became obvious. The technique has since been used to synthesize a range of useful new proteins.

Born in Blackpool, Smith was educated at Manchester University, where he graduated in biochemistry. After obtaining his PhD in 1956, he flew to Vancouver, Canada, where he worked as a post-doctoral fellow at the University of British Columbia. In 1961, he was offered a position on the Fisheries Research Board of Canada, but went back to work at the university in 1966. A popular man, he was later appointed professor of biochemistry, and subsequently director of the biotechnology laboratory.

Smith Theobald 1859–1934. US microbiologist and immunologist who demonstrated the involvement of a bacillus (a rod-shaped bacteria) in the onset of tuberculosis (TB), a respiratory disease.

Continuing the previous work of French physician Jean Louis Villemin, Smith made a careful morphological comparison 1896 of tuberculosis bacilli extracted from humans and cows and showed that there were differences. He also demonstrated that they show marked differences in virulence for the ox and rabbit. He proved that there are two distinct types of mammalian tubercle bacillus, one human and the other bovine. It was later proved that the bovine bacillus is transmissible to humans.

Smith also worked on vaccines for smallpox and cholera, as well as diphtheria antitoxins (antibodies). He noted that guinea pigs used for the testing of diphtheria antitoxin became acutely ill if test injections were separated by long intervals. This reaction was later discovered to be caused by hypersensitiveness, or anaphylaxis, by Charles Richet. He also went on to devise ultrasensitive methods for the detection of certain types of bacteria in drinking water, milk, and sewage.

Smith was born in Albany, New York, where he trained as a doctor at Albany Medical College. He went on to become an eminent bacteriologist with research laboratories at Harvard Medical School and the Rockefeller Institute in New York.

Smith William 1769–1839. English geologist who produced the first geological maps of England and Wales, setting the pattern for geological cartography. Often called the founder of stratigraphical geology, he determined the succession of English strata across the whole country, from the Carboniferous up to the Cretaceous. He also established their fossil specimens.

Working as a canal engineer, he observed while supervising excavations that different beds of rock could be identified by their fossils, and so established the basis of stratigraphy.

Smith was born in Churchill, Oxfordshire, and became a drainage expert, a canal surveyor, and a mining prospector.

Smith was not the first geologist to recognize the principles of stratigraphy nor the usefulness of type fossils. His primary accomplishment lay in mapping. He began in 1799 but it was not until 1815 that he published *A Delineation of the Strata of England and Wales*, a geological map using a scale of five miles to the inch. Between 1816 and 1824 he published *Strata Identified by Organized Fossils* and *Stratigraphical System of Organized Fossils*, a descriptive catalogue. He also issued various charts and sections, and geological maps of 21 counties.

Smithson James Louis Macie 1765–1829. British chemist and mineralogist whose bequest of $100,000 led to the establishment of the Smithsonian Institution.

Snell George Davies 1903–1996. US geneticist who shared the Nobel Prize for Physiology or Medicine 1980 with Jean Dausset and Baruj Benacerraf for their work to identify histocompatability genes, genes that control the acceptance or rejection of tissue and organ transplants, in mice.

Snell was born in Bradford, Massachusetts and trained at the Harvard University and then became one of the founding members of the Jackson Laboratory in Bar Harbor, Maine 1929. The Jackson Laboratory is a nonprofitmaking, independent laboratory for the investigation of the causes of cancer and other diseases through mouse research.

Snell Willebrord van Roijen 1581–1626. Dutch mathematician and physicist who devised the basic law of refraction, known as ***Snell's law*** 1621.This states that the ratio between the sine of the angle of incidence and the sine of the angle of refraction is constant. He also founded the method of determining distances by triangulation.

Snell was born in Leiden, where he studied and eventually became professor.

Snell developed the method of triangulation in 1615, starting with his house and the spires of nearby churches as reference points. He used a large quadrant over 2 m/7 ft long to determine angles, and by building up a network of triangles, was able to obtain a value for the distance between two towns on the same meridian. From this, Snell made an accurate determination of the radius of the Earth.

The laws describing the reflection of light were well known in antiquity, but the principles governing the refraction of light were little understood. Snell's law was published by French mathematician Descartes in 1637. He expressed the law differently from Snell, but could easily have derived it from Snell's original formulation. Whether Descartes knew of Snell's work or discovered the law independently is not known.

Snyder Solomon Halbert 1938– . US pharmacologist and neuroscientist who has studied the chemistry of the brain, and co-discovered the receptor mechanism for the body's own opiates, the encephalins.

Snyder was born in Washington DC and studied at Georgetown University. From 1965 he worked at the Johns Hopkins Medical School, becoming professor 1970.

In the early 1970s, in collaboration with his research student Candace Pert (1946–), Snyder realized that the very specific effects of synthetic opiates given in small doses suggested that they must bind to highly selective target receptor sites. Using radioactively labelled compounds, they located such receptors in specialized areas of the mammalian brain and from this finding deduced that there might be natural opiate-like substances in the brain that used these sites. These chemicals, the encephalins, were discovered by others shortly afterwards.

Continuing to examine the relationships of chemicals to neural functioning, Snyder has made a particular study of the receptor sites for the benzodiazepine drugs, which are widely used in psychiatry.

Soddy Frederick 1877–1956. English physical chemist who pioneered research into atomic disintegration and coined the term isotope. He was awarded a Nobel prize 1921 for investigating the origin and nature of isotopes.

The displacement law, introduced by Soddy in 1913, explains the changes in atomic mass and atomic number for all the radioactive intermediates in the decay processes.

After his chemical discoveries, Soddy spent some 40 years developing a theory of 'energy economics', which he called 'Cartesian economics'. He argued for the abolition of debt and compound interest, the nationalization of credit, and a new theory of value based on the quantity of energy contained in a thing.

Soddy was born in Eastbourne, Sussex, and studied at Oxford. He worked 1900–02 with physicist Ernest Rutherford at McGill University in Montréal, Canada. Soddy was professor at Aberdeen 1914–19 and at Oxford 1919–36. He opposed military use of atomic power.

Soddy and Rutherford postulated that radioactive decay is an atomic or subatomic process, and formulated a disintegration law. They also predicted that helium should be a decay product of radium, a fact that Soddy proved spectrographically in 1903.

His works include *Chemistry of the Radio-Elements* 1912–14, *The Interpretation of the Atom* 1932, and *The Story of Atomic Energy* 1949.

Solander Daniel Carl 1736–1782. Swedish botanist. In 1768, as assistant to Joseph Banks, he accompanied the explorer James Cook on his first voyage to the S Pacific, during which he made extensive collections of plants.

Solander was born in Norrland and studied under Swedish botanist Linnaeus. In 1771 he became secretary and librarian to Banks and in 1773 became keeper of the natural-history department of the British Museum. Named after him are a genus of Australian plants and a cape at the entrance to Botany Bay.

Solvay Ernest 1838–1922. Belgian industrial chemist who in the 1860s invented the ammonia-soda process, also known as the ***Solvay process***, for making the alkali sodium carbonate. It is a multistage process in which carbon dioxide is generated from limestone and passed through brine saturated with ammonia. Sodium hydrogen carbonate is isolated and heated to yield sodium carbonate. All intermediate by-products are recycled so that the only ultimate by-product is calcium chloride.

Solvay was born near Brussels and had little formal education, but carried out chemical experiments in a small home laboratory. In 1860 he went to work at a gasworks, and there learned about the industrial handling of ammonia both as a gas and as an aqueous solution. Within a year he had discovered and patented the reactions that are the basis of the Solvay process. Trial production failed; for two years Solvay knew he had the chemistry right, but could not solve the considerable problems of chemical engineering. He built his first factories 1863 and 1873.

Solvay soon realized that there was more money to be made from granting licences to other manufacturers than there was in making soda. Throughout the world, the Solvay process replaced the old Leblanc process, which was more expensive and polluting. Solvay became a very rich man and entered politics, becoming a member of the Belgian senate and a minister of state.

Somerville Mary, (born Fairfax) 1780–1872. Scottish scientific writer who produced several widely used textbooks, despite having just one year of formal education. Somerville College, Oxford, is named after her.

Her main works were *Mechanism of the Heavens* 1831 (a translation of French astronomer Pierre Laplace's treatise on celestial mechanics), *On the Connexion of Physical Sciences* 1834, *Physical Geography* 1848, and *On Molecular and Microscopic Science* 1869.

Somerville *Scottish mathematician and astronomer Mary Somerville, who published several scientific expository works and helped popularize science in the early 19th century. She strongly supported the emancipation and education of women, and Somerville College at Oxford is named after her.*

Sommeiller *French engineer Germain Sommeiller designed this drilling machine which was powered by compressed air, the prototype pneumatic drill. It was used during the construction of the Mont Cenis tunnel, which links France and Switzerland, in the latter half of the nineteenth century.*

Sommeiller Germain 1815–1871. French engineer who built the Mont Cenis Tunnel, 12 km/7 mi long, between Switzerland and France. The tunnel was drilled with his invention the pneumatic drill.

Sommerfeld Arnold Johannes Wilhelm 1868–1951. German physicist who demonstrated that difficulties with Niels Bohr's model of the atom, in which electrons move around a central nucleus in circular orbits, could be overcome by supposing that electrons adopt elliptical orbits.

This led him in 1916 to predict a series of spectral lines based on the relativistic effects that would occur with elliptical orbits. Friedrich Paschen (1865–1945) undertook the spectroscopic work required and confirmed Sommerfeld's predictions.

Sommerfeld was born and educated in Königsberg, Prussia (now Kaliningrad, Russia). He was professor at the Mining Academy in Clausthal 1897–1900, at the Technical Institute in Aachen 1900–06, and then moved to Munich University as director of the Institute of Theoretical Physics, specially established for him. Sommerfeld built his institute into a leading centre of physics.

Sommerfeld was active as a theoretician both in physics and engineering, and he produced a four-volume work on the theory of gyroscopes 1897–1910 with mathematician Felix Klein. Sommerfeld's other works include *Atombau und Spektrallinien/Atomic Structure and Spectral Lines* 1919 and *Wellenmechanischer Ergänzungsband/Wave Mechanics* 1929.

Sommerville Duncan MacLaren Young 1879–1934. British mathematician who made significant contributions to the study of non-Euclidean geometry.

Sommerville was born in Beawar, Rajasthan, India, and educated in Scotland, graduating from St Andrews. In 1915 he emigrated to New Zealand as professor at Victoria University College, Wellington.

Sommerville explained how non-Euclidean geometries arise from the use of alternatives to Euclid's postulate of parallels, and showed that both Euclidean and non-Euclidean geometries – such as hyperbolic and elliptic geometries – can be considered as sub-geometries of projective geometry. He stated that projective geometry is the invariant theory associated with the group of linear fractional transformations. He studied the tessellations of Euclidean and non-Euclidean space and showed that, although there are only three regular tessellations in the Euclidean plane, there are five congruent regular polygons of the same kind in the elliptical plane and an infinite number of such patterns in the hyperbolic plane. The variety is even greater if 'semi-regular' networks of regular polygons of different kinds are allowed (because the regular patterns are topologically equivalent to the irregular designs). In his later work on *n*-dimensional geometry, Sommerville generalized his earlier analysis to include 'honeycombs' of polyhedra in three-dimensional spaces and of polytopes in spaces of 4, 5, ..., *n* dimensions – including both Euclidean and non-Euclidean geometries.

Sommerville's interest in crystallography played a significant part in motivating him to investigate repetitive space-filling geometric patterns.

Sorby Henry Clifton 1826–1908. English geologist whose discovery 1863 of the crystalline nature of steel led to the study of metallography. Thin-slicing of hard minerals enabled him to study the constituent minerals microscopically in transmitted light. He later employed the same techniques in the study of iron and steel under stress.

Sorby was born near Sheffield, privately educated, and spent his life as an independent scientific researcher.

In addition to microscopic study, Sorby used a Nicol prism to distinguish the different component minerals by the effect they produced on polarized light.

Sorby also extrapolated from laboratory models and small-scale natural processes in the expectation of explaining vast events in geological history.

He published *On the Microscopical Structure of Crystals* 1858.

Sørensen Søren Peter Lauritz 1868–1939. Danish chemist who in 1909 introduced the concept of using the pH scale as a measure of the acidity of a solution. On Sørensen's scale, still used today, a pH of 7 is neutral; higher numbers represent alkalinity, and lower numbers acidity.

South James 1785–1867. English astronomer who published two catalogues of double stars 1824 and 1826, the former with John Herschel. For this catalogue, they were awarded the Gold Medal of the Astronomical Society 1826. Knighted 1830.

South was born in London. He studied medicine and surgery, but form the age of 31 he devoted his life to astronomy. His marriage in 1816 made him wealthy enough to establish observatories in London and in Paris and equip them with the best telescopes available.

South had an argumentative temperament, and his public criticism of the Royal Society for participating in the decline of the sciences in Britain offended other Fellows.

Spallanzani Lazzaro 1729–1799. Italian biologist. He disproved the theory that microbes spontaneously generate out of rotten food by showing that they would not grow in flasks of broth that had been boiled for 30 minutes and then sealed.

Spallanzani also concluded that the fundamental factor in digestion is the solvent property of gastric juice – a term first used by him. He studied respiration, proving that tissues use oxygen and give off carbon dioxide.

Spallanzani was born in Scandiano, Emilia-Romagna, and studied at Bologna. He was professor at Reggio College 1754–60, at Modena University 1760–69, and from 1769 at Pavia. He was also a priest.

In 1771, while examining a chick embryo, Spallanzani discovered vascular connections between arteries and veins – the first time this had been observed in a warm-blooded animal. He studied the effects of growth on the circulation in chick embryos and tadpoles, and showed that the arterial pulse is caused by sideways pressure on the expansile artery walls from heartbeats transmitted by the bloodstream.

Spallanzani also studied the migration of swallows and eels, the flight of bats, and the electric discharge of torpedo fish. In addition to his biological work, Spallanzani pioneered the science of vulcanology.

Spemann Hans 1869–1941. German embryologist who discovered the phenomenon of embryonic induction – the influence exerted by various regions of an embryo that controls the development of cells into specific organs and tissues. Nobel Prize for Physiology or Medicine 1935.

Spemann was born in Stuttgart and studied at Heidelberg and Würtzburg. In 1908 he was appointed professor at Rostock.

He was director of the Kaiser Wilhelm Institute of Biology in Berlin 1914–19, and professor at Freiburg-im-Breisgau 1919–35.

Spemann carried out his research on newt embryos. He found that embryos split in half at an early stage of development either died or developed into a whole embryo, but if they were divided at a later stage, half-embryos formed. Next he transplanted various embryonic parts to other areas of the embryo and to different embryos, and demonstrated that one area of embryonic tissue influences the development of neighbouring tissues.

In another series of experiments, Spemann found that embryonic tissue from newts always gives rise to newt organs, even when transplanted into a frog embryo, and that frog tissue always develops into frog organs in a newt embryo.

In the course of this work, Spemann pioneered techniques of microsurgery.

Spencer Jones Harold 1890–1960. English astronomer who made a determination of solar parallax, using observations of the asteroid Eros. He also studied the speed of rotation of the Earth, and the motions of the Sun, Moon, and planets. Knighted 1943.

Spencer Jones was born in London and studied at Cambridge. He worked at the Royal Observatory, Greenwich, 1913–23; was His Majesty's Astronomer on the Cape of Good Hope, South Africa, 1923–33; and ended his career as the tenth Astronomer Royal 1933–55.

While at the Cape of Good Hope, Spencer Jones published a catalogue containing the radial velocities of the southern stars, calculated the orbits of a number of spectroscopic binary stars, and made a spectroscopic determination of the constant of aberration. In 1925, he obtained and described a long series of spectra of a nova which had appeared in the constellation of Pictor.

Spencer Jones proved that fluctuations in the observed longitudes of the Sun, Moon, and planets are due not to any peculiarities in their motion, but to fluctuations in the angular velocity of rotation of the Earth. He also investigated the Earth's magnetism and oblateness (that its equatorial diameter is greater than its polar one), and he estimated the mass of the Moon.

Sperry Elmer Ambrose 1860–1930. US engineer who developed various devices using gyroscopes, such as gyrostabilizers (for ships and torpedoes) and gyro-controlled autopilots.

The first gyrostabilizers dated from 1912, and during World War I Sperry designed a pilotless aircraft that could carry up to 450 kg/990 lb of explosives a distance of 160 km/100 mi under gyroscopic control – the first flying bomb. By the mid-1930s ***Sperry autopilots*** were standard equipment on most large ships.

Sperry was born in Cortland County, New York. He set up his own research and development enterprise 1888, then formed a mining-machinery company and, moving to Cleveland, Ohio, in 1893, developed and manufactured trams. He was president of the Sperry Gyroscope Company until 1926.

Sperry's first invention was a generator with characteristics suited to arc lighting. He produced a superior storage battery (accumulator), teaching himself chemistry in the process. He was also active in internal-combustion engine research and developed ways of detecting substandard railway track.

Sperry's gyrostabilizer, patented 1908, mounted the gyro with its axis vertical in the hold of the ship. The axis was free to move in a fore and aft direction, but not from side to side. He used an electric motor to precess the gyro (tilt its rotor) artificially just as the ship began to roll. The gyro responded by exerting a force to one side or the other. Since it was fixed rigidly to the ship in the plane, the ship's roll was largely counteracted.

Sperry introduced many of the concepts now common in control theory, cybernetics, and automation.

Sperry Roger Wolcott 1913–1994. US neurologist who shared the Nobel Prize for Physiology or Medicine with David Hubel and Torsten Wiesel 1981 for his work to elucidate the functions of different parts of the human brain.

Sperry investigated how the nerves in the embryo connect with one another to form the correct synapses (the links between nerve cells which transmit nerve impulses) at the right time. In 1954, he started work on split brains (brains in which the connection between the right hemisphere and the left hemisphere is severed) in animal models in an attempt to understand how different sides of the brain are specialized to perform different functions (cerebral specialization).

He also examined the effect of split brains on behaviour and bodily functions in humans and was the first to show that the left side of the brain is mainly involved in verbal and written skills, logic, and powers of deduction; whereas the right side is more active in visual recognition and artistic expression. His work forms the basis of human developmental neurobiology and psychobiology.

Sperry was born in Hartford, Connecticut and studied psychology at Oberlin College and zoology at the University of Chicago. He worked at Harvard University and Chicago University before moving to the California Institute of Technology 1952 as a professor.

We feel and know that we are eternal.

Benedict Spinoza

Ethics

Spinoza Benedict, or Baruch 1632–1677. Dutch philosopher who believed in a rationalistic pantheism that owed much to Descartes's mathematical appreciation of the universe. Mind and matter are two modes of an infinite substance that he called God or Nature, good and evil being relative. He was a determinist, believing that human action was motivated by self-preservation.

Ethics 1677 is his main work. *A Treatise on Religious and Political Philosophy* 1670 was the only one of his works published during his life, and was attacked by Christians. He was excommunicated by the Jewish community in Amsterdam on charges of heretical thought and practice 1656. He was a lens-grinder by trade.

Spitzer Lyman 1914–1997. US astrophysicist who developed influential theories about the formation of stars and planetary systems.

Spitzer was born in Toledo, Ohio, and studied at Yale, with a year in the UK at Cambridge. He stayed at Yale until 1947, when he moved to Princeton as head of the astronomy department.

Spitzer proposed that only a magnetic field could contain gases at temperatures as high as

100 million degrees, by which point hydrogen gas fuses to form helium, and he devised a figure-of-eight design to describe this field. His model was important to later attempts to bring about the controlled fusion of hydrogen.

Spitzer criticized the theory that our planetary system is the result of a gas cloud or gaseous filaments breaking off from the Sun to become planetary fragments. He showed that a gas would be dispersed into interstellar space long before it had cooled sufficiently to condense into planets.

The more different people have studied different methods of bringing up children the more they have come to the conclusion that what good mothers and fathers instinctively feel like doing for their babies is best after all.

Benjamin Spock

The Common Sense Book of Baby and Child Care

Spock Benjamin McLane 1903– . US paediatrician and writer on child care. His *Common Sense Book of Baby and Child Care* 1946 urged less rigidity in bringing up children than had been advised by previous generations of writers on the subject, but this was misunderstood as advocating permissiveness. He was also active in the peace movement, especially during the Vietnam War.

In his later work he stressed that his common-sense approach had not implied rejecting all discipline, but that his main aim was to give parents the confidence to trust their own judgement rather than rely on books by experts who did not know a particular child.

Sprengel Christian Konrad 1750–1816. German botanist. Writing in 1793, he described the phenomenon of dichogamy, the process whereby stigma and anthers on the same flower ripen at different times and so guarantee cross-fertilization.

Spruce Richard 1817–1893. English botanist who travelled widely in South America. He studied bryophytes (mosses and liverworts) collecting many new specimens in the Pyrenees.

In 1849, he went to South America, where he spent 15 years studying the Amazon valley. During this time he sent 7,000 specimens back to England. In 1860, he was commissioned by the British government to find Cinchona plants in Ecuador that would be suitable for cultivation in India. These plants were used to produce the quinine required for the treatment of malaria. After exploring the coast of Ecuador and Peru he, retired to Yorkshire 1864, and published *Palmae Amazonicae* and *Hepaticae Amazonicae et Andinae*.

Spruce was born in Ganthorpe near Malton, England. He followed his father into teaching, becoming a schoolmaster first at Haxby and then at the Collegiate School of York. His hobby, however, was botany, and he published several papers on bryophytes (mosses and liverworts) and the local flora. When the school was closed 1844, he took up botany as a career. In 1845 and 1846, he travelled in the Pyrenees, collecting large numbers of specimens of bryophytes that had been previously unknown in the area.

Stahl Franklin William 1929– . US molecular biologist who, with Matthew Meselson, confirmed that replication of the genetic material DNA is semiconservative (that is, the daughter cells each receive one strand of DNA from the original parent cell and one newly replicated strand).

Stahl was born in Boston, Massachusetts, and studied at Harvard and Rochester. In 1970 he became professor at the University of Oregon and a research associate of the Institute of Molecular Biology.

Working at the California Institute of Technology, Meselson and Stahl began experimenting with viruses in 1957, and then carried out their successful experiment with bacteria to prove the semiconservative nature of DNA replication. The concept was first suggested by Francis Crick and James Watson, who pioneered the study of DNA.

In 1961, working with scientists Sidney Brenner and François Jacob, Meselson and Stahl demonstrated that ribosomes require instructions in order to be able to manufacture proteins, and can make different proteins from those normally produced by a particular cell. They also showed that messenger RNA supplies the instructions to the ribosomes.

Stahl has also researched into the genetics of bacteriophages.

Stahl Georg Ernst 1660–1734. German chemist who developed the theory that objects burn because they contain a combustible substance, phlogiston. Substances rich in phlogiston, such as wood, burn almost completely away. Metals, which are low in phlogiston, burn less well. Chemists spent much of the 18th century evaluating Stahl's theories before these were finally proved false by Antoine Lavoisier.

Stahl was born in the principality of Ansbach and studied medicine at Jena. He became a physician to the duke of Saxe-Weimar in 1687. As professor of medicine at Halle 1694–1716, he also lectured in chemistry. From 1716 he was physician to King Frederick I of Prussia.

The phlogiston theory was the first attempt at a rational explanation for combustion and what we would term oxidation. Doubt crept in only when chemists began weighing the products of such reactions, which should always have been lighter, having lost phlogiston. Stahl accounted for observations to the contrary by suggesting that phlogiston was weightless or could even have negative weight.

Stanley Wendell Meredith 1904–1971. US biochemist who crystallized the tobacco mosaic virus (TMV) in 1935. He demonstrated that, despite its crystalline state, TMV remained infectious. Together with John Northrop and James Sumner, Stanley received the 1946 Nobel Prize for Chemistry.

Stark Johannes 1874–1957. German physicist. In 1902 he predicted, correctly, that high-velocity rays of positive ions (canal rays) would demonstrate the Doppler effect, and in 1913 showed that a strong electric field can alter the wavelength of light emitted by atoms (the ***Stark effect***). Nobel Prize for Physics 1919.

Stark modified in 1913 the photo-equivalence law proposed by Albert Einstein in 1906. Now called the ***Stark–Einstein law***, it states that each molecule involved in a photochemical reaction absorbs only one quantum of the radiation that causes the reaction.

Stark was born in Schickenhof, Bavaria, and studied at Munich.

He became professor 1906 at the Technische Hochschule in Hanover and subsequently held other academic posts until 1922, when he attempted to set up a porcelain factory. This scheme failed, largely because of the depressed state of the German economy. Stark joined the Nazi party 1930 and three years later became president of the Reich Physical-Technical Institute and also president of the German Research Association. But his attempts to become an important influence in German physics brought him into conflict with the authorities and he was forced to resign in 1939. After World War II, he was sentenced to four years' imprisonment by a denazification court 1947.

Starling Ernest Henry 1866–1927. English physiologist who, with William Bayliss, discovered secretin and in 1905 coined the word 'hormone'. He formulated ***Starling's law***, which states that the force of the heart's contraction is a function of the length of the muscle fibres. He is considered one of the founders of endocrinology.

Starling and Bayliss researched the nervous mechanisms that control the activities of the organs of the chest and abdomen, and together they discovered the peristaltic wave in the intestine. In 1902 they found the hormone secretin, which is produced by the small intestine of vertebrates and stimulates the pancreas to secrete its digestive juices when acid chyme passes from the stomach into the duodenum. It was the first time that a specific chemical substance had been seen to act as a stimulus for an organ at a distance from its site of origin.

Starling was born and educated in London, and was professor of physiology at University College, London, 1899–1923.

Starling also studied the conditions that cause fluids to leave blood vessels and enter the tissues. In 1896 he demonstrated the ***Starling equilibrium***: the balance between hydrostatic pressure, causing fluids to flow out of the capillary membrane, and osmotic pressure, causing the fluids to be absorbed from the tissues into the capillary.

Stas Jean Servais 1813–1891. Belgian analytical chemist who made the first accurate determinations of atomic weights (relative atomic masses).

Stas was born in Louvain and studied medicine there. He went to Paris 1837 as assistant to French chemist Jean Baptiste Dumas. From 1840 to 1869 Stas was professor at the Ecole Royale Militaire in Brussels, and he advised the Belgian government on military topics. He was also commissioner of the Mint, but disagreed with the monetary policy of the government and retired 1872. He was openly critical of the part played by the Christian church in education.

While he was working with Dumas in Paris, Stas helped to redetermine the atomic weights of oxygen and carbon. In the 1850s and 1860s, Stas measured the atomic weights of many elements, using oxygen = 16 as a standard. His results discredited English physicist William Prout's hypothesis that all atomic weights are whole numbers, provided the foundation for the work of Dmitri Mendeleyev and others on the periodic system, and remained the standard of accuracy for 50 years.

Staudinger Hermann 1881–1965. German organic chemist, founder of macromolecular chemistry, who carried out pioneering research into the structure of albumen and cellulose. Nobel prize 1953.

To measure the high molecular weights of polymers he devised a relationship, now known as ***Staudinger's law***, between the viscosity of polymer solutions and their molecular weight.

Staudinger was born in Worms, Hesse, and studied at a number of German universities. He became professor 1908 at the Technische Hochschule in Karlsruhe, moved to Zürich, Switzerland, 1912, and from 1926 was at the University of Freiburg-im-Breisgau, where in 1940 he was made director of the Chemical Laboratory and Research Institute for Macromolecular Chemistry.

He devised a new and simple synthesis of isoprene (the monomer for the production of the synthetic rubber polyisoprene) in 1910.

Most chemists thought that polymers were disorderly conglomerates of small molecules, but from 1926 Staudinger put forward the view that polymers are giant molecules held together with ordinary chemical bonds. To give credence to the theory, he made chemical changes to polymers that left their molecular weights almost unchanged; for example, he hydrogenated rubber to produce a saturated hydrocarbon polymer.

In his book *Macromolekulare Chemie und Biologie* 1947, Staudinger anticipated the molecular biology of the future.

Stebbins George Ledyard 1906– . US botanist and plant geneticist who was the first scientist to apply neo-Darwinism to plants in his *Variation and Evolution in Plants* 1950. With Ernest B Babcock, he developed a technique for doubling the chromosome number of a plant and producing polyploids (plants possessing three or more sets of chromosomes) artificially.

Stebbins was born in New York and studied biology at Harvard University, graduating with a PhD 1931. He was a lecturer at Colgate University in Hamilton, New York. In 1950, he went to the University of California, where he founded the genetics department at the Davis campus, and remained at the University of California until 1973. Stebbins has written several books on evolution including: *Processes of Organic Evolution* 1966, *Flowering Plants: Evolution Above the Species Level* 1974, and *Evolution* 1977 with Theodosius Dobzhansky, Francisco Ayala, and J Valentine.

Stebbins Joel 1878–1966. US astronomer, the first to develop the technique of electric photometry in the study of stars.

Stebbins was born in Omaha, Nebraska, and studied at the universities of Nebraska, Wisconsin, and California. He was professor at the University of Illinois 1913–22 and at Wisconsin 1922–48, as well as director of the Washburn Observatory. He continued research at the Lick Observatory until 1958.

In 1906 Stebbins began attempting to use electronic methods in photometry. At first only the brightest objects in the sky (such as the Moon) could be studied in this manner. From 1909 to 1925 he devoted much of his time to improving the photoelectric cell and using it to study the light curves of eclipsing binary stars. As the sensitivity of the device was increased, he could observe variations in the light of cooler stars.

During the 1930s Stebbins applied photoelectric research to the nature and distribution of interstellar dust and its effects on the transmission of stellar light. He analysed the degree of reddening of the light of hot stars and of globular clusters. His discoveries contributed to an understanding of the structure and size of our Galaxy.

Steele Edward John 1948– . Australian immunologist whose research into the inheritance of immunity has lent a certain amount of support to the Lamarckian theory of the inheritance of acquired characteristics, thus challenging modern theories of heredity and evolution.

Steele was born in Darwin, Northern Territory, and educated at the University of Adelaide. He carried out research at the Ontario Cancer Institute, Canada, 1977–80, and then moved to the UK to work at the Medical Research Council in Harrow, Middlesex.

Steele found that mice that have been made immune to certain antigens can pass on this acquired immunity to first and second generations of their offspring.

Steenstrup Johannes Iapetus Smith 1813–1897. Danish zoologist who is considered to be one of the founders of archaeology after his studies of fossilized plant remains in Danish peat bogs.

Steenstrup was born in Vang, Denmark and despite his lack of a University education and degree went on to teach at the Soro/ Academy. In 1842, he published a paper on the alternation of sexual and asexual generations in certain animal species which led to his promotion to professor of zoology at the University of Copenhagen 1846. He also worked on the cephalopod group of invertebrates, including octopus and squid, and other marine invertebrates. He took up the study of fossils 1950. He is known to have corresponded with Charles Darwin, although he did not agree with his theory of evolution.

Stefan Josef 1835–1893. Austrian physicist who established one of the basic laws of heat radiation in 1879, since known as the ***Stefan–Boltzmann law***. This states that the heat radiated by a hot body is proportional to the fourth power of its absolute temperature.

Stefan was born in Klagenfurt and studied at Vienna, becoming professor there 1863 and director of the Institute for Experimental Physics 1866.

Stefan deduced his radiation law from experiments done by Irish physicist John Tyndall with a platinum wire. From his law, Stefan was able to make the first accurate determination of the surface temperature of the Sun, obtaining a value of approximately 6,000°C/11,000°F.

In 1884 Ludwig Boltzmann, a former student of Stefan's, gave a theoretical explanation of Stefan's law based on thermodynamic principles and kinetic theory. Boltzmann pointed out that it held only for perfect black bodies, and Stefan had been able to derive the law because platinum approximates to a black body.

Stein William Howard 1911–1980. US biochemist who determined the amino acid sequence of the enzyme ribonuclease. He worked closely with Stanford Moore to describe the amino acid composition of bovine pancreatic ribonuclease (1954–56), which complemented previous structural studies by Anfinsen on the enzyme. Stein, Moore, and Anfinsen shared the Nobel Prize for Chemistry 1972.

Stein was born in New York City. He was a promising student, graduating in chemistry from Harvard in 1933 and obtaining his PhD in biochemistry from Columbia University in 1938. He was appointed professor of biochemistry at the Rockefeller Institute in 1954.

Steinberger Jack 1921– . US physicist who, with Melvin Schwartz and Leon Lederman, used neutrinos to study elementary particles and the discovered the muon neutrino. Nobel Prize for Physics 1988.

Steinberger, Lederman, and Schwartz produced the world's first beam of neutrinos at the Brookhaven National Laboratory, Long Island, New York 1961, using a large proton accelerator as a source of neutrinos. A huge detector weighing 10 tonnes caught a small number of neutrinos out of the 100,000 billion neutrinos that flowed from the accelerator during the experiment. A 13 m/ 43 ft thick steel wall, built from a scrapped battleship, shielded the detector from unwanted particles, such as cosmic rays. Using the neutrino beam produced by this apparatus, Steinberger, Lederman, and Schwartz investigated the weak nuclear force and the quark structure of matter. They also found a new type of neutrino, called the muon neutrino, which is paired with and can be transformed into the muon particle. The other, previously recognized, neutrino is similarly paired with the electron and is called the electron neutrino.

Steinberger was born in Bad Kissingen, Germany. He left Germany 1934 as a Jewish refugee and later studied chemistry at the University of Chicago. In 1942 he joined the radiation laboratory at the Massachusetts Institute of Technology (MIT). After the war, he studied physics at the University of Chicago under Enrico Fermi, Edward Teller, and others. After brief spells at the Institute for Advanced Study in Princeton, and the University of California at Berkeley, he moved to Columbia University, New York 1950. From 1968, Steinberger worked at CERN, the European particle physics laboratory near Geneva, Switzerland. In 1986 he retired from CERN and became part-time professor at the Scuola Normale Superiore in Pisa, Italy.

Steiner Jakob 1796–1863. Swiss mathematician, the founder of modern synthetic, or projective, geometry. He discovered the ***Steiner surface*** (also called the Roman surface), which has a double infinity of conic sections on it, and the Steiner theorem.

Steiner was born at Utzenstorf, near Bern, and did not learn to read and write until the age of 14. After training as a teacher in Germany, he was admitted to the University of Berlin 1822. By 1825 he was teaching at the university and in 1834 a professorship of geometry was created for him, which he held for the rest of his life.

His first published paper, which appeared in 1826, contained his discovery of the geometrical transformation known as inversion geometry.

The ***Steiner theorem*** states that two pencils (collections of geometric objects) by which a conic is projected from two of its points are projectively related.

In the ***Steiner–Poncelet theorem***, an extension of work done by French mathematician Jean Poncelet in 1822, Steiner proved that any Euclidean figure could be generated using only a straight rule if the plane of construction had a circle with its centre marked on it already.

His most important work is *Systematische Entwicklung der Abhängigkeit geometrischer Gestalten von Einander* 1832.

Steinmetz Charles Proteus 1865–1923. US engineer who formulated the ***Steinmetz hysteresis law*** in 1891, which describes the dissipation of energy that occurs when a system is subject to an alternating magnetic force.

He worked on the design of alternating current transmission and from 1894 to his death served as consulting engineer to General Electric.

Steno Nicolaus 1638–1686. Latinized form of Niels Steensen. Danish anatomist and naturalist, one of the founders of stratigraphy. To illustrate his ideas, Steno sketched what are probably the earliest geological sections.

Steno was born in Copenhagen and studied medicine in Leiden, the Netherlands. In 1666 he was appointed personal physician to the grand duke of Tuscany, and became royal anatomist in Copenhagen 1672. Returning to Italy, he was ordained a Catholic priest 1675, and gave up science on being appointed vicar-apostolic to N Germany and Scandinavia.

As a physician he discovered ***Steno's duct*** of the parotid (salivary) gland, and investigated the workings of the ovaries. Showing that a pineal gland resembling the human one is found in other creatures, he used this finding to challenge French philosopher René Descartes's claim that the gland was the seat of the human soul.

Steno's examination of quartz crystals disclosed that, despite differences in the shapes, the angle formed by corresponding faces is invariable for a particular mineral. This constancy is known as ***Steno's law***.

Having found fossil teeth far inland closely resembling those of a shark he had dissected, in his *Sample of the Elements of Myology* 1667 Steno championed the organic origin of fossils. On the basis of his palaeontological findings, he set out a view of geological history, contending that sedimentary strata had been deposited in former seas.

Stephan Peter 1943– . English therapeutic immunologist. In 1981, he developed Omnigen, a total cellular extract and serum for treating premature cellular degeneration.

Stephan was born in Middlesborough, Yorkshire (now Cleveland).

He learned about cell therapy from his father, who died in 1964, leaving his son to run his Harley Street clinic at the age of only 21. He studied homoeopathic medicine and graduated in 1970.

The concept of cell therapy is that by injecting healthy cells into the body, the general state of health of the body can be improved. Stephan's treatment involves injecting an organ-specific RNA (nucleic acid copied from the genetic material DNA) to boost the cellular RNA and then injecting tissue-specific antisera. These antisera travel to the defective cells and kill them while at the same time they stimulate the body's immune system, and the healthy cells become active and reproduce.

Stephan's recent work has been concerned with the substance that is formed before the production of the antibody from which he prepares his sera. This substance seems to have beneficial effects which may be used therapeutically.

Stephenson George 1781–1848. English engineer who built the first successful steam locomotive. He also invented a safety lamp independently of Humphrey Davy in 1815. He was appointed engineer of the Stockton and Darlington Railway, the world's first public railway, in 1821, and of the Liverpool and Manchester Railway in 1826. In 1829 he won a prize with his locomotive *Rocket*.

Experimenting with various gradients, Stephenson found that a slope of 1 in 200, common enough on roads, reduced the haulage power of a locomotive by 50% (on a completely even surface, a tractive force of less than 5 kg/11 lb would move a tonne). Friction was virtually independent of speed. It followed that railway gradients should always be as low as possible, and cuttings, tunnels, and embankments were therefore necessary. He also advocated the use of malleable iron rails instead of cast iron. The gauge for the Stockton and Darlington railway was set by Stephenson at 1.4 m/4 ft 8 in, which became the standard gauge for railways in most of the world.

Stephenson was born near Newcastle-upon-Tyne and received no formal education. He worked at a coal mine, servicing the steam pumping engine, and it was there he built his first locomotive in 1814. After the Liverpool and Manchester railway opened 1830, he worked as a consultant engineer to several newly emerging railway companies, all in the north of England or the Midlands.

In his first locomotive, Stephenson introduced a system by which exhaust steam was redirected into the chimney through a blast pipe, bringing in air with it and increasing the draught through the fire. This development made the locomotive truly practical.

With his son Robert, he established locomotive works at Newcastle. The Stockton and Darlington Railway was opened 1825 by Stephenson's engine *Locomotion*, travelling at a top speed of 24 kph/15 mph.

Stephenson was engaged to design the railway from Manchester to Liverpool, but there was an open competition for the most efficient locomotive. Three other engines were entered, but on day of the trials the *Rocket* was the only locomotive ready on time. It weighed 4.2 tonnes, half the weight of *Locomotion*.

Stephenson Robert 1803–1859. English civil engineer who constructed railway bridges such as the high-level bridge at Newcastle-upon-Tyne, England, and the Menai and Conway tubular bridges in Wales. He was the son of George Stephenson.

The successful *Rocket* steam locomotive was built under his direction 1829, as were subsequent improvements to it.

Stephenson was born near Newcastle-upon-Tyne, and began his working life assisting his father in the survey of the Stockton and Darlington Railway in 1821. He managed the locomotive factory his father had established in Newcastle, with a three-year break in South America, superintending some gold and silver mines in

Stephenson, Robert *High level bridge over the River Tyne in Newcastle built by Robert Stephenson between 1846–49. The bridge carried track for the North British Railway. Construction of the Kilsby Tunnel in the Midlands on 8 July 1837. The tunnel was also designed by Robert Stephenson. It carried the London to Birmingham railway.*

Colombia. In 1833 he became engineer for a projected railway from Birmingham to London. The line was completed 1838, and from then on he was engaged on railway work for the rest of his life.

In 1844 construction began, under Stephenson's supervision, of a railway line from Chester to Holyhead. His bridge for the Menai Straits, in which the railway tracks were completely enclosed in parallel iron tubes, was so successful that he adopted the same design for other bridges. One such, the Victoria Bridge over the St Lawrence at Montréal, Canada, built 1854–59, was for many years the longest bridge in the world.

Steptoe Patrick Christopher 1913–1988. English obstetrician who pioneered in vitro fertilization. Steptoe, together with biologist Robert Edwards, was the first to succeed in implanting in the womb an egg fertilized outside the body. The first 'test-tube baby' was born in 1978.

Steptoe developed laparoscopy for exploring the interior of the abdomen without a major operation.

Steptoe was educated at King's College and St George's Hospital Medical School, London. He worked at Oldham General Hospital 1951–78, and from 1969 was director of the Centre for Human Reproduction.

In 1968 Edwards and Steptoe began to collaborate on the fertilization of human eggs. Steptoe treated volunteer patients with a fertility drug to stimulate maturation of the eggs in the ovary. In 1971, Edwards and Steptoe were ready to introduce an eight-celled embryo into the uterus of a volunteer patient who hoped to become pregnant, but their attempts were unsuccessful. In 1977 it was decided to abandon the use of the fertility drug and remove the egg at precisely the right natural stage of maturity; once fertilized, the egg was reimplanted in the mother two days later. This pregnancy came to term and the baby was delivered by Caesarean section.

Stern Otto 1888–1969. German physicist who demonstrated by means of the ***Stern–Gerlach apparatus*** that elementary particles have wavelike properties as well as the properties of matter that had been demonstrated. Nobel Prize for Physics 1943.

Stern was born in Sohrau, Upper Silesia (now Zory in Poland), and studied at a number of German universities. He then worked with Albert Einstein in Prague and Zürich. In 1923 Stern became professor at Hamburg and director of the Institute of Physical Chemistry. With the rise of the Nazis, he emigrated to the USA in 1933 and set up a department for the study of molecular beams at the Carnegie Technical Institute in Pittsburgh.

In 1920 Stern and Walther Gerlach (1899–1979) carried out their experiment, which consisted of passing a narrow beam of silver atoms through a strong magnetic field. Classical theory predicted that this field would cause the beam to broaden, but quantum theory predicted that the beam would split into two separate beams. The result, showing a split beam, was the first clear evidence for space quantization – the phenomenon that, in a magnetic field, certain atoms behave like tiny magnets which can only take up particular orientations with respect to the direction of the field. Stern went on to improve this molecular-beam technique and in 1931 was able to detect the wave nature of particles in the beams.

In 1933, Stern measured the magnetic moment of the proton and the deuteron, and demonstrated that the proton's magnetic moment was 2.5 times greater than expected.

Stevens Nettie Maria 1861–1912. US biologist whose experiments with a species of beetle showed that sex was determined by a specific chromosome. This was the first direct evidence that the units of heredity postulated by Austrian biologist Gregor Mendel were associated with chromosomes.

Stevens was born in Cavendish, Vermont, and worked as a librarian; not until the age of 35 did she begin to study at Stanford University, moving to Bryn Mawr College for her PhD. She spent research periods at marine and zoological laboratories in Europe, and was an associate professor at Bryn Mawr from 1905.

Stevens studied regenerative processes in lower invertebrates.

From there she moved on to working on the development of the roundworm, examining its regenerative properties after exposure to ultraviolet radiation, and showed that even in very early embryonic life, cells were restricted in their regenerative capabilities.

Stevenson Robert 1772–1850. Scottish engineer who built many lighthouses, including the Bell Rock lighthouse 1807–11.

Stevinus Simon *c.* 1548–1620. Flemish scientist who, in physics, developed statics and hydrodynamics; he also introduced decimal notation into Western mathematics.

Stevinus was born in Bruges (now in Belgium). He began work in Antwerp as a clerk and then entered Dutch government service, using his engineering skills to become quartermaster-general to the army. He designed sluices that could be used to flood parts of Holland to defend it from attack.

In statics Stevinus made use of the parallelogram of forces and in dynamics he made a scientific study of pulley systems. In hydrostatics he noted that the pressure exerted by a liquid depends only on its height and is independent of the shape of the vessel containing it. He is supposed to have carried out an experiment (later attributed to Italian physicist Galileo) in which he dropped two unequal weights from a tall building to demonstrate that they fell at the same rate.

Stevinus wrote in the vernacular (a principle he advocated for all scientists). His book on mechanics is *De Beghinselen der Weeghcoust* 1586.

Stieltjes Thomas Jan 1856–1894. Dutch-born French mathematician who contributed greatly to the theory of series and is often called the founder of analytical theory. His analysis of continued fractions, in particular, has had immense influence in the development of mathematics.

Stieltjes was born in Zwolle and studied at the polytechnic in Delft. He worked at the Leiden Observatory 1877–83. In 1884 he became professor of mathematics at Groningen. From 1886 he taught at the University of Toulouse, France.

Stieltjes studied almost all the problems in analysis then known – the theory of ordinary and partial differential equations, Euler's gamma functions, elliptic functions, interpolation theory, and asymptotic series. His researches also raised the mathematical status of discontinuous functions and divergent series.

His memoir 'Recherches sur les fractions continues', completed just before he died and published in two parts (1894 and 1895), was a milestone in mathematical history. Stieltjes was the first mathematician to give a general treatment of continued fractions as part of complex analytical function theory.

Stirling Robert 1790–1878. Scottish inventor of the first practicable hot-air engine 1816. The Stirling engine has a high thermal efficiency and a large number of inherent advantages, such as flexibility in the choice of fuel, that could make it as important as the internal-combustion engine.

Stirling was born in Cloag, Perthshire, and attended Glasgow and Edinburgh universities. He was ordained a Presbyterian minister in 1816, responsible for the parish of Galston, Ayrshire, 1824–76. He also designed and made scientific instruments.

The patent on the first air engine (and related patents until 1840) was taken out jointly with his younger brother James, a mechanical engineer. Their first engine appeared 1818. It had a vertical cylinder about 60 cm/2 ft in diameter, produced about 1.5 kW/2 hp pumping water from a quarry, and ran for two years before the hot sections of the cylinder burned out.

In 1824, the brothers started work on improved engines and in 1843 converted a steam engine at a Dundee factory to operate as a Stirling engine. It is said to have produced 28 kW/37 hp and to have used less coal per unit of power than the steam design it replaced. However, the hot parts burned out continually.

The Stirling-cycle engine differs from the internal-combustion engine in that the working fluid (in Stirling's case, air) remains in the working chambers. The heat is applied from an external source, so anything from wood to nuclear fuel can be used. It also means that combustion can be made to take place under the best conditions, making the control of emissions (pollution) considerably easier. The burning of the fuel is continuous, not intermittent as in an internal-combustion engine, so there is less noise and vibration.

Stirling engines use what is effectively two pistons to push the working fluid between two working spaces. One space is kept at a high temperature by the heat source and the other at a low temperature. Between these two spaces is a regenerator which alternately receives and gives up heat to the working fluid. The pistons are connected to a mechanism which keeps them out of phase (usually by 90°). It is this differential motion that moves the working fluid from one space to the other. On its way to the hot space, the fluid passes through the regenerator, gaining heat. In the hot space it gains more heat and expands, giving power. After the power stroke the fluid is pushed back through the regenerator, where it gives up its residual heat into the cold space and is ready to start the cycle again.

Stock Alfred 1876–1946. German inorganic chemist who prepared many of the hydrides of boron (called boranes). He introduced sensitive tests for mercury and devised improved laboratory techniques for dealing with the metal to minimize the risk of poisoning.

Stock was born in Danzig (now Gdansk, Poland) and studied at Berlin. He was director of the Chemistry Department at the Technische Hochschule in Karlsruhe 1926–36. By 1923 he was suffering from chronic mercury poisoning.

Stock began studying the boron hydrides – general formula B_xH_y – in 1909 at Breslau. In 1912, he devised a high-vacuum method for separating mixtures of them. In the 1960s boron hydrides found their first practical use as additives to rocket fuel.

In 1921, Stock prepared beryllium (scarcely known before in the metallic state) by electrolysing a fused mixture of sodium and beryllium fluorides. This method made beryllium available for industrial use, as in special alloys and glasses and for making windows in X-ray tubes.

Stokes George Gabriel 1819–1903. Irish physicist who studied the viscosity (resistance to relative motion) of fluids. This culminated in ***Stokes' law***, $F = 6\pi \eta r v$, which applies to a force acting on a sphere falling through a liquid, where η is the liquid's viscosity and r and v are the radius and velocity of the sphere. 1st baronet 1889.

In 1852 Stokes gave the first explanation of the phenomenon of fluorescence, a term he coined. He noticed that ultraviolet light was being absorbed and then re-emitted as visible light.

This led him to use fluorescence as a method to study ultraviolet spectra.

Stokes was born in Sligo, Ireland; he studied at Cambridge, where he became professor of mathematics 1849.

Stokes's investigation into fluid dynamics in the late 1840s led him to consider the problem of the ether, the hypothetical medium for the propagation of light waves. He showed that the laws of optics held if the Earth pulled the ether with it in its motion through space, and from this he assumed the ether to be an elastic substance that flowed with the Earth.

Stokes realized in 1854 that the Sun's spectrum is made up of spectra of the elements it contains. He concluded that the dark Fraunhofer lines are the spectral lines of elements absorbing light in the Sun's outer layers.

Stokes William 1804–1878. Irish physician who, with John Cheyne, gave his name to ***Cheyne–Stokes breathing***, or periodic respiration.

Stokes studied clinical medicine at the Meath Hospital, Dublin, and at Edinburgh. He returned to Dublin and held a post at the Meath Hospital.

When Stokes referred to Cheyne's paper on periodic respiration in his book *The Diseases of the Heart and Aorta*, the phenomenon became known by both their names.

Stokes's name was also applied to Stokes–Adams attacks after his paper 'Observations on Some Cases of Permanently Slow Pulse' 1846.

Stopes Marie Charlotte Carmichael 1880–1958. Scottish birth-control campaigner. With her husband H V Roe (1878–1949), an aircraft manufacturer, she founded Britain's first birth-control clinic in London 1921. She wrote plays and verse as well as the best-selling manual *Married Love* 1918, in which she urged women to enjoy sexual intercourse within their marriage, a revolutionary view for the time.

Stopes was born in Edinburgh and studied botany at University College, London, and in Germany at Munich. She taught at the University of Manchester 1905–11, as the first woman to be appointed to the science staff there. Her field was palaeobotanical research into fossil plants and primitive cycads.

Her other works include *Wise Parenthood* 1918 and *Radiant Motherhood* 1921. The Well Woman Centre in Marie Stopes House, London, commemorates her work.

Strabo *c.* 63 BC–AD *c.* 24. Greek geographer and historian who travelled widely to collect first-hand material for his *Geography*.

Strasburger Eduard Adolf 1844–1912. German botanist who discovered that the nucleus of plant cells divides during cell division and clarified the role that chromosomes play in heredity. It had previously been thought that the nucleus disappeared during cell division until Strasburger saw the nucleus of a dividing cell divide.

He coined the terms: 'chloroplast' (the structure in a plant cell where photosynthesis occurs), 'cytoplasm' (the part of a cell that is outside the nucleus), 'haploid' (having a single set of chromosomes in each cell), 'diploid' (having two sets of chromosomes in each cell) and 'nucleoplasm' (or karyoplasm, the substance of the nucleus).

Strasburger was born in Warsaw, Poland and left for Paris 1862 to study at the Sorbonne. Two years later he moved to the University of Bonn, where he studied botany and heard Julius von Sachs lecture at nearby Poppelsdorf. It was while he was in Bonn that he met Nathaniel Pringsheim and Ernst Haeckel. He later went to Jena to become Pringsheim's laboratory assistant. He obtained his doctorate 1866 from Jena University. When he was 27 years old, he was appointed professor and director of the botanical gardens in Jena. In 1873, he and Haeckel undertook an expedition to Egypt and the Red Sea. In 1881, he was made professor of cytology at Bonn University, and he was rector of the university 1891–92.

Street J(abez) C(urry) 1906–1989. US physicist who, with Edward C Stevenson, discovered the muon (an elementary particle) in 1937.

Strömgren Bengt Georg Daniel 1908–1987. Swedish-born Danish astronomer whose 1940 hypothesis about the so-called ***Strömgren spheres*** – zones of ionized hydrogen gas surrounding hot stars embedded in gas clouds – proved fundamental to our understanding of the structure of interstellar material.

Strömgren was born in Göteborg. He studied in Denmark at Copenhagen, where he became professor 1938; in 1940 he succeeded his father Elis Strömgren as director of the observatory there. From 1946 he held mainly US academic posts; he was professor at the University of Chicago 1951–57 and director of both the Yerkes and McDonald observatories, and was a member of the Institute for Advanced Study at Princeton 1957–67.

Some gaseous nebulae that can be observed within our Galaxy are luminous. Strömgren proposed that this light was caused by hot stars within obscuring layers of gas in the nebulae. He suggested that these stars ionize hydrogen gas and that the dimensions of the ionized zone (the Strömgren sphere) depend on both the density of the surrounding gas and the temperature of the star. His calculations of the sizes of these zones have been confirmed by observations.

Strömgren's other work included an analysis of the spectral classification of stars by means of photoelectric photometry, and research into the internal composition of stars.

Struve F(riedrich) G(eorg) W(ilhelm) von 1793–1864. German-born Russian astronomer, a pioneer in the observation of double stars and one of the first to measure stellar parallax, in 1830. He was the founder of a dynasty of astronomers that spanned four generations.

Struve was born in Altona, Schleswig-Holstein. To avoid conscription into the German army, he fled to Estonia and studied at Dorpat (now Tartu). He became professor there 1813, and director of the Dorpat Observatory 1817. In 1839 he became the first director of Pulkovo Observatory near St Petersburg, and was succeeded 1862 by his son Otto Wilhelm Struve.

Struve published a catalogue of about 800 double stars 1822, and instigated an extensive observational programme. The number of such stars known had increased to more than 3,000 by 1827. In addition, Struve described more than 500 multiple stars in a paper 1843.

In 1846 Struve published his observations of the absorption of stellar light in the galactic plane, which he correctly deduced to be caused by the presence of interstellar material.

Struve made significant contributions to geodesy with his survey of Livonia 1816–19 and his measurements of the arc of meridian 1822–27.

Struve (Gustav Wilhelm) Ludwig (Ottovich) von 1858–1920. Russian astronomer, an expert on the occultation of stars during a total lunar eclipse, and on stellar motion.

Struve was born in Pulkovo, near St Petersburg. The son of Otto Wilhelm von Struve, he followed the family tradition by studying astronomy at Dorpat (now Tartu, Estonia). He began his research at Pulkovo Observatory, and visited observatories in many European countries. In 1894 he moved to the University of Kharkov, where he became professor 1897 and director of the observatory. From 1919 he was professor at Tauris University in Simferopol.

Struve was interested in precession and he investigated the whole question of motion within the Solar System. This led him to work on the positions and motions of stars, and to an estimation of the rate of rotation of the Galaxy.

Struve (Karl) Hermann von 1854–1920. Russian astronomer, an expert on Saturn. His other work was largely concerned with features of the Solar System, although he shared the family interest in stellar astronomy.

Struve was born in Pulkovo, near St Petersburg, and studied at Dorpat (now Tartu, Estonia), travelled in Europe and visited major centres of astronomical research. He began his career at Pulkovo Observatory (founded by his grandfather Wilhelm von Struve), becoming its director 1890. In 1895 he moved to Germany, first as professor at Königsberg, and from 1904 as director of the Observatory of Berlin-Babalsberg (the Neubabalsberg Observatory from 1913).

Among the many features of the Solar System studied by Struve were the transit of Venus, the orbits of Mars and Saturn, the satellites (especially Iapetus and Titan) of Saturn, and Jupiter and Neptune. Struve's 1898 paper on the ring system of Saturn formed the basis of much of his later research.

Struve Otto von 1897–1963. Russian-born US astronomer who developed a nebular spectrograph to study interstellar gas clouds. In 1938 he showed that ionized hydrogen is present in interstellar matter. He also determined that the interstellar hydrogen is concentrated in the galactic plane.

Struve was born in Kharkov, where his father Ludwig von Struve was director of the observatory. He served in the Imperial Russian Army on the Turkish front during World War I, then graduated from Kharkov. Conscripted into the counterrevolutionary White Army during the Civil War in 1919, he fled to Turkey in 1920 and emigrated to the USA 1921. He worked at the Yerkes Observatory, becoming its director 1932. He was professor of astrophysics at Chicago

1932–50 and at the University of California at Berkeley 1950–59. He was also the founder of the McDonald Observatory in Texas, and the first director of the National Radioastronomy Observatory at Green Bank, West Virginia. In 1962 he was appointed joint professor of the Institute for Advanced Study at Princeton and the California Institute of Technology.

Struve did early work on stellar rotation and demonstrated the rotation of blue giant stars and the relationship between stellar temperature (and hence spectral type) and speed of rotation. In 1931 he found, as he had anticipated, that stars that spun at a high rate deposited gaseous material around their equators.

Struve believed that the establishment of a planetary system should be thought of as the normal course of events in stellar evolution and not a freak occurrence.

Struve Otto Wilhelm von 1819–1905. Russian astronomer who made an accurate determination of the constant of precession. He discovered about 500 double stars.

Struve was born in Dorpat (now Tartu), Estonia, studied there, and worked at the Pulkovo Observatory, near St Petersburg, from 1839. From 1847 to 1862 he was also a military adviser in St Petersburg. In 1862 he succeeded his father Wilhelm von Struve as director of the observatory.

Struve studied Saturn's rings, discovered a satellite of Uranus, and calculated the mass of Neptune. He also concerned himself with the measurement of stellar parallax, the movement of the Sun through space, and the structure of the universe, although he was among those astronomers who erroneously believed our Galaxy to be the extent of the whole universe.

Sturgeon William 1783–1850. English physicist and inventor who made the first electromagnets. He also invented a galvanometer in 1836.

Sturgeon was born in Whittington, Lancashire, and apprenticed to a shoemaker. He was in the army 1802–20, and in 1824 became a lecturer in science and philosophy at the East India Royal Military College of Addiscombe. In 1832 he was appointed to the lecturing staff of the Adelaide Gallery of Practical Science, and in 1840 he moved to Manchester to become superintendent of the Royal Victoria Gallery of Practical Science. When this failed, he became an itinerant lecturer. In 1836 he established *Annals of Electricity*, the first English-language journal devoted wholly to electricity. It ran until 1843, after which he founded other publications.

In 1828 Sturgeon put into practice the idea of a solenoid, first proposed by French physicist André Ampère, by wrapping wire round an iron core and passing a current through the wire. He found that a magnetic field was formed, which seemed to be concentrated in the iron core and disappeared as soon as the current was switched off. His device was capable of lifting 20 times its own weight.

Sturtevant Alfred Henry 1891–1970. US geneticist who was the first to map the position of genes on a chromosome.

Sturtevant worked with Thomas Morgan, one of the first great US biologists. In 1903, he began his seminal experiments on the genetics of the fruit fly *Drosophila melanogaster*. In May 1910, Sturtevant and Morgan discovered a mutated *Drosophila* with a white eye, instead of the normal red. Sturtevant's first breeding experiments were with white-eyed and red-eyed flies. He also developed other mutations using genotoxic agents such as X-rays.

Using these variant flies, Sturtevant developed methods for mapping gene positions on chromosomes. In 1911, he produced the first gene map ever derived, showing the positioning of five genes on a *Drosophila* X chromosome: white-eyed, vermilion-eyed, rudimentary wings, small wings, and yellow body.

Sturtevant and Morgan determined the gene order along the chromosome by working out the recombination frequencies between the linked genes. During meiosis (cell division that produces reproductive cells with half the number of chromosomes) in a heterozygous, phenotypically normal female, an X chromosome carrying the mutations and a normal X chromosome will rearrange or 'crossover'. Males derived from these females will carry a combination of mutations. Those mutations that are closest together on the chromosome will be seen with the greatest frequency in the phenotypes (traits) of the male flies.

Inheritance of these mutations in flies was not strictly according to Mendel's rules since female flies had to inherit two mutated X chromosomes to demonstrate the phenotype, whereas males only had to acquire a single mutated X chromosome. This made the discovery that the five genes were on the same chromosome striking.

Suess Eduard 1831–1914. Austrian geologist who helped pave the way for the theories of continental drift. He suggested that there had once been a great supercontinent, made up of the present southern continents; this he named Gondwanaland, after a region of India.

Suess was born in London and studied at Prague. He was professor at Vienna from 1861, and served as a member of the Reichstag for 25 years.

As a palaeontologist, Suess investigated the fossil mammals of the Danube Basin. He carried out research into the structure of the Alps, the tectonic geology of Italy, and seismology. The possibility of a former landbridge between N Africa and Europe caught his attention.

In his book *The Face of the Earth* 1885–1909, Suess analysed the physical agencies contributing to the Earth's geographical evolution. He offered an encyclopedic view of crustal movement, the structure and grouping of mountain chains, sunken continents, and the history of the oceans. He also made significant contributions to rewriting the structural geology of each continent.

Sufi, al- 903–986. Persian astronomer whose importance lies in his compilation of a valuable catalogue of 1,018 stars with their approximate positions, magnitudes, and colours.

Sumner James Batcheller 1887–1955. US biochemist. In 1926 he succeeded in crystallizing the enzyme urease and demonstrating its protein nature. For this work Sumner shared the 1946 Nobel Prize for Chemistry with John Northrop and Wendell Stanley.

Sutherland Earl Wilbur Jr 1915–1974. US physiologist, discoverer of cyclic AMP, a chemical 'messenger' made by a special enzyme in the wall of living cells. Many hormones operate by means of this messenger. Nobel Prize for Physiology or Medicine 1971.

Sutherland was born in Burlingame, Kansas, and studied at Washburn College, Topeka, and Washington University Medical School, St Louis. He was director of the Department of Medicine at Western Reserve (now Case Western Reserve) University in Cleveland 1953–63, then moved to Vanderbilt University, Nashville, and in 1973 to the University of Miami Medical School.

Sutherland began studying hormones at Washington under biochemists Carl and Gerty Cori and then spent the 1950s doing research on his own. At that time it was thought that hormones, carried in the bloodstream, activated their target organs directly. Sutherland showed that the key to the process – the activating agent of the organ concerned – is cyclic adenosine 3,5 monophosphate (cyclic AMP). The arrival of a hormone increases the organ's cellular level of cyclic AMP, which in turn triggers or inhibits the cellular activity.

Sutherland Gordon Brims Black McIvor 1907–1980. Scottish physicist who used infrared spectroscopy to study molecular structure. He elucidated the structure of a wide range of substances, from proteins to diamonds. Knighted 1960.

Sutherland was born in Caithness and studied at St Andrews and Cambridge. He was at Cambridge 1934–49, then went to the USA as professor at the University of Michigan. In 1956 he returned to Britain to become director of the National Physical Laboratory. He was master of Emmanuel College, Cambridge, 1964–77.

Working with English physicist William Penney, Sutherland showed that the four atoms of hydrogen peroxide (H_2O_2) do not lie in the same plane but that the molecule's structure resembles a partly opened book, with the oxygen atoms aligned along the spine and the O–H bonds lying across each cover. Later, during World War II, Sutherland and his research group at Cambridge analysed fuel from crashed German aircraft in order to discover their sources of oil. At Michigan he was one of the first to use spectroscopy to study biophysical problems; he also continued his investigations into simpler molecules and crystals.

Sutherland Ivan Edward 1938– . US electronics engineer who pioneered the development of computer graphics, the method by which computers display pictorial (as opposed to alphanumeric) information on screen.

Sutherland was born in Hastings, Nebraska, and studied at the Carnegie Institute of Technology and the Massachusetts Institute of Technology (MIT). In Salt Lake City, Utah, he founded the Evans and Sutherland Computer Corporation and was professor at Utah 1972–76. In 1976 he moved to the California Institute of Technology.

The Sketchpad project was the first system of computer graphics that could be altered by the operator in the course of its use. Sketchpad used complex arrangements of the data fed into the computer to produce representations of the objects in space as well as geometrical detail. Programs could be altered using light pens, which touch the surface of the screen. Sutherland worked on this at MIT 1960–63.

At Utah, Sutherland was engaged in the design of a colour graphics system able to represent fine distinctions of colour as well as accurate perspective. The image could be moved, rotated, made larger or smaller to give a realistic image of the object, rendering the computer suitable for use in engineering and architectural design.

Sutton-Pringle John William 1912–1982. British zoologist who established much of our knowledge of the anatomical mechanisms involved in insect flight.

Sutton-Pringle studied at Cambridge, where he spent his whole academic career.

Most insects have two hindwings and two forewings, and in many species each pair of wings acts in unison. Not all species use both pairs of wings for flight; in the housefly, for example, the hindwings are reduced in size and serve as balancing organs during flight.

Insect flight is achieved by simple up-and-down movements of the wings. In aphids, for example, these wing movements are brought about by the contractions of two separate sets of muscles. When moving through the air, the front edge of the wings remains rigid while the back edge bends. This causes the development of a localized region of high-pressure air behind the insect, which propels the insect forwards. The faster the wing beats, the greater the displacement of the posterior wing edges, the greater the pressure exerted on the insect from behind, and therefore the faster the insect flies.

Svedberg Theodor 1884–1971. Swedish chemist. In 1924 he constructed the first ultracentrifuge, a machine that allowed the rapid separation of particles by mass. This can reveal the presence of contaminants in a sample of a new protein, or distinguish between various long-chain polymers. Nobel Prize for Chemistry 1926.

Svedberg was born near Gävle, studied at Uppsala and spent his career there, as professor 1912–49 and head of the Institute of Nuclear Chemistry 1949–67.

Svedberg prepared a number of new organosols from more than 30 metals. Through an ultramicroscope, he studied the particles in these sols and confirmed Albert Einstein's theories about Brownian movement.

Svedberg discovered that thorium-X crystallizes with lead and barium salts (but not with others), anticipating English chemist Frederick Soddy's demonstration of the existence of isotopes.

Svedberg also investigated, about 1923, the chemistry involved in the formation of latent images in photographic emulsions.

Working on synthetic polymers, Svedberg

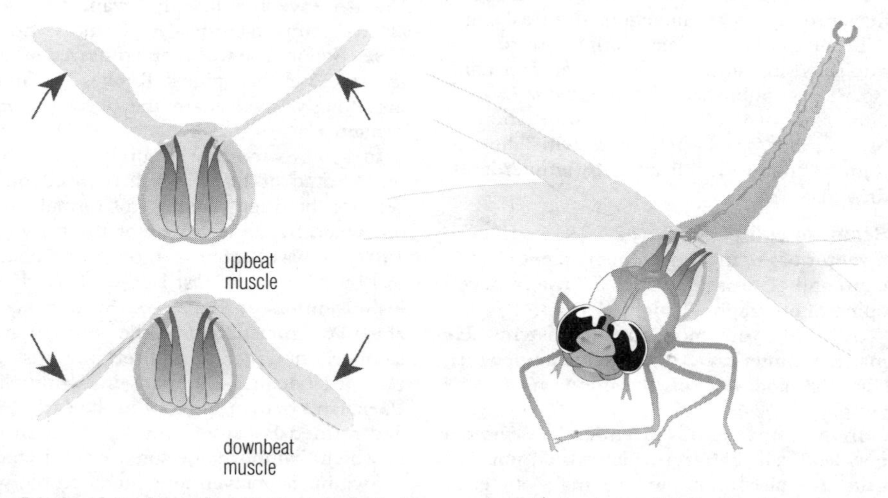

Sutton-Pringle *Sutton-Pringle showed that the rapid wing-beats necessary for flight in insects are achieved by alternate sets of muscles.*

introduced electron microscopy to study natural and regenerated cellulose, X-ray diffraction techniques to investigate cellulose fibres, and electron diffraction to analyse colloidal micelles and crystallites.

Swammerdam Jan 1637–1680. Dutch naturalist who is considered a founder of both comparative anatomy and entomology. Based on their metamorphic development, he classified insects into four main groups, three of which are still used in a modified form in insect classification. He was also probably the first to discover red blood cells when he observed oval particles in frog's blood in 1658.

Swammerdam was born in Amsterdam and studied medicine at Leiden but never practised as a physician. From 1673 he was under the influence of a religious zealot.

Swammerdam accurately described and illustrated the life cycles and anatomies of many insect species, including bees, mayflies, and dragonflies.

Swammerdam also provided a substantial body of new knowledge about vertebrates. He anticipated the discovery of the role of oxygen in respiration by postulating that air contained a volatile element that could pass from the lungs to the heart and then to the muscles, providing the energy for muscle contraction.

In his work on human and mammalian anatomy, Swammerdam discovered valves in the lymphatic system (now called ***Swammerdam valves***). He also investigated the human reproductive system and was one of the first to show that female mammals produce eggs, analogous to birds' eggs.

Swammerdam's manuscripts were not published in full until 1737, when Hermann Boerhaave published *Biblia naturae/Bible of Nature*, a two-volume Latin translation of Swammerdam's Dutch text, with illustrations engraved from Swammerdam's drawings.

Swan Joseph Wilson 1828–1914. English inventor of the incandescent-filament electric lamp and of bromide paper for use in developing photographs. Knighted 1904.

Swan took out more than 70 patents. He made a miner's electric safety lamp which was the ancestor of the modern miner's lamp.

In the course of this invention he devised a new lead cell (battery) which would not spill acid. He also attempted to make an early type of fuel cell.

Swan was born in Sunderland and went to work in a chemical firm. Interested in electric lighting from about 1845, he began making filaments by cutting strips of paper and baking them at high temperatures to produce a carbon fibre. In making the first lamps, he connected the ends of a filament to wire (itself a difficult task), placed the filament in a glass bottle, and attempted to evacuate the air and seal the bottle with a cork. Usually the filament burned away very quickly in the remaining air, blackening the glass at the same time. Only after the invention of the vacuum pump 1865 was Swan able to produce a fairly durable incandescent lamp. For this he made a new type of filament from cotton thread partly dissolved by sulphuric acid. He patented the process in 1880 and began manufacturing lamps. In 1882 US inventor Thomas Edison initiated litigation for patent infringement against Swan, but this was dismissed and the joint company Edison and Swan United Electric Light Company came into being in 1883.

A wet process for producing photographic prints, using a gelatine film impregnated with carbon or other pigment granules and photosensitized using potassium dichromate, was patented by Swan in 1864. This was known as the carbon or autotype process.

Swinburne James 1858–1958. Scottish engineer, a pioneer in electrical engineering and plastics. Baronet 1934.

Swinburne was born in Inverness and apprenticed to a locomotive works in Manchester. During 1881–85 he worked for English inventor Joseph Swan, setting up electric-lamp factories in France and the USA. Swinburne was then employed as assistant to English engineer Rookes Crompton, particularly involved in the development of dynamos.

In 1894 Swinburne set up his own laboratory. Some of his research focused on the reaction between phenol and formaldehyde, but when he came to patent the product in 1907, he was beaten to the idea by just one day by Belgian chemist Leo Baekeland (with his invention of Bakelite). Swinburne was able to obtain a patent on the production of a lacquer, however, and set up his own Damard Lacquer Company in Birmingham. Baekeland bought him out in the early 1920s and formed Bakelite Limited, Great Britain, of which Swinburne became the first chair.

Swinburne's inventions included the watt-hour meter and the 'hedgehog' transformer for stepping up medium-voltage alternating current to high voltages for long-distance power transmission. The words 'motor' and 'stator' are thought to have been coined by him.

Swineshead Richard lived *c.* 1340–1355. British scientist and leading member of a group of natural scientists associated with Merton College, Oxford, who attempted to analyse and quantify the various forms of motion.

Swineshead was known as 'the Calculator'.

Swings Pol(idore) F F 1906–1983. Belgian astrophysicist who used spectroscopy to identify the elements in astronomical bodies, especially comets.

Swings was born in Ransart, near Charleroi, and studied at Liège, where he was professor 1932–75, with several years as visiting professor at a number of US universities.

From his study of cometary atmospheres, he is credited with the discovery of the ***Swings bands*** and the ***Swings effect***. Swings bands are emission lines resulting from the presence of certain atoms of carbon; the Swings effect was discovered with the aid of a slit spectrograph and is attributed to fluorescence resulting partly from solar radiation.

Swings also made spectroscopic studies of interstellar space and investigated stellar rotation, as well as nebulae, novae, and variable stars.

Sydenham Thomas 1624–1689. English physician, the first person to describe measles and to recommend the use of quinine for relieving symptoms of malaria. His original reputation as the 'English Hippocrates' rested upon his belief that careful observation is more useful than speculation. His *Observationes medicae* was published in 1676.

Sylow Ludwig Mejdell 1832–1918. Norwegian mathematician who in 1872 published his fundamental theorem on groups and the type of subgroups now named after him.

Sylow was born in Christiania (now Oslo) and studied at the University of Christiania. For 40 years he taught mathematics in a school in Halden. From 1898 he was professor at Christiania.

Sylow collaborated with Sophus Lie, then professor of mathematics at Christiania, on producing an edition of the works of Norwegian mathematician Niels Abel, published 1881; it was followed by Sylow's edition of Abel's letters in 1901.

Sylvester James Joseph 1814–1897. English mathematician who was one of the discoverers of the theory of algebraic invariants. He coined the term 'matrix' in 1850 to describe a rectangular array of numbers out of which determinants can be formed.

Sylvester was born in London and studied mainly at Cambridge. He became professor of natural philosophy at University College, London, 1837. In 1841 he went to the USA to become professor at the University of Virginia, but resigned 1845 and returned to England. For the next ten years he abandoned academic life, although he took in private pupils, including nursing pioneer Florence Nightingale. In 1844 he joined an insurance company, and in 1850 he became a barrister. In 1855 he returned to academic life, becoming professor of mathematics at the Royal Military Academy in Woolwich, London.

In 1877 he again went to the USA, becoming professor at the newly founded Johns Hopkins University in Baltimore. Returning to the UK, he was from 1883 professor at Oxford.

Sylvester laid the foundations, with English mathematician Arthur Cayley (with whom he did not collaborate), of modern invariant algebra. He also wrote on the nature of roots in quintic equations and on the theory of numbers, especially in partitions and Diophantine analysis.

Symington William 1763–1831. Scottish engineer who built the first successful steamboat. He invented the steam road locomotive in 1787 and a steamboat engine in 1788. His steamboat *Charlotte Dundas* was completed in 1802.

Synge Richard Laurence Millington 1914–1994. British biochemist who improved paper chromatography (a means of separating mixtures) to the point where individual amino acids could be identified. He developed the technique, known as partition chromatography, with his colleague Archer Martin 1944. They shared the 1952 Nobel Prize for Chemistry.

Synge was born in Liverpool and studied at Winchester College and Trinity College, Cambridge. From 1967 until his retirement in 1974 he worked as a biochemist at the Food Research Institute of the Agricultural Research Council in Norwich.

Martin and Synge worked together at Cambridge and at the Wool Industries Research Association in Leeds. Their chromatographic method became an immediate success, widely adopted. It was soon demonstrated that not only the type but the concentration of each amino acid can be determined.

In the early 1940s there were only crude chromatographic techniques available for separating proteins in reasonably large samples; no method existed for the separation of the amino acids that make up proteins. Martin and Synge developed the technique of paper chromatography, using porous filter paper to separate out amino acids using a solvent. A minute quantity of the amino acid solution is placed at the tip of the filter paper; once dry it is dipped (or suspended) in a solvent. As the solvent passes the mixture the amino acids move with it, but they do so at different rates and so become separated. Once dry the paper is sprayed with a developer and the amino acids show up as dark dots. Synge and Martin announced their technique 1944; it was soon being applied to a wide variety of experimental problems.

Paper chromatography is so precise that it can be used to identify amino acid concentration, as well as type, enabling Synge to work out the exact structure of the antibiotic peptide Gramicidin-S, a piece of research important to Frederick Sanger's determination of the structure of insulin 1953.

In 1948 Synge moved to the Rowett Research Institute, Aberdeen, where he remained in charge of protein chemistry until 1967. He then moved to Norwich where he became Honorary Professor of Biology at the University of East Anglia. Synge was very active in the peace movement, and after his retirement in 1976, he became treasurer of the Norwich Peace Council. He died in Norwich.

Discovery consists of seeing what everybody has seen, thinking what nobody has thought.

Albert Szent-Györgyi

The Scientist Speculates ed I G Good, 1962

Szent-Györgyi Albert von Nagyrapolt 1893–1986. Hungarian-born US biochemist who isolated vitamin C and B_2, and studied the chemistry of muscular activity. He was awarded the Nobel Prize for Physiology or Medicine 1937.

In 1928 Szent-Györgyi isolated a substance from the adrenal glands that he named hexuronic acid; he also found it in oranges and paprika, and in 1932 proved it to be vitamin C.

Szent-Györgyi also studied the uptake of oxygen by muscle tissue. In 1940 he isolated two kinds of muscle protein, actin and myosin, and named the combined compound actomyosin. When adenosine triphosphate (ATP) is added to it, a change takes place in the relationship of the two components which results in the contraction of the muscle. When a muscle contracts, myosin and actin move in relation to one another powered by energy released by hydrolysis of ATP and elevated levels of calcium in the muscle cells.

In the 1960s Szent-Györgyi began studying the thymus gland, and isolated several compounds from the thymus that seem to be involved in the control of growth.

Szent-Györgyi was born in Budapest and educated there and at universities elsewhere in Europe and in the USA. He was active in the anti-Nazi underground movement during World War II; after the war he became professor at Budapest. In 1947 he emigrated to the USA, where he became director of the National Institute of Muscle Research at Woods Hole, Massachusetts. From 1975 he was scientific director of the National Foundation for Cancer Research. His last work was *Electronic Biology and Cancer* 1976.

We turned the switch, saw the flashes, watched for ten minutes, then switched everything off and went home. That night I knew the world was headed for sorrow.

Leo Szilard

On taking part in the Columbia University experiment 1939 that confirmed that the atom could be split.

Szilard Leo 1898–1964. Hungarian-born US physicist who, in 1934, was one of the first scientists to realize that nuclear fission, or atom splitting, could lead to a chain reaction releasing enormous amounts of instantaneous energy. He emigrated to the USA in 1938 and there influenced Albert Einstein to advise President Roosevelt to begin the nuclear-arms programme. After World War II he turned his attention to the newly emerging field of molecular biology.

Tabor David 1913– . English physicist who worked mainly in tribology, the study of friction and wear between solid surfaces.

Tabor was born in London and studied there at the Royal College of Science. From 1946 he worked at Cambridge in the Cavendish Laboratory, becoming professor 1973.

Tabor showed that the low friction of Teflon is due not to poor adhesion but to molecular structure, and worked on the self-lubrication of polymers by incorporating surface materials into the polymer itself. A study of the friction of rubber led to the introduction of high-hysteresis rubber into vehicle tyres as means of increasing their skid-resistance.

Tabor researched into the shear properties of molecular films of long-chain organic molecules as an extension of earlier work on the mechanism of boundary lubrication, and showed that the shear strength of these materials rises sharply when they are subjected to high pressure. This sheds light on the mechanism of thin film lubrication.

Tabor's work on the hardness of solids includes an explanation of the indentation hardness of metals, plastic indentation and elastic recovery, the first correlation of hardness behaviour with the creep properties of the material (hot-hardness), a study of scratch hardness, and a simple physical explanation of Mohs' scale (used in the testing of minerals).

The results of his study of the creep of polycrystalline ice have a bearing on the flow of glaciers.

Studying the adhesion of steel to cement, Tabor demonstrated the important role of shear stresses in compacting the cement.

Tagliacozzi Gaspare 1546–1599. Italian surgeon who pioneered plastic surgery. He was the first to repair noses lost in duels or through syphilis. He also repaired ears. His method involved taking flaps of skin from the arm and grafting them into place.

Talbot William Henry Fox 1800–1877. English pioneer of photography. He invented the paper-based calotype process 1841, the first negative/positive method. Talbot made photograms several years before Louis Daguerre's invention was announced.

In 1851 he made instantaneous photographs by electric light and in 1852 photo engravings. *The Pencil of Nature* 1844–46 by Talbot was the first book illustrated with photographs to be published.

Talbot was born in Melbury, Dorset, and studied at Cambridge. He was elected Liberal member of Parliament for Chippenham 1833. During a trip to Italy he tried to capture the images obtained in a camera obscura and by 1835 had succeeded in fixing outlines of objects laid on sensitized paper. Images of his home, Lacock Abbey, Wiltshire, followed. Talbot was also a mathematician and classical scholar, and was one of the first to decipher the cuneiform inscriptions of Nineveh, Assyria.

In Talbot's calotype process, writing paper was coated successively with solutions of silver nitrate and potassium iodide, forming silver iodide. The iodized paper was made more sensitive by brushing with solutions of gallic acid and silver nitrate, and then it was exposed (either moist or dry). The latent image was developed with an application of gallo-silver nitrate solution, and when the image became visible, the paper was warmed for one to two minutes. It was fixed with a solution of potassium bromide (later replaced by sodium hyposulphite). Calotypes did not have the sharp definition of daguerreotypes and were generally considered inferior.

Talbot patented an enlarger in 1843. In the decade to 1851, he took out some patents that contained previously published claims, but in

1852 he cleared the way for amateurs to use processes developed in other countries.

Tamm Igor Yevgenyevich 1895–1971. Russian theoretical physicist who, with Paul Cherenkov and Ilya Frank, explained the blue light (Cherenkov radiation) emitted from water exposed to radioactivity from radium. They showed that the radiation arises when a charged particle travels through a medium (liquid or solid) at greater than the speed of light in the medium. They were able to predict the direction and polarization of the radiation. Nobel Prize for Physics 1958.

Tamm, the son of an engineer, was born in Vladivostok, Russia, and educated at the universities if Edinburgh and Moscow. He taught at the Moscow University 1924–34, and then moved to the Physics Institute of the Academy of Sciences of the USSR in Moscow.

Tansley Arthur George 1871–1955. English botanist, a pioneer in the science of plant ecology. He coordinated a large project to map the vegetation of the British Isles; the results were published in *Types of British Vegetation* 1911. He was also instrumental in the formation of organizations devoted to the study of ecology and the protection of wildlife. Knighted 1950.

Tansley was born in London and studied at Cambridge. He founded the journal *New Phytologist* 1902, remaining its editor for 30 years. As co-founder of the British Ecological Society 1913, he also edited its *Journal of Ecology* 1916–38. From 1923 to 1924 he abandoned botany to study under Sigmund Freud, the founder of psychoanalysis, in Austria. Tansley was professor of botany at Oxford 1927–39, and chair of Nature Conservancy 1949–53.

In *The British Islands and their Vegetation* 1939, Tansley showed how vegetation is affected by soil, climate, the presence of wild and domesticated animals, previous land management, and contemporary human activities. He also reviewed all known accounts of British flora and then linked the two themes, thereby demonstrating which factors are important in influencing the various types of vegetation.

Tartaglia adopted name of Niccolò Fontana *c.* 1499–1557. Italian Renaissance mathematician and physicist who specialized in military problems, topography, and mechanical physics.

Tartaglia was born in Brescia, Lombardy. He was called Tartaglia ('stammerer') because of a speech defect resulting from a wound caused by French soldiers sacking the town when he was 12. Although self-educated, he taught school in Verona 1516–33. He then moved to Venice, where he eventually became professor of mathematics.

Tartaglia solved the problems of calculating the volume of a tetrahedron from the length of its sides, and of inscribing within a triangle three circles tangent to one another.

He delighted in planning the disposition of artillery, surveying the topography in relation to the best means of defence, and in designing fortifications. He also attempted a study of the motion of projectiles, and formulated ***Tartaglia's theorem***: the trajectory of a projectile is a curved line everywhere, and the maximum range at any speed of its projection is obtained with a firing elevation of 45°.

When Tartaglia translated Euclid's *Elements* into Italian 1543, it was the first translation of Euclid into a contemporary European language.

Tatum Edward Lawrie 1909–1975. US microbiologist. For his work on biochemical genetics, he shared the 1958 Nobel Prize for Physiology or Medicine with his co-workers George Beadle and Joshua Lederberg.

Tatum was born in Boulder, Colorado, and studied at the University of Wisconsin. He worked with Beadle at Stanford 1937–41 and with Lederberg at Yale, where he became professor 1946. He ended his career at the Rockefeller Institute for Medical Research from 1957.

Beadle and Tatum used X-rays to cause mutations in bread mould, studying particularly the changes in the enzymes of the various mutant strains. This led them to conclude that for each enzyme there is a corresponding gene. From 1945, with Lederberg, Tatum applied the same technique to bacteria and showed that genetic information can be passed from one bacterium to another. The discovery that a form of sexual reproduction can occur in bacteria led to extensive use of these organisms in genetic research.

Taube Henry 1915– . US chemist who established the basis of inorganic chemistry through his study of the loss or gain of electrons by atoms during chemical reactions. He was awarded a Nobel prize 1983 for his work on electron transference between molecules in chemical reactions.

Taussig Helen Brooke 1898–1986. US cardiologist who developed surgery for 'blue' babies. Such babies never fully develop the shunting mechanism in the circulatory system that allows blood to be oxygenated in the lungs before passing to the rest of the body. The babies are born chronically short of oxygen and usually do not survive without surgery.

Taylor Joseph H 1941– . US radio astronomer who, with Russell Hulse, discovered a new type of pulsar, a binary pulsar. The pulsar had an invisible companion orbiting around it and radiated gravitational waves. This was the first experimental confirmation of the reality of waves predicted by Einstein's general theory of relativity. Nobel Prize for Physics 1993.

Taylor and Hulse discovered the pulsar 1974, when Taylor was working at the Unversity of Massachusetts in Amherst, Massachusetts, USA, and Hulse was a postgraduate student. The two astronomers detected a small regular change in the intervals between the radio pulses emitted by a pulsar called PSR1913+16. This occurred because the pulsar was orbiting another body. The pulses were slowing down in a regular way, losing about 75 microseconds a year due to energy being lost as gravitational waves were emitted. Over the next 18 years, Taylor refined his observations of the binary system's pulses and was able to confirm that the pulses were slowing down at the rate predicted by general relativity.

Taylor is now professor of physics at Princeton University, New Jersey.

Taylor Richard 1929– . Canadian physicist who worked with Jerome I Friedman and Henry Kendall conducting pioneering research into the collision of high-energy electrons with protons and neutrons, which were important in developing the quark model of particle physics. Nobel Prize for Physics 1990.

Taylor, Friedman, and Kendall experimented with bombarding protons (and later, neutrons) with high-energy electrons 1970. They found that the electrons were sometimes scattered through large angles inside the proton. This result was interpreted by James D Bjorken and Richard P Feynman, who suggested that the electrons were hitting hard pointlike objects inside the proton. These objects were soon shown to be quarks.

Taylor was born in Medicine Hat, Alberta, Canada, and educated at the University of Alberta in Edmonton. He moved to Stanford University in California, where he joined the High Energy Physics Laboratory. From 1958 to 1961 he worked in Paris on an accelerator under construction at Orsay. On his return to the United States he worked at the Lawrence Berkeley Laboratory at the University of California for a year and then moved back to Stanford. The next two decades were spent working on various electron scattering experiments, with periods at CERN, the European particle physics laboratory near Geneva, Switzerland, and DESY, the German accelerator laboratory in Hamburg. In 1982 he became Associate Director for Research at Stanford, a post he held until 1986 when he resigned to return to research.

Tebbutt John 1834–1916. Australian astronomer who made detailed observations of the orbits of comets and minor planets over a 50-year period, beginning in 1854, from an observatory he built at his home in Windsor, near Sydney. He published more than 300 scientific papers and was the first person to see the 'Great Comet' of 1861 (later named Tebbutt's comet).

Telford Thomas 1757–1834. Scottish civil engineer who opened up N Scotland by building roads and waterways. He constructed many aqueducts and canals, including the Caledonian canal 1802–23, and erected the Menai road suspension bridge 1819–26, a type of structure scarcely tried previously in England. In Scotland he constructed over 1,600 km/1,000 mi of road and 1,200 bridges, churches, and harbours.

In 1963 the new town of Telford, Shropshire, 32 km/20 mi NW of Birmingham, was named after him.

Telford was born in Westerkirk, Dumfries, and began as a stonemason. Moving to London, he found employment building additions to Somerset House in the Strand under the supervision of architect William

Telford *Two bridges over the Menai Straits. In the foreground is the bridge built by Robert Stephenson for the Chester to Holyhead railway between 1846 and 1850. In the background is the suspension road bridge built by Thomas Telford 1820–1826.*

Chambers. Recognizing his talents, the rich and famous were soon consulting him about their own buildings.

In 1786, Telford was appointed official surveyor to the county of Shropshire. There he built three bridges over the river Severn, among other structures. He also rebuilt many Roman roads to meet the need for faster travel.

As engineer to the Ellesmere Canal Company from 1793, Telford was responsible for the building of aqueducts over the Ceirog and Dee valleys in Wales, using a new method of construction consisting of troughs made from cast-iron plates and fixed in masonry.

Teller Edward 1908– . Hungarian-born US physicist known as the father of the hydrogen bomb. He worked on the fission bomb – the first atomic bomb – 1942–46 (the Manhattan Project) and on the fusion bomb, or H-bomb, 1946–52. In the 1980s he was one of the leading supporters of the Star Wars programme (Strategic Defense Initiative).

He was a key witness against his colleague Robert Oppenheimer at the security hearings 1954. Teller was widely believed to be the model for the leading character in Stanley Kubrick's 1964 film *Dr Strangelove*. It was also Teller who convinced President Reagan of the feasibility of the Star Wars project for militarizing space with fission-bomb-powered X-ray lasers. Millions of dollars were spent before the project was discredited. Teller then suggested 'brilliant pebbles' – thousands of missile-interceptors based in space – and the use of nuclear explosions to prevent asteroids hitting the Earth. He opposed all test-ban treaties.

Two paradoxes are better than one; they may even suggest a solution.

Edward Teller

Teller was born in Budapest and studied there and at various German universities. He left Germany 1933 when the Nazis came to power, and emigrated to the USA 1935, becoming professor at George Washington University, Washington DC. He joined the University of Chicago 1942 to work on atomic fission, and then moved to Los Alamos. By the end of World War II, Teller had designed an H-bomb, and in 1951 he was given responsibility for constructing one. It was successfully tested on Eniwetok Atoll in the Pacific Ocean in 1952. By then, a second nuclear-weapons research facility had opened, the Lawrence-Livermore Laboratory near Berkeley, California. Teller was Livermore's associate director 1954–75, as well as professor at the University of California from 1953.

The original idea of using a fission explosion to ignite a thermonuclear (fusion) explosion in deuterium (heavy hydrogen) came from Italian-born physicist Enrico Fermi. Polish-born mathematician Stanislaw Ulam (1909–1985) then suggested a configuration in which shock waves from the fission explosion would compress and heat the deuterium, causing it to explode. Teller modified this idea to use X-rays from the first explosion, rather than shock waves.

Temin Howard Martin 1934–1994. US virologist concerned with cancer research. For his work on the genetic inheritance of viral elements he shared the 1975 Nobel Prize for Physiology or Medicine with David Baltimore (1938–) and Renato Dulbecco (1914–).

Temin was born in Philadelphia and educated at Swarthmore College, Pennsylvania, and the California Institute of Technology. From 1960 he worked at the University of Wisconsin, becoming professor 1969.

Temin's prizewinning research was on a virus that has a mechanism which incorporates its material into mammalian genes. He discovered that beneficial mutations outside the germ line are naturally selected and that the mechanism adds genetic information from outside into the germ line.

Tereshkova Valentina Vladimirovna 1937– . Soviet cosmonaut, the first woman to fly in space. In June 1963 she made a three-day flight in *Vostok 6*, orbiting the Earth 48 times.

Tesla Nikola 1856–1943. Croatian-born US physicist and electrical engineer who invented fluorescent lighting, the Tesla induction motor 1882–87, and the Tesla coil, and developed the alternating current (AC) electrical supply system.

The ***Tesla coil*** is an air core transformer with the primary and secondary windings tuned in resonance to produce high-frequency, high-voltage electricity. Using this device, Tesla produced an electric spark 40 m/135 ft long in 1899. He also lit more than 200 lamps over a distance of 40 km/25 mi without the use of intervening wires.

Gas-filled tubes are readily energized by high-frequency currents and so lights of this type were easily operated within the field of a large Tesla coil. Tesla soon developed all manner of coils which have since found numerous applications in electrical and electronic devices.

Tesla was born in Croatia of Serbian parents. He emigrated to the USA 1884, and from 1888 was associated with industrialist George Westinghouse, who bought and successfully exploited Tesla's patents, leading to the introduction of alternating current for power transmission. Tesla neglected to patent many of his discoveries and made little profit from them.

Tesla was very interested in the possibility of radio communication and as early as 1897 he demonstrated remote control of two model boats on a pond. He extended this to guided weapons, in particular a remote-control torpedo. In 1900 he began to construct a broadcasting station on Long Island, New York, in the hope of developing 'World Wireless', but lost his funding. He also outlined a scheme for detecting ships at sea, which was later developed as radar. One of his most ambitious ideas was to transmit electricity to anywhere in the world without wires by using the Earth itself as an enormous oscillator.

In his later years, Tesla grew steadily more paranoid. He would not shake hands for fear of contamination by germs, and he was fearful of spherical surfaces. He spent his time on fantastic projects, like the invention of devices for producing death rays or for photographing thoughts on the retina of the eye.

Thales *c.* 624–*c.* 547 BC. Greek philosopher and scientist. He made advances in geometry, predicted an eclipse of the Sun 585 BC, and, as a philosophical materialist, theorized that water was the first principle of all things. He speculated that the Earth floated on water, and so proposed an explanation for earthquakes. He lived in Miletus in Asia Minor.

Thales explained such events as earthquakes in terms of natural phenomena, rather than in the usual terms of activity by

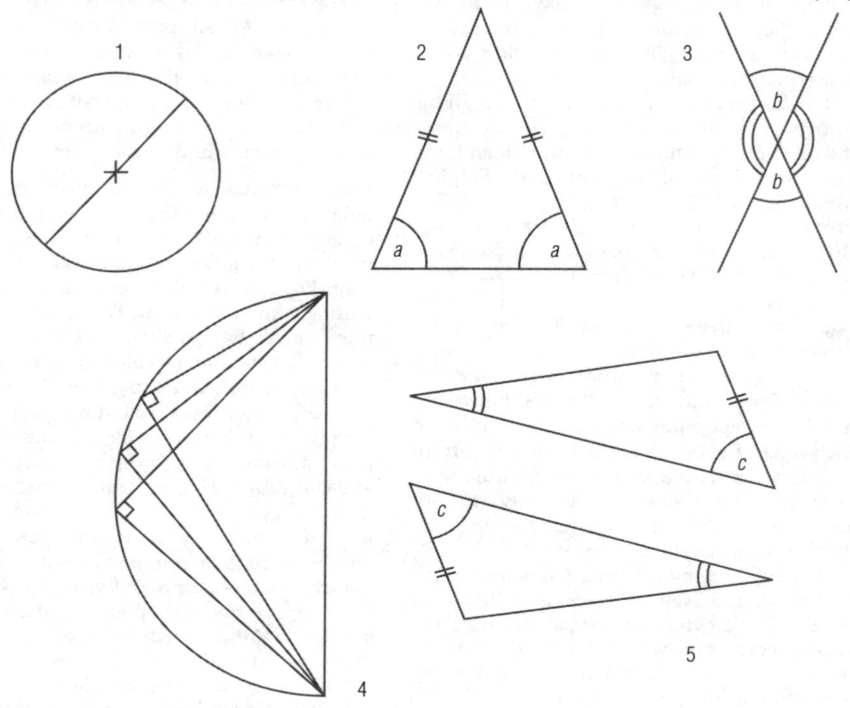

Thales *Some of the basic rules of geometry were laid down by ancient Greek philosopher Thales: (1) a circle is bisected by its diameter; (2) in an isosceles triangle, the two angles opposite the equal sides are themselves equal; (3) when straight lines cross, the opposite angles created are equal; (4) the angle in a semicircle is a right-angle; and (5) two triangles are congruent (the same shape and size) if they have two angles and one side identical.*

the gods. Aristotle records that, when he was reproached for being impractical, Thales, having predicted that weather conditions the next year would be conducive to a large olive harvest, bought up all the olive presses in Miletus and exploited his monopoly to make a large profit.

In five fundamental propositions Thales laid down the foundations on which classical geometry was raised. (1) A circle is bisected by its diameter. (2) In an isosceles triangle the two angles opposite the equal sides are themselves equal to each other. (3) When two straight lines intersect, four angles are produced, the opposite ones being equal. (4) The angle in a semicircle is a right angle. (5) Two triangles are congruent if they have two angles and one side that are respectively equal to each other. He is also said to have introduced the notion of proof by the deductive method.

Theiler Max 1899–1972. South African-born US microbiologist who developed an effective vaccine against yellow fever, which gained him the 1951 Nobel Prize for Physiology or Medicine. He also researched into various other diseases, including Weil's disease and poliomyelitis.

Theiler was born in Pretoria and began his studies in South Africa, completing them in the UK at St Thomas's Hospital and the London School of Hygiene and Tropical Medicine. In 1922 he emigrated to the USA. From 1930 to 1964 he worked at the Rockefeller Institute for Medical Research (now Rockefeller University), New York City, becoming director of the Virus Laboratory there in 1950. He ended his career as professor at Yale.

Theiler used albino mice in his work on yellow fever, and eventually combined the mouse-adapted viral strain with serum from the blood of people who had recovered from yellow fever and injected the mixture into humans. This produced immunity without affecting the kidneys, and was the first safe vaccine against yellow fever.

However, human serum containing antibodies against yellow fever is difficult to obtain. Theiler therefore began culturing the virus in chick embryos, and in 1937 he developed vaccine 17-D, still the main form of protection against yellow fever.

Theodoric of Freiburg *c.* 1250–1310. German scientist and monk. He studied in Paris 1275–77. In his work *De iride/On the Rainbow* he describes how he used a water-filled sphere to simulate a raindrop, and determined that colours are formed in the raindrops and that light is reflected within the drop and can be reflected again, which explains secondary rainbows.

Theon of Smyrna lived *c.* AD 130. Greek astronomer and mathematician. In his celestial mechanics, the planets, Sun, Moon, and the sphere of fixed stars were all set at intervals congruent with an octave. His only surviving work, *Expositio rerum mathematicarum AD legendum Platoneum utilium,* is in two manuscripts, one on mathematics and one on astronomy and astrology.

Theon collated and organized discoveries made by his predecessors, and articulated the interrelationships between arithmetic, geometry, music, and astronomy. The section on mathematics deals with prime, geometrical, and other numbers in the Pythagorean pantheon; the section on music considers instrumental music, mathematical relations between musical intervals, and the harmony of the universe.

The astronomical section is by far the most important. Theon puts forward what was then known about conjunctions, eclipses, occultations, and transits. Other subjects covered include descriptions of eccentric and epicyclic orbits, and estimates of the greatest arcs of Mercury and Venus from the Sun.

Theophrastus *c.* 372–*c.* 287 BC. Greek philosopher, regarded as the founder of botany. A pupil of Aristotle, Theophrastus took over the leadership of his school 323 BC, consolidating its reputation. Of his extensive writings, surviving work is mainly on scientific topics, but includes the *Characters*, a series of caricatures which may have influenced the comic dramatist Menander.

Theophrastus covered most aspects of botany: descriptions of plants, classification, plant distribution, propagation, germination, and cultivation. He distinguished between two main groups of flowering plants – dicotyledons and monocotyledons in modern terms – and between flowering plants and cone-bearing trees (angiosperms and gymnosperms).

Theophrastus was born on Lesvos but studied in Athens at the Academy, which he then headed until his death.

Theophrastus classified plants into trees, shrubs, undershrubs, and herbs. He described and discussed more than 500 species and varieties of plants from lands bordering the Atlantic and Mediterranean. He noted that some flowers bear petals whereas others do not, and observed the different relative positions of the petals and ovary. In his work on propagation and germination, Theophrastus described the various ways in which specific plants and trees can grow: from seeds, from roots, from pieces torn off, from a branch or twig, or from a small piece of cleft wood.

Theremin Leon 1896–1993. Russian inventor of the theremin 1922, a monophonic synthesizer, and of other valve-amplified instruments in the 1930s. Following commercial and public success in the USA and Hollywood, he returned to Russia 1938 to imprisonment and obscurity. After 1945 he continued acoustic research in Moscow and reappeared at a Stockholm electronic music symposium 1990.

The theremin was played without being touched by converting the operator's arm movements into musical tones. The instrument could produce a wide range of sounds and was used to make eerie sound effects for films. He also invented an electronic dance platform, called the terpsitone, in which a dancer's movements were converted into musical tones. Other inventions included a stringless electronic cello, the first syncopated rhythm machine, a colour television system, and a security system which was installed at Sing Sing and Alcatraz Prisons.

Theremin was educated as a physicist and musician. In 1922 he demonstrated his theremin to Lenin and was sent on a tour which included sell-out concerts in Berlin, Paris, London, and New York. He established a studio in New York, where he lived 1927–38. He was then ordered to return to the USSR and sent to a Siberian labour camp during the great Stalin purge. During World War II he worked in a military laboratory, producing a radio-controlled aircraft, tracking systems for ships and submarines, and television systems which are still in use. He also invented a miniature listening device, or 'bug', for the KGB. For this he received the Stalin Prize, First Class, and was allowed to live in Moscow, where he became professor of acoustics at the Moscow Conservatory of Music. He was sacked and his laboratory closed when a chance encounter with a reporter from the *New York Times* resulted in a newspaper article which revealed his earlier imprisonment. He continued his acoustics research, and work aimed at reversing the ageing process, until he died.

Thom René Frédéric 1923– . French mathematician who developed catastrophe theory in 1966. He is a specialist in the fields of differentiable manifolds and topology.

In catastrophe theory, Thom showed that the growth of an organism proceeds by a series of gradual changes that are triggered by, and in turn trigger, large-scale changes or 'catastrophic' jumps. It also has applications in engineering – for example, the gradual strain on the structure of a bridge that can eventually result in a sudden collapse – and has even been extended to economic and psychological events.

Thom was born in Montbéliard in the Vosges and studied in Paris at the Ecole Normale Supérieure. He was professor at Strasbourg 1954–63 and then at the Institute of Advanced Scientific Studies at Bures-sur-Yvette.

Thom formulated a precise series associated with 'space spherical bundles' and demonstrated that the fundamental class of open spherical bundles showed topological invariance and formed a differential geometry. In his work on the theory of forms Thom showed that there are complete homological classes that cannot be the representation of any differential form, and formulated auxiliary spaces, now known as ***Thom spaces***.

In 1956, Thom developed the theory of transversality, and contributed to the examination of singularities of smooth maps. This work laid the ground for his statement of the catastrophe theory, which is, in fact, a model, not yet an explanation. He proposed seven 'elementary catastrophes', which he hoped would be sufficient to describe processes within human experience of space-time

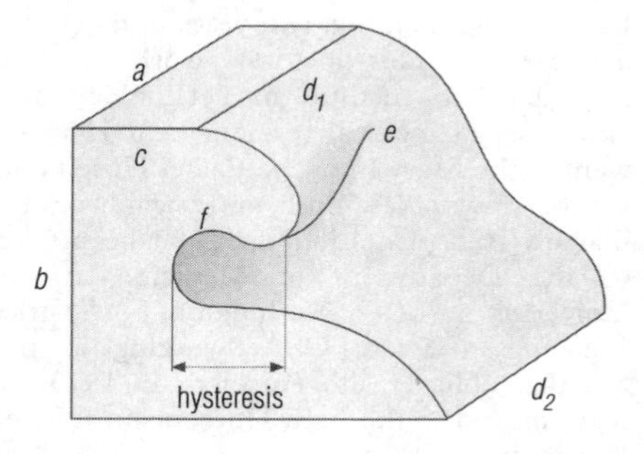

Thom *The cusp is one of seven elementary catastrophes in catastrophe theory: a, b, and c are axes describing interacting factors in a process; d is a folded sheet with an upper zone (d_1) and a lower zone (d_2) describing the behaviour under investigation. At e there is a bifurcation, beyond which (d_1) and (d_2) are separate; f is the intermediate zone.*

dimensions. The seven elementary catastrophes are the fold, cusp, swallow-tail, butterfly, hyperbolic, elliptic, and parabolic. Of these, the cusp is both the simplest and the most useful.

In his book *Stabilité structurelle et morphogenèse/Structural Stability and Morphogenesis* 1972, Thom discusses how the concept of structural stability may be applied to the life sciences. He also introduces the notion of the 'universal unfolding' of a singularity.

Thomas E Donnal 1920– . US scientist who shared the 1990 Nobel Prize for Physiology or Medicine with Joseph E Murray for their discoveries concerning organ and cell transplantation in the treatment of human disease.

Thomas developed the use of bone marrow transplantation to cure leukaemia (cancer of the white blood cells and bone marrow), certain inherited cancers of the bone marrow and some severe blood disorders. The patient is treated with radiation to kill the cancer and then healthy bone marrow cells are given to the patient by infusing it into a blood vessel. Thomas found it was possible to extract about 1 l/1.75 pt of bone marrow from the bones of a healthy donor. In the patient's body, the transplanted bone marrow cells produce new, normal and functioning blood cells. Thomas showed that rejection of the transplanted cells could be avoided by using drugs and if the donor was carefully selected, usually a sibling.

Thomas was born in Texas and educated at the University of Texas in Austin, and the Harvard Medical School, Boston, Massachusetts, receiving the MD degree 1946. There followed an internship, a year of haematology training, two years in the army, a year of post-doctoral work at the Massachusetts Institute of Technology, and two years of medical residency. In 1955 he went to the Mary Imogene Basset Hospital in Cooperstown, New York, and began work on marrow transplantation. In 1963 he moved to the Department of Medicine at the University of Washington, Seattle, Washington State, USA, working in the Seattle Public Health Hospital. In 1975 his team moved to the Fred Hutchinson Cancer Research Center, Seattle.

Thomas Lewis 1913–1993. US physician and academic, best known as a writer of essays on science, especially medicine and biology, for the nonscientist. Essays include 'The Lives of a Cell' 1974 and 'Medusa and the Snail' 1979.

We are built to make mistakes, coded for error.

Lewis Thomas

The Medusa and the Snail
'To Err is Human'

Thomas's father was a doctor and his mother a nurse. As a child in New York, Lewis would often accompany his father on his rounds. Later, he attended Princeton University, from which he graduated to take an internship at the Boston City Hospital 1937. During World War II he served in the navy, carrying out medical experiments in the Pacific. After the war, he continued his medical career; he was dean of the New York University School of Medicine 1966–72 and of the Yale School of Medicine 1972–73, and was president and chief executive of the Memorial Sloan Kettering Cancer Center 1973–80.

Thomas Seth 1785–1859. US clock manufacturer. Establishing his own firm 1812, he became enormously successful in the manufacture of affordable shelf clocks. In 1853 the firm was reorganized as the Seth Thomas Clock Company and continued to prosper into the 20th century.

Born in Wolcott, Connecticut, Thomas trained as a carpenter and cabinetmaker. In 1807 he joined a partnership in Plymouth, Connecticut, for the production of clocks. Seth Thomas Clock Company's headquarters were based in Plymouth Hollow, Connecticut, now renamed Thomaston.

Thomas Sidney Gilchrist 1850–1885. English metallurgist and inventor who, with his cousin Percy Gilchrist, developed a process for removing phosphorus impurities from the iron melted during steel manufacture.

Thomas was born in London and for most of his life worked as a police-court clerk. His deep interest was in industrial chemistry; he experimented systematically at home and attended the laboratories of various chemistry teachers.

The steel produced from phosphoric ores (such as most British, French, German, and Belgian iron ore) was brittle and of little use. Towards the end of 1875 Thomas arrived at a theoretical solution to the problem of how to dephosphorize pig iron when it was loaded into a Bessemer converter. He thought of adding lime, or the chemically similar magnesia or magnesian limestone, to the lining of the converter or furnace. Gilchrist, a chemist at a large ironworks, helped him try out this idea, and Thomas's patent was filed in 1878. The slag that formed as a by-product of this process found use in the developing artificial fertilizer industry.

Thompson Benjamin, original name of Count ◊Rumford, American-born physicist.

Thompson D'Arcy Wentworth 1860–1948. Scottish biologist and classical scholar who interpreted the structure and growth of organisms in terms of the physical forces to which every individual is subjected throughout its life. He also hypothesized, in his book *On Growth and Form* 1917, that the evolution of one species into another results mainly from transformations involving the entire organism.

The concept of an average, the equation to a curve, the description of a froth or cellular tissue, all come within the scope of mathematics for no other reason than that they are summations of more elementary principles or phenomena. Growth and Form are throughout of this composite nature; therefore the laws of mathematics are bound to underlie them, and her methods to be peculiarly fitted to interpret them.

D'Arcy Thompson

On Growth and Form

Thompson was born in Edinburgh. His mother died in childbirth and from the age of three he was brought up by his maternal grandfather who was a veterinary surgeon. In 1877 he went to Edinburgh University to study medicine and in 1880 continued his studies at Trinity College, Cambridge. He graduated 1883 and was appointed professor of biology 1884 at the New University College, Dundee. In 1896 and 1897 he went on expeditions to the Pribilof Islands as a member of a British–American commission on fur-seal hunting in the Bering Sea. He was also one of the British representatives on the International Council for the Exploration of the Sea.

He was awarded a CB 1898 and made a member of the Fishery Board for Scotland. In 1917 he was appointed to the chair of natural history at United College, now a part of St Andrews University. He was elected Fellow of the Royal Society 1916 and awarded both the Darwin medal and the Linnaean Society Gold medal. He was knighted 1937.

In the 1942 revised edition of *On Growth and Form*, Thompson admitted that his evolutionary theory did not adequately account for the cumulative effect of successive small modifications.

Thompson wrote many papers on fisheries and oceanography. He also published works on classical natural history, including *A Glossary of Greek Birds* 1895 and an edition of Aristotle's *Historia animalium* 1910.

Thomson Elihu 1853–1937. US inventor. He founded, with E J Houston (1847–1914), the Thomson–Houston Electric Company 1882, later merging with the Edison Company to form the General Electric Company. He made advances into the nature of the electric arc and invented the first high-frequency dynamo and transformer.

Thomson George Paget 1892–1975. English physicist whose work on interference phenomena in the scattering of electrons by crystals helped to confirm the wavelike nature of particles. He shared a Nobel prize 1937. Knighted 1943.

In the USA, C J Davisson made the same discovery independently, earlier the same year, using a different method.

He was born and educated at Cambridge, the son of physicist J J Thomson. His first professorship was 1922–30 at Aberdeen, moving to Imperial College, London, 1930–52. During World War II, Thomson headed many government committees, including one concerned with atomic weapons.

During the mid-1920s, Thomson carried out a series of experiments hoping to verify French physicist Louis de Broglie's hypothesis that electrons possess duality, acting both as particles and as waves. The experiment involved bombarding very thin metal (aluminium, gold, and platinum) and celluloid foils with a narrow beam of electrons. The beam was scattered into a series of rings. Applying mathematical formulas to measurements of the rings, together with a knowledge of the crystal lattice, Thomson showed in 1927 that all the readings were in complete agreement with de Broglie's theory.

Thomson J(oseph) J(ohn) 1856–1940. English physicist who discovered the electron

1897. His work inaugurated the electrical theory of the atom, and his elucidation of positive rays and their application to an analysis of neon led to the discovery of isotopes. Nobel prize 1906. Knighted 1908.

Using magnetic and electric fields to deflect positive rays, Thomson found in 1912 that ions of neon gas are deflected by different amounts, indicating that they consist of a mixture of ions with different charge-to-mass ratios. English chemist Frederick Soddy had earlier proposed the existence of isotopes and Thomson proved this idea correct when he identified, also in 1912, the isotope neon-22. This work was continued by his student Frederick Aston.

Thomson was born near Manchester and studied there and at Cambridge, where he spent his entire career. As professor of experimental physics 1884–1918, he developed the Cavendish Laboratory into the world's leading centre for subatomic physics. His son was George Paget Thomson.

Investigating cathode rays, Thomson proved that they were particulate and found their charge-to-mass ratio to be constant and with a value nearly 1,000 times smaller than that obtained for hydrogen ions in liquid electrolysis. He also measured the charge of the cathode-ray particles and found it to be the same in the gaseous discharge as in electrolysis. Thus he demonstrated that cathode rays are fundamental, negatively charged particles; the term 'electron' was introduced later.

Thomson, J J *English physicist and discoverer of the electron J J Thomson. In 1876, he won a scholarship to Trinity College, Cambridge, where he eventually succeeded Lord Rayleigh as Cavendish Professor of Experimental Physics. Thomson revolutionized the Cavendish Laboratory, turning it into the greatest research institution in the world – seven of his research assistants went on to win Nobel prizes. Thomson died 1940 and was buried near Newton in the nave of Westminster Abbey.*

The assumption of a state of matter more finely subdivided than the atom of an element is a somewhat startling one.

J J Thomson

1897 Royal Institution Lecture

Thomson James 1822–1892. Northern Irish physicist and engineer who discovered 1849 that the melting point of ice decreases with pressure. He was also an authority on hydrodynamics and invented the vortex waterwheel 1850.

Thomson was born in Belfast, the brother of the future Lord Kelvin. At the age of only ten he began to attend Glasgow University, obtaining an MA in 1839. He held a succession of engineering posts before settling in Belfast in 1851 as a civil engineer. He was professor of civil engineering at Belfast 1857–73 and at Glasgow 1873–89.

The vortex water wheel was a smaller and more efficient turbine than those in use at the time, and it came into wide use. Thomson continued his investigations into whirling fluids, making improvements to pumps, fans, and turbines.

Thomson's discovery about the melting point of ice led him to an understanding of the way in which glaciers flow. He also carried out painstaking studies of the phase relationships of solids, liquids, and gases, and was involved in both geology and meteorology, producing scientific papers on currents and winds.

Thomson William. Irish physicist, see Lord ◊Kelvin.

Thunberg Carl Peter 1743–1828. Swedish botanical explorer who made extensive collections of the flora from Japan and South Africa – 3,000 botanical specimens from South Africa alone. In *Flora japonica* 1784 he identified 21 new genera of plants and hundreds of new species.

Thunberg was born in Jonkoping, Sweden and studied first in Jonkoping and then at Uppsala University, where he was influenced by the work of the eminent naturalist Carolus Linnaeus. He studied natural sciences including botany as well as medicine. In 1770, he graduated with an MD and went to Paris, where he obtained a position on a Dutch merchant ship bound for Japan, setting sail 1772. (The Dutch were the only Europeans permitted access to Japan at that time.) He travelled to Japan via South Africa, where he learnt Dutch and collected 3,000 botanical specimens. In 1775, the ship arrived in Nagasaki, Japan. Thunberg was not permitted to travel in Japan, but he exchanged his Western medical knowledge for specimens that were collected by locals.

In 1776, he travelled to Java and Ceylon before returning to Holland via the Cape Colony. He was made a demonstrator at Uppsala University 1779, but his career was hampered by his disagreements with Linnaeus's son who had replaced him. He was appointed professor of botany in the faculty of medicine.

Tinbergen Niko(laas) 1907–1988. Dutch-born British zoologist who specialized in the study of instinctive behaviour. One of the founders of ethology, the scientific study of animal behaviour in natural surroundings, he shared a Nobel prize 1973 with Konrad Lorenz (with whom he worked on several projects) and Karl von Frisch.

Tinbergen investigated other aspects of animal behaviour, such as learning, and also studied human behaviour, particularly aggression, which he believed to be an inherited instinct that developed when humans changed from being predominantly herbivorous to being hunting carnivores.

Tinbergen was born in The Hague and educated at Leiden, where he became professor 1947. From 1949 he was in the UK at Oxford, and established a school of animal-behaviour studies there.

In *The Study of Instinct* 1951, Tinbergen showed that the aggressive behaviour of the male three-spined stickleback is stimulated by the red coloration on the underside of other males (which develops during the mating season). He also demonstrated that the courtship dance of the male is stimulated by the sight of the swollen belly of a female that is ready to lay eggs.

In *The Herring Gull's World* 1953, Tinbergen described the social behaviour of gulls, emphasizing the importance of stimulus–response processes in territorial behaviour.

Ting Samuel Chao Chung 1936– . US physicist. In 1974 he and his team at the Brookhaven National Laboratory, New York, detected a new subatomic particle, which he named the J particle. It was found to be identical to the Ψ particle discovered in the same year by Burton Richter and his team at the Stanford Linear Accelerator Center, California. Ting and Richter shared the Nobel Prize for Physics 1976. In 1996 Ting was at CERN and the Massachusett's Institute for Technology working on the Alpha Magnetic Spectrometer, a space based antiparticle detector under development.

Tiselius Arne Wilhelm Kaurin 1902–1971. Swedish chemist who developed a powerful method of chemical analysis known as electrophoresis. Electrophoresis is the diffusion of charged particles through a fluid under the influence of an electric field. It can be used to separate molecules of different sizes, which diffuse at different rates. He applied his new techniques to the analysis of animal proteins. Nobel prize 1948.

Tiselius was born in Stockholm and studied at Uppsala, where he spent his career. From 1938 he was director of the Institute of Biochemistry.

Working at Princeton in the USA 1934–35, Tiselius investigated zeolite minerals, and their capacity to retain their crystal structure by exchanging their water of crystallization for other substances. The crystal structure remainins intact even after the water has been removed under vacuum. He studied the optical changes that occur when the dried crystals are rehydrated.

Tiselius first used electrophoresis in the 1920s. In the 1930s he separated the proteins in horse serum and revealed for the first time the existence of three components which he named α-, β-, and γ-globulin. Later he developed new techniques in chromatography.

Tiselius founded the Nobel Symposia, which take place every year in each of the five prize fields to discuss the social, ethical, and other implications of the award-winning work.

Todd Alexander Robertus, Baron Todd 1907–1997. Scottish organic chemist who won a Nobel prize 1957 for his work on the role of nucleic acids in genetics. He also synthesized vitamins B_1, B_{12}, and E. Knighted 1954, Baron 1962.

Todd was born in Glasgow and studied there and in Germany at Frankfurt. He was professor at Manchester 1938–44 and Cambridge 1944–71.

Todd began his work on the synthesis of organic molecules 1934 with vitamin B_1. In the late 1940s and early 1950s he worked on nucleotides; he synthesized adenosine triphosphate (ATP) and adenosine diphosphate (ADP), the key substances in generating energy in the body. He developed new methods for the synthesis of all the major nucleotides and their related co-enzymes, and established in detail the chemical structures of the nucleic acids, such as DNA (deoxyribonucleic acid), the hereditary material of cell nuclei. During the course of this work, which provided the essential basis for further developments in the fields of genetics and of protein synthesis in living cells, Todd also devised an approach to the synthesis of the nucleic acids themselves.

Todt Fritz 1891–1942. German engineer who was responsible for building the first autobahns (German motorways) and, in World War II, the Siegfried Line of defence along Germany's western frontier, and the Atlantic Wall.

Todt's success as minister for road construction led Nazi dictator Hitler to put him in charge of completing the Siegfried Line 1938. His ***Organization Todt***, formed for this task, continued constructing defences on the Atlantic Coast using forced labour until 1944. He was made minister for arms and munitions 1940. In 1942, alarmed at the attrition of equipment on the Eastern Front, he advised Hitler to end the war with the USSR. He was killed in an air crash on the way back from this meeting.

Tolansky Samuel 1907–1973. English physicist who analysed spectra to investigate nuclear spin and magnetic and quadrupole moments. He used multiple-beam interferometry to explore the fine details of surface structure.

Tolansky was born in Newcastle-upon-Tyne and studied mainly at Durham. From 1947 he was professor at London University.

Tolansky made a particular study of the spectrum of mercury. He also studied the hyperfine structure of the spectra of halogen gases such as chlorine and bromine, and of arsenic, iron, copper, and platinum.

During World War II, Tolansky was asked to ascertain the spin of uranium-235, which is the isotope capable of fission in a nuclear chain reaction. Although he had to use samples in which the proportion was only 0.7%, he was fairly successful.

Multiple-beam interferometry can be used to resolve a structure as small as 15 Å (15×10^{-10} m) in height. Tolansky examined the vibration patterns in oscillating quartz crystals and the microtopography of many different crystals, particularly diamonds.

Tombaugh Clyde William 1906– . US astronomer who discovered the planet Pluto 1930.

Tombaugh, born in Streator, Illinois, became an assistant at the Lowell Observatory in Flagstaff, Arizona, in 1929, and photographed the sky in search of an undiscovered but predicted remote planet.

The new planet would be dim, so each photograph could be expected to show anything between 50,000 and 500,000 stars. And, because of its distance from the Earth, any visible motion would be very slight. Tombaugh solved the problem by comparing two photographs of the same part of the sky taken on different days. The photographic plates were focused at a single point and alternately flashed rapidly on to a screen. A planet moving against the background of stars would appear to move back and forth on the screen. Tombaugh found Pluto on 18 Feb 1930, from plates taken three weeks earlier. He continued his search for new planets across the entire sky; his failure to find any placed strict limits on the possible existence of planets beyond Pluto.

Tomonaga Sin-Itiro 1906–1979. Japanese theoretical physicist who developed the theory of quantum electrodynamics (QED). Tomonaga, Richard Feynman, and Julian Schwinger independently developed methods for calculating the interaction between electrons, positrons and photons. The three approaches were essentially the same and QED remains one of the most accurate physical theories known. Nobel Prize for Physics 1965.

Tomonaga developed his quantum theory of the interaction between high-energy sub-atomic particles 1941–43. His central concept was that two sub-atomic particles interact by exchanging a third particle, like footballers throwing the ball to one another. Tomonaga was able to use this idea to produce a theory that was consistent with Einstein's theory of special relativity. In America, Feynman and Schwinger developed the same theory but the three were unaware of each other's work. World War II prevented news of Tomonaga's work from circulating outside Japan until 1947. All three had arrived independently at the same theory.

Tomonaga was born in Tokyo, Japan, and graduated from Kyoto University 1929. He became professor of physics at the University of Tokyo 1941 and President of the University 1956.

Tonegawa Susumu 1939– . Japanese molecular biologist who was awarded the 1987 Nobel Prize for Physiology or Medicine for his discovery that around 1,000 genes are used by the human body to generate billions of different antibodies. Antibodies are molecules which defend the body against infection. When a foreign substances, called an antigen, enters the body, it is attacked and destroyed by antibodies.

When Tonegawa began his work, scientists knew that antibody molecules were produced by genes in white blood cells called B lymphocytes or B cells. However, there were a limited number of genes in the B cells and scientists could not explain how a huge number of different antibodies were produced from so few genes. Tonegawa discovered that the genes were shuffled and recombined in a random manner as a cell grows, producing billions of different gene combinations which could then produce billions of different antibodies.

Tonegawa was born in Nagoya. He was educated at Kyotoa University and the University of California at San Diego, USA, receiving his doctorate 1968. He moved to the Basel Institute of Immunology in Switzerland 1971. In 1981 he was appointed professor at the Center for Cancer Research at the Massachusetts Institute of Technology, Cambridge, Massachusetts, USA.

Torrey John 1796–1873. US botanist who specialized in forms of plant life indigenous to North America and helped to create the Botanical Gardens in New York. He published several books on the botany of America, including *A Catalogue of Plants... within 30 miles of the city of New York* 1819 and *Flora of North America* 1838–43.

Torrey was born in New York and studied medicine at the New York College of Physicians and Surgeons, from where he graduated with his MD 1818. He initially practised medicine in New York but maintained his interests in mineralogy and botany, classifying many plant specimens. In 1824, he was appointed professor of chemistry, mineralogy, and geology at West Point. In 1827, he returned to the New York College of Physicians and Surgeons as the newly appointed professor of chemistry. During this period, he also taught natural history and chemistry at the College of New Jersey (now Princeton).

Torricelli Evangelista 1608–1647. Italian physicist who established the existence of atmospheric pressure and devised the mercury barometer 1644.

In 1643 Torricelli filled a long glass tube, closed at one end, with mercury and inverted it in a dish of mercury. Atmospheric pressure supported a column of mercury about 76 cm/30 in long; the space above the mercury was a vacuum. Noticing that the height of the mercury column varied slightly from day to day, he came to the conclusion that this was a reflection of variations in atmospheric pressure.

Torricelli was born in Faenza and studied at the Sapienza College, Rome. When physicist Galileo read Torricelli's *De motu gravium naturaliter descendentium et proiectorum* 1641, which dealt with movement, he invited him to Florence, where he served as Galileo's secretary for the three months till his death. From 1642 Torricelli was professor of mathematics at Florence.

Tournefort Joseph Pitton de 1656–1708. French botanist who was responsible for defining a genus as a cluster of species and distinguished between the description of a plant and its nomenclature. He wrote *Elemens de botanique* 1694, though he refused to use a microscope to help him with his botanical research.

Tournefort was born in Aix-en-Provence, France and educated by the Jesuits in classics and science. From 1677, he studied natural history and botany at Montpellier University. He decided to travel throughout France and Spain before his appointment as professor at the Jardin du Roi in Paris 1683. His *Schola botanica* lecture notes were published 1689, having been prepared for publication by William Sherad. In 1691, he entered the Acadamie des Sciences. Throughout his life he continued to work as a physician, despite his devotion to botany. He died in Paris 28 November.

Townes Charles Hard 1915– . US physicist who in 1953 designed and constructed the first maser. For this work, he shared the 1964 Nobel prize.

A maser (acronym for *m*icrowave *a*mplification by *s*timulated *e*mission of *r*adiation) is a high-frequency microwave amplifier or oscillator in which the signal to be amplified is used to stimulate unstable atoms into emitting energy at the same frequency. Atoms or molecules are raised to a higher energy level and then allowed to lose this energy by radiation emitted at a precise frequency.

Townes was born in Greenville, South Carolina, and studied there and at Duke and the California Institute of Technology. He was professor at Columbia 1950–61, at the Massachusetts Institute of Technology 1961–67, and from 1967 at the University of California, Berkeley.

Ammonia molecules can occupy only two energy levels and, Townes argued, if a molecule in the high energy level can be made to absorb a photon of a specific frequency, then the molecule should fall to the lower energy level, emitting two photons of the same frequency and producing a coherent beam of single-frequency microwave radiation. Townes had to develop a method (now called population inversion) for separating the relatively scarce high-energy molecules from the more common lower energy ones. He succeeded by using an electric field that focused the high-energy ammonia molecules into a resonator.

In 1958 Townes published a paper that demonstrated the theoretical possibility of producing an optical maser to produce a coherent beam of single-frequency visible light.

Townsend John Sealy Edward 1868–1957. Irish mathematical physicist who studied the kinetics of electrons and ions in gases. He was the first to obtain a value for the charge on the electron, in 1898, and to explain how electric discharges pass through gases. Knighted 1941.

Townsend was born in Galway and studied at Trinity College, Dublin. In 1895 he went to England, initially as a research student at Cambridge together with New Zealander Ernest Rutherford. From 1900 Townsend was professor at Oxford.

Townsend studied the conductivity of gases ionized by the newly discovered X-rays and in 1897 developed a method for producing ionized gases using electrolysis. In 1898, he began the first study of diffusion in gases that had been ionized (or electrified) by means of the so-called ***Townsend discharge*** of a weak current through low-pressure gases. In Townsend's collision theory of ionization, collisions by negative ions (electrons) could induce the formation of secondary ions, thus carrying an electric charge through a gas.

Townsend also studied the electrical conditions that lead to the production of a spark in a gas, and the confusing role played in this by the positive ions that are produced simultaneously with the electrons.

During the 1920s, Townsend was involved with the measurement of the average fraction of energy lost by an electron in a single collision.

Tradescant John 1570–*c.* 1638. English gardener and botanist who travelled widely in Europe and is thought to have introduced the cos lettuce to England from the Greek island of that name. He was appointed gardener to Charles I and was succeeded by his son, John Tradescant the Younger (1608–1662). The younger Tradescant undertook three plant-collecting trips to Virginia in North America.

The Tradescants introduced many new plants to Britain, including the acacia, lilac, and occidental plane. Tradescant senior is generally considered the earliest collector of plants and other natural-history objects.

In 1604 the elder Tradescant became gardener to the earl of Salisbury, who in 1610 for the first time sent him abroad to collect plants. In 1620 he accompanied an official expedition against the North African Barbary pirates and brought back to England gutta-percha and various fruits and seeds. Later, when he became gardener to Charles I, Tradescant set up his own garden and museum in London. In 1624 he published a catalogue of 750 plants grown in his garden.

The Tradescants' collection of specimens formed the nucleus of the Ashmolean Museum in Oxford. Swedish botanist Carolus Linnaeus named the genus *Tradescantia* (the spiderworts) after the younger Tradescant.

Traube Ludwig 1818–1876. German pioneer of experimental pathology who used animals to study the mechanisms of disease. Using this approach, he worked on the pathology of fever, the control and dysfunction of muscles, and various forms of heart disease.

Traube discovered that when the vagus nerve (one of two nerves to the heart that regulate the speed of the pacemaker) was severed, the speed of the pacemaker – consequently heart contractions – could not be slowed. The mechanism for controlling the transport of respiratory gases around the body, therefore, was lost, and breathing was affected as the animal tried to compensate.

Traube was born in Ratibor and educated

at Breslau and then at the University of Berlin.

Travers Morris William 1872–1961. English chemist who, with Scottish chemist William Ramsay, between 1894 and 1908 first identified what were called the inert or noble gases: krypton, xenon, and radon.

Travers was born in London and studied there at University College, where he became professor 1903. He went to Bangalore 1906 as director of the new Indian Institute of Scientists, but returned to Britain at the outbreak of World War I and directed the manufacture of glass at Duroglass Limited. In 1920 he became involved with high-temperature furnaces and fuel technology, including the gasification of coal.

Travers helped Ramsay to determine the properties of the newly discovered gases argon and helium. They also heated minerals and meteorites in the search for further gases, but found none. Then in 1898 they obtained a large quantity of liquid air and subjected it to fractional distillation. Spectral analysis of the least volatile fraction revealed the presence of krypton. They examined the argon fraction for a constituent of lower boiling point, and discovered neon. Finally xenon, occurring as an even less volatile companion to krypton, was identified spectroscopically.

Travers continued his researches in cryogenics and made the first accurate temperature measurements of liquid gases. He also helped to build several experimental liquid air plants in Europe.

Trésaguet Pierre-Marie-Jérôme 1716–1796. French civil engineer who introduced improved methods of road building.

Scottish civil engineer Thomas Telford put the principle into practice when he was surveyor to the county of Shropshire.

Trésaguet was born in Nevers and worked for the Corps des Ponts et Chaussées on civil engineering projects. He eventually rose to inspector general.

Trésaguet realized that lasting improvement to roads could be made only by providing a solid foundation, one that could withstand winter rains and frost, and the effects of traffic. He chose to dig out the roadbed to a depth of about 25 cm/10 in and lay first a course of uniform flat stones, laid on edge to permit drainage. Well hammered in, they provided a solid base on top of which he spread a layer of much smaller stones for a smoother surface. His roads were built 5.4 m/18 ft wide, with a crown that rose 15 cm/6 in above the outside edge.

His method was first used for a main road that ran from Paris to the Spanish border, via Toulouse. It proved so successful that many other countries copied the idea.

Trevithick Richard 1771–1833. English engineer, constructor of a steam road locomotive 1801, the first to carry passengers, and probably the first steam engine to run on rails 1804.

Trevithick also built steamboats, river dredgers, and threshing machines.

Trevithick was born in Illogan, Cornwall. As a boy he was fascinated by mining machinery and the large stationary steam engines that worked the pumps. He made a working model of a steam road locomotive 1797 and went on to build various full-sized engines.

Trevithick's road locomotive *Puffing Devil* made its debut on Christmas Eve 1801, but burned out while he and his friends were celebrating their success at a nearby inn. He then made a larger version which he drove from Cornwall to London the following year, at a top speed of 19 kph/12 mph.

By 1804 he had produced his first railway locomotive, able to haul 10 tonnes and 70 people for 15 km/9.5 mi on rails used by horse-drawn trains at a mine in Wales. He set up in London 1808 giving novelty rides on the engine *Catch-me-who-can*. Then in 1816 he left England for Peru. When he returned, after making and losing a fortune, he found that steam transport had become a thriving concern. Trevithick had been overtaken, and he died a poor man.

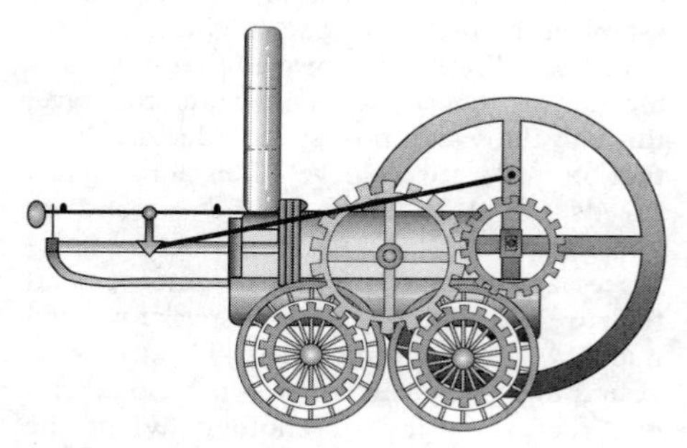

Trevithick *The first steam engine to run on rails was invented by Richard Trevithick 1804. The locomotive was first used on rails designed for horse-drawn wagons, hauling up to 10 metric tons and 70 people to and from the Penydarren ironworks, near Merthyr Tydfil, Wales.*

Trumpler Robert Julius 1886–1956. Swiss-born US astronomer who studied and classified star clusters found in our Galaxy. He also took part in observational tests of the general theory of relativity 1922.

Trumpler was born in Zürich and studied there and in Germany at Göttingen. In 1915 he moved to the USA; he was professor of astronomy at the University of California from 1930.

At the Allegheny Observatory in Pennsylvania, Trumpler noted that galactic star clusters contain an irregular distribution of different classes of stars, and these observations paved the way for later theories about stellar evolution. In 1930 he showed that interstellar material was responsible for obscuring some light from galaxies, which had led to overestimations of their distances from Earth.

Working at the Lick Observatory, near Chicago, he studied the planet Mars, concluding that some of the supposed 'canals' observed by Italian astronomer Giovanni Schiaparelli could be volcanic faults. Trumpler's hypothesis, made in 1924, did not gain real support until the return of the photographs taken by the *Mariner 9* space probe to Mars, more than 50 years later.

Tsiolkovsky Konstantin Eduardovich 1857–1935. Russian scientist who developed the theory of space flight. He published the first practical paper on astronautics 1903, dealing with space travel by rockets using liquid propellants, such as liquid oxygen.

Tsiolkovsky was born in the Spassk district and had little formal education; he was deaf from the age of ten. He never actually constructed a rocket, but his theories and designs were fundamental in helping to establish the reality of space flight.

In 1883 Tsiolkovsky proved that it is feasible for a rocket-propelled craft to travel through the vacuum of space. He calculated that in order to achieve flight into space, speeds of 11.26 km/7 mi per second or 40,232 kph/25,000 mph would be needed – the escape velocity for Earth. Known solid fuels were too heavy, so Tsiolkovsky worked out how to use liquid fuels. He also suggested the 'piggyback' or step principle, with one rocket on top of another. When the lower one was expended, it could be jettisoned (reducing the weight) while the next one fired and took over.

Tswett Mikhail Semyonovich 1872–1919. Italian-born Russian scientist who made an extensive study of plant pigments and developed the technique of chromatography to separate them.

Tswett was born in Asti and studied at Geneva, Switzerland. He worked in Warsaw from 1901, and during World War I organized the evacuation of the Botany Department of the Warsaw Polytechnic Institute to Moscow and Gorky. In 1917 he was appointed professor of botany at Yuriev University (Estonia), but under threat of German invasion had to move once again, to Voronezh.

Tswett showed that green leaves contain more than one type of chlorophyll, and by 1906 he had devised an adsorption method of separating the pigments. He ground up leaves in petroleum ether and let the liquid trickle down a glass tube filled with powdered chalk or alumina. As the mixture seeped downwards, each pigment showed a different degree of readiness to attach itself to the absorbent, and in this way the pigments became separated as different-coloured layers in the tube. Tswett called the new technique chromatography because the result of the analysis was 'written in colour' along the length of the adsorbent column. Eventually he found six different pigments.

Tull Jethro 1674–1741. English agriculturist who about 1701 developed a drill that enabled seeds to be sown mechanically and spaced so that cultivation between rows was possible in the growth period. His chief work, *Horse-Hoeing Husbandry*, was published 1733.

Tull also developed a plough with blades set in such as way that grass and roots were pulled up and left on the surface to dry. Basically the design of a plough is much the same today.

Tull was born in Berkshire, studied at Oxford and qualified as a barrister, but took up farming about 1700.

The seed drill was a revolutionary piece of equipment, designed to incorporate three previously separate actions into one: drilling, sowing, and covering the seeds. The drill consisted of a box capable of delivering the seed in a regulated amount, a hopper mounted above it for holding the seed, and a plough and harrow for cutting the drill (groove in the soil) and turning over the soil to cover the sown seeds.

Turing Alan Mathison 1912–1954. English mathematician and logician. In 1936 he described a 'universal computing machine' that could theoretically be programmed to solve any problem capable of solution by a specially designed machine. This concept, now called the ***Turing machine***, foreshadowed the digital computer.

We do not need to have an infinity of different machines doing different jobs. A single one will suffice. The engineering problem of producing various machines for various jobs is replaced by the office work of 'programming' the universal machine to do these jobs.

Alan Turing

Quoted in A Hodges, *Alan Turing: The Enigma of Intelligence,* 1985

Turing is believed to have been the first to suggest (in 1950) the possibility of machine learning and artificial intelligence. His test for distinguishing between real (human) and simulated (computer) thought is known as the ***Turing test***: with a person in one room and the machine in another, an interrogator in a third room asks questions of both to try to identify them. When the interrogator cannot distinguish between them by questioning, the machine will have reached a state of humanlike intelligence.

Turing was born in London and studied at Cambridge. During World War II he worked on the Ultra project in the team that cracked the German Enigma cipher code. After the war he worked briefly on the project to design the general computer known as the Automatic Computing Engine, or ACE, and was involved in the pioneering computer developed at Manchester University from 1948.

Turing was concerned with mechanistic interpretations of the natural world and attempted to erect a mathematical theory of the chemical basis of organic growth. He was able to formulate and solve complicated differential equations to express certain examples of symmetry in biology and also certain phenomena such as the shapes of brown and black patches on cows.

He committed suicide following a prosecution for a minor homosexual offence.

Turner Charles Henry 1867–1923. US biologist who carried out research into insect behaviour patterns. He was the first to prove that insects can hear and distinguish pitch and that cockroaches learn by trial and error.

Turner was born in Cincinnati, studied there, and went on to teach at schools and colleges. He also wrote nature stories for children and was active in the civil-rights movement in St Louis.

From 1892 until his death, Turner conducted experiments on ants, bees, moths, spiders, and cockroaches. In French literature the turning movement of the ant towards its nest was given the name 'Turner's circling'.

Turner published over 50 papers on neurology, animal behaviour, and invertebrate ecology, including his dissertation 'The homing of ants: an experimental study of ant behaviour' 1907.

Twort Frederick William 1877–1950. English bacteriologist, the original discoverer in 1915 of bacteriophages (often called phages), the relatively large viruses that attack and destroy bacteria. He also researched into Johne's disease, a chronic intestinal infection of cattle.

Twort was born in Camberley, Surrey, and studied medicine at St Thomas's Hospital, London. From 1909 he was superintendent of the Brown Institute, a pathology research centre, and he was also professor of bacteriology at the University of London from 1919.

While working with cultures of *Staphylococcus aureus* (the bacterium that causes the common boil), Twort noticed that colonies of these bacteria were being destroyed. He isolated the substance that produced this effect and found that it was transmitted indefinitely to subsequent generations of the bacterium. He then suggested that the substance was a virus. Twort was unable to continue this work, and the importance of bacteriophages was not recognized until the 1950s.

Twort also discovered that vitamin K is needed by growing leprosy bacteria, which opened a new field of research into the nutritional requirements of microorganisms.

Tyndall John 1820–1893. Irish physicist who 1869 studied the scattering of light by invisibly small suspended particles in colloids. Known as the ***Tyndall effect***, it was first observed with colloidal solutions, in which a beam of light is made visible when it is scattered by minute colloidal particles (whereas a pure solvent does not scatter light). Similar scattering of blue wavelengths of sunlight by particles in the atmosphere makes the sky look blue (beyond the atmosphere, the sky is black).

The mind of man may be compared to a musical instrument with a certain range of notes, beyond which in both directions we have an infinitude of silence.

John Tyndall

Fragments of Science

Tyndall was born in County Carlow and studied at Marburg, Germany. He became professor at the Royal Institution 1853 and was also professor at the Royal School of Mines 1859–68. As superintendent of the Royal Institution from 1867, he did much to popularize science in Britain and also in the USA, where he toured from 1872 to 1873.

Having established that there are dust particles suspended in the air, Tyndall was able to show that the air contains living microorganisms. This confirmed the work of French chemist Louis Pasteur that rejected the spontaneous generation of life, and it also inspired Tyndall to develop methods of sterilizing by heat treatment.

Tyndall also carried out experimental work on the absorption and transmission of heat by gases, especially water vapour and atmospheric gases, which was important in the development of meteorology.

Ulam Stanislaw Marcin 1909–1985. Polish-born US mathematician. He was a member of the Manhattan Project, which produced the first atom bomb, 1943–45, and from 1946 collaborated with Edward Teller on the design of the hydrogen bomb, solving the problem of how to ignite the bomb. All previous designs had collapsed.

Ulam was born and educated in Lvov (now in Ukraine). On the invitation of mathematician John Von Neumann, he emigrated to the USA 1936 and joined the Institute for Advanced Study at Princeton.

Ulugh Beg (Turkish 'great lord') title of Muhammad Taragay 1394–1449. Mongol mathematician and astronomer, ruler of Samarkand from 1409 and of the Mongol Empire from 1447. He built an observatory from which he made very accurate observations of the Sun and planets. He published a set of astronomical tables, called the *Zij of Ulugh Beg*.

Ulugh Beg was born at Sulaniyya in central Asia and brought up at the court of his grandfather Tamerlane. At the age of 15 Ulugh Beg became ruler of the city of Samarkand and the province of Maverannakhr. In 1447 he succeeded his father, Shahrukh, to the throne, but was assassinated two years later in a coup by his son.

In 1420 Ulugh Beg founded an institution of higher learning, or 'madrasa', in Samarkand. It specialized in astronomy and higher mathematics. Four years later he built a three-storey observatory and a sextant. By observing the altitude of the Sun at noon every day, he was able to deduce the Sun's meridianal height, its distance from the zenith, and the inclination of the ecliptic.

The *Zij of Ulugh Beg and his school* is written in Tajik. It consists of a theoretical section and tables of calendar calculations, of trigonometry, and of the positions of planets, as well as a star catalogue of 1,018 stars. This includes 992 stars whose positions Ulugh Beg redetermined with unusual precision.

Urey Harold Clayton 1893–1981. US chemist. In 1932 he isolated heavy water and discovered deuterium, for which he was awarded the 1934 Nobel Prize for Chemistry.

During World War II he was a member of the Manhattan Project, which produced the atomic bomb, and after the war he worked on tritium (another isotope of hydrogen, of mass 3) for use in the hydrogen bomb, but later he advocated nuclear disarmament and world government.

Urey was born in Indiana and educated at Montana State University. He became professor of chemistry at Columbia 1934, and was at Chicago 1945–58.

After deuterium, Urey went on to isolate heavy isotopes of carbon, nitrogen, oxygen, and sulphur. His group provided the basic information for the separation of the fissionable isotope uranium-235 from the much more common uranium-238.

Urey also developed theories about the formation of the Earth.

He thought that the Earth had not been molten at the time when its materials accumulated. In 1952, he suggested that molecules found in its primitive atmosphere could have united spontaneously to give rise to life. The Moon, he believed, had a separate origin from the Earth.

Vail Alfred Lewis 1807–1859. US communications pioneer. A close associate of Samuel Morse 1837, he developed an improved design for the telegraph mechanism, beginning production of the new model 1838. With US Congressional funding for a telegraph line between Washington and Baltimore, Vail renewed his working relationship with Morse 1844.

Vail was born in Morristown, New Jersey, educated at New York University, and worked as a mechanic at his father's iron foundry.

His book *The American Electro Magnetic Telegraph* was published 1845, and he retired from business 1849.

Van Allen James Alfred 1914– . US physicist whose instruments aboard the first US satellite *Explorer 1* 1958 led to the discovery of the Van Allen belts, two zones of intense radiation around the Earth. He pioneered high-altitude research with rockets after World War II.

The ***Van Allen radiation belts*** are two areas of charged particles around the Earth's magnetosphere. The atomic particles come from the upper atmosphere and the solar wind, and are trapped by the Earth's magnetic field. The inner belt lies 1,000–5,000 km/620–3,100 mi above the equator, and contains protons and electrons. The outer belt lies 15,000–25,000 km/9,300–15,500 mi above the equator, but is lower around the magnetic poles. It contains mostly electrons from the solar wind.

Born in Mount Pleasant, Iowa, Van Allen studied at the University of Iowa. He organized and led scientific expeditions to Peru 1949, the Gulf of Alaska 1950, Greenland 1952 and 1957, and Antarctica 1957 to study cosmic radiation. From 1951 he was professor at Iowa. He participated 1953–54 in Project Matterhorn, which was concerned with the study of controlled thermonuclear reactions, and he was responsible for the instrumentation of the first US satellites.

After the end of World War II, Van Allen began utilizing unused German V2 rockets to measure levels of cosmic radiation in the outer atmosphere, the data being radioed back to Earth. He then conceived of rocket-balloons (rockoons), which began to be used in 1952. They consisted of a small rocket which was lifted by means of a balloon into the stratosphere and then fired off.

Van Allen *US physicist James Alfred Van Allen, who discovered the magnetosphere. Using V2 rockets left over from World War II, Van Allen measured cosmic-ray intensities in the upper atmosphere and recorded anomolously high values. When the US government launched the Explorer satellites in 1958, their onboard geiger counters showed that the Earth's magnetic field trapped high-speed charged particles in two doughnut-shaped zones, which were named Van Allen belts or, more commonly, the magnetosphere.*

van de Graaff Robert Jemison 1901–1967. US physicist who from 1929 developed a high-voltage generator, which in its modern form can produce more than a million volts.

The generator consists of a continuous vertical conveyor belt that carries electrostatic charges (resulting from friction) up to a large hollow sphere supported on an insulated stand. The lower end of the belt is earthed, so that charge accumulates on the sphere.

The size of the voltage built up in air depends on the radius of the sphere, but can be increased by enclosing the generator in an inert atmosphere, such as nitrogen.

Van de Graaff was born in Tuscaloosa, Alabama, and studied at the University of Alabama, in France at the Sorbonne, and in the UK at Oxford. He worked at the Massachusetts Institute of Technology 1931–60. In 1946 he set up the High Voltage Engineering Corporation with his collaborator John Trump.

Trump and van de Graaff had modified the generator so that it would produce hard X-rays for use in radiotherapy in treating internal tumours (the first machine was installed in a Boston hospital 1937). In the 1940s they began commercial production, and developed the van de Graaff generator for a wide variety of scientific, medical, and industrial research purposes. The tandem principle of particle acceleration and a new insulating core transformer invented by van de Graaff contributed to these advances.

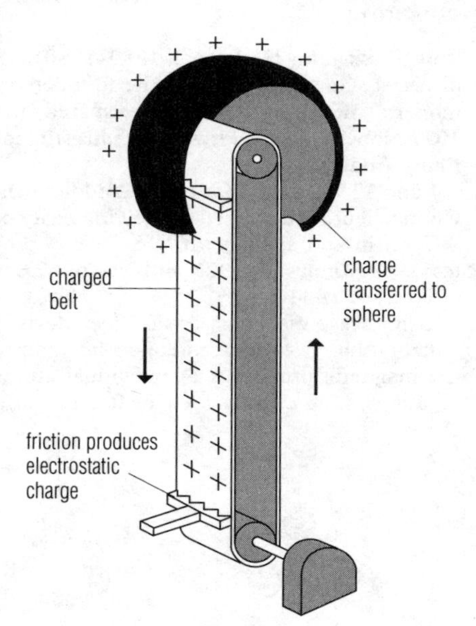

van de Graaff *US physicist Robert Jemison van de Graaff developed this high-powered generator that can produce more than a million volts. Experiments involving charged particles make use of van de Graaff generators as particle accelerators.*

van der Meer Simon 1925– . Dutch physicist who discovered the W and Z elementary particles, the particles responsible for propagating the weak nuclear force. The discovery of these particles confirmed a theory developed in the early 1970s by Steven Weinberg, Sheldon Glashow, Abdus Salam, and others that suggested that the weak nuclear force and the electromagnetic force were linked. Nobel Prize for Physics 1984 (shared with Carlo Rubbia).

The W and Z particles were discovered using the particle accelerators at CERN, the European particle physics laboratory near Geneva, to collide high-energy protons and antiprotons. The collisions produced a shower of particles, like the sparks of a firework, including the W and Z particles. Van der Meer's main contribution to the experiment was the design of a process, called stochastic cooling, which was used to produce large numbers of antiprotons.

Van der Meer was born in The Hague, Netherlands, and educated in engineering at the Technical University in Delft. In 1956 he joined the staff of CERN.

Vandermonde Alexandre-Théophile 1735–1796. French musician and musical theorist who wrote original and influential papers on algebraic equations and determinants.

Vandermonde was born in Paris. He played a part in the founding of the Conservatoire des Arts et des Métiers and served as its director after 1782.

Vandermonde wrote four mathematical papers, all between 1771 and 1773. The first considered the solvability of algebraic equations. He found formulas for solving general quadratic equations, cubic equations, and quartic equations. In addition he found the solution to the equation:

$$x^{11} - 1 = 0$$

and stated, without giving a proof, that:

$$x^{n} - 1 = 0$$

must have a solution where n is a prime number.

The second and third papers are less important; the fourth paper is controversial, and includes the ***Vandermonde determinant***.

Vane John Robert 1927– . British pharmacologist who discovered the wide role of prostaglandins in the human body, produced in response to illness and stress. He shared the 1982 Nobel Prize for Physiology or Medicine with Sune Bergström and Bengt Samuelsson of Sweden.

van't Hoff Jacobus Henricus 1852–1911. Dutch physical chemist. He explained the 'asymmetric' carbon atom occurring in optically active compounds. His greatest work – the concept of chemical affinity as the maximum work obtainable from a reaction – was shown with measurements of osmotic and gas pressures, and reversible electrical cells. He was the first recipient of the Nobel Prize for Chemistry, in 1901.

Van't Hoff was born in Rotterdam and studied at several universities in Europe. He was professor at Amsterdam 1878–96 and at the Prussian Academy of Sciences in Berlin from 1896.

In 1874 van't Hoff postulated that the four valencies of a carbon atom are directed towards the corners of a regular tetrahedron. This allows it to be asymmetric (connected to four different atoms or groups) in certain compounds, and it is these compounds that exhibit optical activity. Van't Hoff ascribed the ability to rotate the plane of polarized light to the asymmetric carbon atom in the molecule, and showed that optical isomers are left- and right-handed forms (mirror images) of the same molecule.

Van't Hoff's first ideas about chemical thermodynamics and affinity were published in 1877, and consolidated in his *Etudes de dynamique chimique* 1884. He applied thermodynamics to chemical equilibria, developing the principles of chemical kinetics and describing a new method of determining the order of a reaction. He deduced the connection between the equilibrium constant and temperature in the form of an equation known as the ***van't Hoff isochore***.

Van Tiegheim Phillipe 1839–1914. French botanist and biologist who defined the plant as having three distinct parts, the stem, the root, and the leaf, and studied the origin and differentiation of each type of plant tissue. His best known research included studies of the gross anatomy of the phanerogams (plants with reproductive organs) and the cryptogams (plants without reproductive organs, such as mosses and ferns).

Van Tiegheim was born in Bailleul, Nord, France. As an orphan, he was brought up by his aunt and uncle and, in 1856, he went to Bailleul College, where he obtained his baccalaureate in science. In 1858, he moved to the Ecole Normale Supérieure, and obtained a position as a teacher of botany and mineralogy there 1861. He went on to work with Louis Pasteur in Paris on the principles of fermentation.

He held a succession of academic posts in France: professor of botany at the Ecole Normale Supérieure 1864, professor and administrator at the Museum of Natural History 1879, and professor of biology at the Ecole Centrale des Arts et Manufactures 1873. From 1885–1912, he was the professor at the Ecole Normale Supérieure des Jeunes Filles and Sèvres, and was professor of plant biology at the Institut Agronomique 1898–1914.

Van Vleck John Hasbrouck 1899–1980. US physicist, considered one of the founders of modern magnetic theory. He shared the 1977 Nobel Prize for Physics with his student Philip Anderson.

Van Vleck was born in Middletown, Connecticut, and studied at the University of Wisconsin and at Harvard. He became professor at Minnesota 1927 and was professor at Harvard 1934–69.

Using wave mechanics, Van Vleck devised a theory that gives an accurate explanation of the magnetic properties of individual atoms in a series of chemical elements. He also

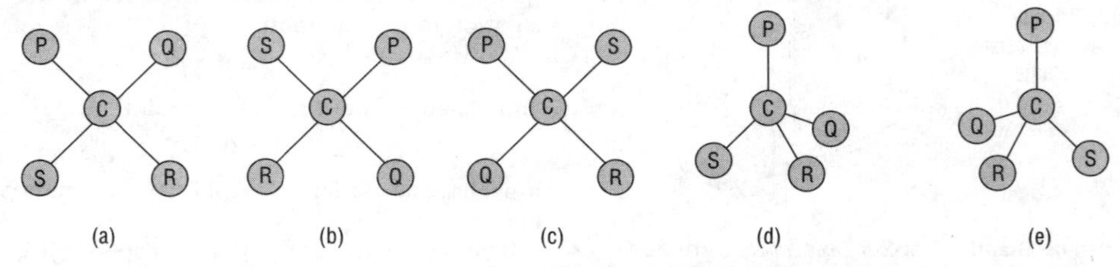

van't Hoff *The three square structures (a), (b), and (c) are all equivalent (rotating (b) 90° anticlockwise gives (a); rotating (c) 180° along the axis P–R also gives (a)). The tetrahedral structures (d) and (e) are mirror images of each other, but one cannot be rotated in any way to make it the same as the other; they represent a pair of (optical) streoisomers.*

introduced the idea of temperature-independent susceptibility in paramagnetic materials – now called ***Van Vleck paramagnetism***. His formulation of the ligand field theory is one of the most useful tools for interpreting the patterns of chemical bonds in complex compounds. This theory explains the magnetic, electrical, and optical properties of many elements and compounds by considering the influences exerted on the electrons in particular atoms by other atoms nearby.

Varmus Harold Elliot 1939– . US molecular biologist who shared the Nobel Prize for Physiology or Medicine 1989 with John Michael Bishop for his demonstration of the presence of oncogenes (cancer-causing genes) in certain viruses and mammalian cells. His work showed that a single viral gene is sufficient to transform a cell into a cancer-like cell, and that this gene is also present in normal mammalian cells. Since the discovery of the first oncogene, many more have been found and used in cancer research.

The Rous sarcoma, or avian sarcoma virus, produces highly malignant tumours in chickens. Varmus and Bishop demonstrated that when the gene *src* is introduced into cells by this virus, it transforms the cell into a cancer-like cell. They also demonstrated, using DNA hybridization experiments, that the *src* gene was present, albeit usually in a quiescent form, in healthy mammalian cells.

Varmus was born in Oceanside, New York and educated at Amherst College, Massachusetts, Harvard University, and Columbia University, where he received an MD. Since 1970 he has worked at the Medical Center of the University of California. He is currently the director of the US National Institutes of Health.

Vaucouleurs Gerard Henri De 1918– . French-born US astronomer; see ◊De Vaucouleurs.

Vauquelin Louis Nicolas 1763–1829. French chemist who worked mainly in the inorganic field, analysing minerals. He discovered the elements chromium (1797) and beryllium.

Vauquelin was born in Saint-André d'Héberôt, Calvados, and was apprenticed to an apothecary. Moving to Paris, he became a laboratory assistant at the Jardin du Roi and was befriended by a professor of chemistry. In 1791 he was made a member of the Academy of Sciences and from that time he helped to edit the journal *Annales de Chimie*, although he left the country for a while during the height of the French Revolution. On his return in 1794 he became professor of chemistry at the Ecole des Mines in Paris, and in 1802 was appointed assayer to the Mint. From 1809 he was professor at the University of Paris. He was elected to the Chamber of Deputies in 1828.

In organic chemistry, Vauquelin also made some significant discoveries. In 1806, working with asparagus, he isolated the amino acid aspargine, the first one to be discovered. He also discovered pectin and malic acid in apples, and isolated camphoric acid and quinic acid.

Vening Meinesz Felix Andries 1887–1966. Dutch geophysicist who originated the method of making very precise gravity measurements in the stable environment of a submarine. The results he obtained were important in the fields of geophysics and geodesy. He was able to discount the model of the Earth's shape that proposed a flattening at the equator.

Vening Meinesz was born in The Hague and studied civil engineering at the Technical University of Delft. Employed by the government to take part in a gravimetric survey of the Netherlands, he took measurements at over 50 sites. He was professor at Utrecht 1927–57 and also at Delft 1938–57. Between 1923 and 1939 he undertook 11 scientific expeditions in submarines.

Vening Meinesz realized that measurements of the Earth's gravitational field could yield indications of the internal features of the Earth. He developed a device requiring the measurement of the mean periods of two pendulums that swing from the same apparatus. The mean of the two periods is not affected by disturbances in the horizontal plane, and so can be used to determine the local gravitational force accurately.

Underwater studies led to the discovery of low-gravity belts in the Indonesian archipelago. Vening Meinesz proposed that these were the result of a downward buckling of the crust causing light sediments to fill the resulting depressions. This is the origin of the concept of the syncline.

Venn John 1834–1923. English logician whose diagram, known as the Venn diagram, is much used in the teaching of elementary mathematics.

The ***Venn diagram*** represents sets and the logical relationships between them. Sets are drawn as circles. An area of overlap between two circles contains elements that

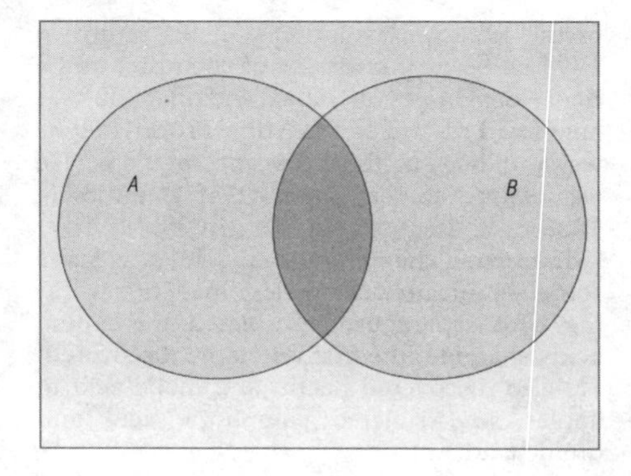

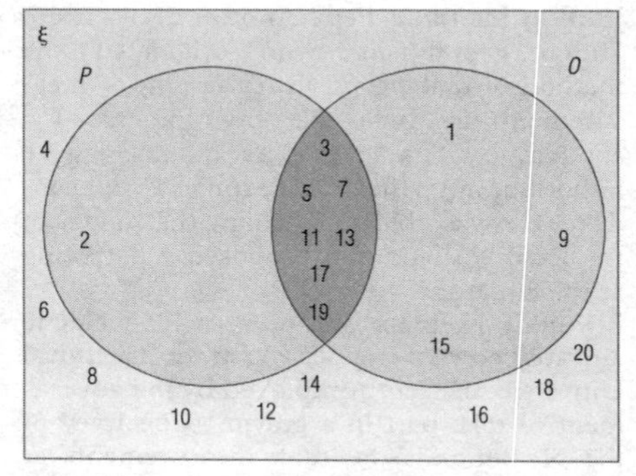

Venn *Venn diagram. Above, a Venn diagram of two intersecting sets; below, a Venn diagram showing the set of whole numbers from 1 to 20 and the subsets P and O of prime and odd numbers, respectively. The intersection of P and O contains all the prime numbers that are also odd.*

are common to both sets, and thus represents a third set. Circles that do not overlap represent sets with no elements in common (disjoint sets).

Venn was born in Hull, Yorkshire, and studied at Cambridge. He became a priest, but abjured his clerical orders 1883. From 1862 he was a Cambridge college lecturer in moral sciences.

The use of geometrical representations to illustrate syllogistic logic was not new. Venn adopted the method of illustrating propositions by means of exclusive and inclusive circles, and added the new device of shading the segments of the circles to represent the possibilities that were excluded by the propositions at issue. Later, he extended his method by proposing a series of circles dividing the plane into compartments, so that each successive circle should intersect all the compartments already existing. This idea, taken up and refined by Charles Dodgson (the writer Lewis Carroll), led to the use of the closed compartment enclosing the whole diagram to define what is now known as the universal set.

Venn published three standard texts: *The Logic of Chance* 1866, *Symbolic Logic* 1881, and *The Principles of Empirical Logic* 1889.

Vernier Pierre *c.* 1580–1637. French engineer and instrumentmaker who invented a means of making very precise measurements with what is now called the vernier scale. The vernier consists of a short divided scale that carries an index or pointer and is slid along a main scale.

Vernier was born in Ornans, near Besançon. He realized the need for a more accurate way of reading angles on the surveying instruments he used in mapmaking whilst working as a military engineer. In 1630 he was appointed to the service of the count of Burgundy, for whom he built fortifications.

In 1631 Vernier published *La construction, l'usage, et les proprietez du quadrant nouveau mathématique/The Construction, Uses and Properties of a New Mathematical Quadrant*, in which he explained his method.

Vesalius Andreas 1514–1564. Belgian physician who revolutionized anatomy. His great innovations were to perform postmortem dissections and to make use of illustrations in teaching anatomy.

The dissections (then illegal) enabled him to discover that Galen's system of medicine was based on fundamental anatomical errors. Vesalius's book *De humani corporis fabrica/On the Structure of the Human Body* 1543, together with the main work of astronomer Copernicus, published in the same year, marked the dawn of modern science.

Vesalius was born in Brussels and studied at Louvain, Paris, and Padua in Italy, where he was professor 1537–42. He became court physician to Charles V, and later to his son Philip II of Spain. On his way back from a pilgrimage to Jerusalem, Vesalius died in a shipwreck off Greece.

Dissatisfied with the instruction he had received, Vesalius resolved to make his own observations, which disagreed with Galen's. For instance, he disproved that men had a rib less than women – a belief that had been widely held until then. He also believed, contrary to Aristotle's theory of the heart being the centre of the mind and emotion, that the brain and the nervous system are the centre.

Between 1539 and 1542 Vesalius prepared his masterpiece, a book that employed

Vesalius *The title page of Andreas Vesalius's 1543 masterpiece on human anatomy* De humani corporis fabrica. *It contained magnificent illustrations of a quality never before seen in a medical book, and described and depicted several organs, such as the thalamus, for the first time. Vesalius's book met with bitter controversy and criticism and led him to abandon anatomy altogether.*

talented artists to provide the anatomical illustrations.

Viète François 1540–1603. French mathematician who developed algebra and its notation. He was the first mathematician to use letters of the alphabet to denote both known and unknown quantities, and is credited with introducing the term 'coefficient' into algebra.

Viète was born in Fontenay-le-Comte in the Poitou region and studied law at Poitiers. In 1570 he moved to Paris and was employed by Charles IX until 1584, when persecution of the Huguenots forced him to flee to Beauvoir-sur-Mer. It was in these years that his most fruitful algebraic research was carried out. On the accession of Henry IV in 1589, Viète returned to the royal service, and deciphered coded messages captured during war with Spain. He was dismissed from the court in 1602.

Viète's mathematical achievements were the result of his interest in cosmology; for example, a table giving the values of six trigonometrical lines based on a method originally used by Egyptian astronomer Ptolemy. Viète was the first person to use the cosine law for plane triangles and he also published the law of tangents.

His works include *Canon mathematicus seu AD triangula* 1579, *In artem analytica isogoge* 1591, and *De aequationum recognitione et emandatione* 1615.

Villemin Jean-Antoine 1827–1892. French military physician who provided early evidence that tuberculosis (TB) is transmitted and caused by an infective agent. Contrary to popular belief, Villemin proved that TB could be passed from humans or cows to rabbits, and from these rabbits, to other rabbits in a series. He also proved that the disease in the rabbit when passed from a cow was considerably worse than when passed from a person.

In 1865, Villemin inoculated a healthy rabbit with TB material obtained from the lung cavity of a human who had died of pulmonary tuberculosis. Another rabbit from the same mother was kept under identical conditions to the inoculated rabbit as a control. The rabbits were killed after three and a half months. The inoculated rabbit's lungs were full of TB lesions, while the control rabbit showed no symptoms of the disease.

He went on to inoculate a rabbit with TB material obtained from the lung of cow. This rabbit became acutely ill after two months. Its organs showed acute TB on post mortem. Villemin then inoculated a series of six rabbits with material obtained from the lesions of rabbit lungs. His experiments proved that TB could be transmitted from animal to animl and between species in a series.

Villemin was born in Prey, Vosges and studied medicine in Strasburg and Paris, where he received his MD 1853. He worked as a general practitioner but ran his own private research laboratory in his spare time. The nature of the infective microorganism was not discovered for another twenty years until the work of the German bacteriologist Robert Koch 1882.

Virchow Rudolf Ludwig Carl 1821–1902. German pathologist, the founder of cellular pathology. Virchow was the first to describe leukaemia (cancer of the blood). In his book

Die cellulare pathologie/Cellular Pathology 1858, he proposed that disease is not due to sudden invasions or changes, but to slow processes in which normal cells give rise to abnormal ones.

There can be no scientific dispute with respect to faith, for science and faith exclude one another.

Rudolf Virchow

Disease, Life and Man, 'On Man'

Virtanen Artturi Ilmari 1895–1973. Finnish chemist who from 1920 made discoveries in agricultural chemistry. Because green fodder tends to ferment and produce a variety of harmful acids, it cannot be preserved for long. Virtanen prevented the process from starting by acidifying the fodder. In this form it lasted longer and remained nutritious. Nobel Prize for Chemistry 1945.

Vogel Hermann Carl 1842–1907. German astronomer who discovered spectroscopic binary stars. By measuring the Doppler effect in the spectral lines of stars to ascertain their velocity, he ended the controversy over the value of Christian Doppler's theory for investigating motion in the universe.

Vogel was born and educated in Leipzig and in 1863 began working at the observatory there. From 1882 he was director of the Potsdam Observatory near Berlin.

Vogel worked intensively on the spectroscopic properties of planets, nebulae, the northern lights, comet III 1871, and the Sun, and examined the spectra of some 4,000 stars. He used spectrophotometry to study Nova Cygni in 1876 and his results provided the first evidence that changes occur in the spectrum of a nova during its fading phase.

Vogel's discovery of spectroscopic binary stars arose from a study of the periodic displacements of the spectral lines of the stars Algol and Spica, eclipsing binary stars whose components could not, at the time, be detected as separate entities by optical means. From his spectrographs, Vogel derived the dimensions of this double star system, the diameter of both components, the orbital velocity of Algol, the total mass of the system and in 1889, he derived the distance between the two component stars from each other.

Vogt Marthe Louise 1903– . Neuropharmacist who was involved in proving that acetylcholine is a neurotransmitter (chemical that transmits impulses between nerve cells) involved in the stimulation of muscles by nerves. Acetylcholine is now known to be one of the principal neurotransmitters that communicates nerve impulses across synapses, the junctions between individual nerves.

In her experiments, Vogt and her co-workers removed sensory nerves from various muscles of cats and dogs, ensuring that only motor neurons could be stimulated. The muscles were then perfused (coated) in a solution (Locke's solution) to prevent the destruction of any neurotransmitters released on stimulation. They were able to consistently detect the release of acetylcholine when the muscle was stimulated. This release ceased immediately when the stimulus stopped.

If the nerve was made to degenerate, stimulation caused muscle contraction but no release of acetylcholine, showing that the chemical originated from the neuron. If the muscle was paralyzed, stimulation caused the production of acetylcholine but no contraction, demonstrating that the acetylcholine was not released as a result of muscle contraction. She concluded that the acetylcholine was associated with the transmission of an impulse from a motor neuron to a muscle.

Volhard Jacob 1834–1910. German chemist who devised various significant methods of organic synthesis, and a method of quantitatively analysing for an element via silver chloride. Bromides can also be determined using this technique.

Volhard was born in Darmstadt and studied at Giessen. He held professorial appointments at three German universities: Munich 1864–79, Erlangen 1879–82, and Halle 1882–1910.

During the 1860s Volhard developed methods for the syntheses of the amino acids sarcosine (*N*-methylaminoethanoic acid) and creatine, and the heterocyclic compound thiophen; he also did research on guanidine and cyanimide.

Volhard's method of preparing halogenated organic acids has become known as the ***Hell–Volhard–Zelinsky reaction***, in which the acid is treated with chlorine or bromine in the presence of phosphorus. The reaction is also useful for syntheses because the substituted halogen atom(s) can easily be replaced by a cyanide group (by treatment with potassium cyanide), which in the presence of an aqueous acid and ethyl alcohol (ethanol) yields the corresponding malonic ester (diethylpropandioate), from which barbiturate drugs can be synthesized.

Volta Alessandro Giuseppe Antonio Anastasio, Count 1745–1827. Italian physicist who invented the first electric cell (the voltaic pile, 1800), the electrophorus (an early electrostatic generator, 1775), and an electroscope.

An electroscope is a device for detecting an electric charge. Volta's version consisted of a vertical conducting (metal) rod ending in two pieces of straw, mounted inside and insulated from an earthed metal case. An electric charge applied to the metal rod made the straws diverge, because they each received a similar charge (positive or negative) and so repelled each other.

In 1776 Volta discovered methane by examining marsh gas found in Lago Maggiore. He then made the first accurate estimate of the proportion of oxygen in the air by exploding air with hydrogen to remove the oxygen. In about 1795, Volta recognized that the vapour pressure of a liquid is independent of the pressure of the atmosphere and depends only on temperature.

Volta was born in Como; he became professor there 1775 and was professor at Pavia 1778–1819. The ***volt*** is named after him.

Volta's electrophorus consisted of a disc made of turpentine, resin, and wax, which was rubbed to give it a negative charge. A plate covered in tin foil was lowered by an insulated handle on to the disc, which induced a positive charge on the lower side of the foil. The negative charge that was likewise induced on the upper surface was removed by touching it to ground the charge, leaving a positive charge on the foil. This process could then be repeated to build up a greater and greater charge. Volta went on to realize from his electrostatic experiments that the quantity of charge produced is proportional to the product of its tension and the capacity of the conductor.

Volta repeated and built on Italian physiologist Luigi Galvani's experiments with metals and the muscles of dead animals, and in 1792, Volta concluded that the source of the electricity was in the junction of two different metals and not, as Galvani thought, in the animals. Volta even succeeded in producing a list of metals in order of their electricity production based on the strength of the sensation they made on his tongue, thereby deriving the electromotive series.

In 1800 Volta described two arrangements of conductors that produced an electric current. One was a pile of silver and zinc discs separated by cardboard moistened with brine, and the other a series of glasses of salty or alkaline water in which bimetallic curved electrodes were dipped. Volta's electric cell was a sensation, for it enabled high electric currents to be produced for the first time.

Volterra Vito 1860–1940. Italian mathematician whose chief work was in the fields of function theory and differential equations. His chief method, hit upon as a young boy, was based on dividing a problem into a small interval of time and assuming one of the variables to be constant during each time period.

Volterra was born in Ancona and studied at Florence and Pisa. He was professor at Pisa 1883–92, Turin 1892–1900, and Rome 1900–31. During World War I he established the Italian Office of War Inventions, where he designed armaments and proposed that helium be used in place of hydrogen in airships. After the war he became increasingly involved in politics, speaking in the Senate and voicing his opposition to the Fascist regime. For his views he was eventually dismissed from his academic post and banned from taking part in any Italian scientific meeting.

At the age of 13, after reading Jules Verne's novel *From the Earth to the Moon*, Volterra became interested in projectile problems and came up with a plausible determination for the trajectory of a spacecraft which had been fired from a gun. His solution was based on the device of breaking time down into small intervals during which it could be assumed that the force was constant. The trajectory could thus be viewed as a series of small parabolic arcs. This was the essence of the argument he developed in detail 40 years later in a series of lectures at the Sorbonne, France.

Volterra contributed especially to the foundation of the theory of functionals, the solution of integral equations with variable limits, and the integration of hyperbolic partial differential equations. His papers on partial differential equations of the early 1890s included the solution of equations for cylindrical waves.

He also brought his knowledge of mathematics to bear on biological matters, constructing a model for population change in which the prey and the predator interact in a continuous manner.

Volterra's main works are *The Theory of Permutable Functions* 1915 and *The Theory of Functionals and of Integral and Integro-differential Equations* 1930.

von Braun Wernher Magnus Maximilian 1912–1977. German rocket engineer who developed military rockets (V1 and V2) during World War II and later worked for the space agency NASA in the USA.

During the 1940s his research team at Peenemünde on the Baltic coast produced the V1 (flying bomb) and supersonic V2 rockets.

In the 1950s von Braun was part of the team that produced rockets for US satellites (the first, *Explorer 1*, was launched early 1958) and early space flights by astronauts.

Von Braun was born in Wirsitz (now in Poland) and studied at Berlin and in Switzerland at Zürich. In 1930 he joined a group of scientists who were experimenting with rockets, and in 1938 he became technical director of the Peenemünde military rocket establishment; he joined the Nazi Party 1940. In the last days of the war in 1945 von Braun and his staff, not wishing to be captured in the Soviet-occupied part of Germany, travelled to the West to surrender to US forces. Soon afterwards von Braun began work at the US Army Ordnance Corps testing grounds at White Sands, New Mexico. In 1952 he became technical director of the army's ballistic-missile programme. He held an administrative post at NASA 1970–72.

von Gesner Konrad, Swiss naturalist. See ◊Gesner, Konrad von.

von Klitzing Klaus 1943– . German physicist who discovered the quantized Hall effect. This effect involves the behaviour of electric currents in thin films of material held in a strong magnetic field at a low temperature. If the magnetic field is varied, the current changes in a step-like fashion, rather than in a smooth regular way. This steplike behaviour indicates that quantum effects are taking place. The quantized Hall effect is one of the few situations where quantum effects can be studied using ordinary macroscopic measurements. The effect is used to make very accurate measurements of electrical voltages. Nobel Prize for Physics 1985.

Von Klitzing was born in Schroda (Posen). He studied physics at the Technical University of Braunschweig and at the University of Würtzburg, gaining his doctorate 1972. He became a professor at the Technical University at München 1980. In 1985 he was appointed director of the Max Planck Institute, Stuttgart.

Von Neumann John, (originally Johann) 1903–1957. Hungarian-born US scientist and mathematician, a pioneer of computer design. He invented his 'rings of operators' (called Von Neumann algebras) in the late 1930s, and also contributed to set theory, game theory, quantum mechanics, cybernetics (with his theory of self-reproducing automata, called ***Von Neumann machines***), and the development of the atomic and hydrogen bombs.

He designed and supervised the construction of the first computer able to use a flexible stored program (named MANIAC-1) at the Institute for Advanced Study at Princeton 1940–1952. This work laid the foundations for the design of all subsequent programmable computers.

It would appear that we have reached the limits of what it is possible to achieve with computer technology, although one should be careful with such statements, as they tend to sound pretty silly in 5 years.

John Von Neumann

Von Neumann was born in Budapest and studied in Germany and Switzerland. In 1930 he emigrated to the USA, where he became professor at Princeton 1931, and from 1933 he was a member of the Institute for Advanced Study there. He also held a number of advisory posts with the US government 1940–54.

Von Neumann's book *The Mathematical Foundations of Quantum Mechanics* 1932 defended mathematically the uncertainty principle of German physicist Werner Heisenberg. In 1944, Von Neumann showed that matrix mechanics and wave mechanics were equivalent.

The monumental *Theory of Games and Economic Behavior* 1944, written with Oskar Morgenstern (1902–1977), laid the foundations for modern game theory.

Von Neumann originated the basic ideas of game theory in 1928 by proving that a quantitative mathematical model could be constructed for determining the best strategy –

the one that, in the long term, would produce maximal gains and minimal losses in any game, even one of chance or one with more than two players. Situations for which this theory found immediate use were in business, warfare, and the social sciences.

Voronoff Serge 1866–1951. Russian physiologist who carried out testicular transplants from 1919 for rejuvenation. He believed that glandular secretions could enhance a person's lifespan and began to graft animal glands into aging humans.

He initially experimented on male rejuvenation by transplanting monkey testicles into human subjects. At the time, the medical profession was against these experiments, which were subsequently shown to be ineffective and unsatisfactory due to unpleasant side effects.

Voronoff was born in Voronezh and educated in Paris. He worked at the College de France at Paris and then moved to Switzerland, where he began to work on the control of human longevity.

Vorontsov-Vel'iaminov Boris Aleksandrovich 1904– . Russian astronomer and astrophysicist. In 1930, independently of Swiss astronomer Robert Trumpler, Vorontsov-Vel'iaminov demonstrated the occurrence of the absorption of stellar light by interstellar dust. As a result, it became possible to determine astronomical distances and, in turn, the size of the universe more accurately.

Vorontsov-Vel'iaminov was professor at Moscow from 1934. Analysing the Hertzsprung–Russell diagram, with reference to the evolution of stars, he made particularly important contributions to the study of the blue–white star sequence, which was the subject of a book he published in 1947.

In 1959 Vorontsov-Vel'iaminov recorded and listed the positions of 350 interacting galaxies clustered so closely that they seem to perturb each other slightly in structure. Besides this catalogue, he compiled a more extensive catalogue of galaxies in 1962, in which he listed and described more than 30,000 examples.

Waddington Conrad Hal 1905–1975 English geneticist and biologist who examined the evolutionary significance of the interrelations between the genetic constitution, characteristics, and environment of an organism. Waddington performed experiments that demonstrated the theory that natural selection is the cause of hereditary change. He was a champion of the principle of 'organic selection', in which environmentally induced changes in somatic (body) cells can result in hereditary changes, not because they affect the hereditary material (DNA) itself, but because they enable the population to survive long enough to allow the accumulation and selection of similar hereditary changes.

In one experiment, Waddington gave the pupae of the fruit fly *Drosophila melanogaster* a heat shock for several hours, resulting in part or all of a vein across the wing being absent (crossveinless). Flies that showed the variation, approximately 40% of the population, were crossed with each other. Matings eventually produced flies that were crossveinless without heat shock; the environmentally produced characteristic had become an inherited characteristic.

Waddington was born in Evesham in the UK and trained originally as a geologist at Cambridge University, graduating 1926. He then turned his attention to palaeontology and finally to embryology and evolution.

Wagner-Jauregg Julius 1857–1940. Austrian neurologist. He received a Nobel prize in 1927 for his work on the use of induced fevers in treating mental illness.

Waite Edgar Ravenswood 1866–1928. Australian ornithologist and zoologist. He was a member of several expeditions into the subantarctic islands, New Guinea, and central Australia which contributed significantly to scientific knowledge of vertebrates. His published work includes more than 200 scientific papers, *Popular Account of Australian Snakes* 1898 and *The Fishes of South Australia* 1923.

Waksman Selman Abraham 1888–1973. US biochemist, born in Ukraine. He coined the word 'antibiotic' for bacteria-killing chemicals derived from microorganisms. Waksman was awarded a Nobel prize in 1952 for the discovery of streptomycin, an antibiotic used against tuberculosis.

Waksman, was professor of soil microbiology at Rutgers University in New Jersey in the USA from 1910.

Wald George 1906– . US biochemist who explored the chemistry of vision. He discovered the role played in night vision by the retinal pigment rhodopsin, and later identified the three primary-colour pigments. Nobel Prize for Physiology or Medicine 1967.

Wald was born and educated in New York. He spent his academic career at Harvard from 1935, becoming professor of biology 1948. In the 1970s he spoke out against the US role in the Vietnam War.

Studying rhodopsin, which occurs in the rods (dim-light receptors) of the retina, Wald discovered in 1933 that this substance consists of the colourless protein opsin in combination with retinal, a yellow carotenoid compound that is the aldehyde of vitamin A. Rhodopsin molecules are split into these two compounds when they are struck by light, and the enzyme alcohol dehydrogenase then further reduces the retinal to form vitamin A. In the dark the process is reversed, but over a period of time some of the retinal is lost. This deficiency has to be made up from vitamin A, and if the body's stores are inadequate, night blindness results.

In the 1950s Wald found the retinal pigments that detect red and yellow-green light, and a few years later the pigment for blue light. All these are related to vitamin A, and in the 1960s he demonstrated that the absence of one or more of them results in colour blindness.

Wallace Alfred Russel 1823–1913. Welsh naturalist who collected animal and plant specimens in South America and SE Asia, and independently arrived at a theory of evolution by natural selection similar to that proposed by Charles Darwin.

In 1858, Wallace wrote an essay outlining his ideas on evolution and sent it to Darwin, who had not yet published his. Together they presented a paper to the Linnaean Society that year. Wallace's section, entitled 'On the Tendency of Varieties to Depart Indefinitely from the Original Type', described the survival of the fittest.

Although both thought that the human race had evolved to its present physical form by natural selection, Wallace was of the opinion that humans' higher mental capabilities had arisen from some 'metabiological' agency.

He defined the ***Wallace line***, a hypothetical line that separates the S Asian (Oriental) and Australian biogeographical regions, each of which has its own distinctive animals. This line follows a deep-water channel that runs between the larger islands of Borneo and Celebes and the smaller ones of Bali and Lombok.

In proportion as physical characteristics become of less importance, mental and moral qualities will have an increasing importance to the well-being of the race. Capacity for acting in concert, for protection of food and shelter; sympathy, which leads all in turn to assist each other; the sense of right, which checks depredation upon our fellows ... all qualities that from earliest appearance must have been for the benefit of each community, and would therefore have become objects of natural selection.

Alfred Wallace

'Origin of human races and the antiquity of man' in *Journal of the Royal Anthropological Society* London 1864 clviii

---

Wallace was born in Usk (now in Gwent). While working as a schoolteacher, he met English naturalist Henry Bates; together they planned a collecting trip to the Amazon, and arrived in South America 1848. When Wallace was returning to the UK 1852, his ship sank and although he survived, all his specimens were lost except those that had been shipped earlier. From 1854 to 1862 he explored the Malay Peninsula and archipelago, from which he collected more than 125,000 specimens, and in 1869–70 he made an expedition to Borneo and Maluku.

Wallace *Welsh naturalist Alfred Russel Wallace was, along with Charles Darwin, one of the pioneers of the theory of evolution. During his expedition to the Malay Archipelago, Wallace became the first European to observe orang-utans in the wild, and it was here that he conceived the notion of the 'survival of the fittest' to explain natural selection.*

Wallace also promoted socialism, campaigning for women's suffrage and land nationalization.

Wallace's works include *A Narrative of Travels on the Amazon and Rio Negro* 1853, *On the Law Which Has Regulated the Introduction of New Species* 1855, *The Malay Archipelago* 1869, *Contributions to the Theory of Natural Selection* 1870; and a pioneering work on zoogeography, *Geographical Distribution of Animals* 1876.

Wallach Otto 1847–1931. German analytic chemist who isolated a new class of compounds, called ***terpenes***, from essential oils (oils extracted from plants and used in medicine, aromatherapy, and perfume). Terpenes are made up of isoprene units containing five carbon atoms, with the general formula $(C_5H_8)_n$. He went on to determine the formula of the most important terpene, limonene, $C_{10}H_{16}$. Nobel Prize for Chemistry in 1910.

Wallach was born in Königsberg, Prussia (now Kaliningrad, Russia) on 27 March 1847. He studied in Berlin and gained a PhD from the University of Göttingen in 1867 for his work on toluene compounds. After briefly working in industry, Wallach moved to Bonn to become the assistant of Friedrich Kekulé.

Wallach was appointed director of the Chemical Institute in Göttingen 1895 and continued to research the molecular structure of terpenes. In 1915, when he retired to enjoy his substantial art collection, he had already published over 120 papers.

Wallis Barnes Neville 1887–1979. British aeronautical engineer who designed the airship R-100, and during World War II perfected the 'bouncing bombs' used by the Royal Air Force Dambusters Squadron to destroy the German Möhne and Eder dams in 1943. He also assisted in the development of the Concorde supersonic airliner and developed the swing wing aircraft. Knighted 1968.

Wallis was born in Derbyshire and trained as a marine engineer. From shipbuilding he turned to the design of airships and then to aeroplanes. He joined the Vickers Company 1911 and worked there as a designer until the end of World War II, moving to the British Aircraft Corporation 1945.

In the G4–31 biplane 1932, Wallis introduced a lattice-work system derived from the wire-netting used to contain the gas bags on the airship R-100. A full geodesic structure was first employed in the monoplane that became the Wellesley bomber. In the Wallis lattice pattern, if one series of members was in tension, the opposite members were in compression, so that the system was stress-balanced in all directions. The Wellesley was responsible for a great technical advance in design in the mid-1930s, which eventually produced the Wellington bomber of World War II.

Wallis John 1616–1703. English mathematician and cleric who made important contributions to the development of algebra and analytical geometry. He was one of the founders of the Royal Society.

Wallis was born in Ashford, Kent, and studied at Cambridge. In 1640 he was ordained in the Church of England. He moved to London 1645 and assisted the Parliamentary side by deciphering captured coded letters during the Civil War. From 1649 he was professor of geometry at Oxford, and in 1658 he was appointed keeper of the university archives. In 1660 Charles II chose him as his royal chaplain. After the revolution of 1688–89, which drove James II from the throne, Wallis was employed by William III as a decipherer.

Wallis also conducted experiments in speech and attempted to teach, with some success, congenitally deaf people to speak. His method was described in his *Grammatica linguae anglicanae* 1652.

Wallis's *Arithmetica infinitorum* 1655 was the most substantial single work on mathematics yet to appear in England. It introduced the symbol ∞ to represent infinity, the germ of the differential calculus, and, by an impressive use of interpolation (the word was Wallis's invention), the value for π. His *Mechanica* 1669–71 was the fullest treatment of the subject then existing, and his *Algebra* 1685 introduced the principles of analogy and continuity into mathematics.

Walton Ernest Thomas Sinton 1903–1995. Irish physicist who collaborated with John Cockcroft on investigating the structure of the atom. In 1932 they succeeded in splitting the atom; for this experiment they shared the 1951 Nobel Prize for Physics.

Walton and Cockcroft built the first successful particle accelerator. This used an arrangement of condensers to produce a beam of protons and was completed in 1932.

Walton was born in County Waterford and studied at Trinity College, Dublin, and 1927–34 at the Cavendish Laboratory in Cambridge, England. He returned to Trinity and was professor there 1947–74.

Using the proton beam to bombard lithium, Walton and Cockcroft observed the production of large quantities of alpha particles, showing that the lithium nuclei had captured the protons and formed unstable beryllium nuclei which instantaneously decayed into two alpha particles travelling in opposite directions. They detected these alpha particles with a fluorescent screen. Later they investigated the transmutation of other light elements using proton beams, and also deuterons (nuclei of deuterium) derived from heavy water.

Wang An 1920–1990. Chinese-born US engineer, founder of Wang Laboratories 1951, one of the world's largest computer companies in the 1970s. In 1948 he invented the computer memory core, the most common device used for storing computer data before the invention of the integrated circuit (chip).

Wang emigrated to the USA 1945. He developed his own company with the $500,000 he received from IBM from the sale of his patent. One of his early contracts was the first electronic scoreboard, installed at New York's Shea Stadium. His company took off in 1964 with the introduction of a desktop calculator. Later, Wang switched with great success to the newly emerging market for word-processing systems based on cheap silicon chips, turning Wang Laboratories into a multibillion-dollar company. But with the advent of the personal computer, the company fell behind. Wang Laboratories made a loss of $400 million 1989.

Wankel Felix 1902–1988. German engineer who developed by 1956 the rotary engine that bears his name.

The ***Wankel engine*** consists of a figure-eight chamber with a triangular rotor. In one revolution the rotor successively isolates various parts of the chamber, allowing for the intake of fuel and air, compression, ignition; expansion and exhaust to take place. While this driving cycle is taking place relative to one face of the rotor, two more are taking place relative to the other two faces; as a result the crankshaft rotates three times for every turn of the rotor. Thus the Wankel engine produces more power for its weight than the more conventional Otto and diesel engines. There are no separate valves, because the induction and exhaust of vapours is controlled by the movement of the rotor.

Wankel was born in Luhran. In 1927 he became a partner in an engineering works before opening his own research establishment. Later he carried out work for the German Air Ministry. At the end of World War II he began to work for a number of German motor manufacturers at the Technische Entwicklungstelle in Lindow; he was made its director 1960.

During the 1930s Wankel carried out a systematic investigation of internal-combustion engines, particularly rotary engines. The German motor firm NSU sponsored the development of his engine with a view to its possible use in motorcycles. Eventually he rearranged his early designs and produced a successful prototype of a practical engine in 1956.

Wankel engines are easily connected together in pairs. They have few moving parts compared with an ordinary motorcar engine; there are no piston rods or camshafts. The saving in engine weight means that slightly less power is required from engines of this type when they are used in cars. Companies throughout the world have bought the rights to manufacture and use the Wankel engine.

Warburg Otto Heinrich 1883–1970. German biochemist who in 1923 devised a manometer (pressure gauge) sensitive enough to measure oxygen uptake of respiring tissue. By measuring the rate at which cells absorb oxygen under differing conditions, he was able to show that enzymes called cytochromes enable cells to process oxygen. Nobel Prize for Physiology or Medicine 1931.

Later he discovered the mechanism of the conversion of light energy to chemical energy that occurs in photosynthesis. He also demonstrated that cancerous cells absorb less oxygen than normal cells.

Warburg was born in Freiburg-im-Breisgau and studied at Berlin and Heidelberg. In 1913 he went to the Kaiser Wilhelm (later Max Planck) Institute for Cell Physiology in Berlin, becoming a professor

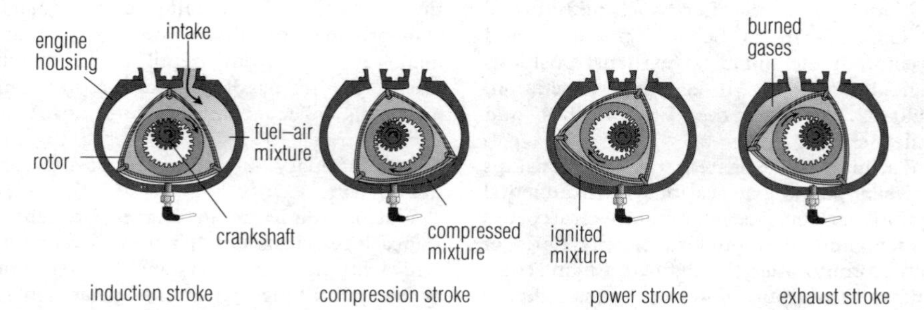

Wankel *The rotary Wankel engine uses the same four stages as the four-stroke Otto cycle: induction, compression, combustion, and exhaust.*

there in 1918 and its director in 1931. In 1941 Warburg, being part-Jewish, was removed from his post but such was his international prestige that he was soon reinstated. In 1944 he was nominated for a second Nobel prize but Nazi rules prevented him from accepting the award.

Warburg discovered that in both charcoal systems and living cells, the uptake of oxygen is inhibited by the presence of cyanide or hydrogen sulphide, which combine with heavy metals.

He also showed that, in the dark, carbon monoxide inhibits the respiration of yeast but does not do so in the light. He was aware that heavy metals form complexes with carbon monoxide and that the iron complex is dissociated by light, which provided further evidence for the existence of an iron-containing respiratory enzyme. He then investigated the efficiency of light in overcoming the carbon monoxide inhibition of respiration, and determined the photochemical absorption spectrum of the respiratory enzyme, which proved to be a haemoprotein (a protein with an iron-containing group) similar to haemoglobin; he called it iron oxygenase.

Warming Johannes Eugenius Bülow 1841–1924. Danish botanist whose pioneering studies of the relationships between plants and their natural environments established plant ecology as a new discipline within botany.

Warming was born on the island of Mandø, Denmark. He studied at Copenhagen and at German universities. While still a student, he spent 1863–66 at Lagoa Santa, Brazil, where he undertook a study of tropical vegetation. In 1882 he became professor at the Royal Institute of Technology in Stockholm, Sweden. He went on an expedition to Greenland in 1884 and to Norway in 1885, after which he returned to Copenhagen to become professor and director of the botanical gardens, positions he held until 1911. His last major expedition, 1890–92, was to the West Indies and Venezuela.

Warming investigated the relationships between plants and various environmental conditions, such as light, temperature, and rainfall, and attempted to classify types of plant communities (he defined a plant community as a group of several species that is subject to the same environmental conditions, which he called ecological factors). In *Plantesamfund/Oecology of Plants* 1895 he also formulated a programme for future research into the subject.

Wassermann August Paul von 1866–1925. German professor of medicine. In 1907 he discovered a diagnostic blood test for syphilis, known as the ***Wassermann reaction***.

Waterston John James 1811–1883. Scottish physicist who first formulated the essential features of the kinetic theory of gases 1843–45. He also estimated the temperature of the Sun 1857.

Waterston was born and educated in Edinburgh. He moved to London 1833 to do surveying for the railways, then took a job in the Hydrographers' Department of the Admiralty. In 1839 he went to India as teacher of the East India Company's cadets in Bombay, returning to Edinburgh 1857 to devote all his efforts to research. His work was repeatedly rejected or ignored, causing him to withdraw from the scientific community.

Waterston's first scientific paper, published when he was only 19, concerned a model which he proposed might explain gravitational force without the necessity for postulating an effect that operated at great distances.

In 1843, Waterston wrote a book on the nervous system in which he attempted to apply molecular theory to physiology. It included several fundamental features of the kinetic theory of gases, among them the idea that temperature and pressure are related to the motion of molecules. He formulated kinetic theory more fully in a paper submitted to the Royal Society in 1845, but this was turned down, delaying progress by about 15 years.

Waterston wrote other papers on sound, capillarity, latent heat, and various aspects of astronomy.

Watson David Meredith Seares 1886–1973. English embryologist and palaeobiologist who provided the first evidence that mammals evolved from reptiles. From the fossilized remains of primitive reptiles and mammals collected on trips to South Africa and Australia 1911–14, he pieced together the evolutionary line linking reptiles to early mammals.

Watson was born in Higher Broughton, Lancashire. He graduated from Manchester University in chemistry and geology, but began publishing papers on palaeobiology while he was still an undergraduate.

With the outbreak of World War I, he cut short this work to return to England to join the RAF. From 1921 until his retirement in 1965 he was based at University College, London. In 1952 he was Alexander Agassiz visiting professor at Harvard. He was elected to the Royal Society 1922, and was awarded the Darwin medal 1942 and the Linnaeus medal 1949.

Watson James Dewey 1928– . US biologist whose research on the molecular structure of DNA and the genetic code, in collaboration with Francis Crick, earned him a shared Nobel prize in 1962. Based on earlier works, they were able to show that DNA formed a double helix of two spiral strands held together by base pairs.

Crick and Watson published their work on the proposed structure of DNA in 1953, and explained how genetic information could be coded.

Watson was born in Chicago and studied there and at Indiana. He initially specialized in viruses but shifted to molecular biology and in 1951 he went to the Cavendish Laboratory at Cambridge University, where he performed the work on DNA with Crick. In 1953 Watson returned to the USA. He became professor at Harvard 1961 and director of the Cold Spring Harbor Laboratory of Quantitative Biology 1968, and was head of the US government's Human Genome Project 1989–92.

Watson *US biologist James Dewey Watson who shared the Nobel Prize for Physiology or Medicine with Francis Crick and Maurice Wilkins in 1962. Watson and Crick determined the structure of the genetic material DNA at the Cavendish Laboratory in Cambridge, England. The characters and work underlying the discovery are described in Watson's book* The Double Helix.

Crick and Watson envisaged DNA replication occurring by a parting of the two strands of the double helix, each organic base thus exposed linking with a nucleotide (from the free nucleotides within a cell) bearing the complementary base. Thus two complete DNA molecules would eventually be formed by this step-by-step linking of nucleotides, with each of the new DNA molecules comprising one strand from the original DNA and one new strand.

It is necessary to be slightly underemployed if you want to do something significant.

James Watson

The Eighth Day of Creation

Watson-Watt Robert Alexander 1892–1973. Scottish physicist who developed a forerunner of radar. He proposed in 1935 a method of radiolocation of aircraft – a key factor in the Allied victory over German aircraft in World War II. Knighted 1942.

Watson-Watt was born in Brechin, Angus, and educated at the University of St Andrews. He spent a long career in government service (1915–52).

Watson-Watt patented in 1919 a device concerned with radiolocation by means of short-wave radio waves. By 1935 he had made it possible to follow an aeroplane by the radio-wave reflections it sent back. By 1938, radar stations were in operation, and during the Battle of Britain in 1940 radar made it possible for the British to detect incoming German aircraft as easily by night as by day, and in all weathers including fog. Early in 1943 microwave aircraft-interceptor radars were operational, ending night-bombing raids on Britain.

The first radar sets specifically designed for airborne surface-vessel detection were flown in 1943. Before the USA joined the war, Watson-Watt went there to advise on the setting-up of radar systems.

Watt James 1736–1819. Scottish engineer who developed the steam engine in the 1760s.

He made Thomas Newcomen's steam engine vastly more efficient by cooling the used steam in a condenser separate from the main cylinder.

Watt *Scottish engineer and inventor James Watt shown repairing a small working model of Newcomen's steam engine in his workshop at Glasgow University. Watt noticed several defects in Newcomen's engine and built his own, which incorporated a separate condenser. Watt's improvements to the design increased the efficiency of the engine by a factor of three and provided a practical way of producing power for British industry during the Industrial Revolution.*

Steam engines incorporating governors, sun-and-planet gears, and other devices of his invention were successfully built by him in partnership with Matthew Boulton and were vital to the Industrial Revolution. Watt also devised the horsepower as a description of an engine's rate of working.

Watt was born in Greenock (now in Strathclyde) and trained as an instrument-maker. Between 1767 and 1774, he made his living as a canal surveyor. In 1775 Boulton and Watt went into partnership and manufactured Watt's engines at the Soho Foundry, near Birmingham. In 1782 Watt improved his machine by making it drive on both the forward and backward strokes of the piston, and a sun-and-planet gear produced rotary motion. This highly adaptable engine was quickly adopted by cotton and woollen mills.

Watt also invented artistic instruments and a chemical copying process for documents.

Weatherall David John 1933– . English physician and prominent exponent of the role of genetics in modern medicine. Weatherall researched the genetic basis for the thalassaemias, a group of inherited anaemias that result from faulty haemoglobin synthesis.

These conditions are particularly prevalent in people living near the Mediterranean Sea, some 20% of the population of some parts of Italy being carriers of the disease. The geographic distribution of some thalassaemia genes parallels that of malaria, which suggests that heterozygotes may benefit from the presence of the mutation, as in the sickle-cell disease. These discoveries have had a major impact on molecular biology, medicine, and genetics since they have provided striking illustrations of the consequences of mutations in human disease.

Weatherall graduated from the University of Liverpool 1954 and held various junior medical jobs before taking up a career in medical research at Johns Hopkins Medical School 1960. He then moved back to England to take up a teaching post at Liverpool University and was made professor of haematology there 1971. Since then, he has been appointed professor of clinical medicine 1974 and then Regius professor of medicine 1992 at Oxford University. He is also the director of the Institute for Molecular Medicine in Oxford. He was elected a Fellow of the Royal Society 1974 and knighted 1987 for his contribution to the genetics of human disease.

Weber Ernst Heinrich 1795–1878. German anatomist and physiologist. He applied hydrodynamics to study blood circulation, and formulated ***Weber's law***, relating response to stimulus.

Weber's law (also known as the Weber–Fechner law) states that sensation is proportional to the logarithm of the stimulus. It is the basis of the scales used to measure the loudness of sounds.

In 1825, with his brother Wilhelm Weber, he made the first experimental study of interference in sound.

Weber Heinrich 1842–1913. German mathematician whose chief work was in the fields of algebra and number theory. He demonstrated Norwegian mathematician Niels Abel's theorem in its most general form.

Weber was born and educated in Heidelberg, where he became professor 1869. He then taught at a number of institutions in Germany and Switzerland.

Weber proved German mathematician Leopold Kronecker's theorem that the absolute Abelian fields are cyclotomic; that is, that they are derived from the rational numbers by the adjunction of roots of unity.

Weber's work in such subjects as heat, electricity, and electrolytic dissociation was contained in his *Die partiellen Differentialgleichungen der mathematischen Physik* 1900–01, which was essentially a reworking of, and a commentary upon, a book of the same title based on lectures given by Bernhard Riemann and written by Karl Hattendorff.

Weber's *Lehrbuch der Algebra* 1896 became a standard text.

Weber Wilhelm Eduard 1804–1891. German physicist who studied magnetism and electricity. Working with mathematician Karl Gauss, he made sensitive magnetometers to measure magnetic fields, and instruments to measure direct and alternating currents. He also built an electric telegraph. The SI unit of magnetic flux, the ***weber***, is named after him.

Weber defined an electromagnetic unit for electric current which was applied to measurements of current made by the deflection of the magnetic needle of a galvanometer. In 1846, he developed the electrodynamometer, in which a current causes a coil suspended within another coil to turn when a current is passed through both. In 1852, Weber defined the absolute unit of electrical resistance.

Weber was born in Wittenberg, the brother of Ernst Weber, and studied at Halle. In 1831 he became professor at Göttingen; he lost this post after making a political protest 1837, but was reinstated 1849.

At Göttingen, he built a 3-km/2-mi telegraph to connect the physics laboratory with the astronomical observatory where Gauss worked, and this was the first practical telegraph to operate anywhere in the world. Gauss and Weber organized a network of observation stations 1836–41 to correlate measurements of terrestrial magnetism made around the world.

Weber put forward in 1871 the view that atoms contain positive charges that are surrounded by rotating negative particles and that the application of an electric potential to a conductor causes the negative particles to migrate from one atom to another. He also provided similar explanations of thermal conduction and thermoelectricity.

Wedderburn Joseph Henry Maclagan 1882–1948. Scottish mathematician who opened new lines of thought in the subject of mathematical fields and who had a deep influence on the development of modern algebra.

Wedderburn was born in Forfar and studied at Edinburgh and Chicago, USA. He taught in the USA at Princeton 1909–45, though during World War I he saw active duty in France as a soldier in the British army.

The first paper that Wedderburn published, 'Theorem on Finite Algebra' 1905, was a milestone in algebraic history. By introducing new methods, he showed that it was possible to arrive at total understanding of the structure of semi-simple algebras using hyper-complex numbers as well as real or complex numbers.

Wedderburn went on to derive the two theorems to which his name has become attached. The first was contained in his paper 'On Hyper-Complex Numbers' 1907, in which he demonstrated that a simple algebra consists of matrices of a given degree with elements taken from a division of algebra.

The first Wedderburn theorem states that 'if the algebra is a finite division algebra (that is, that it has only a finite number of elements and always permits division by a non-zero element), then the multiplication law must be commutative, so that the algebra is actually a finite field'.

Wedderburn's second theorem states that a central-simple algebra is isomorphic to the algebra of all $n \times n$ algebras. He arrived at it by an investigation of skew fields with a finite number of elements.

His discovery that every field with a finite number of elements is commutative under multiplication led to a complete classification of all semi-simple algebras with a finite number of elements.

Wedgwood Josiah 1730–1795. English pottery manufacturer. He set up business in Staffordshire in the early 1760s to produce his agateware as well as unglazed blue or green stoneware (jasper) decorated with white neo-Classical designs, using pigments of his own invention.

Wedgwood was born in Burslem, Staffordshire, and worked in the family pottery. Eventually he set up in business on his own at the Ivy House Factory in Burslem, and there he perfected cream-colonial earthenware, which became known as queen's ware because of the interest and patronage of Queen Charlotte in 1765. In 1768 he expanded the company into the Brick House Bell Works Factory. He then built the Etruria Factory, using his engineering skills in the design of machinery and the high-temperature beehive-shaped kilns, which were more than 4 m/12 ft wide.

Wegener Alfred Lothar 1880–1930. German meteorologist and geophysicist whose theory of continental drift, expounded in *Origin of Continents and Oceans* 1915, was

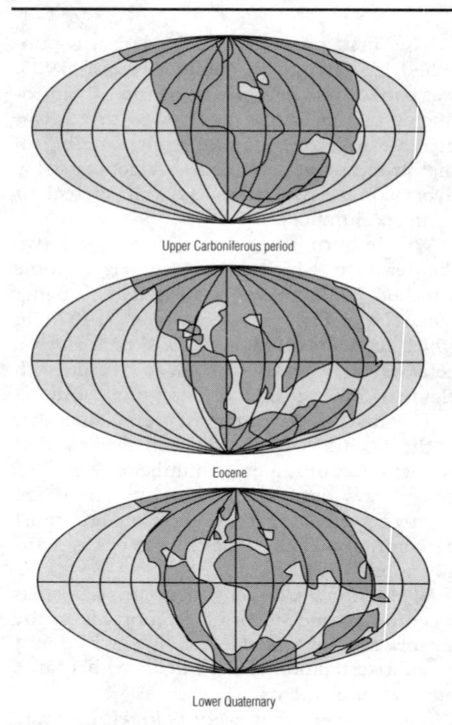

Wegener: *continental drift*

originally known as 'Wegener's hypothesis'. His ideas can now be explained in terms of plate tectonics, the idea that the Earth's crust consists of a number of plates, all moving with respect to one another.

Wegener was born in Berlin and studied at Heidelberg, Innsbruck, and Berlin. From 1924 he was professor of meteorology and geophysics at Graz, Austria. He completed three expeditions to Greenland and died on a fourth.

Wegener supposed that a united supercontinent, Pangaea, had existed in the Mesozoic. This had developed numerous fractures and had drifted apart some 200 million years ago. During the Cretaceous, South America and Africa had largely been split, but not until the end of the Quaternary had North America and Europe finally separated; the same was true of the break between South America and Antarctica. Australia had been severed from Antarctica during the Eocene.

If it turns out that sense and meaning are now becoming evident in the whole history of the Earth's development, why should we hesitate to toss the old views overboard?

Alfred Wegener

Alfred Wegener, the Father of Continental Drift

Weick Fred 1899–1993. US aeronautical engineer. His designs include the Erco Ercoupe 1936, remarkable for its safety and simplicity of design, and the Piper Cherokee in the 1950s, the first cheap all-metal aircraft, of which over 30,000 have since been built.

He developed the fully enclosed engine cowling that streamlined and reduced the drag of cumbersome radial aero engines, and a novel form of undercarriage that he called 'tricycle landing gear', which has since become the standard configuration. Impressed by the safety of the Erco Ercoupe, the US Civil Aeronautics Authority reduced the time for a trainee pilot to fly solo from eight to five hours when flying an Ercoupe.

Weick became interested in aviation as a child. On graduating from the University of Illinois as a mechanical engineer, he took a job converting surplus bombers into mail planes. He made his first solo flight 1923 – a 100-mile journey in bad weather – but it was not until 1939 that he bothered to obtain a pilot's licence. In 1923 he joined the Bureau of Aeronautics in Washington DC, working on propeller technology. He was invited to run the propeller research wind tunnel of the National Advisory Committee for Aeronautics (NACA), the forerunner of NASA. Whilst working for NACA, he developed the streamlined cowl and continued his pioneering work on propeller design. His interest in aircraft stability and control led him to develop a safe, simple aeroplane for private owners – the Weick W-1, which first flew 1934. In 1936 he was invited by the Engineering and Research Corporation (Erco) of Washington DC to develop a commercial version of the W-1, and the Ercoupe was born.

Weierstrass Karl Theodor Wilhelm 1815–1897. German mathematician who deepened and broadened the understanding of functions. He demonstrated in 1871 that there exist continuous functions in an interval which have derivatives nowhere in the interval.

Weierstrass was born in Ostenfelde, Westphalia, and trained as a teacher. From 1856 he was professor at the Royal Polytechnic School in Berlin as well as the university.

Weierstrass's breakthrough came with a paper 1854 that solved the inversion of hyperelliptic integrals. He did much to clarify basic concepts such as 'function', 'derivative', and 'maximum'. His development of the modern theory of functions was described in his *Abhandlungen aus der Funktionlehre* 1886, a text derived chiefly from his students' lecture notes. In the 1890s Weierstrass planned the publication of his life's work, again to be compiled from lecture notes. Two volumes were published before his death and five more appeared during the next three decades.

Weil André 1906– . French mathematician who worked on number theory and group theory, and contributed to the generalization of algebraic geometry. He was a founder member of a secretive group that published mathematical papers under the pseudonym Nicolas Bourbaki.

Weil was born and educated in Paris, the brother of political writer Simone Weil. In 1930 he went to India for two years as professor at Aligarh Muslim University. Returning to France, he took up a similar post at Strasbourg University. In 1940 he emigrated, and was professor at the University of Saõ Paolo, Brazil, 1945–47. He moved to the USA and the University of Chicago 1947–58, and then transferred to the Institute for Advanced Study at Princeton.

In 1929, Weil extended some work by Henri Poincaré; this resulted in the postulation of what is now called the ***Mordell–Weil theorem***, which is closely connected to the theory of Diophantine equations.

Weil worked on quadratic forms with algebraic coefficients and extended Austrian mathematician Emil Artin's work on the theory of quadratic number fields.

Weil's chief work was *Foundations of Algebraic Geometry* 1946.

Weinberg Steven 1933– . US physicist who in 1967 demonstrated, together with Abdus Salam, that the weak nuclear force and the electromagnetic force (two of the fundamental forces of nature) are variations of a single underlying force, now called the electroweak force. Weinberg and Salam shared a Nobel prize in 1979.

Weinberg and Salam's theory involved the prediction of a new interaction, the neutral current (discovered in 1973), which required the presence of charm, a type of quark.

Weismann August Friedrich Leopold 1834–1914. German biologist, one of the founders of genetics. He postulated that every living organism contains a special hereditary substance, the 'germ plasm', and in 1892 he proposed that changes to the body do not in turn cause an alteration of the genetic material.

This 'central dogma' of biology remains of vital importance to biologists supporting the Darwinian theory of evolution. If the genetic material can be altered only by chance mutation and recombination, then the Lamarckian view that acquired bodily changes can subsequently be inherited becomes obsolete.

Weismann was born in Frankfurt-am-Main and studied medicine at Göttingen. From 1863 he taught at Freiburg; persuading the university to build a zoological institute and museum, he became its director. Failing eyesight forced him to turn from microscopy to theoretical work in the 1860s.

Weismann realized that the germ plasm controls the development of every part of the organism and is transmitted from one generation to the next in an unbroken line of descent. Since repeated mixing of the germ plasm at fertilization would lead to a progressive increase in the amount of hereditary material, he predicted that there must be a type of nuclear division.

Weizsäcker Carl Friedrich von 1912– . German theoretical physicist who investigated the way in which energy is generated in the cores of stars. He also developed a theory that planetary systems were formed as a natural by-product of stellar evolution, condensing from vortices of gas.

Weizsäcker was born in Kiel and studied at Leipzig. He was professor at Strasbourg 1942–44. During World War II, he was a member of the research team investigating the feasibility of constructing nuclear weapons and harnessing nuclear energy, although he did not want his team to develop a weapon for the Nazi government. In 1946 he became director of a department in the Max Planck Institute of Physics in Göttingen. He was professor of philosophy at Hamburg 1957–69, and in 1970 he became a director of the Max Planck Institute.

In 1938, Weizsäcker and German-born physicist Hans Bethe independently proposed the same theory of stellar evolution, which accounted both for the very high temperatures in stellar cores and for the production of ionizing and particulate radiation by stars. They proposed that hydrogen atoms fused to form helium via a proton–proton chain reaction.

Welch William Henry 1850–1934. US pathologist and bacteriologist who discovered the microorganism *Clostridium welchii*, which causes gas-gangrene. He advanced the sciences of pathology and bacteriology in the USA, especially at his own university in New York, with his high-calibre teaching and research.

Welch was born in Norfolk, Connecticut and studied medicine at Yale and at several German universities. He held the chair of pathology at Johns Hopkins University from its foundation 1884 until 1918. Upon his retirement, he went on to become the founding director of the Schools of Hygiene and Public Health and the History of Medicine at Johns Hopkins.

Weller Thomas Huckle 1915– . US virologist who shared the Nobel Prize for Physiology or Medicine 1954 with John Enders and Frederick Robbins for their work on the polio virus. He developed new ways of growing and studying the poliomyelitis virus in culture, groundwork for the later development of a vaccine. Weller went on to study the role of a single virus in chickenpox and shingles and to show that a single microorganism was involved in German measles.

Weller was born in Ann Arbor, Michigan and graduated from the University of Michigan in zoology 1936. After the end of World War II, he went to work at the Boston Children's Hospital with Enders and Robbins. In 1954, he was elected to a chair in tropical public health at Harvard and retired from this post 27 years later.

Wells Horace 1815–1848. US dentist who discovered nitrous oxide anaesthesia and, in 1844, was the first to use the gas in dentistry.

Wells was born in Hartford, Vermont. He set up a dental practice in Hartford, Connecticut, in partnership with William Morton, who later pioneered the use of ether as an anaesthetic. In 1844, while watching an exhibition of the effects of laughing gas (nitrous oxide) staged by a travelling show, Wells observed that the gas induced anaesthesia. He tried it first on himself and then used it to perform tooth extractions on his patients. In 1845 Wells went to Boston where, with the help of Morton (then no longer his partner) and others, he arranged to give a demonstration at the Massachusetts General Hospital. However, the patient cried out and, although the patient later claimed to have felt no pain, the audience believed that the demonstration had failed.

After this, Wells gave up his dental practice and became a travelling sales representative, selling canaries and then showerbaths in Connecticut. Once Morton had given a successful demonstration of ether anaesthesia, Wells went to Paris 1847 to try to establish his priority in using anaesthesia. At about this time he also began experimenting on himself with nitrous oxide, ether, and various other intoxicating chemicals; as a result he became addicted to chloroform. He committed suicide in prison.

Welsbach Carl Auer, Baron von Welsbach 1858–1929. Austrian chemist and engineer who discovered two rare-earth elements and invented the incandescent gas mantle and a lighter flint.

Auer was born in Vienna and studied in Germany at Heidelberg.

He showed that didymium, previously thought to be an element, actually consisted of two very similar but different elements: praseodymium and neodymium. He also found that another rare earth element, cerium, added as its nitrate salt to a cylindrical fabric impregnated with thorium nitrate, produced a fragile mantle that glowed with white incandescence when heated in a gas flame. The 'Welsbach mantle' was patented 1885.

Most lighter flints consist of Welsbach's invention *Mitschmetall*, a pyrophoric mixture containing about 50% cerium, 25% lanthanum, 15% neodymium, and 10% other rare metals and iron. When it is struck or scraped, it produces hot metal sparks. *Mitschmetall* is also used as a deoxidizer in vacuum tubes and as an alloying agent for magnesium.

Wenner-Gren Axel Leonard 1881–1961. Swedish industrialist who founded the Electrolux Company 1921, manufacturing electrical appliances, and developed a monorail system.

Wenner-Gren was born in Uddevalla, educated in Germany, and began his career working for the Swedish Electric Lamp Company, where he eventually became a majority shareholder. In 1921 he founded the Electrolux Company to manufacture vacuum cleaners and, later, refrigerators. He then acquired one of country's largest wood-pulp mills and the Bofors munition works. From the profits, he donated a large sum for the foundation of an institute for the development of scientific research in Sweden, the Wenner-Gren Foundation for Nordic Cooperation and Research.

Wenner-Gren's monorail, the Alweg line, consisted of a concrete beam carried on concrete supports. The cars straddled the beam on rubber-tyred wheels, and there were also horizontal wheels in two rows on each side of the beam. The system was used, for example, for the 13.3-km/8.3-mi line in Japan from Tokyo to Haneda Airport.

Werner Abraham Gottlob 1749–1817. German geologist, one of the first to classify minerals systematically. He also developed the later discarded theory of neptunism – that the Earth was initially covered by water, with every mineral in suspension; as the water receded, layers of rocks 'crystallized'.

Werner was born in Silesia and studied at the Mining School at Freiberg, Saxony, and the University of Leipzig. From 1775 he taught at the Freiberg Akademie.

Werner's geology was particularly important for establishing a physically based stratigraphy, grounded on precise mineralogical knowledge. He linked the order of the strata to the history of the Earth, and related studies of mineralogy and strata.

Werner Alfred 1866–1919. French-born Swiss chemist. He was awarded a Nobel prize in 1913 for his work on valency theory, which gave rise to the concept of coordinate bonds and coordination compounds.

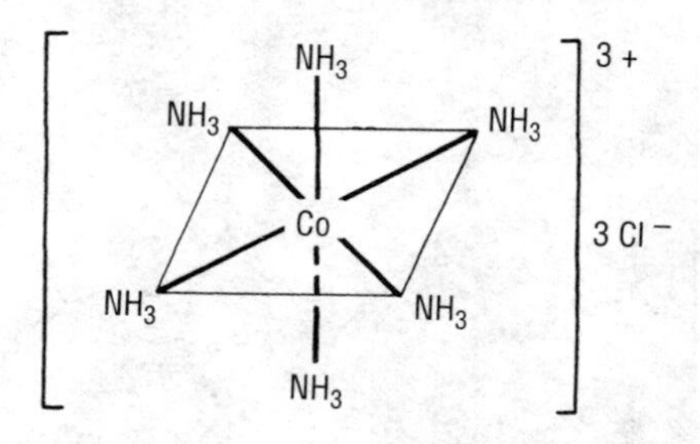

Werner *Hexamminocobalt chloride, $[Co(NH_3)_6]\ Cl_3$, with a coordination number of 6.*

Werner demonstrated that different three-dimensional arrangements of atoms in inorganic compounds gives rise to optical isomerism (the rotation of polarized light in opposite directions by molecules that contain the same atoms but are mirror images of each other).

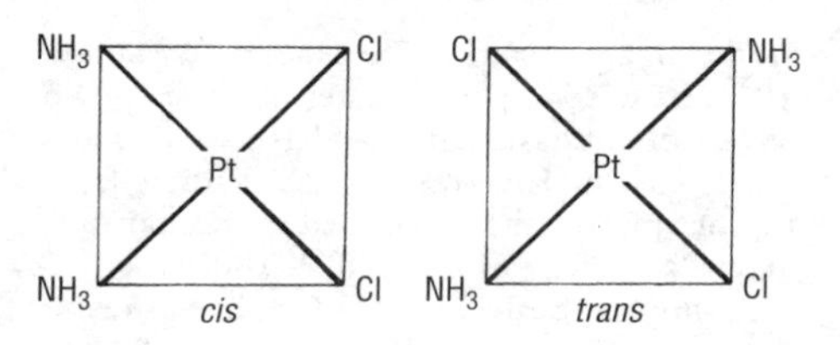

Werner *Stereoisomers of diamminoplatinum chloride, $Pt(NH_4)_2CL_2$ each with a coordination number of 4.*

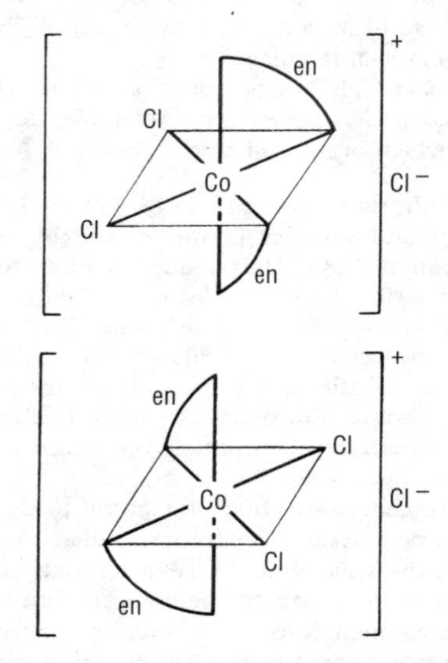

Werner *A pair of optically isomeric coordination compounds: mirror images of cis- $[Co\ (en)_2Cl_2{}^{+}Cl^{-}$, each with a coordination number of 6.*

Werner was born in Mulhouse, Alsace, and studied in Switzerland at the Zürich Polytechnic, becoming professor there 1895.

In addition to ionic and covalent bonds, Werner proposed the existence of a set of coordination bonds resulting from an attractive force from the centre of an atom acting uniformly in all directions. The number of groups, or ligands, that can thus be bonded to the central atom depends on its coordination number and determines the structure (geometry) of the resulting molecules. Neutral ligands (such as ammonia and water) leave the central atom's ionic charge unchanged; ionic ligands (such as chloride or cyanide ion) alter the central charge accordingly.

Wernicke Carl 1848–1905. German neurologist and psychiatrist. He is known for his study of aphasia. In *The Aphasic Syndrome* 1874, he described what later became known as sensory aphasia (that is, defects in, or loss of, speech and expression) as distinct from motor aphasia, first described by French surgeon Paul Pierre Broca (1824–1880).

Although both forms of aphasia result from brain damage, Wernicke found that the locus of the damage differed, sensory aphasia being induced by lesions to the left temporal lobe, motor aphasia by lesions to the left posterior frontal lobe. He used the differential clinical features of the two aphasias to formulate a general theory of the neural bases of language.

Wernicke also described a form of encephalopathy induced by thiamine deficiency which bears his name.

Westinghouse George 1846–1914. US inventor and founder of the Westinghouse Corporation 1886. He patented a powerful air brake for trains 1869, which allowed trains to run more safely with greater loads at higher speeds. In the 1880s he turned his attention to the generation of electricity. Unlike Thomas Edison, Westinghouse introduced alternating current (AC) into his power stations.

Westinghouse was born in Central Bridge, New York, and ran away from school at 15 to fight in the Civil War. In 1865 he took out the first of more than 400 patents. He formed various companies to manufacture his inventions, several based in Pittsburgh and nearby in Turtle Creek Valley, where in 1889 Westinghouse built a model town for his workers. During 1907–08 a series of financial crises and takeovers caused him to lose control of the Westinghouse Industries.

Westinghouse helped to standardize railway components, including the development of a completely new signalling system. He also developed a system of gas mains. In the 1880s Westinghouse got his engineers to design equipment suitable for a new high-tension (voltage) AC system. He also secured the services of Croatian physicist Nikola Tesla. In 1895 the Westinghouse Electric Company harnessed Niagara Falls to generate electricity for the lights and trams of the nearby town of Buffalo.

Resentful that AC current was chosen as the standard for domestic electricity supply, Edison, who supported DC current transmission, coined the term 'Westinghoused' to describe the fate of someone who had been executed by electric chair.

Weyl Hermann 1885–1955. German mathematician and mathematical physicist who studied mainly topological space and the geometry of Bernhard Riemann, but also quantum mechanics and number theory.

Weyl was born in Elmshorn, near Hamburg, and studied at Göttingen. He was professor at the Technische Hochschule in Zürich, Switzerland, 1913–30, and then at Göttingen, but the unfavourable political climate of Nazi Germany prompted him to move to the Institute for Advanced Study at Princeton, USA.

As a colleague of Albert Einstein during 1913, Weyl became interested in the developing general theory of relativity and came to believe (erroneously) that he had found a way to a grand unification of gravitation and electromagnetism. Weyl was able to anticipate the nonconservation of parity, which has since been found to be characteristic of weak interactions between leptons (a class of subatomic particles).

Weyl's most important work on Riemann surfaces was the definition of the complex manifold of the first dimension, which has been important in all later work on the theory of both complex and of differential manifolds.

Wheatstone *English physicist Charles Wheatstone who, as well as performing important experimental work in the fields of acoustics and electricity, was also an eminent inventor. Wheatstone introduced the term microphone and invented the concertina, kaleidoscope and the first electric telegraph. Ironically, he did not actually invent the Wheatstone bridge (an instrument for determining resistances in a circuit), but instead played a major role in aplying the device to new physical problems.*

Weyl's works include *Raum-Zeit-Materie/Space-Time-Matter* 1918. He also published works on philosophy, logic, and the history of mathematics.

Wheatstone Charles 1802–1875. English physicist and inventor. With William Cooke, he patented a railway telegraph 1837, and, developing an idea of Samuel Christie (1784–1865), devised the ***Wheatstone bridge***, an electrical network for measuring resistance. He also invented the concertina. Knighted 1868.

In 1834 Wheatstone made the first determination of the velocity of electricity along a wire. He also improved on early versions of the dynamo so that current was generated continuously.

Wheatstone was born in Gloucester and joined the family business making musical instruments. His work in acoustics led to his appointment as professor of experimental physics at King's College, London, 1834, a position he retained for the rest of his life.

In 1827 Wheatstone invented a device called the kaleidophone, which visually demonstrated the vibration of sounding surfaces by causing an illuminated spot to vibrate and produce curves by the persistence of vision. He went on to investigate the transmission of sound in instruments and discovered modes of vibration in air columns 1832 and vibrating plates 1833.

Wheatstone showed in 1835 that spectra produced by spark discharges from metal electrodes have different lines and colours formed by different electrodes. He predicted correctly that with development, spectroscopy would become a technique for the analysis of elements.

In 1860, he demonstrated how the visual combination of two similar pictures in a stereoscope gives an illusion of three dimensions.

Man is the interpreter of nature, science the right interpretation.

William Whewell

The Philosophy of the Inductive Sciences 1837

Whewell William 1794–1866. British physicist and philosopher who coined the term 'scientist' along with such words as 'Eocene' and 'Miocene', 'electrode', 'cathode', and 'anode'. He produced two works of great scholarship, *The History of the Inductive Sciences* 1837 and *The Philosophy of the Inductive Sciences* 1840.

Most of his career was connected with Cambridge University, where he became the Master of Trinity College.

Whipple Fred Lawrence 1906– . US astronomer whose hypothesis in 1949 that the nucleus of a comet is like a dirty snowball was confirmed 1986 by space-probe studies of Halley's comet.

Whipple was born in Red Oak, Iowa, and studied at the University of California. He was professor at Harvard 1950–77 and director of the Smithsonian Astrophysics Observatory 1955–73.

In addition to discovering six new comets, Whipple proposed that the nucleus of a comet consisted of a frozen mass of water, ammonia, methane, and other hydrogen compounds together with silicates, dust, and other materials. As the comet's orbit brought it nearer to the Sun, solar radiation would cause the frozen material to evaporate, thus producing a large amount of silicate dust which would form the comet's tail.

Whipple also worked on ascertaining cometary orbits and defining the relationship between comets and meteors. In the 1950s he became active in the programme to devise effective means of tracking artificial satellites.

Whipple George Hoyt 1878–1976. US physiologist whose research interest concerned the formation of haemoglobin in the blood. He showed that anaemic dogs, kept under restricted diets, responded well to a liver regime, and that their haemoglobin quickly regenerated. This work led to a cure for pernicious anaemia. He shared the 1934 Nobel Prize for Physiology or Medicine with George Minot (1885–1950) and William Murphy (1892–1987).

White Gilbert 1720–1793. English naturalist and cleric. He was the author of *The Natural History and Antiquities of Selborne* 1789, which records the flora and fauna of an area of Hampshire.

White studied at Oxford. Although assigned to parishes elsewhere, he chose to live in his native Selborne.

White's book is based on a diary of his observations and on letters to two naturalist friends over a period of about 20 years. Elegantly written, *The Natural History* contains descriptions of rural life and acute observations of a wide variety of natural-history subjects, such as the migration of swallows, the recognition of three distinct species of British leaf warblers, and the identification of the harvest mouse and the noctule bat as British species.

White also wrote *Calendar of Flora and the Garden* 1765, an account of observations he made in his garden in 1751, and *Naturalist's Journal* (begun in 1768).

Whitehead Alfred North 1861–1947. English philosopher and mathematician. In his 'theory of organism', he attempted a synthesis of metaphysics and science. His works include *Principia Mathematica* 1910–13 (with Bertrand Russell), *The Concept of Nature* 1920, and *Adventures of Ideas* 1933.

Whitehead's research in mathematics involved a highly original attempt – incorporating the principles of logic – to create an extension of ordinary algebra to universal algebra (*A Treatise of Universal Algebra* 1898), and a meticulous re-examination of the relativity theory of Albert Einstein.

A science which hesitates to forget its founders is lost.

Alfred Whitehead

Attributed remark

Whitehead was born in Ramsgate, Kent, and studied at Cambridge. He was professor of applied mathematics at London University 1914–24 and professor of philosophy at Harvard University, USA, 1924–37.

At the International Congress of Philosophy in 1900, Whitehead and Russell heard Italian mathematician Giuseppe Peano describe the method by which he had arrived at his axioms concerning the natural numbers, and they spent the next ten years on their project to deduce mathematics from logic in a general and fundamental way.

Other works include *Principles of Natural Knowledge* 1919, *Science and the Modern World* 1925, and *Process and Reality* 1929.

Whitehead John Henry Constantine 1904–1960. British mathematician who studied the more abstract areas of differential geometry, and of algebraic and geometrical topology.

Whitehead was born in Madras, India, and studied at Oxford, where he became professor 1947.

Whitehead's early research was on differential geometry, and the application of differential calculus and differential equations to the study of geometrical figures.

Some of his most significant work was in the study of knots. In geometry a knot is a two-dimensional representation of a three-dimensional curve that because of its dimensional reduction (by topological distortion) appears to have nodes (to loop onto itself).

Whitehead wrote a textbook together with Oswald Veblen (1880–1960), *Foundations of Differential Geometry* 1932.

Whitehead Robert 1823–1905. English engineer who invented the self-propelled torpedo 1866.

He devised methods of accurately firing torpedoes either above or below water from the fastest ships, no matter what the speed or bearing of the target.

Whitehead developed the torpedo for the Austrian Empire. Typically it was 4 m/13 ft long, could carry a 9-kg/20-lb dynamite warhead, and by 1889 had a speed of 29 knots. It was powered by compressed air and had a balancing mechanism and, later, gyroscopic controls. In 1876 he developed a servomotor, which controlled the steering gear and gave the torpedo a truer path through the water.

Whitehead was born in Bolton, Lancashire, and apprenticed to an engineering company. In 1847 he set up his own business in Milan (then part of Austria), moving to Trieste 1848 and Fiume 1856.

Whitehead designed pumps for draining part of the Lombardy marshes and made improvements to silk-weaving looms. From 1856, in Fiume, he built naval marine engines; he designed and built the engines of the ironclad warship *Ferdinand Max*.

Whitney Eli 1765–1825. US inventor who in 1794 patented the cotton gin, a device for separating cotton fibre from its seeds. Also a manufacturer of firearms, he created a standardization system that was the precursor of the assembly line.

Whitney's cotton gin had a wooden cylinder bearing rows of spikes set 1.3 cm/0.5 in apart, which extended between the bars of a grid set so closely together that only the cotton lint (and not the seeds) could pass through. A revolving brush cleaned the cotton off the spikes and the seeds fell into another compartment. The machine was hand-cranked, and one gin could produce about 23 kg/50 lb of cleaned cotton per day – a 50-fold increase in a worker's output.

Whitney was born in Westborough,

Massachusetts, and studied law at Yale. His cotton gin was so easy to copy and manufacture that he eventually gave up defending his patent and in 1798 turned to the manufacture of firearms in New Haven, Connecticut. He used machine tools to make arms with fully interchangeable parts, and introduced division of labour and mass production.

In 1818 he made a small milling machine, with a power-driven table that moved horizontally beneath and at right angles to a rotating cutter.

Whitten-Brown Arthur 1886–1948. British aviator. After serving in World War I, he took part in the first nonstop flight across the Atlantic as navigator to Captain John Alcock 1919. KBE 1919.

Whittle Frank 1907–1996. British engineer who patented the basic design for the turbojet engine 1930. In the Royal Air Force he worked on jet propulsion 1937–46. In May 1941 the Gloster E 28/39 aircraft first flew with the Whittle jet engine. Both the German (first operational jet planes) and the US jet aircraft were built using his principles. Knighted 1948.

Whittle was born in Coventry and joined the RAF as an apprentice, later training as a fighter pilot. He had the idea for a jet engine 1928 but could not persuade the Air Ministry of its potential until 1935, when he formed the Power Jets Company. He retired from the RAF with the rank of air commodore 1948 and took up a university appointment in the USA.

Whitworth Joseph 1803–1887. English engineer who established new standards of accuracy in the production of machine tools and precision measuring instruments. He devised standard gauges and screw threads, and introduced new methods of making gun barrels. Baronet 1869.

Whitworth was born in Stockport, Cheshire, and left school at 14. He worked for English engineer Henry Maudslay in his London workshops 1825–33, then moved to Manchester and set up in business as a toolmaker. From this, a large factory developed. He was concerned with the training, lives, and leisure time of his workers, and donated large sums to educational organizations.

Whitworth brought standardization to his company and the engineering industry as a whole by developing means of measuring to tolerances never before possible, so that shafts, bearings, gears, and screws could be interchanged. The Whitworth company produced many machines for cutting, shaping, and planing. Whitworth also designed a knitting machine and a horse-drawn mechanical roadsweeper.

At the Whitworth works, guns of all sizes were produced, and Whitworth supervised many experiments to investigate the forces acting on the breech and barrel of a gun. He also made advances in the design of rifling for the barrels of small-calibre weapons.

Wickham Henry 1846–1928. British planter who founded the rubber plantations of Sri Lanka and Malaysia, and broke the monopoly in rubber production then held by Brazil. He collected rubber seeds from Brazil, where they grew naturally, cultivated them at Kew Gardens, Surrey, and re-exported them to the Far East.

Wieland Heinrich Otto 1877–1957. German organic chemist who determining the structures of steroids and related compounds. He also studied other natural compounds, such as alkaloids and pterins, and contributed to the investigation of biological oxidation. Nobel prize 1927.

Wieland was born in Pforzheim in the Black Forest and attended several universities. He spent most of his career at the University and Technische Hochschule of Munich. During World War I he researched into chemical warfare.

In 1912 Wieland showed that bile acids have similar structures to that of cholesterol. Later he worked out what he thought was the basic skeleton of a steroid molecule (for which he was awarded the Nobel prize), but it was found to be incorrect. In 1932 he and his co-workers produced a somewhat modified structure, which is still accepted today.

Wieland did other work with the bile acids, demonstrating their role in converting fats into water-soluble cholic acids (a key process in digestion). He determined the structures of, and synthesized many, toadstool poisons, such as phalloidine from the deadly *Amanita* fungus. He also began research into the composition and synthesis of pterins, the pigments that give the colour to butterflies' wings.

Wieland proved experimentally that biological oxidation (the process within living tissues by which food substances such as glucose are converted to carbon dioxide and energy) was in fact a catalytic dehydrogenation. This was in direct opposition to the findings of Otto Warburg, who had shown that biological oxidation was an addition of oxygen, and the controversy sparked debate and research. In the end both dehydrogenation and oxidation were shown to occur.

Wien Wilhelm Carl Werner Otto Fritz Franz 1864–1928. German physicist who studied radiation and established the principle, since known as ***Wien's law***, that the wavelength at which the radiation from an idealized radiating body is most intense is inversely proportional to the body's absolute temperature. (That is, the hotter the body, the shorter the wavelength.) For this and other work on radiation, he was awarded the 1911 Nobel Prize for Physics.

Wiener Norbert 1894–1964. US mathematician, credited with the establishment of the science of cybernetics in his book *Cybernetics* 1948. In mathematics, he laid the foundation of the study of stochastic processes (those dependent on random events), particularly Brownian motion.

Wiener was born in Columbia, Missouri, and received his PhD from Harvard at the age of 19. He then went to Europe to study under leading mathematicians (Bertrand Russell at Cambridge, England, and David Hilbert at Göttingen, Germany). From 1919 he taught at the Massachusetts Institute of Technology, becoming professor 1932.

Wiener devoted much of his efforts to methodology, developing mathematical approaches that could usefully be applied to continuously changing processes.

During World War II, Wiener worked on the control of anti-aircraft guns (which required him to consider factors such as the machinery itself, the gunner, and the unpredictable evasive action on the part of the target's pilot), on filtering 'noise' from useful information for radar, and on coding and decoding. His investigations stimulated his interest in information transfer and processes such as information feedback.

Wieschaus Eric F 1947– . US geneticist who shared the 1995 Nobel Prize for Physiology or Medicine with Edward Lewis and Christiane Nüsslein-Volhard for their discoveries concerning the genes that control early embryonic development.

Inspired by the work of Lewis, Wieschaus and Nüsslein, at the European Molecular Biology Laboratory in Heidelberg, Germany, set out in 1978 to identify the clusters of genes which determine the development of the individual body parts of fruit flies. They fed adult flies chemicals that cause mutations in the gentic DNA. They then bred the flies and looked for abnormal embryos. By checking back to the parent flies, they were able to pinpoint the genes responsible for different mutations. By 1980 they had identified 139 developmental genes, which fell into several classes. One class caused whole body segments to be missing; another class caused defects in every second body segment. Similar principles are now known to control the development of other species, including humans.

Wiesel Torsten N 1924– . Swedish neuroscientist who shared the 1981 Nobel Prize for Physiology or Medicine with Roger W Sperry and David H Hubel for work on the brain. Wiesel and Hubel studied the brain processes underlying sight.

The signal that the eye sends to the brain can be regarded as a code which only the brain can interpret. Wiesel and Hubel succeeded in breaking the code. They did this by tapping the signals from the nerve cells in the various cell layers of the brain. They were able to show how various parts of the image on the retina of the eye are read out and interpreted by the brain. They also showed that the ability to interpret messages from the retina develop soon after birth.

Wiesel was born in Uppsala. In 1954, he began his research career at the Karolinska Institute, Uppsala, undertaking basic neurophysiological studies. The following year he was invited to the United States as a postdoctoral fellow at the Wilmer Institute, Johns Hopkins Medical School, Baltimore, Maryland. Hubel joined the laboratory in 1958, beginning a collaboration which lasted 20 years. Wiesel and Hubel moved with their professor, Stephen Kuffler, to the Harvard Medical School 1959. In 1973 Wiesel was appointed head of the department of neurobiology at Harvard.

Wigglesworth Vincent Brian 1899–1994. English entomologist whose research covered many areas of insect physiology, especially the role of hormones in growth and metamorphosis. Knighted 1964.

Wigglesworth was born in Kirkham, Lancashire, and studied at Cambridge. During World War I he served in France, and then qualified in medicine at St Thomas's Hospital, London. He was director of the Agricultural Research Council Unit of Insect Physiology at Cambridge 1943–67, as well as professor of biology at Cambridge 1952–66.

Wigglesworth's work on insect metamorphosis was carried out mainly on the bloodsucking insect *Rhodnius prolixus*. He demonstrated that the hormones responsible for growth and moulting, and for preventing the development of adult characteristics until the insect larva is fully grown, are produced in specific areas of the brain.

Wigglesworth investigated many other aspects of insect anatomy and physiology, including the mechanisms involved in hatching; the mode of action of adhesive organs in walking; the role of the outer waxy layer on insects' bodies in preventing water loss; the respiration of insect eggs; insect sense organs and their use in orientation; and the functions of insect blood cells.

His book *The Principles of Insect Physiology* 1939 became the standard text on insect physiology.

Wigner Eugene Paul 1902–1995. Hungarian-born US physicist who introduced the notion of parity, or symmetry theory, into nuclear physics, showing that all nuclear processes should be indistinguishable from their mirror images. For this and other work on nuclear structure, he shared the 1963 Nobel Prize for Physics.

The ***Wigner effect*** is a rapid rise in temperature in a nuclear reactor pile when, under particle bombardment, such materials as graphite deform, swell, then suddenly release large amounts of energy. This was the cause of the fire at the British Windscale plant 1957.

Educated at the Lutheran Gymnasium in Budapest, Wigner took up postgraduate studies in Berlin where he was present at Albert Einstein's seminars in the 1920s. He emigrated to the USA in 1930, and became a US citizen 1937.

He was one of the scientists who persuaded President Roosevelt to commit the USA to developing the atom bomb. In 1960, he was awarded the Atoms for Peace Award, in recognition of his vigorous support for the peaceful use of atomic energy. He taught as a professor of mathematics at Princeton University for 40 years until his retirement 1971.

Wilcox Stephen 1830–1893. US inventor who, with Herman Babcock, designed a steam-tube boiler which was developed into one of the most efficient sources of high-pressure steam.

Wilcox was born in Westerley, Rhode Island. After leaving school he went to work on improving old machines and inventing new ones, such as a hot-air engine for operating fog signals. In about 1856 Wilcox patented a steam-tube boiler which was not entirely successful. Ten years later, with his boyhood friend Babcock, he designed an improved safety water-tube boiler, and they formed a company to manufacture this. Throughout his inventing career Wilcox acquired nearly 50 patents.

The Babcock–Wilcox boiler, patented 1867, had straight tubes inclined to the horizontal and connected together at their ends, through which the hot water gradually rose by convection. The firebox surrounded the tubes to give rapid heating, and there was a reservoir of hot water above the firebox and tubes, with steam above the water. These steam engines were used in the first American electricity generating stations and played an important part in the subsequent development of electric lighting.

Wilkes Maurice Vincent 1913– . English mathematician who led the team at Cambridge University that built the EDSAC (electronic delay storage automatic calculator) 1949, one of the earliest of the British electronic computers.

Wilkes was born in Dudley and studied at Cambridge. During World War II he became involved with the development of radar.

He was director of the Cambridge Mathematical Laboratory 1946–80.

In the late 1940s Wilkes and his team began to build the EDSAC. At the time, electronic computers were in their infancy. Wilkes chose the serial mode, in which the information in the computer is processed in sequence (and not several parts at once, as in the parallel type). This design incorporated mercury delay lines (developed at the Massachusetts Institute of Technology, USA) as the elements of the memory.

In May 1949 the EDSAC ran its first program and became the first delay-line computer in the world. From early 1950 it offered a regular computing facility to the members of Cambridge University, the first general-purpose computer service. Much time was spent by the research group on programming and on the compilation of a library of programs. The EDSAC was in operation until 1958.

EDSAC II came into service 1957. This was a parallel-processing machine and the delay line was abandoned in favour of magnetic storage methods.

Wilkins Maurice Hugh Frederick 1916– . New Zealand-born British molecular biologist. In 1962 he shared the Nobel Prize for Physiology or Medicine with Francis Crick and James Watson for his work on the molecular structure of nucleic acids, particularly DNA, using X-ray diffraction.

Wilkins began his career as a physicist working on luminescence and phosphorescence, radar, and the separation of uranium isotopes, and worked in the USA during World War II on the development of the atomic bomb. After the war he turned his attention from nuclear physics to molecular biology, and studied the genetic effects of ultrasonic waves, nucleic acids, and viruses by using ultraviolet light.

Wilkins was born in Pongaroa and studied in the UK at Cambridge. He became professor of biophysics at London 1970.

Studying the X-ray diffraction pattern of DNA, he discovered that the molecule has a double helical structure and passed on his findings to Crick and Watson.

Wilkinson Geoffrey 1921–1996. English inorganic chemist who shared a Nobel prize 1973 for his pioneering work on the organometallic compounds of the transition metals. Knighted 1976.

Wilkinson was born near Manchester and studied at Imperial College, University of London. He held numerous posts in North America, and was assistant professor at Harvard 1951–56. He then moved back to Imperial College, where he was professor of inorganic chemistry, retiring 1988.

Wilkinson's Nobel-prizewinning work was done with US chemist R B Woodward. An organometallic molecule consists of a metal atom sandwiched between carbon rings. The synthetic compound they were investigating, ferrocene, turned out to have a single iron atom sandwiched between two five-sided carbon rings; materials were later created with other metals and four-, six-, seven-, and eight-membered carbon rings.

Wilks Samuel Stanley 1906–1964. US statistician whose work in data analysis enabled him to formulate methods of deriving valid information from small samples. He also concentrated on the developments and applications of techniques for the analysis of variance.

Wilks was born in Little Elm, Texas. He studied architecture at Texas and statistics at Iowa. He spent his career at Princeton, becoming professor 1944.

Wilks's investigations of the analysis of variance were devoted especially to multivariate analysis. Two of his most original contributions were the Wilks criterion and his multiple correlation coefficient.

The US College Entrance Examination Board, which carries out educational tests, found his assistance invaluable in analysing their results. Seeking also to apply these methods to industrial problems, Wilks did fundamental work in the establishment of the theory of statistical tolerance.

Williams Frederic Calland 1911–1977. English electrical and electronics engineer who developed cathode-ray-tube storage devices used in many early computers.

He took part in building the first stored-program computer 1948. Knighted 1976.

Williams was born near Stockport, Cheshire, and studied at Manchester and Oxford. He was professor at Manchester from 1946.

During World War II, Williams played a major part in the development of radar and allied devices, and in the design of the feedback systems known as servomechanisms. Visiting the Massachusetts Institute of Technology, USA, he learned of attempts to use cathode-ray tubes to store information as dots on the screen, and in 1946 began to develop this system in the UK. The phosphor in the tubes allowed an image to persist for only a fraction of a second, and at first he transferred information to and from two tubes; later he designed the appropriate circuitry to repeat the dots in one tube so that they would persist indefinitely.

Together with M H A Newman, Williams built a computer which began operation in June 1948. After modification, the machine went into production with Ferranti Limited.

In the 1950s Williams began work on electrical machines, principally induction motors and induction-excited alternators. During his later years he worked on an automatic transmission for motor vehicles.

Williamson Alexander William 1824–1904. English organic chemist who made significant discoveries concerning alcohols and ethers, catalysis, and reversible reactions. He was the first to explain the action of a catalyst in terms of the formation of an intermediate compound.

Williamson was born in London and studied at several European universities. He was professor at University College, London, 1849–87.

Williamson was the first to make 'mixed' ethers, with two different alkyl groups, and his method is still known as the Williamson synthesis. It involves treating an alkoxide with an alkyl halide (haloalkane).

Some of the reactions of alcohols and ethers are reversible (that is, the products of a reaction may recombine to form the reactants), a phenomenon first noted and described by Williamson in the early 1850s. If the rate of the forward reaction is the same as that of the reverse reaction, all compounds in the process coexist and the system is said to be in 'dynamic equilibrium' (a term also introduced by Williamson).

Williamson William Crawford 1816–1895. English botanist, surgeon, zoologist, and palaeontologist who was regarded as one of the founders of modern palaeobotany. His research included work on deep-sea deposits, protozoans (single-celled animals), and cryptogams (plants that grow from spores, such as algae, ferns, or mosses). He showed that not all plant fossils containing secondary wood were necessarily spermatophytes (seed plants), but that some were spore-bearing.

Williamson was born in Scarborough, Yorkshire and was apprenticed to an apothecary 1832, although he studied natural history in his spare time, especially fossilized plant remains. He trained as a doctor at University College, London and returned to Manchester to practise medicine, where he was made curator of the Manchester Natural History Museum 1835. He became the professor of natural history, anatomy, and physiology at Owens College in Manchester and professor of botany there 1880.

The significance of plant fossils was not appreciated in the 19th century, however. After 40 years of teaching and dedicated research at Owens College, he was refused a pension upon retirement.

Willis Thomas 1621–1673. English physician and a founding member of the Royal Society who contributed much to our knowledge of the nervous system and cardiovascular system. He carried out a detailed study of the circulation of the brain, including the cerebral arterial circle, which he discovered under the base of the brain, that bears his name.

Willis wrote extensively about many diseases, including the mental disorders hypochondria, hysteria, and melancholia. He rediscovered the sweetness of urine excreted from patients with diabetes mellitus (which had originally been identified by physicians in India around 400 BC).

Willstätter Richard 1872–1942. German organic chemist who investigated plant pigments – such as chlorophyll – and alkaloids, determining the structure of cocaine, tropine, and atropine. Nobel prize 1915.

Willstätter was born in Karlsruhe and studied at the Munich Technische Hochschule. In 1905 he became professor at the Technische Hochschule in Zürich, Switzerland; he worked at the Kaiser Wilhelm Institute in Berlin 1912–16. In 1916 he became professor at Munich, but resigned in 1925 because of mounting anti-Semitism. At the start of World War II in 1939 he left Germany for Switzerland.

Willstätter showed that chlorophyll is made up of four components: two green ones, chlorophyll *a* and *b*, and two yellow ones, carotene and xanthophyll, and established the ratio in which they occur. In order to separate the complex substances, he redeveloped the technique of chromatography.

Willstätter also worked on quinones and, by following English chemist William Perkin's method of oxidizing aniline (phenylamine) with chromic acid, determined the structure of the dyestuff aniline black. Later he studied enzymes and catalytic hydrogenation, particularly in the presence of oxygen. He worked on the degradation of cellulose, investigated fermentation, and pioneered the use of hydrogels for absorption.

Wilson Charles Thomson Rees 1869–1959. Scottish physicist who in 1911 invented the Wilson cloud chamber, an apparatus for studying subatomic particles. He shared a Nobel prize 1927.

A cloud chamber consists of a vessel fitted with a piston and filled with air or other gas, supersaturated with water vapour. When the volume of the vessel is suddenly expanded by moving the piston outwards, the vapour cools and a cloud of tiny droplets forms on any nuclei, dust, or ions present. As fast-moving ionizing particles collide with the air or gas molecules, they show as visible tracks.

Wilson was born near Edinburgh and studied at Manchester and Cambridge. He was professor at Cambridge 1925–34.

Wilson originally devised the cloud chamber to simulate clouds in his laboratory. From 1895 to 1899, he carried out many experiments and established that the

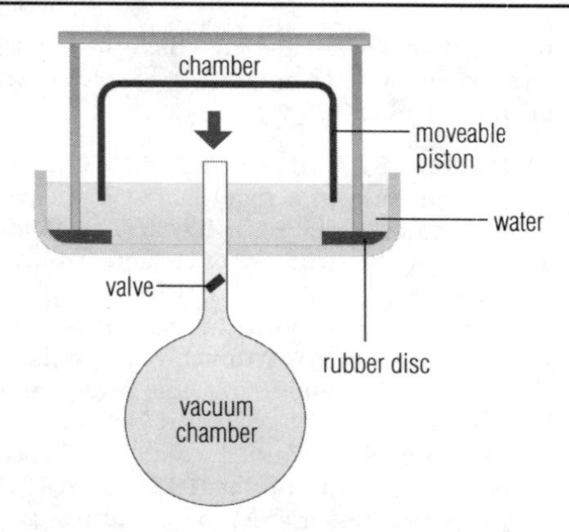

Wilson *The cloud chamber devised by C T R Wilson was the first instrument to detect the tracks of atomic particles. It consisted originally of a cylindrical glass chamber fitted with a hollow piston, which was connected, via a valve, to a large evacuated flask. The piston falls rapidly when the valve is opened, and water vapour condenses along the tracks of any particles in the chamber.*

nucleation of droplets can take place in the absence of dust particles. He also demonstrated that once the gas is supersaturated with water vapour, nucleation can occur and is greatly improved by exposure to X-rays. This showed that ions are the nucleation sites on which water droplets form in the absence of dust.

Wilson realized 1910 that the cloud chamber could possibly show the track of a charged particle moving through it, and applying a magnetic field to the chamber would cause the track to curve, giving a measure of the charge and mass of the particle. The adapted Wilson cloud chamber immediately became vital to the study of radioactivity.

During 1900–10, Wilson studied electrical conduction in dust-free air by means of a gold-leaf electroscope he devised.

Wilson Edmund Beecher 1856–1939. US zoologist and embryologist who is considered one of the founders of modern genetics. His research programme focussed on the way vertebrate cells divide. He was one of the first to describe the process of meiosis, the division of gametes (eggs and sperm) in the sexual organs during embryonic and adult life. He also investigated the role of chromosomes in heredity, particularly the sex chromosomes X and Y.

Wilson was born in Geneva, Illinois and studied medicine at Yale University and Johns Hopkins University. His research was crucial to the development of the field of genetics and in 1921 he was appointed professor of zoology at Columbia University in recognition of his contribution to the understanding of the cellular mechanisms of heredity.

Wilson Edward Osborne 1929– . US zoologist whose books have stimulated interest in biogeography, the study of the distribution of species, and sociobiology, the evolution of behaviour. He is a world authority on ants.

Wilson was born in Birmingham, Alabama. His works include *Sociobiology: The New Synthesis* 1975, *On Human Nature* 1978, and *The Diversity of Life* 1992.

Wilson John Tuzo 1908–1993. Canadian geologist and geophysicist who established and brought about a general understanding of the concept of plate tectonics.

Born in Ottawa, Wilson studied geology and physics – an original combination that led directly to the development of the science of geophysics – at the University of Toronto, and obtained his doctoral degree at Princeton University, New Jersey, USA. His particular interest was the movement of the continents across the Earth's surface – then a poorly understood and not widely accepted concept known as 'continental drift'. He spent 28 years as professor of geophysics at the University of Toronto, retiring 1974 just as interest in plate tectonics was developing worldwide. From then on he was the director-general of the Ontario Science Centre and later the chancellor of York University, Toronto, finally retiring 1987. In 1957 he was the president of the International Union of Geodesy and Geophysics – the most senior administrative post in the field.

Wilson's great strength was in education. He pioneered hands-on interactive museum exhibits, and could explain complex subjects like the movement of continents, the spreading of ocean floors, and the creation of island chains by using astonishingly simple models. He was an active outdoor man, leading expeditions into the remote north of Canada, and he made the first ascent of Mount Hague in Montana, USA, 1935.

The Wilson Range in Antarctica is named after him.

Wilson Robert Woodrow 1936– . US radio astronomer who, with Arno Penzias, in 1964–65 detected the cosmic background radiation, which is thought to represent a residue of the primordial Big Bang. He and Penzias shared the 1978 Nobel Prize for Physics for their work on microwave radiation.

Wilson was born in Houston, Texas, and studied at Rice University and the California Institute of Technology. In 1963 he joined the Bell Telephone Laboratories in New Jersey; he was made head of the radiophysics department in 1976.

In 1964, Wilson and Penzias tested a radiotelescope and receiver system for the Bell Telephone Laboratories with the intention of tracking down all possible sources of static that were causing interference in satellite communications. They found a high level of isotropic background radiation at a wavelength of 7.3 cm/2.9 in, with a temperature of 3.1 K (–276°C/–529°F). This radiation was a hundred times more powerful than any that could be accounted for on the basis of any known sources.

Unable to explain this signal, Wilson and Penzias contacted Princeton University, where it was immediately realized that their findings confirmed predictions of residual microwave radiation from the beginning of the universe.

Windaus Adolf Otto Reinhold 1876–1959. German chemist who was awarded the Nobel Prize for Chemistry 1928 for his research on the structure of cholesterol, its relationship to vitamin D, and his discovery that steroids are precursors of vitamins. He also linked the roles of sunlight and vitamin D in the prevention of rickets.

Windaus discovered that the precursor (provitamin) of vitamin D is a steroid, ergosterol, a fat-soluble impurity of cholesterol. In the 1920s, he demonstrated that ergosterol converts to vitamin D in the presence of sunlight, providing an explanation for why both cod-liver oil (which contains vitamin D) and sunlight help to prevent rickets.

Windaus was born on 26 December in Berlin. He began reading medicine at the Universities of Berlin and Freiburg in 1895. However, under the guidance of Emil Fischer he gave up medicine in favour of chemistry and, in 1899, completed a doctoral thesis on the cardiac poisons of digitalis. He was appointed professor at Freiburg University, then moved to become professor of medical chemistry at the University of Innsbruck, and finally settled at the University of Göttingen. His study concentrated on the relationship between cholesterol and vitamin D. In the 1930s Windaus continued his study of natural products, such as vitamin B and colchicine, and retired in 1944.

Winsor Frederick Albert 1763–1830. German inventor, one of the pioneers of gas lighting in the UK and France.

Winsor was born in Brunswick. He went to Britain before 1799 and became interested in the technology and economics of fuels. In 1802 he went to Paris to investigate the 'thermo-lamp' which French engineer Philippe Lebon had patented in 1799. Returning to Britain, he started a gasworks and in 1807 lit one side of Pall Mall, London, with gas lamps. In 1804–09 he was granted various patents for gas furnaces and purifiers.

His application to Parliament for a charter for the Light and Heat Company having failed, Winsor once more moved to France, but in Paris his company made little progress and was liquidated 1819.

The distilling retort Winsor used consisted of an iron pot with a fitted lid. The lid had a pipe in the centre leading to the conical condensing vessel, which was compartmented inside with perforated divisions to spread the gas to purify it of hydrogen sulphide and ammonia. The device was not very successful, and the gas being burned was impure and emitted a pungent smell.

Winsor published *Description of the Thermo-lamp Invented by Lebon of Paris* 1802, *Analogy between Animal and Vegetable Life, Demonstrating the Beneficial Application of the Patent Light Stoves to all Green and Hot Houses* 1807, and others.

Withering William 1741–1799. English physician, botanist, and mineralogist who investigated the drug digitalis (from the foxglove plant), which he initially used as a diuretic to treat oedema.

Withering was born in Wellington, Shropshire, and studied at the Edinburgh Medical School. From 1775 he had a practice in Birmingham, where he met prominent contemporary scientists; he later became physician at Birmingham General Hospital. Because he publicly expressed his sympathies with the French Revolution, in 1791 his house was attacked by a mob.

Withering began studying *Digitalis purpurea* in 1775, after noting its use in traditional herbal remedies. He worked out precise dosages of dried foxglove leaves for oedema, and also suggested the possible use of the drug in the treatment of heart disease. It is now widely used for treating heart failure.

Withering published *Account of the Foxglove* 1785.

His *Botanical Arrangement* 1776, based on the system of Swedish botanist Linnaeus, became a standard work, and his activities in geology are remembered through the mineral ore ***witherite*** (barium carbonate), which was named after him.

Wittgenstein Ludwig Josef Johann 1889–1951. Austrian philosopher. *Tractatus Logico-Philosophicus* 1922 postulated the 'picture theory' of language: that words represent things according to social agreement. He subsequently rejected this idea, and developed the idea that usage was more important than convention.

If a lion could talk, we could not understand him.

Ludwig Wittgenstein

Philosophical Investigations

The picture theory said that it must be possible to break down a sentence into 'atomic propositions' whose elements stand for elements of the real world. After he rejected this idea, his later philosophy developed a quite different, anthropological view of language: words are used according to different rules in a variety of human activities – different 'language games' are played with them. The traditional philosophical problems arise through the assumption that words (like 'exist' in the sentence 'Physical objects do not really exist') carry a fixed meaning with them, independent of context.

Wittgenstein was born in Vienna and studied in the UK at Cambridge, where he taught in the 1930s and 1940s, becoming professor 1939. His *Philosophical Investigations* 1953 and *On Certainty* 1969 were published posthumously.

Wittig Georg 1897–1987. German chemist whose method of synthesizing olefins (alkenes) from carbonyl compounds is a reaction often termed the ***Wittig synthesis***. For this achievement he shared the 1979 Nobel Prize for Chemistry.

Wittig was born in Berlin and studied at Kassel and Marburg. He was professor at Freiburg 1937–44, Tübingen 1944–56, and Heidelberg 1956–67.

In the Wittig reaction, which he first demonstrated 1954, a carbonyl compound (aldehyde or ketone) reacts with an organic phosphorus compound, an alkylidenetriphenylphosphorane, $(C_6H_5)_3P{=}CR_2$, where R is a hydrogen atom or an organic radical. The alkylidene group ($=CR_2$) of the reagent reacts with the oxygen atom of the carbonyl group to form a hydrocarbon with a double bond, an olefin (alkene). In general:

$$(C_6H_5)_3P{=}CR_2 + R_2`CO \rightarrow$$
$$(C_6H_5)_3PO + R_2C{=}CR_2$$

The reaction is widely used in organic synthesis, for example to make squalene (the synthetic precursor of cholesterol) and vitamin D_3.

Wöhler Friedrich 1800–1882. German chemist who in 1828 became the first person to synthesize an organic compound (urea) from an inorganic compound (ammonium cyanate). He also devised a method 1827 that isolated the metals aluminium, beryllium, yttrium, and titanium from their ores.

$$NH_4OCN \xrightarrow{\text{heat}} O{=}C(NH_2)_2$$

ammonium cyanate — urea (carbamide)

Wöhler was born near Frankfurt-am-Main and studied at German universities and in Stockholm, Sweden, under Jöns Berzelius. He was professor at Göttingen from 1836.

Until Wöhler's landmark synthesis of urea there had been a basic misconception in scientific thinking that the chemical changes undergone by substances in living organisms were not governed by the same laws as inanimate substances.

Organic chemistry just now is enough to drive one mad. It gives one the impression of a primeval, tropical forest full of the most remarkable things, a monstrous and boundless thicket, with no way of escape, into which one may well dread to enter.

Friedrich Wöhler

Letter to Berzelius 28 January 1835

Wöhler worked with German chemist Justus von Liebig on a number of important investigations. In 1830, they proved the polymerism of cyanates and fulminates; in 1837, they investigated uric acid and its derivatives.

Wöhler discovered quinone (cyclohexadiene-1,4-dione), hydroquinone or quinol (benzene-1,4-diol), and calcium carbide. He isolated boron and silicon.

Wolf Maximilian Franz Joseph Cornelius 1863–1932. German astronomer who developed new photographic methods for observational astronomy. He discovered several new nebulae, both within the Milky Way and outside our Galaxy; more than 200 asteroids; and in 1883 a comet, which now bears his name.

Wolf was born and educated in Heidelberg, where he spent most of his career, becoming professor 1893. He used a small private observatory 1885–96, and then became the director of a new observatory at Königstuhl, near Heidelberg, built at his instigation.

Wolf was the first to use time-lapse photography in astronomy, a technique he used for detecting asteroids. In 1903 he discovered the first of the so-called Trojan satellites (number 588, later named Achilles), whose orbits are in precise synchrony with that of Jupiter's; they form a gravitationally stable configuration between Jupiter and the Sun. This kind of triangular three-bodied system had been analysed and predicted theoretically by Joseph Lagrange in the 1770s.

Independently of US astronomer Edward Barnard, Wolf discovered that the dark 'voids' in the Milky Way are in fact nebulae which are obscured by vast quantities of dust, and he studied their spectral characteristics and distribution.

Wolf also was the first to observe Halley's comet when it approached the Earth in 1909.

Wolff Christian 1679–1754. German philosopher, mathematician, and scientist who invented the terms 'cosmology' and 'teleology'. He was science adviser to Peter the Great of Russia 1716–25.

Wolff worked in many fields, including theology, psychology, botany, and physics. His philosophy was influenced by Gottfried Leibniz and scholasticism. His numerous works include *Vernunftige Gedanken von Gott, der Welt und der Seele der Menschen/Rational Ideas on God, the World and the Soul of Man* 1720.

He was professor of mathematics at Halle 1707–23 and professor of mathematics and philosophy at Marburg 1723–40.

Wolff Heinz Siegfried 1928– . German-born British biomedical engineer who worked on high-technology instruments and the application of technology to medicine.

Born in Berlin, Wolff left Germany for the UK as a boy. He studied physiology at London and then worked for the Medical Research Council (MRC) at the National Institute for Medical Research, where he specialized in the development of instrumentation. Formerly director of the Biomedical Division of the MRC's Clinical Research Centre, he then became director of the Institute for Bioengineering at Brunel University, Uxbridge, in the UK.

Wolff's interests range from the invention of new high technology instruments to the widespread and sensible application of technology to the problems of the elderly and the disabled. He believes that small, specialized pieces of equipment that can be worked by doctors and nurses may be preferable to large centralized units to which patients have to go for tests or treatment. Machines should be simple to use, should show when they are not working properly, and should be designed so that they can be repaired on the spot by the operator.

Wolff Kaspar Friedrich 1733–1794. German surgeon and physiologist who is regarded as the founder of embryology. He introduced the idea that initially unspecialized cells later differentiate to produce the separate organs and systems of the plant or animal body.

Wolff was born in Berlin and studied at Halle and the Berlin Medical School. In 1766 he accepted an invitation from Catherine II of Russia to take the post of Academician for anatomy and physiology in St Petersburg. He remained there until his death.

Wolff produced his revolutionary work *Theoria generationis* in 1759. Until that time it was generally believed that each living organism develops from an exact miniature of the adult within the seed or sperm – the so-called preformation or homunculus theory. In fact, Wolff's view that plants and animals are composed of cells was still a subject of controversy, and his findings were largely ignored for more than 50 years.

His name is also associated with, among other parts of the anatomy, the Wolffian body, a structure in an animal embryo that eventually develops into the kidney.

Wollaston William Hyde 1766–1828. English chemist and physicist who discovered in 1804 how to make malleable platinum. He went on to discover the new elements palladium 1804 and rhodium 1805. He also contributed to optics through the invention of a number of ingenious and still useful measuring instruments.

Wollaston was born in East Dereham, Norfolk, and studied medicine at Cambridge.

Wollaston initiated the technique of powder metallurgy when working with platinum. Using aqua regia (a mixture of concentrated nitric and hydrochloric acids), he dissolved the platinum from crude platina, a mixed platinum–iridium ore. He then prepared ammonium platinichloride, which he decomposed by heating to yield fine grains of platinum metal. The grains were worked using heat, pressure, and hammering to form sheets, which he sold to industrial chemists. He donated much of the profits to various scientific societies to help finance their researches.

In 1807 he developed the *camera lucida*, which was to inspire William Fox Talbot to his discoveries in photography.

A supporter of John Dalton's atomic theory, Wollaston suggested 1808 that a knowledge of the arrangements of atoms in three dimensions would be a great leap forward.

As a member of a Royal Commission in 1819, Wollaston was instrumental in the rejection of the decimal system of weights and measures.

Woodger Joseph Henry 1894–1981. English biologist who attempted to provide biology with a systematic and logical foundation on which observations, theories, and methods could be based.

Woodger was born in Great Yarmouth, Norfolk, and studied at London. He taught at the Middlesex Hospital Medical School from 1924 and became professor there 1947.

Woodger developed the idea that one of the characteristics of a living system is the organization of its substance, and that this order is of a hierarchical nature. Thus the components of an organism can be classified on a scale of increasing size and complexity: molecular, macromolecular, cell components, cells, tissues, organs, and organisms. Each class exhibits specifically new modes of behaviour, which cannot be interpreted as being merely additive phenomena from the previous class. Living matter shows not only spatial hierarchical order but also divisional hierarchies (each cell or group of cells has a parent cell), and he showed that many difficulties in biological theory arose originally through viewing an organism as a series of components ordered in space but not in time.

Woodger's works include the textbook *Elementary Morphology and Physiology* 1924, *Biological Principles: a critical study* 1929, and *The Technique of Theory Construction* 1939.

Woodward Robert Burns 1917–1979. US chemist who worked on synthesizing a large number of complex molecules. These included quinine 1944, cholesterol 1951, chlorophyll 1960, and vitamin B_{12} 1971. Nobel prize 1965.

Woodward was born in Boston and studied at the Massachusetts Institute of Technology. He worked throughout his career at Harvard, becoming professor 1950.

In 1947 Woodward worked out the structure of penicillin and two years later that of strychnine. In the early 1950s, he began to synthesize steroids, and in 1954 he synthesized the poisonous alkaloid strychnine and lysergic acid, the basis of the hallucinogenic drug LSD. In 1956 he made reserpine, the first of the tranquillizing drugs, Turning his attention again to antibiotics, he and his coworkers produced a tetracycline 1962 and cephalosporin C in 1965. The synthesis of vitamin B_{12} was made in collaboration with Swiss chemists and took ten years.

Woolley Richard van der Riet 1906–1986. English astronomer whose work included observational and theoretical astrophysics, stellar dynamics and the dynamics of the Galaxy. Knighted 1963.

Woolley was born in Weymouth, Dorset, and studied at Cape Town, South Africa, and at Cambridge. He was director of the Commonwealth Observatory at Mount Stromlo, Canberra, Australia, 1939–55, and Astronomer Royal at Herstmonceux, Sussex, England, 1956–70.

The observatory at Mount Stromlo was devoted mainly to solar physics, and Woolley devoted much of his time there to the study of photospheric convection, emission spectra of the chromosphere, and the solar corona. He pioneered the observation of monochromatic magnitudes and constructed colour magnitude arrays for globular clusters.

During the 15 years that he spent as Astronomer Royal, his personal interests were globular clusters, the evolution of galactic orbits, improvements of radial velocities, and a re-evaluation of RR Lyrae luminosities.

Wright Almroth Edward 1861–1947. English bacteriologist who developed a vaccine against typhoid fever. He established a new discipline within medicine, that of therapeutic immunization by vaccination, which was aimed at treating microbial diseases rather than preventing them. Knighted 1906.

Wright was born near Richmond, Yorkshire, and studied at Trinity College, Dublin, Ireland, and at German universities. He was professor of pathology at the Army Medical School 1892–1902, and at St Mary's Hospital in London 1902–46. In 1911, Wright went to South Africa, where he introduced prophylactic inoculation against pneumonia for workers in the Rand gold mines. On returning to England, he was appointed director of the Department of Bacteriology of the newly founded Medical Research Committee (later Council).

By 1896 Wright had succeeded in developing an effective antityphoid vaccine, which he prepared from killed typhoid bacilli. Preliminary trials of the vaccine on troops of the Indian Army proved its effectiveness and the vaccine was subsequently used successfully among the British soldiers in the Boer War. He also originated vaccines against enteric tuberculosis and pneumonia.

Wright proved that the human bloodstream contains bacteriotropins (opsonins) in the serum and that these substances can destroy bacteria by phagocytosis.

Wright Louis Tompkins 1891–1952. US physician, surgeon, and civil-rights leader who specialized in head injuries and fractures, but also venereal disease and cancer.

Wright was born in LaGrange, Georgia, and studied at Clark University, Atlanta, and at Harvard Medical School. From 1919 he worked at the Harlem Hospital, New York, rising to director of surgery 1942 and director of the medical board of Harlem Hospitals 1948. He was the first black doctor to be appointed to a municipal hospital position in New York City. In 1929 he also became the first black police surgeon in the history of the city. Wright was chair of the board of directors of the National Association for the Advancement of Colored People for 17 years.

Wright originated the intradermal method of vaccination against smallpox 1918. He devised a brace for neck fractures, a blade for the treatment of fractures of the knee joint, and a plate made of tantalum for the repair of recurrent hernias. He also became an authority on the venereal disease lymphogranuloma venereum, and was the first physician to experiment with the antibiotics Aureomycin and Terramycin. In 1948, Wright moved into the field of cancer research and studied the effectiveness of chemotherapeutic drugs.

Wright Thomas 1711–1786. English astronomer and teacher. He was the first to propose that the Milky Way is a layer of stars, including the Sun, in his *An Original Theory or New Hypothesis of the Universe* 1750.

Wright was born near Durham and apprenticed to a clockmaker. He taught mathematics and lectured on popular scientific subjects.

In his book, Wright described the Milky Way as a flattened disc, in which the Sun does not occupy a central position. Furthermore, he stated that nebulae lie outside the Milky Way.

These views were more than 150 years ahead of their time.
However, he believed that the centre of the system was occupied by a divine presence.

Wright's other work included thoughts on the particulate nature of the rings of Saturn, and reflections on such diverse fields as architecture and reincarnation.

Wright Sewall 1889–1988. US geneticist and statistician. During the 1920s he helped modernize Charles Darwin's theory of evolution, using statistics to model the behaviour of populations of genes.

Wright's work on genetic drift centred on a phenomenon occurring in small isolated colonies where the chance disappearance of some types of gene leads to evolution without the influence of natural selection.

The airplane stays up because it doesn't have the time to fall.

Orville Wright

Wright brothers Orville (1871–1948) and Wilbur (1867–1912) US inventors; brothers who pioneered piloted, powered flight. Inspired by Otto Lilienthal's gliding, they perfected their piloted glider 1902. In 1903 they built a powered machine, a 12-hp 341-kg/750-lb plane, and became the first to make a successful powered flight, near Kitty Hawk, North Carolina. Orville flew 36.6 m/120 ft in 12 sec; Wilbur, 260 m/852 ft in 59 sec.

Orville was born in Dayton, Ohio, and

Wright, Wilbur *US aviation pioneer Wilbur Wright who, with his younger brother Orville, became the first to achieve powered flight in 1903. After the first historic flight, the brothers improved the design and by 1909 were flying public demonstrations in Europe. During litigation over a patent dispute with a competitor in 1912, Wilbur Wright caught typhoid fever and died.*

Wilbur in Indiana. They built their piloted glider while running a bicycle business in Dayton. In 1905 the Wrights offered their aeroplane design to the US War Department, though they did not exhibit the planes publicly until 1908–09, when they took them to Europe. From 1909 they manufactured aeroplanes, but Orville sold his interest in the Wright Company 1915, after Wilbur's death.

In 1899 the Wrights flew a large kite with controls for warping the wings to achieve control of direction and stability. These were the forerunners of ailerons. They then built a small wind tunnel and tested various wing designs and cambers, compiling the first accurate tables of lift and drag.

Wurtz Charles Adolphe 1817–1884. French organic chemist who discovered ethylamine, the first organic derivative of ammonia, 1849 and ethylene glycol (1,2-ethanediol) 1856.

Wurtz was born in Wolfisheim, near Strasbourg, and studied at Strasbourg. In 1844 he moved to Paris and worked at the Sorbonne, where he became professor 1874. He also held public office as mayor of the Seventh Arrondissement of Paris and as a senator.

Wurtz initially worked on the oxides and oxyacids of phosphorus; in 1846 he discovered phosphorus oxychloride ($POCl_3$). He later turned to organic chemistry.

In 1855, Wurtz discovered a method of producing paraffin hydrocarbons (alkanes) using alkyl halides (haloalkanes) and sodium in ether. The method was named the ***Wurtz reaction***.

Wurtz discovered aldol (3-hydroxybutanal) while investigating the polymerization of acetaldehyde (ethanal), devised a method of making esters from alkyl halides, and in 1867, with German chemist Friedrich Kekulé, synthesized phenol from benzene.

Wynne-Edwards Vero Copner 1906–1997. English zoologist who argued that animal behaviour is often altruistic and that animals will behave for the good of the group, even if this entails individual sacrifice. His study *Animal Dispersal in Relation to Social Behaviour* was published 1962.

The theory that animals are genetically programmed to behave for the good of the species has since fallen into disrepute. From this dispute grew a new interpretation of animal behaviour, seen in the work of biologist E O Wilson.

Yalow Rosalyn Sussman 1921– . US physicist who developed radioimmunoassay (RIA), a technique for detecting minute quantities of hormones present in the blood. It can be used to discover a range of hormones produced in the hypothalamic region of the brain. She shared the Nobel Prize for Physiology or Medicine 1977.

Sussman was born in New York and studied at Hunter College and at the University of Illinois. In the 1940s she started working in the Radioisotope Unit of the Veterans Administration Hospital in the Bronx, New York, with a medical doctor, Sol Berson. When he died 1972, Yalow was appointed director of the laboratory.

To measure the concentration of a natural hormone, a solution containing a known amount of the radioisotope-labelled form of the hormone and its antibody is prepared. When a solution containing the natural hormone is added to the first solution, some of the labelled hormone is displaced from the hormone–antibody complex. The fraction of labelled hormone displaced is proportional to the amount of the natural hormone (which is unknown).

Yang Chen Ning 1922– . Chinese-born US physicist who, with with Tsung Dao Lee, overthrew a long-standing principle in theoretical physics called the conservation of parity. This principle states that physical laws are the same in all mirror-image systems. The work of Yang and Lee showed that the weak nuclear force caused particles called K-mesons to decay in a way which violated parity conservation. Nobel Prize for Physics 1957.

Yang also established the basis of modern quantum field theory. In 1954, with Robert Mills, he elaborated the methods needed to describe elementary particles by using mathematical entities called gauge-invariant fields. He also made contributions to statistical mechanics.

Yang was born in Hefei, China, and was the son of a mathematics professor. He was educated at the National Southwest Associated University in Kunming. In 1945 he travelled to the United States to study under Edward Teller and Enrico Fermi at the University of Chicago. He received his doctorate 1948. In 1949 he went to the Institute for Advanced Study at Princeton, New Jersey, where he worked on the theory of elementary particles and the weak nuclear force. In 1951 he and Lee started to study the unusual decay pattern of K-mesons. By 1956, they had concluded that the parity was not conserved in the decay. This result was quickly confirmed experimentally by Chien-Shung Wu, a physicist at Columbia University, New York. Yang was made professor of physics at Princeton 1955. In 1966 he was appointed Director for Theoretical Physics at the University of New York at Stony Brook.

Yanofsky Charles 1925– . US geneticist who demonstrated a linear correspondence between the sequence of bases along the DNA of a gene and the amino acid sequence of the gene's protein product.

He examined numerous strains of the bacterium *E. coli* that carry mutations in the gene for tryptophan synthetase (the enzyme that catalyses the final step in the production of the amino acid tryptophan). The positions of the mutations in the tryptophan synthetase gene were mapped by measuring frequencies of recombination (the 'shuffling' of genetic material that increases variation in the offspring). The position and nature of the change in each mutant tryptophan synthetase protein was then identified. The order of the mutations in the gene proved to correspond to the amino acid sequence of the gene's protein product. These results were the first proof that genes and their polypeptide products correspond.

Yanofsky was born in New York City and studied biology at New York City College and Yale University. He has been professor of biology at Stanford University in California since 1961.

Yersin Alexandre Emile Jean 1863–1943. Swiss bacteriologist who discovered the bubonic plague bacillus in Hong Kong 1894 and prepared a serum against it. The bacillus was discovered independently, in the same epidemic, by Japanese bacteriologist Shibasaburō Kitasato, who published his results before Yersin did.

Young Charles Augustus 1834–1908. US astronomer who made some of the first spectroscopic investigations of the Sun. He was the first person to observe the spectrum of the solar corona.

Young was born in Hanover, New Hampshire, and studied there at Dartmouth College. He was professor at the Western Reserve College, Hudson, Ohio, 1856–66; at Dartmouth College 1866–77; and at Princeton 1877–1905.

Young discovered a layer in the solar atmosphere in which the dark hues of the Sun's spectrum are momentarily reversed at the moment of a total solar eclipse. He published a series of papers relating his spectroscopic observations of the solar chromosphere, solar prominences, and sunspots. He also compiled a catalogue of bright spectral lines in the Sun and used these to measure its rotational velocity.

Young wrote several best-selling textbooks: *General Astronomy* 1888, *Lessons in Astronomy* 1891, and *Manual of Astronomy* 1902.

Young John Watts 1930– . US astronaut. His first flight was on *Gemini 3* 1965. He landed on the Moon with *Apollo 16* 1972, and was commander of the first flight of the space shuttle *Columbia* 1981.

Young was born in San Francisco. He became a NASA astronaut in 1962, and chief of the Astronaut Office in 1975. He flew on *Apollo 10* in 1969 and commanded the ninth space-shuttle flight in 1983.

Young J(ohn) Z(achary) 1907– . English zoologist who discovered and studied the giant nerve fibres in squids, contributing greatly to knowledge of nerve structure and function. He also did research on the central nervous system of octopuses, demonstrating that memory stores are located in the brain.

Young was born in Bristol and studied at Oxford and the zoological station in Naples, Italy. He set up a unit at Oxford to study nerve regeneration in mammals. In 1945 he became the first nonmedical scientist in Britain to hold a professorship in anatomy, at London.

Young discovered that certain nerve fibres of squids are about 100 times the diameter of mammalian neurons and are covered with a relatively thin myelin sheath (unlike mammalian nerve fibres, which have thick sheaths). These properties make them easy to experiment on and to obtain intracellular nerve material.

Turning his attention to the central nervous system, Young showed that octopuses can learn to discriminate between different orientations of the same object. When presented with horizontal and vertical rectangles, for example, the octopuses attacked one but avoided the other. He also demonstrated that octopuses can learn to recognize objects by touch. In addition, Young proposed a model to explain the processes involved in memory.

Young published the textbooks *The Life of Vertebrates* 1950 and *The Life of Mammals* 1957.

Young Thomas 1773–1829. British physicist, physician, and Egyptologist who revived the wave theory of light and identified the phenomenon of interference in 1801. Interference is the phenomenon of two or more wave motions interacting and combining to produce a wave of larger or smaller amplitude. He also established many important concepts in mechanics.

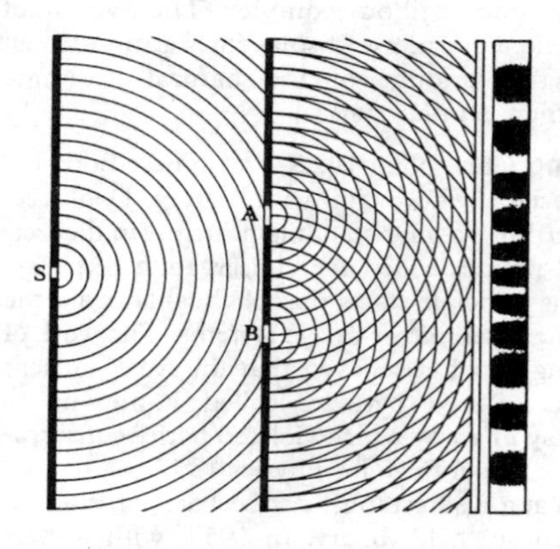

Young, Thomas *Young demonstrated the interface of light by passing monochromatic light through a narrow slit S (he originally used a small hole) and then letting it pass through a pair of closely spaced slits (A and B). Reinforcement and cancellation of the wave trains as they reached the screen produced characteristic interference fringes of alternate light and dark bands.*

In 1793, Young recognized that focusing of the eye (accommodation) is achieved by a change of shape in the lens of the eye, the lens being composed of muscle fibres. He also showed that astigmatism is due to irregular curvature of the cornea. In 1801, he became the first to recognize that colour sensation is due to the presence in the retina of structures that respond to the three colours red, green, and violet.

Young was born in Milverton, Somerset. A child prodigy, he had learned most European and many ancient languages by the age of 20. He studied medicine in London and at Edinburgh and Göttingen, Germany. He was professor of natural philosophy at the Royal Institution 1801–03 and worked as a physician at St George's Hospital, London, from 1811.

Young assumed that light waves are propagated in a similar way to sound waves, and proposed that different colours consist of different frequencies. He obtained experimental proof for the principle of interference by passing light through extremely narrow openings and observing the interference patterns produced.

In mechanics, Young was the first to use the terms 'energy' for the product of the mass of a body with the square of its velocity and 'labour expended' for the product of the force exerted on a body 'with the distance through which it moved'. He also stated that these two products are proportional to each other. He introduced an absolute measurement in elasticity, now known as ***Young's modulus***.

From 1815 onwards, Young published papers on Egyptology; his account of the Rosetta stone played a crucial role in the stone's eventual decipherment.

Yukawa Hideki 1907–1981. Japanese physicist. In 1935 he discovered the strong nuclear force that binds protons and neutrons together in the atomic nucleus, and predicted the existence of the subatomic particle called the meson. Nobel prize 1949.

Yukawa was born and educated in Kyoto and spent his career at Kyoto University, becoming professor 1939 and director of the university's newly created Research Institute for Fundamental Physics from 1953.

Yukawa's theory of nuclear forces postulated the existence of a nuclear 'exchange force' that counteracted the mutual repulsion of the protons and therefore held the nucleus together. He predicted that this exchange force would involve the transfer of a particle (the existence of which was then unknown), and calculated the range of the force and the mass of the hypothetical particle, which would be radioactive, with an extremely short half-life. The muon, or μ meson, discovered 1936, fitted part of the description, and the pion, or π meson, discovered 1947, fitted all of it.

In 1936 Yukawa predicted that a nucleus could absorb one of the innermost orbiting electrons and that this would be equivalent to emitting a positron. These innermost electrons belong to the K electron shell, and this process of electron absorption by the nucleus is known as K capture.

Zassenhaus Hans Julius 1912– . German mathematician who mainly studied group theory and number theory. ***Zassenhaus groups*** form part of the basis for the contemporary development of finite group theory.

Zassenhaus was born in Koblenz and studied at Hamburg. He was professor at Montréal, Canada, 1949–59. Transferring to the USA, he finally became professor at Ohio State 1964.

Group theory is the study of systems in which the product of any two members of a system results in another member of the same system (for example, even numbers). Zassenhaus's most significant results were obtained in his investigations into finite groups. In group extensions he postulated with Issai Schur (1875–1941) what has become known as the Schur–Zassenhaus theory.

Zassenhaus also made contributions to the study of Lie algebras, the geometry of numbers, and applied mathematics.

Zeeman Erik Christopher 1925– . British mathematician concerned with catastrophe theory and its applications to the physical, biological, and behavioural sciences, and with dynamical systems. For example, he has constructed catastrophe models of the heartbeat and the propagation of nerve impulses.

Catastrophe theory is a mathematical method for dealing with discontinuous and divergent phenomena. French mathematician René Thom showed in his book *Structural Stability and Morphogenesis* 1972 that the growth of an organism proceeds by a series of gradual changes that are triggered by, and in turn trigger, large-scale changes or 'catastrophic' jumps. It can be applied to any situations where gradually changing forces lead to abrupt changes in behaviour.

Zeeman studied at Cambridge. In 1964 he became professor at Warwick, where he founded and became director of the Mathematics Research Centre.

... mathematics is a natural and a fundamental language. It may well be that it's a property of human beings, that only human beings can think maths. But I think it's probably true that any intelligence in the universe would have this language as well. So maybe it's even greater than – no, not greater than, but more universal than – the human race.

Erik Zeeman

L Wolpert and A Richards, *A Passion for Science*, 1988

Zeeman's interest in catastrophe theory arose from his study of topology and brain modelling. If two elementary catastrophes (two conflicting behavioural drives, for example) are plotted as axes on the horizontal plane – called the control surface – and the complementary result (the resulting behaviour) is plotted on a third axis perpendicular to the first two, resultant points can be plotted for the entire control surface, and when connected they form a surface of their own. Catastrophe theory reveals that in the middle of the surface is a pleat, which becomes narrower towards the back. For Zeeman, all the points on the surface represent the most probable behaviour, with the exception of those on the pleated middle part, which represent the least likely behaviour. At the edge of the pleat, the sheet on which the behaviour points have been travelling folds under and is wiped out. The behaviour state falls to the bottom sheet of the graph and there is a sudden change in behaviour.

Zeeman Pieter 1865–1943. Dutch physicist who discovered 1896 that when light from certain elements, such as sodium or lithium (when heated), is passed through a spectroscope in the presence of a strong magnetic field, the spectrum splits into a number of distinct lines. His discovery, known as the ***Zeeman effect***, won him a share of the 1902 Nobel Prize for Physics with Hendrik Lorentz.

Zeeman was born in Zonnemaire, Zeeland, and studied under Lorentz at Leiden. He was professor at Amsterdam 1900–35, and from 1923 director of a new laboratory named the Zeeman Laboratory.

Lorentz proposed that light is caused by the vibration of electrons and suggested that imposing a magnetic field on light would result in a splitting of spectral lines by varying the wavelengths of the lines. Using a sodium flare between the poles of a powerful electromagnet and producing spectra with a large concave diffraction grating, Zeeman was able to detect a broadening of the spectral lines when the current was activated. In 1897 he refined the experiment and was successful in resolving the broadening of the narrow blue-green spectral line of cadmium produced in a vacuum discharge into a triplet of three component lines.

Zeeman's attention turned to the velocity of light in moving media and he was able to show that the results were in agreement with the theory of relativity.

Zel'dovich Yakov Borisovich 1914–1987. Soviet astrophysicist, originally a specialist in nuclear physics. His cosmological theories led to more accurate determinations of the abundance of helium in older stars.

Zel'dovich was born in Minsk, studied at Leningrad (now St Petersburg), and in 1931 began work at the Soviet Academy of Sciences. During World War II he contributed research towards the war effort. He later worked at the Institute of Cosmic Research at the Space Research Institute of the Soviet Academy of Sciences in Moscow.

During the 1930s, Zel'dovich was involved in a research programme aimed at discovering the mechanism of oxidation of nitrogen during an explosion. He also wrote about the chemical reactions of explosions, the subsequent generation of shockwaves, and the related subjects of gas dynamics and flame propagation.

Zel'dovich participated in early work on the mechanism of fission during the radioactive decay of uranium. In the 1950s he developed an interest in cosmology and in such subjects as quark annihilation and neutrino detection. In 1967 he proposed that in its initial stages the universe was uniform in all directions, but that as it has expanded, this isotropy has diminished.

Zeppelin Ferdinand Adolf August Heinrich, Count von Zeppelin 1838–1917. German airship pioneer. His first airship was built and tested 1900. During World War I a number of ***zeppelins*** bombed England. They were also used for luxury passenger transport but the construction of hydrogen-filled airships with rigid keels was abandoned after several disasters in the 1920s and 1930s. Zeppelin also helped to pioneer large multi-engine bomber planes.

Zeppelin was born in Constance, Baden, and entered the army 1858. In the 1860s he joined an expedition to North America to explore the sources of the Mississippi River, and in 1870 at Fort Snelling, Minnesota, he made his first ascent in a (military) balloon. Zeppelin rose to the rank of brigadier general before retiring from the army 1891. He devoted himself to the study of aeronautics, and by 1906 he had developed a practical airship. The German government then subsidized him with a National Zeppelin Fund.

In World War I zeppelins were used extensively for air raids on Britain and France 1915–16 but the large size and slow speed of the zeppelin made it a relatively easy target, and it remained in use only as a supply transport.

The early airships – the LZ-5, for example – had chain-driven propellers and, in the stern, multiple rudders and elevators. The main principle of Zeppelin's invention was streamlining the all-over envelope, inside which separate hydrogen-filled gasbags were raised inside a steel skeleton. In some craft, as in the LZ-1, a balancing rod ran the whole length of the ship below the envelope, and could be used for horizontal trimming.

Zermelo Ernst Friedrich Ferdinand 1871–1953. German mathematician who made important contributions to the development of set theory, particularly in developing the axiomatic set theory that now bears his name.

Zermelo was born in Berlin and studied at Halle and Freiburg. In 1905 he became professor at Göttingen, eventually moving to Freiburg, where he resigned his post 1935 in protest at the Nazi regime, but was reinstated 1946.

In 1900 Zermelo provided an ingenious proof to the well-ordering theorem, which states that every set can be well ordered (that is, can be arranged in a series in which each subclass – not being null – has a first term). He said that a relation $a < b$ (a comes before b) can be introduced such that for any two statements a and b, either $a = b$, or $a < b$ or $b < a$. If there are three elements a, b, and c, then if a b and $b < c$, then $a < c$. This gave rise to the Zermelo axiom that every class can be well ordered.

In 1904 Zermelo defined the ***axiom of choice***, the use of which had previously been unrecognized in mathematical reasoning. The first formulations of axioms for set theory – an axiom system for German mathematician Georg Cantor's theory of sets – were made by Zermelo in 1908.

Zernike Frits 1888–1966. Dutch physicist who developed the phase-contrast microscope 1935. Earlier microscopes allowed many specimens to be examined only after they had been transformed by heavy staining and other treatment. The phase-contrast microscope allowed living cells to be directly observed by making use of the difference in refractive indices between specimens and medium. Nobel Prize for Physics 1953.

Ziegler Karl 1898–1973. German organic chemist. In 1963 he shared the Nobel Prize for Chemistry with Giulio Natta of Italy for his work on the chemistry and technology of large polymers. He combined simple molecules of the gas ethylene (ethene) into the long-chain plastic polyethylene (polyethene).

Ziegler and Natta discovered 1953 a family of stereo-specific catalysts capable of introducing an exact and regular structure to various polymers. This discovery formed the basis of nearly all later developments in synthetic plastics, fibres, rubbers, and films derived from such olefins as ethylene (ethene) and butadiene (but-1,2:3,4-diene).

Ziegler was born near Kassel and studied at Marburg. From 1943 he was director of the Kaiser Wilhelm (later Max Planck) Institute for Coal Research in Mülheim.

In 1933, Ziegler discovered a method of making compounds that contain large rings of carbon atoms. Later he carried out research on the organic compounds of aluminium. Using electrochemical techniques, he prepared various other metal alkyls from the aluminium ones, including tetraethyl lead, which was used as an additive to petrol.

In 1953 Ziegler found that organometallic compounds mixed with certain heavy metals polymerize ethylene at atmospheric pressure to produce a linear polymer of high molecular weight (relative molecular mass) and with valuable properties, such as high melting point.

Zsigmondy Richard Adolf 1865–1929. Austrian-born German chemist who devised and built an ultramicroscope in 1903. The microscope's illumination was placed at right angles to the axis. (In a conventional microscope the light source is placed parallel to the instrument's axis.) Zsigmondy's arrangement made it possible to observe particles with a diameter of 10-millionth of a millimetre. Nobel Prize for Chemistry 1925.

Zsigmondy was born in Vienna and studied at Munich. He stayed in Germany, becoming professor at Göttingen 1908. Working at the Glass Manufacturing Company in Jena 1897–1900, Zsigmondy became concerned with coloured and turbid glasses and he invented a type of milk glass. This aroused his interest in colloids, because it is colloidal inclusions that give glass its colour or opacity. His belief that the suspended particles in gold sols are kept apart by electric charges was generally accepted, and the sols became model systems for much of his later work on colloids.

Using the ultramicroscope Zsigmondy was able to count the number of particles in a given volume and indirectly estimate their sizes. He showed that colour changes in sols reflect changes in particle size caused by coagulation when salts are added, and that the addition of agents such as gelatin stabilizes the colloid by inhibiting coagulation.

Zuckerman Solly. Baron Zuckerman of Burnham Thorpe 1904–1993. South African-born British zoologist, educationalist, and establishment figure. He did extensive research on primates, publishing a number of books that became classics in their field, including *The Social Life of Monkeys and Apes* 1932 and *Functional Affinities of Man, Monkeys and Apes* 1933. He was chief scientific adviser to the British government 1964–71

Born in Cape Town, where he was demonstrator in anatomy at the university, Zuckerman came to London in the 1920s and soon established himself as a leading anatomist with the Zoological Society. He joined the faculty of Oxford University 1934 and during World War II, as a government scientific adviser, investigated the biological effects of bomb blasts. He was professor of anatomy at Birmingham University 1946–68, and was created a peer 1971. As chief scientific adviser to the government during Harold Wilson's premiership, he had his own office within the Cabinet Office, with direct access to the prime minister himself. He published his autobiography *From Apes to Warlords* 1978.

Zwicky Fritz 1898–1974. Swiss astronomer who predicted the existence of neutron stars 1934. He discovered 18 supernovae and determined that cosmic rays originate in them.

Zwicky observed that most galaxies occur in clusters, each of which contains several thousand galaxies. He made spectroscopic studies of the Virgo and Coma Berenices clusters and calculated that the distribution of galaxies in the Coma Berenices cluster was statistically similar to the distribution of molecules in a gas when its temperature is at equilibrium. Beginning 1936, he compiled a catalogue of galaxies and galaxy clusters in which he listed 10,000 clusters.

Zwicky was born in Varna, Bulgaria, but his parents were Swiss and he retained his Swiss nationality throughout his life. He studied at the Federal Institute of Technology in Zürich, then moved to the USA 1925 to join the California Institute of Technology, where he spent his whole career.

Zwicky was among the first to suggest that there is a relationship between supernovae and neutron stars. He suggested in the early 1930s that the outer layers of a star that explodes as a supernova leave a core that collapses upon itself as a result of gravitational forces.

Zworykin Vladimir Kosma 1889–1982. Russian-born US electronics engineer who invented a television camera tube and developed the electron microscope.

Zworykin was born in Murom, Russia, and studied at the St Petersburg Institute of Technology and at the Collège de France in Paris. During World War I he worked as a radio officer in Russia. In 1919 he emigrated to the USA. He joined the Westinghouse corporation in Pittsburgh, Pennsylvania, and in 1923 took out a patent for the iconoscope (his TV camera tube), followed a year later by the kinescope (a TV receiver tube). In 1929 he demonstrated an improved electronic television system and became director of electronic research for the Radio Corporation of America (RCA), where he rose to vice president 1947.

The iconoscope tube uses an electron beam to scan the charge pattern on a signal plate, which corresponds to the pattern of light and dark of an image focused on it by a lens. Zworykin's inventions also included an early form of electric eye and an electronic image tube sensitive to infrared light, which was the basis for World War II inventions for seeing in the dark. In 1957 he patented a device that uses ultraviolet light and television to throw a colour picture of living cells on a screen, which opened up new prospects for biological investigation.

He worked with James Hillier (1915–) on the development of the electron microscope, a device Hillier constructed while a member of Zworykin's research group at RCA.

The electron microscope uses a beam of electrons to form a magnified image of a specimen. Useful magnifications of 1 million times can be obtained, which is sufficiently large to disclose a disarranged cluster of atoms in the lattice of a crystal. This is because the effective wavelength of an electron beam is several thousand times shorter than that of light.

Appendices

Astronomy: chronology

2300 BC Chinese astronomers made their earliest observations.
2000 Babylonian priests made their first observational records.
1900 Stonehenge was constructed: first phase.
434 Anaxagoras claimed the Sun is made up of hot rock.
365 The Chinese observed the satellites of Jupiter with the naked eye.
3rd century Aristarchus argued that the Sun is the centre of the Solar System.
2nd century AD Ptolemy's complicated Earth-centred system was promulgated, which dominated the astronomy of the Middle Ages.
1543 Copernicus revived the ideas of Aristarchus in *De Revolutionibus*.
1608 Hans Lippershey invented the telescope, which was first used by Galileo 1609.
1609 Johannes Kepler's first two laws of planetary motion were published (the third appeared 1619).
1632 The world's first official observatory was established in Leiden in the Netherlands.
1633 Galileo's theories were condemned by the Inquisition.
1675 The Royal Greenwich Observatory was founded in England.
1687 Isaac Newton's *Principia* was published, including his 'law of universal gravitation'.
1705 Edmund Halley correctly predicted that the comet that had passed the Earth in 1682 would return in 1758; the comet was later to be known by his name.
1781 William Herschel discovered Uranus and recognized stellar systems beyond our Galaxy.
1796 Pierre Laplace elaborated his theory of the origin of the Solar System.
1801 Giuseppe Piazzi discovered the first asteroid, Ceres.
1814 Joseph von Fraunhofer first studied absorption lines in the solar spectrum.
1846 Neptune was identified by Johann Galle, following predictions by John Adams and Urbain Leverrier.
1859 Gustav Kirchhoff explained dark lines in the Sun's spectrum.
1887 The earliest photographic star charts were produced.
1889 Edward Barnard took the first photographs of the Milky Way.
1908 Fragment of comet fell at Tunguska, Siberia.
1920 Arthur Eddington began the study of interstellar matter.
1923 Edwin Hubble proved that the galaxies are systems independent of the Milky Way, and by 1930 had confirmed the concept of an expanding universe.
1930 The planet Pluto was discovered by Clyde Tombaugh at the Lowell Observatory, Arizona, USA.
1931 Karl Jansky founded radio astronomy.
1945 Radar contact with the Moon was established by Z Bay of Hungary and the US Army Signal Corps Laboratory.
1948 The 5-m/200-in Hale reflector telescope was installed at Mount Palomar, California, USA.
1957 The Jodrell Bank telescope dish in England was completed.
1957 The first Sputnik satellite (USSR) opened the age of space observation.
1962 The first X-ray source was discovered in Scorpius.
1963 The first quasar was discovered.
1967 The first pulsar was discovered by Jocelyn Bell (now Bell Burnell) and Antony Hewish.
1969 The first crewed Moon landing was made by US astronauts.
1976 A 6-m/240-in reflector telescope was installed at Mount Semirodniki, USSR.
1977 Uranus was discovered to have rings.
1977 The spacecraft *Voyager 1* and *2* were launched, passing Jupiter and Saturn 1979–1981.
1978 The spacecraft *Pioneer Venus 1* and 2 reached Venus.
1978 A satellite of Pluto, Charon, was discovered by James Christy of the US Naval Observatory.
1986 Halley's comet returned. *Voyager 2* flew past Uranus and discovered six new moons.
1987 Supernova SN1987A flared up, becoming the first supernova to be visible to the naked eye since 1604.
The 4.2 m/165 in William Herschel Telescope on La Palma, Canary Islands, and the James Clerk Maxwell Telescope on Mauna Kea, Hawaii, began operation.
1988 The most distant individual star was recorded – a supernova, 5 billion light years away, in the AC118 cluster of galaxies.
1989 *Voyager 2* flew by Neptune and discovered eight moons and three rings.
1990 Hubble Space Telescope was launched into orbit by the US space shuttle.
1991 The space probe *Galileo* flew past the asteroid Gaspra, approaching it to within 26,000 km/16,200 mi.
1992 COBE satellite detected ripples from the Big Bang that mark the first stage in the formation of galaxies.
1995 Fragments of comet Shoemaker–Levy struck Jupiter.

Biochemistry: chronology

c.1830 Johannes Müller discovered proteins.
1833 Anselme Payen and J F Persoz first isolated an enzyme.
1862 Haemoglobin was first crystallized.
1869 The genetic material DNA (deoxyribonucleic acid) was discovered by Friedrich Mieschler
1899 Emil Fischer postulated the 'lock-and-key' hypothesis to explain the specificity of enzyme action.
1913 Leonor Michaelis and M L Menten developed a mathematical equation describing the rate of enzyme-catalysed reactions
1915 The hormone thyroxine was first isolated from thyroid gland tissue.
1920 The chromosome theory of heredity was postulated by Thomas H Morgan; growth hormone was discovered by Herbert McLean Evans and J A Long.
1921 Insulin was first isolated from the pancreas by Frederick Banting and Charles Best.
1926 Insulin was obtained in pure crystalline form.
1927 Thyroxine was first synthesized.
1928 Alexander Fleming discovered penicillin.
1931 Paul Karrer deduced the structure of retinol (vitamin A); vitamin D compounds were obtained in crystalline form by Adolf Windaus and Askew, independently of each other.
1932 Charles Glen King isolated ascorbic acid (vitamin C).
1933 Tadeus Reichstein synthesized ascorbic acid.
1935 Richard Kuhn and Karrer established the structure of riboflavin (vitamin B_2).
1936 Robert Williams established the structure of thiamine (vitamin B_1); biotin was isolated by Kogl and Tonnis.
1937 Niacin was isolated and identified by Conrad Arnold Elvehjem.
1938 Pyridoxine (vitamin B_6) was isolated in pure crystalline form.
1939 The structure of pyridoxine was determined by Kuhn.
1940 Hans Krebs proposed the Krebs (citric acid) cycle; Hickman isolated retinol in pure crystalline form; Williams established the structure of pantothenic acid; biotin was identified by Albert Szent-Györgyi, Vincent Du Vigneaud, and co-workers.
1941 Penicillin was isolated and characterized by Howard Florey and Ernst Chain.
1943 The role of DNA in genetic inheritance was first demonstrated by Oswald Avery, Colin MacLeod, and Maclyn McCarty.
1950 The basic components of DNA were established by Erwin Chargaff; the alpha-helical structure of proteins was established by Linus Pauling and R B Corey.
1953 James Watson and Francis Crick determined the molecular structure of DNA.
1956 Mahlon Hoagland and Paul Zamecnick discovered transfer RNA (ribonucleic acid); mechanisms for the biosynthesis of RNA and DNA were discovered by Arthur Kornberg and Severo Ochoa.
1957 Interferon was discovered by Alick Isaacs and Jean Lindemann.
1958 The structure of RNA was determined.
1960 Messenger RNA was discovered by Sydney Brenner and François Jacob.
1961 Marshall Nirenberg and Ochoa determined the chemical nature of the genetic code.
1965 Insulin was first synthesized.
1966 The immobilization of enzymes was achieved by Chibata.
1968 Brain hormones were discovered by Roger Guillemin and Andrew Schally.
1975 J Hughes and Hans Kosterlitz discovered encephalins.
1976 Guillemin discovered endorphins.
1977 J Baxter determined the genetic code for human growth hormone.
1978 Human insulin was first produced by genetic engineering.
1979 The biosynthetic production of human growth hormone was announced by Howard Goodman and J Baxter of the University of California, and by D V Goeddel and Seeburg of Genentech.
1982 Louis Chedid and Michael Sela developed the first synthesized vaccine.
1983 The first commercially available product of genetic engineering (Humulin) was launched.
1985 Alec Jeffreys devised genetic fingerprinting.
1993 UK researchers introduced a healthy version of the gene for cystic fibrosis into the lungs of mice with induced cystic fibrosis, restoring normal function.

Biology: chronology

c.500 BC	First studies of the structure and behaviour of animals, by the Greek Alcmaeon of Croton.
c.450	Hippocrates of Kos undertook the first detailed studies of human anatomy.
c.350	Aristotle laid down the basic philosophy of the biological sciences and outlined a theory of evolution.
c.300	Theophrastus carried out the first detailed studies of plants.
c. AD 175	Galen established the basic principles of anatomy and physiology.
c.1500	Leonardo da Vinci studied human anatomy to improve his drawing ability and produced detailed anatomical drawings.
1628	William Harvey described the circulation of the blood and the function of the heart as a pump.
1665	Robert Hooke used a microscope to describe the cellular structure of plants.
1672	Marcelle Malphigi undertook the first studies in embryology by describing the development of a chicken egg.
1677	Anton van Leeuwenhoek greatly improved the microscope and used it to describe spermatozoa as well as many microorganisms.
1736	Carolus Linnaeus published his systematic classification of plants, so establishing taxonomy.
1768–79	James Cook's voyages of discovery in the Pacific revealed an undreamed-of diversity of living species, prompting the development of theories to explain their origin.
1796	Edward Jenner established the practice of vaccination against smallpox, laying the foundations for theories of antibodies and immune reactions.
1809	Jean-Baptiste Lamarck advocated a theory of evolution through inheritance of acquired characteristics.
1839	Theodor Schwann proposed that all living matter is made up of cells.
1857	Louis Pasteur established that microorganisms are responsible for fermentation, creating the discipline of microbiology.
1859	Charles Darwin published *On the Origin of Species,* expounding his theory of the evolution of species by natural selection.
1866	Gregor Mendel pioneered the study of inheritance with his experiments on peas, but achieved little recognition.
1883	August Weismann proposed his theory of the continuity of the germ plasm.
1900	Mendel's work was rediscovered and the science of genetics founded.
1935	Konrad Lorenz published the first of many major studies of animal behaviour, which founded the discipline of ethology.
1953	James Watson and Francis Crick described the molecular structure of the genetic material, DNA.
1964	William Hamilton recognized the importance of inclusive fitness, so paving the way for the development of sociobiology.
1975	Discovery of endogenous opiates (the brain's own painkillers) opened up a new phase in the study of brain chemistry.
1976	Har Gobind Khorana and his colleagues constructed the first artificial gene to function naturally when inserted into a bacterial cell, a major step in genetic engineering.
1982	Gene databases were established at Heidelberg, Germany, for the European Molecular Biology Laboratory, and at Los Alamos, USA, for the US National Laboratories.
1985	The first human cancer gene, retinoblastoma, was isolated by researchers at the Massachusetts Eye and Ear Infirmary and the Whitehead Institute, Massachusetts.
1988	The Human Genome Organization (HUGO) was established in Washington DC with the aim of mapping the complete sequence of DNA.
1991	Biosphere 2, an experiment that attempted to reproduce the world's biosphere in miniature within a sealed glass dome, was launched in Arizona, USA.
1992	Researchers at the University of California, USA, stimulated the multiplication of isolated brain cells of mice, overturning the axiom that mammalian brains cannot produce replacement cells once birth has taken place.
1994	Scientists from Pakistan and the USA unearthed a 50-million-year-old fossil whale with hind legs that would have enabled it to walk on land.
1995	A new phylum, the Cycliophora, was discovered, containing a single species *Symbion pandora* (a lobster parasite).

Chemistry: chronology

c.3000 BC Egyptians were producing bronze – an alloy of copper and tin.

c.450 BC Greek philosopher Empedocles proposed that all substances are made up of a combination of four elements – earth, air, fire, and water – an idea that was developed by Plato and Aristotle and persisted for over 2,000 years.

c.400 BC Greek philosopher Democritus theorized that matter consists ultimately of tiny, indivisible particles, *atomos.*

AD 1 Gold, silver, copper, lead, iron, tin, and mercury were known.

200 The techniques of solution, filtration, and distillation were known.

7th–17th centuries Chemistry was dominated by alchemy, the attempt to transform nonprecious metals such as lead and copper into gold. Though misguided, it led to the discovery of many new chemicals and techniques, such as sublimation and distillation.

12th century Alcohol was first distilled in Europe.

1242 Gunpowder introduced to Europe from the Far East.

1620 Scientific method of reasoning expounded by Francis Bacon in his *Novum Organum*

1650 Leyden University in the Netherlands set up the first chemistry laboratory.

1661 Robert Boyle defined an element as any substance that cannot be broken down into still simpler substances and asserted that matter is composed of 'corpuscles' (atoms) of various sorts and sizes, capable of arranging themselves into groups, each of which constitutes a chemical substance.

1662 Boyle described the inverse relationship between the volume and pressure of a fixed mass of gas (Boyle's law).

1697 Georg Stahl proposed the erroneous theory that substances burn because they are rich in a certain substance, called phlogiston.

1755 Joseph Black discovered carbon dioxide.

1774 Joseph Priestley discovered oxygen, which he called 'dephlogisticated air'. Antoine Lavoisier demonstrated his law of conservation of mass.

1777 Lavoisier showed air to be made up of a mixture of gases, and showed that one of these – oxygen – is the substance necessary for combustion (burning) and rusting to take place.

1781 Henry Cavendish showed water to be a compound.

1792 Alessandra Volta demonstrated the electrochemical series.

1807 Humphry Davy passed electric current through molten compounds (the process of electrolysis) in order to isolate elements, such as potassium, that had never been separated by chemical means. Jöns Berzelius proposed that chemicals produced by living creatures should be termed 'organic'.

1808 John Dalton published his atomic theory, which states that every element consists of similar indivisible particles – called atoms – which differ from the atoms of other elements in their mass; he also drew up a list of relative atomic masses. Joseph Gay-Lussac announced that the volumes of gases that combine chemically with one another are in simple ratios.

1811 Publication of Amedeo Avogadro's hypothesis on the relation between the volume and number of molecules of a gas, and its temperature and pressure.

1813–14 Berzelius devised the chemical symbols and formulae still used to represent elements and compounds.

1828 Franz Wöhler converted ammonium cyanate into urea – the first synthesis of an organic compound from an inorganic substance.

1832–33 Michael Faraday expounded the laws of electrolysis, and adopted the term 'ion' for the particles believed to be responsible for carrying current.

1846 Thomas Graham expounded his law of diffusion.

1853 Robert Bunsen invented the Bunsen burner.

1858 Stanislao Cannizzaro differentiated between atomic and molecular weights (masses).

1861 Organic chemistry was defined by German chemist Friedrich Kekulé as the chemistry of carbon compounds.

1864 John Newlands devised the first periodic table of the elements.

1869 Dmitri Mendeleyev expounded his periodic table of the elements (based on atomic mass), leaving gaps for elements that had not yet been discovered.

1874 Jacobus van't Hoff suggested that the four bonds of carbon are arranged

Chemistry: chronology (continued)

	tetrahedrally, and that carbon compounds can therefore be three-dimensional and asymmetric.
1884	Swedish chemist Svante Arrhenius suggested that electrolytes (solutions or molten compounds that conduct electricity) dissociate into ions, atoms or groups of atoms that carry a positive or negative charge.
1894	William Ramsey and Lord Rayleigh discovered the first inert gases, argon.
1897	The electron was discovered by J J Thomson.
1901	Mikhail Tsvet invented paper chromatography as a means of separating pigments.
1909	Sören Sörensen devised the pH scale of acidity.
1912	Max von Laue showed crystals to be composed of regular, repeating arrays of atoms by studying the patterns in which they diffract X-rays.
1913–14	Henry Moseley equated the atomic number of an element with the positive charge on its nuclei, and drew up the periodic table, based on atomic number, that is used today.
1916	Gilbert Newton Lewis explained covalent bonding between atoms as a sharing of electrons.
1927	Nevil Sidgwick published his theory of valency, based on the numbers of electrons in the outer shells of the reacting atoms.
1930	Electrophoresis, which separates particles in suspension in an electric field, was invented by Arne Tiselius.
1932	Deuterium (heavy hydrogen), an isotope of hydrogen, was discovered by Harold Urey.
1940	Edwin McMillan and Philip Abelson showed that new elements with a higher atomic number than uranium can be formed by bombarding uranium with neutrons, and synthesized the first transuranic element, neptunium.
1942	Plutonium was first synthesized by Glenn T Seaborg and Edwin McMillan.
1950	Derek Barton deduced that some properties of organic compounds are affected by the orientation of their functional groups (the study of which became known as conformational analysis).
1954	Einsteinium and fermium were synthesized.
1955	Ilya Prigogine described the thermodynamics of irreversible processes (the transformations of energy that take place in, for example, many reactions within living cells).
1962	Neil Bartlett prepared the first compound of an inert gas, xenon hexafluoroplatinate; it was previously believed that inert gases could not take part in a chemical reaction.
1965	Robert B Woodward synthesized complex organic compounds.
1981	Quantum mechanics applied to predict course of chemical reactions by US chemist Roald Hoffmann and Kenichi Fukui of Japan.
1982	Element 109, unnilennium, synthesized.
1985	Fullerenes, a new class of carbon solids made up of closed cages of carbon atoms, were discovered by Harold Kroto and David Walton at the University of Sussex, England.
1987	US chemists Donald Cram and Charles Pederson, and Jean-Marie Lehn of France created artificial molecules that mimic the vital chemical reactions of life processes.
1990	Jean-Marie Lehn, Ulrich Koert, and Margaret Harding reported the synthesis of a new class of compounds, called nucleohelicates, that mimic the double helical structure of DNA, turned inside out.
1993	US chemists synthesized rapamycin, one of a group of complex, naturally occurring antibiotics and immunosuppressants.
1994	Elements 110 (ununnilium) and 111 (unununium) discovered at the GSI heavy-ion cyclotron, Darmstadt, Germany.
1996	Element 112 discovered as the GSI heavy-ion cyclotron, Darmstadt, Germany.

Earth science: chronology

1735 English lawyer George Hadley described the circulation of the atmosphere as large-scale convection currents centred on the equator.

1743 Christopher Packe produced the first geological map, of S England.

1744 The first map produced on modern surveying principles was produced by César-François Cassini in France.

1745 In Russia, Mikhail Vasilievich Lomonosov published a catalogue of over 3,000 minerals.

1746 A French expedition to Lapland proved the Earth to be flattened at the poles.

1760 Lomonosov explained the formation of icebergs. John Mitchell proposed that earthquakes are produced when one layer of rock rubs against another.

1766 The fossilized bones of a huge animal (later called *Mosasaurus*) were found in a quarry near the river Meuse, the Netherlands.

1776 James Keir suggested that some rocks, such as those making up the Giant's Causeway in Ireland, may have formed as molten material that cooled and then crystallized.

1779 French naturalist Comte George de Buffon speculated that the Earth may be much older than the 6,000 years suggested by the Bible.

1785 Scottish geologist James Hutton proposed the theory of uniformitarianism: all geological features are the result of processes that are at work today, acting over long periods of time.

1786 German–Swiss Johann von Carpentier described the European ice age.

1793 Jean Baptiste Lamarck argued that fossils are the remains of once-living animals and plants.

1794 William Smith produced the first large-scale geological maps of England.

1795 In France, Georges Cuvier identified the fossil bones discovered in the Netherlands in 1766 as being those of a reptile, now extinct.

1804 French physicists Jean Biot and Joseph Gay-Lussac studied the atmosphere from a hot-air balloon.

1809 The first geological survey of the eastern USA was produced by William Maclure.

1815 In England, William Smith showed how rock strata (layers) can be identified on the basis of the fossils found in them.

1822 Mary Ann Mantell discovered on the English coast the first fossil to be recognized as that of a dinosaur (an iguanodon). In Germany, Friedrich Mohs introduced a scale for specifying mineral hardness.

1825 Cuvier proposed his theory of catastrophes as the cause of the extinction of large groups of animals.

1830 Scottish geologist Charles Lyell published the first volume of *The Principles of Geology*, which described the Earth as being several hundred million years old.

1839 In the USA, Louis Agassiz described the motion and laying down of glaciers, confirming the reality of the ice ages.

1842 English palaeontologist Richard Owen coined the name 'dinosaur' for the reptiles, now extinct, that lived about 175 million years ago.

1846 Irish physicist William Thomson (Lord Kelvin) estimated, using the temperature of the Earth, that the Earth is 100 million years old.

1850 US naval officer Matthew Fontaine Maury mapped the Atlantic Ocean, noting that it is deeper near its edges than at the centre.

1852 Edward Sabine in Ireland showed a link between sunspot activity and changes in the Earth's magnetic field.

1853 James Coffin described the three major wind bands that girdle each hemisphere.

1854 English astronomer George Airy calculated the mass of the Earth by measuring gravity at the top and bottom of a coal mine.

1859 Edwin Drake drilled the world's first oil well at Titusville, Pennsylvania, USA.

1872 The beginning of the world's first major oceanographic expedition, the four-year voyage of the *Challenger*.

1882 Scottish physicist Balfour Stewart postulated the existence of the ionosphere (the ionized layer of the outer atmosphere) to account for differences in the Earth's magnetic field.

1884 German meteorologist Vladimir Köppen introduced a classification of the world's temperature zones.

1890 English geologist Arthur Holmes used radioactivity to date rocks, establishing the Earth to be 4.6 billion years old.

1895 In the USA, Jeanette Picard launched the first balloon to be used for stratospheric research.

1896 Swedish chemist Svante Arrhenius discovered a link between the amount of carbon dioxide in the atmosphere and the global temperature.

Earth science: chronology (continued)

1897 Norwegian-US meteorologist Jacob Bjerknes and his father Vilhelm developed the mathematical theory of weather forecasting.

1902 British physicist Oliver Heaviside and US engineer Arthur Edwin Kennelly predicted the existence of an electrified layer in the atmosphere that reflects radio waves. In France, Léon Teisserenc discovered layers of different temperatures in the atmosphere, which he called the troposphere and stratosphere.

1906 Richard Dixon Oldham proved the Earth to have a molten core by studying seismic waves.

1909 Yugoslav physicist Andrija Mohorovičić discovered a discontinuity in the Earth's crust, about 30 km/18 mi below the surface, that forms the boundary between the crust and the mantle.

1912 In Germany, Alfred Wegener proposed the theory of continental drift and the existence of a supercontinent, Pangaea, in the distant past.

1913 French physicist Charles Fabry discovered the ozone layer in the upper atmosphere.

1914 German-US geologist Beno Gutenberg discovered the discontinuity that marks the boundary between the Earth's mantle and the outer core.

1922 British meteorologist Lewis Fry Richardson developed a method of numerical weather forecasting.

1925 A German expedition discovered the Mid-Atlantic Ridge by means of sonar. Edward Appleton discovered a layer of the atmosphere that reflects radio waves; it was later named after him.

1929 By studying the magnetism of rocks, Japanese geologist Motonori Matuyama showed that the Earth's magnetic field reverses direction from time to time.

1935 US seismologist Charles Francis Richter established a scale for measuring the magnitude of earthquakes.

1936 Danish seismologist Inge Lehmann postulated the existence of a solid inner core of the Earth from the study of seismic waves.

1939 In Germany, Walter Maurice Elsasser proposed that eddy currents in the molten iron core cause the Earth's magnetism.

1950 Hungarian-US mathematician John Von Neumann made the first 24-hour weather forecast by computer.

1956 US geologists Bruce Charles Heezen and Maurice Ewing discovered a global network of oceanic ridges and rifts that divide the Earth's surface into plates.

1958 Using rockets, US physicist James Van Allen discovered a belt of radiation (later named after him) around the Earth.

1960 The world's first weather satellite, *TIROS 1*, was launched. US geologist Harry Hammond Hess showed that the sea floor spreads out from ocean ridges and descends back into the mantle at deep-sea trenches.

1963 British geophysicists Fred Vine and Drummond Matthews analysed the magnetism of rocks in the Atlantic Ocean floor and found conclusive proof of seafloor spreading.

1985 A British expedition to the Antarctic discovered a hole in the ozone layer above the South Pole.

1991 A borehole in the Kola Peninsula in Arctic Russia, begun in the 1970s, reached a depth of 12,261 m/40,240 ft (where the temperature was found to be 210°C/410°F). It is expected to reach a depth of 15,000 m/49,000 ft by 1995.

Mathematics: chronology

c.2500 BC The people of Mesopotamia (now Iraq) developed a positional numbering (place-value) system, in which the value of a digit depends on its position in a number.

c.2000 Mesopotamian mathematicians solved quadratic equations (algebraic equations in which the highest power of a variable is 2).

876 A symbol for zero was used for the first time, in India.

c.550 Greek mathematician Pythagoras formulated a theorem relating the lengths of the sides of a right-angled triangle. The theorem was already known by earlier mathematicians in China, Mesopotamia, and Egypt.

c.450 Hipparcos of Metapontum discovered that some numbers are irrational (cannot be expressed as the ratio of two integers).

300 Euclid laid out the laws of geometry in his book *Elements*, which was to remain a standard text for 2,000 years.

c.230 Eratosthenes developed a method for finding all prime numbers.

c.100 Chinese mathematicians began using negative numbers.

c.190 Chinese mathematicians used powers of 10 to express magnitudes.

c. AD 210 Diophantus of Alexandria wrote the first book on algebra.

c.600 A decimal number system was developed in India.

829 Persian mathematician Muhammad ibn-Musa al-Khwarizmi published a work on algebra that made use of the decimal number system.

1202 Italian mathematician Leonardo Fibonacci studied the sequence of numbers (1, 1, 2, 3, 5, 8, 13, 21, ...) in which each number is the sum of the two preceding ones.

1550 In Germany, Rheticus published trigonometrical tables that simplified calculations involving triangles.

1614 Scottish mathematician John Napier invented logarithms, which enable lengthy calculations involving multiplication and division to be carried out by addition and subtraction.

1623 Wilhelm Schickard invented the mechanical calculating machine.

1637 French mathematician and philosopher René Descartes introduced coordinate geometry.

1654 In France, Blaise Pascal and Pierre de Fermat developed probability theory.

1666 Isaac Newton developed differential calculus, a method of calculating rates of change.

1675 German mathematician Gottfried Wilhelm Leibniz introduced the modern notation for integral calculus, a method of calculating volumes.

1679 Leibniz introduced binary arithmetic, in which only two symbols are used to represent all numbers.

1684 Leibniz published the first account of differential calculus.

1718 Jakob Bernoulli in Switzerland published his work on the calculus of variations (the study of functions that are close to their minimum or maximum values).

1746 In France, Jean le Rond d'Alembert developed the theory of complex numbers.

1747 D'Alembert used partial differential equations in mathematical physics.

1798 Norwegian mathematician Caspar Wessel introduced the vector representation of complex numbers.

1799 Karl Friedrich Gauss of Germany proved the fundamental theorem of algebra: the number of solutions of an algebraic equation is the same as the exponent of the highest term.

1810 In France, Jean Baptiste Joseph Fourier published his method of representing functions by a series of trigonometric functions.

1812 French mathematician Pierre Simon Laplace published the first complete account of probability theory.

1822 In the UK, Charles Babbage began construction of the first mechanical computer, the difference machine, a device for calculating logarithms and trigonometric functions.

1827 Gauss introduced differential geometry, in which small features of curves are described by analytical methods.

1829 In Russia, Nikolai Ivanonvich Lobachevsky developed hyperbolic geometry, in which a plane is regarded as part of a hyperbolic surface, shaped like a saddle. In France, Evariste Galois introduced the theory of groups (collections whose members obey certain simple rules of addition and multiplication).

Mathematics: chronology (continued)

1844	French mathematician Joseph Liouville found the first transcendental number, which cannot be expressed as an algebraic equation with rational coefficients. In Germany, Hermann Grassmann studied vectors with more than three dimensions.
1854	George Boole in the UK published his system of symbolic logic, now called Boolean algebra.
1858	English mathematician Arthur Cayley developed calculations using ordered tables called matrices.
1865	August Ferdinand Möbius in Germany described how a strip of paper can have only one side and one edge.
1892	German mathematician Georg Cantor showed that there are different kinds of infinity and studied transfinite numbers.
1895	Jules Henri Poincaré published the first paper on topology, often called 'the geometry of rubber sheets'.
1931	In the USA, Austrian-born mathematician Kurt Gödel proved that any formal system strong enough to include the laws of arithmetic is either incomplete or inconsistent.
1937	English mathematician Alan Turing published the mathematical theory of computing.
1944	John Von Neumann and Oscar Morgenstern developed game theory in the USA.
1945	The first general purpose, fully electronic digital computer, ENIAC (electronic numerator, integrator, analyser, and computer), was built at the University of Pennsylvania, USA.
1961	Meteorologist Edward Lorenz at the Massachusetts Institute of Technology, USA, discovered a mathematical system with chaotic behaviour, leading to a new branch of mathematics – chaos theory.
1962	Benoit Mandelbrot in the USA invented fractal images, using a computer that repeats the same mathematical pattern over and over again.
1975	US mathematician Mitchell Feigenbaum discovered a new fundamental constant (approximately 4.669201609103), which plays an important role in chaos theory.
1980	Mathematicians worldwide completed the classification of all finite and simple groups, a task that took over a hundred mathematicians more than 35 years to complete and whose results took up more than 14,000 pages in mathematical journals.
1989	A team of US computer mathematicians at Amdahl Corporation, California, discovered the highest known prime number (it contains 65,087 digits).
1993	British mathematician Andrew Wiles published a 1,000-page proof of Fermat's last theorem, one of the most baffling challenges in pure mathematics.
1994	Wiles's proof was rejected but he submitted a revised and successful version in October.

Medicine: chronology

c.400 BC	Hippocrates recognized that disease had natural causes.
c. AD 200	Galen consolidated the work of the Alexandrian doctors.
1543	Andreas Vesalius gave the first accurate account of the human body.
1628	William Harvey discovered the circulation of the blood.
1768	John Hunter began the foundation of experimental and surgical pathology.
1785	Digitalis was used to treat heart disease; the active ingredient was isolated 1904.
1798	Edward Jenner published his work on vaccination.
1877	Patrick Manson studied animal carriers of infectious diseases.
1882	Robert Koch isolated the bacillus responsible for tuberculosis.
1884	Edwin Krebs isolated the diphtheria bacillus.
1885	Louis Pasteur produced a vaccine against rabies.
1890	Joseph Lister demonstrated antiseptic surgery.
1895	Wilhelm Röntgen discovered X-rays.
1897	Martinus Beijerinck discovered viruses.
1899	Felix Hoffman developed aspirin; Sigmund Freud founded psychiatry.
1900	Karl Landsteiner identified the first three blood groups, later designated A, B, and O.
1910	Paul Ehrlich developed the first specific antibacterial agent, Salvarsan, a cure for syphilis.
1922	Insulin was first used to treat diabetes.
1928	Alexander Fleming discovered penicillin.
1932	Gerhard Domagk discovered the first antibacterial sulphonamide drug, Prontosil.
1937	Electro-convulsive therapy (ECT) was developed.
1940s	Lithium treatment for manic-depressive illness was developed.
1950s	Antidepressant drugs and beta-blockers for heart disease were developed. Manipulation of the molecules of synthetic chemicals became the main source of new drugs. Peter Medawar studied the body's tolerance of transplanted organs and skin grafts.
1950	Proof of a link between cigarette smoking and lung cancer was established.
1953	Francis Crick and James Watson announced the structure of DNA. Jonas Salk developed a vaccine against polio.
1958	Ian Donald pioneered diagnostic ultrasound.
1960s	A new generation of minor tranquillizers called benzodiazepines was developed.
1967	Christiaan Barnard performed the first human heart-transplant operation.
1971	Viroids, disease-causing organisms even smaller than viruses, were isolated outside the living body.
1972	The CAT scan, pioneered by Godfrey Hounsfield, was first used to image the human brain.
1975	César Milstein developed monoclonal antibodies.
1978	World's first 'test-tube baby' was born in the UK.
1980s	AIDS (acquired immune-deficiency syndrome) was first recognized in the USA. Barbara McClintock's discovery of the transposable gene was recognized.
1980	The World Health Organization reported the eradication of smallpox.
1983	The virus responsible for AIDS, now known as human immunodeficiency virus (HIV), was identified by Luc Montagnier at the Institut Pasteur, Paris; Robert Gallo at the National Cancer Institute, Maryland, USA discovered the virus independently 1984.
1984	The first vaccine against leprosy was developed.
1987	The world's longest-surviving heart-transplant patient died in France, 18 years after his operation.
1989	Grafts of fetal brain tissue were first used to treat Parkinson's disease.
1990	Gene for maleness discovered by UK researchers.
1991	First successful use of gene therapy (to treat severe combined immune deficiency) was reported in the USA.
1993	First trials of gene therapy against cystic fibrosis took place in the USA.
1994	Gene that triggers breast cancer identified. *BRCA1* is responsible for about 50% of cases of inherited breast cancer.
1995	Tuberculosis was responsible for more than a quarter of deaths in developing countries.

Physics: chronology

c.400 BC	The first 'atomic' theory was put forward by Democritus.
c.250	Archimedes' principle of buoyancy was established.
AD 1600	Magnetism was described by William Gilbert.
1608	Hans Lippershey invented the refracting telescope.
c.1610	The principle of falling bodies descending to earth at the same speed was established by Galileo.
1642	The principles of hydraulics were put forward by Blaise Pascal.
1643	The mercury barometer was invented by Evangelista Torricelli.
1656	The pendulum clock was invented by Christiaan Huygens.
1662	Boyle's law concerning the behaviour of gases was established by Robert Boyle.
c.1665	Isaac Newton put forward the law of gravity, stating that the Earth exerts a constant force on falling bodies.
1690	The wave theory of light was propounded by Christiaan Huygens.
1704	The corpuscular theory of light was put forward by Isaac Newton.
1714	The mercury thermometer was invented by Daniel Fahrenheit.
1764	Specific and latent heats were described by Joseph Black.
1771	The link between nerve action and electricity was discovered by Luigi Galvani.
c.1787	Charles's law relating the pressure, volume, and temperature of a gas was established by Jacques Charles.
1795	The metric system was adopted in France.
1798	The link between heat and friction was discovered by Benjamin Rumford.
1800	Alessandro Volta invented the Voltaic cell.
1801	Interference of light was discovered by Thomas Young.
1808	The 'modern' atomic theory was propounded by John Dalton.
1811	Avogadro's hypothesis relating volumes and numbers of molecules of gases was proposed by Amedeo Avogadro.
1814	Fraunhofer lines in the solar spectrum were mapped by Joseph von Fraunhofer.
1815	Refraction of light was explained by Augustin Fresnel.
1819	The discovery of electromagnetism was made by Hans Oersted.
1821	The dynamo principle was described by Michael Faraday; the thermocouple was discovered by Thomas Seebeck.
1822	The laws of electrodynamics were established by André Ampère.
1824	Thermodynamics as a branch of physics was proposed by Sadi Carnot.
1827	Ohm's law of electrical resistance was established by Georg Ohm; Brownian movement resulting from molecular vibrations was observed by Robert Brown.
1829	The law of gaseous diffusion was established by Thomas Graham.
1831	Electromagnetic induction was discovered by Faraday.
1834	Faraday discovered self-induction.
1842	The principle of conservation of energy was observed by Julius von Mayer.
c.1847	The mechanical equivalent of heat was described by James Joule.
1849	A measurement of speed of light was put forward by French physicist Armand Fizeau (1819–1896).
1851	The rotation of the Earth was demonstrated by Jean Foucault.
1858	The mirror galvanometer, an instrument for measuring small electric currents, was invented by William Thomson (Lord Kelvin).
1859	Spectrographic analysis was made by Robert Bunsen and Gustav Kirchhoff.
1861	Osmosis was discovered.
1873	Light was conceived as electromagnetic radiation by James Maxwell.
1877	A theory of sound as vibrations in an elastic medium was propounded by John Rayleigh.
1880	Piezoelectricity was discovered by Pierre Curie.
1887	The existence of radio waves was predicted by Heinrich Hertz.
1895	X-rays were discovered by Wilhelm Röntgen.
1896	The discovery of radioactivity was made by Antoine Becquerel.
1897	Joseph Thomson discovered the electron.
1899	Ernest Rutherford discovered alpha and beta rays.
1900	Quantum theory was propounded by Max Planck; the discovery of gamma rays was made by French physicist Paul-Ulrich Villard (1860–1934).

Physics: chronology (continued)

1902	Oliver Heaviside discovered the ionosphere.
1904	The theory of radioactivity was put forward by Rutherford and Frederick Soddy.
1905	Albert Einstein propounded his special theory of relativity.
1908	The Geiger counter was invented by Hans Geiger and Rutherford.
1911	The discovery of the atomic nucleus was made by Rutherford.
1913	The orbiting electron atomic theory was propounded by Danish physicist Niels Bohr.
1915	X-ray crystallography was discovered by William and Lawrence Bragg.
1916	Einstein put forward his general theory of relativity; mass spectrography was discovered by William Aston.
1924	Edward Appleton made his study of the Heaviside layer.
1926	Wave mechanics was introduced by Erwin Schrödinger.
1927	The uncertainty principle of quantum physics was established by Werner Heisenberg.
1931	The cyclotron was developed by Ernest Lawrence.
1932	The discovery of the neutron was made by James Chadwick; the electron microscope was developed by Vladimir Zworykin.
1933	The positron, the antiparticle of the electron, was discovered by Carl Anderson.
1934	Artificial radioactivity was developed by Frédéric and Irène Joliot-Curie.
1939	The discovery of nuclear fission was made by Otto Hahn and Fritz Strassmann.
1942	The first controlled nuclear chain reaction was achieved by Enrico Fermi.
1956	The neutrino, an elementary particle, was discovered by Clyde Cowan and Fred Reines.
1960	The Mössbauer effect of atom emissions was discovered by Rudolf Mössbauer; the first laser and the first maser were developed by US physicist Theodore Maiman (1927–).
1964	Murray Gell-Mann and George Zweig discovered the quark.
1967	Jocelyn Bell (now Bell Burnell) and Antony Hewish discovered pulsars (rapidly rotating neutron stars that emit pulses of energy).
1971	The theory of superconductivity was announced, where electrical resistance in some metals vanishes above absolute zero.
1979	The discovery of the asymmetry of elementary particles was made by US physicists James W Cronin and L Fitch.
1982	The discovery of processes involved in the evolution of stars was made by Subrahmanyan Chandrasekhar and William Fowler.
1983	Evidence of the existence of weakons (W and Z particles) was confirmed at CERN, validating the link between the weak nuclear force and the electromagnetic force.
1986	The first high-temperature superconductor was discovered, able to conduct electricity without resistance at a temperature of –238°C/–396°F.
1989	CERN's Large Electron Positron Collider (LEP), a particle accelerator with a circumference of 27 km/16.8 mi, came into operation.
1991	LEP experiments demonstrated the existence of three generations of elementary particles, each with two quarks and two leptons.
1995	Top quark discovered at Fermilab, the US particle-physics laboratory, near Chicago, USA. US researchers announce the discovery of a material which is superconducting at the temperature of liquid nitrogen – a much higher temperature than previously achieved.

Discovery of the elements

date	element (symbol)	discoverer
	arsenic (As)	
	bismuth (Bi)	
	carbon (C)	
	copper (Cu)	
	gold (Au)	
	iron (Fe)	
	lead (Pb)	
	mercury (Hg)	
	silver (Ag)	
	sulphur (S)	
	tin (Sn)	
	zinc (Zn)	
1557	platinum (Pt)	Julius Scaliger
1674	phosphorus (P)	Hennig Brand
1730	cobalt (Co)	Georg Brandt
1751	nickel (Ni)	Axel Cronstedt
1755	magnesium (Mg)	Joseph Black (oxide isolated by Humphry Davy 1808) (pure form isolated by Antoine-Alexandre-Brutus Bussy 1828)
1766	hydrogen (H)	Henry Cavendish
1771	fluorine (F)	Karl Scheele (isolated by Henri Moissan 1886)
1772	nitrogen (N)	Daniel Rutherford
1774	chlorine (Cl)	Karl Scheele
	manganese (Mn)	Johann Gottlieb Gahn
	oxygen (O)	Joseph Priestley and Karl Scheele, independently of each other
1781	molybdenum (Mo)	named by Karl Scheele (isolated by Peter Jacob Hjelm 1782)
1782	tellurium (Te)	Franz Müller
1783	tungsten (W)	isolated by Juan José Elhuyar and Fausto Elhuyar
1789	uranium (U)	Martin Klaproth (isolated by Eugène Péligot 1841)
	zirconium (Zr)	Martin Klaproth
1790	titanium (Ti)	William Gregor
1794	yttrium (Y)	Johan Gadolin
1797	chromium (Cr)	Louis-Nicolas Vauquelin
1798	beryllium (Be)	Louis-Nicolas Vauquelin (isolated by Friedrich Wöhler and Antoine-Alexandre-Brutus Bussy 1828)
1801	vanadium (V)	Andrés del Rio (disputed), or Nils Sefström 1830
	niobium (Nb)	Charles Hatchett
1802	tantalum (Ta)	Anders Ekeberg
1804	cerium (Ce)	Jöns Berzelius and Wilhelm Hisinger, and independently by Martin Klaproth
	iridium (Ir)	Smithson Tennant
	osmium (Os)	Smithson Tennant
	palladium (Pd)	William Wollaston
	rhodium (Rh)	William Wollaston
1807	potassium (K)	Humphry Davy
	sodium (Na)	Humphry Davy
1808	barium (Ba)	Humphry Davy
	boron (B)	Humphry Davy, and independently by Joseph Gay-Lussac and Louis-Jacques Thénard
	calcium (Ca)	Humphry Davy
	strontium (Sr)	Humphry Davy
1811	iodine (I)	Bernard Courtois
1817	cadmium (Cd)	Friedrich Strohmeyer
	lithium (Li)	Johan Arfwedson
	selenium (Se)	Jöns Berzelius
1823	silicon (Si)	Jöns Berzelius
1824	aluminium (Al)	Hans Oersted (also attributed to Friedrich Wöhler 1827)
1826	bromine (Br)	Antoine-Jérôme Balard
1827	ruthenium (Ru)	G W Osann (isolated by Karl Klaus 1844)
1828	thorium (Th)	Jöns Berzelius
1839	lanthanum (La)	Carl Mosander
1843	erbium (Er)	Carl Mosander
1843	terbium (Tb)	Carl Mosander
1860	caesium (Cs)	Robert Bunsen and Gustav Kirchhoff
1861	rubidium (Rb)	Robert Bunsen and Gustav Kirchhoff
	thallium (Tl)	William Crookes (isolated by William Crookes and Claude August Lamy, independently of each other 1862)
1863	indium (In)	Ferdinand Reich and Hieronymus Richter
1868	helium (He)	Pierre Janssen
1875	gallium (Ga)	Paul Lecoq de Boisbaudran
1876	scandium (Sc)	Lars Nilson
1878	ytterbium (Yb)	Jean Charles de Marignac
1879	holmium (Ho)	Per Cleve
	samarium (Sm)	Paul Lecoq de Boisbaudran
	thulium (Tm)	Per Cleve
1885	neodymium (Nd)	Carl von Welsbach
	praseodymium (Pr)	Carl von Welsbach
1886	dysprosium (Dy)	Paul Lecoq de Boisbaudran
	gadolinium (Gd)	Paul Lecoq de Boisbaudran
	germanium (Ge)	Clemens Winkler
1894	argon (Ar)	John Rayleigh and William Ramsay
1898	krypton (Kr)	William Ramsay and Morris Travers
	neon (Ne)	William Ramsay and Morris Travers
	polonium (Po)	Marie and Pierre Curie
	radium (Ra)	Marie Curie
	xenon (Xe)	William Ramsay and Morris Travers

Discovery of the elements: (continued)

date	element (symbol)	discoverer
1899	actinium (Ac)	André Debierne
1900	radon (Rn)	Friedrich Dorn
1901	europium (Eu)	Eugène Demarçay
1907	lutetium (Lu)	Georges Urbain and Carl von Welsbach independently of each other
1913	protactinium (Pa)	Kasimir Fajans and O Göhring
	hafnium (Hf)	Dirk Coster and Georg von Hevesy
1925	rhenium (Re)	Walter Noddack, Ida Tacke, and Otto Berg
1937	technetium (Tc)	Carlo Perrier and Emilio Segrè
1939	francium (Fr)	Marguérite Perey
1940	astatine (At)	Dale R Corson, K R MacKenzie, and Emilio Segrè
	neptunium (Np)	Edwin McMillan and Philip Abelson
	plutonium (Pu)	Glenn Seaborg, Edwin McMillan, Joseph Kennedy, and Arthur Wahl
1944	americium (Am)	Glenn Seaborg, Ralph James, Leon Morgan, and Albert Ghiorso
	curium (Cm)	Glenn Seaborg, Ralph James, and Albert Ghiorso
1945	promethium (Pm)	J A Marinsky, Lawrence Glendenin, and Charles Coryell
1949	berkelium (Bk)	Glenn Seaborg, Stanley Thompson, and Albert Ghiorso
1950	californium (Cf)	Glenn Seaborg, Stanley Thompson, Kenneth Street Jr, and Albert Ghiorso
1952	einsteinium (Es)	Albert Ghiorso and co-workers
1955	fermium (Fm)	Albert Ghiorso and co-workers
	mendelevium (Md)	Albert Ghiorso, Bernard G Harvey, Gregory Choppin, Stanley Thompson, and Glenn Seaborg
1958	nobelium (No)	Albert Ghiorso, Torbjørn Sikkeland, J R Walton, and Glenn Seaborg
1961	lawrencium (Lr)	Albert Ghiorso, Torbjørn Sikkeland, Almon Larsh, and Robert Latimer
1964	dubnium (Db)	claimed by Soviet scientist Georgii Flerov and co-workers (disputed by US workers)
1967	unnilpentium (Unp)	claimed by Georgii Flerov and co-workers (disputed by US workers)
1969	dubnium (Db)	claimed by US scientist Albert Ghiorso and co-workers (disputed by Soviet workers)
1970	unnilpentium (Unp)	claimed by Albert Ghiorso and co-workers (disputed by Soviet workers)
1974	rutherfordium (Rf)	claimed by Georgii Flerov and co-workers, and, independently, by Albert Ghiorso and co-workers
1976	unnilseptium (Uns)	Georgii Flerov and Yuri Oganessian (confirmed by German scientist Peter Armbruster and co-workers)
1982	unnilennium (Une)	Peter Armbruster and co-workers
1984	unniloctium (Uno)	Peter Armbruster and co-workers
1994	ununnilium (Uun)	team at GSI heavy-ion cyclotron, Darmstadt, Germany
1994	unununium (Uuu)	team at GSI heavy-ion cyclotron, Darmstadt, Germany
1996	ununduoium (Uud)	team at GSI heavy-ion cyclotron, Darmstadt, Germany

Nobel Prize for Chemistry

1901 Jacobus van't Hoff (Netherlands): laws of chemical dynamics and osmotic pressure
1902 Emil Fischer (Germany): sugar and purine syntheses
1903 Svante Arrhenius (Sweden): electrolytic theory of dissociation
1904 William Ramsay (UK): inert gases in air and their locations in the periodic table
1905 Adolf von Baeyer (Germany): organic dyes and hydroaromatic compounds
1906 Henri Moissan (France): isolation of fluorine and adoption of electric furnace
1907 Eduard Buchner (Germany): biochemical researches and discovery of cell-free fermentation
1908 Ernest Rutherford (New Zealand): atomic disintegration, and the chemistry of radioactive substances
1909 Wilhelm Ostwald (Germany): catalysis, and principles of equilibria and rates of reaction
1910 Otto Wallach (Germany): alicyclic compounds
1911 Marie Curie (Poland): discovery of radium and polonium, and the isolation and study of radium
1912 Victor Grignard (France): discovery of Grignard reagent; Paul Sabatier (France): catalytic hydrogenation of organic compounds
1913 Alfred Werner (Switzerland): bonding of atoms within molecules
1914 Theodore Richards (USA): accurate determination of the atomic masses of many elements
1915 Richard Willstäter (Germany): research into plant pigments, especially chlorophyll
1916–17 no award
1918 Fritz Haber (Germany): synthesis of ammonia from its elements
1919 no award
1920 Walther Nernst (Germany): work on thermochemistry
1921 Frederick Soddy (UK): work on radioactive substances, especially isotopes
1922 Francis Aston (UK): mass spectrometry of isotopes of radioactive elements, and enunciation of the whole-number rule
1923 Fritz Pregl (Austria): microanalysis of organic substances
1924 no award
1925 Richard Zsigmondy (Austria): heterogeneity of colloids
1926 Theodor Svedberg (Sweden): investigation of dispersed systems
1927 Heinrich Wieland (Germany): constitution of bile acids and related substances
1928 Adolf Windaus (Germany): constitution of sterols and related vitamins
1929 Arthur Harden (UK) and Hans von Euler-Chelpin (Germany): fermentation of sugar, and fermentative enzymes
1930 Hans Fischer (Germany): analysis of haem (the iron-bearing group in haemoglobin) and chlorophyll, and the synthesis of haemin (a compound of haem)
1931 Carl Bosch (Germany) and Friedrich Bergius (Germany): invention and development of chemical high-pressure methods
1932 Irving Langmuir (USA): surface chemistry
1933 no award
1934 Harold Urey (USA): discovery of deuterium (heavy hydrogen)
1935 Irène and Frédéric Joliot-Curie (France): synthesis of new radioactive elements
1936 Peter Debye (Netherlands): work on molecular structures by investigation of dipole moments and the diffraction of X-rays and electrons in gases
1937 Norman Haworth (UK): work on carbohydrates and ascorbic acid (vitamin C); Paul Karrer (Switzerland): work on carotenoids, flavins, retinol (vitamin A) and riboflavin (vitamin B_2)
1938 Richard Kuhn (Austria): carotenoids and vitamins
1939 Adolf Butenandt (Germany): work on sex hormones; Leopold Ruzicka (Switzerland) polymethylenes and higher terpenes
1940–42 no prizes awarded
1943 Georg von Hevesy (Sweden): use of isotopes as tracers in chemical processes
1944 Otto Hahn (Germany): discovery of nuclear fission
1945 Artturi Virtanen (Finland): agriculture and nutrition, especially fodder preservation
1946 James Sumner (USA): crystallization of enzymes; John Northrop (USA) and Wendell Stanley (USA): preparation of pure enzymes and virus proteins
1947 Robert Robinson (UK): biologically important plant products, especially alkaloids
1948 Arne Tiselius (Sweden): electrophoresis and adsorption analysis, and discoveries concerning serum proteins
1949 William Giauque (USA): chemical thermodynamics, especially at very low temperatures
1950 Otto Diels (Germany) and Kurt Alder (Germany): discovery and development of diene synthesis
1951 Edwin McMillan (USA) and Glenn Seaborg (USA): chemistry of transuranic elements
1952 Archer Martin (UK) and Richard Synge (UK): invention of partition chromatography
1953 Hermann Staudinger (West Germany): discoveries in macromolecular chemistry
1954 Linus Pauling (USA): nature of chemical bonds, especially in complex substances
1955 Vincent Du Vigneaud (USA): investigations into biochemically important sulphur

compounds, and the first synthesis of a polypeptide hormone

1956 Cyril Hinshelwood (UK) and Nikoly Semenov (USSR): mechanism of chemical reactions

1957 Alexander Todd (UK): nucleotides and nucleotide coenzymes

1958 Frederick Sanger (UK): structure of proteins, especially insulin

1959 Jaroslav Heyrovsky (Czechoslovakia): polarographic methods of chemical analysis

1960 Willard Libby (USA): radiocarbon dating in archaeology, geology, and geography

1961 Melvin Calvin (USA): assimilation of carbon dioxide by plants

1962 Max Perutz (UK) and John Kendrew (UK): structures of globular proteins

1963 Karl Ziegler (West Germany) and Giulio Natta (Italy): chemistry and technology of high polymers

1964 Dorothy Crowfoot Hodgkin (UK): crystallographic determination of the structures of biochemical compounds, notably penicillin and cyanocobalamin (vitamin B_{12})

1965 Robert Woodward (USA): organic synthesis

1966 Robert Mulliken (USA): molecular orbital theory of chemical bonds and structures

1967 Manfred Eigen (West Germany), Ronald Norrish (UK), and George Porter (UK): investigation of rapid chemical reactions by means of very short pulses of energy

1968 Lars Onsager (USA): discovery of reciprocal relations, fundamental for the thermodynamics of irreversible processes

1969 Derek Barton (UK) and Odd Hassel (Norway): concept and applications of conformation

1970 Luis Frederico Leloir (Argentina): discovery of sugar nucleotides and their role in carbohydrate biosynthesis

1971 Gerhard Herzberg (Canada): electronic structure and geometry of molecules, particularly free radicals

1972 Christian Anfinsen (USA), Stanford Moore (USA), and William Stein (USA): amino-acid structure and biological activity of the enzyme ribonuclease

1973 Ernst Fischer (West Germany) and Geoffrey Wilkinson (UK): chemistry of organometallic sandwich compounds

1974 Paul Flory (USA): physical chemistry of macromolecules

1975 John Cornforth (Australia): stereochemistry of enzyme-catalysed reactions; Vladimir Prelog (Yugoslavia): stereochemistry of organic molecules and their reactions

1976 William N Lipscomb (USA): structure and chemical bonding of boranes (compounds of boron and hydrogen)

1977 Ilya Prigogine (USSR): thermodynamics of irreversible and dissipative processes

1978 Peter Mitchell (UK): biological energy transfer and chemiosmotic theory

1979 Herbert Brown (USA) and Georg Wittig (West Germany): use of boron and phosphorus compounds, respectively, in organic syntheses

1980 Paul Berg (USA): biochemistry of nucleic acids, especially recombinant-DNA; Walter Gilbert (USA) and Frederick Sanger (UK): base sequences in nucleic acids

1981 Kenichi Fukui (Japan) and Roald Hoffmann (USA): theories concerning chemical reactions

1982 Aaron Klug (UK): crystallographic electron microscopy: structure of biologically important nucleic-acid–protein complexes

1983 Henry Taube (USA): electron-transfer reactions in inorganic chemical reactions

1984 Bruce Merrifield (USA): chemical syntheses on a solid matrix

1985 Herbert A Hauptman (USA) and Jerome Karle (USA): methods of determining crystal structures

1986 Dudley Herschbach (USA), Yuan Lee (USA), and John Polanyi (Canada): dynamics of chemical elementary processes

1987 Donald Cram (USA), Jean-Marie Lehn (France), and Charles Pedersen (USA): molecules with highly selective structure-specific interactions

1988 Johann Deisenhofer (West Germany), Robert Huber (West Germany), and Hartmut Michel (West Germany): three-dimensional structure of the reaction centre of photosynthesis

1989 Sydney Altman (USA) and Thomas Cech (USA): discovery of catalytic function of RNA

1990 Elias James Corey (USA): new methods of synthesizing chemical compounds

1991 Richard R Ernst (Switzerland): improvements in the technology of nuclear magnetic resonance (NMR) imaging

1992 Rudolph A Marcus (USA): theoretical discoveries relating to reduction and oxidation reactions

1993 Kary Mullis (USA): invention of the polymerase chain reaction technique for amplifying DNA; Michael Smith (Canada): development of techniques for splicing foreign genetic segments into an organism's DNA in order to modify the proteins produced

1994 George A Olah (USA): development of technique for examining hydrocarbon molecules

1995 Sherwood Roland (USA), Mario Molina (Mexico), and Paul Crutzen (Netherlands): explaining the mechanisms of the ozone layer

1996 Harold Kroto (UK), Robert Curl (USA) and Richard Smalley (USA): discovery of the structure of "buckminster fullerene" ("buckyballs" for short) a form of carbon composed of 60 atoms.

Nobel Prize for Physiology or Medicine

1901 Emil von Behring (Germany): discovery that the body produces antitoxins, and development of serum therapy for diseases such as diphtheria
1902 Ronald Ross (UK): role of the *Anopheles* mosquito in transmitting malaria
1903 Niels Finsen (Denmark): use of ultraviolet light to treat skin diseases
1904 Ivan Pavlov (Russia): physiology of digestion
1905 Robert Koch (Germany): investigations and discoveries in relation to tuberculosis
1906 Camillo Golgi (Italy) and Santiago Ramón y Cajal (Spain): fine structure of nervous system
1907 Charles Laveran (France): discovery that certain protozoa can cause disease
1908 Ilya Mechnikov (Russia) and Paul Ehrlich (Germany): work on immunity
1909 Emil Kocher (Switzerland): physiology, pathology, and surgery of the thyroid gland
1910 Albrecht Kossel (Germany): study of cell proteins and nucleic acids
1911 Allvar Gullstrand (Sweden): refraction of light through the different components of the eye
1912 Alexis Carrel (USA): techniques for connecting severed blood vessels and transplanting organs
1913 Charles Richet (France): allergic responses
1914 Robert Bárány (Austria): physiology and pathology of the equilibrium organs of the inner ear
1915–18 no award
1919 Jules Bordet (Belgium): work on immunity
1920 August Krogh (Denmark): discovery of mechanism regulating the dilation and constriction of blood capillaries
1921 no award
1922 Archibald Hill (UK): production of heat in contracting muscle; Otto Meyerhof (Germany): relationship between oxygen consumption and metabolism of lactic acid in muscle
1923 Frederick Banting (Canada) and John Macleod (UK): discovery and isolation of the hormone insulin
1924 Willem Einthoven (Netherlands): invention of the electrocardiograph
1925 no award
1926 Johannes Fibiger (Denmark): discovery of a parasite *Spiroptera carcinoma* that causes cancer
1927 Julius Wagner-Jauregg (Austria): use of induced malarial fever to treat paralysis caused by mental deterioration
1928 Charles Nicolle (France): role of the body louse in transmitting typhus
1929 Christiaan Eijkman (Netherlands): discovery of a cure for beriberi, a vitamin-deficiency disease; Frederick Hopkins (UK): discovery of trace substances, now known as vitamins, that stimulate growth
1930 Karl Landsteiner (USA): discovery of human blood groups
1931 Otto Warburg (Germany): discovery of respiratory enzymes that enable cells to process oxygen
1932 Charles Sherrington (UK) and Edgar Adrian (UK): function of neurons (nerve cells)
1933 Thomas Morgan (USA): role of chromosomes in heredity
1934 George Whipple (USA), George Minot (USA), and William Murphy (USA): treatment of pernicious anaemia by increasing the amount of liver in the diet
1935 Hans Spemann (Germany): organizer effect in embryonic development
1936 Henry Dale (UK) and Otto Loewi (Germany): chemical transmission of nerve impulses
1937 Albert Szent-Györgyi (Hungary): investigation of biological oxidation processes and of the action of ascorbic acid (vitamin C)
1938 Corneille Heymans (Belgium): mechanisms regulating respiration
1939 Gerhard Domagk (Germany): discovery of the first antibacterial sulphonamide drug
1940–42 no award
1943 Carl Dam (Denmark): discovery of vitamin K; Edward Doisy (USA): chemical nature of vitamin K
1944 Joseph Erlanger (USA) and Herbert Gasser (USA): transmission of impulses by nerve fibres
1945 Alexander Fleming (UK): discovery of the bactericidal effect of penicillin; Ernst Chain (UK) and Howard Florey (Australia): isolation of penicillin and its development as an antibiotic drug
1946 Hermann Muller (USA): discovery that X-ray irradiation can cause mutation
1947 Carl Cori (USA) and Gerty Cori (USA): production and breakdown of glycogen (animal starch); Bernardo Houssay (Argentina): function of the pituitary gland in sugar metabolism

Nobel Prize for Physiology or Medicine (continued)

1948 Paul Müller (Switzerland): discovery of the first synthetic contact insecticide DDT

1949 Walter Hess (Switzerland): mapping areas of the midbrain that control the activities of certain body organs; Antonio Egas Moniz (Portugal): therapeutic value of prefrontal leucotomy in certain psychoses

1950 Edward Kendall (USA), Tadeus Reichstein (Poland), and Philip Hench (USA): structure and biological effects of hormones of the adrenal cortex

1951 Max Theiler (South Africa): discovery of a vaccine against yellow fever

1952 Selman Waksman (USA): discovery of streptomycin, the first antibiotic effective against tuberculosis

1953 Hans Krebs (UK): discovery of the Krebs cycle; Fritz Lipmann (USA): discovery of coenzyme A, a nonprotein compound that acts in conjunction with enzymes to catalyse metabolic reactions leading up to the Krebs cycle

1954 John Enders (USA), Thomas Weller (USA), and Frederick Robbins (USA): cultivation of the polio virus in the laboratory

1955 Hugo Theorell (Sweden): nature and action of oxidation enzymes

1956 André Cournand (USA), Werner Forssmann (Germany), and Dickinson Richards Jr (USA): technique for passing a catheter into the heart for diagnostic purposes

1957 Daniel Bovet (Switzerland): discovery of synthetic drugs used as muscle relaxants in anaesthesia

1958 George Beadle (USA) and Edward Tatum (USA): discovery that genes regulate precise chemical effects; Joshua Lederberg (USA): genetic recombination and the organization of bacterial genetic material

1959 Severo Ochoa (USA) and Arthur Kornberg (USA): discovery of enzymes that catalyse the formation of RNA (ribonucleic acid) and DNA (deoxyribonucleic acid)

1960 Macfarlane Burnet (Australia) and Peter Medawar (UK): acquired immunological tolerance of transplanted tissues

1961 Georg von Békésy (USA): investigations into the mechanism of hearing within the cochlea of the inner ear

1962 Francis Crick (UK), James Watson (USA), and Maurice Wilkins (UK): discovery of the double-helical structure of DNA and of the significance of this structure in the replication and transfer of genetic information

1963 John Eccles (Australia), Alan Hodgkin (UK), and Andrew Huxley (UK): ionic mechanisms involved in the communication or inhibition of impulses across neuron (nerve cell) membranes

1964 Konrad Bloch (USA) and Feodor Lynen (West Germany): cholesterol and fatty-acid metabolism

1965 François Jacob (France), André Lwoff (France), and Jacques Monod (France): genetic control of enzyme and virus synthesis

1966 Peyton Rous (USA): discovery of tumour-inducing viruses; Charles Huggins (USA): hormonal treatment of prostatic cancer

1967 Ragnar Granit (Sweden), Haldan Hartline (USA), and George Wald (USA): physiology and chemistry of vision

1968 Robert Holley (USA), Har Gobind Khorana (USA), and Marshall Nirenberg (USA): interpretation of genetic code and its function in protein synthesis

1969 Max Delbruck (USA), Alfred Hershey (USA), and Salvador Luria (USA): replication mechanism and genetic structure of viruses

1970 Bernard Katz (UK), Ulf von Euler (Austria), and Julius Axelrod (USA): storage, release, and inactivation of neurotransmitters

1971 Earl Sutherland (USA): discovery of cyclic AMP, a chemical messenger that plays a role in the action of many hormones

1972 Gerald Edelman (USA) and Rodney Porter (UK): chemical structure of antibodies

1973 Karl von Frisch (Austria), Konrad Lorenz (Austria), and Nikolaas Tinbergen (Netherlands): animal behaviour patterns

1974 Albert Claude (USA), Christian de Duve (Belgium), and George Palade (USA): structural and functional organization of the cell

1975 David Baltimore (USA), Renato Dulbecco (USA), and Howard Temin (USA): interactions between tumour-inducing viruses and the genetic material of the cell

1976 Baruch Blumberg (USA) and Carleton Gajdusek (USA): new mechanisms for the origin and transmission of infectious diseases

1977 Roger Guillemin (USA) and Andrew Schally (USA): discovery of hormones produced by the hypothalamus region of

Nobel Prize for Physiology or Medicine (continued)

	the brain; Rosalyn Yalow (USA): radioimmunoassay techniques by which minute quantities of hormone may be detected
1978	Werner Arber (Switzerland), Daniel Nathans (USA), and Hamilton Smith (USA): discovery of restriction enzymes and their application to molecular genetics
1979	Allan Cormack (USA) and Godfrey Hounsfield (UK): development of the CAT scan
1980	Baruj Benacerraf (USA), Jean Dausset (France), and George Snell (USA): genetically determined structures on the cell surface that regulate immunological reactions
1981	Roger Sperry (USA): functional specialization of the brain's cerebral hemispheres; David Hubel (USA) and Torsten Wiesel (Sweden): visual perception
1982	Sune Bergström (Sweden), Bengt Samuelson (Sweden), and John Vane (UK): discovery of prostaglandins and related biologically reactive substances
1983	Barbara McClintock (USA): discovery of mobile genetic elements
1984	Niels Jerne (Denmark), Georges Köhler (West Germany), and César Milstein (UK): work on immunity and discovery of a technique for producing highly specific, monoclonal antibodies
1985	Michael Brown (USA) and Joseph L Goldstein (USA): regulation of cholesterol metabolism
1986	Stanley Cohen (USA) and Rita Levi-Montalcini (Italy): discovery of factors that promote the growth of nerve and epidermal cells
1987	Susumu Tonegawa (Japan): process by which genes alter to produce a range of different antibodies
1988	James Black (UK), Gertrude Elion (USA), and George Hitchings (USA): principles governing the design of new drug treatment
1989	Michael Bishop (USA) and Harold Varmus (USA): discovery of oncogenes, genes carried by viruses that can trigger cancerous growth in normal cells
1990	Joseph Murray (USA) and Donnall Thomas (USA): pioneering work in organ and cell transplants
1991	Erwin Neher (Germany) and Bert Sakmann (Germany): discovery of how gatelike structures (ion channels) regulate the flow of ions into and out of cells
1992	Edmund Fisher (USA) and Erwin Krebs (USA): isolating and describing the action of the enzyme responsible for reversible protein phosphorylation, a major biological control mechanism
1993	Phillip Sharp (USA) and Richard Roberts (UK): discovery of split genes (genes interrupted by nonsense segments of DNA)
1994	Alfred Gilman (USA) and Martin Rodbell (USA): discovery of a family of proteins (G proteins) that translate messages – in the form of hormones or other chemical signals – into action inside cells
1995	Edward B Lewis and Eric F Wieschaus (USA), and Christiane Nüsslein-Volhard (Germany): discovery of genes which control the early stages of the body's development
1996	Peter Doherty (Australia) and Rolf Zinkernagel (Switzerland): discovery of how the immune systrem recognizes and kills virus-infected cells.

Nobel Prize for Physics

1901 Wilhelm Röntgen (Germany): discovery of X-rays
1902 Hendrik Lorentz (Netherlands) and Pieter Zeeman (Netherlands): influence of magnetism on radiation phenomena
1903 Antoine Becquerel (France): discovery of spontaneous radioactivity; Pierre Curie (France) and Marie Curie (Poland): researches on radiation phenomena
1904 John Strutt (Lord Rayleigh, UK): densities of gases and discovery of argon
1905 Philipp von Lenard (Germany): work on cathode rays
1906 Joseph J Thomson (UK): theoretical and experimental work on the conduction of electricity by gases
1907 Albert Michelson (USA): measurement of the speed of light through the design and application of precise optical instruments such as the interferometer
1908 Gabriel Lippmann (France): photographic reproduction of colours by interference
1909 Guglielmo Marconi (Italy) and Karl Braun (Germany): development of wireless telegraphy
1910 Johannes van der Waals (Netherlands): equation describing the physical behaviour of gases and liquids
1911 Wilhelm Wien (Germany): laws governing radiation of heat
1912 Nils Dalen (Sweden): invention of light-controlled valves, which allow lighthouses and buoys to operate automatically
1913 Heike Kamerlingh Onnes (Netherlands): studies of properties of matter at low temperatures
1914 Max von Laue (Germany): discovery of diffraction of X-rays by crystals
1915 William Bragg (UK) and Lawrence Bragg (UK): X-ray analysis of crystal structures
1916 no award
1917 Charles Barkla (UK): discovery of characteristic X-ray emission of the elements
1918 Max Planck (Germany): formulation of quantum theory
1919 Johannes Stark (Germany): discovery of Doppler effect in rays of positive ions, and splitting of spectral lines in electric fields
1920 Charles Guillaume (Switzerland): precision measurements through anomalies in nickel–steel alloys
1921 Albert Einstein (Switzerland): theoretical physics, especially law of photoelectric effect
1922 Niels Bohr (Denmark): structure of atoms and radiation emanating from them
1923 Robert Millikan (USA): discovery of the electric charge of an electron, and study of the photoelectric effect
1924 Karl Siegbahn (Sweden): X-ray spectroscopy
1925 James Franck (USA) and Gustav Hertz (Germany): laws governing the impact of an electron upon an atom
1926 Jean Perrin (France): confirmation of the discontinuous structure of matter
1927 Arthur Compton (USA): transfer of energy from electromagnetic radiation to a particle; Charles Wilson (UK): invention of the Wilson cloud chamber, by which the movement of electrically charged particles may be tracked
1928 Owen Richardson (UK): thermionic phenomena and associated law
1929 Louis Victor de Broglie (France): discovery of wavelike nature of electrons
1930 Venkata Raman (India): discovery of the scattering of single-wavelength light when it is passed through a transparent substance
1931 no award
1932 Werner Heisenberg (Germany): creation of quantum mechanics
1933 Erwin Schrödinger (Austria) and Paul Dirac (UK): development of quantum mechanics
1934 no award
1935 James Chadwick (UK): discovery of the neutron
1936 Victor Hess (Austria): discovery of cosmic radiation; Carl Anderson (USA): discovery of the positron
1937 Clinton Davisson (USA) and George Thomson (UK): diffraction of electrons by crystals
1938 Enrico Fermi (USA): use of neutron irradiation to produce new elements, and discovery of nuclear reactions induced by slow neutrons
1939 Ernest O Lawrence (USA): invention and development of cyclotron, and production of artificial radioactive elements
1940–42 no award
1943 Otto Stern (Germany): molecular-ray method of investigating elementary particles, and discovery of magnetic moment of proton
1944 Isidor Isaac Rabi (USA): resonance method of recording the magnetic properties of atomic nuclei

Nobel Prize for Physics (continued)

1945	Wolfgang Pauli (Austria): discovery of the exclusion principle
1946	Percy Bridgman (USA): development of high-pressure physics
1947	Edward Appleton (UK): physics of the upper atmosphere
1948	Patrick Blackett (UK): application of the Wilson cloud chamber to nuclear physics and cosmic radiation
1949	Hideki Yukawa (Japan): theoretical work predicting existence of mesons
1950	Cecil Powell (UK): use of photographic emulsion to study nuclear processes, and discovery of pions (pi mesons)
1951	John Cockcroft (UK) and Ernest Walton (Ireland): transmutation of atomic nuclei by means of accelerated subatomic particles
1952	Felix Bloch (USA) and Edward Purcell (USA): precise nuclear-magnetic measurements
1953	Frits Zernike (Netherlands): invention of phase-contrast microscope
1954	Max Born (Germany): statistical interpretation of wave function in quantum mechanics; Walther Bothe (Germany): coincidence method of detecting the emission of electrons
1955	Willis Lamb (USA): structure of hydrogen spectrum; Polykarp Kusch (USA): determination of magnetic moment of the electron
1956	William Shockley (USA), John Bardeen (USA), and Walter Houser Brattain (USA): study of semiconductors, and discovery of transistor effect
1957	Yang Chen Ning (USA) and Lee Tsung-Dao (China): investigations of weak interactions between elementary particles
1958	Pavel Cherenkov (USSR), Ilya Frank (USSR), and Igor Tamm (USA): discovery and interpretation of Cherenkov radiation
1959	Emilio Segrè (Italy) and Owen Chamberlain (USA): discovery of the antiproton
1960	Donald Glaser (USA): invention of the bubble chamber
1961	Robert Hofstadter (USA): scattering of electrons in atomic nuclei, and structure of protons and neutrons; Rudolf Mössbauer (Germany): resonance absorption of gamma radiation
1962	Lev Landau (USSR): theories of condensed matter, especially liquid helium
1963	Eugene Wigner (USA): discovery and application of symmetry principles in atomic physics; Maria Goeppert-Mayer (USA) and Hans Jensen (Germany): discovery of the shell-like structure of atomic nuclei
1964	Charles Townes (USA), Nikolai Basov (USSR), and Aleksandr Prokhorov (USSR): quantum electronics leading to construction of oscillators and amplifiers based on maser–laser principle
1965	Sin-Itiro Tomonaga (Japan), Julian Schwinger (USA), and Richard Feynman (USA): quantum electrodynamics
1966	Alfred Kastler (France): development of optical pumping, whereby atoms are raised to higher energy levels by illumination
1967	Hans Bethe (USA): theory of nuclear reactions, and discoveries concerning production of energy in stars
1968	Luis Alvarez (USA): elementary-particle physics, and discovery of resonance states, using hydrogen bubble chamber and data analysis
1969	Murray Gell-Mann (USA): classification of elementary particles, and study of their interactions
1970	Hannes Alfvén (Sweden): magnetohydrodynamics and its applications in plasma physics; Louis Néel (France): antiferromagnetism and ferromagnetism in solid-state physics
1971	Dennis Gabor (UK): invention and development of holography
1972	John Bardeen (USA), Leon Cooper (USA), and John Robert Schrieffer (USA): theory of superconductivity
1973	Leo Esaki (Japan) and Ivar Giaver (USA): tunnelling phenomena in semiconductors and superconductors; Brian Josephson (UK): theoretical predictions of the properties of a supercurrent through a tunnel barrier
1974	Martin Ryle (UK) and Antony Hewish (UK): development of radio astronomy, particularly aperture-synthesis technique, and the discovery of pulsars
1975	Aage Bohr (Denmark), Ben Mottelson (Denmark), and James Rainwater (USA): discovery of connection between collective motion and particle motion in atomic nuclei, and development of theory of nuclear structure
1976	Burton Richter (USA) and Samuel Ting (USA): discovery of the psi meson
1977	Philip Anderson (USA), Nevill Mott (UK), and John Van Vleck (USA): electronic structure of magnetic and disordered systems

Nobel Prize for Physics (continued)

1978 Pyotr Kapitza (USSR): low-temperature physics; Arno Penzias (Germany) and Robert Wilson (USA): discovery of cosmic background radiation

1979 Sheldon Glashow (USA), Abdus Salam (Pakistan), and Steven Weinberg (USA): unified theory of weak and electromagnetic fundamental forces, and prediction of the existence of the weak neutral current

1980 James W Cronin (USA) and Val Fitch (USA): violations of fundamental symmetry principles in the decay of neutral kaon mesons

1981 Nicolaas Bloemergen (USA) and Arthur Schawlow (USA): development of laser spectroscopy; Kai Siegbahn (Sweden): high-resolution electron spectroscopy

1982 Kenneth Wilson (USA): theory for critical phenomena in connection with phase transitions

1983 Subrahmanyan Chandrasekhar (USA): theoretical studies of physical processes in connection with structure and evolution of stars; William Fowler (USA): nuclear reactions involved in the formation of chemical elements in the universe

1984 Carlo Rubbia (Italy) and Simon van der Meer (Netherlands): contributions to the discovery of the W and Z particles (weakons)

1985 Klaus von Klitzing (Germany): discovery of the quantized Hall effect

1986 Erns Ruska (Germany): electron optics, and design of the first electron microscope; Gerd Binnig (Germany) and Heinrich Rohrer (Switzerland): design of scanning tunnelling microscope

1987 Georg Bednorz (Germany) and Alex Müller (Switzerland): superconductivity in ceramic materials

1988 Leon M Lederman (USA), Melvin Schwartz (USA), and Jack Steinberger (Germany): neutrino-beam method, and demonstration of the doublet structure of leptons through discovery of muon neutrino

1989 Norman Ramsey (USA): measurement techniques leading to discovery of caesium atomic clock; Hans Dehmelt (USA) and Wolfgang Paul (Germany): ion-trap method for isolating single atoms

1990 Jerome Friedman (USA), Henry Kendall (USA), and Richard Taylor (Canada): experiments demonstrating that protons and neutrons are made up of quarks

1991 Pierre-Gilles de Gennes (France): work on disordered systems including polymers and liquid crystals; development of mathematical methods for studying the behaviour of molecules in a liquid on the verge of solidifying

1992 Georges Charpak (Poland): invention and development of detectors used in high-energy physics

1993 Joseph Taylor (USA) and Russell Hulse (USA): discovery of first binary pulsar (confirming the existence of gravitational waves)

1994 Clifford G Shull (USA) and Bertram N Brockhouse (Canada): development of technique known as 'neutron scattering' which led to advances in semiconductor technology

1995 Frederick Reines (USA): discovery of the neutrino; Martin L Perl (USA): discovery of the tau lepton

1996 David Lee (USA), Robert Richardson (USA) and Douglas Osheroff (USA): discovery of the "superfluid" properties of helium at temperatures close to absolute zero.

Fields medal (mathematics)

awarded every four years by the International Congress of Mathematicians

1936 Lars Ahlfors, Finland; Jesse Douglas, USA
1950 Atle Selberg, USA; Laurent Schwartz, France
1954 Kunihiko Kodaira, USA; Jean-Pierre Serre, France
1958 Klaus Roth, UK; René Thom, France
1962 Lars Hörmander, Sweden; John Milnor, USA
1966 Michael Atiyah, UK; Paul J Cohen, USA; Alexander Grothendieck, France; Stephen Smale, USA
1970 Alan Baker, UK; Heisuke Hironaka, USA; Sergei Novikov, USSR; John G Thompson, USA
1974 Enrico Bombieri, Italy; David Mumford, USA
1978 Pierre Deligne, Belgium; Charles Fefferman, USA; G A Margulis, USSR; Daniel Quillen, USA
1982 Alain Connes, France; William Thurston, USA; S T Yau, USA
1986 Simon Donaldson, UK; Gerd Faltings, West Germany; Michael Freedman, USA
1990 Vladimir Drinfeld, USSR; Vaughan F R Jones, USA; Shigefumi Mori, Japan; Edward Witten, USA
1994 L J Bourgain, USA/France; P-L Lions, France; J-C Yoccoz, France; E I Zelmanov, USA